T0213214

Lecture Notes in Computer Science 14071

Founding Editors

Gerhard Goos
Juris Hartmanis

The series Lecture Notes in Computer Science (LNCS), including its subseries Lecture Notes in Artificial Intelligence (LNAI) and Lecture Notes in Bioinformatics (LNBI), has established itself as a medium for the publication of new developments in computer science and information technology research, teaching, and education.

LNCS enjoys close cooperation with the computer science R & D community, the series counts many renowned academics among its volume editors and paper authors, and collaborates with prestigious societies. Its mission is to serve this international community by providing an invaluable service, mainly focused on the publication of conference and workshop proceedings and postproceedings. LNCS commenced publication in 1973.

Frank Nielsen · Frédéric Barbaresco
Editors

Geometric Science of Information

6th International Conference, GSI 2023
St. Malo, France, August 30 – September 1, 2023
Proceedings, Part I

 Springer

Editors
Frank Nielsen ⓘ
Sony Computer Science Laboratories Inc.
Tokyo, Japan

Frédéric Barbaresco ⓘ
THALES Land and Air Systems
Meudon, France

ISSN 0302-9743   ISSN 1611-3349 (electronic)
Lecture Notes in Computer Science
ISBN 978-3-031-38270-3   ISBN 978-3-031-38271-0 (eBook)
https://doi.org/10.1007/978-3-031-38271-0

This Springer imprint is published by the registered company Springer Nature Switzerland AG
The registered company address is: Gewerbestrasse 11, 6330 Cham, Switzerland

6th Geometric Science of Information Conference (GSI'23): From Classical To Quantum Information Geometry

Saint-Malo, France, Pierre Louis Moreau de Maupertuis' Birthplace

We are celebrating the 10th anniversary of the launch of the GSI conferences cycle, which were initiated in 2013. As for GSI'13, GSI'15, GSI'17, GSI'19 and GSI'21 (https://franknielsen.github.io/GSI/), the objective of this 6th edition of the SEE GSI conference, hosted in Saint-Malo, birthplace of Pierre Louis Moreau de Maupertuis, is to bring together pure and applied mathematicians and engineers with a common interest in geometric tools and their applications for information analysis. GSI emphasizes the active participation of young researchers to discuss emerging areas of collaborative research on the topic of "Geometric Science of Information and its Applications". In 2023, GSI's main theme was "FROM CLASSICAL TO QUANTUM INFORMATION GEOMETRY", and the conference took place at the Palais du Grand Large, in Saint-Malo, France.

The GSI conference cycle was initiated by the Brillouin Seminar Team as early as 2009 (http://repmus.ircam.fr/brillouin/home). The GSI'21 event was motivated by the continuity of the first initiative, launched in 2013 (https://web2.see.asso.fr/gsi2013), at Mines ParisTech, consolidated in 2015 (https://web2.see.asso.fr/gsi2015) at Ecole Polytechnique, and opened to new communities in 2017 (https://web2.see.asso.fr/gsi2017) at Mines ParisTech, 2019 (https://web2.see.asso.fr/gsi2019) at ENAC Toulouse and 2021 (https://web2.see.asso.fr/gsi2021) at Sorbonne University. We mention that in 2011, we organized an Indo-French workshop on the topic of "Matrix Information Geometry" (https://www.lix.polytechnique.fr/~nielsen/MIG/) that yielded an edited book in 2013, and in 2017, collaborated at a CIRM seminar in Luminy on the event TGSI'17 "Topological & Geometrical Structures of Information" (https://fconferences.cirm-math.fr/1680.html).

GSI satellite events were organized in 2019 and 2020 as FGSI'19 "Foundation of Geometric Structures of Information" in Montpellier (https://fgsi2019.sciencesconf.org/) and Les Houches Seminar SPIGL'20 "Joint Structures and Common Foundations of Statistical Physics, Information Geometry and Inference for Learning" (https://franknielsen.github.io/SPIG-LesHouches2020/).

The technical program of GSI'23 covered all the main topics and highlights in the domain of the "Geometric Science of Information" including information geometry manifolds of structured data/information and their advanced applications. These Springer LNCS proceedings consist solely of original research papers that have been carefully single-blind peer-reviewed by at least two or three experts. 125 of 161 submissions were accepted for this volume. Accepted contributions were revised before acceptance.

Like GSI'13, GSI'15, GSI'17, GSI'19, and GSI'21, GSI'23 addresses inter-relations between different mathematical domains such as shape spaces (geometric statistics on manifolds and Lie groups, deformations in shape space, ...), probability/optimization and algorithms on manifolds (structured matrix manifolds, structured data/information, ...), relational and discrete metric spaces (graph metrics, distance geometry, relational analysis,...), computational and Hessian information geometry, geometric structures in thermodynamics and statistical physics, algebraic/infinite-dimensional/Banach information manifolds, divergence geometry, tensor-valued morphology, optimal transport theory, manifold and topology learning, ... and applications such as geometries of audio-processing, inverse problems and signal/image processing. GSI'23 topics were enriched with contributions from Lie Group Machine Learning, Harmonic Analysis on Lie Groups, Geometric Deep Learning, Geometry of Hamiltonian Monte Carlo, Geometric & (Poly)Symplectic Integrators, Contact Geometry & Hamiltonian Control, Geometric and structure-preserving discretizations, Probability Density Estimation & Sampling in High Dimension, Geometry of Graphs and Networks and Geometry in Neuroscience & Cognitive Sciences.

At the turn of the century, new and fruitful interactions were discovered between several branches of science: Information Sciences (information theory, digital communications, statistical signal processing), Mathematics (group theory, geometry and topology, probability, statistics, sheaf theory, ...) and Physics (geometric mechanics, thermodynamics, statistical physics, quantum mechanics, ...). The GSI biannual international conference cycle is an effort to discover joint mathematical structures to all these disciplines by elaboration of a "General Theory of Information" embracing physics science, information science, and cognitive science in a global scheme.

The GSI'23 conference was structured in 25 sessions of more than 120 papers and a poster session:

- **Geometry and Machine Learning**

  - **Geometric Green Learning** - Alice Barbara Tumpach, Diarra Fall & Guillaume Charpiat
  - **Neurogeometry Meets Geometric Deep Learning** - Remco Duits & Erik Bekkers, Alessandro Sarti
  - **Divergences in Statistics & Machine Learning** - Michel Broniatowski & Wolfgang Stummer

- **Divergences and Computational Information Geometry**

  - **Computational Information Geometry and Divergences** - Frank Nielsen & Olivier Rioul
  - **Statistical Manifolds and Hessian Information Geometry** - Michel Nguiffo Boyom

- **Statistics, Topology and Shape Spaces**

  - **Statistics, Information and Topology** - Pierre Baudot & Grégoire Seargeant-Perthuis
  - **Information Theory and Statistics** - Olivier Rioul
  - **Statistical Shape Analysis and more Non-Euclidean Statistics** - Stephan Huckemann & Xavier Pennec

- **Probability and Statistics on Manifolds** - Cyrus Mostajeran
- **Computing Geometry & Algebraic Statistics** - Eliana Duarte & Elias Tsigaridas

- **Geometry & Mechanics**

  - **Geometric and Analytical Aspects of Quantization and Non-Commutative Harmonic Analysis on Lie Groups** - Pierre Bieliavsky & Jean-Pierre Gazeau
  - **Deep Learning: Methods, Analysis and Applications to Mechanical Systems** - Elena Celledoni, James Jackaman, Davide Murari and Brynjulf Owren
  - **Stochastic Geometric Mechanics** - Ana Bela Cruzeiro & Jean-Claude Zambrini
  - **Geometric Mechanics** - Géry de Saxcé & Zdravko Terze
  - **New trends in Nonholonomic Systems** - Manuel de Leon & Leonardo Colombo

- **Geometry, Learning Dynamics & Thermodynamics**

  - **Symplectic Structures of Heat & Information Geometry** - Frédéric Barbaresco & Pierre Bieliavsky
  - **Geometric Methods in Mechanics and Thermodynamics** - François Gay-Balmaz & Hiroaki Yoshimura
  - **Fluid Mechanics and Symmetry** - François Gay-Balmaz & Cesare Tronci
  - **Learning of Dynamic Processes** - Lyudmila Grigoryeva

- **Quantum Information Geometry**

  - **The Geometry of Quantum States** - Florio M. Ciaglia
  - **Integrable Systems and Information Geometry (From Classical to Quantum)** - Jean-Pierre Francoise, Daisuke Tarama

- **Geometry & Biological Structures**

  - **Neurogeometry** - Alessandro Sarti, Giovanna Citti & Giovanni Petri
  - **Bio-Molecular Structure Determination by Geometric Approaches** - Antonio Mucherino
  - **Geometric Features Extraction in Medical Imaging** - Stéphanie Jehan-Besson & Patrick Clarysse

- **Geometry & Applications**

  - **Applied Geometric Learning** - Pierre-Yves Lagrave, Santiago Velasco-Forero & Teodora Petrisor

June 2023

Frank Nielsen
Frédéric Barbaresco

The original version of the book was revised: the book was inadvertently published with a typo in the frontmatter. This has been corrected. The correction to the book is available at https://doi.org/10.1007/978-3-031-38271-0_62

Organization

Conference Co-chairs

Frank Nielsen   Sony Computer Science Laboratories Inc., Japan
Frédéric Barbaresco   Thales Land & Air Systems, France

Local Organizing Committee

SEE Groupe Régional GRAND OUEST
Christophe Laot, SEE & IMT Atlantique, France
Alain Alcaras, SEE & THALES SIX, France
Jacques Claverie, SEE & CREC St-Cyr Cöetquidan, France
Palais du Grand Large Team, Saint-Malo

Secretariat

Imene Ahmed   SEE, France

Scientific Committee

Bijan Afsari   Johns Hopkins University, USA
Pierre-Antoine Absil   Université Catholique de Louvain, Belgium
Jesus Angulo   Mines ParisTech, France
Nihat Ay   Max Planck Institute, Germany
Simone Azeglio   ENS Paris, France
Frédéric Barbaresco   Thales Land & Air Systems, France
Pierre Baudot   Median Technologies, France
Daniel Bennequin   Paris-Diderot University, France
Pierre Bieliavsky   Université Catholique de Louvain, Belgium
Michel Boyom   Montpellier University, France
Goffredo Chirco   University of Naples Federico II, Italy
Florio M. Ciaglia   Max Planck Institute, Germany
Nicolas Couellan   ENAC, France
Ana Bela Ferreira Cruzeiro Universidade de Lisboa, Portugal
Ariana Di Bernardo   ENS Paris, France

Zdravko Terze	University of Zagreb, Croatia
Alain Trouvé	Ecole Normale Supérieure, France
Alice Barbara Tumpach	Université de Lille, France
Hông Vân Lê	Institute of Mathematics of Czech Academy of Sciences, Czech Republic
Geert Verdoolaege	Ghent University, Belgium
Hiroaki Yoshimura	Waseda University, Japan
Jean Claude Zambrini	Universidade de Lisboa, Portugal
Jun Zhang	University of Michigan, Ann Arbor, USA

GSI'23 Keynote Speakers

Information Theory with Kernel Methods

Francis Bach

Inria, Ecole Normale Supérieure

Abstract. Estimating and computing entropies of probability distributions are key computational tasks throughout data science. In many situations, the underlying distributions are only known through the expectation of some feature vectors, which has led to a series of works within kernel methods. In this talk, I will explore the particular situation where the feature vector is a rank-one positive definite matrix, and show how the associated expectations (a covariance matrix) can be used with information divergences from quantum information theory to draw direct links with the classical notions of Shannon entropies.

Reference

1. Francis, B.: Information theory with kernel methods. To appear IEEE Trans. Inf. Theor (2022). https://arxiv.org/pdf/2202.08545

From Alan Turing to Contact Geometry: Towards a "Fluid Computer"

Eva Miranda

Universitat Politècnica de Catalunya and Centre de Recerca Matemàtica

Abstract. Is hydrodynamics capable of performing computations? (Moore 1991) Can a mechanical system (including a fluid flow) simulate a universal Turing machine? (Tao, 2016) Etnyre and Ghrist unveiled a mirror between contact geometry and fluid dynamics reflecting Reeb vector fields as Beltrami vector fields. With the aid of this mirror, we can answer in the positive the questions raised by Moore and Tao. This is a recent result that mixes up techniques from Alan Turing with modern Geometry (contact geometry) to construct a "Fluid computer" in dimension 3. This construction shows, in particular, the existence of undecidable fluid paths. I will also explain applications of this mirror to the detection of escape trajectories in Celestial Mechanics (for which I'll need to extend the mirror to a singular set-up). This mirror allows us to construct a tunnel connecting problems in Celestial Mechanics and Fluid Dynamics.

References

1. Robert, C., Eva, M., Daniel, P.-S., Francisco,.P.: Constructing turing complete euler flows in dimension 3. Proc. Natl. Acad. Sci. USA **118**(19), 9. Paper No. e2026818118 (2021)
2. Etnyre, J., Ghrist, R.: Contact topology and hydrodynamics: I. Beltrami fields and the Seifert conjecture. Nonlinearity **13**, 441 (2000)
3. Miranda, E., Oms, C., Peralta-Salas, D.: On the singular Weinstein conjecture and the existence of escape orbits for b-Beltrami fields. Commun. Contemp. Math. **24**(7), 25. Paper No. 2150076 (2022)
4. Tao, T.: Finite time blowup for an averaged three-dimensional Navier–Stokes equation. J. Am. Math. Soc. **29**, 601–674 (2016)
5. Turing, A.: On computable numbers, with an application to the entscheidungsproblem. Proc. London Math. Soc. **s2–42**(1), 230–265 (1937). DOI:10.1112/plms/s2-42.1.230 ISSN 0024-6115

Transverse Poisson Structures to Adjoint Orbits in a Complex Semi-simple Lie Algebra

Hervé Sabourin

Director for Strategic projects of the Réseau Figure® (network of 31 universities)
Former Regional Director of the A.U.F. (Agence Universitaire de la Francophonie)
for the Middle East
Former Vice-President of the University of Poitiers, France

Abstract. The notion of transverse Poisson structure has been introduced by Alan Weinstein stating in his famous splitting theorem that any Poisson Manifold M is, in the neighbourhood of each point m, the product of a symplectic manifold, the symplectic leaf S at m, and a submanifold N which can be endowed with a structure of Poisson manifold of rank 0 at m. N is called a transverse slice at M of S. When M is the dual of a complex Lie algebra g equipped with its standard Lie-Poisson structure, we know that the symplectic leaf through x is the coadjoint G. x of the adjoint Lie group G of g. Moreover, there is a natural way to describe the transverse slice to the coadjoint orbit and, using a canonical system of linear coordinates $(q1, \ldots, qk)$, it follows that the coefficients of the transverse Poisson structure are rational in $(q1, \ldots, qk)$. Then, one can wonder for which cases that structure is polynomial. Nice answers have been given when g is semi-simple, taking advantage of the explicit machinery of semi-simple Lie algebras. One shows that a general adjoint orbit can be reduced to the case of a nilpotent orbit where the transverse Poisson structure can be expressed in terms of quasihomogeneous polynomials. In particular, in the case of the subregular nilpotent orbit the Poisson structure is given by a determinantal formula and is entirely determined by the singular variety of nilpotent elements of the slice.

References

1. Sabourin, H.: Sur la structure transverse à une orbite nilpotente adjointe. Canad. J. Math. **57**(4), 750–770 (2005)
2. Sabourin, H.: Orbites nilpotentes sphériques et représentations unipotentes associées : Le cas SL(n). Represent. Theor. **9**, 468–506 (2005)
3. Sabourin, H.: Mémoire d'HDR, Quelques aspects de la méthode des orbites en théorie de Lie, Décembre (2005)
4. Damianou, P., Sabourin, H., Vanhaecke, P.: Transverse poisson structures to adjoint orbits in semi-simple Lie algebras, Pacific J. Math. **232**, 111–139 (2007)

5. Sabourin, H., Damianou, P., Vanhaecke, P.: Transverse poisson structures: the sub-regular and the minimal orbits, differential geometry and its applications. Proc. Conf. Honour Leonhard Euler, Olomouc, August (2007)
6. Sabourin, H., Damianou, P., Vanhaecke, P.: Nilpotent orbits in simple Lie algebras and their transverse poisson structures. Am. Inst. Phys. Conf. Proc. Ser. **1023**, 148–152 (2008)

Statistics Methods for Medical Image Processing and Reconstruction

Diarra Fall

Institut Denis Poisson, UMR CNRS, Université d'Orléans & Université de Tours, France

Abstract. In this talk we will see how statistical methods, from the simplest to the most advanced ones, can be used to address various problems in medical image processing and reconstruction for different imaging modalities. Image reconstruction allows the images in question to be obtained, while image processing (on the already reconstructed images) aims at extracting some information of interest. We will review several statistical methods (mainly Bayesian) to address various problems of this type.

Keywords: Image processing · Image reconstruction · Statistics · Frequentist · Bayesian · Parametrics · Nonparametrics

References

1. Fall, M.D., Dobigeon, N., Auzou, P.: A bayesian estimation formulation to voxel-based lesion symptom mapping. In: Proceedings of European Signal Processing Conference (EUSIPCO), Belgrade, Serbia, September (2022)
2. Fall, M.D.: Bayesian nonparametrics and biostatistics: the case of PET imaging. Int. J. Biostat. (2019)
3. Fall, M.D., Lavau, E., Auzou, P.: Voxel-based lesion-symptom mapping: a nonparametric bayesian approach. In: Proceedings of IEEE International Conference on Acoustics, Speech and Signl Processing (ICASSP) (2018)

Algebraic Statistics and Gibbs Manifolds

Bernd Sturmfels

MPI-MiS Leipzig, Germany

Abstract. Gibbs manifolds are images of affine spaces of symmetric matrices under the exponential map. They arise in applications such as optimization, statistics and quantum physics, where they extend the ubiquitous role of toric geometry. The Gibbs variety is the zero locus of all polynomials that vanish on the Gibbs manifold. This lecture gives an introduction to these objects from the perspective of Algebraic Statistics.

References

1. Pavlov, D., Sturmfels, B., Telen, S.: Gibbs manifolds. arXiv:2211.15490
2. Sturmfels, B., Telen, S., Vialard, F.-X., von Renesse, M.: Toric geometry of entropic regularization. arXiv:2202.01571
3. Sullivant, S.: Algebraic Statistics. graduate studies in mathematics, Am. Math. Soc. Providence, RI, 194 (2018)
4. Huh, J., Sturmfels, B.: Likelihood geometry, in combinatorial algebraic geometry. In: Conca, A., et al. Lecture Notes in Mathematics, vol. 2108, Springer, pp. 63–117 (2014)
5. Geiger, D., Meek, C., Sturmfels, B.: On the toric algebra of graphical models, Annal. Stat. **34**, 1463–1492 (2006)

Learning of Dynamic Processes

Juan-Pablo Ortega

Head, Division of Mathematical Sciences, Associate Chair (Faculty), School of Physical and Mathematical Sciences, Nanyang Technological University, Singapore

Abstract. The last decade has seen the emergence of learning techniques that use the computational power of dynamical systems for information processing. Some of those paradigms are based on architectures that are partially randomly generated and require a relatively cheap training effort, which makes them ideal in many applications. The need for a mathematical understanding of the working principles underlying this approach, collectively known as Reservoir Computing, has led to the construction of new techniques that put together well-known results in systems theory and dynamics with others coming from approximation and statistical learning theory. In recent times, this combination has allowed Reservoir Computing to be elevated to the realm of provable machine learning paradigms and, as we will see in this talk, it also hints at various connections with kernel maps, structure-preserving algorithms, and physics-inspired learning.

References

1. Gonon, L., Grigoryeva, L., Ortega, J.-P.: Approximation bounds for random neural networks and reservoir systems. To appear in The Annals of Applied Probability. Paper (2022)
2. Cuchiero, C., Gonon, L., Grigoryeva, L., Ortega, J.-P., Teichmann, J.: Expressive power of randomized signature. NeurIPS. Paper (2021)
3. Cuchiero, C., Gonon, L., Grigoryeva, L., Ortega, J.-P., Teichmann, J.: Discrete-time signatures and randomness in reservoir computing. IEEE Trans. Neural Netw. Learn. Syst. **33**(11), 6321–6330. Paper (2021)
4. Gonon, L., Ortega, J.-P.: Fading memory echo state networks are universal. Neural Netw. **138**, 10–13. Paper (2021)
5. Gonon, L., Grigoryeva, L., Ortega, J.-P.: Risk bounds for reservoir computing. J. Mach. Learn. Res. **21**(240), 1–61. Paper (2020)
6. Gonon, L., Ortega, J.-P.: Reservoir computing universality with stochastic inputs. IEEE Trans. Neural Netw. Learn. Syst. **31**(1), 100–112. Paper (2020)
7. Grigoryeva, L., Ortega, J.-P.: Differentiable reservoir computing. J. Mach. Learn. Res. **20**(179), 1–62. Paper (2019)

8. Grigoryeva, L., Ortega, J.-P.: Echo state networks are universal. Neural Netw. **108**, 495–508. Paper (2018)
9. Grigoryeva, L., Ortega, J.-P.: Universal discrete-time reservoir computers with stochastic inputs and linear readouts using non-homogeneous state-affine systems. J. Mach. Learn. Res. **19**(24), 1–40. Paper (2018)

Pierre Louis Moreau de Maupertuis, King's Musketeer Lieutenant of Science and Son of a Saint-Malo Corsaire

« Héros de la physique, Argonautes nouveaux/Qui franchissez les monts, qui traversez les eaux/Dont le travail immense et l'exacte mesure/De la Terre étonnée ont fixé la figure./Dévoilez ces ressorts, qui font la pesanteur./Vous connaissez les lois qu'établit son auteur. » ... [Heroes of physics, new Argonauts/Who cross the mountains, who cross the waters/Whose immense work and the exact measure/Of the astonished Earth fixed the figure./Reveal these springs, which make gravity./You know the laws established by its author.] - Voltaire on Pierre Louis Moreau de Maupertuis

Son of René Moreau de Maupertuis (1664–1746) a corsair and ship owner from Saint-Malo, director of the Compagnie des Indes and knighted by Louis XIV, Maupertuis was offered a cavalry regiment at the age of twenty. His father, with whom he had a very close relationship, thus opened the doors of the gray musketeers to him, of which he became lieutenant. Between 1718 and 1721, Maupertuis devoted himself to a military career, first joining the company of gray musketeers, then a cavalry regiment in Lille, without abandoning his studies. In 1718, Maupertuis entered the gray musketeers, writes Formey in his Éloge (1760), but he carried there the love of study, and above all the taste for geometry. However, his profession as a soldier was not to last long and at the end of 1721, the learned Malouin finally and permanently went to Paris, as he could not last long in the idleness of the state of a former military officer in time of peace, and soon he took leave of it. This moment marks the official entry of Maupertuis into Parisian intellectual life, halfway between the literary cafés and the benches of the Academy. He nevertheless preferred to abandon this military career to devote himself to the study of mathematics, an orientation crowned in 1723 by his appointment as a member of the Academy of Sciences.

He then published various works of mechanics and astronomy. In 1728, Maupertuis visited London, a trip which marked a decisive turning point in his career. Elected associate member of the Royal Society, he discovered Newton's ideas, in particular universal attraction, of which he was to become an ardent propagandist in France, which D'Alembert, in the Discourse preliminary to the Encyclopedia, did not miss. Academician at 25, Pierre-Louis Moreau de Maupertuis led a perilous expedition to Lapland to verify Newton's theory and became famous as "the man who flattened the earth". Called by Frederick II to direct the Berlin Academy of Sciences, he was as comfortable in the royal courts as in the Parisian salons.

The rejection of the Newtonian approach, as well as the distrust of the Cartesian approach, led Maupertuis to the elaboration of a cosmology different from both the finalism of some and the anti-finalism of others. It is a cosmology that cannot be attributed to any particular tradition, and that must rather be read as an independent and creative elaboration. All of Maupertuis' cosmology is based on a physical principle which he was the first to formulate, namely the principle of least action, the novelty and generality of which he underlines on several occasions.

His "principle of least action" constitutes an essential contribution to physics to this day, a fundamental principle in classical mechanics. It states that the motion of a particle between two points in a conservative system is such that the action integral, defined as the integral of the Lagrangian over the time interval of motion, is minimized. Maupertuis' principle was renewed by the Cartan-Poincaré Integral Invariant in the field of geometric mechanics. In geometric mechanics, the motion of a mechanical system is described in terms of differential forms on a configuration manifold and the Cartan-Poincaré integral invariant is associated with a particular differential form called the symplectic form, which encodes the dynamics of the system. The integral invariant is defined as the integral of the symplectic form over a closed loop in the configuration manifold. More recently, Maupertuis' principle has been extended more recently by Jean-Marie Souriau through Maxwell's principle with the hypothesis that the exterior derivative of the Lagrange 2-form of a general dynamical system vanishes. For systems of material points, Maxwell's principle allows us, under certain conditions, to define a Lagrangian and to show that the Lagrange form is nothing else than the exterior derivative of the Cartan form, in the study of the calculus of variations. Without denying the importance of the principle of least action nor the usefulness of these formalisms, Jean-Marie Souriau declares that Maupertuis' principle and least action principle seem to him less fundamental than Maxwell's principle. His viewpoint seems to him justified because the existence of a Lagrangian is ensured only locally, and because there exist important systems, such as those made of particles with spin, to which Maxwell's principle applies while they have not a globally defined Lagrangian. Jean-Marie Souriau has also geometrized Noether's theorem (algebraic theorem proving that we can associate invariants to symmetries) with "moment map" (components of moment map are Noether's invariants).

« La lumière ne pouvant aller tout-à-la fois par le chemin le plus court, et par celui du temps le plus prompt ... ne suit-elle aucun des deux, elle prend une route qui a un avantage plus réel : le chemin qu'elle tient est celui par lequel la quantité d'action est la moindre. » [Since light cannot go both by the shortest path and by that of the quickest time... if it does not follow either of the two, it takes a route which has a more real advantage: the path that it holds is that by which the quantity of action is least.] - Maupertuis 1744.

ALEAE GEOMETRIA – BLAISE PASCAL's 400th Birthday

We celebrate in 2023 Blaise Pascal's 400th birthday. GSI'23 motto is "ALEA GEOME-TRIA".

In 1654, Blaise Pascal submitted a paper to « Celeberrimae matheseos Academiae Parisiensi » entitled « ALEAE GEOMETRIA : De compositione aleae in ludis ipsi subjectis »

- « … et sic matheseos demonstrationes cum aleae incertitudine jugendo, et quae contraria videntur conciliando, ab utraque nominationem suam accipiens, stupendum hunc titulum jure sibi arrogat: **Aleae Geometria** »
- « … par l'union ainsi réalisée entre les démonstrations des mathématiques et l'incertitude du hasard, et par la conciliation entre les contraires apparents, elle peut tirer son nom de part et d'autre et s'arroger à bon droit ce titre étonnant: **Géométrie du Hasard** »
- « … by the union thus achieved between the demonstrations of mathematics and the uncertainty of chance, and by the conciliation between apparent opposites, it can take its name from both sides and arrogate to right this amazing title: **Geometry of Chance** »

Blaise Pascal had a multi-disciplinary approach of Science, and has developed 4 topics directly related to GSI'23:*

- **Blaise Pascal and COMPUTER:** Pascaline marks the beginning of the development of mechanical calculus in Europe, followed by Charles Babbage analytical machine from 1834 to 1837, a programmable calculating machine combining the inventions of Blaise Pascal and Jacquard's machine, with instructions written on perforated cards.
- **Blaise Pascal and PROBABILITY:** The "calculation of probabilities" began in a correspondence between Blaise Pascal and Pierre Fermat. In 1654, Blaise Pascal submitted a short paper to "Celeberrimae matheseos Academiae Parisiensi" with the title "Aleae Geometria" (Geometry of Chance), that was the seminal paper founding Probability as a new discipline in Science.
- **Blaise Pascal and THERMODYNAMICS:** Pascal's Experiment in the Puy de Dôme to Test the Relation between Atmospheric Pressure and Altitude. In 1647, Blaise Pascal suggests to raise Torricelli's mercury barometer at the top of the Puy de Dome Mountain (France) in order to test the "weight of air" assumption.
- **Blaise Pascal and DUALITY:** Pascal's Hexagrammum Mysticum Theorem, and its dual Brianchon's Theorem. In 1639 Blaise Pascal discovered, at age sixteen, the famous hexagon theorem, also developed in "Essay pour les Coniques", printed in 1640, declaring his intention of writing a treatise on conics in which he would derive the major theorems of Apollonius from his new theorem.

The GSI'23 Conference is Dedicated to the Memory of Mademoiselle Paulette Libermann, Geometer Student of Elie Cartan and André Lichnerowicz, PhD Student of Charles Ehresmann and Familiar with the Emerald Coast of French Brittany

Paulette Libermann died on July 10, 2007 in Montrouge near Paris. Admitted to the entrance examination to the Ecole Normale Supérieure de Sèvres in 1938, she was a pupil of Elie Cartan and André Lichnerowicz. Paulette Libermann was able to learn about mathematical research under the direction of Elie Cartan, and was a faithful friend of the Cartan family. After her aggregation, she was appointed to Strasbourg and rubbed shoulders with Georges Reeb, René Thom and Jean-Louis Koszul. She prepared a thesis under the direction of Charles Ehresmann, defended in 1953. She was the first ENS Sèvres woman to hold a doctorate in mathematics. She was then appointed professor at the University of Rennes and after at the Faculty of Sciences of the University of Paris in 1966. She began to collaborate with Charles-Michel Marle in 1967. She led a seminar with Charles Ehresmann until his death in 1979, and then alone until 1990. In her thesis, entitled "On the problem of equivalence of regular infinitesimal structures", she studied the symplectic manifolds provided with two transverse Lagrangian foliations and showed the existence, on the leaves of these foliations, of a canonical flat connection. Later, Dazord and Molino, in the South-Rhodanian geometry seminar, introduced the notion of Libermann foliation, linked to Stefan foliations and Haefliger Γ-structures. Paulette Libermann also deepened the importance of the foliations of a symplectic manifold which she called "simplectically complete", such as the Poisson bracket of two functions, locally defined, constant on each leaf, that is also constant on each leaf. She proved that this property is equivalent to the existence of a Poisson structure on the space of leaves, such that the canonical projection is a Poisson map, and also equivalent to the complete integrability of the subbundle symplectically orthogonal to the bundle tangent to the leaves. She wrote a famous book with Professor Charles-Michel Marle, "Symplectic Geometry and Analytical Mechanics". Professor Charles-Michel Marle told us that Miss Paulette Libermann had bought an apartment in Dinard and spent her summers just in front of Saint-Malo, and so was familiar with the emerald coast of French Brittany.

GSI'23 Sponsors

THALES (https://www.thalesgroup.com/en) and European Horizon CaLIGOLA (https://site.unibo.it/caligola/en) were both PLATINIUM SPONSORS of the SEE GSI'23 conference.

Contents – Part I

Statistics, Information and Topology

Information Theory and Statistics

Statistical Shape Analysis and more Non-Euclidean Statistics

Probability and Statistics on Manifolds

Computing Geometry and Algebraic Statistics

Geometric and Analytical Aspects of Quantization and Non-Commutative Harmonic Analysis on Lie Groups

Deep Learning: Methods, Analysis and Applications to Mechanical Systems

Contents – Part II

Fluid Mechanics and Symmetry

Integrable Systems and Information Geometry (From Classical to Quantum)

Neurogeometry

Geometric Green Learning

Three Methods to Put a Riemannian Metric on Shape Space

Alice Barbora Tumpach[1,2]([✉]) and Stephen C. Preston[3]

[1] Institut CNRS Pauli, Vienna, Austria
`alice-barbora.tumpach@univ-lille.fr`
[2] University of Lille, Lille, France
[3] Brooklyn College and CUNY Graduate Center, New York, USA
`stephen.preston@brooklyn.cuny.edu`

Abstract. In many applications, one is interested in the shape of an object, like the contour of a bone or the trajectory of joints of a tennis player, irrespective of the way these shapes are parameterized. However for analysis of these shape spaces, it is sometimes useful to have a parameterization at hand, in particular if one is interested in deforming shapes. The purpose of the paper is to examine three different methods that one can follow to endow shape spaces with a Riemannian metric that is measuring deformations in a parameterization independent way. The first is via Riemannian submersion on a quotient; the second is via isometric immersion on a particular slice; and the third is an alternative method that allows for an arbitrarily chosen complement to the vertical space and a metric degenerate along the fibers, which we call the gauge-invariant metric. This allows some additional flexibility in applications, as we describe.

Keywords: Shape space · Geometric green learning · Geometric invariants

1 Introduction, Motivation, and a Simple Example

In this paper we will describe three ways to think about geometry on a quotient space of a trivial principal bundle, with application to shape space. The first is the standard approach via quotients by a group and a Riemannian submersion. The second is by considering a particular global section of the bundle and inducing a metric by isometric immersion. The third is newer and consists of specifying a normal bundle complementary to the vertical bundle, projecting the metric onto the normal bundle, and taking the quotient of the resulting degenerate metric; we refer to this as the gauge-invariant approach. We will begin with some motivations about our main concern of shape space before presenting an explicit example in finite dimensions to fix ideas. Then we describe the three basic methods as (I), (II), and (III), and finally we discuss how to get from one to another and the meaning of gauge invariance.

Supported by FWF grant I 5015-N, Institut CNRS Pauli and University of Lille.

F. Nielsen and F. Barbaresco (Eds.): GSI 2023, LNCS 14071, pp. 3–11, 2023.
https://doi.org/10.1007/978-3-031-38271-0_1

Group G	Elements of an orbit under G	one preferred element in the orbit	another choice of preferred element
$\mathbb{R}^2$ acting by translations	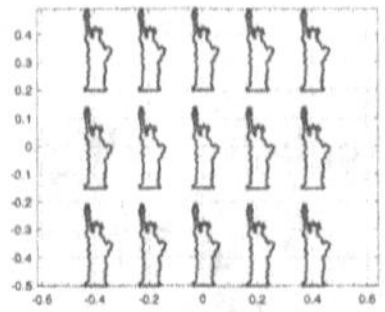	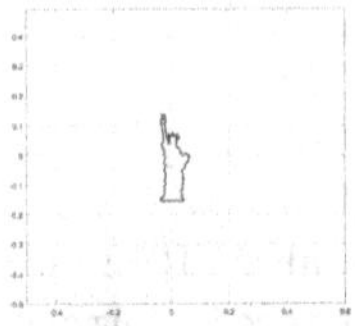 centroid at origin	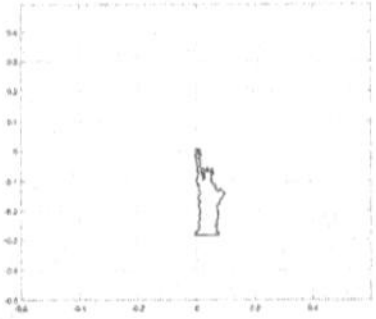 curve starting at origin
SO(2) acting by rotations		 axes of approximating ellipse aligned	 tangent vector at starting point horizontal
$\mathbb{R}^+$ acting by scaling	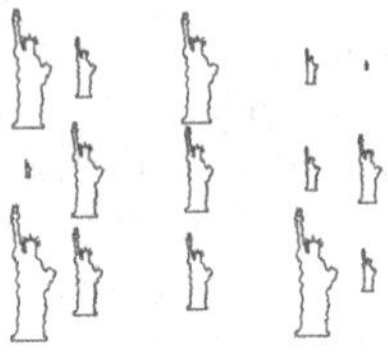	 length = 1	 enclosed area =1

Fig. 1. Examples of group actions on 2D simple closed curves and different choices of sections of the corresponding fiber bundle.

In order to explain the ideas of the present paper in a simple way, we will consider the contour of the Statue of Liberty appearing in different layouts in Fig. 1. Imagine a camera that scans a photo of the Statue and needs to recognize it regardless of how the photo is held (at any distance, position, or rotation angle). If we require the photo to be held perpendicular to the camera lens, then the transformations are rotations, translations, and rescalings. These are all shape-preserving and we would like the scanner to be able to detect the shape independently of them. We can handle this in two ways:

1. either one groups together the photos that are transformations of each other,
2. or one specifies a preferred choice of position in space and/or scale as a representative.

The first option consists of considering the *orbit* of the photo under the group; the second option consists of considering a preferred *section* of the quotient space of curves modulo the group. These two different ways of thinking about shapes modulo a given group of transformations are illustrated in Fig. 1. In the second column we show some examples of the photos in the same orbit under the action of the group. In the third and fourth columns, a representative of this orbit is singled out.

Then, when considering contours of objects, another group acting by shape-preserving transformations is the group of reparameterizations of the contour. Making the analysis invariant by the group of reparameterizations is a much more difficult problem than that in the previous paragraph, but it has the same essential nature, and is the main source of motivation for us.

Corresponding mathematical objects. The mathematical picture to start with is the following: the group of shape-preserving transformations G is acting on the space of curves or surfaces $\mathcal{F}$ and the shape space $\mathcal{S}$ that retain just the informations that we need is the *quotient space* $\mathcal{S} := \mathcal{F}/G$. The map that sends a curve to its orbit under the group G is called the *canonical projection* and will be denoted be $p : \mathcal{F} \to \mathcal{S}$. The *orbit* of a element $f \in \mathcal{F}$ will also be denoted by $[f] \in \mathcal{F}/\mathcal{S}$, in particular $p(f) = [f]$ for any $f \in \mathcal{F}$. The triple $(p, \mathcal{F}, \mathcal{S})$ is a particular example of fiber bundle attached to a smooth action of a group on a manifold.

When we specify which procedure we follow to choose a representation of each orbit, one is selecting a preferred *section* of the fiber bundle $p : \mathcal{F} \to \mathcal{S}$. A global section of the fiber bundle $p : \mathcal{F} \to \mathcal{S}$ is a smooth application $s : \mathcal{S} \to \mathcal{F}$, such that $p \circ s([f]) = [f]$ for any $[f] \in \mathcal{S}$. There is one-to-one correspondance between the shape space $\mathcal{S}$ and the range of s. Defining a global section of $p : \mathcal{F} \to \mathcal{S}$ is in fact defining a way to choose a preferred element in the fiber $p^{-1}([f])$ over $[f]$. In the case of the group of reparameterizations, it consists of singling out a preferred parameterization of each oriented shape.

Why do we care about the distinction? Depending on the representation of shape space as a quotient space or as a preferred section, shape analysis may give different results. Very often, curves or surfaces are centered and scaled as a pre-processing step. However, the procedure to center or scale the shapes may influence further analysis. For instance, a Statue of Liberty whose contour has a fractal behaviour will appear very small if scaling variability is taken care of by fixing the length of the curve to 1 and will seem visually very different to analogous statues with smooth boundaries.

Example 1. We begin with the simplest nontrivial example of the three methods we have in mind for producing a metric on the quotient space by a group action, given a metric on the full space. Here our full space will be the Heisenberg group $\mathcal{F} \cong \mathbb{R}^3$ with the left-invariant metric

$$ds^2 = dx^2 + dy^2 + (dz - y\,dx)^2 \tag{1}$$

on it, while the group action is vertical translation in the z-direction by a real number, generated by the flow of the vector field $\xi = \partial_z$. Hence the group is $G = \mathbb{R}$ under addition, and the quotient space is $\mathcal{S} = \mathbb{R}^2$ with projection $p(x, y, z) = (x, y)$. We will denote by $\{e_1, e_2\}$ the canonical basis of $\mathbb{R}^2$. The metric (1) is invariant under this action since none of the components depend on z.

(I) At every point the field ξ is vertical since it projects to zero. Horizontal vectors are those orthogonal to this in the metric (1), and the horizontal bundle is spanned by the fields $h_1 = \partial_x + y\,\partial_z$ and $h_2 = \partial_y$. This basis is special since $Dp(h_1) = e_1$ and $Dp(h_2) = e_2$, so we get the usual basis on the quotient. If the inner product on $\{e_1, e_2\}$ comes from the inner product on $\{h_1, h_2\}$, the result is the Riemannian submersion quotient metric

$$ds^2 = du^2 + dv^2. \tag{2}$$

(II) The second way to get a natural metric on the quotient $\mathcal{S}$ is to embed it back into $\mathcal{F}$ by a section $s\colon \mathcal{S} \to \mathcal{F}$, so that $p \circ s$ is the identity. All such sections are given by the graph of a function $(x, y, z) = s(u, v) = (u, v, \psi(u, v))$ for some $\psi\colon \mathbb{R}^2 \to \mathbb{R}$, a choice of a particular representative $z = \psi(u, v)$ in the equivalence class $\pi^{-1}(u, v)$. The image of s is a submanifold of $\mathcal{F}$ which we denote by $\mathcal{A}$, and it inherits the isometric immersion metric

$$ds^2 = du^2 + dv^2 + \left[\psi_v\,dv + (\psi_u - v)\,du\right]^2. \tag{3}$$

(III) The third way to get a metric is to declare that movement in the z direction will be "free," and only movement transverse to the vertical direction will have some cost. This corresponds to specifying a space of normal vectors along each fiber (arbitrary except that it is transverse to the tangent vectors ∂_z). Any normal bundle is generated by the span of vector fields of the form $n_1 = \partial_x + \varphi_1(x, y)\,\partial_z$ and $n_2 = \partial_y + \varphi_2(x, y)\,\partial_z$ for some functions $\varphi_1, \varphi_2\colon \mathbb{R}^2 \to \mathbb{R}$ independent of z to ensure G-invariance. Again this basis is special since $\pi_*(n_1) = e_1$ and $\pi_*(n_2) = e_2$. To measure movement only in the normal direction, we define $g_{GI}(U, V) = g_{\mathcal{F}}\big(p_N(U), p_N(V)\big)$ for any vectors U and V, where p_N is the projection onto the normal bundle parallel to the vertical direction. This results in the degenerate metric

$$ds^2 = dx^2 + dy^2 + \left[\varphi_2\,dy + (\varphi_1 - y)\,dx\right]^2. \tag{4}$$

This formula then induces a nondegenerate quotient metric on the quotient $\mathbb{R}^2$.

It is clear that (III) is the most general choice, and that both (I) and (II) are special cases. The metric (4) matches (2) when the normal bundle coincides with the horizontal bundle, and otherwise is strictly larger. Meanwhile the immersion metric (II) in (3) is strictly larger than (I), and no choice of ψ will reproduce it since ψ would have to satisfy $\psi_v = 0$ and $\psi_u = v$. This is a failure of integrability, see Sect. 3.

2   Different Methods to Endow a Quotient with a Riemannian Metric

In this section, we will suppose that we have at our disposal a Riemannian metric $g_{\mathcal{F}}$ on the space of curves or surfaces $\mathcal{F}$ we are interested in, and that this metric is invariant under a group of shape-preserving transformations G.

In other words, G preserves the metric $g_{\mathcal{F}}$, i.e. G acts by isometries on $\mathcal{F}$. We will explain three different ways to endow the quotient space $\mathcal{S} := \mathcal{F}/G$ with a Riemannian metric.

The action of a group G on a space of curves or surfaces $\mathcal{F}$ will be denoted by a dot. For instance, if G is the group of translations acting on curves in $\mathbb{R}^2$, $g \cdot F = F + C$ where C is the constant function given by the coordinates of the vector of translation defined by the translation g. When G is the group of reparameterizations, then $g \cdot F := F \circ g^{-1}$.

(I) Quotient Riemannian metric. The first way to endow the quotient space $\mathcal{S} := \mathcal{F}/G$ with a Riemannian metric is through the *quotient Riemannian metric*. We recall the following classical Theorem of Riemannian geometry [5,11].

Theorem 1 (Riemannian submersion Theorem). *Let $\mathcal{F}$ be a manifold endowed with a Riemannian metric $g_{\mathcal{F}}$, and G a Lie group acting on $\mathcal{F}$ in such a way that $\mathcal{F}/G$ is a smooth manifold. Suppose $g_{\mathcal{F}}$ is G-invariant and $T_F\mathcal{F}$ splits into the direct sum of the tangent space to the fiber and its orthogonal complement, i.e.,*

$$g_{\mathcal{F}}(X,Y) = g_{\mathcal{F}}(g \cdot X, g \cdot Y), \forall X, Y \in T\mathcal{F}, \forall g \in G,$$
$$T_F\mathcal{F} = \mathrm{Ker}(dp)_F \oplus \mathrm{Ker}(dp)_F^{\perp}, \forall F \in \mathcal{F}, \tag{5}$$

then there exists a unique Riemannian metric $g_{1,\mathcal{S}}$ on the quotient space $\mathcal{S} = \mathcal{F}/G$ such that the canonical projection $p : \mathcal{F} \to \mathcal{S}$ is a Riemannian submersion, i.e. such that $dp : \mathrm{Ker}(dp)^{\perp} \to T\mathcal{S}$ is an isometry.

In this Theorem, the space $\mathrm{Hor} := \mathrm{Ker}(dp)^{\perp}$ is called the *horizontal space* because it is defined as the orthogonal with respect to $g_{\mathcal{F}}$ of the *vertical space* $\mathrm{Ver} := \mathrm{Ker}(dp)$ (traditionally the fibers of a fiber bundle are depicted vertically). Condition (5) is added in order to deal with the infinite-dimensional case where, for weak Riemannian metrics, this identity is not automatic.

One way to understand the Riemannian submersion Theorem is the following: first, in order to define a Riemannian metric on the quotient space, one looks for a subbundle of $T\mathcal{F}$ which is in bijection with $T\mathcal{S}$. Since the vertical space is killed by the projection, the transverse space to the vertical space given by the orthogonal complement is a candidate. The restriction of the Riemannian metric on it defines uniquely a Riemannian metric on the quotient.

(II) Riemannian metric induced on a smooth section. Now suppose that we have chosen a preferred smooth section $s : \mathcal{S} \to \mathcal{F}$ of the fiber bundle $p : \mathcal{F} \to \mathcal{S} = \mathcal{F}/G$, for instance the space of arc-length parameterized curves in the case where G is the group of orientation-preserving reparameterizations, or the space of centered curves when G is the group of translations. The smoothness assumption means that the range of s is a smooth manifold of $\mathcal{F}$, like the space of arc-length parameterized curves in the space of parameterized curves. We will denote it by $\mathcal{A} := s(\mathcal{S})$. By construction, there is a isomorphism between $\mathcal{S}$ and $\mathcal{A}$ which one can use to endow the quotient space $\mathcal{S}$ with the induced Riemannian structure on $\mathcal{A}$ by $\mathcal{F}$.

Theorem 2 (Riemannian immersion Theorem). *Given a smooth section* $s : \mathcal{S} \to \mathcal{F}$, *there exists a unique Riemannian metric* $g_{\mathcal{A}}$ *on* $\mathcal{A} := s(\mathcal{S})$ *such that the inclusion* $\iota : \mathcal{A} \hookrightarrow \mathcal{F}$ *is an isometry. Using the isomorphism* $s : \mathcal{S} \to \mathcal{A}$, *there exists a unique Riemannian metric* $g_{2,\mathcal{S}}$ *on* $\mathcal{S}$ *such that* $s : \mathcal{S} \to \mathcal{F}$ *is an isometry.*

(III) Gauge invariant metric. Here we suppose that we have a vector bundle Nor over $\mathcal{F}$ which is a G-invariant subbundle of $T\mathcal{F}$ transverse to the vertical bundle Ver := Ker(dp). Using any G-invariant metric $g_{\mathcal{F}}$ on $\mathcal{F}$, one can define a G-invariant metric g_{GI} on $\mathcal{F}$ that is degenerate along the fiber of the projection $p : \mathcal{F} \to \mathcal{S}$. We will explain the meaning of "gauge invariance" later.

Theorem 3. *Let* $g_{\mathcal{F}}$ *be a* G-*invariant metric on* $\mathcal{F}$ *and* Nor $\subset T\mathcal{F}$ *be a* G-*invariant subbundle of* $T\mathcal{F}$ *such that*

$$T_F\mathcal{F} = \mathrm{Ker}(dp)_F \oplus \mathrm{Nor}_F, \forall F \in \mathcal{F}. \tag{6}$$

There exists a unique metric g_{GI} *on* $T\mathcal{F}$ *which coincides with* $g_{\mathcal{F}}$ *on* Nor *and is degenerate exactly along the vertical fibers of* $p : \mathcal{F} \to \mathcal{S}$. *It induces a Riemannian metric* $g_{3,\mathcal{S}}$ *on shape space* $\mathcal{S}$ *such that* $dp :$ Nor $\to T\mathcal{S}$ *is an isometry.*

Since we want the inner product to be the same in g_{GI} as in $g_{\mathcal{F}}$ when the vectors are normal, and zero if either vector is vertical, we define g_{GI} by simply projecting an arbitrary vector onto the normal bundle:

$$g_{GI}(X,Y) = g_{\mathcal{F}}\big(p_{\mathrm{Nor}}(X), p_{\mathrm{Nor}}(Y)\big), \tag{7}$$

where $p_N : T_F\mathcal{F} \to$ Nor is the projection onto the normal bundle parallel to the vertical space. This is nondegenerate on the quotient since the projection onto the quotient is an isomorphism when restricted to the normal bundle.

Remark 1. In the case where Nor = Hor, the Riemannian metric $g_{3,\mathcal{S}}$ coincides with the quotient metric $g_{1,\mathcal{S}}$. Another choice of G-invariant complement to the vertical space will give another Riemannian metric on the quotient space.

Example 2. The main example for shape space consists of the elastic metric first defined in [4] on the space of planar curves $\mathcal{F} = \{F : [0,1] \to \mathbb{R}^2\}$ by the formula

$$g_{\mathcal{F}}^{a,b}(h_1, h_2) = \int_0^1 \left[a(D_s h_1, \boldsymbol{t})(D_s h_2, \boldsymbol{t}) + b(D_s h_1, \boldsymbol{n})(D_s h_2, \boldsymbol{n}) \right] ds,$$

$$F \in \mathcal{F}, \; h_i \in T_F\mathcal{F}, ds = \|F'(t)\| dt, \; D_s h(t) = \frac{\dot{h}(t)}{\|\dot{F}(t)\|}, \; \boldsymbol{t} = \frac{\dot{F}}{\|\dot{F}\|}, \; \boldsymbol{n} = \boldsymbol{t}^{\perp}. \tag{8}$$

See [1] for a recent survey of its properties. We will follow [6,10] below.

Our group G is the (orientation-preserving) reparameterizations of all these curves, since we only care about the image $F[0,1]$, and the shape space is the quotient $\mathcal{F}/G$. At any $F \in \mathcal{F}$ the vertical space is Ver $= \{m\boldsymbol{t} \,|\, m : [0,1] \to \mathbb{R}\}$. A natural section $s : \mathcal{S} \to \mathcal{A} \subset \mathcal{F}$ comes from parameterizing all curves proportional to arc length. The tangent space $T\mathcal{A}$ to the space of arc-length parameterized curves is the space of vector fields w along F such that $w' \cdot \boldsymbol{t} = 0$. The horizontal bundle in the metric (8) is given at each F by the space of vector fields w along F such that $\frac{d}{dt}(w' \cdot t) - \frac{b}{a}\kappa(w' \cdot n) = 0$ for all $t \in [0,1]$, where κ is the curvature function. Hence computing the projections requires solving an ODE.

A much simpler normal bundle is obtained by just taking the pointwise normal, i.e., using $\mathrm{Nor} := \{\Phi n \mid \Phi \colon [0,1] \to \mathbb{R}\}$, where now the tangential and normal projections can be computed without solving an ODE.

Instead of parameterizing by arc length we can choose other special parameterizations to get the section; for example using speed proportional to the curvature of the shape as in [9]. An example of application to action recognition is given in [3]. Similar metrics can be defined on surfaces in $\mathbb{R}^3$ to get two-dimensional shape spaces; see for example [7,8].

Remark 2. In the infinite-dimensional case, it is not always possible to find a complement to the vertical space $\mathrm{Ker}(dp)$ as in (6). An example of this phenomenon is provided by shape spaces of non-linear flags (see [2]). In this case, one has to work with the quotient vector spaces $T_F \mathcal{F} / \mathrm{Ker}(dp)_F$. See [11] for this more general case.

3 Relationships of the 3 Methods and Gauge Invariance

3.1 Converting Between (I), (II), and (III)

We have seen in Example 1 that in some cases the three metrics coincide when we start with the *same* base metric $g_{\mathcal{F}}$, but typically they do not. However if we allow the metric on $\mathcal{F}$ to change, we can convert any metric of the form (I), (II), or (III) into a metric of the other forms. Here we demonstrate how to do it.

(I) $\Rightarrow$ (III). If we start with a quotient Riemannian submersion metric arising from $g_{\mathcal{F}}$, how do we get a gauge-invariant metric? We simply define the normal bundle Nor to be the horizontal bundle Hor of vectors orthogonal in $g_{\mathcal{F}}$ to the vertical bundle, and use the projection p_{Nor} as in (7). The new metric g_{GI} on $\mathcal{F}$ will be degenerate but will produce the same metric on the quotient.

(II) $\Rightarrow$ (III). If we start with a section s that embeds the quotient $\mathcal{S}$ into a submanifold $\mathcal{A}$ of $\mathcal{F}$, how do we obtain a gauge-invariant metric? Here we define the normal bundle Nor to be the tangent bundle of the $\mathcal{A}$ and proceed as in (7). Again the new degenerate metric on $\mathcal{F}$ will agree with the induced metric on $\mathcal{A}$ (and in particular be nondegenerate there).

(I) $\Rightarrow$ (II), (III)$\Rightarrow$(II) As in Example 1, a given normal bundle (in particular a horizontal bundle from a metric) may not be the tangent bundle of any manifold due to failure of integrability; hence there may not be any way to express a particular instance of (I) or (III) as a version of (II) with the horizontal space equal to the tangent space of a particular section. The Frobenius integrability condition for the bundle, which is equivalent to the curvature of the corresponding connection vanishing, has to be satisfied in order to build a section tangent to the horizontal space. If one does not require the section to be horizontal, we may proceed as in [10] to pull-back the metric from the quotient to the particular section.

(III) ⇒ **(I)**, **(II)**⇒**(I)**. Both these cases come from the same metric $g_{\mathcal{F}}$, and as in Example 1 they may not coincide. However if we are willing to consider a *different* metric on $\mathcal{F}$, then we can go from method (III) to method (I) and still have the same metric on the quotient. Given a G-invariant normal bundle Nor and a G-invariant metric $g_{2,\mathcal{F}}$ which generates the degenerate metric g_{GI} on $\mathcal{F}$ and a metric $g_{\mathcal{S}}$ by method (III) via (7), we define a new metric $g_{1,\mathcal{F}}$ by choosing any G-invariant Riemannian metric on the tangent bundle Ver to the fiber and by declaring that Ver and Nor are orthogonal. By construction, the subbundle Nor is the horizontal bundle in this metric, and we are in case (I). The same construction works to go from (II) to (I), using the above to get from (II) to (III).

3.2   The Meaning of Gauge Invariance

In geometry on shape space, we are often interested primarily in finding minimizing paths between shapes, and a common algorithm is to construct an initial path between shapes and shorten it by some method. Paths are typically easy to construct in $\mathcal{F}$ and difficult to construct directly in the quotient space $\mathcal{S}$. A section $s\colon \mathcal{S} \to \mathcal{F}$ makes this simpler, but especially if the image $\mathcal{A} = s(\mathcal{S})$ is not flat, it can be difficult to keep the shortening constrained on that submanifold. Our motivating example is when the shapes are parameterized by arc length as in [10], with a path-straightening or gradient descent algorithm based on the metric $g_{\mathcal{F}}$: the optimal reduction gives intermediate curves that are typically no longer parameterized by arc length, and in fact the parameterizations can become degenerate. As such we may want to apply the group action of reparameterization independently on each of the intermediate shapes to avoid this breakdown.

If G is a group acting on a space $\mathcal{F}$, define the *gauge group* as the group $\mathcal{G} := \{g : [0,1] \to G\}$ of paths in G acting on the space of paths $\gamma\colon [0,1] \to \mathcal{F}$ in $\mathcal{F}$ by the obvious formula $(g \cdot \gamma)(t) = g(t) \cdot \gamma(t)$, i.e., pointwise action in the t parameter. We would like the metric on $\mathcal{F}$ to have the property that lengths of paths are invariant under this action, which essentially allows us to change the section s "on the fly" if it's convenient. Since this involves pushing the path in the direction of the fibers, it is intuitively clear that the metric will need to be degenerate in those directions. We have the following proposition, whose proof we defer to [11].

Proposition 1. *The length of a path $[\gamma]$ in $\mathcal{S}$ measured with the quotient metric $g_{1,\mathcal{S}}$ is equal to the length of any lift γ of $[\gamma]$ in $\mathcal{F}$ measured with the gauge invariant metric g_{GI}.*

Remark 3. The length of γ measured with the metric $g_{\mathcal{F}}$ differs from the length of $[\gamma]$ unless γ is also horizontal. In this case, $\gamma = g_0 \cdot \gamma_0$ for a fixed $g_0 \in G$.

4   Conclusion

We have shown how to view the problem of constructing a Riemannian metric on a principal bundle quotient such as shape space in three different ways. Any desired metric on the quotient can be viewed as any one of the three depending on what is computationally convenient. The gauge-invariant method (III) is the most general and flexible, capturing the other two more familiar methods as special cases; it has the advantage that it is convenient for shape space computations without needing to work on the difficult shape space explicitly.

References

1. Bauer, M., Charon, N., Klassen, E., Kurtek, S., Needham, T., Pierron, T.: Elastic metrics on spaces of Euclidean curves: theory and algorithms (2022). arXiv:2209.09862v1
2. Ciuclea, I., Tumpach, A.B., Vizman, C.: Shape spaces of non-linear flags. To appear in GSI23
3. Drira, H., Tumpach, A.B., Daoudi, M.: Gauge invariant framework for trajectories analysis. In: DIFFCV Workshop 2015 (2015). https://doi.org/10.5244/C.29. DIFFCV.6. GIFcurve.pdf
4. Mio, W., Srivastava, A., Joshi, S.H.: On shape of plane elastic curves. Int. J. Comput. Vision **73**, 307–324 (2007)
5. Lang, S.: Fundamentals of Differential Geometry. Springer, New York (1999). https://doi.org/10.1007/978-1-4612-0541-8
6. Preston, S.C.: The geometry of whips. Ann. Global Anal. Geometry **41**(3) (2012)
7. Tumpach, A.B.: Gauge invariance of degenerate Riemannian metrics. Not. Amer. Math. Soc. (2016). http://www.math.univ-lille1.fr/~tumpach/Site/researchfiles/ Noticesfull.pdf
8. Tumpach, A.B., Drira, H., Daoudi, M., Srivastava, A.: Gauge invariant framework for shape analysis of surfaces. IEEE Trans. Pattern Anal. Mach. Intell. **38**(1) (2016)
9. Tumpach, A.B.: On canonical parameterizations of $2D$-shapes To appear in GSI23
10. Tumpach, A.B., Preston, S.C.: Quotient elastic metrics on the manifold of arc-length parameterized plane curves. J. Geom. Mech. **9**(2), 227–256 (2017)
11. Tumpach, A.B., Preston, S.C.: Riemannian metrics on shape spaces: comparison of different constructions. (in preparation)

Riemannian Locally Linear Embedding with Application to Kendall Shape Spaces

Elodie Maignant[1,2]($\boxtimes$) (iD), Alain Trouvé[2], and Xavier Pennec[1] (iD)

[1] Université Côte d'Azur and Inria, Epione Team, Nice, France
elodie.maignant@inria.fr
[2] Centre Borelli, ENS Paris Saclay, Gif-sur-Yvette, France

Abstract. Locally Linear Embedding is a dimensionality reduction method which relies on the conservation of barycentric alignments of neighbour points. It has been designed to learn the intrinsic structure of a set of points of a Euclidean space lying close to some submanifold. In this paper, we propose to generalise the method to manifold-valued data, that is a set of points lying close to some submanifold of a given manifold in which the points are modelled. We demonstrate our algorithm on some examples in Kendall shape spaces.

Keywords: Locally Linear Embedding · Optimisation on Quotient Manifolds · Shape Spaces

1 Introduction

Dimensionality reduction is a critical issue when it comes to data analysis on complex structures. Especially, the data modelled in the context of shape analysis – for example protein conformations or anatomical shapes – are by nature high-dimensional data. Common tools for dimensionality reduction have been originally designed for data described in a Euclidean space. However, objects like shapes are rather naturally described in a manifold. As an example, Kendall manifolds [3] encode the idea that two configurations of points – e.g. two protein conformations – should be compared independently of the coordinate system they are written in. We refer to such data as manifold-valued data. A first approach to process manifold-valued data then consists in embedding them in a larger Euclidean space – or equivalently to work extrinsically. This approach has two main drawbacks. First of all, it ignores the structural information contained in the manifold model, which then may not be well recovered in areas of low sampling density. Moreover, there might be a significant gap in dimensionality between the intrinsic and the extrinsic model in some cases. The manifold of unparameterised curves [5,8] illustrates well this second point as the extrinsic and the intrinsic descriptions differ by the removal of parametrisations – diffeomorphisms – which is an infinite dimensional space. Thus, when the data are modelled in a known manifold, it is relevant to look for a generalisation of existing tools for vector-valued data to manifold-valued data. Locally Linear Embedding (LLE) has been introduced by Roweis and Saul in [7] as a nonlinear

© The Author(s), under exclusive license to Springer Nature Switzerland AG 2023
F. Nielsen and F. Barbaresco (Eds.): GSI 2023, LNCS 14071, pp. 12–20, 2023.
https://doi.org/10.1007/978-3-031-38271-0_2

dimensionality reduction tool. Given a set of points of a vector space sampled from some underlying submanifold of lower dimension, the method leverages the locally linear assumption to characterise each data point as a weighted barycentre of the other points nearby and then embed them in a low-dimensional vector space accordingly. Essentially, both the weights and the embedding are written as solutions of a least square problem such that the algorithm is straightforward to write and implement. LLE differs significantly from other dimension reduction methods, firstly as it relies on an intrinsic description of the data which is local – unlike PCA for example – and secondly as it implements a criterion which is affine rather than metric – as opposed to distance-based methods like Multi-Dimensional Scaling (MDS) or Isomap. Therefore, because LLE preserves local affine relationships rather than distances, we expect it to be able to retrieve different information from the data. While there has already been a consequent work about extending PCA [1] and MDS methods to manifold-valued data, LLE has not been yet generalised to our knowledge. In this paper, we propose a new Riemannian formulation of LLE which we refer to as Riemannian Locally Embedding (RLLE) and we detail an algorithm for the weights estimation. We illustrate our method on two examples in Kendall shape spaces and we evaluate RLLE performance in this setting with respect to the LLE one.

2   Riemannian Locally Linear Embedding

In this section, we recall the algorithm implemented by Locally Linear Embedding (LLE). Since it relies on barycentric coordinates, we extend their definition in order to generalise the method to Riemannian manifolds.

2.1   Outline of Locally Linear Embedding

Consider n points $x_1, \ldots, x_n \in \mathbb{R}^m$ sampled from some underlying submanifold of lower dimension d. Then at a sufficiently local scale, the points should lie close to a linear subspace. Under this assumption, each point can be written as a linear combination of its neighbours up to some residual error. Following this observation, LLE implements two main steps. First, compute the approximate barycentric coordinates of x_i with respect to its k nearest neighbours, that is for all i, find w_{ij} which solve

$$\min_{w_{i1}, \ldots, w_{in} \in \mathbb{R}} \quad \left\| x_i - \sum_{j=1}^{n} w_{ij} x_j \right\|^2 \tag{1}$$
$$\text{subject to} \quad \sum_j w_{ij} = 1$$

such that $w_{ij} = 0$ if x_j is not one of the k nearest neighbours of x_i. This amounts to solving n linear system $\in \mathbb{R}^k$. Then second step consists of finding n new points $y_1, \ldots, y_n \in \mathbb{R}^d$ which best retrieve the weights estimated in Problem 1, that is solve

$$\min_{y_1, \ldots, y_n \in \mathbb{R}^d} \quad \sum_{i=1}^{n} \left\| y_i - \sum_{j=1}^{n} w_{ij} y_j \right\|^2. \tag{2}$$

Problem 2 is equivalent to an eigenvalue decomposition problem.

2.2 Riemannian Barycentric Coordinates

In order to generalise the method to manifold-valued data, we need to rewrite Problem 1 in a general non-Euclidean setting. Precisely, we need to extend the definition of a barycentre and barycentric coordinates. Intuitively, the weighted barycentre of a set of points is the point which minimises the weighted sum of squared distances. Let M denote a Riemannian manifold. We define the following.

Definition 1 (Riemannian barycentric coordinates). *A point $x \in M$ has barycentric coordinates $w_1, \dots, w_n$ with respect to $x_1, \dots, x_n \in M$ if*

$$\sum_{i=1}^{n} w_i \log_x(x_i) = 0. \tag{3}$$

where $\log_x$ denotes the logarithm map of M at x. A point x which satisfies the previous is called a weighted barycentre of $x_1, \dots, x_n \in M$ affected with the weights $w_1, \dots, w_n$.

Equation 3 can be interpreted as a first order condition on the minimisation of the weighted sum of squared distances. We can check that both descriptions coincide with the usual definition of barycentric coordinates in a Euclidean setting. Indeed, for $M = \mathbb{R}^m$ and $x, y \in M$, the logarithm map of M at x is simply $\log_x(y) = y - x$. Therefore, according to the previous definition, the barycentre of $x_1, \dots, x_n \in M$ affected with the weights $w_1, \dots, w_n$ is the point x which satisfies $x = \sum_i w_i x_i$. For more details, we refer the reader to [6]. Let us now rewrite Problem 1 for points of a Riemannian manifold.

2.3 Towards Riemannian Locally Linear Embedding

The generalisation of Problem 1 is not straightforward as the previous definition is an implicit definition and does not allow to write a weighted barycentre in closed-form except in the Euclidean case. Instead, one can introduce an auxiliary variable $\widehat{x}_i \in M$ satisfying Eq. 3 and generalise Problem 1 as a constrained optimisation problem on manifolds. More explicitly, for $x_1, \dots x_n \in M$, solve

$$\min_{\substack{\widehat{x}_i \in M, \\ w_{i1}, \dots, w_{in} \in \mathbb{R}}} \quad d_M(x_i, \widehat{x}_i)^2$$
$$\text{subject to} \quad \sum_j w_{ij} \log_{\widehat{x}_i}(x_j) = 0 \tag{4}$$
$$\sum_j w_{ij} = 1.$$

It is not trivial however to solve this problem in practice. Especially, the constraint $\sum_j w_{ij} \log_{\widehat{x}_i}(x_j) = 0$ lies in the tangent space $T_{\widehat{x}_i} M$ which depends itself on the value of $\widehat{x}_i$ which we wish to optimise. Rather, we propose to look at the equivalent translated problem

$$\min_{\substack{\widehat{x}_i \in M, \\ w_{i1}, \dots, w_{in} \in \mathbb{R}}} \quad d_M(x_i, \widehat{x}_i)^2$$
$$\text{subject to} \quad \sum_j w_{ij} P_{\widehat{x}_i, x_i}\left(\log_{\widehat{x}_i}(x_j)\right) = 0 \tag{5}$$
$$\sum_j w_{ij} = 1.$$

where $P_{\widehat{x}_i, x_i}$ denotes the parallel transport map of M along the geodesic joining $\widehat{x}_i$ and x_i. Since the parallel transport is an isometric map, Problems 4 and 5 are equivalent. In this new formulation however, the constraint lies in the tangent space at x_i, which is independent of the optimisation state.

Riemannian Locally Linear Embedding (RLLE) implements two steps according to the same scheme as LLE. First, the reconstruction step consists in estimating the weights w_{ij} solving Problem 5. Then, the embedding step consists in computing the points y_i solving Problem 2.

3   Algorithm and Implementation

In this section, we provide one possible algorithm to solve the optimisation Problem 2 and describe the implementation of our method. We detail the algorithm for Kendall shape spaces.

3.1   Tangent Space Formulation of the Optimisation Problem

In its current formulation, Problem 5 can be solved using Lagrangian methods for constrained optimisation on manifolds. However, it can be also be formulated alternatively as a vector-valued optimisation problem. Precisely, we keep track of the estimate $\widehat{x}_i$ with the tangent vector $v_i \in T_{x_i} M$ such that

$$\exp_{x_i}(v_i) = \widehat{x}_i \tag{6}$$

where $\exp_{x_i}$ denotes the exponential map of M at x_i. Additionally, we set

$$u_{ij} = P_{\widehat{x}_i, x_i}\left(\log_{\widehat{x}_i}(x_j)\right). \tag{7}$$

We derive the following optimisation problem

$$\min_{\substack{u_{i1},\ldots,u_{in} \in T_{x_i} M, \\ v_i \in T_{x_i} M, \\ w_{i1},\ldots,w_{in} \in \mathbb{R}}} \quad \|v_i\|^2$$

$$\text{subject to} \quad \sum_j w_{ij} u_{ij} = 0 \tag{8}$$

$$\sum_j w_{ij} = 1$$

$$\exp_{[\exp_{x_i}(v_i)]}\left(P_{x_i,[\exp_{x_i}(v_i)]}(u_{ij})\right) = x_j \quad (\forall j).$$

Problem 8 is a priori a vector-valued optimisation problem on the product space $(T_{x_i} M)^n \times T_{x_i} M \times \mathbb{R}^n$. In fact, since the weight w_{ij} is set to be 0 whenever x_j is not a neighbour of x_i, then the correct search space is $(T_{x_i} M)^k \times T_{x_i} M \times \mathbb{R}^k$ where k is the number of neighbours. Now, provided that a basis of $T_{x_i} M^n$ can been explicitly computed, then the search space is the Euclidean space $\mathbb{R}^{mk+m+k}$, where m is the dimension of M, and the optimisation task is performed using standard Lagrangian methods implemented in most libraries. As a reference, we use the SLSQP solver from scipy. Note that in practice, the complexity of the algorithm strongly depends on whether one knows the exponential map and the parallel transport in closed-form as we will discuss later in the paper. In any case, it requires an implementation of both methods which is compatible with automatic differentiation.

Note that another possible way to solve Problem 5 would be to address it directly as a Riemannian constrained optimisation problem using specific optimisation tools like the ones implemented in the Manopt library.

3.2   Riemannian Locally Linear Embedding for Quotient Manifolds

In what follows, we propose to detail the algorithm in the concrete case of Kendall shape spaces [3]. We recall that Kendall shape spaces carry a quotient structure. We first describe the algorithm for a general quotient manifold and then give an explicit formulation for Kendall shape spaces. For there does not always exist an explicit description for quotient objects, computations in quotient spaces are generally performed in the top space. It is also often more comfortable. The main motivation of this subsection is to show how to perform the previous optimisation task in the top space.

The setting is the following. Let again M be a Riemannian manifold and let G be a group acting on M. Given $x_1, \ldots, x_n \in M$, we want to solve Problem 5 for the corresponding data points $\pi(x_1), \ldots, \pi(x_n) \in M/G$, where $\pi : M \to M/G$ is the canonical quotient map. Assume that π is a Riemannian submersion. The vertical space of M at a point x, denoted by $\mathrm{Ver}_x M$, is defined by

$$\mathrm{Ver}_x M = \ker d_x \pi. \tag{9}$$

The tangent space of M at x admits an orthogonal decomposition

$$T_x M = \mathrm{Ver}_x M \oplus \mathrm{Hor}_x M \tag{10}$$

and $\mathrm{Hor}_x M$ is called the horizontal subspace of M at x. A central property is that the tangent space of M/G at a point $\pi(x)$ identifies with the horizontal space of M at x through the tangent map $d\pi$. Moreover, geodesics of M/G correspond exactly to the projection by π of horizontal geodesics of M, that is geodesics spanned by a horizontal vector. Additionally, we define the following

Definition 2 (Horizontal parallel transport). *Let γ be a horizontal curve in M. Then we say that the vector field $t \to v(t)$ is the horizontal parallel transport of a horizontal vector v along γ if it is horizontal and if its projection to the tangent bundle of M/G is the parallel transport of $d_x\pi(v)$ along $\pi(\gamma)$. We denote the horizontal transport map of M from a point x to a point y by $P_{x,y}^H$.*

Now let us go back to the algorithm. Since the tangent map $d\pi$ allows to identify the tangent spaces of M/G and the horizontal spaces of M, we can lift up Problem 8 to the top manifold M

$$\min_{\substack{g_{i1},\dots,g_{in}\in G \\ u_{i1},\dots,u_{in}\in \mathrm{Hor}_{x_i} M, \\ v_i\in \mathrm{Hor}_{x_i} M, \\ w_{i1},\dots,w_{in}\in\mathbb{R}}} \|v_i\|^2$$

$$\text{subject to} \quad \sum_j w_{ij}u_{ij} = 0 \tag{11}$$

$$\sum_j w_{ij} = 1$$

$$g_{ij}\cdot \exp_{[\exp_{x_i}(v_i)]}\left(P^H_{x_i,[\exp_{x_i}(v_i)]}(u_{ij})\right) = x_j \quad (\forall j).$$

The optimisation variables g_{ij} are the elements of the group G lifting up the equality constraint

$$\pi\left(\exp_{[\exp_{x_i}(v_i)]}\left(P^H_{x_i,[\exp_{x_i}(v_i)]}(u_{ij})\right)\right) = \pi(x_j)$$

to the top space. If G is a matrix Lie group, then it identifies to its Lie algebra $\mathfrak{g}$ through the matrix exponential such that the search space can still be written as a vector space.

3.3   Implementation in Kendall Shape Spaces

The implementation of the algorithm for Kendall shape spaces requires to compute the horizontal spaces, the exponential map and the horizontal parallel transport map. The Kendall shape space Σ_p^q is defined as the quotient

$$\Sigma_p^q = \mathcal{S}_p^q/SO(p) \tag{12}$$

where $\mathcal{S}_p^q = \{x \in M(p,q) \mid \sum x_i = 0 \text{ and } \|x\| = 1\}$ is referred as the pre-shape space [3]. The shape space Σ_p^q describes the possible configurations of a set of points independently of any similarity transformation of the ambient space. The space $\mathcal{S}_p^q$ can be understood as the hypersphere of $\mathbb{R}^{p(q-1)}$ and its exponential map is given by

$$\exp_x(v) = \cos(\|v\|)x + \sin(\|v\|)\frac{v}{\|v\|}. \tag{13}$$

The horizontal subspace at x is described as

$$\mathrm{Hor}_x M = \left\{v \in M(p,q) \mid \sum x_i = 0 \text{ and } vx^t = xv^t \text{ and } \langle x, v\rangle = 0\right\}.$$

where $\langle \cdot, \cdot \rangle$ denotes the usual Frobenius scalar product. The horizontal parallel transport can be computed as the solution of a first-order differential equation as described in [4]. Its implementation has already been discussed in the previous work [2] and is available in the library geomstats. Finally, the group of rotations $SO(p)$ is a Lie group and its Lie algebra is $\mathfrak{so}(p) = \mathrm{Skew}(p)$.

4   Benchmark Experiment

In this section, we illustrate RLLE on two examples in the shape space Σ_3^3 and we compare its performance to LLE performance. We then discuss more generally the advantages of our method depending on the type of data.

4.1   A Swiss Roll Example in the Shape Space Σ_3^3

We run two experiments in the Kendall shape space Σ_3^3, for which we have an isometric embedding in a hemisphere and therefore a way to visually evaluate the embedding computed by either method. We first illustrate our algorithm on a set of data points generated from a mixture of normal distributions on the hemisphere (Fig. 1). We compare the ability of LLE and RLLE to embed such a set in the plane. Our implementation of LLE performs first a Procrustean alignment step before solving Problem 1. Precisely, it aligns each neighbour x_j onto the point x_i by applying the optimal rotation. Note that without this alignment step, the performance of LLE drops significantly. Then we demonstrate numerically the performance of RLLE on an example derived from the "Swiss Roll" data set. Explicitly, given a set of shapes sampled along a logarithmic spiral curve in Σ_3^3 (Fig. 2), we evaluate the one-dimensional parametrisation computed by LLE and RLLE with respect to the one given by the arc-length.

LLE   RLLE   Exact

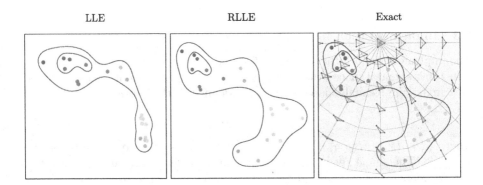

Fig. 1. First experiment. The points are equally sampled from two normal distributions in the hemisphere, flattened into a disk for visualisation. For both methods, the neighbour graph has been generated with the Riemannian distance. The number of neighbours k is chosen to optimise the performance of each method: $k = 9$ for LLE and $k = 10$ for RLLE. Since the methods are by definition invariant by affine transformation, we align each embedding onto the flattened data set. We observe that LLE is able to retrieve very local alignments (blue) but fails at a more global scale (red). This was to be expected as the Euclidean distance approximates the Riemannian distance locally. On the other hand, RLLE is able to retrieve the alignments at every scale. (Color figure online)

4.2   Computational Complexity

RLLE shares the main drawback of intrinsic manifold learning methods: it is computationally quite expensive. Let us detail this point. We mainly focus on the reconstruction step as the embedding step is common to the LLE method. First, the search space is a space of dimension $km + m + k$, where m is the dimension

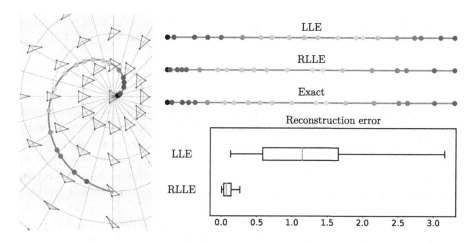

Fig. 2. Second Experiment. The points are sampled along a spiral curve. We fix $k = 3$ for both methods. We illustrate the experiment for one sample in the left and upper-right sub-figures. Both methods are ordering the points correctly. Relative distances however are better preserved by RLLE. We estimate the absolute error between the exact embedding and the one computed by RLLE (resp. LLE). The experiment is run 100 times. We summarise the performance of each method in a box plot. We observe that RLLE performs significantly better and moreover is more stable.

of the manifold M and k is the number of neighbours. For the Kendall shape space of parameters p and q, we have $m = p(q - \frac{1}{2}(p-1)) - 1$. Then the number of optimisation problems to solve is n. Finally, we need to take into account the computational cost of the exponential and the parallel transport methods. In the case of Kendall shape spaces, while the first one is free, the latter performs in $\mathcal{O}(2sp^3 + pq)$ as detailed in [2], where s is the number of integration steps. Each evaluation of the constraint – and so each step of the optimisation task – costs the same. Finally, our methods implements a SQP method to solve each optimisation problem. As a comparison, the solution of Problem 1 is equivalent to a matrix inversion of dimension k. Therefore, LLE performs in roughly $\mathcal{O}(nk^3)$, such that its complexity does not depend on the dimension m of the data.

4.3 Further Discussion

Given the computational cost of RLLE, it is important to understand for which type of data it is particularly suited. Typically, the locally linear assumption made by LLE may be valid for large and well-concentrated data sets. In these cases, both methods should perform the same. Moreover, non-local methods like PCA or its manifold generalisations provide a correct estimation whenever the data are sufficiently concentrated. Finally, LLE and RLLE seem of particular use in cases where a distance-based method does not perform well. These remarks suggest that RLLE is more specifically designed for small sample size data sets with large dispersion, and provides an embedding which might allow to observe more informative patterns than the ones characterised by the distance only.

5　Conclusion

As for now, RLLE has been implemented for Kendall shape spaces only. We are contemplating a general implementation of the method into the library geomstats for various manifolds and quotient manifolds – for example the space of unlabelled graphs. Applications to real data sets will also be developed in future works. Especially, we wish to further investigate the analysis of protein conformations using Kendall's framework following one of our previous works.

Acknowledgements. We thank G-stats team for the fruitful discussions. This work was supported by the ERC grant #786854 G-Statistics from the European Research Council under the European Union's Horizon 2020 research and innovation program and by the French government through the 3IA Côte d'Azur Investments ANR-19-P3IA-0002 managed by the National Research Agency.

References

1. Fletcher, P.T., Lu, C., Pizer, S.M., Joshi, S.: Principal geodesic analysis for the study of nonlinear statistics of shape. IEEE Trans. Med. Imaging **23**(8), 995–1005 (2004)
2. Guigui, N., Maignant, E., Trouvé, A., Pennec, X.: Parallel transport on Kendall shape spaces. In: Nielsen, F., Barbaresco, F. (eds.) GSI 2021. LNCS, vol. 12829, pp. 103–110. Springer, Cham (2021). https://doi.org/10.1007/978-3-030-80209-7_12
3. Kendall, D.G.: Shape manifolds, procrustean metrics, and complex projective spaces. Bull. Lond. Math. Soc. **16**(2), 81–121 (1984)
4. Kim, K.R., Dryden, I.L., Le, H., Severn, K.E.: Smoothing splines on Riemannian manifolds, with applications to 3D shape space. J. Roy. Stat. Soc. Ser. B (Stat. Methodol.) **83**(1), 108–132 (2021)
5. Lahiri, S., Robinson, D., Klassen, E.: Precise matching of PL curves in $\mathbb{R}^N$ in the square root velocity framework. arXiv preprint arXiv:1501.00577 (2015)
6. Pennec, X.: Barycentric subspace analysis on manifolds. Ann. Stat. **46**(6A), 2711–2746 (2018)
7. Roweis, S.T., Saul, L.K.: Nonlinear dimensionality reduction by locally linear embedding. Science **290**(5500), 2323–2326 (2000)
8. Tumpach, A.B., Preston, S.C.: Quotient elastic metrics on the manifold of arc-length parameterized plane loops. arXiv preprint arXiv:1601.06139 (2016)

A Product Shape Manifold Approach for Optimizing Piecewise-Smooth Shapes

Lidiya Pryymak[1]([✉]), Tim Suchan[2], and Kathrin Welker[1]

[1] TU Bergakademie Freiberg, Akademiestr. 6, 09599 Freiberg, Germany
{lidiya.pryymak,kathrin.welker}@math.tu-freiberg.de
[2] Helmut Schmidt University, Holstenhofweg 85, 22043 Hamburg, Germany
suchan@hsu-hh.de

Abstract. Spaces where each element describes a shape, so-called shape spaces, are of particular interest in shape optimization and its applications. Theory and algorithms in shape optimization are often based on techniques from differential geometry. Challenges arise when an application demands a non-smooth shape, which is commonly-encountered as an optimal shape for fluid-mechanical problems. In order to avoid the restriction to infinitely-smooth shapes of a commonly-used shape space, we construct a space containing shapes in $\mathbb{R}^2$ that can be identified with a Riemannian product manifold but at the same time admits piecewise-smooth curves as elements. We combine the new product manifold with an approach for optimizing multiple non-intersecting shapes. For the newly-defined shapes, adjustments are made in the known shape optimization definitions and algorithms to ensure their usability in applications. Numerical results regarding a fluid-mechanical problem constrained by the Navier-Stokes equations, where the viscous energy dissipation is minimized, show its applicability.

Keywords: shape optimization · Riemannian manifolds · product manifolds · piecewise-smooth shapes · Navier-Stokes equations

1 Introduction

Shape optimization is commonly-applied in engineering in order to optimize shapes w.r.t. to an objective functional that relies on the solution of a partial differential equation (PDE). The PDE is required to model the underlying physical phenomenon, e.g. elastic displacements due to loadings or fluid movement due to pressure differences. Different methods are available for the shape optimization, however we focus on gradient-based techniques on shape spaces.

An ideal shape space would enable the usage of classical optimization methods like gradient descent algorithms. Since this is usually not the case, it is

This work has been partly supported by the German Research Foundation (DFG) within the priority program SPP 1962 under contract number WE 6629/1-1 and by the state of Hamburg (Germany) within the Landesforschungsförderung under project SENSUS with project number LFF-GK11.

desirable to define a shape u to be an element of a Riemannian manifold. An important example of a smooth[1] manifold allowing a Riemannian structure is the shape space

$$B_e := B_e(S^1, \mathbb{R}^2) := \text{Emb}(S^1, \mathbb{R}^2)/\text{Diff}(S^1).$$

An element of B_e is a smooth simple closed curve in $\mathbb{R}^2$. The space was briefly investigated in [11]. The existence of Riemannian metrics, geodesics or, more generally, the differential geometric structure of B_e (cf., e.g. [10,11]) reveals many possibilities like the computation of the shape gradient in shape optimization (cf., e.g. [15]). However, since an element of B_e is a smooth curve in $\mathbb{R}^2$, the shape space is in general not sufficient to carry out optimization algorithms on piecewise-smooth shapes, which are often encountered as an optimal shape for fluid-mechanical problems, see e.g. [14] for a prominent example. In particular, we are interested in shapes with kinks. Such piecewise-smooth shapes are generally not elements of a shape space that provides the desired geometrical properties for applications in shape optimization. Some effort has been put into constructing a shape space that contains non-smooth shapes, however so far only a diffeological space structure could be found, cf. e.g. [19,20]. A further issue for many applications in shape optimization [1,5,8], such as the electrical impedance tomography, is to consider multi-shapes. A first approach for optimizing smooth multi-shapes has been presented in [6].

In this paper, we aim to construct a novel shape space holding a Riemannian structure for optimizing piecewise-smooth multi-shapes. The structure of the paper is as follows: In Sect. 2, we extend the findings related to multi-shapes in [6] to a novel shape space considering piecewise-smooth shapes. Hereby, we use the fact that the space of simple, open curves

$$B_e([0,1], \mathbb{R}^2) := \text{Emb}([0,1], \mathbb{R}^2)/\text{Diff}([0,1])$$

is a smooth manifold as well (cf. [9]) and interpret a closed curve with kinks as a glued-together curve of smooth, open curves, i.e. elements of $B_e([0,1], \mathbb{R}^2)$. Moreover, we derive a shape optimization procedure on the novel shape space. In Sect. 3, we apply the presented optimization technique to a shape optimization problem constrained by Navier-Stokes equations and present numerical results.

2 Product Space for Optimizing Piecewise-Smooth Shapes

In this section, we aim to construct a gradient descent algorithm for optimizing piecewise-smooth multi-shapes, e.g. the multi-shape $u = (u_1, u_2)$ from Fig. 1. In Sect. 2.1, we therefore introduce a novel shape space which has the structure of a Riemannian product manifold. An optimization algorithm on the novel shape space is formulated in Sect. 2.2.

---

[1] Throughout this paper, the term smooth shall refer to infinite differentiability.

2.1   Product Shape Space

In the following, we introduce a novel shape manifold, whose structure will later be used to optimize piecewise-smooth shapes. The construction of the novel shape space is based on a Riemannian product manifold. Therefore, we first investigate the structure of product manifolds.

We define $(\mathcal{U}_i, G^i)$ to be Riemannian manifolds equipped with the Riemannian metrics G^i for all $i = 1, \ldots, N \in \mathbb{N}$. The Riemannian metric G^i at the point $p \in \mathcal{U}_i$ will be denoted by

$$G_p^i(\cdot, \cdot) \colon T_p\mathcal{U}_i \times T_p\mathcal{U}_i \to \mathbb{R},$$

where $T_p\mathcal{U}_i$ denotes the tangent space at a point $p \in \mathcal{U}_i$. We then define the product manifold as

$$\mathcal{U}^N := \mathcal{U}_1 \times \ldots \times \mathcal{U}_N = \prod_{i=1}^N \mathcal{U}_i.$$

As shown in [6], for the tangent space of product manifolds it holds

$$T_{\tilde{u}}\mathcal{U}^N \cong T_{\tilde{u}_1}\mathcal{U}_1 \times \cdots \times T_{\tilde{u}_N}\mathcal{U}_N.$$

Moreover, a product metric can be defined as

$$\mathcal{G}^N = \sum_{i=1}^N \pi_i^* G^i, \tag{1}$$

where π_i^* are the pushforwards associated with canonical projections. It is obvious to use the space B_e defined in Sect. 1 to construct a specific product shape space. An issue arises for non-smooth shapes, e.g. the shape u_1 from Fig. 1. To fix this issue, we now introduce the new multi-shape space for s shapes built on the Riemannian product manifold $\mathcal{U}^N$.

Definition 1. *Let $(\mathcal{U}_i, G^i)$ be Riemannian manifolds equipped with Riemannian metrics G^i for all $i = 1, \ldots, N$. Moreover, $\mathcal{U}^N := \prod_{i=1}^N \mathcal{U}_i$. For $s \in \mathbb{N}$, we define the s-dimensional shape space on $\mathcal{U}^N$ by*

$$M_s(\mathcal{U}^N) := \{u = (u_1, \ldots, u_s) \mid u_j \in \prod_{l=k_j}^{k_j+n_j-1} \mathcal{U}_l, \sum_{j=1}^s n_j = N \text{ and}$$

$$k_1 = 1, k_{j+1} = k_j + n_j \ \forall j = 1, \ldots, s-1\}.$$

With Definition 1, an element in $M_s(\mathcal{U}^N)$ is defined as a group of s shapes $u_1, \ldots, u_s$, where each shape u_j is an element of the product of n_j smooth manifolds. For $\mathcal{U}_l = B_e([0,1], \mathbb{R}^2)$ for $l = 1, \ldots, 12$ and $\mathcal{U}_{13} = B_e(S^1, \mathbb{R}^2)$, we can define the shapes presented in Fig. 1 by $(u_1, u_2) \in M_2(\mathcal{U}^{13})$, where $u_1 \in \prod_{l=1}^{12} \mathcal{U}_l$ and $u_2 \in \mathcal{U}_{13}$.

For applications of Definition 1 in shape optimization problems, it is of great interest to look at the tangent space of $M_s(\mathcal{U}^N)$. Since any element $u = (u_1, \ldots, u_s) \in M_s(\mathcal{U}^N)$ can be understood as an element $\tilde{u} = (\tilde{u}_1, \ldots, \tilde{u}_N) \in \mathcal{U}^N$, we set $T_u M_s(\mathcal{U}^N) = T_{\tilde{u}} \mathcal{U}^N$ and

$$G_u(\varphi, \psi) = G_{\tilde{u}}(\varphi, \psi) \; \forall \varphi, \psi \in T_u M_s(\mathcal{U}^N) = T_{\tilde{u}} \mathcal{U}^N.$$

Next, we consider shape optimization problems, i.e. we investigate so-called shape functionals. A shape functional on $M_s(\mathcal{U}^N)$ is given by $j \colon M_s(\mathcal{U}^N) \to \mathbb{R}$, $u \mapsto j(u)$. In the following paragraph, we investigate solution techniques for shape optimization problems, i.e. for problems of the form

$$\min_{u \in M_s(\mathcal{U}^N)} j(u). \tag{2}$$

2.2 Optimization Technique on $M_s(\mathcal{U}^N)$ for Optimizing Piecewise-Smooth Shapes

A theoretical framework for shape optimization depending on multi-shapes is presented in [6], where the optimization variable can be represented as a multi-shape belonging to a product shape space. Among other things, a multi-pushforward and multi-shape gradient are defined; however, each shape is assumed to be an element of one shape space. In contrast, Definition 1 also allows that a shape itself is represented by a product shape space. Therefore, we need to adapt the findings in [6] to our setting.

To derive a gradient descent algorithm for a shape optimization problem as in (2), we need a definition for differentiating a shape functional mapping from a smooth manifold to $\mathbb{R}$. For smooth manifolds, this is achieved using a pushforward.

Definition 2. *For each shape $u \in M_s(\mathcal{U}^N)$, the multi-pushforward of a shape functional $j \colon M_s(\mathcal{U}^N) \to \mathbb{R}$ is given by the map*

$$(j_*)_u \colon T_u M_s(\mathcal{U}^N) \to \mathbb{R}, \quad \varphi \mapsto \frac{d}{dt} j(\varphi(t))_{t=0} = (j \circ \varphi)'(0).$$

Thanks to the multi-pushforward, we can define the so-called multi-shape gradient, which is required for optimization algorithms.

Definition 3. *The multi-shape gradient for a shape functional $j \colon M_s(\mathcal{U}^N) \to \mathbb{R}$ at the point $u \in M_s(\mathcal{U}^N)$ is given by $\psi \in T_u M_s(\mathcal{U}^N)$ satisfying*

$$\mathcal{G}_u^N(\psi, \varphi) = (j_*)_u \varphi \quad \forall \varphi \in T_{\tilde{u}} M_s(\mathcal{U}^N).$$

We are now able to formulate a gradient descent algorithm on $M_s(\mathcal{U}^N)$ similar to the one presented in [6]. For updating the multi-shape u in each iteration, the multi-exponential map

$$\exp_u^N \colon T_u M_s(\mathcal{U}^N) \to M_s(\mathcal{U}^N), \quad \varphi = (\varphi_1, \ldots, \varphi_N) \mapsto (\exp_{\tilde{u}_1} \varphi_1, \ldots, \exp_{\tilde{u}_N} \varphi_N)$$

Algorithm 1. Gradient descent algorithm on $M_s(\mathcal{U}^N)$ with Armijo backtracking line search to solve (2)

Input: Initial shape $u = (u_1, \ldots, u_s) = (\tilde{u}_1, \ldots, \tilde{u}_N) = \tilde{u}$, constants for Armijo backtracking and $\epsilon > 0$ for break condition

1: **while** $\|v\|_{\mathcal{G}^N} > \epsilon$ **do**
2:     Compute the multi-shape gradient v with respect to $\mathcal{G}^N$
3:     Compute stepsize α with Armijo backtracking
4:     $u \leftarrow \exp_u^N(-\alpha v)$
5: **end while**

---

is used. The algorithm is depicted in Algorithm 1.

So far, we have derived an optimization algorithm on $M_s(\mathcal{U}^N)$, i.e. an algorithm for optimizing a non-intersecting group of shapes, where each shape is an element of a product manifold with a varying number of factor spaces. With the main goal of this section in mind, we need to further restrict the choice of shapes in $M_s(\mathcal{U}^N)$ to glued-together piecewise-smooth shapes: We assume that $\mathcal{U}_i$ is either $B_e(S^1, \mathbb{R}^2)$ or $B_e([0,1], \mathbb{R}^2)$. Moreover, we assume that each shape $(u_1, \ldots, u_s)$ is closed, where $u = (u_1, \ldots, u_s)$ is chosen from $M_s(\mathcal{U}^N)$. By that we mean that if a shape is $u_j \in \prod_{l=k_j}^{k_j+n_j-1} \mathcal{U}_l$, then either

$$n_j = 1 \text{ and } \mathcal{U}_{k_j} = B_e(S^1, \mathbb{R}^2)$$

or

$$\mathcal{U}_l = B_e([0,1], \mathbb{R}^2) \ \forall l = k_j, \ldots, k_j + n_j - 1 \text{ and for}$$
$$u_j = (u_{k_j}, \ldots, u_{k_j+n_j-1}), \text{ it holds that}$$
$$u_{k_j+h}(1) = u_{k_j+h+1}(0) \ \forall h = 0, \ldots, n_j - 2 \text{ and } u_{k_j}(0) = u_{k_j+n_j-1}(1).$$

Finally, we want to address another important issue in shape optimization algorithms: the development of kinks in smooth shapes over the course of the optimization. If we view a smooth initial shape, e.g. u_2 from Fig. 1, as an element in $B_e(S^1, \mathbb{R}^2)$ no kinks can arise during the optimization of the shape. An approach to fix this issue for applications, where the developments of kinks in shapes is desired, is to approximate a smooth shape with elements of $B_e([0,1], \mathbb{R}^2)$. A simple but sufficient choice is using initially straight lines connecting locations of possible kinks. In this manner, the multi-shape $u = (u_1, u_2)$ from Fig. 1 would be an element of

$M_2(\mathcal{U}^{12+l_1+l_2})$, where $l_1, l_2 \in \mathbb{N}$ and

$$u_1 \in \prod_{l=1}^{12+l_1} \mathcal{U}_l = B_e([0,1], \mathbb{R}^2)^{12+l_1}, \ u_2 \in \prod_{l=13+l_1}^{12+l_1+l_2} \mathcal{U}_l = B_e([0,1], \mathbb{R}^2)^{l_2}. \tag{3}$$

3   Application to Navier-Stokes Flow

In the following, we apply Algorithm 1 to a shape optimization problem constrained by steady-state Navier-Stokes equations and geometrical constraints. In Sect. 3.1, we briefly describe the numerical implementation of Algorithm 1. Afterwards, we formulate the optimization problem that will be considered for the numerical studies in Sect. 3.2, and finally, in Sect. 3.3, we describe the numerical results.

3.1   Adjustments of Algorithm 1 for Numerical Computations

In order to ensure the numerical applicability of Algorithm 1, adjustments must be made. We define the space $\mathcal{W} := \{W \in H^1(D_u, \mathbb{R}^2) | W = 0 \text{ on } \partial D_u \setminus u\}$, and similarly to [6], we use an optimization approach based on partial shape derivatives, together with the Steklov-Poincaré metric in Eq. (1). The Steklov-Poincaré metric is defined in [16] and yields $\mathcal{G}^i(V|_u, W|_u) = a(V, W)$ with a symmetric and coercive bilinear form $a \colon \mathcal{W} \times \mathcal{W}$. We replace the multi-shape gradient with the mesh deformation $V \in \mathcal{W}$, which is obtained by replacing the multi-pushforward with the multi-shape derivative[2] in Definition 3. A common choice for the bilinear form when using the Steklov-Poincaré metric is linear elasticity

$$\int_{D_u} \varepsilon(V) : C : \varepsilon(W) \, \mathrm{d}x = \mathrm{d}j(u)[W] \quad \forall W \in \mathcal{W}, \tag{4}$$

where $\varepsilon(V) = \operatorname{sym}\operatorname{grad}(V)$ and C describes the linear elasticity tensor and $A : B$ is the standard Frobenius inner product and $\mathrm{d}j(u)[W]$ denotes the shape derivative of j at u in direction W. Due to the equivalence of the Steklov-Poincaré metric and the bilinear form a, we replace the $\mathcal{G}^N$-norm in the stopping criterion of the Algorithm 1 with the H^1-norm in D_u. Finally, since the exponential map used in Algorithm 1 is an expensive operation, it is common to replace it by a so-called retraction. In our computations, we use the retraction introduced in [17].

3.2   Model Formulation

We consider the problem

$$\min_{u \in M_s(\mathcal{U}^N)} j(u) := \min_{u \in M_s(\mathcal{U}^N)} \int_{D_u} \frac{\mu}{2} \nabla v : \nabla v \, \mathrm{d}x, \tag{5}$$

where we constrain the optimization problem by the Navier-Stokes equations and choose $M_s(\mathcal{U}^N)$ as in (3). The state is denoted as $y = (v, p)$ for which the Navier-Stokes equations can be found in standard literature and will be omitted here for brevity. The material constants dynamic viscosity and density are defined as $\mu =$

---

[2] We refer to [6] for the definition and details about the multi-shape derivative.

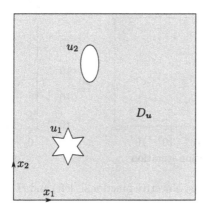

Fig. 1. Sketch of two shapes u_1, u_2 surrounded by a domain $D_u \subset \mathbb{R}^2$.

1.81 and $\rho = 1.2 \cdot 10^5$, respectively. We choose homogenous Dirichlet boundary conditions on the top and bottom boundary as well as on both shapes. The right boundary is modelled as homogenous Neumann, and the left boundary has the inhomogenous Dirichlet boundary condition $v = (-0.08421\, x_2\, (x_2 - 1), 0)^\top$. We choose the hold-all domain $D = (0,1)^2$, in which two shapes u_1 and u_2 are embedded as shown in Fig. 1 with barycenters at $(0.3, 0.3)^\top$ and $(0.45, 0.75)^\top$, respectively.

Additional geometrical constraints are required in order to avoid trivial solutions, see e.g. [12,13], which are implemented as inequality constraints with an Augmented Lagrange approach as described in [18]. We restrict the area of each shape $\text{vol}(u_i)$ to be at 100% initial area. Further, the barycenter $\text{bary}(u_1)$ is constrained to stay between $(-0.03, -0.05)^\top$ and $(0.04, 0.03)^\top$ of the initial position in x and y direction, respectively, and the barycenter $\text{bary}(u_2)$ to stay between $(-0.075, -0.02)^\top$ and $(0.02, 0.05)^\top$ of the initial position.

3.3   Numerical Results

The computational domain is discretized with 3512 nodes and 7026 triangular elements using Gmsh [7] with standard Taylor-Hood elements. An automatic remesher is available in case the mesh quality deteriorates below a threshold. The optimization is performed in FEniCS 2019.1.0 [2]. We use a Newton solver and solve the linearized system of equations using MUMPS 5.5.1 [3,4]. Armijo backtracking is performed as described in Algorithm 1 with $\tilde{\alpha} = 0.0125$, $\sigma = 10^{-4}$ and $\tilde{\rho} = \frac{1}{10}$. The stopping criterion for each gradient descent is reached when the H^1-norm of the mesh deformation is at or below 10^{-4}. The objective functional and the H^1-norm of the mesh deformation over the course of the optimization are shown in Fig. 2 and the magnitude of the fluid velocity in the computational domain before, during, and after optimization can be found in Fig. 3. The optimized shapes can be seen in Fig. 3 on the right. Over the course of the optimization we observe a reduction of the objective functional by approximately

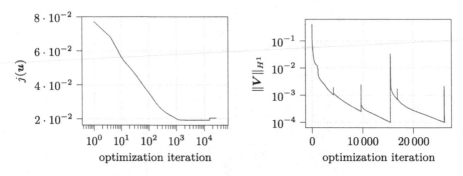

Fig. 2. Optimization results: objective functional (left) and H^1-norm of the mesh deformation (right).

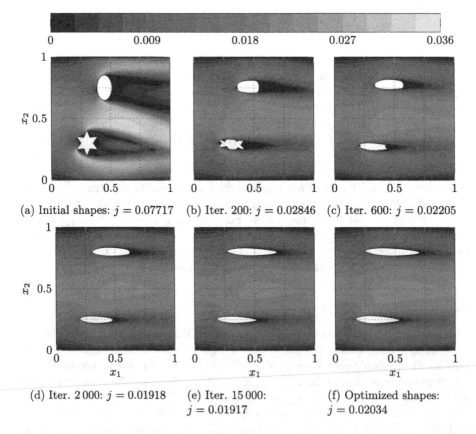

(a) Initial shapes: $j = 0.07717$ (b) Iter. 200: $j = 0.02846$ (c) Iter. 600: $j = 0.02205$

(d) Iter. 2 000: $j = 0.01918$ (e) Iter. 15 000:   (f) Optimized shapes:
                                    $j = 0.01917$ $j = 0.02034$

Fig. 3. Fluid velocity magnitude at different stages of the optimization. Figure 3f has an increased objective functional value in comparison to Fig. 3d and 3e, however it fulfills the geometrical constraints while the others do not yet.

74%. The norm of the mesh deformation shows an exponential decrease, similar to a classical gradient descent algorithm. The peaks are caused by remeshing or by the adjustment of Augmented Lagrange parameters. Initially, the optimizer is mainly concerned with obtaining a approximate optimized shape, see Fig. 3b–3d, while the exact fulfillment of geometrical constraints is less relevant. The later stages optimize small features like the leading and trailing edge of the shape, see Fig. 3e, any suboptimal kinks that were still remaining are removed, and in Fig. 3f the geometrical constraints are fulfilled with an infeasibility of below 10^{-6} after $k = 7$ Augmented Lagrange iterations.

4 Conclusion

A novel shape space $M_s(\mathcal{U}^N)$ that provides both, a Riemannian structure and a possibility to consider glued-together shapes (in particular, shapes with kinks) is introduced. Additionally, an optimization algorithm, based on findings from [6], is formulated. The new algorithm is applied to solve an optimization problem constrained by the Navier-Stokes equations with additional geometrical inequality constraints, where we have observed a strong reduction of the objective functional and convergence of the gradient descent on $M_s(\mathcal{U}^N)$ similar to a classical gradient descent algorithm. Forthcoming research should involve an investigation of the development of the shapes' overlaps (glued-together points) over the course of the optimization. Moreover, convergence statements need to be investigated.

References

1. Albuquerque, Y.F., Laurain, A., Sturm, K.: A shape optimization approach for electrical impedance tomography with point measurements. Inverse Probl. **36**(9) (2020). https://doi.org/10.1088/1361-6420/ab9f87
2. Alnæs, M., et al.: The FEniCS project version 1.5. Arch. Numer. Softw. **3**(100) (2015). https://doi.org/10.11588/ans.2015.100.20553
3. Amestoy, P.R., Duff, I.S., Koster, J., L'Excellent, J.Y.: A fully asynchronous multifrontal solver using distributed dynamic scheduling. SIAM J. Matrix Anal. Appl. **23**(1), 15–41 (2001). https://doi.org/10.1137/S0895479899358194
4. Amestoy, P.R., Guermouche, A., L'Excellent, J.Y., Pralet, S.: Hybrid scheduling for the parallel solution of linear systems. Parallel Comput. **32**(2), 136–156 (2006). https://doi.org/10.1016/j.parco.2005.07.004
5. Cheney, M., Isaacson, D., Newell, J.C.: Electrical impedance tomography. SIAM Rev. Soc. Ind. Appl. Math. **41**(1), 85–101 (1999). https://doi.org/10.1137/S0036144598333613
6. Geiersbach, C., Loayza-Romero, E., Welker, K.: PDE-constrained shape optimization: towards product shape spaces and stochastic models. In: Chen, K., Schönlieb, C.B., Tai, X.C., Younes, L. (eds.) Handbook of Mathematical Models and Algorithms in Computer Vision and Imaging, pp. 1–46. Springer, Cham (2022). https://doi.org/10.1007/978-3-030-03009-4_120-1

7. Geuzaine, C., Remacle, J.F.: Gmsh: a 3-D finite element mesh generator with built-in pre- and post-processing facilities. Int. J. Numer. Methods Eng. **79**(11), 1309–1331 (2009). https://doi.org/10.1002/nme.2579

8. Kwon, O., Woo, E.J., Yoon, J.R., Seo, J.K.: Magnetic resonance electrical impedance tomography (MREIT): simulation study of j-substitution algorithm. IEEE Trans. Biomed. Eng. **49**(2), 160–167 (2002). https://doi.org/10.1109/10.979355

9. Michor, P.W.: Manifolds of Differentiable Mappings, vol. 3. Shiva Mathematics Series (1980). https://www.mat.univie.ac.at/~michor/manifolds_of_differentiable_mappings.pdf

10. Michor, P.W., Mumford, D.: An overview of the Riemannian metrics on spaces of curves using the Hamiltonian approach. Appl. Comput. Harmon. Anal. **23**(1), 74–113 (2007). https://doi.org/10.1016/j.acha.2006.07.004

11. Michor, P.W., Mumford, D.B.: Riemannian geometries on spaces of plane curves. J. Eur. Math. Soc. **8**, 1–48 (2006). https://doi.org/10.4171/JEMS/37

12. Mohammadi, B., Pironneau, O.: Applied Shape Optimization for Fluids. Oxford University Press (2009). https://doi.org/10.1093/acprof:oso/9780199546909.001.0001

13. Müller, P.M., Kühl, N., Siebenborn, M., Deckelnick, K., Hinze, M., Rung, T.: A novel p-harmonic descent approach applied to fluid dynamic shape optimization. Struct. Multidiscip. Optim. **64**(6), 3489–3503 (2021). https://doi.org/10.1007/s00158-021-03030-x

14. Pironneau, O.: On optimum profiles in Stokes flow. J. Fluid Mech. **59**(1), 117–128 (1973). https://doi.org/10.1017/s002211207300145x

15. Schulz, V.H.: A Riemannian view on shape optimization. Found. Comput. Math. **14**(3), 483–501 (2014). https://doi.org/10.1007/s10208-014-9200-5

16. Schulz, V.H., Siebenborn, M., Welker, K.: Efficient PDE constrained shape optimization based on Steklov-Poincaré-type metrics. SIAM J. Optim. **26**(4), 2800–2819 (2016). https://doi.org/10.1137/15m1029369

17. Schulz, V.H., Welker, K.: On optimization transfer operators in shape spaces. In: Schulz, V., Seck, D. (eds.) Shape Optimization, Homogenization and Optimal Control, pp. 259–275. Springer, Cham (2018). https://doi.org/10.1007/978-3-319-90469-6_13

18. Steck, D.: Lagrange multiplier methods for constrained optimization and variational problems in Banach spaces. Ph.D. thesis, Universität Würzburg (2018). https://opus.bibliothek.uni-wuerzburg.de/frontdoor/index/index/year/2018/docId/17444

19. Welker, K.: Efficient PDE constrained shape optimization in shape spaces. Ph.D. thesis, Universität Trier (2016). https://doi.org/10.25353/ubtr-xxxx-6575-788c/

20. Welker, K.: Suitable spaces for shape optimization. Appl. Math. Optim. **84**(1), 869–902 (2021). https://doi.org/10.1007/s00245-021-09788-2

On Canonical Parameterizations of 2D-Shapes

Alice Barbora Tumpach[1,2]($\boxtimes$) (iD)

[1] Institut CNRS Pauli, Oskar-Morgenstern-Platz 1, 1090 Vienna, Austria
[2] University of Lille, Cité scientifique, 59650 Villeneuve d'Ascq, France
`alice-barbora.tumpach@univ-lille.fr`
`http://math.univ-lille1.fr/~tumpach/Site/home.html`

Abstract. This paper is devoted to the study of unparameterized simple curves in the plane. We propose diverse canonical parameterization of a 2D-curve. For instance, the arc-length parameterization is canonical, but we consider other natural parameterizations like the parameterization proportional to the curvature of the curve. Both aforementionned parameterizations are very natural and correspond to a natural physical movement: the arc-length parameterization corresponds to travelling along the curve at constant speed, whereas parameterization proportional to curvature corresponds to a constant-speed moving frame. Since the curvature function of a curve is a geometric invariant of the unparameterized curve, a parameterization using the curvature function is a canonical parameterization. The main idea is that to any physically meaningful strictly increasing function is associated a natural parameterization of 2D-curves, which gives an optimal sampling, and which can be used to compare unparameterized curves in an efficient and pertinent way. An application to point correspondence in medical imaging is given.

Keywords: Canonical parameterization · Geometric Green Learning · shape space

1 Introduction

Curves in $\mathbb{R}^2$ appear in many applications: in shape recognition as outline of an object, in radar detection as the signature of a signal, as trajectories of cars etc. There are two main features of the curve: the route and the speed profile. In this paper, we are only interested in the route drawn by the curve and we will called it the unparameterized curve. An unparameterized curve can be parameterized in multiple ways, and the chosen parameterization selects the speed at which the curve is traversed. Hence a curve can be travelled with many different speed profiles, like a car can travel with different speeds (not necessarily constant) along a given road. The choice of a speed profile is called a parameterization of the curve. It may be physically meaningful or not. For instance, depending on

Supported by FWF grant I 5015-N, Institut CNRS Pauli, Vienna, Austria, and University of Lille, France.

applications, there may not be any relevant parameterization of the contour of the statue of Liberty depicted in Fig. 1. In this paper, we propose various very natural parameterization of 2D-curves. They are based on the curvature, which together with the arc-length measure form a complete set of geometric invariants or descriptors of the unparameterized curves.

2   Different Parameterizations of 2D-shapes

2.1   Arc-Length Parameterization and Signed Curvature

By 2D-shape, we mean the shape drawn by a parameterized curve in the plane. It is the ordered set of points visited by the curve. The shapes of two curves are identical if one can reparameterize one curve into the other (using a continuous increasing function). Any rectifiable planar curve admits a canonical parameterization, its *arc-length parameterization*, which draws the same shape, but with constant unit speed. The set of 2D-shapes can be therefore identified with the set of arc-length parameterized curves, which is not a vector subspace, but rather an infinite-dimensional submanifold of the space of parameterized curves (see [5]).

It may be difficult to compute an explicit formula of the arc-length parameterization of a given rectifiable curve. Fortunately, when working with a computer, one do not need it. One neither need a concrete parameterization of the curve to depict it, a sample of points on the curve is enough. To draw the statue of Liberty as in Fig. 1 left, one just needs a finite ordered set of points (the red stars). The discrete version of an arc-length parameterized curve is a uniformly sampled curve, i.e. an ordered set of equally distant points (for the euclidean metric). Resampling a curve uniformly is immediate using some appropriate interpolation function like the matlab function *spline* (the second picture in Fig. 1 shows a uniform resampling of the statue of liberty).

Consider the set of 2D simple closed curves, such as the contour of Elie Cartan's head in Fig. 2. After the choice of a starting point and a direction, there is a unique way to travel the curve at unit speed. In Fig. 2, we have drawn the velocity vector near the glasses of Elie Cartan, as well as the unit normal vector which is obtained from the unit tangent vector by a rotation of $+\frac{\pi}{2}$. These two vectors form an orthonormal basis, i.e. an element (modulo the choice of a basis of $\mathbb{R}^2$) of the Lie group $SO(2)$, which is characterized by a rotation angle. The rate of variation of this rotation angle is called the signed curvature of the curve. For instance, when moving along the external outline of the glasses, this curvature equals the inverse of the radius of the glasses. We have depicted the curvature function κ of Elie Cartan's head in Fig. 3, first line, when the parameter $s \in [0; 1]$ on the horizontal axis is proportional to arc-length, and such that the entire contour of Elie Cartan's head is travelled when the parameter reaches 1. It corresponds to a uniform sampling of the contour. The curvature function is also depicted when parameterized by two other canonical parameters, namely by the curvature-length parameter (second line) and the curvarc-length parameter (third line).

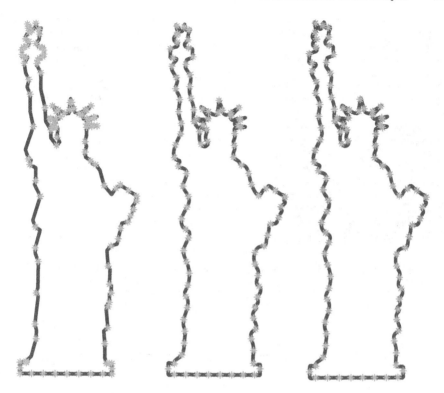

Fig. 1. The statue of Liberty (left), a uniform resampling using Matlab function spline (middle), a reconstruction of the statue using its discrete curvature (right). (Color figure online)

A discrete version of an arc-length parameterized curve is an equilateral polygon. To draw an equilateral polygon, one just need to know the length of the edges, the position of the first edge, and the angles between two successive edges. The sequence of turning-angles is the discrete version of the curvature and defines a equilateral polygon modulo rotation and translation. In Fig. 1, right, we have reconstructed the statue of Liberty using the discrete curvature.

In order to interpolate between two parameterized curves, it is easier when the domains of the parameter coincide. For this reason we will always consider curves parameterized with a parameter in $[0; 1]$. A natural parameterization is then the parameterization proportional to arc-length. It is obtained from the parameterization by arc-length by dividing the arc-length parameter by the length of the curve L. The corresponding curvature function is also defined on $[0; 1]$ and is obtained from the curvature function parameterized by arc-length by compressing the x-axis by a factor L. To recover the initial curve from the curvature function associated to the parameter $s \in [0; 1]$ proportional to arc-length, one only need to know the length of the curve.

Fig. 2. Elie Cartan and the moving frame associated to the contour of his head.

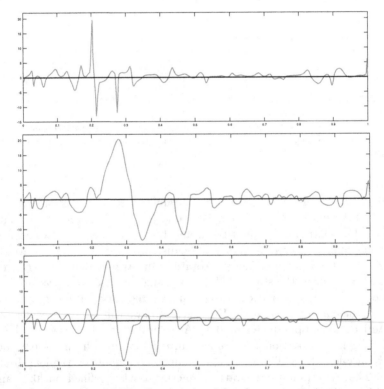

Fig. 3. Signed curvature of Elie Cartan's head for the parameterization proportional to arc-length (first line), proportional to the curvature-length (second line), and proportional to the curvarc length (third line).

2.2 Parameterization Proportional to Curvature-Length

In the same spirit as the scale space of T. Lindeberg [2], and the curvature scale space of Mackworth and Farcin Mokhtarian [3], we now define another very natural parameterization space of 2D curves. Its relies on the fact that the integral of the absolute value of the curvature κ is an increasing function on the interval $[0; 1]$, strictly increasing when there are no flat pieces. In that case the function

$$r(s) = \frac{\int_0^s |\kappa(s)| ds}{\int_0^1 |\kappa(s)| ds} \tag{1}$$

(where κ denotes the curvature of the curve) belongs to the group of orientation preserving diffeomorphisms of the parameter space $[0; 1]$, denoted by $\mathrm{Diff}^+([0; 1])$. Note that its inverse $s(r)$ can be computed graphically using the fact that its graph is the symmetric of the graph of $r(s)$ with respect to $y = x$. The contour of Elie Cartan's head can be reparameterized using the parameter $r \in [0; 1]$ instead of the parameter $s \in [0; 1]$. In Fig. 4 upper left, we have depicted the graph of the function $s \mapsto r(s)$. A uniform sampling with respect to the parameter r is obtain by uniformly sampling the vertical-axis (this is materialized by the green equidistributed horizontal lines) and resampling Elie Cartan's head at the sequence of values of the s-parameter given by the abscissa of the corresponding points on the graph of r (where a green line hits the graph of r a red vertical line materializes the corresponding abscissa). One sees that this reparameterization naturally increases the number of points where the 2D contour is the most curved, and decreases the number of points on nearly flat pieces of the contour. For a given number of points, it gives an optimal way to store the information contained in the contour. The quantity

$$C = L \int_0^1 |\kappa(s)| ds, \tag{2}$$

where $s \in [0; 1]$ is proportional to arc-length, is called the *total curvature-length* of the curve. It is the length of the curve drawn in SO(2) by the moving frame associated with the arc-length parameterized curve. For this reason we call this parameterization the *parameterization proportional to curvature-length*. In the right picture of Fig. 4, we show the corresponding resampling of the contour of Elie Cartan's head.

This resampling can naturally be adapted in the case of flat pieces resulting in a sampling where there is no points between two points on the curve joint by a straight line. In the left picture of Fig. 5, we have depicted a sampling of the statue of Liberty proportional to curvature-length. Note that there are no points on the base of the statue. The corresponding parameterization has the advantage of concentrating on the pieces of the contour that are very complex, i.e. where there is a lot of curvature, and not distributing points on the flat pieces which are easy to reconstruct (connecting two points by a straight line is easy, but drawing the moustache of Elie Cartan is harder and needs more information).

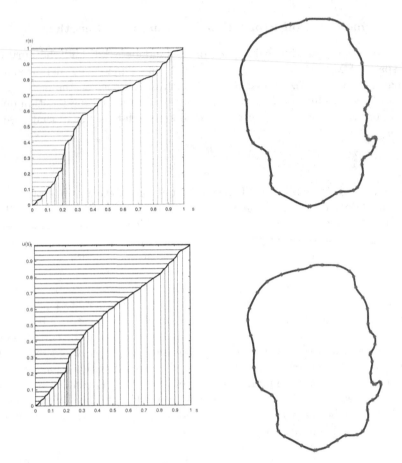

Fig. 4. First line: Integral of the (renormalized) absolute value of the curvature (left), and corresponding resampling of Elie's Cartan head (right). Second line: Integral of the (renormalized) curvarc length (left), and corresponding resampling of Elie's Cartan head (right). (Color figure online)

The drawback of using the parameterization proportional to curvature-length is that one can not reconstruct the flat pieces of a shape without knowing their lengths (remember that the parameterization proportional to curvature-length put no point at all on flat pieces). For this reason we propose a parameterization intermediate between arc-length parameterization and curvature-length parameterization. We call it *curvarc-length parameterization*.

2.3   Curvarc-Length Parameterization

In order to define the curvarc-length parameterization, we consider the triple $(P(s), v(s), n(s))$, where $P(s)$ is the point of the shape parameterized proportionally to arc-length with $s \in [0;1]$, $v(s)$ and $n(s)$ the corresponding unit tangent vector and unit normal vector respectively. It defines an element of the group of rigid motions of $\mathbb{R}^2$, called the special Euclidean group and denoted by $\mathrm{SE}(2) := \mathbb{R}^2 \rtimes \mathrm{SO}(2)$. The point $P(s)$ corresponds to the translation part of the rigid motion, it is the vector of translation needed to move the origin to the point of the curve corresponding to the parameter value s. The moving frame $O(s)$ defined by $\boldsymbol{v}(s)$ and $\boldsymbol{n}(s)$ is the rotation part of the rigid motion. One has the following equations:

$$\frac{dP}{ds} = L\boldsymbol{v}(s) \quad \text{and} \quad O(s)^{-1}\frac{d}{ds}O(s) = \begin{pmatrix} 0 & -\kappa(s) \\ \kappa(s) & 0 \end{pmatrix}, \tag{3}$$

where L is the length of the curve. Endow $\mathrm{SE}(2) := \mathbb{R}^2 \rtimes \mathrm{SO}(2)$ with the structure of a Riemannian manifold, product of the plane and the Lie group $\mathrm{SO}(2) \simeq \mathbb{S}^1$. Then the norm of the tangent vector to the curve $s \mapsto (P(s), \boldsymbol{v}(s), \boldsymbol{n}(s))$ is $L + |\kappa(s)|$. Therefore the length of the $\mathrm{SE}(2)$-valued curve is $L + \int_0^1 |\kappa(s)|ds = L + \frac{C}{L}$. We call it the total curvarc-length. It follows that the following function

$$u(s) = \frac{\int_0^s (L + |\kappa(s)|)ds}{\int_0^1 (L + |\kappa(s)|)ds} \tag{4}$$

defines a reparameterization of $[0; 1]$. More generally, one can use the following canonical parameter to reparameterize a curve in a canonical way:

$$u_\lambda(s) = \frac{\int_0^s L\lambda + |\kappa(s)|)ds}{\int_0^1 L\lambda + |\kappa(s)|)ds}, \tag{5}$$

where s is the arc-length parameter. In Fig. 5 we show the resulting sampling of the Statue of Liberty for different values of λ. Note that for $\lambda = 0$, one recovers the curvature-length parameterization (1), for $\lambda = 1$ one obtains the curvarc-length parameterization (4), and when $\lambda \to +\infty$ the parameterization tends to the arc-length parameterization.

3   Application to Medical Imaging: Parameterization of Bones

In the analysis of diseases like Rheumatoid Arthritis, one uses X-ray scans to evaluate how the disease analogous the bones. One effect of Rheumatoid Arthritis is erosion of bones, another is joint shrinking [1]. In order to measure joint space, one has to solve a point correspondence problem. For this, one uses landmarks along the contours of bones as in Fig. 6. These landmarks have to be placed at the same anatomical positions for every patient. In Fig. 7 they are placed using a method by Hans Henrik Thodberg [4], based on minimum description length which minimizes the description of a Principal Component Analysis model capturing the variability of the landmark positions. For instance in Fig. 7 left, the landmark number 56 should always be in the middle of the head of the bone

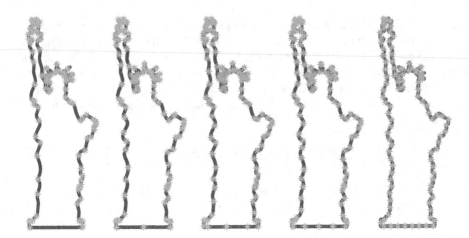

Fig. 5. Resampling of the statue of Liberty using Eq. (5) for (from left to right) $\lambda = 0$ (curvature-length parameterization); $\lambda = 0.3$; $\lambda = 1$(curvarc-length parameterization); $\lambda = 2$; $\lambda = 100$. The parameterization tends to arc-length parameterization when $\lambda \to \infty$.

because it is used to measure the width between two adjacent bones in order to detect rheumatoid arthritis.

Although the method by Hans Henrik Thodberg gives good results, it is computationally expensive. In this paper we propose to recover similar results with a quicker algorithm. It is based on the fact that any geometrically meaningful parameterization of a contour can be expressed using the arc-length measure and the curvature of the contour, which are the only geometric invariants of a 2D-curve (modulo translation and rotation). It follows that the parameterization calculated by Thodberg's algorithm should be recovered as a parameterization expressed using arc-length and curvature. We investigate a 2 parameter family of parameterizations defined by

$$u(s) = \frac{\int_0^s (c * L + |\kappa(s)|^\lambda) ds}{\int_0^1 (c * L + |\kappa(s)|^\lambda) ds} \tag{6}$$

where c and λ are positive parameters and where L is the length of the curve and κ its curvature function. We recover an analogous parameterization to the one given by Thodberg's algorithm with $c = 1$ and $\lambda = 7$ at real-time speed (gain of 2 order of magnitude). Hence, instead of running Thodberg's algorithm on new samples (which takes many minutes on a Mac M1), one can use the optimal parameterization function (6) to place landmarks on the bones at real time speed (Fig. 8).

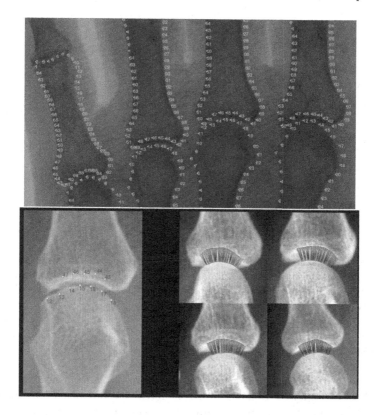

Fig. 6. Landmarks on bones used to measure joint space (courtesy of [1]).

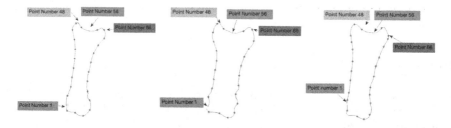

Fig. 7. Point correspondence on three different bones using the method of [4].

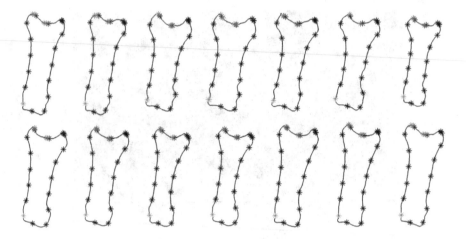

Fig. 8. 14 bones parameterized by Thodberg's algorithm on one hand and the parameterization defined by (6) with $c = 1$ and $\lambda = 7$ on the other hand (the two parameterizations are superposed). The colored points corresponds to points labelled 1, 48, 56, 66. They overlap for the two methods.

4 Conclusion

We proposed diverse canonical parameterization of 2D-contours, which are expressed using arc-length and curvature of curves. The curvature-length parameterization and the curvarc-length parameterization are very natural examples, since they corresponds to a constant-speed moving frame in $SO(2)$ and $SE(2)$. We present an application to the point matching problem in medical imaging consisting of automatically labeling key points along the contour of bones. We recover a parameterization analogous to that proposed by Thodberg at real-time speed. Since we have a family of two-parameter parameterizations, fine-tuning can be applied to our results to further improve the point matching.

References

1. Langs, G., Peloschek, P., Bischof, H., Kainberger, F.: Automatic quantification of joint space narrowing and erosions in Rheumatoid Arthritis. IEEE Trans. Med. Imaging **28**(1) (2009)
2. Lindeberg, T.: Image matching using generalized scale-space interest points. J. Math. Imaging Vis. **52**(1), 3–36 (2015)
3. Mokhtarian, F., Mackworth, A.K.: A theory of multiscale, curvature-based shape representation for planar curves. IEEE Trans. Pattern Anal. Mach. Intell. **14**(8), 789–805 (1992)
4. Taylor, C., Noble, J.A. (eds.): IPMI 2003. LNCS, vol. 2732. Springer, Heidelberg (2003). https://doi.org/10.1007/b11820
5. Tumpach, A.B., Preston, S.C.: Quotient elastic metrics on the manifold of arc-length parameterized plane curves. J. Geom. Mech. **9**(2), 227–256 (2017)

Shape Spaces of Nonlinear Flags

Ioana Ciuclea[1]([✉]) [ID], Alice Barbora Tumpach[2,3] [ID], and Cornelia Vizman[1] [ID]

[1] West University of Timişoara, bld. Vasile Pârvan no. 4, 300223 Timişoara, Romania
ioana.ciuclea@e-uvt.ro, cornelia.vizman@e-uvt.ro
[2] Institut CNRS Pauli, Oskar-Morgenstern-Platz 1, 1090 Vienna, Austria
alice-barbora.tumpach@univ-lille.fr
[3] University of Lille, Cité scientifique, 59650 Villeneuve d'Ascq, France
http://math.univ-lille1.fr/~tumpach/Site/home.html

Abstract. The shape space considered in this article consists of surfaces embedded in $\mathbb{R}^3$, that are decorated with curves. It is a special case of the Fréchet manifolds of nonlinear flags, i.e. nested submanifolds of a fixed type. The gauge invariant elastic metric on the shape space of surfaces involves the mean curvature and the normal deformation, i.e. the sum and the difference of the principal curvatures κ_1, κ_2. The proposed gauge invariant elastic metrics on the space of surfaces decorated with curves involve, in addition, the geodesic and normal curvatures κ_g, κ_n of the curve on the surface, as well as the geodesic torsion τ_g.

More precisely, we show that, with the help of the Euclidean metric, the tangent space at (C, Σ) can be identified with $C^\infty(C) \times C^\infty(\Sigma)$ and the gauge invariant elastic metrics form a 6-parameter family:

$$\mathcal{G}_{(C,\Sigma)}(h_1, h_2) = a_1 \int_C (h_1\kappa_g + h_2|_C \kappa_n)^2 d\ell \quad + a_2 \int_\Sigma (h_2)^2 (\kappa_1 - \kappa_2)^2 dA$$

$$+ b_1 \int_C (D_s h_1 - h_2|_C \tau_g)^2 d\ell \quad + b_2 \int_\Sigma (h_2)^2 (\kappa_1 + \kappa_2)^2 dA$$

$$+ c_1 \int_C (D_s(h_2|_C) + h_1\tau_g)^2 d\ell \quad + c_2 \int_\Sigma |\nabla h_2|^2 dA,$$

where $h_1 \in C^\infty(C), h_2 \in C^\infty(\Sigma)$.

Keywords: Shape space · Geometric Green Learning · Gauge invariant elastic metrics

1 Introduction

In this paper we use the elastic metrics on parameterized curves ([5,10]) and parameterized surfaces ([4]) in order to endow the shape space of surfaces decorated with curves with a family of Riemannian metrics. This shape space of

Alice Barbora Tumpach is supported by FWF grant I 5015-N, Institut CNRS Pauli, and University of Lille. The first and the third authors are supported by a grant of the Romanian Ministry of Education and Research, CNCS-UEFISCDI, project number PN-III-P4-ID-PCE-2020-2888, within PNCDI III.

F. Nielsen and F. Barbaresco (Eds.): GSI 2023, LNCS 14071, pp. 41–50, 2023.
https://doi.org/10.1007/978-3-031-38271-0_5

decorated surfaces is an example of Fréchet manifold of nonlinear flags, studied in [3]. These consist of nested submanifolds of an ambient manifold M, of a fixed type $S_1 \xrightarrow{\iota_1} S_2 \xrightarrow{\iota_2} \cdots \to S_r$. In our case the ambient manifold is $\mathbb{R}^3$ and the type is the embedding of the equator into the sphere: $\mathbb{S}^1 \xrightarrow{\iota} \mathbb{S}^2$.

We emphasize that we do not use the quotient elastic metrics on curves and surfaces, but rather their gauge invariant relatives (see [8,9]). Indeed, following [6,7] for surfaces in $\mathbb{R}^3$, and [2] for curves in $\mathbb{R}^3$, we construct degenerate metrics on parameterized curves and surfaces by first projecting an arbitrary variation of a given curve or surface onto the space of vector fields perpendicular to the curve or surface (for the Euclidean product of $\mathbb{R}^3$) and then applying the elastic metric on this component. By construction, vector fields tangent to the curve or surface will have vanishing norms, leading to a degenerate metric on pre-shape space. However, since the degeneracy is exactly along the tangent space to the orbit of the reparameterization group, these degenerate metrics define Riemannian (i.e. non-degenerate) metrics on shapes spaces of curves and surfaces (as in Theorem 3 in [8]). The following advantages of this procedure can be mentioned:
• there is no need to compute a complicated horizontal space in order to define a Riemannian metric on shape space
• the length of paths in shape space equals the length of any of their lifts for the corresponding degenerate metric, a property called *gauge invariance* in [6,7].
• the resulting Riemannian metric on shape spaces can be easily expressed in terms of geometric invariants of curves and surfaces, leading to expressions that are completely independant of parameterizations.

In this paper, we use the gauge invariant (degenerate) metrics on parameterized curves and surfaces obtained from the elastic metrics via the procedure described above in order to define Riemannian metrics on shape spaces of curves and surfaces. Then we embed the shape space of nonlinear flags consisting of surfaces decorated with curves into the Cartesian product of the shape space of curves in $\mathbb{R}^3$ with the shape space of surfaces in $\mathbb{R}^3$. The Riemannian metric obtained on the shape space of nonlinear flags can be made explicit thanks to a precise description of its tangent space (Theorem 1) and thanks to the geometric expressions of the metrics used on curves and surfaces, leading to a 6-parameter family of natural Riemannian metrics (Theorem 2).

2   Shape Spaces of Decorated Surfaces as Manifolds of Nonlinear Flags

We will consider the *shape space* of nonlinear flags consisting of pairs (C, Σ) such that C is a curve on the surface Σ embedded in $\mathbb{R}^3$. We will restrict our attention to surfaces of genus 0, and simple curves (the complement to the curve in the surface has only two connected components), but our construction can be extended without substantial changes to surfaces of genus g and to a finite number of curves. The general setting for the Fréchet manifolds on nonlinear flags of a given type $S_1 \xrightarrow{\iota_1} S_2 \xrightarrow{\iota_2} \cdots \to S_r$ can be found in [3]. Our type is the embedding ι of the unit circle $\mathbb{S}^1$ in the unit sphere $\mathbb{S}^2$ as the equator, thus the shape space of nonlinear flags is in this case $\mathcal{F} := \text{Flag}_{\mathbb{S}^1 \overset{\iota}{\hookrightarrow} \mathbb{S}^2}(\mathbb{R}^3)$. Examples of elements $(C, \Sigma) \in \mathcal{F}$ are given in Fig. 1.

Fig. 1. Examples of elements in the shape space of nonlinear Flags: a black belt on a human body (left), the model nonlinear flag consisting of the equator (in white) on the sphere, and a black collar on a cat (right)

We will be interested in deforming flags. To this aim, we will represent a general flag (C, Σ) using an embedding $F : \mathbb{S}^2 \to \mathbb{R}^3$ such that the image of the restriction $F \circ \iota$ of F to the equator is C. The pair $(F \circ \iota, F)$ is also called a *parameterization* of the flag (C, Σ). The space of parametrized flags is

$$\mathcal{P} := \{F : \mathbb{S}^2 \to \mathbb{R}^3, F \text{ is an embedding}\}.$$

It is called the *pre-shape space* of flags since objects with same shape but different parameterizations correspond to different points in $\mathcal{P}$. The set $\mathcal{P}$ is a manifold, as an open subset of the linear space $\mathcal{C}^\infty(\mathbb{S}^2, \mathbb{R}^3)$ of smooth functions from $\mathbb{S}^2$ to $\mathbb{R}^3$. The tangent space to $\mathcal{P}$ at F, denoted by $T_F\mathcal{P}$, is therefore just $\mathcal{C}^\infty(\mathbb{S}^2, \mathbb{R}^3)$.

There is a natural projection π from the space of parameterized flags $\mathcal{P}$ onto the space of flags $\mathcal{F}$ given by

$$\pi(F) = ((F \circ \iota)(\mathbb{S}^1), F(\mathbb{S}^2)). \tag{1}$$

Since we are only interested in unparameterized nonlinear flags, we would like to identify pairs of parameterized curves and surfaces that can be related through reparameterization. The reparametrization group G is the group of diffeomorphisms γ of $\mathbb{S}^2$ which restrict to a diffeomorphism of the equator $\iota : \mathbb{S}^1 \hookrightarrow \mathbb{S}^2$:

$$G = \{\gamma \in \text{Diff}(\mathbb{S}^2) : \gamma \circ \iota = \iota \circ \bar{\gamma} \text{ for some } \bar{\gamma} \in \text{Diff}(\mathbb{S}^1)\}.$$

The group G is an infinite-dimensional Fréchet Lie group whose Lie algebra is the space of vector fields on $\mathbb{S}^2$ whose restriction to the equator is tangent to the equator. The right action of G on $\mathcal{P}$ is given by $F \cdot \gamma := F \circ \gamma$. It's a principal action for the principal G-bundle $\pi : \mathcal{P} \to \mathcal{F}$.

The elements in $\mathcal{P}$ obtained by following a fixed parameterized flag $F \in \mathcal{P}$ when acted on by all elements of G is called the *G-orbit* of F or the *equivalence class* of F under the action of G, and will be denoted by $[F]$. The orbit of $F \in \mathcal{P}$ is characterized by the surface $\Sigma := F(\mathbb{S}^2)$ and the curve $C := (F \circ \iota)(\mathbb{S}^1)$, hence $\pi(F) = (C, \Sigma)$ (see (1)). The elements in the orbit $[F] = \{F \circ \gamma, \text{ for } \gamma \in G\}$ are all possible parameterizations of (C, Σ) of the form $(F \circ \iota, F)$. For instance in Fig. 2 one can see some parameterized hands with bracelets that are elements of the same orbit. The set of orbits of $\mathcal{P}$ under the group G is called the *quotient space* and will be denoted by $\mathcal{P}/G$.

Fig. 2. Examples of elements in the same orbit under the group of reparameterizations: a hand with a white bracelet presented under different parameterizations.

Proposition 1. *The shape space $\mathcal{F}$ is isomorphic to the quotient space of the pre-shape space $\mathcal{P}$ by the shape-preserving group $G = \mathrm{Diff}(\mathbb{S}^2; \iota)$:*

$$\mathcal{F} = \mathcal{P}/G.$$

The shape space $\mathcal{F} = \mathcal{P}/G$ is a smooth manifold and the canonical projection $\pi : \mathcal{P} \to \mathcal{F}$, $F \mapsto [F]$ is a submersion (see for instance [3]). The kernel of the differential of this projection is called the *vertical space*. It is the tangent space to the orbit of $F \in \mathcal{P}$ under the action of the group G.

Proposition 2. *The vertical space Ver_F of π at some embedding $F \in \mathcal{P}$ is the space of vector fields $X_F \in \mathcal{C}^\infty(\mathbb{S}^2, \mathbb{R}^3)$ such that the deformation vector field $X_F \circ F^{-1}$ is tangent to the surface $\Sigma := F(\mathbb{S}^2)$ and such that the restriction of $X_F \circ F^{-1}$ to the curve $C := F \circ \iota(\mathbb{S}^1)$ is tangent to C.*

The normal bundle Nor is the vector bundle over the pre-shape space $\mathcal{P}$, whose fiber over an embedding F is the quotient vector space

$$\mathrm{Nor}_F := T_F\mathcal{P}/\mathrm{Ver}_F.$$

Proposition 3. *The right action of G on $\mathcal{P}$ induces an action on $T\mathcal{P}$ which preserves the vertical bundle, hence it descends to an action on Nor by vector bundle homomorphisms. The quotient bundle Nor/G can be identified with the tangent bundle $T\mathcal{F}$.*

Consider a nonlinear flag (C, Σ). Let us denote by ν the unit normal vector field on the oriented surface Σ, and by t the unit vector field tangent to the oriented curve C. Set $n := \nu \times t$ the unit normal to the curve C contained in the tangent space to the surface Σ. The triple (t, n, ν) is an orthonormal frame along C, called the *Darboux frame*. We will denote by $\langle \cdot, \cdot \rangle$ the Euclidian scalar product on $\mathbb{R}^3$.

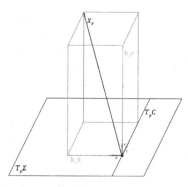

Fig. 3. Deformation vector field and Darboux frame (t, n, ν), where ν is the unit normal vector field on the oriented surface σ, t is the unit vector field tangent to the oriented curve C and n $:= \nu \times t$ is the unit normal to the curve C contained in the tangent space to the surface.

Theorem 1. *Let F be a parameterization of (C, Σ). Consider the linear surjective map*

$$\Psi_F : T_F \mathcal{P} \simeq \mathcal{C}^\infty(\mathbb{S}^2, \mathbb{R}^3) \to \mathcal{C}^\infty(C) \times \mathcal{C}^\infty(\Sigma), \tag{2}$$

which maps $X_F \in T_F \mathcal{P}$ to (h_1, h_2) defined by

$$h_1 := \langle (X_F \circ \iota) \circ (F \circ \iota)^{-1}, n \rangle \in \mathcal{C}^\infty(C),$$
$$h_2 := \langle X_F \circ F^{-1}, \nu \rangle \in \mathcal{C}^\infty(\Sigma).$$

Then the kernel of Ψ_F is the vertical subspace Ver_F, hence Ψ_F defines a map from the quotient space $\mathrm{Nor}_F = T_F \mathcal{P} / \mathrm{Ver}_F$ into $\mathcal{C}^\infty(C) \times \mathcal{C}^\infty(\Sigma)$. The resulting bundle map Ψ is G-invariant providing an isomorphism between the tangent space $T_{(C, \Sigma)} \mathcal{F}$ and $\mathcal{C}^\infty(C) \times \mathcal{C}^\infty(\Sigma)$.

Proof. Consider X_F such that $\Psi_F(X_F) = 0$. Since $h_2 = 0$, $X_F \circ F^{-1}$ is a vector field tangent to Σ. Since $h_1 = 0$, the restriction of $X_F \circ F^{-1}$ to the curve C, given by $(X_F \circ \iota) \circ (F \circ \iota)^{-1}$, is tangent to Σ and orthogonal to n, hence it is tangent to C. Thus, by Proposition 2, the kernel of Ψ is exactly Ver_F.

Let us show that Ψ is G-invariant, i.e. that for $\gamma \in G$,

$$\Psi_F(X_F) = \Psi_{F \circ \gamma}(X_F \circ \gamma). \tag{3}$$

One has $\pi(F \circ \gamma) = \pi(F) = (C, \Sigma)$. Moreover the normal vector fields $\nu : \Sigma \to \mathbb{R}^3$ and n $: C \to \mathbb{R}^3$ do not depend on the parameterizations of Σ and C. For $\gamma \in G$, we have

$$(X_F \circ \gamma) \circ (F \circ \gamma)^{-1} = X_F \circ F^{-1}$$

as deformation vector fields on Σ. On the other hand, using the fact that $\gamma \circ \iota = \iota \circ \bar{\gamma}$ for some $\bar{\gamma} \in \mathrm{Diff}(\mathbb{S}^1)$, we get

$$(X_F \circ \gamma \circ \iota) \circ (F \circ \gamma \circ \iota)^{-1} = (X_F \circ \iota) \circ (F \circ \iota)^{-1}$$

as deformation vector fields on C. The invariance property (3) follows.

The projection $\pi \; : \mathcal{P} \to \mathcal{F} = \mathcal{P}/G$ is a principal G-bundle, hence the G-action preserves the vertical bundle Ver. It induces a well-defined G-action on Nor : for $\gamma \in G$, the class $[X_F] \in \mathrm{Nor}_F$ is mapped to the class $[X_F \circ \gamma] \in \mathrm{Nor}_{F \circ \gamma}$. By G-invariance of Ψ, we get a well-defined map on Nor $/G$ which maps isomorphically the tangent space $T_{(C,\Sigma)}\mathcal{F}$ into $\mathcal{C}^\infty(C) \times \mathcal{C}^\infty(\Sigma)$.

Remark 1. The construction in Theorem 1 also works for complete nonlinear flags of length k in $\mathbb{R}^{k+1}$. These are nested submanifolds $N_1 \subset ... \subset N_k \subset \mathbb{R}^{k+1}$ with $\dim N_i = i$. In this case, the tangent space at a complete nonlinear flag can be identified with $C^\infty(N_1) \times ... \times C^\infty(N_k)$.

Remark 2. For the pre-shape space of embedded surfaces, there is a natural section of the projection $T_F\mathcal{P} \to \mathcal{C}^\infty(\Sigma)$, given by variations that are in the direction of the normal vector field ν to the surface Σ. In this case, the Euclidean metric on $\mathbb{R}^3$ induces a connection on the principal $\mathrm{Diff}(\mathbb{S}^2)$-bundle $\mathcal{P} \to \mathcal{S}$, where $\mathcal{S}$ denotes the shape space of surfaces. The problem of finding similar principal connections for the G-bundle $\mathcal{P} \to \mathcal{F}$ will be addressed in [1].

Remark 3. As in Proposition 2.9 in [3], the shape space $\mathcal{F}$ can be seen as a homogeneous space for the compactly supported diffeomorphism group $\mathrm{Diff}_c(\mathbb{R}^3)$, with origin the nonlinear flag $\mathbb{S}^1 \subset \mathbb{S}^2$ in $\mathbb{R}^3$.

3   Riemannian Metrics on Shape Spaces of Nonlinear Flags

As in [6] and [7], we endow the pre-shape space of parameterized curves and surfaces with a family of gauge invariant metrics which descend to a family of Riemannian metric on shape spaces of curves and surfaces. The construction is explained in Subsect. 3.1. The Riemannian metrics on parameterized curves and surfaces used in this construction are the elastic metrics given in Subsect. 3.2. The expression of the Riemannian metrics obtained on nonlinear flags in terms of the geometric invariants of curves and surfaces is given in Subsect. 3.3.

3.1   Procedure to Construct the Riemannian Metrics

In order to construct a Riemannian metric on the space $\mathcal{F}$ of nonlinear flags, we proceed as follows:

1. we embed our shape space $\mathcal{F}$ of surfaces decorated with curves in $\mathcal{S}_1 \times \mathcal{S}_2$, with $\mathcal{S}_1$ the shape space of curves and $\mathcal{S}_2$ the shape space of surfaces.
2. we choose a family $g^{a,b}$ of $\mathrm{Diff}^+(\mathbb{S}^1)$-invariant metrics on the space of parameterized curves $\mathcal{P}_1$ (Eq. (6))
3. the family $g^{a,b}$ defines a family of Riemannian metrics on the shape space of curves $\mathcal{S}_1$ by restricting to the normal variations of curves
4. we choose a family $g^{a',b',c'}$ of $\mathrm{Diff}^+(\mathbb{S}^2)$-invariant metrics on the space of parameterized surfaces $\mathcal{P}_2$ (Eq. (7)).

5. the family $g^{a',b',c'}$ defines a family of Riemannian metrics on the shape space of surfaces $\mathcal{S}_2$ by restricting to the normal variations of surfaces

6. the product of these metrics is then restricted to $\mathcal{F}$ using the characterization of the tangent space to $\mathcal{F}$ given in Theorem 1.

Remark 4. An equivalent procedure is to pull back to $\mathcal{P}$, via $F \mapsto (F \circ \iota, F) \in \mathcal{P}_1 \times \mathcal{P}_2$, the sum of the gauge invariant elastic metrics on the preshape space $\mathcal{P}_1$ for curves and $\mathcal{P}_2$ for surfaces. The result is gauge invariant under G, so it descends to a Riemannian metric on the shape space $\mathcal{F}$.

$$
\begin{array}{ccc}
F \in \mathcal{P} \longhookrightarrow & \mathcal{P}_1 \times \mathcal{P}_2 \ni (F \circ \iota, F) \\
{\scriptstyle G}\Big\downarrow & \Big\downarrow {\scriptstyle \mathrm{Diff}(\mathbb{S}^1) \times \mathrm{Diff}(\mathbb{S}^2)} \\
(C, \Sigma) \in \mathcal{F} \longhookrightarrow & \mathcal{S}_1 \times \mathcal{S}_2 \ni (C, \Sigma).
\end{array}
\tag{4}
$$

3.2 Elastic Metrics on Manifolds of Parameterized Curves and Surfaces

The family of metrics measuring deformations of curves that we will use is the family of $\mathrm{Diff}^+(\mathbb{S}^1)$-invariant elastic metrics on parameterized curves $\mathcal{P}_1$ in [5]:

$$
g_f^{a,b}(h_1, h_2) = \int_C \left[a(D_s h_1^{\|})(D_s h_2^{\|}) + b(D_s h_1^\perp)(D_s h_2^\perp) \right] d\ell,
\tag{5}
$$

where $f \in \mathcal{P}_1$ is a parameterization of the curve C, $h_i \in T_f\mathcal{P}_1$ are tangent vectors to the space of parameterized curves, $d\ell = \|\dot{f}(t)\| dt$, $D_s h(t) = \frac{\dot{h}(t)}{\|\dot{f}(t)\|}$ is the arc-length derivative of the variation h, $D_s h^{\|} = \langle D_s h, \mathrm{t} \rangle$ is the component along the unit tangent vector field $\mathrm{t} = \frac{\dot{f}}{\|\dot{f}\|}$ to the curve, $D_s h^\perp = D_s h - \langle D_s h, \mathrm{t} \rangle \mathrm{t}$ is the component orthogonal to the tangent vector t. Here the a-term measures streching of the curve, while the b-term measures its bending. Note that this metric is degenerate: it has a kernel induced by translations of curves.

Let δf denote a perturbation of a parametrized curve $f : \mathbb{S}^1 \to \mathbb{R}^3$, and let $(\delta r, \delta t)$ denote the corresponding variation of the speed $r := \|\dot{f}(t)\|$ and of the unit tangent vector field t. It is easy to check that the squared norm of δf for the metric (5) reads:

$$
\mathcal{E}'_f(\delta f) := g_f^{a,b}(\delta f, \delta f) = a \int_{\mathbb{S}^1} \left(\frac{\delta r}{r} \right)^2 d\ell + b \int_{\mathbb{S}^1} |\delta t|^2 d\ell.
\tag{6}
$$

The family of metrics measuring deformations of surfaces that we will use is the family of $\mathrm{Diff}^+(\mathbb{S}^2)$-invariant metrics introduced in [4] and called elastic metrics. Let δF denote a perturbation of a parametrized surface F, and let $(\delta g, \delta \nu)$ denote the corresponding perturbation of the induced metric $g = F^*\langle \cdot, \cdot \rangle_{\mathbb{R}^3}$ and of the unit normal vector field ν. Then the squared norm of δF, namely $g_F^{a',b',c'}(\delta F, \delta F)$, is:

$$
\mathcal{E}''_F(\delta F) = a' \int_{\mathbb{S}^2} \mathrm{Tr}((g^{-1}\delta g)_0)^2 dA + b' \int_{\mathbb{S}^2} (\mathrm{Tr}(g^{-1}\delta g))^2 dA + c' \int_{\mathbb{S}^2} |\delta \nu|^2 dA
\tag{7}
$$

where B_0 is the traceless part of a 2×2-matrix B defined as $B_0 = B - \frac{\mathrm{Tr}(B)}{2} I_{2 \times 2}$. The a'-term measures area-preserving changes in the induced metric g, the b'-term measures changes in the area of patches, and the c'-term measures bending. Similarly to the case of curves, this metric also has kernel induced by translations.

3.3 Geometric Expression of the Riemannian Metrics on Manifolds of Decorated Surfaces

In this subsection we restrict the reparametrization invariant metrics (6) and (7) to normal variations. This allows us to express them with the help of the principal curvatures κ_1 and κ_2 of the surface, geodesic and normal curvatures κ_g, κ_n of the curve on the surface, as well as its geodesic torsion τ_g. We recall the identities involving $(\mathrm{t}, \mathrm{n}, \nu)$, the Darboux frame:

$$\dot{\mathrm{t}} = r(\kappa_g \,\mathrm{n} + \kappa_n \nu), \quad \dot{\mathrm{n}} = r(-\kappa_g \,\mathrm{t} + \tau_g \nu), \quad \dot{\nu} = r(-\kappa_n \,\mathrm{t} - \tau_g \,\mathrm{n}). \tag{8}$$

For functions h on the curve we will use the arc-length derivative $D_s h = \dot{h}/r$, because it is invariant under reparametrizations.

Moreover, we split the b-term in (6) into two terms in order to put different weights on the variations along ν and n. This leads to the following result :

Theorem 2. *The gauge invariant elastic metrics for parameterized curves respectively surfaces lead to a 6-parameter family of Riemannian metrics on the shape space of embedded surfaces decorated with curves:*

$$\mathcal{G}_{(C,\Sigma)}(h_1, h_2) = a_1 \int_C (h_1 \kappa_g + h_2|_C \kappa_n)^2 d\ell \quad + a_2 \int_\Sigma (h_2)^2 (\kappa_1 - \kappa_2)^2 dA$$

$$+ b_1 \int_C (D_s h_1 - h_2|_C \tau_g)^2 d\ell \quad + b_2 \int_\Sigma (h_2)^2 (\kappa_1 + \kappa_2)^2 dA$$

$$+ c_1 \int_C (D_s(h_2|_C) + h_1 \tau_g)^2 d\ell \quad + c_2 \int_\Sigma |\nabla h_2|^2 dA, \tag{9}$$

for $h_1 \in \mathcal{C}^\infty(C)$ and $h_2 \in \mathcal{C}^\infty(\Sigma)$.

In the proof we will use the following two lemmas:

Lemma 1. *Given the normal variation $\delta f = h_1 \,\mathrm{n} + h_2|_C \nu$ of the parametrized curve $f = F \circ \iota$ on the parametrized surface F, the variation of the speed r and of the unit tangent vector field t are*

$$\delta r = -r(h_1 \kappa_g + h_2|_C \kappa_n), \quad \delta \mathrm{t} = (\tfrac{1}{r}\dot{h}_1 - h_2|_C \tau_g)\,\mathrm{n} + (\tfrac{1}{r}\dot{h}_2|_C + h_1 \tau_g)\nu$$

Proof. For $f_\varepsilon = f + \varepsilon(h_1 \,\mathrm{n} + h_2 \nu) + O(\varepsilon^2)$, we get

$$r_\varepsilon^2 = r^2 - 2\varepsilon r^2 (h_1 \kappa_g + h_2|_C \kappa_n) + O(\varepsilon^2)$$

using the well known identities (8). Thus $2r\delta r = -2r^2(h_1\kappa_g + h_2|_C\kappa_n)$, hence the first identity. We use it in the computation of the variation of the unit tangent goes as follows:

$$\delta\, \mathbf{t} = \tfrac{1}{r}\delta \dot{f} - \tfrac{\delta r}{r}\mathbf{t} = \tfrac{1}{r}(-r(h_1\kappa_g + h_2|_C\kappa_n))\mathbf{t} + (\dot{h}_1 - rh_2|_C\tau_g)\mathbf{n}$$
$$+ (\dot{h}_2|_C + rh_1\tau_g)\nu) + (h_1\kappa_g + h_2|_C\kappa_n)\mathbf{t} = (\tfrac{\dot{h}_1}{r} - h_2|_C\tau_g)\mathbf{n} + (\tfrac{\dot{h}_2|_C}{r} + h_1\tau_g)\nu,$$

hence the second identity.

Lemma 2. *Given the normal variation $\delta F = h\nu$ of the parametrized surface F, the variation of the unit normal vector field ν and the gradient of h with respect to the induced metric on the surface have the same norm.*

Proof. Let (u,v) denote coordinates on $\mathbb{S}^2$ and let F_u, F_v denote the partial derivatives of F (and similarly for h). Then, as in [6], we get the variation

$$\delta\nu = -(h_u, h_v)g^{-1}(F_u, F_v)^\top.$$

On the other hand $\nabla h = g^{-1}(h_u, h_v)^\top$. Now we compute

$$|\delta\nu|^2 = (h_u, h_v)g^{-1}(F_u, F_v)^\top(F_u, F_v)g^{-1}(h_u, h_v)^\top$$
$$= (h_u, h_v)g^{-1}(h_u, h_v)^\top = (\nabla h)^\top g\nabla h = |\nabla h|^2,$$

using the fact that $g = (F_u, F_v)^\top(F_u, F_v)$.

Proof (of Theorem (2)). Let $(\mathbf{t}, \mathbf{n}, \nu)$ be the Darboux frame along the curve $C \subset \Sigma$. The normal vector field $h_1\mathbf{n} + (h_2|_C)\nu$ to the curve C encodes the variation of the curve which doesn't leave the surface Σ. Using the Lemma 1, we obtain the following expression for the elastic metric (6) restricted to this normal variation:

$$\mathcal{E}_F'(h_1\mathbf{n} + (h_2|_C)\nu) = a\int_C (h_1\kappa_g + h_2|_C\kappa_n)^2 d\ell$$
$$+ b\int_C \left((D_s h_1 - h_2|_C\tau_g)^2 + (D_s(h_2|_C) + h_1\tau_g)^2\right) d\ell.$$

We will split the b-term in two parts, thus obtaining a 3-parameter family of metrics, namely

$$a_1\int_C (h_1\kappa_g + h_2|_C\kappa_n)^2 d\ell + b_1\int_C (D_s h_1 - h_2|_C\tau_g)^2 d\ell + c_1\int_C (D_s(h_2|_C) + h_1\tau_g)^2 d\ell.$$

The normal vector field $h_2\nu$ to the surface Σ encodes the variation of the surface. Using Eqn. (12) in [6] the elastic metric (7) restricted to this normal variation is given by the following geometric expression :

$$\mathcal{E}_F''(h_2\nu) = 2a\int_\Sigma (h_2)^2(\kappa_1 - \kappa_2)^2 dA + 4b\int_\Sigma (h_2)^2(\kappa_1 + \kappa_2)^2 dA + c\int_\Sigma |\nabla h_2|^2 dA.$$

Here we use the fact that $g^{-1}\delta g = -2h_2L$, where L is the shape operator of the surface, as well as the identity $|\delta\nu| = |\nabla h_2|$, where ∇ denotes the gradient with respect to the induced metric on the surface, by Lemma 2.

Renaming the parameters and adding to this elastic metric for the surface the elastic metric for the curve on the surface obtained above leads to the 6-parameter family of elastic metrics for the shape space $\mathcal{F}$.

Remark 5. Assuming that the functions h_1, h_2 are constant along the curve C, the b_1-term becomes $\int_C (h_2|_C \tau_g)^2 d\ell$ and encodes the variation of the curve normal to the surface (variation together with the surface) while the c_1 term becomes $\int_C (h_1\tau_g)^2 d\ell$ and encodes the normal variation of the curve inside the surface.

4   Conclusion

In this paper, we identify the tangent spaces to nonlinear flags consisting of surfaces of genus zero decorated with a simple curve. We use gauge invariant metrics on parameterized curves and surfaces to endow the space of nonlinear flags with a family of Riemannian metrics, whose expression is given in terms of geometric invariants of curves and surfaces.

References

1. Ciuclea, I., Tumpach, A.B., Vizman, C.: Connections on principal bundles of embeddings (in progress)
2. Drira, H., Tumpach, A.B., Daoudi, M.: Gauge invariant framework for trajectories analysis. In: DIFFCV Workshop (2015). https://doi.org/10.5244/C.29.DIFFCV.6
3. Haller, S., Vizman, C.: Nonlinear flag manifolds as coadjoint orbits. Ann. Glob. Anal. Geom. **58**(4), 385–413 (2020). https://doi.org/10.1007/s10455-020-09725-6
4. Jermyn, I.H., Kurtek, S., Klassen, E., Srivastava, A.: Elastic shape matching of parameterized surfaces using square root normal fields. In: Fitzgibbon, A., Lazebnik, S., Perona, P., Sato, Y., Schmid, C. (eds.) ECCV 2012. LNCS, vol. 7576, pp. 804–817. Springer, Heidelberg (2012). https://doi.org/10.1007/978-3-642-33715-4_58
5. Mio, W., Srivastava, A., Joshi, S.H.: On shape of plane elastic curves. Int. J. Comput. Vision **73**(3), 307–324 (2007)
6. Tumpach, A.B., Drira, H., Daoudi, M., Srivastava, A.: Gauge invariant framework for shape analysis of surfaces. IEEE Trans Pattern Anal. Mach. Intell. **38**(1) (2016)
7. Tumpach, A.B.: Gauge Invariance of degenerate Riemannian metrics, Notices of American Mathematical Society (April 2016)
8. Tumpach, A.B., Preston, S.: Three methods to put a Riemannian metric on shape spaces, GSI23
9. Tumpach, A.B., Preston, S.C.: Riemannian metrics on shape spaces: comparison of different constructions (in progress)
10. Srivastava, A., Klassen, E., Joshi, S.H., Jermyn, I.H.: Shape analysis of elastic curves in euclidean spaces. IEEE Trans. Pattern Anal. Mach. Intell. **33**(7), 1415–1428 (2011)

Neurogeometry Meets Geometric Deep
Learning

A Neurogeometric Stereo Model
for Individuation of 3D Perceptual Units

Maria Virginia Bolelli[1,2,4(✉)], Giovanna Citti[1,2], Alessandro Sarti[1,2],
and Steven Zucker[3]

[1] Department of mathematics, University of Bologna, Bologna, Italy
giovanna.citti@unibo.it
[2] Laboratoire CAMS, CNRS-EHESS, Paris, France
alessandro.sarti@ehess.fr
[3] Departments of Computer Science and Biomedical Engineering, Yale University,
New Haven, CT, USA
steven.zucker@yale.edu
[4] Laboratoire des Signaux et Systèmes, Université Paris-Saclay, CentraleSupélec,
Gif-sur-Yvette, France
maria-virginia.bolelli@centralesupelec.fr

Abstract. We present a neurogeometric model for stereo vision and individuation of 3D perceptual units. We first model the space of position and orientation of 3D curves in the visual scene as a sub-Riemannian structure. Horizontal curves in this setting express good continuation principles in 3D. Starting from the equation of neural activity we apply harmonic analysis techniques in the sub-Riemannian structure to solve the correspondence problem and find 3D percepts.

Keywords: Neurogeometry · Stereo vision · 3D perceptual units · 3D good continuation

1 Introduction

We propose here a neurogeometrical model of stereo vision, in order to describe the ability of the visual system to infer the three-dimensionality of a visual scene from the pair of images projected respectively on the left and right retina.

The first differential models of the visual cortex, devoted to the description of monocular vision, have been proposed by Hoffmann [16] and Koenderink-van Doorn [19]. Results were unified under the name of neurogeometry by Petitot and Tondut [23], who related psychophysical experiments of Field, Hayes and Hess [14] with the contact geometry introduced by Hoffmann [16] and the stochastic approach of Mumford [21]. The functional architecture of the visual cortex has been described through sub-Riemannian metrics by Citti and Sarti [8] and through Frenet frames by Zucker [29], and after that a large litterature was developed.

The geometric optics of stereo vision has been proposed by Faugeras in [13] and a differential model for stereo was proposed by Zucker [29]. A sub-Riemannian structure of 3D space has been introduced by Duits et al. in [11,12]

© The Author(s), under exclusive license to Springer Nature Switzerland AG 2023
F. Nielsen and F. Barbaresco (Eds.): GSI 2023, LNCS 14071, pp. 53–62, 2023.
https://doi.org/10.1007/978-3-031-38271-0_6

and [24] for 3D image processing. Our model, first introduced in [4], generalizes these models introducing a sub-Riemannian geometry for stereo vision: it is presented in Sect. 3. In particular, we will focus on association fields, introduced in 2D by Field, Hayes and Hess in [14] and modeled in [8,23] and [5]. We will extend this approach to neural connectivity with integral curves and justify psychophysical experiments on perceptual organization of oriented elements in $\mathbb{R}^3$ ([9,15,17]).

The main contribution with respect to [4] is the constitution of 3D percepts, presented in Sect. 4. We start from the model of interactions between neural populations proposed by Bressloff-Cowan ([6]) and we modify the integro-differential equation they propose with the connectivity kernel obtained as fundamental solution of a sub-Riemannian Fokker Planck. Then, we generalize the stability analysis proposed by [6] for hallucinations, by [26] for emergence of percepts, and we show that in this case they correspond to 3D perceptual units.

2 The Stereo Problem

The stereo problem deals with the reconstruction of the three-dimensional visual scene starting from its perspective projection through left C_L and right C_R optical centers on the two eyes. The setting of the problem involves classical triangulation instruments (e.g. [13]), and the main issue is to couple in a correct way the correspondent left $Q_L = (x_L, y)$ and right $Q_R = (x_R, y)$ points on the parallel retinal planes ($y = y_L = y_R$), in order to project them back into the environment space to obtain $Q = (r_1, r_2, r_3) \in \mathbb{R}^3$, see Fig. 1,(a). This goes under the name of *stereo correspondence*.

The main clues for solving the correspondence are the slight differences in the two projected images, namely the disparities. Our main focus will be on horizontal positional disparity $d := (x_L - x_R)/2$, which introduces the set of *cyclopean coordinates* (x, y, d), together with the mean position $x := (x_L + x_R)/2$. Since binocularly driven neurons in the primary visual cortex, which perform the binocular integration, receive input from monocular (orientation selective) cells, we will choose as additional variables the orientations on left and right monocular structures: θ_L and θ_R; but we will not consider orientation disparity, because it does not seem to be coded directly in the visual cortex, see for example [7].

2.1 The Monocular Model for Orientation-Selective Cells

The hypercolumnar structure selective for orientation of monocular left and right simple cells in V1 (denoted respectively $i = L, R$) can be modeled in term of a fiber bundle, with base $(x_i, y) \in \mathbb{R}^2$ identified with the retinal plane, (see [23]) and fiber $\theta_i \in \mathbb{R}/2\pi\mathbb{Z} \equiv \mathbb{S}^1$. The response $O_i(x_i, y, \theta_i)$ of these cells to a visual signal on the retina $I(x_i, y)$ is quantified in terms of a function $\varphi(x_i, y, \theta_i)$, called receptive profile RP and well described by Gabor filters, see Fig. 1,(b). Following the work of Citti and Sarti [8], the action of these RPs induces a choice of contact

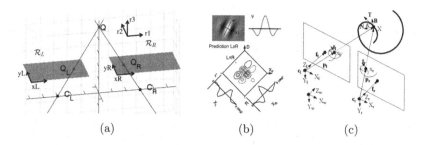

(a)   (b)   (c)

Fig. 1. (a) Stereo geometry. (b) Above: Gabor filter: model of 2D receptive profile, its 1D section. Below: binocular receptive profile (image adapted from [2]) (c) The Zucker model (image adapted from [20]).

form on the whole space:

$$\omega_{\theta_i} = -\sin\theta_i dx_i + \cos\theta_i dy. \tag{1}$$

The visual signal propagates in this cortical structure along integral curves of vector fields lying in the kernel of this contact form.

2.2 Models of Binocular Cells and Stereo Vision

Ohzawa et al. in [2] found that binocular simple cells in V1 perform a non-linear integration of left and right monocular cells, displayed in Fig. 1, (b). They proposed the binocular energy model (BEM), which characterize the binocular output through an interaction term O_B, product of left O_L and right O_R monocular outputs:

$$O_B = O_R O_L. \tag{2}$$

The mathematical model for stereo vision built by Zucker et al. in [1,20] is based on neural mechanisms of selectivity to position, orientations and curvatures of the visual stimulus and it is expressed via instruments of Frenet differential geometry. The connections between binocular neurons are described by helices whose spiral develops along the depth axis, encoding simultaneously position and orientation disparities. The model is illustrated in Fig. 1, (c).

3 A Sub-Riemannian Model for Stereo Vision

In this section we present the biologically-inspired model proposed in [4].

3.1 The Fiber Bundle of Binocular Cells

The binocular structure is based on monocular ones and it is equipped with a symmetry that involves the left and right structures, allowing the use of cyclopean coordinates (x, y, d) defined in Sect. 2. The set of binocular cells will be expressed a fiber bundle with base $\mathcal{B} = \mathbb{R}^2$ the cyclopean retina of coordinates (x, y). The structure of the fiber is $\mathcal{F} = \mathbb{R} \times \mathbb{S}^1 \times \mathbb{S}^1$, with coordinates $(d, \theta_L, \theta_R) \in \mathcal{F}$. Schematic representation is provided in Fig. 2, (a) and (b).

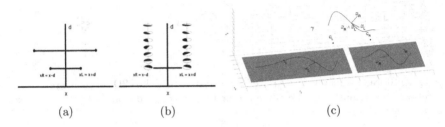

(a)   (b)   (c)

Fig. 2. Binocular cell structure and 3D reconstruction. (a) and (b) schematically represent the binocular fiber bundle in 2D: we visualized a 1D restriction to the direction x of the basis, the fiber of disparity d in (a) and the fiber of orientations θ_L and θ_R in (b). (c) describes reconstruction of a 3D curve from its projections. The normal to the curves γ_L and γ_R on retinal planes are identified by the 1-forms ω_{θ_L} and ω_{θ_R}. The wedge product $\tilde{\omega}_{\theta_L} \wedge \tilde{\omega}_{\theta_R}$ of their 3D counterpart identify the tangent vector to the 3D corresponding curve $\gamma : \mathbb{R} \to \mathbb{R}^3$.

3.2   Compatibility with Stereo Triangulation

We can introduce a 2-form starting from the monocular structures that embodies the binocular energy model, since Eq. (2) can be written in terms of monocular left and right RPs, see [4, Eq. (18),(49)], obtaining the following result.

Proposition 1. *The binocular interaction term O_B of (2) can be recast as wedge product of the two monocular 1-forms ω_{θ_L} and ω_{θ_R} defined in (1):*

$$O_B = \omega_{\theta_R} \wedge \omega_{\theta_L}. \tag{3}$$

It is possible to extend the monocular 1-forms ω_{θ_L} and ω_{θ_R} on retinal planes to $\tilde{\omega}_{\theta_L}$ and $\tilde{\omega}_{\theta_R}$ 1-forms in $\mathbb{R}^3$ and obtaining $\tilde{\omega}_{\theta_R} \wedge \tilde{\omega}_{\theta_L}$. Through the Hodge duality this 2-form identifies a vector that can be interpreted as the direction of the tangent to a potential 3D curve in the scene, see Fig. 2 (c).

So, binocular cells couple positions, identified with points in $\mathbb{R}^3$, and orientations in $\mathbb{S}^2$, identified with three-dimensional unitary tangent vectors. To solve the stereo problem the visual system must take into account suitable types of connections ([27]). It is therefore natural to introduce the perceptual space via the manifold $\mathcal{M} = \mathbb{R}^3_{(r_1, r_2, r_3)} \rtimes \mathbb{S}^2_{(\theta, \varphi)}$, and look for appropriate curves in $\mathcal{M}$.

3.3   Stereo Sub-Riemannian Geometry

The sub-Riemannian structure on $\mathcal{M}$ can be expressed locally using the chart $\theta \in (0, 2\pi), \varphi \in (0, \pi)$ by considering an orthonormal frame $\{Y_3, Y_\theta, Y_\varphi\}$, where:

$$Y_3 = \cos\theta \sin\varphi \partial_1 + \sin\theta \sin\varphi \partial_2 + \cos\varphi \partial_3, \quad Y_\theta = \frac{1}{\sin\varphi}\partial_\theta, \quad Y_\varphi = \partial_\varphi. \tag{4}$$

The vector field Y_3 encodes the tangent of the stimulus, Y_φ involves orientation in the depth direction, while Y_θ involves orientation on the fronto-parallel plane. We take here into account that contour detectability systematically changed with the degree to which they are oriented in depth, see [18]. Indeed the vector Y_θ is not defined for $\varphi = 0$, meaning that we do not perceive correctly contours which are completely oriented in the depth direction. The vector fields satisfy the Hörmander condition since the whole space is spanned at every point by the vectors $\{Y_3, Y_\theta, Y_\varphi\}$ and their commutators.

Remark 1. As noted by Duits and Franken in [12], the space $\mathbb{R}^3 \rtimes \mathbb{S}^2$ can be identified with the quotient $SE(3)/\{0_{|\mathbb{R}^3}\} \times SO(2)$. Different sections have different invariance properties; in [24], the authors provide a section which preserves isotropy in the spherical tangent plane and give the same role to all the angular variables [11, Thm.1 and Thm.4].

Integral curves with constant coefficients in the local orthonormal frame (4) are defined by the differential equation:

$$\dot{\Gamma}(t) = \vec{Y}_{3,\Gamma(t)} + c_1 \vec{Y}_{\theta,\Gamma(t)} + c_2 \vec{Y}_{\varphi,\Gamma(t)} \quad c_1, c_2 \in \mathbb{R}. \tag{5}$$

These curves, displayed in Fig. 3 (a), can be thought of in terms of trajectories in $\mathbb{R}^3$ describing a movement in the Y_3 direction, and by varying the coefficients c_1 and c_2 in $\mathbb{R}$, they can twist and bend in all space directions. Formally, the amount of "twisting and bending" in space is measured by curvature k and torsion τ, which in this setting read as: $k = \sqrt{c_1^2 + c_2^2}$, and $\tau = -c_1 \cotan \varphi$.

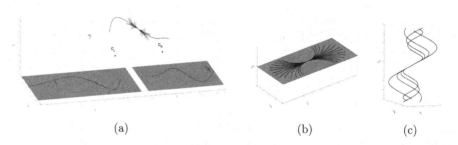

(a)   (b)   (c)

Fig. 3. Different families of integral curves (5). (a) General fan of integral curves described by Eq. (5) with varying c_1 and c_2 in $\mathbb{R}$, enveloping a curve $\gamma \in \mathbb{R}^3$. (b) Arc of circles for constant $\varphi = \pi/2$. (c) $r3$-helices for constant $\varphi = \pi/3$.

The model is then compatible with the previous models of [8] of monocular vision, since if $c_1 = 0$ or $\varphi = \pi/2$ then $\Gamma(t)_{|\mathbb{R}^3}$ is a piece of circle (Fig. 3, (b)). In addition it is compatible with the results of [1], based on properties of curvature, since if $\varphi = \varphi_0$ with $\varphi_0 \neq \pi/2$, then $\Gamma(t)_{|\mathbb{R}^3}$ is a $r3$-helix. The main difference is that curvature is an extracted feature in [1], while it is coded in connectivity in our model.

3.4   Good Continuation in 3D and Stereo Association Fields

The family (5) model neural connectivity (see [4]) and it can be related to the geometric relationships deriving from psychophysical experiments on perceptual organization of oriented elements in $\mathbb{R}^3$, the basis of the Gestalt law of good continuation ([28]). This generalizes the 2D concept introduced by Field Hayes and Hess in [14] (Fig. 4, (a)) of an association field in 3D.

The geometrical affinities between orientations under which a pair of position-orientation elements in $\mathbb{R}^3 \rtimes \mathbb{S}^2$ are perceived as connected in a 3D scene, have been determined by [17] with the theory of 3D relatability. Curves that are suitable to connect these 3D relatable points have the properties of being smooth and monotonic [9,15], extending good continuity/ regularity in depth. Moreover, the strength of the relatable edges in co-planar planes with the initial edge must meet the relations of the bi-dimensional association fields [17].

The family of integral curves (5) locally connects the association fan generated by 3D relatability geometry (Fig. 4, (b)), satisfying smoothness, monotonicity and compatibility with 2D association fields.

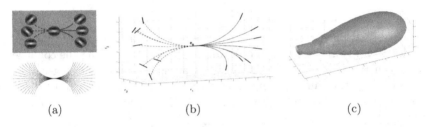

| (a) | (b) | (c) |

Fig. 4. Display of connectivity. (a) Field Hayes and Hess association field (top) and 2D integral curves of the Citti-Sarti model [8] (bottom). (b) Fan of 3D relatable points connected by integral curves (5).(c) Iso-surface in $\mathbb{R}^3$ of probability density (7) associated to the curves (5).

4   Constitution of 3D Visual Perceptual Units

Integral curves model the good continuation law, playing a fundamental role within the problem of perceptual grouping, individuating 3D visual units.

4.1   Sub-Riemannian Fokker Plank Equation and Connectivity Kernel

The emergence of 3D visual percepts derives from interactions between binocular cells: according to the Gestalt law of good continuation, entities described by similar local orientations are more likely to belong to the same perceptual unit.

Following [3,25], we suppose that the signal starting from a binocular neuron $\xi \in \mathbb{R}^3 \rtimes \mathbb{S}^2$ evolves following the stochastic process described by the SDE:

$$d\Gamma(t) = Y_{\mathbb{R}^3,\Gamma(t)}dt + \lambda(Y_{\theta,\Gamma(t)}, Y_{\varphi,\Gamma(t)})dB(t), \quad \lambda \in \mathbb{R}, \tag{6}$$

with $B(t)$ 2-dimensional Brownian motion. The probability of interaction between points ξ and $\xi' \in \mathcal{M}$, has a (time-independent) density:

$$\{J_\lambda(\xi, \xi')\}_{\lambda \in \mathbb{R}}, \tag{7}$$

whose iso-surfaces in $\mathbb{R}^3$ are displayed in Fig. 4, (c). This probability density coincides with the (time-integrated) fundamental solution of the forward Kolmogorov differential equation associated to (6) with operator $\mathcal{L} = -Y_3 + \lambda(Y_\theta^2 + Y_\varphi^2)$ written in terms of the chosen vector fields (4). Analytical approximation of the fundamental solutions have been provided in [12,24], and numerical approximation with Fourier methods and Monte-Carlo simulations in [10]. We implement here the latter, following the approach presented in [3], since it is more physiological being based on the stochastic integral curves.

Remark 2. The authors in [24] have shown that the space $\mathbb{R}^3 \rtimes \mathbb{S}^2$ can be identified with a section of $SE(3)$ where kernels have symmetry properties with respect to the group law, and all angles have the same role. In our model, 3D association field fan depends on the choice of the vector fields, which is not invariant, due to the different meaning of the considered orientations. Nevertheless, we expect the kernel to preserve invariance. A comparison between the two approaches based on parametrix method will be provided in a future paper.

4.2 From Neural Activity to 3D Perceptual Units

The kernels (7) are implemented as facilitation patterns to define the evolution in time $t \mapsto a(\xi, t)$ of the activity of the neural population at $\xi \in \mathcal{M}$. This activity is usually modeled through a mean field equation, see [6]:

$$\partial_t a(\xi, t) = -a(\xi, t) + \sigma\left(\int_{\mathcal{M}} J(\xi, \xi') a(\xi', t) d\xi' + h(\xi, t) \right), \tag{8}$$

where h is the feedforward input, σ is a sigmoidal function and J a symmetrization of (7). When the input h is constant over a subset Ω of $\mathcal{M}$ and zero elsewhere, it has been proved in [26] that the domain of Eq. (8) reduces to Ω since the population activity is negligible in the complementary set $\mathcal{M} \setminus \Omega$.

We extend the stability analysis around a stationary state a_1 proposed by [6] for hallucination and [26] for perceptual units. A perturbation u, difference between two solutions $a - a_1$, satisfies the eigenvalue problem associated to the linearized time independent operator

$$\int_\Omega J(\xi, \xi') u(\xi', t) d\xi' = \frac{1}{\mu} u(\xi, t) \tag{9}$$

where $\mu = \sigma'(0)$. As shown in [26] for the 2D case, the eigenvectors represent the perceptual units, and the eigenvalues their salience. The whole process is strictly linked with spectral clustering and dimensionality reduction results ([22]).

4.3 The Proposed Model for the Correspondence Problem

The model can be described as follows. We start from two rectified stereo images. We couple all possible corresponding points (left and right retinal points with the same abscissa coordinate): this lifts retinal points in points $\xi_i \in \Omega$ generating also false matches, i.e. points that do not belong to the original stimulus. We call affinity matrix the kernel J evaluated on every couple of lifted points $\xi_i, \xi_j \in \Omega$: $\mathbf{J}_{i,j} := J(\xi_i, \xi_j)$. Spectral analysis on $\mathbf{J}$ individuates 3D perceptual units, and solves the stereo correspondence. In this process false matches are eliminated since the similarity measure introduced by the kernel $\mathbf{J}$ groups elements satisfying the good-continuation constraints.

4.4 Numerical Experiments

We develop the ideas illustrated so far by numerical examples; the main steps of the algorithm are summarized in Table 1.

Table 1. Recovering 3D visual percepts starting from rectified stereo images.

0	Gabor filtering the left and right retinal images to obtain for every point (x_i, y_i) its corresponding orientation θ_i for $i = L, R$
1	Recover the domain $\Omega \subset \mathbb{R}^3 \rtimes \mathbb{S}^2$, $\xi_k \in \Omega$, $k = 1, \ldots n$, from the coupling of retinal images by inverting perspective projections.
2	Call affinity matrix $\mathbf{J}$ the discretization of the kernel J: $\mathbf{J}_{ij} := J(\xi_i, \xi_j)$.
3	Solve the eigenvalue problem $\mathbf{J}a = \iota a$.
4	Find the q largest eigenvalues $\{\iota_i\}_{i=1}^q$ and the associated eigenvectors $\{a_i\}_{i=1}^q$.
5	For $k = 1, \ldots, n$ assign the point ξ_k to the clustered labeled by $\max_i\{a_i(k)\}_{i=1}^q$.
6	Join together the clusters with less than Q elements.

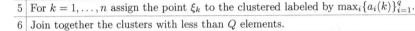

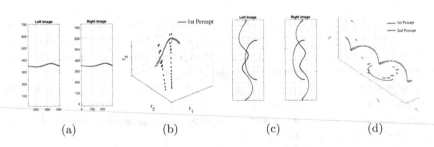

| (a) | (b) | (c) | (d) |

Fig. 5. (a) Stereo images of a 3D curve. (b) Lifting of the stimulus in $\mathbb{R}^3 \rtimes \mathbb{S}^2$: points clustered together are marked by the same color (one main red colored 3D percept); black points do not belong to any cluster. (c) Stereo images of a 3D helix and arc of a circle. (d) Lifting of (c) in $\mathbb{R}^3 \rtimes \mathbb{S}^2$: two main clusters (red and blue) correctly segment into two perceptual units the 3D visual scene. (Color figure online)

The model is first tested on synthetic stereo images of 3D curves (Fig. 5 (a),(c)), and perceptual units are correctly recovered (Fig. 5 (c),(d)).

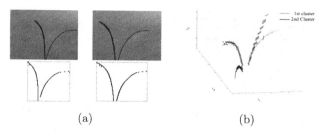

(a)   (b)

Fig. 6. (a) Top: couple of natural images. Bottom: Gabor filtering to recover position and orientation in retinal planes. (b) The application of the algorithm defined in Table 1 individuates the two 3D perceptual units (red and blue points). (Color figure online)

A second test is performed on a natural image: we pre-process the images via Gabor filtering, to recover position and orientation on the two retinas, and then we apply the model. Results are illustrated in Fig. 6.

Acknowledgement. MVB, GC, AS were supported by GHAIA project, H2020 MSCA RISE n. 777622 and by NGEU-MUR-NRRP, project MNESYS (PE0000006) (DN. 1553 11.10.2022). SWZ was supported by NIH Grant EY031059 and NSF CRCNS Grant 1822598.

References

1. Alibhai, S., Zucker, S.W.: Contour-based correspondence for stereo. In: Vernon, D. (ed.) ECCV 2000. LNCS, vol. 1842, pp. 314–330. Springer, Heidelberg (2000). https://doi.org/10.1007/3-540-45054-8_21
2. Anzai, A., Ohzawa, I., Freeman, R.: Neural mechanisms for processing binocular information i. simple cells. J. Neurophysiol. **82**(2), 891–908 (1999)
3. Barbieri, D., Citti, G., Cocci, G., Sarti, A.: A cortical-inspired geometry for contour perception and motion integration. J. Math. Imaging Vis. **49**(3), 511–529 (2014)
4. Bolelli, M.V., Citti, G., Sarti, A., Zucker, S.W.: Good continuation in 3D: the neurogeometry of stereo vision. arXiv preprint arXiv:2301.04542 (2023)
5. Boscain, U., Duits, R., Rossi, F., Sachkov, Y.: Curve cuspless reconstruction via sub-riemannian geometry. ESAIM: Control, Optimisat, Calculus Variations **20**, 748–770 (2014)
6. Bressloff, P.C., Cowan, J.D.: The functional geometry of local and horizontal connections in a model of V1. J. Physiol. Paris **97**(2), 221–236 (2003)
7. Bridge, H., Cumming, B.: Responses of macaque V1 neurons to binocular orientation differences. J. Neurosci. **21**(18), 7293–7302 (2001)
8. Citti, G., Sarti, A.: A cortical based model of perceptual completion in the roto-translation space. J. Math. Imaging Vis. **24**(3), 307–326 (2006)
9. Deas, L.M., Wilcox, L.M.: Perceptual grouping via binocular disparity: The impact of stereoscopic good continuation. J. Vis. **15**(11), 11 (2015)
10. Duits, R., Bekkers, E.J., Mashtakov, A.: Fourier transform on the homogeneous space of 3D positions and orientations for exact solutions to linear pdes. Entropy **21**(1), 38 (2019)
11. Duits, R., Dela Haije, T., Creusen, E., Ghosh, A.: Morphological and linear scale spaces for fiber enhancement in dw-mri. J. Mathematical Imaging Vis. **46**, 326–368 (2013)

12. Duits, R., Franken, E.: Left-invariant diffusions on the space of positions and orientations and their application to crossing-preserving smoothing of HARDI images. IJCV (2011)

13. Faugeras, O.: Three-dimensional computer vision: a geometric viewpoint. MIT press (1993)

14. Field, D.J., Hayes, A., Hess, R.F.: Contour integration by the human visual system: Evidence for a local "association field." Vision Res. **33**(2), 173–193 (1993)

15. Hess, R.F., Hayes, A., Kingdom, F.A.A.: Integrating contours within and through depth. Vision. Res. **37**(6), 691–696 (1997)

16. Hoffman, W.C.: The visual cortex is a contact bundle. Appl. Math. Comput. **32**(2), 137–167 (1989)

17. Kellman, P.J., Garrigan, P., Shipley, T.F., Yin, C., Machado, L.: 3-d interpolation in object perception: Evidence from an objective performance paradigm. J. Exp. Psychol. Hum. Percept. Perform. **31**(3), 558–583 (2005)

18. Khuu, S.K., Honson, V., Kim, J.: The perception of three-dimensional contours and the effect of luminance polarity and color change on their detection. J. Vis. (2016)

19. Koenderink, J.J., van Doorn, A.J.: Representation of local geometry in the visual system. Biol. Cybern. **55**(6), 367–375 (1987)

20. Li, G., Zucker, S.W.: Contextual inference in contour-based stereo correspondence. IJCV **69**(1), 59–75 (2006)

21. Mumford, D.: Elastica and computer vision. In: Algebraic geometry and its applications, pp. 491–506. Springer (1994). https://doi.org/10.1007/978-1-4612-2628-4_31

22. Perona, P., Freeman, W.: A factorization approach to grouping. In: Burkhardt, H., Neumann, B. (eds.) ECCV 1998. LNCS, vol. 1406, pp. 655–670. Springer, Heidelberg (1998). https://doi.org/10.1007/BFb0055696

23. Petitot, J., Tondut, Y.: Vers une neurogéométrie. Fibrations corticales, structures de contact et contours subjectifs modaux. Math. Scien. Humain. **145**, 5–101 (1999)

24. Portegies, J.M., Duits, R.: New exact and numerical solutions of the (convection-)diffusion kernels on SE(3). Differential Geom. Appl. **53**, 182–219 (2017)

25. Sanguinetti, G., Citti, G., Sarti, A.: A model of natural image edge co-occurrence in the rototranslation group. J. Vis. (2010)

26. Sarti, A., Citti, G.: The constitution of visual perceptual units in the functional architecture of V1. J. Comput. Neurosci. (2015)

27. Scholl, B., Tepohl, C., Ryan, M.A., Thomas, C.I., Kamasawa, N., Fitzpatrick, D.: A binocular synaptic network supports interocular response alignment in visual cortical neurons. Neuron **110**(9), 1573–1584 (2022)

28. Wagemans, J., et al.: A century of gestalt psychology in visual perception: I perceptual grouping and figure-ground organization. Psych. Bull. **138**(6), 1172 (2012)

29. Zucker, S.W.: Differential geometry from the Frenet point of view: boundary detection, stereo, texture and color. In: Handbook of Mathematical Models in Computer Vision, pp. 357–373. Springer (2006). https://doi.org/10.1007/0-387-28831-7_22

Functional Properties of PDE-Based Group Equivariant Convolutional Neural Networks

Gautam Pai$^{(\boxtimes)}$, Gijs Bellaard, Bart M. N. Smets, and Remco Duits

Center for Analysis, Scientific Computing and Applications, (Group: Geometric Learning and Differential Geometry), Department of Mathematics and Computer Science, Eindhoven University of Technology, Eindhoven, The Netherlands
{g.pai,g.bellaard,b.m.n.smets,r.duits}@tue.nl

Abstract. We build on the recently introduced PDE-G-CNN framework, which proposed the concept of non-linear morphological convolutions that are motivated by solving HJB-PDEs on lifted homogeneous spaces such as the homogeneous space of 2D positions and orientations isomorphic to $G = SE(2)$. PDE-G-CNNs generalize G-CNNs and are provably equivariant to actions of the roto-translation group $SE(2)$. Moreover, PDE-G-CNNs automate geometric image processing via orientation scores and allow for a meaningful geometric interpretation.

In this article, we show various functional properties of these networks:

(1.) PDE-G-CNNs satisfy crucial geometric and algebraic symmetries: they are semiring quasilinear, equivariant, invariant under time scaling, isometric, and are solved by semiring group convolutions.
(2.) PDE-G-CNNs exhibit a high degree of data efficiency: even under limited availability of training data they show a distinct gain in performance and generalize to unseen test cases from different datasets.
(3.) PDE-G-CNNs are extendable to well-known convolutional architectures. We explore a UNet variant of PDE-G-CNNs which has a new *equivariant* U-Net structure with PDE-based morphological convolutions.
We verify the properties and show favorable results on various datasets.

Keywords: Group Equivariant Convolutional Neural Networks · PDE-Based Image Processing · Lie Groups · Semirings · Riemannian Geometry

1 Introduction and Background

Convolutional Neural Networks (CNNs) are ubiquitous in modern day computer vision, and are considered standard computational architectures in most data-dependent image processing tasks. Recently, there has been a great focus on designing neural networks that must be *invariant* or, more generally, *equivariant* to generic geometric transformations. CNNs are structurally equivariant to the group of translations applied on the image function and this is a key reason for

© The Author(s), under exclusive license to Springer Nature Switzerland AG 2023
F. Nielsen and F. Barbaresco (Eds.): GSI 2023, LNCS 14071, pp. 63–72, 2023.
https://doi.org/10.1007/978-3-031-38271-0_7

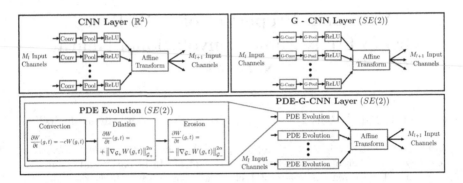

Fig. 1. A schematic for a CNN, G-CNN and the PDE-G-CNN Layer in a deep neural network. CNNs are typically processed with $\mathbb{R}^2$ convolutions, whereas G-CNNs work with linear G-convolutions in lifted space $SE(2)$. Elements in $SE(2)$ are denoted by $g = (x, y, \theta)$. In PDE-G-CNNs only the convection vector c and the metric parameters $\mathcal{G}_+, \mathcal{G}_-$ are *learned* and they lead to kernels that are used for non-linear *morphological* convolutions that solve the respective Erosion and Dilation PDE's in lifted space $SE(2)$.

their success and widespread use. However, CNNs are not invariant/equivariant to general and more challenging transformations, like, e.g. rotations or affine transformations. This led to the development of Group Equivariant Convolutional Neural Networks or G-CNNs [3,8,20,23]. G-CNNs do not waste network capacities and hard-code symmetries in the neural network. However, despite the impact made by enforcing geometry in network architectures, CNNs and G-CNNs still have some shortcomings:

1. They lack geometric interpretability in their action. The kernels themselves lack geometric structure and fail to relate to cortical association fields [7,15]
2. Overall, CNNs and G-CNNs have high network complexity,
3. Training these architectures requires vast amounts of well-annotated, clean data - that is often sparsely available in applications (e.g. medical imaging).

In [21], PDE-based group convolutional neural networks (PDE-G-CNNs) were introduced. A typical PDE-G-CNN layer employs numerical solvers for specifically chosen non-linear evolution PDE's. These PDE's are not arbitrary and have been explored previously in the domain of geometric image analysis [7,11] and yield theoretically interpretable solutions: they train sparse sets of association fields from neurogeometry as shown in [4, App-B & Ch. 1]. The dominant PDEs in a PDE-G-CNN are convection, dilation and erosion. These three PDE's correspond to the common notions of shifting, max pooling, and min pooling over Riemannian balls respectively. They are solved by re-samplings and so-called *morphological* convolutions. For a conceptual schematic of the 3 types of convolutional networks, see Fig. 1.

Despite the theoretical development of PDE-G-CNNs [4,5,21], some important functional properties of these networks have not been reported in prior work. In this paper we focus on three important aspects:

1. In Sect. 2, we theoretically enumerate a number of geometric and algebraic properties desirable by a neural network and show that PDE-G-CNNs satisfy

them. For e.g. PDE-G-CNNs consist solely of *quasilinear* (and PDE-solving) layers w.r.t. a semiring R, that can either be a linear or tropical semiring. Tropical semirings have been shown to be useful for machine learning [9,14, 18], and we report that they are particularly valuable in PDE-G-CNNs.

2. In Sect. 3, we perform experimental evaluations of CNNs, G-CNNs and PDE-G-CNNs by varying the amount of data used for training. We evaluate comparable versions of these architectures and show that PDE-G-CNNs allow for strong training data reduction without performance decrease. *PDE-G-CNNs exhibit a significantly improved training data efficiency.*

3. Motivated by the success of convolutional U-Nets [17], we build a PDE based *equivariant* PDE-G-UNet and show its applicability in vessel segmentation.

2 Geometric and Algebraic Properties of PDE-G-CNNs

In this section, we provide a concise theory on PDE-G-CNNs with Lie group domain $G = SE(2)$ and semiring co-domain R. In Theorem 1 we list geometric properties of PDE-G-CNNs, and end with an explanation on the practical significance of these properties in geometric deep learning.

Set $G = SE(2)$. It sets the *Lie group domain* of functions in PDE-G-CNNs. Recall $G = \mathbb{R}^2 \rtimes SO(2) \equiv \mathbb{R}^2 \rtimes S^1 \equiv \mathbb{R}^2 \rtimes \mathbb{R}/(2\pi\mathbb{Z})$ and write $g = (x, y, \theta) = (\mathbf{x}, \theta) \in G$. The group product is $g_1 g_2 = (\mathbf{x}_1, \theta_1)(\mathbf{x}_2, \theta_2) = (\mathbf{x}_1 + R_{\theta_1}\mathbf{x}_2, \theta_1 + \theta_2 \text{ Mod } 2\pi)$.

Set semiring $R = (R, \oplus, \otimes, \mathbb{0}, \mathbb{1})$. It sets the *Semiring co-domain* of functions in PDE-G-CNNs. It is either the linear semiring $L = (\mathbb{R}, +, \cdot, 0, 1)$ or a tropical semiring: $T_- = (\mathbb{R} \cup \{\infty\}, \min, +, \infty, 0)$, $T_+ = (\mathbb{R} \cup \{-\infty\}, \max, +, -\infty, 0)$. If $R = L$ we sometimes extend to $\mathbb{C}$.

We set $\mathcal{M}_R(G)$ as a set of functions on G associated to the semiring R. For $R = L$ the set $\mathcal{M}_L(G)$ is the set of square integrable functions w.r.t. the Haarmeasure μ_G on G. For $R = T_-$ the set $\mathcal{M}_{T_-}(G)$ is the set of upper semicontinuous functions on G bounded from below. For $R = T_+$ the set $\mathcal{M}_{T_+}(G)$ is the set of lower semicontinuous functions on G bounded from above.

We define the integral I_R associated to that semiring:

$$I_L(f) = \int_G f(g) \, d\mu_G(g), \ I_{T_-}(f) = \inf_{g \in G} f(g), \ I_{T_+}(f) = \sup_{g \in G} f(g). \quad (1)$$

Write $I_R(f) =: \bigoplus_{g \in G} f(g)$. Define the semimodule, i.e. a 'R-linear vector space':

$$H_R^G = \{f : G \to R \mid \delta_R(f, \mathbb{0}) < \infty, f \in \mathcal{M}_R(G)\} / \sim \quad (2)$$

with a partition of equivalence classes w.r.t. to the equivalence relation:

$$f \sim g \Leftrightarrow \delta_R(f, g) = 0 \text{ with } \delta_R(f, g) := \epsilon_R \bigoplus_{x \in G} \epsilon_R \, \rho(f(x), g(x)), \quad (3)$$

with $\epsilon_L = 1$, $\epsilon_{T_-} = -1$, $\epsilon_{T_+} = 1$, and we assume the continuous function $\rho_R : R \times R \to \mathbb{R}_{\geq 0}$ is a metric up to a monotonic transformation. Then denote $\mathcal{H}_R^G$ as the completion of the space H_R^G. On $\mathcal{H}_R^G$ we define the semiring group convolution ⊛ by

$$(f_1 \circledast f_2)(g) = \bigoplus_{h \in G} f_1(h^{-1}g) \otimes f_2(h), \qquad g \in G. \tag{4}$$

For $R = L$ we set $\rho_L(a,b) = |a - b|^2$ and have $\delta_L(f,g) = \|f - g\|_{\mathbb{L}_2(G)}^2$ and $\mathcal{H}_L^G = \mathbb{L}_2(G)$. Equation (4) is then a linear group convolution.

For $R = T_-$ we set $\rho_{T_-}(a,b) = |e^{-a} - e^{-b}|$ and $\mathcal{H}_{T_-}^G$ is the closure of the semi-module $H_{T_-}^G$. Equation (4) is then a 'morphological group convolution' [21]. For $R = T_+$ we set $\rho_{T_+}(a,b) = |e^a - e^b|$ and $\mathcal{H}_{T_+}^G$ is the closure of the semi-module $H_{T_+}^G$. As sets one has $\mathbb{L}_\infty(G) \cap C(G) = \mathcal{H}_{T_+}^G \cap \mathcal{H}_{T_-}^G$ as $\mathcal{H}_{T_+}^G = -\mathcal{H}_{T_-}^G$ and the metrics relate similarly by $\delta_{T_-}(f,g) = \delta_{T_+}(-f,-g)$ and

$$0 \le \psi \in C(G) \Rightarrow \delta_{T_+}(\psi, -\infty) = e^{\|\psi\|_{\mathbb{L}_\infty(G)}}, \delta_{T_-}(-\psi, \infty) = e^{\|\psi\|_{\mathbb{L}_\infty(G)}}. \tag{5}$$

A PDE-G-CNN consist of PDE-layers, Fig. 1, each layer corresponds to a choice of semiring R and a corresponding PDE system that is solved on $G \times \mathbb{R}^+$:

$$\boxed{\begin{aligned} & L : \partial_t W = -cW, \text{ or: } \partial_t W = -i^q(-\Delta_{\mathcal{G}_0})^{\frac{\alpha}{2}}W \text{ with } q \in \{0,1\}, \\ & T_- : \partial_t W = -\frac{1}{\alpha}\|\nabla_{\mathcal{G}_-}W\|^\alpha, \ T_+ : \partial_t W = \frac{1}{\alpha}\|\nabla_{\mathcal{G}_+}W\|^\alpha, \end{aligned}} \tag{6}$$

for all $t \ge 0$, always with $\alpha \in (1,2]$ and with initial condition $W|_{t=0} = f \in \mathcal{H}_R^G$. Here f is input from the previous layer and $W|_{t=1} \in \mathcal{H}_R^G$ is the output. Vector field c on G is left-invariant and transports along exponential curves in G. The PDEs are quasilinear w.r.t. the indicated semiring R.

In (6) the gradient $\nabla_{\mathcal{G}}$ and Laplacian $\Delta_{\mathcal{G}}$ are indexed by a left-invariant metric tensor field $\mathcal{G} \in \{\mathcal{G}_0, \mathcal{G}_+, \mathcal{G}_-\}$. Then by left-invariance $\mathcal{G} = \sum_{i,j=1}^3 g_{ij} \omega^i \otimes \omega^j$ has constant coefficients $[g_{ij}] \in \mathbb{R}^{3 \times 3}$ w.r.t. left-invariant (dual) basis:

$$\begin{aligned} \mathcal{A}_1 &= \cos\theta \, \partial_x + \sin\theta \, \partial_y, & \mathcal{A}_2 &= -\sin\theta \, \partial_x + \cos\theta \, \partial_y, & \mathcal{A}_3 &= \partial_\theta, \\ \omega^1 &= \cos\theta \, dx + \sin\theta \, dy, & \omega^2 &= -\sin\theta \, dx + \cos\theta \, dy, & \omega^3 &= d\theta. \end{aligned} \tag{7}$$

with $\langle \omega^i, \mathcal{A}_j \rangle = \delta_j^i$. Similarly $c = \sum_{i=1}^3 c^i \mathcal{A}_i$ for constant $\boldsymbol{c} = (c^1, c^2, c^3) \in \mathbb{R}^3$.

Remark 1. (**Network Parameters of PDE-G-CNNs**)
The network parameters in a PDE-G-CNN are given by all convection vectors c, all symmetric positive definite(SPD) matrices $[\mathcal{G}] = [g_{ij}]$ of all $\{\mathcal{G}_0, \mathcal{G}_+, \mathcal{G}_-\}$, and linear-combination weights of each layer [21].

Remark 2. (**Hyper-parameters in PDE-G-CNNs**)
Parameter α is constrained to $\alpha \in (1,2]$. For $R = T_-, T_+$ it regulates soft min/max pooling over balls [4, Prop. 1]. For $R = L$ it connects the Poisson and Gaussian semigroup. Parameter $q \in \{0,1\}$ switches from fractional diffusion (classical PDE-G-CNNs) to fractional Schrödinger PDEs (quantum PDE-G-CNNs).

Remark 3. (**Well-posed Solutions of PDE-G-CNNs**)
The solutions of PDE-systems (6) are for $R = L$ strongly continuous semigroups [24] on G, and for $R = T_-$ as viscosity solutions (Lax-Oleinik solutions) of dynamic HJB-systems on Riemannian manifold $(G, \mathcal{G}_\pm)$ [4, Prop. 1].

Theorem 1. *Let $\Phi_t : \mathcal{H}_R^G \to \mathcal{H}_R^G$ be the PDE-solver of the equivariant PDE-evolution (6) on $G = SE(2)$ with semiring structure $R = (R, \oplus, \otimes, 0, 1) \in \{L, T_-, T_+\}$ on the co-domain. Then $\Phi_t = \Phi_t^{[\mathcal{G}], \mathbf{c}}$ parameterized by symmetric positive definite(SPD) $[\mathcal{G}]$ or[1] by $\mathbf{c}$ satisfies:*

$$a)\ \Phi_t \circ \Phi_s = \Phi_{t+s},\ b)\ \exists_{k_t^{\mathcal{G}} \in \mathcal{H}_R^G} : \Phi_t f = k_t^{\mathcal{G}} \circledast f,\ c)\ \lim_{t \downarrow 0} \Phi_t f = f,$$

$$d)\ \forall_{SPD\ [\mathcal{G}]\ or\ \mathbf{c} \in \mathbb{R}^3} \exists_{\psi \in C^1(\mathbb{R}), \psi' > 0} : \Phi_t^{[\mathcal{G}], \mathbf{c}} = \Phi_1^{(\psi(t))^{-2}[\mathcal{G}], t\mathbf{c}}, \tag{8}$$

$$e)\ \Phi_t \circ \mathcal{L}_g = \mathcal{L}_g \circ \Phi_t,\ where\ \mathcal{L}_g f(h) = f(g^{-1}h),$$

$$f)\ \Phi_t(\alpha \otimes f_1 \oplus \beta \otimes f_2) = \alpha \otimes \Phi_t(f_1) \oplus \beta \otimes \Phi_t(f_2), \tag{9}$$

$$g)\ I_R(k_t^{\mathcal{G}}) = 1\ and\ I_R(\Phi_t f) = I_R(f),$$

for all $t, s \geq 0$, $g, h \in G$, $\alpha, \beta \in R$, $f, f_1, f_2 \in \mathcal{H}_R^G$. If we moreover impose h) $\forall_{f \in H_R} \forall_{t \geq 0} : \delta_R(f, 0) = \delta_R(\Phi_t f, 0)$, then this discards the (fractional) diffusion in the linear semiring $R = L$ setting.

If we moreover impose i) $k_t > 0$ then this discards the Schrödinger PDE if $R = L$ and it discards the dilation PDEs of $R = T_+$.

Proof. **Item a):** follows from well-defined evolution PDEs (Remarks 3), yielding well-known semigroup property in the linear setting [24] and in the tropical setting [2, Thm. 2.1, (ii)]. See Remark 4 for a short insight on this.
Item b): for $R = L$ the linear evolutions (6) are solved by linear group convolution with a probability kernel derived in [11]. For $R = T_-$ the nonlinear evolutions (6) are solved by a 'Lax-Oleinik formula' that gives the viscosity solution that [4, Prop. 1] is a tropical group convolution with a kernel $k_t \in \mathcal{H}_{T_-}^G$

$$R = T_- \Rightarrow k_t^{\mathcal{G}}(g) = \frac{t}{\beta} \left(\frac{d_{\mathcal{G}}(g,e)}{t} \right)^\beta, \frac{1}{\alpha} + \frac{1}{\beta} = 1 \text{ for } t > 0, \tag{10}$$

with $d_{\mathcal{G}}$ the Riemannian distance of $\mathcal{G}$. Case $R = T_+$ follows from case T_- by $\Phi_t^{T_+}(f) = -\Phi_t^{T_-}(-f)$, $\max(x) = -\min(-x)$. In all cases the solutions are

$$(\Phi_t f)(g) = (k_t^{\mathcal{G}} \circledast f)(g) = \bigoplus_{h \in G} k_t^{\mathcal{G}}(h^{-1}g) \otimes f(h) \tag{11}$$

with a kernel that by a) satisfies $k_{t+s}^{\mathcal{G}} = k_t^{\mathcal{G}} \circledast k_t^{\mathcal{G}}$ for all $t, s \geq 0$.
Item c): strong continuity is well-known: for $R = L, q = 0$ see [24], for $R = L, q = 1$ see [1] also on G [10, Thm. 2]. For $R \in \{T_-, T_+\}$ see [13].

---

[1] Convections and dilations/diffusion do not commute on the non-commutative G, and the desirable order is first convection and then dilation/diffusion [4, App. B, Prop. 3].

Item d): the linear convection PDE in (6) is solved by: $W(\cdot, t) = e^{-tc}f = \mathcal{R}_{e^{-tc|_e}}f$ with unit element $\mathbf{e} \in G$, $\mathcal{R}_g U(h) := U(hg)$, and $\Phi_t^{tc} = e^{-tc} = (e^{-c})^t = \Phi_t^{c}$. For the other PDE cases one has: $\|\nabla_{s^{-2}\mathcal{G}}f\| = s\|\nabla_{\mathcal{G}}f\|$, and $\Delta_{s^{-2}\mathcal{G}}f = s^2\Delta_{\mathcal{G}}f$. Thereby $\forall_{t,s>0}\Phi_t^{s^{-2}[\mathcal{G}]} = \Phi_{s^\alpha t}^{[\mathcal{G}]}$. In particular: $\Phi_1^{\Psi(t)^{-2}[\mathcal{G}]} = \Phi_t^{[\mathcal{G}]}$, with $\Psi(t) = t^{1/\alpha}$.

Item e): Semiring group convolution commutes with $\mathcal{L}_q$: $(k_t^{\mathcal{G}} \circledast \mathcal{L}_q f)(g) = \bigoplus_{h \in G} \kappa_t^{\mathcal{G}}(h^{-1}g) \otimes f(q^{-1}h) = \bigoplus_{v \in G} \kappa_t^{\mathcal{G}}(v^{-1}q^{-1}g) \otimes f(v) = (\mathcal{L}_q(k_t^{\mathcal{G}} \circledast f))(g)$, $v = q^{-1}h$.

Item f): integration $f \mapsto I(f) = \bigoplus_{g \in G} f(g)$ is a continuous quasi-linear operator and so is the pointwise multiplication $f(g) \mapsto k_t^{\mathcal{G}}(h^{-1}g) \otimes f(g)$. Thereby Φ_t (11) is a concatenation of quasi-linear operators and thereby again quasi-linear.

Item g): for $R = L$, $\mathbb{1} = 1$ and $\alpha = 2$ one has $I_R(k_t^{\mathcal{G}}) = \mathbb{1}$: this holds by exact formulas [11]. For $1 < \alpha \leq 2$ the α-th power of the spectrum in the orthogonal spectral decomposition does not affect 0 in the generator spectrum. For $R = T_-, T_+$ it follows directly by inspection of the kernel (10). By integration and quasi-linearity (item f), (11) one has $I_R(\Phi_t f) = I_R(f)$.

Item h): we must show $\delta_R(f, \mathbb{0}) = \delta_R(\Phi_t f, \mathbb{0})$ for all $f \in \mathcal{H}_R^G$:
1) For $R = L$ we have an $\mathbb{L}_2(G)$-isometry $\Leftrightarrow$ the unbounded linear generator $Qf = -cf - i^q |\Delta_{\mathcal{G}_0}|^{\frac{\alpha}{2}}f$ is skew-adjoint: $(e^{tQ})^* = (e^{tQ})^{-1} \Leftrightarrow Q^* = -Q \Leftrightarrow q = 1$.
2) For $R = T_+, T_-$ this follows by g) as $\delta_{T_+}(f, \mathbb{0}) = e^{I_{T_+}(f)}, \delta_{T_-}(f, \mathbb{0}) = e^{-I_{T_-}(f)}$.

Item i): One has $\max\{x\} = -\min\{-x\}$ and $\Phi_t^{T_+}(f) = -\Phi_t^{T_-}(-f)$. So for T_+ the convolution kernels are minus the positive erosion kernels of T_- so they are discarded. For $R = L$ the Schrödinger kernels $q = 1$ follow from the positive diffusion kernels $q = 0$ by $t \mapsto it$ so they are discarded. $\qquad\square$

Remark 4. **(Semigroup Property for $R = T_-$)**
Rewrite (10) as $k_t^{\mathcal{G}}(g) = k_t^{1D}(d_{\mathcal{G}}(g, e))$ with $k_t^{1D}(x) = \frac{t}{\beta}\left|\frac{x}{t}\right|^\beta$, $\frac{1}{\alpha} + \frac{1}{\beta} = 1$. Then

$$\forall_{g \in G}\forall_{s,t \geq 0} : (k_t \circledast k_s)(g) = (k_t^{1D} \circledast_{\mathbb{R}} k_s^{1D})(d_{\mathcal{G}}(g, e)) = k_{t+s}^{1D}(d_{\mathcal{G}}(g, e)) = k_{t+s}(g).$$

Above, the first equality follows by the triangular inequality for $d_{\mathcal{G}}$ and monotony of $k_t^{1D}(\cdot)$ and by realizing that the infimum/minimum over G can be replaced by the *subset* reached by any minimizing geodesic connecting e and g and the semigroup property for morphological scale spaces on $\mathbb{R}$ [19] (that follows by Fenchel transform on $\mathbb{R}$). The final equality is due to the definition of k_t^{1D} and (10). Then $\Phi_{t+s}(f) = k_{t+s}^{\mathcal{G}} \circledast f = k_t^{\mathcal{G}} \circledast (k_s^{\mathcal{G}} \circledast f) = (\Phi_t \circ \Phi_s)(f)$ for all $f \in \mathcal{H}_{T_-}^G$.

Practical Implications: Item a) is called the *semigroup property*. We call b) the *quasi-linear convolution property* allows for fast parallel GPU-code [21] of PDE-G-CNNs. Item c) is called the *strong continuity property* implies well-defined evolution solutions that continuously depend on all $t \geq 0$ via a), item d) is the *scaling equivariance property*; and avoids scale biases in PDE-G-CNNs, item e) yields *equivariance* and item f) is called the *quasi-linearity property*. Item g) yields *normalisation* (of feature maps). Item h) is called the *isometry property* and yields PDE-G-CNNs without fractional diffusion regularisation [12]. The isometry property avoids contractions when the depth of the network

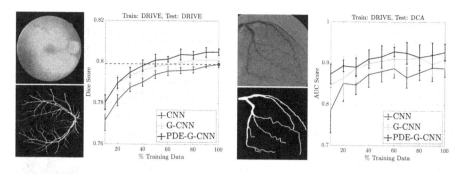

Fig. 2. We compare a 6-layer: CNN (25662 parameters), G-CNN (24632 parameters), and a PDE-G-CNN (2560 parameters) architectures with varying amounts of training data. All networks are trained *only* on DRIVE, and their performances on the test data are plotted as a function of % input training data. PDE-G-CNNs exhibit superior generalization for intra (DRIVE) and inter (DCA) datasets, especially for limited training data. The dashed line shows that PDE-G-CNNs outperform CNNs with just ∼45% training data and ∼10x fewer parameters. The images and their ground truths are representatives of the respective datasets (i.e. DRIVE and DCA).

increases (hindering classification performances). Item i) of *positive kernels* is questionable: The Schrödinger equation gives rise to new Quantum PDE-G-CNNs, providing optimal transport [16] in PDE-G-CNNs.

3 Data Efficiency of PDE-G-CNNs

We evaluate the performance of comparable architectures of a CNN, G-CNN and PDE-G-CNNs by reducing the amount of training data. We compare a 6-layer CNN with 25662 parameters, a 6-layer G-CNN with 24632 parameters and a 6-layer PDE-G-CNN with 2560 parameters. For the G-CNN and PDE-G-CNN we discretize with 8 orientations (45° per orientation) and kernels of size $5 \times 5 \times 5$. We experiment with the vessel segmentation task on the well-known DRIVE [22] and DCA1 [6] datasets. We make a split of 75% images for training and 25% for testing. The networks are trained in overlapping patches of dimension 64×64 with a patch overlap of 16. We randomly shuffle the training patches in the DRIVE dataset and progressively compile 10% to 100% of the total, and use this reduced input for the training of all the networks. In order to avoid possible bias, we run each training 10 times for 60 epochs using randomization for compiling the reduced training data. In Fig. 2, we compare all networks by plotting the mean and variance of the performance metrics as a function of the % input training data.

In addition to reporting the best test error on DRIVE, we apply the *same* best-trained networks to the *unseen and untrained* images in the DCA1 dataset [6] which comprises 134 X-ray angiograms of coronary arteries with ground truth segmentation's (see the example in the right of Fig. 2). All images are preprocessed with the adaptive histogram normalization framework of [25].

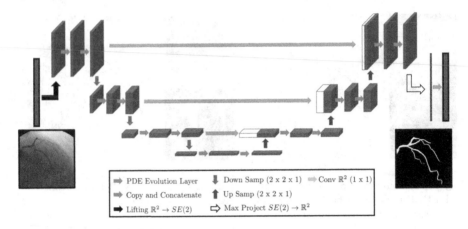

Fig. 3. We compile a 3-stage PDE-G-UNet using PDE-based group convolutions in the following order: Convection-Dilation-Erosion (as shown in Fig. 1). We up and downsample only spatially (2×2) and keeping all orientations ($\times 1$). We compare with an equivalent 3-stage U-Net using standard CNNs [17] in Table 1. PDE-G-UNets show competitive performance, with an order of magnitude lesser parameters. The images and their ground truths are representatives of the DCA1 dataset [6].

The plots in Fig. 2 demonstrate that despite a significant reduction in network complexity, PDE-G-CNNs have a considerable benefit in training with limited data and generalize well to unseen examples. Conceptually, these results highlight the benefits of equivariant PDE action in lifted spaces. In contrast to learning everything from data alone, PDE-G-CNNs enable the inclusion of geometric inductive bias into the architectures which becomes valuable, especially in scenarios of limited training data.

4   Equivariant U-Nets with PDE-G-CNNs

We demonstrate a wider applicability of PDE-based group convolutions by constructing a PDE-based U-Net [17]. We replace regular convolutions (CNNs) in $\mathbb{R}^2$ with the convection PDEs and PDE-based morphological convolutions in lifted space $SE(2)$.

Table 1. Comparing UNets

Architecture	Dice	AUC	Params
6 Layer CNN	0.75	0.90	2.5×10^4
6 Layer PDE-G-CNN	0.76	0.92	2.5×10^3
U-Net (3 stage) [17]	0.77	0.93	6.2×10^6
PDE-G-UNet (3 stage)	0.79	0.92	3.1×10^4

We implement the spatial up and downsampling of the feature maps (of the lifted domain $SE(2)$) using nearest neighbor interpolation. The hierarchical PDE-G-UNet structure is combined with a mandatory *equivariant* lifting and projection layer for the processing of input and output images respectively in $\mathbb{R}^2$. See Fig. 3 for an overview of the proposed architecture. We discretize the PDE-G-UNet with 8 orientations and use kernels of size ($5 \times 5 \times 5$). We report an evaluation by imputing full images of size 320×320 from the DCA1 dataset [6] for all the architectures. We partition the total data into a train-validation-test split of 67%–8%–25% respectively. We compare two 3-stage U-Nets trained for 1000 epochs, 5 times and randomly shuffle the training and validation datasets each iteration. As a reference, we also report the performance of a 6-Layer fully convolutional CNN and an equivalent PDE-G-CNN. In Table 1, we report the mean values of the segmentation metrics for all architectures. We see a favorable performance despite a ∼200x reduction in network parameters.

5 Conclusion and Future Work

In this paper, we theoretically enumerate some crucial geometric and algebraic properties of PDE-G-CNNs. Practically, we report that they have lower network complexity, yield a training data reduction and are extendable to more complex architectures. Nevertheless, processing of feature maps in lifted spaces with non-linear evolution PDE's demands considerable memory and associated computation. We will aim for faster computation and sparsification of feature maps for PDE-G-CNNs in future work.

Acknowledgement. We gratefully acknowledge the Dutch Foundation of Science NWO for funding of VICI 2020 Exact Sciences (Duits, Geometric learning for Image Analysis VI.C. 202-031). The git repository containing the vanilla PDE-G-CNN implementations can be found at: https://gitlab.com/bsmetsjr/lietorch.

References

1. Aguado Lopez, J., et al.: Semigroup theory in quantum mechanics. In: Escuela-Taller de Análisis Funcional, vol. VIII (2018)
2. Balogh, Z.M., Engulatov, A., Hunziker, L., Maasalo, O.E.: Functional inequalities and Hamilton-Jacobi equations in geodesic spaces. Potential Anal. **36**(2), 317–337 (2012)
3. Bekkers, E.J., Lafarge, M.W., Veta, M., Eppenhof, K.A.J., Pluim, J.P.W., Duits, R.: Roto-translation covariant convolutional networks for medical image analysis. In: Frangi, A.F., Schnabel, J.A., Davatzikos, C., Alberola-López, C., Fichtinger, G. (eds.) MICCAI 2018. LNCS, vol. 11070, pp. 440–448. Springer, Cham (2018). https://doi.org/10.1007/978-3-030-00928-1_50
4. Bellaard, G., Bon, D.L., Pai, G., Smets, B.M., Duits, R.: Analysis of (sub-) Riemannian PDE-G-CNNs. J. Math. Imaging Vis., 1–25 (2023). https://doi.org/10.1007/s10851-023-01147-w
5. Bellaard, G., Pai, G., Bescos, J.O., Duits, R.: Geometric adaptations of PDE-G-CNNs. In: Calatroni, L., Donatelli, M., Morigi, S., Prato, M., Santacesaria, M. (eds.) Scale Space and Variational Methods in Computer Vision, SSVM 2023. LNCS, vol. 14009, pp. 538–550. Springer, Cham (2023). https://doi.org/10.1007/978-3-031-31975-4_41
6. Cervantes-Sanchez, F., Cruz-Aceves, I., Hernandez-Aguirre, A., Hernandez-Gonzalez, M.A., Solorio-Meza, S.E.: Automatic segmentation of coronary arteries in X-ray angiograms using multiscale analysis and artificial neural networks. Appl. Sci. **9**(24), 5507 (2019)

7. Citti, G., Sarti, A.: A cortical based model of perceptional completion in the roto-translation space. JMIV **24**(3), 307–326 (2006)
8. Cohen, T.S., Weiler, M., Kicanaoglu, B., Welling, M.: Gauge equivariant convolutional networks and the icosahedral CNN. In: Chaudhuri, K., Salakhutdinov, R. (eds.) ICML, pp. 1321–1330. PMLR (2019)
9. Davidson, J.L., Hummer, F.: Morphology neural networks: an introduction with applications. Circ. Syst. Sig. Process. **12**(2), 177–210 (1993)
10. Duits, R., Florack, L., de Graaf, J., et al.: On the axioms of scale space theory. JMIV **20**, 267–298 (2004). https://doi.org/10.1023/B:JMIV.0000024043.96722.aa
11. Duits, R., Franken, E.: Left-invariant parabolic evolution equations on $SE(2)$ and contour enhancement via orientation scores. QAM-AMS **68**, 255–331 (2010)
12. Duits, R., Smets, B., Bekkers, E., Portegies, J.: Equivariant deep learning via morphological and linear scale space PDEs on the space of positions and orientations. In: Elmoataz, A., Fadili, J., Quéau, Y., Rabin, J., Simon, L. (eds.) SSVM 2021. LNCS, vol. 12679, pp. 27–39. Springer, Cham (2021). https://doi.org/10.1007/978-3-030-75549-2_3
13. Happ, L.: Lax-Oleinik semi-group and weak KAM solutions. https://www.mathi.uni-heidelberg.de/~gbenedetti/13_Happ_Talk.pdf
14. Maragos, P., Charisopoulos, V., Theodosis, E.: Tropical geometry and machine learning. Proc. IEEE **109**(5), 728–755 (2021)
15. Petitot, J.: The neurogeometry of pinwheels as a sub-Riemannian contact structure. J. Physiol. Paris **97**, 265–309 (2003)
16. Renesse, M.: An optimal transport view on Schrödinger's equation, pp. 1–11. arXiv (2009). https://arxiv.org/pdf/0804.4621.pdf
17. Ronneberger, O., Fischer, P., Brox, T.: U-Net: convolutional networks for biomedical image segmentation. In: Navab, N., Hornegger, J., Wells, W.M., Frangi, A.F. (eds.) MICCAI 2015. LNCS, vol. 9351, pp. 234–241. Springer, Cham (2015). https://doi.org/10.1007/978-3-319-24574-4_28
18. Sangalli, M., Blusseau, S., Velasco-Forero, S., Angulo, J.: Scale equivariant neural networks with morphological scale-spaces. In: Lindblad, J., Malmberg, F., Sladoje, N. (eds.) DGMM 2021. LNCS, vol. 12708, pp. 483–495. Springer, Cham (2021). https://doi.org/10.1007/978-3-030-76657-3_35
19. Schmidt, M., Weickert, J.: Morphological counterparts of linear shift-invariant scale-spaces. J. Math. Imaging Vis. **56**(2), 352–366 (2016)
20. Sifre, L., Mallat, S.: Rotation, scaling and deformation invariant scattering for texture discrimination. In: IEEE-CVPR, pp. 1233–1240 (2013)
21. Smets, B., Portegies, J., Bekkers, E.J., Duits, R.: PDE-based group equivariant convolutional neural networks. JMIV **65**, 1–31 (2022). https://doi.org/10.1007/s10851-022-01114-x
22. Staal, J., Abràmoff, M.D., Niemeijer, M., Viergever, M.A., Van Ginneken, B.: Ridge-based vessel segmentation in color images of the retina. IEEE-TMI **23**(4), 501–509 (2004)
23. Weiler, M., Cesa, G.: General E(2)-equivariant steerable CNNs. In: Advances in Neural Information Processing Systems, pp. 14334–14345 (2019)
24. Yosida, K.: Functional Analysis. CM, vol. 123. Springer, Heidelberg (1995). https://doi.org/10.1007/978-3-642-61859-8
25. Zuiderveld, K.: Contrast limited adaptive histogram equalization. In: Graphics Gems, pp. 474–485 (1994)

Continuous Kendall Shape Variational Autoencoders

Sharvaree Vadgama[1]($\boxtimes$), Jakub M. Tomczak[2], and Erik Bekkers[1]

[1] Amsterdam Machine Learning Lab, University of Amsterdam,
Amsterdam, The Netherlands
s.p.vadgama@uva.nl
[2] Generative AI Group, Eindhoven University of Technology, Eindhoven,
The Netherlands

Abstract. We present an approach for unsupervised learning of geometrically meaningful representations via *equivariant* variational autoencoders (VAEs) with *hyperspherical* latent representations. The equivariant encoder/decoder ensures that these latents are geometrically meaningful and grounded in the input space. Mapping these geometry-grounded latents to hyperspheres allows us to interpret them as points in a Kendall shape space. This paper extends the recent *Kendall-shape VAE* paradigm by Vadgama et al. by providing a general definition of Kendall shapes in terms of group representations to allow for more flexible modeling of KS-VAEs. We show that learning with generalized Kendall shapes, instead of landmark-based shapes, improves representation capacity.

Keywords: Generative models · Kendall shape spaces · shape spaces · equivariance · Variational Autoencoders · continuous landmarks

1 Introduction

Variational Autoencoders (VAEs) are a class of probabilistic generative models [11,16] that allows for unsupervised learning of compressed representations. Amongst other likelihood-based generative frameworks [17], VAEs give information-rich latent spaces, can be parameterized by neural networks, and are easy to train; VAEs are a flexible and expressive class of generative models.

In VAEs, disentangling the latent space into an invariant and non-invariant part for any given transformation is a challenging but often desirable task. In most cases, in order to make latent spaces invariant, the models need to be trained separately for each transformation class, and thus generalization can be difficult. However, adding geometric structure to the latent space has been found to enhance the generation capacity of VAEs [4]. It can lead to improved disentanglement if the geometry is designed accordingly [5]. When it comes to tasks such as domain generalization [7] and data compression [14], it is crucial to have a clear understanding of how the latent space is encoded.

In this paper, we focus on an equivariant VAE framework, called the Kendall Shape VAEs (KS-VAEs) [18] as it combines both equivariant representations as well as well-structured hyperspherical latent representations. We extend [18]

F. Nielsen and F. Barbaresco (Eds.): GSI 2023, LNCS 14071, pp. 73–81, 2023.
https://doi.org/10.1007/978-3-031-38271-0_8

by introducing generalized Kendall shape spaces. Through this framework, we alleviate the constraint of KS-VAEs, which could represent only anti-symmetric continuous shapes, and present that band-limited landmarks are more suitable to model symmetric shapes.

2   Background

As a preliminary to KS-VAEs, we give a brief introduction to VAEs as an unsupervised representation learning paradigm, as well as its extension to hyperspherical VAEs.

VAEs assume the generative process of observations x from an unobserved latent z through a latent-conditional generator $p_\theta(x|z)$ and prior $p_\theta(z)$ where θ refers to network parameters. When we parameterize the joint distribution by a neural network, the marginalization over the latent variables to obtain data evidence $p_\theta(x)$ for likelihood maximization is untractable. So instead, our objective is to maximize the evidence lower bound (ELBO), using an approximate posterior $q_\phi(z|x)$, with ϕ being network parameters. The ELBO is given by [11]

$$\mathcal{L}(\theta, \phi) = \mathbb{E}_{z \sim q_\phi(z|x)} \left[\log p_\theta(x|z) - D_{KL}[q_\phi(z|x)||p_\theta(z)] \right].$$

Optimizing the ELBO allows us to train the *encoder* (approx. posterior $q_\phi(z|x)$) to infer compressed representations z of the input x in an unsupervised manner.

Hyperspherical VAEs (SVAEs): In the original VAE, the prior and posterior are both defined by a normal distribution. In SVAE [5], a von Mises Fisher (vMF) distribution is the natural choice for a distribution on a sphere, as it is the stationary distribution of a convection-diffusion process on the hypersphere S^{m-1}, just like the normal is on $\mathbb{R}^m$. The probability density function of the vMF distribution for a random unit vector $z \in S^{m-1}$ is defined as

$$q(z|\mu, \kappa) = \mathcal{C}_m(\kappa) \exp(\kappa \mu^T z), \quad \text{with} \quad \mathcal{C}_m(\kappa) = \frac{\kappa^{m/2-1}}{(2\pi)^{m/2} \mathcal{I}_{m/2-1}(\kappa)},$$

$||\mu||^2 = 1$, κ is a concentration parameter and $\mathcal{I}_\nu$ denoting a modified Bessel function of the first kind at order ν. The Kullback-Leibler divergence can be analytically computed between a vMF distribution $vMF(z|\mu, \kappa)$ and a uniform distribution on S^{m-1} $U(x)$ via

$$KL(vMF(\mu, \kappa)||U(S^{m-1})) = \kappa \frac{\mathcal{I}_{m/2}(k)}{\mathcal{I}_{m/2-1}(k)} + \log \mathcal{C}_m(\kappa) - \log \frac{2(\pi^{m/2})^{-1}}{\Gamma(m/2)}.$$

3   Generalized Kendall Shape Space

Kendall Shape VAEs, as proposed in [18], provide a framework for encoding image data into geometrically meaningful latent representations, which in [18] is given the interpretation of neural ideograms (learned geometric symbols). More specifically, an equivariant encoder is used to encode images into Kendall shapes whose landmarks through an equivariant design follow the same transformation laws of $SE(m)$, which is a special Euclidean group of dimension m.

Kendall defined shapes based on the idea that a shape is a translation, scale, and rotation invariant quantity [9]. More precisely, he defined pre-shapes as k labeled points in Euclidean space $\mathbb{R}^m$. Two configurations of k labeled points are then regarded as equivalent if their pre-shapes can be transformed into the other by a rotation about a shared centroid. The quotient structure was extended in [10] by defining shape space as the quotient of the space of landmark configurations by the group of translations, scale, and rotations. Following this rationale, we generalize the definition of a shape space in terms of group representations, taking the original definition as a special case.

3.1 Preliminaries (Group Representations)

We consider the group $SO(m)$ of rotations in $\mathbb{R}^m$. Let $\rho : SO(m) \to GL(V)$ denote a representation of $SO(m)$. I.e., ρ describes transformations on elements in vector space V that are parameterized by rotations $\mathbf{R} \in SO(m)$, following the group structure via $\rho(\mathbf{R})\rho(\mathbf{R}')v = \rho(\mathbf{R}\mathbf{R}')v$ for all $\mathbf{R}, \mathbf{R}' \in SO(m)$ and all $v \in V$. Note, V does not have to be $\mathbb{R}^m$ but can be any vector space that can be transformed by the group action of $SO(m)$ such as e.g. $\mathbb{L}_2(S^{m-1})$, the space of square-integrable functions on S^{m-1}. With $\overline{\rho}_l$, we denote the irreducible representations of $SO(m)$, i.e., the rotation matrices (Wigner-D matrices for $m = 3$) of frequency l. With $\rho_{\mathcal{L}}$ we denote the left regular representation of $SO(m)$ on functions on homogeneous spaces, i.e., $(\rho_{\mathcal{L}}(\mathbf{R})f)(x) = f(\mathbf{R}^{-1}x)$. With V_ρ we denote the vector space associated with group representation ρ. We assume each vector space V_ρ is equipped with a norm $\|\cdot\|$ and assume ρ to be a unitary representation, thus satisfying $\forall_{\mathbf{R} \in SO(m)} : \|\rho(\mathbf{R})v\| = \|v\|$. Product spaces $V_\rho = V_{\rho_1} \times \cdots \times V_{\rho_k}$ inherit the norm from the subspaces via $\|\mathbf{x}\|^2 = \sum_{i=1}^k \|\mathbf{x}_i\|^2$, with $\mathbf{x} = \mathbf{x}_1 \oplus \cdots \oplus \mathbf{x}_k \in V_\rho$.

3.2 Generalized Kendall Shape Space

We define a shape space as the equivalence class of vectors (pre-shapes) that lie in the same orbit generated by the associated representation ρ, as given below.

Definition 1 (Pre-shape space). *A pre-shape space S_m^ρ is defined by a block-diagonal representation $\rho = \rho_1 \oplus \cdots \oplus \rho_k$ of $SO(m)$, or equivalently by its associated vector space $V_\rho = V_{\rho_1} \times \cdots \times V_{\rho_k}$, as*

$$S_m^\rho = \{\mathbf{x} \in V_\rho \mid \|\mathbf{x}\| = 1\}. \tag{1}$$

The sub-vectors $\mathbf{x}_i \in V_{\rho_i}$ in $\mathbf{x} = \mathbf{x}_1 \oplus \cdots \oplus \mathbf{x}_k$ are called landmarks; S_m^ρ is identified with the hypersphere $S^{d_\rho - 1}$ with $d_\rho = \dim V_\rho$.

Definition 2 (Shape space). *A shape is defined as the equivalence class of pre-shapes $x \in S_m^\rho$, denoted with $[x]$, by*

$$[x] = \{y \in S_m^\rho \mid y \sim x\},$$

in which $x \sim y$ iff $\exists_{\mathbf{R} \in SO(m)} : y = \rho(\mathbf{R})x$. A shape space then is the space of equivalence classes of pre-shapes, denoted as the quotient space $S_m^\rho / \sim$.

The classic Kendall shape space as defined in [6] is a shape space $S_m^\rho / \sim$ with representation $\rho(\mathbf{R}) = \oplus_k \rho_1(\mathbf{R}) = \oplus_k \mathbf{R}$. A Kendall shape thus consists of k m-dimensional landmarks $\mathbf{x}_i \in \mathbb{R}^m$, i.e., the sub-vectors in $\mathbf{x} = \mathbf{x}_1 \oplus \cdots \oplus \mathbf{x}_k \in (\mathbb{R}^m)^k$ that simply transform by the usual rotation matrices. Kendall shapes further have the property that the landmarks are centered, i.e., $\sum_{i=1}^k \mathbf{x}_i = \mathbf{0}$.

Remark 1. In our general formulation of shape spaces, we omit the constraint that the landmarks sum to 0. In some special cases, this constraint can be enforced, but in a general sense, it is not natural because the landmarks do not have to literally correspond to spatial landmarks in $\mathbb{R}^m$. Our definition of pre-shape is independent of how it is obtained. When encoding a shape from a point cloud, it is common to center and normalize it to unit length to obtain translation and scale invariance. However, we do not assume how the shape is obtained, allowing for direct rescaling of the point cloud. In this case, the resulting shape loses translation invariance, and a shifted point cloud is considered a different shape. Therefore, if centering is not included in the encoding process, translation can be seen as an additional degree of freedom to describe a shape.

3.3   The Procrustes Problem, Pose, and Canonicalization

The geometry of pre-shapes is equivalent to that of hyperspheres, which allows for convenient (and analytic) geometric analysis such as the computation of distances between shapes. However, the quotient structure has the practical disadvantage that in order to compare shapes, one has to align them (as much as possible) and compare the pre-shapes in each equivalence class that is closest. This matching problem is called Procrustes analysis and involves minimizing the distance between pre-shapes over translation, reflection, rotation, and scaling.

Since often it is possible to assign a unique pose to a given shape, Procrustes's problem boils down to simply aligning their poses. Specifically, we define the pose $g = (\mathbf{t}, s, \mathbf{R})$ of a generalized shape as a translation $\mathbf{t}$, scale s, and rotation $\mathbf{R}$ in the group $SIM(m)$.

Definition 3 *(Geometric pose, canonical shape). A pre-shape $x \in [x_0] \in S_m^\rho / \sim$ can be equipped with a pose $g = (\mathbf{t}, s, \mathbf{R}) \in SIM(m)$, such $x = \rho(\mathbf{R})x_0$. I.e., the rotation part $\mathbf{R}$ of the pose describes the rotation of a **canonical shape** x_0 to pre-shape x. Any pre-shape lies in the $SO(m)$-orbit of a canonical shape.*

Remark 2. As noted in Remark 1, we consider shapes as equivalence classes over rotations and do not explicitly take translation and scale into account. In applications it might be beneficial to keep track of such attributes via an equivariant pose extractor, that could be used to canonicalize an vector and obtain a fully $SIM(m)$ invariant shape extractor. Here, we only aim for rotation invariance.

If one has access to a pose, the Procrustes problem is significantly simplified by simply aligning any shape y to x via $\rho(\mathbf{R}_x)\rho(\mathbf{R}_y)^{-1}y$. In our work, we use an equivariant architecture to simultaneously predict a pre-shape x and a rotational pose $\mathbf{R}$, which allows us to map any predicted pre-shape to its canonical pose $x_0 = \rho(\mathbf{R}^{-1})x$. This is a practically useful way of representing an equivalence class of shapes to a single representative. In recent deep learning literature, such a mapping is called a **canonicalization** function [8] and has the practical benefit that canonicalized representations are invariant and do not require specialized equivariant architectures in down-stream tasks. In our KS-VAE setting, we work with neural networks that are translation and rotation equivariant by design, through $SE(2)$ equivariant group convolutions [1], and the data does not exhibit scale variations. In principle, one could make use of $SIM(m)$ equivariant architectures [12] to also take scale into account.

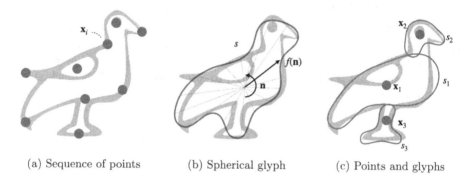

(a) Sequence of points   (b) Spherical glyph   (c) Points and glyphs

Fig. 1. Generalization of classical Kendall shapes.

3.4   Continuous Shapes: Spherical Glyphs as Landmarks

In equivariant deep learning literature, it is common to consider feature fields as a bundle of "fibers" that transform by actions of $SO(m)$ [20]. Features in such fibers are understood to carry directional information and implicitly encode for feature signals on $SO(m)$ at each location in space. That is, these signals can be mapped to a set of Fourier-like coefficients relative to a basis of irreducible representations of $SO(m)$ [13].

Many practical equivariant deep learning libraries explicitly assign to points in space either a signal on S^{m-1} or $SO(m)$, or a vector of Fourier coefficients that can be turned into such a signal via an inverse Fourier transform. That is, regular group convolutional networks encode features fields on $SE(m)$ which allows for considering $SO(m)$ signals at each fixed location, i.e., $\mathbf{f_x}(\mathbf{R}) := f(\mathbf{x}, \mathbf{R})$. Steerable group equivariant architectures encode for fields of Fourier coefficients that can be mapped to signals on $SO(m)$ via $\mathbf{f_x}(\mathbf{R}) := \mathcal{F}^{-1}[\hat{f}(\mathbf{x})]$. Such signals transform via the regular representations $\rho_{\mathcal{L}}$ of $SO(m)$. When designing equivariant encoders/decoders for the purpose of KS-VAEs, it is thus of practical value to understand that such fibers can directly be used as landmarks in a generalized Kendall shape space. We will refer to shape spaces with $\rho = \oplus_{i=1}^{k} \rho_{\mathcal{L}}$ as **continuous Kendall shape spaces**.

It is natural to think of spherical signals as geometric shapes through so-called glyph visualizations [15] for their use in diffusion MRI or [2] for the sake of visualizing feature fields in equivariant graph NNs. Concretely, a spherical signal $f : S^{m-1} \rightarrow \mathbb{R}$ can be turned into its geometric form as *spherical glyph s* via

$$s = \{f(\mathbf{n})\mathbf{n} \mid \mathbf{n} \in S^{m-1}\}.$$

Using such spherical glyphs (equivalently spherical signals) as landmarks thus allows the construction of fully continuous shapes, see Fig. 1b.

3.5   Neural Ideograms

Our general definition of shapes allows for the mixing of landmarks of different types, which conceptually corresponds to defining shapes in terms of geometric primitives. Alluding to the idea of neural ideograms of [18], one could mix multiple glyphs (and their centers) as in Fig. 1c, to form abstract symbols much alike the ancient (hiero)glyphs, however, now to be defined by a neural network.

4   Method

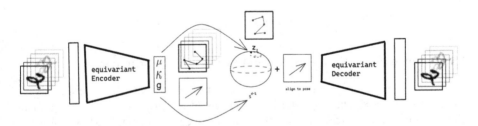

Fig. 2. *Kendall Shape VAE pipeline.* The equivariant encoder encodes the equivalence class of a transformation (here rotations), extracts a pose, and gives a μ and κ. A latent variable is sampled from vMF(μ, κ), and along with the extracted pose, the decoder gives the output reconstruction.

The KS-VAE consists of an encoder Enc : $\mathbb{L}_2(\mathbb{R}^m) \rightarrow V_\rho \times \mathbb{R}^+ \times SO(m)$ that models the approximate posterior distribution $q(z|x) = VMF(z|\mu, \kappa)$, through the prediction of $\mu \in V_\rho$, $kappa \in \mathbb{R}^+$ and a pose $\mathbf{R} \in SO(m)$. As such, it predicts the mean $\mu \in V_\rho$ in an equivarian manner, and the predicted concentration parameter $V_{\bar{\rho}_0} \ni \tilde{\kappa} \mapsto \kappa = e^{\tilde{\kappa}} \in \mathbb{R}^+$ invariantly. The mean μ is obtained as a vector in V_ρ that transforms via ρ, and which we normalize to obtain $\mu \in S_m^\rho$. We emphasize that *the inference of μ should be equivariant if we want the landmarks in S_m^k to be geometrically meaningful* with respect to the content in the input image. In our 2D setting, the pose $\mathbf{R}_\theta$ is obtained by predicting a vector $\mathbf{n} \in V_{\bar{\rho}_1}$ of which $\theta(\mathbf{n}) = \arctan(n_y, n_x)$ parametrizes the rotation angle (Fig. 2).

5   Experiments

In this section, we perform experiments with continuous KS-VAE and compare them with standard KS-VAE. For completeness, we show the Kendall shape spaces in a pure autoencoder form as well (KS-AE) as well as an autoencoder in the continuous case (continuous KS-AE). We compare these models with vanilla VAEs, hyperspherical VAEs as well as their autoencoder counterparts in an unsupervised way, without any prior knowledge of the equivalence classes or the poses of every instance.

All the models use the same architecture for the encoder/decoder, except for the type of convolutions used. The baselines AE and S-AE use normal (non-rotationally equivariant) convolutions. The KS-VAEs use roto-translation equivariant convolutions with fibers that transform via regular representations of the discrete subgroup $SO(2, N) = C_N$ of cyclic permutations. The input images are padded with zeros at the bottom and right to obtain images of size 29×29.

For the equivariant encoder and decoder we used the interpolation-based library se2cnn from [1], but recommend escnn from [3,19] for extensions to 3D. The encoder shrinks the image to a single pixel using multiple group convolutions (without spatial padding). What remains at this pixel is a signal over $SO(2, N)$. Our networks are based on a discretization of $N = 9$. We can decompose these signals into their irreps via a Fourier transform to obtain vectors that transform via different types of irreducible representations $\bar{\rho}_l$ when needed. We then use the following options for the Kendall shape space.

- **KS-VAE**: The encoder predicts k landmarks in $\mathbb{R}^2 \equiv V_{\bar{\rho}_1}$ which are normalized to unit length. The pre-shape space is thus S^{2k-1}.
- **KS-VAE continuous**: The encoder predicts k signals over $SO(2, N)$ which are band-limited to maximum frequency $\lfloor N/2 \rfloor$ and the mean is subtracted. We use $N = 9$ and thus effectively encode to 8-dimensional landmarks, which when normalized gives points in a S^{8k-1}.

For quantitative analysis, we present the results of reconstruction loss with different models for the MNIST dataset, where the training dataset is not rotated and the testing data is randomly rotated. Here, we compare our model with vanilla autoencoder (AE), hyperspherical autoencoder (S-AE) as well as the discrete case of Kendall shape autoencoder (KS-AE) and similarity with their variational counterparts. All the models are trained for 50 epochs.

Table 1. Comparing reconstruction losses of different models

Model	Latent dim	# of landmarks	loss
AE	8	–	99.89 ± 1.32
S-AE	8	–	91.03 ± .020
KS-AE	8	4	89.77 ± .041
KS-AE continuous	8	1	88.06 ± .022
VAE	8	–	124.3 ± .031
S-VAE	8	–	110.3 ± .048
KS-VAE	8	4	84.60 ± .012
KS-VAE continuous	8	1	80.07 ± .057

6   Conclusion

Our approach mixes the advantages of equivariance with that of hyperspherical latent variable models, which combined gives us Kendall shape latent representations. The advantage is supported by results in Table 1, which show that hyperspherical outperform plain VAEs, which in turn are outperformed by equivariant hyperspherical VAEs (KS-VAEs). Our main contribution is the formulation of continuous Kendall shape space in a group theoretical language that is compatible with the state of the art in geometric deep learning. Although not explored in this paper, previous work [18] has shown that it is possible to visualize the learned latent shapes. We believe it is an exciting direction for future work to explore the potential of using the Kendall shape framework to build interactive and interpretable systems for the visualization of features through neural ideograms.

References

1. Bekkers, E.J., et al.: Roto-translation covariant convolutional networks for medical image analysis. In: MICCAI (2018). https://github.com/ebekkers/se2cnn
2. Brandstetter, J., Hesselink, R., van der Pol, E., Bekkers, E., Welling, M.: Geometric and physical quantities improve E(3) equivariant message passing. In: ICLR (2022)
3. Cesa, G., Lang, L., Weiler, M.: A program to build E(N)-equivariant steerable CNNs. In: ICLR (2022)
4. Chadebec, C., Allassonnière, S.: Data augmentation with variational autoencoders and manifold sampling (2021). https://doi.org/10.48550/ARXIV.2103.13751
5. Davidson, T.R., Falorsi, L., De Cao, N., Kipf, T., Tomczak, J.M.: Hyperspherical variational auto-encoders. In: 34th Conference on UAI (2018)
6. Guigui, N., Maignant, E., Trouvé, A., Pennec, X.: Parallel transport on Kendall shape spaces (2021). https://doi.org/10.48550/ARXIV.2103.04611
7. Ilse, M., Tomczak, J.M., Louizos, C., Welling, M.: DIVA: domain invariant variational autoencoder (2020). https://openreview.net/forum?id=rJxotpNYPS

8. Kaba, S.O., Mondal, A.K., Zhang, Y., Bengio, Y., Ravanbakhsh, S.: Equivariance with learned canonicalization functions (2022)
9. Kendall, D., Barden, D.M., Carne, T.K.: Shape and Shape Theory (1999)
10. Kendall, W.S., Last, G., Molchanov, I.S.: New perspectives in stochastic geometry. Oberwolfach Rep. **5**(4), 2655–2702 (2009)
11. Kingma, D.P., Welling, M.: Auto-encoding variational Bayes. arXiv preprint arXiv:1312.6114 (2013)
12. Knigge, D.M., Romero, D.W., Bekkers, E.J.: Exploiting redundancy: separable group convolutional networks on lie groups. In: ICML (2022)
13. Kondor, R., Trivedi, S.: On the generalization of equivariance and convolution in neural networks to the action of compact groups. In: ICML (2018)
14. Kuzina, A., Pratik, K., Massoli, F.V., Behboodi, A.: Equivariant priors for compressed sensing with unknown orientation. In: Proceedings of the 39th ICML (2022)
15. Peeters, T.H., Prckovska, V., van Almsick, M., Vilanova, A., ter Haar Romeny, B.M.: Fast and sleek glyph rendering for interactive HARDI data exploration. In: 2009 IEEE Pacific Visualization Symposium, pp. 153–160. IEEE (2009)
16. Rezende, D.J., Mohamed, S., Wierstra, D.: Stochastic backpropagation and approximate inference in deep generative models (2014)
17. Tomczak, J.M.: Deep Generative Modeling. Springer, Cham (2022). https://doi.org/10.1007/978-3-030-93158-2
18. Vadgama, S., Tomczak, J.M., Bekkers, E.J.: Kendall shape-VAE: learning shapes in a generative framework. In: NeurReps Workshop, NeurIPS 2022 (2022)
19. Weiler, M., Cesa, G.: General E(2)-Equivariant Steerable CNNs. In: Conference on Neural Information Processing Systems (NeurIPS) (2019)
20. Weiler, M., Forré, P., Verlinde, E., Welling, M.: Coordinate independent convolutional networks-isometry and gauge equivariant convolutions on Riemannian manifolds (2021)

Can Generalised Divergences Help for Invariant Neural Networks?

Santiago Velasco-Forero[✉]

MINES Paris-PSL-Research University-Centre de Morphologie Mathématique,
Paris, France
Santiago.Velasco@Mines-Paristech.fr

Abstract. We consider a framework including multiple augmentation regularisation by generalised divergences to induce invariance for non-group transformations during training of convolutional neural networks. Experiments on supervised classification of images at different scales not considered during training illustrate that our proposed method performs better than classical data augmentation.

1 Introduction

Deep neural networks are the primary model for learning functions from data, in different tasks ranging from classification to generation. Convolutional neural networks (CNNs) have become a widely used method across multiple domains. The *translation* equivariance of convolutions is one of the key aspects to their success [23]. This equivariance is induced by applying the same convolutional filter to each area of an image producing learned weights that are independent of the location. This mechanism is called *weight sharing*. Ideally, CNNs should perform equally well regardless of input scale, rotation or reflections. Numerous attempts have been made to address using the formalism of *group-convolutions* [12], *steerable filters* [8], *moving frames* [31], *wavelet* [2], *partial differential equations* [36], *Gaussian filters* [26], *Elementary Symmetric Polynomials* [27] among others. Despite all these recent advances, it is still unclear what is the most adequate way to adapt these methods for the case of more general transformations that cannot be considered as a group [30,38]. The most commonly used solution is to take advantage of *data augmentation*, where the inputs are randomly transformed during training to induce an output (which is) insensible to some given transformations [34]. But still data augmentation implies neither equivariance nor invariance. An more elaborated path would be to apply the *weight sharing* mechanism for each discretisation of the transformation, followed by an integration in the sense of [22]. In this paper, we study the use of contrastive based regularisation on a set of transformations during training. Surprisingly, our proposition presents the best performance considering the power of generalisation outside the interval of values where the transformation has been sampled during training. This phenomenon is illustrated in the case of supervised classification on aerial images and traffic signs at different scales.

F. Nielsen and F. Barbaresco (Eds.): GSI 2023, LNCS 14071, pp. 82–90, 2023.
https://doi.org/10.1007/978-3-031-38271-0_9

2 Proposition

2.1 Motivation

Data augmentation is nowadays one of the main components of the design of efficient training for deep learning models. Initially proposed to improve over-sampling on class-imbalanced datasets [9] or to prevent overfitting when the model contains more parameters than training points [24]. Recent research has shown its interest in increasing generalization ability especially when augmentations yield samples that are diverse [15]. We restrict our study to augmentations which act on a single sample and do not modify labels, this means that we do not consider *mixup augmentations* [40]. Namely, we study augmentations which can be written as $(t(\mathbf{x}), y)$, where $(\mathbf{x}, y)$ denotes an input-label pair, and $t \in \mathcal{T}$ is a random transformation sampled from a set of possible transformations $\mathcal{T}$. Let f denotes a projection from the input space to a *latent space*. The latent space is said to be *invariant* to $\mathcal{T}$ if for any input $\mathbf{x}$ and any $t \in \mathcal{T}$, $f(t(\mathbf{x})) = f(\mathbf{x})$. Practitioners recommend to use *data augmentation* to induce invariance *by training* [3]. Usually, data augmentation consists of randomly applying an element of the set of $\mathcal{T}$ during training. An alternative to data augmentation is possible when $\mathcal{T}$ is a group. One can construct an invariant function $f_\theta(\mathbf{x}; \eta)$ from a non-invariant function $g_\theta(\mathbf{x})$ by integrating over all the group actions. This concept is referred to as *insensitivity* [37], *soft-invariance* [4], or *deformation stability* [6]. The special case in which there exist a subgroup H where the computation can be reduced to summing over H is called *Reynolds design* [29]. For topological groups, there is a non-zero, translation invariant measure called *Haar measure* that can be used to define invariant convolution on a group [20,28] or invariance by integration of kernels [17,25]. An alternative to define an invariant function, is to use composition of *equivariant* functions followed by an invariant pooling in the *Geometric Deep Learning Blueprint* [6]. In both cases, the invariance is defined *by structure* [3] which can be seen as a constraint in the model that one is learning.

However, in many applications, the transformations under study is not a group, so that the above arguments are not easily generalizable. In this paper we are interested in using data augmentation to induce invariance during training in deep learning models for the case where the set $\mathcal{T}$ is not a group. The idea is to include a regularisation term that takes into account K realisations of the transformation family, i.e, in the loss function, we will include $\text{Loss}_{\mathcal{T}}(\mathbf{x}, t_1(\mathbf{x}), t_2(\mathbf{x})), \ldots, t_K(\mathbf{x})))$ where the t_i denotes a random value of transformations $\mathcal{T}$. To apply this method we do not need any requirements on the transformations $\mathcal{T}$.

2.2 Related Work

The idea of using multiple random augmentations (K is our case) during training is also found in the following methods:

Semi-supervised Learning. In a semisupervised case, [41] proposed to learn a classifier penalised for quick changes in its predictions.

Self-training. Self-training also known as decision-directed or self-taught learning machine, is one of the earliest approach in semi-supervised learning [14,32]. The idea of these approaches is to start by learning a supervised classifier on the labelled training set, and then, at each iteration, the classifier selects a part of the unlabelled data and assigns pseudo-labels to them using the classifier's predictions. These pseudo-labeled examples are considered as additional labeled examples in the following iterations. The function loss includes a trade-off term to balance the influence of pseudo labels.

Self-supervised Learning. Most of these works are placed in a *joint-embedding framework* [10,18], where augmented views (usually two) are generated from a source image. These two views are then projected to an encoder, giving representations, and then through a projection back to an embedding space. Finally, a loss minimises the distance between the embeddings, i.e. makes them invariant to the augmentations, and is combined with a regularisation loss to spread embeddings in space.

Data Augmentation Regularisations. A negative aspect of data augmentation has been illustrated in [13] which is the slow down of training speed and a minimal effect on the variance of the model. The idea of using multiple augmentation per image in the same minibatch has been used to solve that problem, and it has been used to improve at the same time the classifier's generalisation performance [13]. This simple modification computes an average of the minibatch on different augmentations that asymptotically approaches a Reynolds operator [29] when the number of considered augmentations gets as large as possible. Recently, [5,39] proposed the use of a regularisation term for multiple augmentations, which is the mechanism that we will evaluate in this paper.

2.3   Supervised Regularisation by Generalised Divergences

We propose to use multiple data augmentations in the target transformation, and use a generalised divergence as a regularisation term. The idea follows those presented in [11], and is contrary to the usual mechanism of *data augmentation*, where the network is trained to classify in the same class each of the augmentations, but never considers a term related to the divergence produced by the transformation. Since we use K augmentations, we must consider a divergence from multiple probability distributions, which is called *generalised divergences*. We consider the classical framework of training deep learning models from N

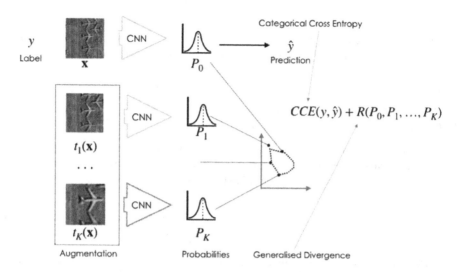

Fig. 1. Scheme of our proposition. We propose to use a regularisation that considers multiple realisations of the transformation family, this regularisation uses generalised divergences. Since you want to evaluate the invariance of a classification problem, the model uses only the classification of the original image (not of the transformations). The probability distributions are obtained in the output of a softmax layer.

samples $\{\mathbf{x}_1, \mathbf{x}_2, \ldots, \mathbf{x}_N\}$ and as objective minimising the following loss function:

$$\text{Loss}(\mathbf{x}, y) = \sum_i^N \text{Loss}_{class}(y_i, \hat{y}_i) + \alpha \sum_i^N \text{Loss}_{\mathcal{T}}(\mathbf{x}_i, t_0(\mathbf{x}_i), t_1(\mathbf{x}_i), \ldots, t_K(\mathbf{x}_i)),$$

$$(1)$$

where $\hat{y}_i$ denotes the prediction of the model, and y_i the ground-truth class of the i-th sample $\mathbf{x}_i$ (Fig. 1).

The first term is a supervised classification term, and the second term $\text{Loss}_{\mathcal{T}}$ is the main interest of our proposition. We propose to use statistical divergences to compare the outputs produced by model f applied to the original data $\mathbf{x}$ and $K + 1$ random augmentations of $\mathbf{x}$, i.e. $\{\mathbf{x}_i, t_0(\mathbf{x}_i), t_1(\mathbf{x}_i), \ldots, t_K(\mathbf{x}_i)\}$. In our supervised case, we use the last layer of the model f, which is usually a sum-one layer (softmax) indicating the probability of belonging to a given class. For two probability distributions P, Q, the most renowned statistical divergence rooted in information theory [21] is the *Kullback-Leibler divergence*,

$$D_{KL}(P\|Q) = \sum P(x) \log \left(\frac{P(x)}{Q(x)} \right).$$

Defining divergence between more than two distributions has been studied for many authors called often *generalised divergences* or *dissimilarity coefficient* in [33]. Let $K \geq 1$ be a fixed natural number. Each generalised divergence R that we consider here, satisfies the following properties:

1. $R(P_0, P_1, P_2, \ldots, P_K) \leq 0$
2. $R(P_0, P_1, P_2, \ldots, P_K) = 0$ whenever $P_0 = P_1 = \ldots = P_K$
3. R is invariant to permutation of input components.

These three properties are important for the minimisation of this divergence to induce the invariance during training in the case we are studying. Accordingly, we consider the following two generalised divergences, the *Average Divergence* [33]

$$R_1(P_0, P_1, P_2, \ldots, P_K) = \frac{1}{K(K+1)} \sum_{i,j=0,i \neq j}^{K} D_{KL}(P_i \| P_j) \qquad (2)$$

the *Information radius* [35] which is the generalised mean of the Rényi's divergences between each of the P_i's and the generalised mean of all the P_i's,

$$R_2(P_0, P_1, P_2, \ldots, P_K) = \frac{1}{K+1} \sum_{i}^{K} D_{KL}((K+1)^{-1} \sum_{j}^{K} P_j \| P_i) \qquad (3)$$

In the following section, we compare the use of (1), considering as $\mathsf{Loss}_\mathcal{T}$ the average divergence in (2) or the information radius (3) (Fig. 2).

3   Experiments

In this experimental section, we have followed the training protocol presented in [1] on two datasets, *Aerial* and *Traffic Signs*, which contains images 64×64 RGB-color images on 48 different scales. The objective is to obtain a scale and translation invariant model for supervised classification on nine (resp. 16) classes on

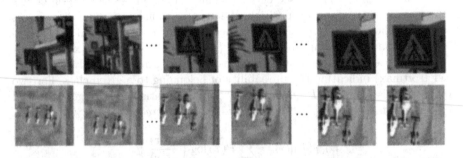

Fig. 2. Examples of images at different parameter transformation in the two considered datasets. From left to right: Scale 1, 3, 23, 25, 45 and 47. Training is done considering only images of intermediate scales (17 to 32) in both training and validation. Evaluation is performed on both small (0 to 16) and large (33–48) scales. In first row: An example of *Traffic Sign* dataset. In second row: An example of *Aerial* dataset.

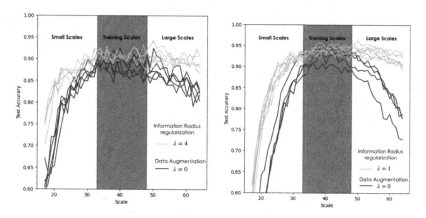

Fig. 3. Detailed plots of scale generalisation on *Mid2Rest* scenario in *Aerial* datasets (Left) and *Traffic Sign* dataset (Right). Five repetitions of the training is illustrated per method. Our proposition performs clearly better than classical data augmentation.

Table 1. Results of the Generalised Divergence in Aerial and Traffic Sign dataset on scales non-considered during training. A visual comparison of results are shown in Fig. 3

Aerial Method	λ	Small Scales test acc.	±std	Large Scales test acc.	±std
Data Aug.	0.0	0.776	±0.014	0.845	±0.008
Av. Div.(2)	0.5	0.854	±0.013	0.889	±0.011
	1.0	0.852	±0.016	0.888	±0.009
	1.5	**0.858**	± 0.013	**0.889**	±0.011
	2.0	0.841	±0.028	0.880	±0.020
	2.5	0.846	±0.015	0.881	±0.009
	3.0	0.834	±0.023	0.877	±0.018
	3.5	0.833	±0.008	0.873	±0.017
	4.0	0.822	±0.021	0.864	±0.016
	5.0	0.828	±0.016	0.872	±0.022
	10.0	0.824	±0.019	0.865	±0.012
Inf. Rad.(3)	0.5	0.845	±0.016	0.878	±0.017
	1.0	0.847	±0.012	0.885	±0.011
	1.5	0.853	±0.010	0.881	±0.009
	2.0	0.853	±0.009	0.884	±0.011
	2.5	0.850	±0.014	**0.887**	±0.005
	3.0	**0.859**	±0.013	0.885	±0.010
	3.5	0.841	±0.008	0.886	±0.009
	4.0	0.839	±0.018	0.877	±0.016
	5.0	0.845	±0.017	0.883	±0.010
	10.0	0.829	±0.014	0.870	±0.011

Traffic Sign Method	λ	Small Scales test acc.	±std	Large Scales test acc.	±std
Data Aug.	0.0	0.721	± 0.021	0.824	±0.024
Av. Div. (2)	0.5	0.821	±0.014	0.898	±0.020
	1.0	**0.829**	±0.012	**0.921**	±0.018
	1.5	0.820	±0.015	0.902	±0.012
	2.0	0.806	±0.027	0.886	±0.025
	2.5	0.797	±0.044	0.883	±0.034
	3.0	0.789	±0.025	0.882	±0.018
	3.5	0.754	±0.026	0.842	±0.028
	4.0	0.743	±0.026	0.832	±0.042
Inf. Rad.(3)	0.5	0.806	±0.012	0.908	±0.017
	1.0	0.825	±0.022	**0.921**	±0.016
	1.5	0.829	±0.014	0.918	±0.015
	2.0	**0.830**	±0.016	0.918	±0.007
	2.5	0.815	±0.020	0.906	±0.016
	3.0	0.828	±0.013	0.909	±0.018
	3.5	0.818	±0.019	0.898	±0.014
	4.0	0.823	±0.020	0.917	±0.016

Aerial (resp. *Traffic Signs*) dataset. Keen readers are referred to [1,16,19,30,38] for a deeper understanding of different propositions for scale invariant convolutional networks. Following [1] the model is a CNN with two layers using categorical cross-entropy as a supervised term in (1). An example per dataset at different scales is shown in Fig. 2. The models are trained on the middle interval of the transformation parameterisation and the performance of the models are evaluated outside this interval. This is called *Mid2Rest* scenario in [1]. The value of K in (3) and (2) has been set equal to three in our experiments. Quantitative comparison of results are found in Table 1 for both considered datasets. The reported result is the average and standard deviation of performance on the scales and images that were not considered during training. On the considered datasets, the information radius (3) presents better results in terms of performance over the unseen scales, with respect to both the average divergence (2), and the classical data augmentation method. Finally, for a better illustration of the difficulty of the task, the best value of the lambda and regularisation function is compared with the data augmentation in five random training runs, and compared across the different scales for the two databases in Fig. 3.

4   Conclusions

In this paper we present a proposal for the use of regularisation from multiple data augmentation with generalised divergences. Quantitative results show the interest of our method in the case of generalisation to scales that have not been considered during training. Future studies may include the study of multiparametric transformations, as these are used to avoid overfitting in large neural networks. Additionally, generalised divergence considering barycenters for probability distributions in [7] seems a promising direction to generalise the results of this article.

Acknowledgements. This work was granted access to the Jean Zay supercomputer under the allocation 2022-AD011012212R2.

References

1. Altstidl, T., et al.: Just a matter of scale? reevaluating scale equivariance in convolutional neural networks. arXiv preprint arXiv:2211.10288 (2022)
2. Andén, J., Mallat, S.: Deep scattering spectrum. IEEE Trans. Signal Process. **62**(16), 4114–4128 (2014)
3. Barnard, E., Casasent, D.: Invariance and neural nets. IEEE Trans. Neural Netw. **2**(5), 498–508 (1991)
4. Benton, G., Finzi, M., Izmailov, P., Wilson, A.G.: Learning invariances in neural networks from training data. Adv. Neural. Inf. Process. Syst. **33**, 17605–17616 (2020)
5. Botev, A., Bauer, M., De, S.: Regularising for invariance to data augmentation improves supervised learning. arXiv preprint arXiv:2203.03304 (2022)
6. Bronstein, M.M., Bruna, J., Cohen, T., Veličković, P.: Geometric deep learning: Grids, groups, graphs, geodesics, and gauges. arXiv preprint arXiv:2104.13478 (2021)
7. Cazelles, E., Tobar, F., Fontbona, J.: A novel notion of barycenter for probability distributions based on optimal weak mass transport. Adv. Neural. Inf. Process. Syst. **34**, 13575–13586 (2021)

8. Cesa, G., Lang, L., Weiler, M.: A program to build E(N)-equivariant steerable CNNs. In: International Conference on Learning Representations (2022)
9. Chawla, N.V., Bowyer, K.W., Hall, L.O., Kegelmeyer, W.P.: Smote: synthetic minority over-sampling technique. J. Artif. Intell. Res. **16**, 321–357 (2002)
10. Chen, T., Kornblith, S., Norouzi, M., Hinton, G.: A simple framework for contrastive learning of visual representations. In: International Conference on Machine Learning, pp. 1597–1607. PMLR (2020)
11. Chen, W., Tian, L., Fan, L., Wang, Y.: Augmentation invariant training. In: Proceedings of the IEEE/CVF International Conference on Computer Vision Workshops, pp. 0–0 (2019)
12. Cohen, T., Welling, M.: Group equivariant convolutional networks. In: International Conference on Machine Learning, pp. 2990–2999. PMLR (2016)
13. Fort, S., Brock, A., Pascanu, R., De, S., Smith, S.L.: Drawing multiple augmentation samples per image during training efficiently decreases test error. arXiv preprint arXiv:2105.13343 (2021)
14. Fralick, S.: Learning to recognize patterns without a teacher. IEEE Trans. Inf. Theory **13**(1), 57–64 (1967)
15. Geiping, J., Goldblum, M., Somepalli, G., Shwartz-Ziv, R., Goldstein, T., Wilson, A.G.: How much data are augmentations worth? an investigation into scaling laws, invariance, and implicit regularization. arXiv preprint arXiv:2210.06441 (2022)
16. Girshick, R., Donahue, J., Darrell, T., Malik, J.: Rich feature hierarchies for accurate object detection and semantic segmentation. In: Proceedings of the IEEE Conference on Computer Vision and Pattern Recognition, pp. 580–587 (2014)
17. Haasdonk, B., Vossen, A., Burkhardt, H.: Invariance in kernel methods by Haar-Integration kernels. In: Kalviainen, H., Parkkinen, J., Kaarna, A. (eds.) SCIA 2005. LNCS, vol. 3540, pp. 841–851. Springer, Heidelberg (2005). https://doi.org/10.1007/11499145_85
18. He, K., Fan, H., Wu, Y., Xie, S., Girshick, R.: Momentum contrast for unsupervised visual representation learning. In: Proceedings of the IEEE/CVF Conference on Computer Vision and Pattern Recognition, pp. 9729–9738 (2020)
19. Kokkinos, I.: Pushing the boundaries of boundary detection using deep learning. arXiv preprint arXiv:1511.07386 (2015)
20. Kondor, R., Trivedi, S.: On the generalization of equivariance and convolution in neural networks to the action of compact groups. In: International Conference on Machine Learning, pp. 2747–2755. PMLR (2018)
21. Kullback, S.: Information theory and statistics. Courier Corporation (1997)
22. Laptev, D., Savinov, N., Buhmann, J.M., Pollefeys, M.: Ti-pooling: transformation-invariant pooling for feature learning in convolutional neural networks. In: Proceedings of the IEEE Conference on Computer Vision and Pattern Recognition, pp. 289–297 (2016)
23. LeCun, Y., Bengio, Y., Hinton, G.: Deep learning. Nature **521**(7553), 436–444 (2015)
24. LeCun, Y., Bottou, L., Bengio, Y., Haffner, P.: Gradient-based learning applied to document recognition. Proc. IEEE **86**(11), 2278–2324 (1998)
25. Mroueh, Y., Voinea, S., Poggio, T.A.: Learning with group invariant features: A kernel perspective. In: Advances in Neural Information Processing Systems, vol. 28 (2015)
26. Penaud-Polge, V., Velasco-Forero, S., Angulo, J.: Fully trainable Gaussian derivative convolutional layer. In: 2022 IEEE International Conference on Image Processing (ICIP), pp. 2421–2425. IEEE (2022)

27. Penaud-Polge, V., Velasco-Forero, S., Angulo, J.: Genharris-resnet: A rotation invariant neural network based on elementary symmetric polynomials. In: Scale Space and Variational Methods in Computer Vision, pp. 149–161. Springer International Publishing, Cham (2023). https://doi.org/10.1007/978-3-031-31975-4_12

28. Procesi, C.: Lie groups: an approach through invariants and representations, vol. 115. Springer (2007). https://doi.org/10.1007/978-0-387-28929-8

29. Rota, G.C.: Reynolds operators. In: Proceedings of Symposia in Applied Mathematics. vol. 16, pp. 70–83. American Mathematical Society Providence, RI (1964)

30. Sangalli, M., Blusseau, S., Velasco-Forero, S., Angulo, J.: Scale equivariant neural networks with morphological scale-spaces. In: Lindblad, J., Malmberg, F., Sladoje, N. (eds.) DGMM 2021. LNCS, vol. 12708, pp. 483–495. Springer, Cham (2021). https://doi.org/10.1007/978-3-030-76657-3_35

31. Sangalli, M., Blusseau, S., Velasco-Forero, S., Angulo, J.: Moving frame net: SE (3)-equivariant network for volumes. In: NeurIPS Workshop on Symmetry and Geometry in Neural Representations, pp. 81–97. PMLR (2023)

32. Scudder, H.: Adaptive communication receivers. IEEE Trans. Inf. Theory **11**(2), 167–174 (1965)

33. Sgarro, A.: Informational divergence and the dissimilarity of probability distributions. Calcolo **18**(3), 293–302 (1981)

34. Shorten, C., Khoshgoftaar, T.M.: A survey on image data augmentation for deep learning. J. big data **6**(1), 1–48 (2019)

35. Sibson, R.: Information radius. Zeitschrift für Wahrscheinlichkeitstheorie und verwandte Gebiete **14**(2), 149–160 (1969)

36. Smets, B.M., Portegies, J., Bekkers, E.J., Duits, R.: Pde-based group equivariant convolutional neural networks. J. Math. Imag. Vision **65**(1), 209–239 (2023)

37. Wilk, M.v.d., Bauer, M., John, S., Hensman, J.: Learning invariances using the marginal likelihood. In: Proceedings of the 32nd International Conference on Neural Information Processing Systems, pp. 9960–9970 (2018)

38. Worrall, D., Welling, M.: Deep scale-spaces: Equivariance over scale. In: Advances in Neural Information Processing Systems, vol. 32 (2019)

39. Yang, S., Dong, Y., Ward, R., Dhillon, I.S., Sanghavi, S., Lei, Q.: Sample efficiency of data augmentation consistency regularization. In: AISTATS (2023)

40. Zhang, H., Cissé, M., Dauphin, Y.N., Lopez-Paz, D.: mixup: Beyond empirical risk minimization. In: ICLR (2018)

41. Zhu, X., Ghahramani, Z., Lafferty, J.D.: Semi-supervised learning using Gaussian fields and harmonic functions. In: Proceedings of the 20th International conference on Machine learning (ICML-03), pp. 912–919 (2003)

Group Equivariant Sparse Coding

Christian Shewmake[1,2,3(✉)], Nina Miolane[3], and Bruno Olshausen[1,2]

[1] University of California Berkeley, Berkeley, CA 94720, USA
[2] The Redwood Center for Theoretical Neuroscience, Berkeley, CA 94720, USA
[3] University of California Santa Barbara, Santa Barbara, CA 93106, USA
shewmake@berkeley.edu

Abstract. We describe a sparse coding model of visual cortex that encodes image transformations in an equivariant and hierarchical manner. The model consists of a group-equivariant convolutional layer with internal recurrent connections that implement sparse coding through neural population attractor dynamics, consistent with the architecture of visual cortex. The layers can be stacked hierarchically by introducing recurrent connections between them. The hierarchical structure enables rich bottom-up and top-down information flows, hypothesized to underlie the visual system's ability for perceptual inference. The model's equivariant representations are demonstrated on time-varying visual scenes.

Keywords: Equivariance · Sparse coding · Generative models

1 Introduction

Brains have the remarkable ability to build internal models from sensory data for reasoning, learning, and prediction to guide actions in dynamic environments. Central to this is the problem of *representation*—i.e., how do neural systems construct internal representations of the world? In the Bayesian view, this requires a generative model mapping from a latent state space to observations, along with a mechanism for inferring latent states from sensory data. Thus, understanding the causal structure of the natural world is essential for forming internal representations. But what is the causal structure of the natural world? Natural images contain complex *transformation groups* that act both on objects and their parts. Variations in object pose, articulation of its parts, even lighting and color changes, can be described by the actions of groups. Additionally, these variations are hierarchical in nature: scenes are composed of objects, objects are composed of parts in relative poses, and so on down to low-level image features. A transformation at the level of an object propagates down the compositional hierarchy, transforming each of its component parts correspondingly. Finally, object parts and sub-parts can undergo their own independent transformations. These variations carry important information for understanding and meaningfully interacting with the world. Thus, a rich compositional hierarchy that is *compatible with group actions* is essential for forming visual representations (Fig. 1).

F. Nielsen and F. Barbaresco (Eds.): GSI 2023, LNCS 14071, pp. 91–101, 2023.
https://doi.org/10.1007/978-3-031-38271-0_10

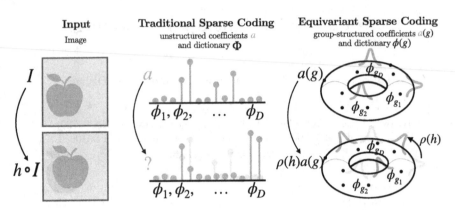

Fig. 1. Traditional vs Equivariant sparse coding as image I is transformed by action h.

Our contribution. We establish a novel Bayesian model for forming representations of visual scenes with equivariant hierarchical part-whole relations by proposing a group-equivariant extension of hierarchical *sparse coding* [7].

2   Background: Sparse Coding for Visual Representations

Sparse coding was originally proposed as a model for how neurons in primary visual cortex represent image data coming from the retina. In contrast to the feedforward cascade of linear filtering followed by point-wise nonlinearity commonly utilized in deep learning architectures, sparse coding uses *recurrent dynamics* to infer a sparse representation of images in terms of a learned dictionary of image features. When trained on natural images, the learned dictionary resembles the oriented, localized, and bandpass receptive fields of neurons in primary visual cortex (area V1) [7].

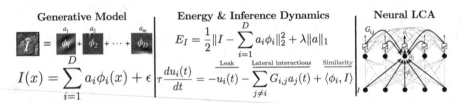

Fig. 2. (left) Generative model, (center) Energy function where $\| \cdot \|_2$ denotes the Euclidean norm, $\| \cdot \|_1$ denotes the ℓ_1 norm, and λ is a regularization parameter controlling the sparsity of a. u_i is the internal state of neuron i, $G_{i,j} = \langle \phi_i, \phi_j \rangle$ models neuronal interactions, and $a(t) = \sigma(u(t))$, where σ is a nonlinearity. (right) LCA circuit model

Generative Model. Sparse coding assumes that natural images are described by a linear generative model with an overcomplete dictionary and additive Gaussian noise $\epsilon(x)$ [7], shown in Fig. 2 (left). Here, the image I is represented as a function $I : X \to \mathbb{R}$, specifically as a vector in the space $L_2(X)$ of square-integrable functions with compact support $X \subset \mathbb{R}^2$. Computationally, the support is discretized as an image patch with n pixels, so that $I \in \mathbb{R}^n$. The dictionary Φ comprises D elements: $\Phi = \{\phi_1, \ldots, \phi_D\}$, with each $\phi_i \in L_2(X)$, for $i \in \{1, ..., D\}$. The size of the dictionary D is typically chosen to be overcomplete, i.e. larger than the image patch dimension n. The coefficients $a = [a_1, \ldots, a_D] \in \mathbb{R}^D$ form the *representation of image I*.

Energy & Inference Dynamics. Given a dataset, sparse coding attempts to find a dictionary Φ and a latent representation $a \in \mathbb{R}^D$ for each image in the dataset such that, in expectation, neural activations are maximally sparse and independent. Sparsity is promoted through the use of an i.i.d.prior over a with scale parameter λ, with the form of the prior chosen to be peaked at zero with heavy tails compared to a Gaussian (typically Laplacian). Finding the optimal representation a is accomplished by maximizing the posterior distribution $P(a|I, \Phi)$ via minimization of the energy function E_I in Fig. 2 (center).

One particularly effective method for minimizing E_I with a clear cortical circuit implementation is the *Locally Competitive Algorithm* (LCA) [9]. In LCA, inference is carried out via the temporal dynamics of a population of D neurons. Each neuron is associated with a dictionary element i, and its internal state is represented by a coefficient $u_i(t)$. The evolution of the neural population state is governed by the dynamics specified in Fig. 2 (center). The gram matrix, $G_{i,j} = \langle \phi_i, \phi_j \rangle$, specifies the interaction between neuron i and j. In neurobiological terms, this corresponds to the excitatory and inhibitory interactions mediated by horizontal connections among V1 neurons. The notation $\langle ., . \rangle$ refers to the inner-product between functions in $L_2(X)$, $\langle \phi_i, \phi_j \rangle = \int_X \phi_i(x)\phi_j(x)dx$. The activations, interpreted as instantaneous neural firing rates, are given by a nonlinearity applied to the internal state: $a_j(t) = \sigma(u_j(t))$, with $\sigma(u) = \frac{u - \alpha\lambda}{1 + e^{-\gamma(u - \lambda)}}$, similar to a smoothed ReLU function with threshold λ and hyperparameters α and γ. At equilibrium, the latent representation of image I is given by $\hat{a} = \arg\min_a E_I$.

Dictionary Learning. The dictionary Φ is adapted to the statistics of the data by minimizing the same energy function E_I averaged over the dataset. This is accomplished by alternating gradient descent on E. Given a current dictionary Φ, along with a batch of images and their inferred latent representations, $\hat{a}$, the dictionary is updated with one gradient step of E with respect to Φ, averaged over the batch.

3   Group Equivariant Sparse Coding

Missing in the current formulation of sparse coding is the mathematical structure to support reasoning about hierarchical object transformations in visual scenes. This limits its utility in both unsupervised learning and mechanistic models of visual cortex. Here we address this problem by explicitly incorporating group equivariant and hierarchical structure into the sparse coding model. Prior work has explored imposing topological relations between dictionary elements by establishing implicit neighborhood relations during training through co-activation penalties [6], or explicitly coupling steerable pairs or n-tuples of dictionary elements [8]. More recent work in Geometric Deep Learning (GDL) has introduced several group equivariant architectures, for example through the use of group convolutions [3,4]. However, these models are feedforward, lacking mechanisms for hierarchical inference or rich top-down and bottom up flows. Aside from [1], these models lack mechanisms for hierarchical part-whole relations.

We explore the implications of inheriting the dictionary's geometric structure through *group actions*. In particular, we propose a model in which each dictionary element is generated by an action of g on a canonical dictionary element, as shown in Fig. 3 (right). For example, the group G of 2D rotations acts on the 2D domain X, inducing a natural action on the space of images in $L_2(X)$ defined over X. We refer the interested reader to [5] for mathematical details on groups and group actions.

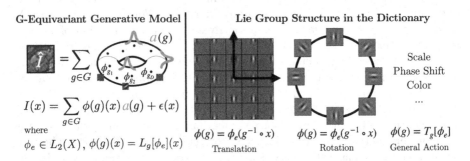

G-Equivariant Generative Model

$$I(x) = \sum_{g \in G} \phi(g)(x)\, a(g) + \epsilon(x)$$

where
$$\phi_e \in L_2(X),\ \phi(g)(x) = L_g[\phi_e](x)$$

Lie Group Structure in the Dictionary

Scale
Phase Shift
Color
...

$$\phi(g) = \phi_e(g^{-1} \circ x) \qquad \phi(g) = \phi_e(g^{-1} \circ x) \qquad \phi(g) = T_g[\phi_e]$$
Translation   Rotation   General Action

Fig. 3. (left) Geometric generative model, (right) Lie group actions relate dictionary elements. Here, e is the identity element of G, and the canonical dictionary element is $\phi_e \in L_2(X)$. Additionally, L is a linear group action of G in the space of functions on the domain $L_h[f](x) = f(h^{-1}x)$.

3.1   Geometric Generative Model

This perspective enables us to rewrite the sparse coding generative model as:

$$I(x) = \sum_{g \in G} \phi(g)(x)\, a(g) + \epsilon(x), \tag{1}$$

where both the dictionary elements $\phi(g)$ and the scalar coefficients $a(g)$ are indexed with group elements, i.e. "coordinates" in G. In other words, images are (1) generated by linear combinations of dictionary elements ϕ, where (2) each dictionary element has an explicit coordinate g in the group. The latent representation a is now a scalar field over the group, $a : G \rightarrow \mathbb{R}$, illustrated in Figs. 3 (left) and 1 (right). Intuitively, this perspective gives an explicit geometric interpretation of both the dictionary Φ and latent representation a in sparse coding, and thus a route toward modeling transformations which was implicit in the unstructured vector representation.

3.2   Geometric Inference and LCA

The geometric perspective of sparse coding above allows us to rewrite the LCA dynamics. Specifically, each neuron is now associated with a group element g, with internal state $u(g)$. The LCA dynamics are typically computationally expensive to compute due to the prohibitive size of the neural interaction matrix $G_{g,h} = \langle \phi(g), \phi(h) \rangle$. However this term can now be written as a group convolution with a ϕ_e-dependent kernel w, leading to a **symmetric, local wiring rule** between neurons and efficient computation during inference that is readily parallelized on GPUs. Hence, we propose a new, provably equivariant inference method—*Geometric LCA*, where $*$ denotes group convolution:

$$\dot{u}(g)(t) = -u(g)(t) - [w * a(t)](g) + \langle \phi(g), I \rangle \qquad (2)$$

Box 1. Isometry and the Derivation of Geometric LCA

Lemma 1 Consider a function $f \in L_2(X)$ and a dictionary element $\phi(g) \in L_2(X)$ indexed by $g \in G$. If the action of $h \in G$ is isometric on the domain X, then, $\forall h \in G$, we have:

$$\langle L_h[\phi(g)], f \rangle = \langle \phi(g), L_{h^{-1}}[f] \rangle$$

Proof. We have: $\langle L_h[\phi(g)], f \rangle$

$$= \int_X L_h[\phi(g)](x)f(x)dx$$

$$= \int_X L_{hg}[\phi_e](x)f(x)dx \quad \text{by def. of } \phi(g)$$

$$= \int_X \phi_e((hg)^{-1}x)f(x)dx \quad \text{by def. of } L$$

$$= \int_X \phi_e(g^{-1}h^{-1}x)f(x)dx$$

$x \leftarrow hx$, h action isometric: $d(hx) = dx$

$$= \int_X \phi_e(g^{-1}x)f(hx)dx$$

$$= \int_X \phi(g)(x)L_{h^{-1}}[f](x)dx$$

$$= \langle \phi(g), L_{h^{-1}}[f] \rangle.$$

This last step leads to the following LCA dynamics.

Proposition 1: Geometric LCA The LCA dynamics have the following geometric formulation

$$\tau \dot{u}(g)(t) = -u(g)(t) - [w * a(t)](g) + \langle \phi(g), I \rangle$$

where $*$ denotes a group convolution.

Proof. Consider the interaction term in the LCA dynamics: $\sum_{h \in G, \; h \neq g} G_{g,h} a(h)(t)$

$$= \sum_{h \in G, \; h \neq g} \langle \phi(g), \phi(h) \rangle a(h)(t) \quad \text{by def. of } G$$

$$= \sum_{h \in G, \; h \neq g} \langle L_h^{-1}[\phi(g)], \phi_e \rangle a(h)(t) \quad \text{Lemma 1}$$

$$= \sum_{h \in G, \; h \neq g} \langle \phi(h^{-1}g), \phi_e \rangle a(h)(t)$$

$$= \sum_{h \in G, \; h \neq g} w(g^{-1}h)a(h)(t)$$

$$= [w * a(t)](g),$$

where we define $w(g) := \langle \phi(g^{-1}), \phi_e \rangle$ for $g \neq e$ and $w(e) = 0$.

Equivariance of Inference and LCA. Next, we demonstrate that the solutions $I \to a$ obtained from the LCA dynamics are equivariant. First, we say a map $\psi : X \to Z$ is *equivariant* to a group G if $\psi(L_g x) = L'_g \psi(x)$ $\forall g \in G$, with L_g, L'_g representations of G on X and Z respectively. For clarity of exposition, L is defined as a group action of G on the space $L_2(X)$ via domain transformations $L_g[f](x) = f(g^{-1}x)$, and the action L' of G is defined on the space $L_2(G)$ of square integrable functions from G to $\mathbb{R}$, defined as:

$$L'_h(a)(g) = a(h^{-1}g), \quad \forall g \in G, \ \forall h \in G, \ \forall a \in L_2(G),$$

where $h^{-1}g$ refers to the group composition of two group elements. First, we show that solutions of the ordinary differential equation (ODE) defining the LCA dynamics exist and are unique. Consider the initial value problem below, where f denotes the LCA dynamics:

$$\text{ODE}(I) : \begin{cases} \dot{u}(g,t) = f(u(g,t), I) & \forall g \in G, t \in \mathbb{R}_+, \\ u(g,0) = 0 & \forall g \in G. \end{cases} \tag{3}$$

Proposition 1 (Existence and Uniqueness of LCA Solutions). *Given an image I, the solution of* $\text{ODE}(I)$ *exists and is unique. We denote it with u^I.*

Proof. The Cauchy-Lipschitz theorem (Picard-Lindelöf theorem) states that the initial value problem defined by $ODE(I)$ has a unique solution if the function f is (i) continuous in t and (ii) Lipschitz continuous in u, where:

$$f(u(g,t), I) = \frac{1}{\tau}\left(-u(g)(t) - [w * a(t)](g) + \langle \phi(g), I \rangle\right) \tag{4}$$

The continuity in t stems from the fact that a and u are continuous. We prove that $f(u, I)$ is Lipschitz continuous in u, i.e. that $\frac{\partial f}{\partial u}(u, I)$ is bounded. Observe that the derivatives of the first and third terms are bounded. The second term is a convolution composed with a smooth, ReLU-like nonlinearity. As convolutions are bounded linear operators, the question reduces to whether derivative of the nonlinearity $\frac{\partial \sigma}{\partial u}$ is bounded, which indeed holds. Thus solutions exist and are unique. Using this fact, we show that the solution of the dynamics transforms equivariantly with image transformations. Let u^I be the unique solution of $\text{ODE}(I)$. Similarly, let $u^{L_h[I]}$ be the unique solution of:

$$\text{ODE}(L_h[I]) : \begin{cases} \dot{u}(g,t) = f(u(g,t), L_h[I]) & \forall g \in G, t \in \mathbb{R}_+, \\ u(g,0) = 0 & \forall g \in G. \end{cases}$$

Proposition 2: Equivariance of LCA Inference Dynamics Take $h \in G$. The solutions of the LCA dynamics $ODE(I)$ and $ODE(L_h[I])$ are related by $u^{L_h I} = L'_h(u^I)$. Since $a(g) = \sigma(u(g))$, it follows that: $a^{L_h I} = L'_h(a^I)$.

Proof Take $h \in G$ and define $v(g,t) := u^I(h^{-1}g, t), \forall g, \forall t$. We show v verifies $ODE(L_h[I])$. First, we verify that v satisfies the initial conditions: $v(g,0) = u^I(h^{-1}g, 0) = 0, \ \forall g \in G$. Next, we verify that v satisfies $ODE(L_h[I]) \ \forall g, \forall t$.

$$\tau \dot{v}(g,t) = \frac{\partial}{\partial t}[\tau u^I(h^{-1}g, t)] \quad \text{(definition of } v\text{)}$$

$$= -u^I(h^{-1}g, t) - \sum_{g' \in G} w((h^{-1}g)^{-1}g') \cdot \sigma(u(g')) + \langle \phi(h^{-1}g), I \rangle$$

$$= -v(g,t) - \sum_{g' \in G} w(g^{-1}hg') \cdot \sigma(u(g')) + \langle \phi(g), L_h[I] \rangle \quad \text{(Lemma 1)}$$

$$= -v(g,t) - \sum_{g' \in G} w(g^{-1}g') \cdot \sigma(u(h^{-1}g')) + \langle \phi(g), L_h[I] \rangle \quad (g' \leftarrow h^{-1}g')$$

$$= -v(g,t) - \sum_{g' \in G} w(g^{-1}g') \cdot \sigma(v(g')) + \langle \phi(g), L_h[I] \rangle \quad \text{(definition of } v\text{)}$$

$$= f(v(g,t), L_h[I]) \quad \text{(definition of ODE } (L_h[I])\text{)}.$$

Thus, v is a solution of $ODE(L_h[I])$, and, by uniqueness, $v(g,t) = u^{L_h[I]}(g,t) \ \forall g, \forall t$. Therefore, $u^I(h^{-1}g, t) = u^{L_h[I]}(g,t) \ \forall g, \forall t$, and $a^{L_h I} = L'_h(a^I)$ as well. Thus, the LCA inference dynamics are equivariant to global image transformations.

3.3 Equivariances of the Generative Model

Here, we show that the generative model, that is, the function $f : a \rightarrow I$ that maps coefficients to images, is also equivariant. There are three types of equivariance important for representing transformations in natural scenes: global, part/local, and hierarchical. Here we define these three types and prove that the generative model is indeed equivariant in these important ways.

Global Equivariance. Traditionally, work on group equivariant neural networks (e.g. GCNNs [3,4]) has focused on global equivariance, i.e. equivariance to a group action L on the domain of the input function. In Box 2, we show that the geometric form of the sparse coding model is globally equivariant. However, transformations of natural scenes typically involve actions on objects and parts at different levels of the hierarchy. That is, transformations of an object at a higher level of the hierarchy should propagate down compatibly with its parts. In the context of equivariant sparse coding, the generative model explicitly decomposes the scene into primitive parts—the first-level dictionary elements. That is, if an image I is composed of M objects $I_1, ..., I_M$ then

$$I(x) = I_1(x) + ... + I_M(x)$$
$$= \sum_{g \in G} \phi(g)(x) a_1(g) + ... + \sum_{g \in G} \phi(g)(x) a_M(g)$$
$$= \sum_{g \in G} \phi(g)(x) (a_1(g) + ... + a_M(g)).$$

In the context of this generative model, we can define two additional notions that are essential for natural scene decompositions—*local* and *hierarchical* equivariance. We prove that the generative model is indeed equivariant in these two additional important ways.

Local Equivariance. Using the decomposition above, we define *local actions* of G on the space of images $L_2(X)$ as:

$$L_h^{(1)}[I] = L_h[I_1] + ... + I_M, \quad \forall h \in G, I \in L_2(X)$$

$L^{(1)}$ only acts on image part 1, represented by image I_1, and likewise on image part m via $L^{(m)}$. We now prove that these local actions are indeed group actions. *Proof* $L^{(1)}$ is a group action. The proof for $L^{(2)}$ follows.

(i) *Identity* : $L_e^{(1)}[I] = I$.

(ii) *Closure* : $L_{h'h}^{(1)}[I] = L_{h'h}[I_1] + ... + I_M = L_{h'}[L_h[I_1]] + ... + I_M = L_{h'}^{(1)}\left[L_h^{(1)}[I]\right]$.

Here, we have used the definition of $L^{(1)}$ and the fact that L is a group action. Similarly, we can define local actions $L'^{(m)}$ on the space $L_2(G)$ of coefficients a_m corresponding to image part I_m. By the linearity of the generative model f, a local action in the space of coefficients yields a local action in the image space, as shown in Box 2, Proposition 3.

Hierarchical Equivariance. The properties of global and local equivariance naturally give rise to the hierarchical equivariance of the new generative model. In other words, when a transformation is applied at the level of an object I (e.g. the whole scene), transformations propagate down compatibly to its parts (e.g. $I_1, ..., I_M$). This hierarchical transformation is directly reflected in actions on the latent coefficients for an object a and its parts $a_1, ..., a_M$. See Box 2, Proposition 5. Thus, a hierarchy of transformations in the scene is equivalent to a hierarchy of transformations in the internal neural representation.

Box 2. Generative Model: Global, Local, & Hierarchical Equivariance

Proposition 2: Global The generative model in Eq. 1, $I = f(a)$ is globally G-equivariant, i.e. for all $h \in G$, we have:

$$f(L'_h(a)) = L_h(f(a)) \quad (5)$$

Proof Take $h \in G$. We have:

$$L_h[f(a)] = L_h \left[\sum_{g \in G} \phi(g)a(g) \right]$$

$$= \sum_{g \in G} L_h \left[L_g[\phi_e] \right] a(g)$$

$$= \sum_{g \in G} \phi(hg)a(g)$$

$$= \sum_{'\in G} \phi(g)a(h^{-1}g) \quad (g \leftarrow h^{-1}g)$$

$$= \sum_{g \in G} \phi(g)L'_h(a)(g).$$

Thus the model is globally G-equivariant.

Proposition 3: Part/Local Consider the linear model f, where $a = a_1 + \dots + a_M$. f is locally G-equivariant, i.e. $\forall h \in G$, for $m \in \{1, 2, \dots, M\}$

$$f \left(L'^{(m)}_h(a) \right) = L^{(m)}_h[f(a)] \quad (6)$$

Proof We have

$$f \left(L'^{(1)}_h(a) \right) = f \left(L'_h(a_1) + \dots + a_M \right)$$

$$= f \left(L'_h(a_1) \right) + \dots + f(a_M)$$

$$= f \left(L'_h(a_1) \right) + \dots + I_M$$

$$= L_h[I_1] + \dots + I_M \quad \text{(by 5)}$$

$$= L^{(1)}[I] \quad \text{(definition of } L^{(1)}\text{)}.$$

Shown for $m = 1$, this property holds for all m, thus the model is locally G-equivariant.

Proposition 4: Hierarchical Consider the linear model f, where $a = a_1 + \dots + a_M$. For all $h, h' \in G$, $m \in \{1, 2, \dots, M\}$ we have:

$$f \left(L'_h \left(L'^{(m)}_{h'}(a) \right) \right) = L_h \left[L^{(m)}_{h'}[f(a)] \right]$$

Proof Directly from global and local cases:

$$f \left(L'_h \left(L'^{(m)}_{h'}(a) \right) \right) = L_h \left[f \left(L'^{(m)}_{h'}(a) \right) \right]$$

$$= L_h \left[L^{(m)}_{h'}[f(a)] \right].$$

Thus, f is *hierarchically* G-equivariant.

3.4   Constructing a Hierarchical Generative Model

Finally, the equivariant sparse coding layers can be composed hierarchically, where first-level activations are describable in terms of second-level activations over arrangements of parts.

$$I(x) = \sum_{g \in G} \phi_0(g)a_0(g) + \epsilon(x), \quad a_0(g) = \sum_{k=1}^{K} \sum_{g \in G} \phi_1^k(g)a_1^k(g) + \epsilon(g) \quad (5)$$

Defining $\hat{I} = \sum_{g \in G} \phi_0(g)a_0(g)$ and $\hat{a}_0 = \sum_{k=1}^{K} \sum_{g \in G} \phi_1^k(g)a_1^k(g)$, the energy [2] and geometric LCA inference dynamics are given by

$$E = \frac{1}{2}||I - \hat{I}||_2^2 + \lambda_0 C(a_0) + \frac{1}{2}||a_0 - \hat{a}_0||_2^2 + \lambda_1 C(a_1)$$

$$\tau_0 \dot{u}_0(g) = -u_0(g) - [w_0 * a_0](g) + \langle \phi_0(g), I \rangle + \hat{a}_0(g)$$

$$\tau_1 \dot{u}_1^k(g) = -u_1^k(g) - [w_1^k * a_1^k](g) + a_1^k(g) + \langle \phi_1^k(g), a_0 \rangle$$

4 Experiments

To evaluate and characterize the behavior of the proposed hierarchical, equivariant sparse coding model, we construct a synthetic dataset of scenes containing (1) objects comprised of lower-level parts where (2) parts and wholes are transformed via group actions. We do this by specifying a group G, constructing the dictionary elements at each level of the hierarchy, and then sampling from the generative model. For the first layer dictionary, we construct an overcomplete dictionary of Gabor functions generated by acting on a canonical Gabor template with a discrete sampling of the group of translations G. The mother Gabor ϕ_e is shown in Fig. 4. We construct $K = 2$ canonical second-layer dictionary elements ϕ_1^1, ϕ_1^2 from arrangements of parts at the preceding level of representation. Next, we generate the "orbit" of each template by again sampling from the group of translations G. The templates and selected dictionary elements are shown in Fig. 4. We then generate a dataset of images by sampling from the generative model. In particular, we create a sequence of frames in which objects present in the scene undergo different translations. The resulting images, inferred latents, and reconstructions are shown in Fig. 4. Note that the latent variables are *sparse* and *transform equivariantly*, as stated in the proofs.

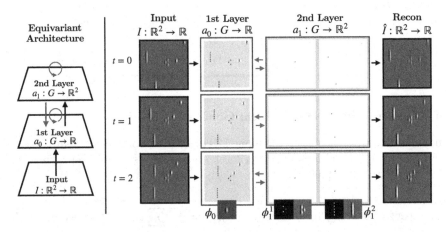

Fig. 4. Figure: (left) a two-layer translation-equivariant architecture with recurrent connections within and between layers, (right) experimental results demonstrating that the neural dynamics converge to a sparse, hierarchical representation of the scene which transforms equivariantly in time with the input video. Column 1: input video frames, Column 2: first layer gabor coefficient map displayed with sparse equivariant activations, Columns 3&4: two second layer "object" coefficient maps displayed with sparse equivariant activations

5 Discussion

By incorporating group structure, we have derived a new sparse coding model that is equivariant in its response to image transformations, both within a layer and across multiple layers stacked in a hierarchy. We believe this is an important step toward developing a hierarchical, probabilistic model of visual cortex capable of performing perceptual inference (e.g. object recognition) on natural scenes. Surprisingly, the network architecture has the same functional form as the neural attractor model of Kechen Zhang [10], suggesting new circuit mechanisms in visual cortex for top-down steering, motion computation, and disparity estimation that could be done in the sparse code domain. Of relevance to deep learning, this new structure enables inference to be implemented efficiently on GPUs as (1) a feed-forward group convolution followed by (2) iterative lateral interaction dynamics implemented by group convolutions between dictionary elements.

Acknowledgements. The authors thank their helpful colleagues at the Redwood Center and Bioshape Lab. CS acknowledges support from the NIH NEI Training Grant T32EY007043.

References

1. Bekkers, E.J., Lafarge, M.W., Veta, M., Eppenhof, K.A.J., Pluim, J.P.W., Duits, R.: Roto-translation covariant convolutional networks for medical image analysis. In: Frangi, A.F., Schnabel, J.A., Davatzikos, C., Alberola-López, C., Fichtinger, G. (eds.) MICCAI 2018. LNCS, vol. 11070, pp. 440–448. Springer, Cham (2018). https://doi.org/10.1007/978-3-030-00928-1_50
2. Boutin, V., Franciosini, A., Ruffier, F., Perrinet, L.: Effect of top-down connections in hierarchical sparse coding. Neural Computation **32**(11), 2279–2309 (Nov 2020). https://doi.org/10.1162/neco_a_01325
3. Bronstein, M.M., Bruna, J., Cohen, T., Veličković, P.: Geometric deep learning: Grids, groups, graphs, geodesics, and gauges. arXiv preprint arXiv:2104.13478 (2021)
4. Cohen, T., Welling, M.: Group equivariant convolutional networks. In: International Conference on Machine Learning, pp. 2990–2999. PMLR (2016)
5. Hall, B.C.: Lie groups, lie algebras, and representations. In: Quantum Theory for Mathematicians, pp. 333–366. Springer (2013). https://doi.org/10.1007/978-3-319-13467-3
6. Hyvärinen, A., Hurri, J., Väyrynen, J.: Bubbles: a unifying framework for low-level statistical properties of natural image sequences. JOSA **20**(7), 1237–1252 (2003). https://doi.org/10.1364/josaa.20.001237
7. Olshausen, B.A., Field, D.J.: Sparse coding with an overcomplete basis set: A strategy employed by v1? Vision. Res. **37**(23), 3311–3325 (1997)
8. Paiton, D.M., Shepard, S., Chan, K.H.R., Olshausen, B.A.: Subspace locally competitive algorithms. In: Proceedings of the Neuro-inspired Computational Elements Workshop, pp. 1–8 (2020)
9. Rozell, C.J., Johnson, D.H., Baraniuk, R.G., Olshausen, B.A.: Sparse coding via thresholding and local competition in neural circuits. Neural Comput. **20**(10), 2526–2563 (2008)
10. Zhang, K.: Representation of spatial orientation by the intrinsic dynamics of the head-direction cell ensemble: A theory. J. Neurosci. **16**(6), 2112–2126 (1996)

Divergences in Statistics and Machine Learning

On a Cornerstone of Bare-Simulation Distance/Divergence Optimization

Michel Broniatowski[1] and Wolfgang Stummer[2]([⊠])

[1] LPSM, Sorbonne Université, 4 place Jussieu, 75252 Paris, France
`michel.broniatowski@sorbonne-universite.fr`
[2] Department of Mathematics, University of Erlangen–Nürnberg, Cauerstrasse 11, 91058 Erlangen, Germany
`stummer@math.fau.de`

Abstract. In information theory—as well as in the adjacent fields of statistics, geometry, machine learning and artificial intelligence—it is important to solve high-dimensional optimization problems on directed distances (divergences), under very non-restrictive (e.g. non-convex) constraints. Such a task can be comfortably achieved by the new *dimension-free bare (pure) simulation* method of [6,7]. In the present paper, we give some new insightful details on one cornerstone of this approach.

Keywords: φ−divergence · High-dimensional optimization

1 Directed Distances, Divergences

As usual, a divergence is a function $D : \mathbb{R}^K \times \mathbb{R}^K \mapsto \mathbb{R}$ with the following properties: $D(\mathbf{Q}, \mathbf{P}) \geq 0$ for $K-$dimensional vectors $\mathbf{Q}, \mathbf{P} \in \mathbb{R}^K$, and $D(\mathbf{Q}, \mathbf{P}) = 0$ iff $\mathbf{Q} = \mathbf{P}$. Since in general, $D(\mathbf{Q}, \mathbf{P}) \neq D(\mathbf{P}, \mathbf{Q})$ and the triangle inequality is not satisfied, $D(\mathbf{Q}, \mathbf{P})$ can be interpreted as *directed distance* from $\mathbf{Q}$ to $\mathbf{P}$; accordingly, the divergences D can be connected to geometric issues in various different ways, see e.g. the detailed discussion in Sect. 1.5 of [5], and [15]. Typically, a divergence D is generated by some function φ. For the latter, here we fundamentally require:

(G1) $\varphi :] - \infty, \infty[\to [0, \infty]$ is lower semicontinuous and convex;
(G2) $\varphi(1) = 0$;
(G3) the effective domain $dom(\varphi) := \{t \in \mathbb{R} : \varphi(t) < \infty\}$ has interior $int(dom(\varphi))$ of the form $int(dom(\varphi)) =]a, b[$ for some $-\infty \leq a < 1 < b \leq \infty$;[1]
(G4') φ is strictly convex in a neighborhood $]t_-^{sc}, t_+^{sc}[\subseteq]a, b[$ of one $(t_-^{sc} < 1 < t_+^{sc})$.

---

[1] notice that (G3) follows from (G1), (G2) and the requirement that $int(dom(\varphi))$ is non-empty.

© The Author(s), under exclusive license to Springer Nature Switzerland AG 2023
F. Nielsen and F. Barbaresco (Eds.): GSI 2023, LNCS 14071, pp. 105–116, 2023.
https://doi.org/10.1007/978-3-031-38271-0_11

Also, we set $\varphi(a) := \lim_{t \downarrow a} \varphi(t)$ and $\varphi(b) := \lim_{t \uparrow b} \varphi(t)$. The class of all functions φ with (G1), (G2), (G3) and (G4') will be denoted by $\widetilde{\varUpsilon}(]a, b[)$. For $\varphi \in \widetilde{\varUpsilon}(]a, b[)$, $\mathbf{P} := (p_1, .., p_K) \in \mathbb{R}_{>0}^K := \{\mathbf{R} := (r_1, \ldots, r_K) \in \mathbb{R}^K : r_i > 0 \text{ for all } i = 1, \ldots, K\}$ and $\mathbf{Q} := (q_1, \ldots, q_K) \in \boldsymbol{\Omega} \subset \mathbb{R}^K$, we define as directed distance the *generalized $\varphi-$divergence* (generalized Csiszár-Ali-Silvey-Morimoto divergence)

$$D_\varphi(\mathbf{Q}, \mathbf{P}) := \sum_{k=1}^K p_k \cdot \varphi\left(\frac{q_k}{p_k}\right);$$

for a comprehensive technical treatment, see e.g. [4]. Comprehensive overviews on these important (generalized) $\varphi-$divergences are given in e.g. [5,6,12,18], and the references therein. Notice that the total variation distance $D_{\varphi_{TV}}(\mathbf{Q}, \mathbf{P}) = \sum_{k=1}^K p_k \cdot \varphi_{TV}\left(\frac{q_k}{p_k}\right) = \sum_{k=1}^K |p_k - q_k|$ with $\varphi_{TV}(t) := |t - 1|$ is not covered here.

2 Optimization and Bare Simulation Solution Method

Problem 1. *For pregiven* $\varphi \in \widetilde{\varUpsilon}(]a, b[)$, *positive-entries vector* $\mathbf{P} := (p_1, .., p_K) \in \mathbb{R}_{>0}^K$ *(or from some subset thereof), and subset* $\boldsymbol{\Omega} \subset \mathbb{R}^K$ *with regularity properties*

$$cl(\boldsymbol{\Omega}) = cl\left(int\left(\boldsymbol{\Omega}\right)\right), \qquad int\left(\boldsymbol{\Omega}\right) \neq \emptyset, \tag{1}$$

find

$$\inf_{\mathbf{Q} \in \boldsymbol{\Omega}} D_\varphi(\mathbf{Q}, \mathbf{P}), \tag{2}$$

provided that

$$\inf_{\mathbf{Q} \in \boldsymbol{\Omega}} D_\varphi(\mathbf{Q}, \mathbf{P}) < \infty. \tag{3}$$

Remark 2. *(a) In case that a minimizer* $\mathbf{Q}^*$ *of (2) exists—i.e.* $\mathbf{Q}^*$ *lies in* $\boldsymbol{\Omega}$ *and satisfies* $D_\varphi(\mathbf{Q}^*, \mathbf{P}) = \inf_{\mathbf{Q} \in \boldsymbol{\Omega}} D_\varphi(\mathbf{Q}, \mathbf{P})$—*then this* $\mathbf{Q}^*$ *can be interpreted as an information projection of* $\mathbf{P}$ *on* $\boldsymbol{\Omega}$. *For the related context of probability measures, information projections have been addressed e.g. in [8,9] for the subcase of the Kullback-Leibler information divergence (also called* $I-$divergence), *see also e.g. [13] for a nice exposition in connection with the differential geometry of probability model parameters and general* φ.
(b) When $\boldsymbol{\Omega}$ *is not closed but merely satisfies (1), then the infimum in (2) may not be reached in* $\boldsymbol{\Omega}$ *although being finite; whenever, additionally,* $\boldsymbol{\Omega}$ *is a closed set then a minimizer* $\mathbf{Q}^* \in \boldsymbol{\Omega}$ *exists. In the subsetup where* $\boldsymbol{\Omega}$ *is a closed convex set and* $int(\boldsymbol{\Omega}) \neq \emptyset$, *(1) is satisfied and the minimizer* $\mathbf{Q}^* \in \boldsymbol{\Omega}$ *in (2) is attained and even unique. When* $\boldsymbol{\Omega}$ *is open and satisfies (1), then the infimum in (2) exists but is generally reached at some generalized projection of* $\mathbf{P}$ *on* $\boldsymbol{\Omega}$ *(see [9] for the Kullback-Leibler divergence case of probability measures, which extends to any* $\varphi-$divergence *in our framework). However, in this paper we only deal with finding the infimum/minimum in (2) (rather than a corresponding minimizer).*
(c) Our approach is predestined for non- or semiparametric models. For instance, (1) is valid for appropriate tubular neighborhoods of parametric models or for more general non-parametric settings such as e.g. shape constraints.

According to our work [6], the above-mentioned Problem 1 can be solved by a new dimension-free precise *bare simulation (BS) method* to be explained in the following, where—(only) for the sake of brevity—we assume the additional constraint $\sum_{i=1}^{K} p_i = 1$ for the rest of this paper. We first suppose

Condition 3. *The divergence generator φ in (2) satisfies (G1) to (G4') (i.e. $\varphi \in \tilde{\Upsilon}(]a,b[)$) and additionally there holds the representation*

$$\varphi(t) = \sup_{z \in \mathbb{R}} \left(z \cdot t - \log \int_{\mathbb{R}} e^{z \cdot y} d\zeta(y) \right), \qquad t \in \mathbb{R}, \tag{4}$$

for some probability measure ζ on the real line $\mathbb{R}$ such that the function $z \mapsto MGF_\zeta(z) := \int_{\mathbb{R}} e^{z \cdot y} d\zeta(y)$ is finite on some open interval containing zero[2].

The class of all functions φ which satisfy this *cornerstone Condition* 3 will be denoted by $\Upsilon(]a,b[)$. By means of this, for each fixed $\varphi \in \Upsilon(]a,b[)$ we construct in [6] a sequence $(\boldsymbol{\xi}_n^{\mathbf{W}})_{n \in \mathbb{N}}$ of $\mathbb{R}^K$−valued random variables/random vectors (on an auxiliary probability space $(\mathfrak{X}, \mathscr{A}, \Pi)$) as follows: for any $n \in \mathbb{N}$ and any $k \in \{1, \ldots, K\}$, let $n_k := \lfloor n \cdot p_k \rfloor$ where $\lfloor x \rfloor$ denotes the integer part of x. Thus one has $\lim_{n \to \infty} \frac{n_k}{n} = p_k$. Moreover, we assume that $n \in \mathbb{N}$ is large enough, namely $n \geq \max_{k \in \{1, \ldots, K\}} \frac{1}{p_k}$, and decompose the set $\{1, \ldots, n\}$ of all integers from 1 to n into the following disjoint blocks: $I_1^{(n)} := \{1, \ldots, n_1\}$, $I_2^{(n)} := \{n_1 + 1, \ldots, n_1 + n_2\}$, and so on until the last block $I_K^{(n)} := \{\sum_{k=1}^{K-1} n_k + 1, \ldots, n\}$ which therefore contains all integers from $n_1 + \ldots + n_{K-1} + 1$ to n. Clearly, $I_k^{(n)}$ has $n_k \geq 1$ elements (i.e. $\text{card}(I_k^{(n)}) = n_k$ where $\text{card}(A)$ denotes the number of elements in a set A) for all $k \in \{1, \ldots, K-1\}$, and the last block $I_K^{(n)}$ has $n - \sum_{k=1}^{K-1} n_k \geq 1$ elements which anyhow satisfies $\lim_{n \to \infty} \text{card}(I_K^{(n)})/n = p_K$. Furthermore, consider a vector $\mathbf{W} := (W_1, \ldots, W_n)$ where the W_i's are i.i.d. copies of the random variable W whose distribution is associated with the divergence-generator φ through (4), in the sense that $\Pi[W \in \cdot] = \zeta[\cdot]$. We group the W_i's according to the above-mentioned blocks and sum them up blockwise, in order to build the following $K-$ component random vector

$$\boldsymbol{\xi}_n^{\mathbf{W}} := \left(\frac{1}{n} \sum_{i \in I_1^{(n)}} W_i, \ldots, \frac{1}{n} \sum_{i \in I_K^{(n)}} W_i \right).$$

For such a context, in [6] we obtain the following solution of Problem 1:

---

[2] this implies that $\int_{\mathbb{R}} y \, d\zeta(y) = 1$ (cf. (G11i) below) and that ζ has light tails.

Theorem 4. *Under Condition 3, there holds the "bare-simulation (BS) minimizability"*

$$\inf_{\mathbf{Q}\in\Omega} D_\varphi(\mathbf{Q},\mathbf{P}) = -\lim_{n\to\infty} \frac{1}{n} \log \Pi\big[\boldsymbol{\xi}_n^{\mathbf{W}} \in \Omega\big]$$

for any $\Omega \subset \mathbb{R}^K$ *with regularity properties* (1) *and finiteness property* (3).

In [6], we also give versions for the case $\sum_{i=1}^K p_i \neq 1$ (and hence, φ needs to be transformed), versions for constraint sets Ω in the probability simplex, as well as numerous solved cases. An extension of this bare simulation method to *arbitrary* divergences $D(\mathbf{Q},\mathbf{P})$ (e.g. Bregman distances) can be found in [7].

Theorem 4 provides our principle for the *approximation* of the solution of the deterministic optimization problem (2). Indeed, by replacing the involved limit by its finite counterpart, we deduce for given large n

$$\inf_{\mathbf{Q}\in\Omega} D_\varphi(\mathbf{Q},\mathbf{P}) \approx -\frac{1}{n} \log \Pi\big[\boldsymbol{\xi}_n^{W} \in \Omega\big] ; \qquad (5)$$

it remains to estimate the right-hand side of (5). The latter can be performed either by a *naive estimator* of the frequency of those replications of $\boldsymbol{\xi}_n^{\mathbf{W}}$ which hit Ω, or more efficiently by some improved estimator, see [6,7] for details.

We can also deduce that the *rate* of convergence in Theorem 4 is $O\left(n^{-1}\cdot\log n\right)$. Indeed, since the K components in $\boldsymbol{\xi}_n^{\mathbf{W}}$ are independent, each of them being a weighted empirical mean of independent summands, we can combine the proof of Theorem 8 in [6] (with the special choice $\sum_{i=1}^K p_i = 1$) with e.g. Theorem 11.1 in [14] in order to derive the existence of two constants $c_1 > 0$, $c_2 > 0$ such that

$$c_1 \cdot n^{-K/2} \cdot e^{-n\cdot\inf_{\mathbf{Q}\in\mathbf{B}} D_\varphi(\mathbf{Q},\mathbf{P})} \leq \Pi\big[\boldsymbol{\xi}_n^{\mathbf{W}} \in \mathbf{B}\big] \leq c_2 \cdot n^{-K/2} \cdot e^{-n\cdot\inf_{\mathbf{Q}\in\mathbf{B}} D_\varphi(\mathbf{Q},\mathbf{P})}$$

for any hyper-rectangle $\mathbf{B}$ in $\mathbb{R}^K$, which entails the above-mentioned order of the convergence in Theorem 4. A complete development of such bounds together with the corresponding tuning of the number of replicates in order to approximate $\Pi\big[\boldsymbol{\xi}_n^{\mathbf{W}} \in \mathbf{B}\big]$ for fixed n is postponed to forthcoming work.

3 Finding/Constructing the Distribution of the Weights

As seen above, in our bare-simulation-optimization context it is important to verify the cornerstone Condition 3. For this, in Theorem 22 of [6] we have developed one *special method* (see (7),(8) below) of constructing "good candidates" φ and ζ for Condition 3, and for those we have also given some additional conditions in order to fully verify Condition 3. As a new contribution, let us now give some *more general view* on the verification of Condition 3, delivering also insights and details additional to our investigations [6]. Subsequently, we discuss the following direction: starting from a concrete optimization problem (2) with pregiven φ satisfying $G(1),(G2),(G3),(G4')$, one would first like to verify whether the representability (4) holds, without finding the corresponding ζ explicitly. For this sake, we first present some fundamental properties of all $\varphi \in \Upsilon(]a,b[)$:

Proposition 5. *Let φ satisfy Condition 3 (i.e. $\varphi \in \Upsilon(]a, b[))$. Then the following assertions hold:*

(G4) φ *is strictly convex **only** in a non-empty neighborhood* $]t_-^{sc}, t_+^{sc}[\subseteq]a, b[$ *of one ($t_-^{sc} < 1 < t_+^{sc}$);*

(G5) φ *is continuously differentiable on* $]a, b[$ *(i.e. $\varphi \in C^1(]a, b[)$;*

(G6) φ *is infinitly differentiable on* $]t_-^{sc}, t_+^{sc}[$ *(i.e. $\varphi \in C^\infty(]t_-^{sc}, t_+^{sc}[)$, and hence, $\varphi'(1) = 0$, $\varphi''(t) > 0$ for all $t \in]t_-^{sc}, t_+^{sc}[$;*

notice that the left-hand second derivative and the right-hand second derivative of φ may not coincide at t_-^{sc} respectively at t_+^{sc} (i.e. possible non-second-differentiability at these two points);

(G7) if $a > -\infty$, then $a = t_-^{sc}$;

if $a = -\infty$, then either $t_-^{sc} = -\infty$ or $\varphi(t) = \varphi(t_-^{sc}) + \varphi'(t_-^{sc}) \cdot (t - t_-^{sc})$ for all $t \in]-\infty, t_-^{sc}[$ (affine-linearity); notice that $\varphi'(t_-^{sc}) < 0$;

(G8) if $b < \infty$, then $b = t_+^{sc}$;

if $b = \infty$, then either $t_+^{sc} = \infty$ or $\varphi(t) = \varphi(t_+^{sc}) + \varphi'(t_+^{sc}) \cdot (t - t_+^{sc})$ for all $t \in]t_+^{sc}, \infty[$ (affine-linearity); notice that $\varphi'(t_+^{sc}) > 0$;

(G9) the Fenchel-Legendre transform (also called convex conjugate) of φ

$$\varphi^*(z) = \sup_{t \in \mathbb{R}} (z \cdot t - \varphi(t)) = \sup_{t \in]a, b[} (z \cdot t - \varphi(t)), \qquad z \in \mathbb{R}, \qquad (6)$$

has the following properties:

(G9i) $int(dom(\varphi^*)) =]\lambda_-, \lambda_+[$, where $dom(\varphi^*) := \{z \in \mathbb{R} : -\infty < \varphi^*(z) < \infty\}$,

$\lambda_- := \inf_{t \in]a, b[} \varphi'(t) = \lim_{t \downarrow a} \varphi'(t) =: \varphi'(a) < 0$ *and*
$\lambda_+ := \sup_{t \in]a, b[} \varphi'(t) = \lim_{t \uparrow b} \varphi'(t) =: \varphi'(b) > 0$;

(G9ii) if $a > -\infty$, then

— $\lambda_- = -\infty$;

— the function $z \mapsto e^{-a \cdot z + \varphi^*(z)} =: M(z)$ is absolutely monotone on $]-\infty, 0[$, i.e. all derivatives exist and satisfy $\frac{\partial^k}{\partial z^k} M(z) \geq 0$ ($k \in \mathbb{N}_0$, $z \in]-\infty, 0[$);

— $\lim_{z \to 0-} M(z) = 1$;

(G9iii) if $b < \infty$, then

— $\lambda_+ = \infty$;

— the function $z \mapsto e^{b \cdot z + \varphi^*(-z)} =: M(z)$ is absolutely monotone on $]-\infty, 0[$;

— $\lim_{z \to 0-} M(z) = 1$;

(G9iv) if $a = -\infty$ and $b = \infty$, then

— the function $z \mapsto e^{\varphi^*(z)} =: M(z)$ is exponentially convex on $]\lambda_-, \lambda_+[$, i.e. $M(\cdot)$ is continuous and satisfies

$$\sum_{i,j=1}^n c_i \cdot c_j \cdot M\left(\frac{z_i + z_j}{2}\right) \geq 0 \qquad \text{for all } n \in \mathbb{N}, c_i, c_j \in \mathbb{R} \text{ and } z_i, z_j \in]\lambda_-, \lambda_+[;$$

notice that exponential-convexity is stronger than the usual log-convexity.

— $\lim_{z \to 0-} M(z) = 1$;

(G10) *the endpoints of* $int(dom(\varphi)) =\,]a,b[$ *have the following important "functioning" for the underlying probability distribution* ζ *(cf. (4)) respectively of an associated random variable* W *with* $\zeta[\cdot] := \sqcap[W \in \cdot]$:

(G10i) $a = \inf supp(\zeta) = \inf supp(W)$, $b = \sup supp(\zeta) = \sup supp(W)$, *where* $supp(\zeta)$ *respectively* $supp(W)$ *denotes the support of* ζ *respectively* W; *consequently,* $]a,b[\,= int(conv(supp(\zeta))) = int(conv(supp(W)))$ *where* $conv(A)$ *denotes the convex hull of a set* A;

(G10ii) *if* $a > -\infty$, *then* $\varphi(a) = -\log \zeta[\{a\}] = -\log \sqcap[W = a]$; *consequently,*
$a = \min supp(\zeta) = \min supp(W)$ *if and only if* $\zeta[\{a\}] = \sqcap[W = a] > 0$ *if and only if* $\varphi(a) < \infty$ *if and only if* $a \in dom(\varphi)$;

(G10iii) *if* $b < \infty$, *then* $\varphi(b) = -\log \zeta[\{b\}] = -\log \sqcap[W = b]$; *consequently,*
$b = \max supp(\zeta) = \max supp(W)$ *if and only if* $\zeta[\{b\}] = \sqcap[W = b] > 0$ *if and only if* $\varphi(b) < \infty$ *if and only if* $b \in dom(\varphi)$.

(G11) *the first two derivatives of* φ *at the point* 1 *have the following important "functioning" for* ζ *respectively* W:

(G11i) $1 = \varphi'^{-1}(0) = \int_{\mathbb{R}} y \, d\zeta(y) = E_{\sqcap}[W]$ *where* $\varphi'^{-1}(\cdot)$ *denotes the inverse of the first derivative* $\varphi'(\cdot)$ *of* $\varphi(\cdot)$,

(G11ii) $\frac{1}{\varphi''(1)} = \int_{\mathbb{R}} \left(y - \int_{\mathbb{R}} \tilde{y} \, d\zeta(\tilde{y})\right)^2 d\zeta(y) = E_{\sqcap}[W^2] - (E_{\sqcap}[W])^2 = Var_{\sqcap}[W]$; *thus, scaling* $\tilde{c} \cdot \varphi$ *($\tilde{c} > 0$) does not change the mean* 1 *but the variance of* W.

Notice that (G4) is stronger than (G4'). The proof of Proposition 5 will be given in Sect. 4. The properties (G1) to (G9iv) constitute *necessary* (shape-geometric) conditions for a pregiven function φ to belong to $\Upsilon(]a,b[))$; accordingly, these should be verified first, in concrete situations where one aims to optimize D_φ through Theorem 4. For the *sufficiency*, we obtain

Proposition 6. *Suppose that* $\varphi :]-\infty,\infty[\mapsto [0,\infty]$ *satisfies (G1) to (G8), and recall the notations in (G9i). Then,* $\varphi \in \Upsilon(]a,b[)$ *if one of the following three conditions holds:*

(a) $a > -\infty$, $\lambda_- = -\infty$, *and* $z \mapsto e^{-a \cdot z + \varphi^*(z)}$ *is absolutely monotone on* $]-\infty, 0[$,

(b) $b < \infty$, $\lambda_+ = \infty$, *and* $z \mapsto e^{b \cdot z + \varphi^*(-z)}$ *is absolutely monotone on* $]-\infty, 0[$,

(c) $a = -\infty$, $b = \infty$, *and* $z \mapsto e^{\varphi^*(z)}$ *is exponentially convex on* $]\lambda_-, \lambda_+[$.

The proof of Proposition 6 will be given in Sect. 4; notice that—due to (G9ii) respectively (G9iii)—the two cases "$a > -\infty$, $\lambda_- > -\infty$" respectively "$b < \infty$, $\lambda_+ < \infty$" can not appear. As far as applicability is concerned, it is well known that, in general, verifying absolute monotonicity is typically more comfortable than verifying exponential convexity. Hence, the boundary points a and b of

$int(dom(\varphi))$ play an important role, also (in case that one starts from a pregiven φ) because they indicate the support of the desired ζ to be searched for; for the latter task, one typically should know the explicit form of the Fenchel-Legendre transform φ^* which can sometimes be hard to determine. This hardness issue also applies for the reverse direction of starting from a concrete probability distribution ζ with light tails, computing its log-moment-generating function $z \mapsto \Lambda_\zeta(z) := \log MGF_\zeta(z)$ and the corresponding Fenchel-Legendre transform Λ_ζ^* which is nothing but the associated divergence generator φ. However, as already indicated above, in Theorem 22 of [6] we have developed a comfortable method to considerably ease these problems, as follows (where for the sake of brevity we only present a special case thereof): we start from a function $F :]-\infty, \infty] \mapsto [-\infty, \infty]$ with $int(dom(F)) =]a_F, b_F[$ for some $-\infty \le a_F < 1 < b_F \le \infty$, which is smooth (infinitely continuously differentiable) and strictly increasing on $]a_F, b_F[$, and which satisfies $F(1) = 0$. From this, we construct $]\lambda_-, \lambda_+[:= int(Range(F))$, $]t_-^{sc}, t_+^{sc}[:=]a_F, b_F[$, $]a, b[$ with $a := t_-^{sc} \cdot 1_{\{-\infty\}}(\lambda_-) - \infty \cdot 1_{]-\infty,0[}(\lambda_-)$ and $b := t_+^{sc} \cdot 1_{\{\infty\}}(\lambda_+) + \infty \cdot 1_{]0,\infty[}(\lambda_+)$, as well as the following two functions $\varphi :]-\infty, \infty[\mapsto [0, \infty]$ and $\Lambda :]-\infty, \infty[\mapsto [-\infty, \infty]$ by

$$\varphi(t) := \begin{cases} t \cdot F(t) - \int_0^{F(t)} F^{-1}(u)\, du \ \in [0, \infty[, & \text{if } t \in]t_-^{sc}, t_+^{sc}[, \\ t_-^{sc} \cdot F(t_-^{sc}) - \int_0^{F(t_-^{sc})} F^{-1}(u)\, du \ \in]0, \infty], & \text{if } t = t_-^{sc} > -\infty, \\ t_+^{sc} \cdot F(t_+^{sc}) - \int_0^{F(t_+^{sc})} F^{-1}(u)\, du \ \in]0, \infty], & \text{if } t = t_+^{sc} < \infty, \\ \varphi(t_-^{sc}) + \lambda_- \cdot (t - t_-^{sc}) \ \in]0, \infty], & \text{if } t_-^{sc} > -\infty \text{ and } t \in]-\infty, t_-^{sc}[, \\ \varphi(t_+^{sc}) + \lambda_+ \cdot (t - t_+^{sc}) \ \in]0, \infty], & \text{if } t_+^{sc} < \infty \text{ and } t \in]t_+^{sc}, \infty[, \\ \infty, & \text{else,} \end{cases} \quad (7)$$

$$\Lambda(z) := \begin{cases} \int_0^z F^{-1}(u)\, du \ \in]-\infty, \infty[, & \text{if } z \in]\lambda_-, \lambda_+[, \\ \int_0^{\lambda_-} F^{-1}(u)\, du \ \in [-\infty, \infty], & \text{if } z = \lambda_- > -\infty, \\ \int_0^{\lambda_+} F^{-1}(u)\, du \ \in [-\infty, \infty], & \text{if } z = \lambda_+ < \infty. \\ \infty, & \text{else.} \end{cases} \quad (8)$$

For this construction, we have shown in [6] that $\varphi(t) = \sup_{z \in \mathbb{R}} (z \cdot t - \Lambda(z))$ for all $t \in \mathbb{R}$ and that $z \mapsto \exp(\Lambda(z))$ is a "good candidate" for a moment generating function of a probability distribution ζ, and hence for the representability (4). Additionally, we can also straightforwardly show that φ satisfies (G1) to (G8), and consequently, Proposition 6 generalizes the Proposition 24 of [6]. Numerous examples for the applicability of (7),(8) are given in [6].

4   Proofs

Proof of Proposition 5. Suppose that φ satisfies (G1) and (G3). Moreover, recall the required representability (4); the involved Laplace-Stieltjes transform

$$\mathbb{R} \ni z \mapsto MGF_\zeta(z) := \int_{\mathbb{R}} e^{z \cdot y}\, d\zeta(y) = E_\Pi[e^{z \cdot W}] \quad (9)$$

of a probability measure ζ on the real line $\mathbb{R}$ respectively of an associated random variable W (with $\zeta[\cdot] := \Pi[W \in \cdot]$) has the following fundamental properties, according to well-known general theory:

(M1) MGF_ζ takes values in $]0, \infty]$;

(M2) $dom(MGF_\zeta)$ is an interval which contains 0 and may be degenerated or even the whole real line; correspondingly, we denote its interior by $]\lambda_-, \lambda_+[:= int(dom(MGF_\zeta))$ which may be the empty set (in case that $dom(MGF_\zeta) = \{0\}$, i.e. $\lambda_- = \lambda_+ = 0$); clearly, there holds $\lambda_- \in [-\infty, 0]$ and $\lambda_+ \in [0, \infty]$;

(M3) MGF_ζ is continuous on $dom(MGF_\zeta)$ and lower semicontinuous on $\mathbb{R}$;

(M4) if $\lambda_- \neq \lambda_+$ then MGF_ζ is real analytic and thus infinitely differentiable on $]\lambda_-, \lambda_+[$;

(M5) if MGF_ζ is finite in a neighborhood of zero, i.e. $0 \in]\lambda_-, \lambda_+[$ with $\lambda_- < \lambda_+$, then for all $k \in \mathbb{N}_0$ the $k-$th moment of ζ respectively W exists and is finite and can be computed in terms of the $k-$th derivative $MGF_\zeta^{(k)}$ as

$$MGF_\zeta^{(k)}(0) = \int_\mathbb{R} y^k \, d\zeta(y) = E_\Pi[W^k],$$

which, by the way, then allows the interpretation of MGF_ζ as "moment generating function of ζ resp. W"; since in Condition 3 we assume $0 \in]\lambda_-, \lambda_+[$, we have used the abbreviation MGF_ζ (rather than LST_ζ) in (9);

(M6) if $\lambda_- \neq \lambda_+$, then MGF_ζ is strictly convex on $]\lambda_-, \lambda_+[$.

Hence, the logarithm of the Laplace-Stieltjes transform

$$z \mapsto \Lambda_\zeta(z) := \log MGF_\zeta(z) := \log \int_\mathbb{R} e^{z \cdot y} \, d\zeta(y) = \log E_\Pi[e^{z \cdot W}] \qquad (10)$$

(which in case of $0 \in]\lambda_-, \lambda_+[$ can be interpreted as cumulant generating function) "carries over" (M1) to (M6), which partially can be even refined:

(C1) Λ_ζ takes values in $] - \infty, \infty]$;

(C2) $dom(\Lambda_\zeta) = dom(MGF_\zeta)$ and thus $int(dom(\Lambda_\zeta)) =]\lambda_-, \lambda_+[$;

(C3) Λ_ζ is continuous on $dom(\Lambda_\zeta)$ and lower semicontinuous on $\mathbb{R}$;

(C4) if $\lambda_- \neq \lambda_+$, then λ_ζ is infinitely differentiable on $]\lambda_-, \lambda_+[$;

(C5) if $0 \in]\lambda_-, \lambda_+[$, then

$$\Lambda_\zeta(0) = 0, \quad \Lambda_\zeta'(0) = \int_\mathbb{R} y \, d\zeta(y) = E_\Pi[W], \qquad (11)$$

$$\Lambda_\zeta''(0) = \int_\mathbb{R} \left(y - \int_\mathbb{R} \tilde{y} \, d\zeta(\tilde{y})\right)^2 d\zeta(y) = E_\Pi[W^2] - (E_\Pi[W])^2 = Var_\Pi[W]; \quad (12)$$

(C6) under the assumption $\lambda_- \neq \lambda_+$ there holds: Λ_ζ is strictly convex on $]\lambda_-, \lambda_+[$ if and only if ζ is not a one-point distribution (Dirac mass) if and only if W is not a.s. constant; otherwise, Λ_ζ is linear;

(C7) under the assumption that ζ is not a one-point distribution (Dirac mass)—with the notations $a := \inf supp(\zeta) = \inf supp(W)$, $b := \sup supp(\zeta) = \sup supp(W)$, $t^{sc}_- := \inf\{\Lambda'_\zeta(z) : z \in]\lambda_-, \lambda_+[\} = \lim_{z \downarrow \lambda_-} \Lambda'_\zeta(z)$ and $t^{sc}_+ := \sup\{\Lambda'_\zeta(z) : z \in]\lambda_-, \lambda_+[\} = \lim_{z \uparrow \lambda_+} \Lambda'_\zeta(z)$—one gets the following:

(C7i) $]t^{sc}_-, t^{sc}_+[\subseteq]a, b[$;

(C7ii) if $a > -\infty$, then $\lambda_- = -\infty$, $t^{sc}_- = \lim_{z \to -\infty} \Lambda'_\zeta(z) = \lim_{z \to -\infty} \frac{\Lambda_\zeta(z)}{z} = a$;

(C7iii) if $b < \infty$, then $\lambda_+ = \infty$, $t^{sc}_+ = \lim_{z \to \infty} \Lambda'_\zeta(z) = \lim_{z \to \infty} \frac{\Lambda_\zeta(z)}{z} = b$;

(C7iv) if $a = -\infty$ and $\lambda_- = -\infty$, then $t^{sc}_- = \lim_{z \to -\infty} \Lambda'_\zeta(z) = -\infty = a$;

(C7v) if $b = \infty$ and $\lambda_+ = \infty$, then $t^{sc}_+ = \lim_{z \to \infty} \Lambda'_\zeta(z) = \infty = b$;

(C7vi) if $\lambda_- \in]-\infty, 0[$ and $t^{sc}_- > -\infty$, then $a = -\infty$, $\Lambda_\zeta(\lambda_-) \in]-\infty, \infty[$, $\Lambda_\zeta(z) = \infty$ for all $z < \lambda_-$, $\Lambda'_\zeta(\lambda_-) \in]-\infty, \infty[$;

(C7vii) if $\lambda_+ \in]0, \infty[$ and $t^{sc}_+ < \infty$, then $b = \infty$, $\Lambda_\zeta(\lambda_+) \in]-\infty, \infty[$, $\Lambda_\zeta(z) = \infty$ for all $z > \lambda_+$, $\Lambda'_\zeta(\lambda_+) \in]-\infty, \infty[$;

(C7viii) if $\lambda_- \in]-\infty, 0[$ and $t^{sc}_- = -\infty$, then $a = -\infty$;

(C7ix) if $\lambda_+ \in]0, \infty[$ and $t^{sc}_+ = \infty$, then $b = \infty$.

Notice that (C7ii) to (C7ix) cover all possible constellations. For a proof of (C7ii) to (C7vii) as well as further details, see e.g. Section 9.1 in [3]. By contradiction, (C7viii) follows from (C7ii) and (C7ix) follows from (C7iii). Moreover, (C7i) is a consequence of (C7ii) to (C7ix). As a side remark, notice that (C6) refines (M6). To proceed with our proof of Proposition 5, due to the requirement (4) one has

$$\varphi(t) = \sup_{z \in \mathbb{R}} (z \cdot t - \Lambda_\zeta(z)) =: \Lambda^*_\zeta(t), \qquad t \in \mathbb{R}, \tag{13}$$

i.e. the divergence generator φ must be equal to the Fenchel-Legendre transform Λ^*_ζ of a cumulant generating function Λ_ζ of some probability distribution ζ, such that $\lambda_- < 0 < \lambda_+$ holds. Moreover, φ should satisfy $\varphi(1) = 0$ (cf. (G2)), and should be finite as well as strictly convex in a non-empty neighborhood $]t^{sc}_-, t^{sc}_+[$ of 1 (cf. (G4')). The latter *rules out* that ζ is any one-point distribution (Dirac distribution), say $\zeta = \delta_{y_0}$ for some $y_0 \in \mathbb{R}$, since in such a situation one gets $\Lambda_\zeta(z) = z \cdot y_0$, and thus $\varphi(t) = \Lambda^*_\zeta(t) = 0$ for $t = y_0$ and $\varphi(t) = \Lambda^*_\zeta(t) = \infty$ for all $t \in \mathbb{R}\backslash\{y_0\}$ (even in the case $y_0 = 1$ for which $\varphi(1) = 0$ is satisfied). Consequently, Λ_ζ is strictly convex on $]\lambda_-, \lambda_+[= int(dom(\Lambda_\zeta))$ (cf. (C6)) and (C7) applies. Clearly, by continuity one gets

$$\Lambda^*_\zeta(t) = \sup_{z \in]\lambda_-, \lambda_+[} (t \cdot z - \Lambda_\zeta(z)), \qquad t \in \mathbb{R}. \tag{14}$$

For $t \in]t^{sc}_-, t^{sc}_+[$, the optimization problem (14) can be solved explicitly by the well-known "pure/original" Legendre transform, namely

$$\Lambda^*_\zeta(t) = t \cdot \Lambda'^{-1}_\zeta(t) - \Lambda_\zeta\left(\Lambda'^{-1}_\zeta(t)\right), \qquad t \in]t^{sc}_-, t^{sc}_+[. \tag{15}$$

Let us inspect the further cases $t \le t^{sc}_-$. In the contexts of (C7iv) and (C7viii), this is obsolete since $t^{sc}_- = a = -\infty$. For (C7ii), where $t^{sc}_- = a > -\infty$, one can show $\Lambda^*_\zeta(a) = -\log \zeta[\{a\}] = -\log \Pi[W = a]$ which together with (13) proves (G10ii); moreover, $\Lambda^*_\zeta(t) = \infty$ for all $t < a$ (see e.g. Section 9.1 of [3]). In the setup (C7vi), where $t^{sc}_- > a = -\infty$ it is clear that $\Lambda^*_\zeta(t^{sc}_-) = t^{sc}_- \cdot \Lambda'^{-1}_\zeta(t^{sc}_-) - \Lambda_\zeta(\Lambda'^{-1}_\zeta(t^{sc}_-)) = t^{sc}_- \cdot \lambda_- - \Lambda_\zeta(\lambda_-)$ and

$$\Lambda^*_\zeta(t) = t \cdot \lambda_- - \Lambda_\zeta(\lambda_-) = \Lambda^*_\zeta(t^{sc}_-) + \lambda_- \cdot (t - t^{sc}_-) \quad \text{for all } t \in \,] - \infty, t^{sc}_-[. \quad (16)$$

As far as the cases $t \ge t^{sc}_+$ is concerned, in the situations of (C7v) and (C7ix), this is obsolete since $t^{sc}_+ = b = \infty$. For (C7iii), where $t^{sc}_+ = b < \infty$, one can show $\Lambda^*_\zeta(b) = -\log \zeta[\{b\}] = -\log \Pi[W = b]$ which together with (13) proves (G10iii); moreover, $\Lambda^*_\zeta(t) = \infty$ for all $t > b$ (see e.g. Sect. 9.1 of [3]). In the setup (C7vii), where $t^{sc}_+ < b = \infty$ it is clear that $\Lambda^*_\zeta(t^{sc}_+) = t^{sc}_+ \cdot \Lambda'^{-1}_\zeta(t^{sc}_+) - \Lambda_\zeta(\Lambda'^{-1}_\zeta(t^{sc}_+)) = t^{sc}_+ \cdot \lambda_+ - \Lambda_\zeta(\lambda_+)$ and

$$\Lambda^*_\zeta(t) = t \cdot \lambda_+ - \Lambda_\zeta(\lambda_+) = \Lambda^*_\zeta(t^{sc}_+) + \lambda_+ \cdot (t - t^{sc}_+) \quad \text{for all } t \in \,]t^{sc}_+, \infty[. \quad (17)$$

As a side effect, we have thus also proved (G10i) (notice that in (C7) we have started with a, b to be the endpoints of the support of ζ respectively W, in contrast to (G3) of the definition of $\widetilde{\Upsilon}(]a, b[)$ where a, b are defined as the endpoints of the effective domain of φ). To proceed, from (13) and (15) we obtain

$$\varphi'(t) = (\Lambda^*_\zeta)'(t) = \Lambda'^{-1}_\zeta(t), \quad \varphi''(t) = (\Lambda^*_\zeta)''(t) = \frac{1}{\Lambda''_\zeta(\Lambda'^{-1}_\zeta(t))} > 0, \quad t \in \,]t^{sc}_-, t^{sc}_+[, \quad (18)$$

which—together with the investigations below (15)—provides (G4) and (G5); moreover, (G6) is immediate since the infinite differentiability is straightforward and $\varphi'(1) = 0$ because we have required both the nonnegativity of φ and (G2) (cf. the definition of $\widetilde{\Upsilon}(]a, b[)$). The property (G7) follows from (C7ii), (C7iv), (C7viii), (13), (16) and $\varphi'(t^{sc}_-) = \Lambda'^{-1}_\zeta(t^{sc}_-) = \lambda_-$. Analogously, we get (G8) from (C7iii), (C7v), (C7ix), (13), (17) and $\varphi'(t^{sc}_+) = \Lambda'^{-1}_\zeta(t^{sc}_+) = \lambda_+$.

Let us continue with (G9). By applying the general theory of double Fenchel-Legendre transforms (bi-conjugates), (6) turns into

$$\varphi^*(z) = \Lambda_\zeta(z), \quad z \in \mathbb{R}, \quad (19)$$

which deduces (G9i). The properties (G9ii), (G9iii) and (G9iv) follow from the following theorem and the discussion thereafter.

Theorem 7. (a) Let $M :] - \infty, 0] \mapsto]0, \infty[$ be continuous on $] - \infty, 0]$ with $M(0) = 1$. Then one has

M is absolutely monotone on $] - \infty, 0[\iff$
$\exists$ unique prob. distr. $\widetilde{\widetilde{\zeta}}$ on $[0, \infty[$ s.t. $M(z) = \int_0^\infty e^{z \cdot y} d\widetilde{\widetilde{\zeta}}(y)$ for all $z \in \,] - \infty, 0]$.

(b) *Let I be an open interval which contains 0, and $M : I \mapsto [0, \infty[$ be continuous with $M(0) = 1$. Then one gets*

M *is exponentially convex* $\Longleftrightarrow$

$\exists$ *unique prob. distr.* $\widetilde{\zeta}$ *on* $] - \infty, \infty[$ *s.t.* $M(z) = \displaystyle\int_{-\infty}^{\infty} e^{z \cdot y} d\widetilde{\zeta}(y)$ *for all $z \in I$.*

Assertion (a) of Theorem 7 is known as (probability-version of) *Bernstein's theorem* [2] (see e.g. also [16]), whereas assertion (b) is known as (probability-version of) *Widder's theorem* [19] (see e.g. also [1,10,11,17,20]). From (10), (19), (G9i) and Theorem 7(b), the first item in (G9iv) follows immediately by using the choice $I =]\lambda_-, \lambda_+[$. Under the additional knowledge $a > -\infty$ (and consequently $\lambda_- = -\infty$) employed together with (G10i) and thus $\sqcap[W \geq a] = \zeta[[a, \infty[] = 1$, one arrives at

$$e^{\varphi^*(z) - a \cdot z} = \int_a^{\infty} e^{z \cdot (y-a)} \, d\zeta(y) = \int_0^{\infty} e^{z \cdot \widetilde{y}} \, d\widetilde{\zeta}(\widetilde{y}) = E_{\sqcap}[e^{z \cdot (W-a)}], \quad z \in]-\infty, \lambda_+[, \quad (20)$$

where the probability distribution $\widetilde{\zeta}[\cdot] := \zeta[\cdot + a]$ is the $a-$shifted companion of ζ; recall that $\lambda_+ > 0$. Put in other words, $\sqcap[\widetilde{W} \in \cdot] = \widetilde{\zeta}[\cdot]$ is the probability distribution of the (a.s.) nonnegative random variable $\widetilde{W} := W - a$. Similarly, if $\varphi \in \Upsilon(]a, b[)$ and $b < \infty$ (and hence $\lambda_+ = \infty$), one can derive from (G10i) and its consequence $\sqcap[W \leq b] = \zeta[] - \infty, b]] = 1$ that

$$e^{\varphi^*(-z) + b \cdot z} = \int_{-\infty}^b e^{z \cdot (b-y)} \, d\zeta(y) = \int_0^{\infty} e^{z \cdot \widetilde{y}} \, d\widetilde{\widetilde{\zeta}}(\widetilde{y}) = E_{\sqcap}[e^{z \cdot (b-W)}], \quad z \in] - \infty, -\lambda_-[,$$
(21)

where $-\lambda_- > 0$ and $\widetilde{\widetilde{\zeta}}[\cdot] := \zeta[b - \cdot]$ is the mirrored$-b-$shifted companion of $\zeta[\cdot]$. This means that $\sqcap[\widetilde{W} \in \cdot] = \widetilde{\widetilde{\zeta}}[\cdot]$ is the probability distribution of the (a.s.) nonnegative random variable $\widetilde{W} := b - W$. By using this, Theorem 7(a) together with (20) (respectively (21)) implies the second item of (G9ii) (respectively of (G9iii)). Finally, we obtain (G11i) and (G11ii) from (18), (11) and (12). ∎

Proof of Proposition 6. The assertions follow straightforwardly from Theorem 7, (10), (19), (20), (21), (18) (and the discussion thereafter) as well as (M5). ∎

Acknowledgements. We are grateful to the four reviewers for their comments and very useful suggestions. W. Stummer is grateful to the Sorbonne Université Paris for its multiple partial financial support and especially to the LPSM for its multiple great hospitality. M. Broniatowski thanks very much the FAU Erlangen-Nürnberg for its partial financial support and hospitality.

References

1. Akhiezer, N.I.: The Classical Moment Problem and Some Related Questions in Analysis. Oliver & Boyd, Edinburgh (1965)

2. Bernstein, S.: Sur les fonctions absolument monotones. Acta Math. **52**, 1–66 (1929)
3. Borovkov, A.A.: Probability Theory. Springer, London (2013). https://doi.org/10.1007/978-1-4471-5201-9
4. Broniatowski, M., Stummer, W.: Some universal insights on divergences for statistics, machine learning and artificial intelligence. In: Nielsen, F. (ed.) Geometric Structures of Information. SCT, pp. 149–211. Springer, Cham (2019). https://doi.org/10.1007/978-3-030-02520-5_8
5. Broniatowski, M., Stummer, W.: A unifying framework for some directed distances in statistics. In: Nielsen, F., Rao, A.S.R.S., Rao, C.R. (eds.) Geometry and Statistics. Handbook of Statistics, vol. 46, pp. 145–223. Academic Press, Cambrigde (2022)
6. Broniatowski, M., Stummer, W.: A precise bare simulation approach to the minimization of some distances. I. Foundations. IEEE Trans. Inf. Theory **69**(5), 3062–3120 (2023)
7. Broniatowski, M., Stummer, W.: A precise bare simulation approach to the minimization of some distances. II. Further foundations (2023)
8. Csiszár, I.: I-divergence geometry of probability distributions and minimization problems. Ann. Prob. **3**(1), 146–158 (1975)
9. Csiszár, I.: Sanov property, generalized I-projection and a conditional limit theorem. Ann. Prob. **12**(3), 768–793 (1984)
10. Jaksetic, J., Pecaric, J.: Exponential convexity method. J. Convex Anal. **20**(1), 181–197 (2013)
11. Kotelina, N.O., Pevny, A.B.: Exponential convexity and total positivity. Siberian Electr. Math. Rep. **17**, 802–806 (2020). https://doi.org/10.33048/semi.2020.17.057
12. Liese, F., Vajda, I.: Convex Statistical Distances. Teubner, Leipzig (1987)
13. Nielsen, F.: What is an information geometry? Not. AMS **65**(3), 321–324 (2018)
14. Rassoul-Agha, F., Seppäläinen, T.: A course on Large Deviations with an Introduction to Gibbs Measures. Graduate Studies in Mathematics, vol. 162. American Mathematical Society, Providence RI (2015)
15. Roensch, B., Stummer, W.: 3D insights to some divergences for robust statistics and machine learning. In: Nielsen, F., Barbaresco, F. (eds.) GSI 2017. LNCS, vol. 10589, pp. 460–469. Springer, Cham (2017). https://doi.org/10.1007/978-3-319-68445-1_54
16. Schilling, R.L., Song, R., Vondracek, Z.: Bernstein Functions, 2nd edn. de Gruyter, Berlin (2012)
17. Shucker, D.S.: Extensions and generalizations of a theorem of Widder and of the theory of symmetric local semigroups. J. Funct. Anal. **58**, 291–309 (1984)
18. Vajda, I.: Theory of Statistical Inference and Information. Kluwer, Dordrecht (1989)
19. Widder, D.V.: Necessary and sufficient conditions for the representation of a function by a doubly infinite Laplace integral. Bull. Amer. Math. Soc. **40**(4), 321–326 (1934)
20. Widder, D.V.: The Laplace Transform. Princeton University Press, Princeton (1941)

Extensive Entropy Functionals and Non-ergodic Random Walks

Valérie Girardin[1] and Philippe Regnault[2]([⊠])

[1] Normandie Université, UNICAEN, CNRS, LMNO, 14000 Caen, France
valerie.girardin@unicaen.fr
[2] Université de Reims Champagne-Ardenne, CNRS, LMR,
BP 1039, 51687 Reims cedex 2, France
philippe.regnault@univ-reims.fr

Abstract. According to Tsallis' seminal book on complex systems: "an entropy of a system is extensive if, for a large number n of its elements, the entropy is (asymptotically) proportional to n". According to whether the focus is on the system or on the entropy, an entropy is extensive for a given system or a system is extensive for a given entropy. Yet, exhibiting the right classes of random sequences that are extensive for the right entropy is far from being trivial, and is mostly a new area for generalized entropies. This paper aims at giving some examples or classes of random walks that are extensive for Tsallis entropy.

Keywords: extensivity · complex systems · phi-entropy · random walks · stochastic process

1 Phi-Entropy Functionals and Extensivity

In the most classical information theory, the sources, identified to random sequences, are assumed to be ergodic or stationary. For such sources, the Asymptotic Equipartition Property (AEP) holds, stating that Shannon entropy asymptotically increases linearly with the number of elements of the source, a consequence of the strong additivity of Shannon entropy; see [3] for precise statements of AEPs for various types of sources. For more complex, non-ergodic systems, this asymptotics can be highly non linear, requiring to investigate alternative behaviors or to consider other entropy functionals.

The φ-entropy functionals (also called trace entropies) have now been widely used and studied in numerous scientific fields. The φ-entropy of a random variable X with finite or countable state space E and distribution P_X is defined as $\mathbb{S}_\varphi(X) = \sum_{x \in E} \varphi(P_X(x))$, with φ some smooth function. Classical examples include Shannon with $\varphi(x) = -x \log(x)$, Taneja with $\varphi(x) = -x^s \log(x)$, and Tsallis with $\mathbb{T}_s(X) = \frac{1}{s-1}[1 - \Lambda(X; s)]$, where $\Lambda(X; s) = \sum_{x \in E} P_X(x)^s$ is the so-called Dirichlet series associated to X. Here, we will focus on Tsallis entropy, and suppose that $s > 0$.

Extensivity of a complex systems is introduced in [10] as follows: *"an entropy of a system is extensive if, for a large number n of its elements (probabilistically independent or not), the entropy is (asymptotically) proportional to n".*

F. Nielsen and F. Barbaresco (Eds.): GSI 2023, LNCS 14071, pp. 117–124, 2023.
https://doi.org/10.1007/978-3-031-38271-0_12

Precisely, a φ-entropy is extensive for a random sequence $\mathbf{X} = (X_n)_{n \in \mathbb{N}^*}$, with $X_{1:n} = (X_1, \ldots, X_n)$, if some $c > 0$ exists such that $\mathbb{S}_\varphi(X_{1:n}) \sim_{n \to \infty} c.n$; the constant c is the φ-entropy rate of the sequence. Intuitively, all variables contribute equally to the global information of the sequence, an appealing property in connection with the AEP in the theory of stochastic processes and complex systems; see, e.g., [2]. Extensivity is a two-way relationship of compatibility between an entropy functional and a complex system: indeed, the entropy is extensive for a given system or the system is extensive for a given entropy, according to whether the focus is on the system or on the entropy. Yet, exhibiting the right class of random sequences that are extensive for the right entropy is far from being trivial, and is mostly a new area for generalized entropies. This paper, as a first step, aims at giving some examples or classes of random walks that are extensive for Tsallis entropy, widely in use in complex systems theory; see [10].

For ergodic systems, Shannon entropy is well-known to be extensive while Tsallis entropy in non-extensive; see e.g. [7]. More generally, [4] establishes that Shannon entropy is the unique extensive φ-entropy for a large class of random sequences called quasi-power (QP) sequences (see definition given by (2) below), among the class of the so-called quasi-power-log (QPL) entropies introduced in [1], satisfying

$$\varphi(x) \sim_0 a x^s (\log x)^\delta + b, \tag{1}$$

for some $a, b \in \mathbb{R}$, $s > 0$, $\delta \in \{0, 1\}$. QPL entropies are considered in [9, Eq. (6.60), p356] and [1] as the simplest expression of generalized entropies for studying the asymptotic behavior of entropy for random sequences, on which the present paper focuses. Indeed, the asymptotic behavior of the marginal QPL entropy of a random sequence is closely linked to the behavior of its Dirichlet series, characterized for QP sequences by the quasi-power property

$$\Lambda(X_{1:n}; s) = c(s)\lambda(s)^{n-1} + R_n, \quad s > \sigma_0, \tag{2}$$

where $0 < \sigma_0 < 1$, c and λ are strictly positive analytic functions, λ is strictly decreasing and $\lambda(1) = c(1) = 1$, and R_n is an analytic function such that $|R_n(s)| = O(\rho(s)^n \lambda(s)^n)$ for some $\rho(s) \in]0, 1[$. Thanks to Perron-Frobenius theorem, the QP property is satisfied by ergodic Markov chains, including independent and identically distributed (i.i.d.) sequences. It is also satisfied by a large variety of dynamic systems, including continuous fraction expansions; see [11].

In another perspective on the characterization of the asymptotic behavior of entropy, [9] studies uniformly distributed systems, in which each X_n is drawn from a uniform distribution on a state space that may depend on n; see also [5] and the references therein. The entropies are classified according to the two parameters $0 < c \le 1$ and $d \in \mathbb{R}$ given by

$$c = 1 - \lim_n \frac{\Omega(n)}{n\,\Omega'(n)}, \qquad d = \lim_n \log \Omega(n) \left[\frac{\Omega(n)}{n\,\Omega'(n)} + c - 1 \right], \tag{3}$$

depending only on the asymptotics of the size $\Omega(n) = |E(1:n)|$ of the state space $E_{1:n}$ of $X_{1:n}$. In the context of [5], the asymptotic behavior of $\Omega(n)$ is assumed to be a smooth expression of n, e.g., n^β with $\beta > 0$; then, $\Omega'(n)$ denotes the derivative of this expression at n. The asymptotic classification includes QPL entropies plus Quasi Power Exponential (QPE) entropies, given by $\varphi(x) \sim_0 ax^s \exp(-\gamma x) + b$, with $a, b \in \mathbb{R}^*$, $s \in \mathbb{R}$, and $\gamma \in \mathbb{R}_+^*$, that are all asymptotically equivalent to the Tsallis one. Linear combination of such cases may also be considered, but are asymptotically dominated by one of the terms. Therefore, the present paper will focus exclusively on the asymptotic behavior of QPL entropies, for which $(c, d) = (s, \delta)$ in (1); see [1] and [9, Table 6.2].

All in all, in [5,9] and the references therein, one can identify the following– non exhaustive– types of growth, attached with class tags linked to the type of maximum entropic distribution:

$\Omega(n) \sim n^b$ is power-law leading to $c = 1 - 1/b, d = 0$, and Tsallis entropy;

$\Omega(n) \sim L^{n^{1-\beta}}$ is sub-exponential and leads to $c = 1$, $d = 1 - 1/\beta$, with $0 < \beta < 1$;

$\Omega(n) \sim \exp(\ell n)$ is exponential with $c = d = 1$, and hence is extensive for Shannon entropy;

$\Omega(n) \sim \exp(\ell n^g)$ is stretched exponential with $c = 1, d = 1/g$, with $g > 1$, and extensive for QPE entropies, asymptotically equal to Tsallis.

The paper aims at showing through examples that various simple systems are extensive for Tsallis entropy, by using the growth rate of both the size of the state space and the behavior of the Dirichlet series. This amounts to using the physics approach in [9] to supplement and clarify the mathematics approach in [1,4]–and other works along the same lines. The approach developed in [9] focuses on the complex systems and the induced maximum entropy distribution, and involves random sequences only via the size of the state space, while we are here interested in entropy as a function of a random sequence. Indeed, we focus on the random variables, together with their distributions, involved in the – asymptotic – behavior of a system and its entropy, as reflected in the Dirichlet series.

Section 2 begins by considering classical random walks, non-extensive for Tsallis entropy, but constituting a good starting point for constructing extensive ones. Then some examples of Tsallis-extensive systems are given in the context of complex systems, in terms of restricted or autocorrelated random walks. Still, the conditions on these systems appear to be difficult to express simply in terms of statistical inference, construction, and simulation, of random sequences. Therefore, the framework is broadened in Sect. 3 by considering non identically distributed increments, that is delayed random walks. Tuning the marginal distributions of the increments leads to Tsallis extensive sequences, with explicit probabilistic conditions allowing for the effective construction of such systems. Precisely, the main result, Theorem 1, gives a procedure for building random walks that are Tsallis-extensive, through an opening to non-uniform systems.

2   Random Walks

Let $\mathbf{X} = (X_n)_{n \in \mathbb{N}^*}$ be a sequence of independent random variables such that, for each $n \in \mathbb{N}^*$, X_n takes values in a finite or countable subset $E(n)$ of $\mathbb{Z}^{\mathbb{N}}$.

Let $\mathbf{W} = (W_n)_{n \in \mathbb{N}^*}$ be the random walk on $\mathbb{Z}^{\mathbb{N}}$ associated to the increments $\mathbf{X}$ through $W_n = \sum_{k=1}^{n} X_n$, for $n \in \mathbb{N}^*$.

We will derive the asymptotic behavior of the classical and extended random walks thanks to the following properties satisfied by the Dirichlet series; see, e.g., [8].

Properties 1 *Let X be a discrete random variable taking values in E. Let $\mathcal{E} = |E|$ denote the number of states, possibly infinite. Let $s > 0$. Then:*

1. *$\Lambda(X, 1) = 1$, and if X is deterministic, then $\Lambda(X, s) = 1$ too for all s.*
2. *$\log m_X < \Lambda(X; s) < \log \mathcal{E}$, where m_X is the number of modes of X.*
3. *$s \mapsto \Lambda(X; s)$ is a smooth decreasing function.*
4. *If $X_1, \ldots, X_n$ are independent variables, then $\Lambda(X_{1:n}; s) = \prod_{k=1}^{n} \Lambda(X_k; s)$.*

Classical isotropic random walks (W_n) on $\mathbb{Z}^I$ are associated to sequences $\mathbf{X}$ of i.i.d. random variables with common uniform distribution, say $X_n \sim \mathcal{U}(E)$, on $E = \{\pm e_i, i \in [\![1, I]\!]\}$, with $I \in \mathbb{N}^*$, where the e_i are the canonical vectors of $\mathbb{R}^I$. Property 1.4 and the i.i.d. assumption yield

$$\Lambda(W_{1:n}; s) = \Lambda(X_{1:n}; s) = \prod_{k=1}^{n} \Lambda(X_k; s), \tag{4}$$

that is to say $\Lambda(W_{1:n}; s) = (2I)^n$, so that $\mathbb{S}(W_n) = n \log 2I$ and $\mathbb{T}_s(W_n) = \frac{1}{s-1} \left[1 - (2I)^{(1-s)n} \right]$, and hence Shannon is extensive while Tsallis is exponential.

Note that (4) still holds for non identically distributed random variables. Clearly, alternative choices for $\Lambda(X_k; s)$ yield alternative behaviors for $\Lambda(X_{1:n}; s)$ and hence for Tsallis entropy. Let us give two examples, where the state space of X_n grows with n.

Example 1

1. The state space of X_n is linearly expanding if $X_n \sim \mathcal{U}(\{\pm e_i, 1 \leq i \leq n\})$, since then $\mathcal{E}(1 : n) = |E(n)| = 2n$ and $\Omega(n) = 2^n.n!$. We compute $\Lambda(W_n; s) = (2^n n!)^{1-s}$, $\mathbb{S}(W_n) \sim_\infty n \log n$, and $\mathbb{T}_s(W_n) = \frac{1}{s-1} \left[1 - (2^n n!)^{1-s} \right]$, making the random walk $\mathbf{W}$ over-extensive for both Shannon and Tsallis.
2. The state space of X_n is exponentially expanding if $X_n \sim \mathcal{U}(\{\pm 1\}^n)$, since then $\mathcal{E}(n) = 2^n$, and $\Omega(n) = 2^{n(n+1)}$, a stretched exponential growth, and leads to a QPE entropy with $c = 1$, $d = 1/2$, asymptotically equal to Tsallis. We compute $\Lambda(W_n; s) = 2^{(1-s)n(n+1)/2}$, and $\mathbb{S}(W_n) \sim_\infty \log 2n^2/2$ and $\mathbb{T}_s(W_n) = \frac{1}{s-1} \left[1 - 2^{(1-s)n(n+1)/2} \right]$.

Both (4) and Examples 1 show that the marginal Tsallis entropy of random walks with such inflating state spaces increases at least exponentially fast. To obtain extensive sequences for Tsallis entropy in this way would require the state spaces to contract, which is impossible. The approach of [5,6] with either restricted state spaces or autocorrelated random variables next presented will pave the way to possible solutions.

The following restricted binary random walks, with $E = \{0, 1\}$, are heuristically described in [5]. If, asymptotically, the proportion of 1 is the same whichever be the length of the sequence, then W_n/n converges to a constant limit $L \in (0; 1)$, and $\Omega(n) = L^n$, with exponential growth, and hence $\mathbf{W}$ is extensive for Shannon entropy.

If W_n goes to infinity slower than $L.n$, its growth is sub-extensive for Shannon, and over-extensive otherwise. Such behaviors induce to restrict in some way the number of either 0 or 1 that the system can produce in n steps. For a power law growth, W_n converges to a constant $g > 0$ and $\Omega(n) \sim n^g$, leading to extensivity for Tsallis entropy with $s = 1 - 1/g$; see [7]. A rigorous presentation of such a sequence will be obtained in Example 5 below.

Further, autocorrelated random walks are considered in [6]; see also [7]. Suppose here that $E(n) = \{-1, 1\}$. In the classical symmetric uncorrelated RW, $\mathbb{P}(X_m = -1) = \mathbb{P}(X_m = 1) = 1/2$, $\mathbb{E} X_m = 0$ and $\mathbb{E} X_n X_m = \delta_{nm}$. Then $\Omega(n) = 2^n$ and hence $(c, d) = (1, 1)$ leads to extensivity for Shannon entropy, as seen above. Suppose now that the X_n are correlated random variables, with $\mathbb{E} X_n X_m = 1$ if $\alpha n^\gamma (\log n)^\beta < z \leq \alpha m^\gamma (\log m)^\beta$ and 0 otherwise, for some fixed integer z and real numbers α, β, γ. Taking $\gamma = 0$ and $\beta \neq 0$ leads to extensivity for Tsallis entropy. [6] conjectures that all choices of (γ, β) lead to all choices of (c, d).

Instead of autocorrelated RW, the somewhat less artificial (sic, [6]) ageing RW can be considered, with $X_n = \eta_n X_{n-1}$ where (η_m) is a sequence of binary random variables taking values ± 1 ; see [6] and [9, Chapter 6]. The ensuing (c, d) depends on the distribution of η_{m+1} conditional on the the number of $0 \leq m \leq n$ such that $\eta_m = 1$. A suitable choice leads for instance to the stretched exponential growth and extensivity for a QPE entropy, asymptotically equal to Tsallis.

Applied systems involving Tsallis entropy are given in [5,9]. For instance, spin systems with a constant network connectivity lead to extensivity for Shannon entropy, while random networks growing with constant connectedness require Tsallis entropy; see [5]. See also [9, p371] for a social network model leading to Tsallis entropy.

Still, both restricted and autocorrelated systems are difficult to express in terms of the behavior, statistical inference or simulation of random variables. The delayed RW that we finally propose in Sect. 3 will be more tractable in these perspectives.

3   Delayed Random Walks

A super diffusive random walk model in porous media is considered in [5]. Each time a direction is drawn, $\lfloor n^\beta \rfloor$ steps occur in this direction before another is drawn, where $\beta \in [0, 1[$ is fixed. More precisely, a first direction X_0 is chosen at random between two possibilities. Then, the $\lfloor 2^\beta \rfloor$ following steps equal X_1 : $X_1 \cdots = X_{\lfloor 2^\beta \rfloor} = X_0$. At time $\lfloor 2^\beta \rfloor$, again a direction is chosen at random and repeated for the following $\lfloor 3^\beta \rfloor$ steps, and so on. The number of random choices after n steps, of order $n^{1-\beta}$, decreases in time, and hence $\Omega(n) \simeq 2^{n^{1-\beta}}$, and $(c,d) = (1, 1/(1-\beta))$; Shannon entropy is no more extensive.

This example leads to the notion of delayed random walks, that we will develop here in order to construct classes of random sequences that are extensive for Tsallis entropy. Precisely, we will say that $\mathbf{W}$ is a delayed random walk (DRW) if for identified indices $n \in D \subseteq \mathbb{N}^*$, the behavior of W_n is deterministic conditionally to $W_{1:n-1}$. In other words, all X_n are deterministic for these n.

Let us first give three examples where we assume that the random increments X_n, for $n \in R = \mathbb{N}^* \backslash D$, are drawn uniformly in a finite set E with cardinal $\mathcal{E}$.

Example 2

1. A constant delay $\kappa \in \mathbb{N}^*$ between random steps leads to $R = \kappa \mathbb{N}^*$ and $\Omega(n) = \mathcal{E}^{\lfloor \frac{n}{\kappa} \rfloor}$, an exponential growth leading to Shannon entropy. We compute $\Lambda(W_{1:n}; s) = \mathcal{E}^{\lfloor \frac{n}{\kappa} \rfloor (1-s)}$.

2. A linearly increasing delay, say $R = \{1 + n(n+1)/2, n \in \mathbb{N}^*\}$, leads to $\Omega(n) = \mathcal{E}^{\lfloor (-1+\sqrt{1+8n})/2 \rfloor}$, a stretched exponential growth leading to a QPE entropy, asymptotically equal to Tsallis. We compute $\Lambda(W_{1:n}; s) = \mathcal{E}^{\lfloor (-1+\sqrt{1+8n})/2 \rfloor (1-s)}$, and $\mathbb{T}_s(W_n) = \frac{1}{s-1}\left[1 - \mathcal{E}^{\lfloor (-1+\sqrt{1+8n})/2 \rfloor (1-s)}\right]$.

3. An exponentially increasing delay, say $R = \{2^n, n \in \mathbb{N}^*\}$, leads to $\Omega(n) = \mathcal{E}^{\lfloor \log_2 n \rfloor}$, a power-law growth leading to Tsallis entropy. We compute $\Lambda(W_{1:n}; s) = \mathcal{E}^{\lfloor \log_2 n \rfloor (1-s)}$, and $\mathbb{T}_s(W_{1:n}) = \frac{1}{s-1}\left[1 - \mathcal{E}^{\lfloor \log_2 n \rfloor (1-s)}\right]$, from which we immediately derive that

$$\frac{1}{s-1}(1 - n^{(1-s)\ln(\mathcal{E})/\ln(2)} \mathcal{E}^{s-1}) < \mathbb{T}_s(W_{1:n}) \leq \frac{1}{s-1}(1 - n^{(1-s)\ln(\mathcal{E})/\ln(2)}).$$

In other words, $\mathbb{T}_s(W_{1:n})$ essentially increases as a power of n. For random increments occurring at times of order $2^{n(1-s)}$ instead of 2^n and if $\mathcal{E} = 2$, we similarly derive that $W_{1:n}$ is extensive for $\mathbb{T}_s$; this will be rigorously stated in Example 3 below.

Examples 1 and 2 illustrate how the Dirichlet series of DRW are affected by state space expansion and delays. On the one hand, the Dirichlet series increase with the expansion of the system while on the other hand, the faster the delay lengths increase between random increments, the slower the Dirichlet series and $\Omega(n)$ increase. More generally, one can generate–theoretically–any prescribed asymptotic behavior for the Dirichlet series and $\Omega(n)$ by suitably balancing between the introduction of delays and the ability to control the Dirichlet series of the random increments.

Precisely, Properties 1.1 and 1.4 yield the following relation between the Dirichlet series of the DRW and the Dirichlet series $l_n = \Lambda(X_{r_n}; s)$ of the increments,

$$\Lambda(W_{1:n}; s) = \prod_{k=1}^{\widetilde{k}} l_k, \quad \widetilde{k} = \max\{k : r_k \leq n\}. \tag{5}$$

Let us now exhibit different types DRW that are either strictly extensive for Tsallis entropy, such that $\lim \mathbb{T}_s(W_n)/n$ exists and is not zero, or weakly extensive in the sense that both $\liminf_n \frac{1}{n}\mathbb{T}_s(W_n)$ and $\limsup_n \frac{1}{n}\mathbb{T}_s(W_n)$ exist and are not zero.

Theorem 1. *Let $s \in (0;1)$. Let $(l_n)_{n\in\mathbb{N}^*}$ be a real sequence such that $l_n > 1$ and $\lfloor \prod_{k=1}^n l_k \rfloor \geq n$ for all n. Let $\mathbf{W} = (W_n)_{n\in\mathbb{N}^*}$ be the DRW associated to increments $(X_n)_{n\in\mathbb{N}^*}$ and delays $r_n = \max\{\lfloor \prod_{k=1}^n l_k \rfloor, r_{n-1}+1\}$, where $l_n = \Lambda(X_{r_n}; s)$. Then $\limsup_{n\to\infty} \frac{1}{n}\mathbb{T}_s(W_{1:n}) = 1/(1-s)$.*

Moreover, if l_n converges to $L \geq 1$, then $\liminf_{n\to\infty} \frac{1}{n}\mathbb{T}_s(W_{1:n}) = 1/(1-s)L$, and $\mathbf{W}$ is weakly extensive for $\mathbb{T}_s$. If l_n converges to $L = 1$, then the extensivity is strict.

Proof. Assume that the sequence $(\lfloor \prod_{k=1}^n l_k \rfloor)$ is strictly increasing so that $r_n = \lfloor \prod_{k=1}^n l_k \rfloor$. Otherwise, simply discard the first components of $\mathbf{W}$ to fit this assumption.

We compute using (5),

$$\Lambda(W_{1:n}; s) = \begin{cases} \prod_{i=1}^k l_i & \text{if } n = r_k, \\ \prod_{i=1}^{k-1} l_i & \text{if } r_{k-1} < n < r_k, \end{cases}$$

that is piecewise-constant and increasing with respect to n. Its supremum limit is obtained for the subsequence $\Lambda(W_{1:r_n}; s) = \prod_{k=1}^n l_n$, $k \in \mathbb{N}^*$. Since $r_n = \lfloor \prod_{k=1}^n l_k \rfloor$, we have $r_n \leq \Lambda(W_{1:r_n}; s) \leq r_n + 1$, so that

$$\frac{1}{1-s}\frac{r_n - 1}{r_n} \leq \frac{1}{r_n}\mathbb{T}_s(W_{1:r_n}) \leq \frac{1}{1-s},$$

and the limsup result holds.

Similarly, the infimum limit exists and is obtained for the subsequence $(W_{1:r_n+1})$ as soon as l_n converges (to $L \geq 1$), which finishes the proof. $\square$

Note that Theorem 1 is based on the existence of a random variable X whose Dirichlet series $\Lambda(X; s)$ takes any prescribed value $\ell > 1$. Thanks to Property 1.2, this can be achieved in various ways, by choosing X in a parametric model with state space E and number of modes m_X as soon as $\ell \in (\log m_X; \log|E|)$; see [8]. Tuning the parameters of the distribution leads to specific values for which $\Lambda(X; s) = \ell$. See Example 4 below for a Bernoulli model, where $l \in (1;2)$.

The following example illustrates how to generate simple random sequences that are weakly extensive for Tsallis entropy by suitably introducing delays. Still, the infimum and supremum limits cannot be equal, hindering strict extensivity.

Example 3. Let $s \in (0,1)$. Let $\mathbf{W}$ be a DRW with exponential delays of order 2^{1-s}, say $r_n = \max\{\lfloor 2^{n(1-s)} \rfloor, r_{n-1}+1\}$ for $n \geq 2$, with $r_1 = 1$. Random increments X_{r_n} are drawn according to a uniform distribution $\mathcal{U}(\{-1,1\})$ so that $l_n = \Lambda(X_{r_n}; s) = 2^{1-s}$.

Then, Theorem 1 yields $\Omega(n) \sim 2^{\lfloor \frac{1}{1-s}\log_2 n \rfloor}$, and

$$\liminf_{n\to\infty} \frac{1}{n}\mathbb{T}_s(W_n) = \frac{1}{1-s}2^{s-1}, \quad \limsup_{n\to\infty} \frac{1}{n}\mathbb{T}_s(W_n) = \frac{1}{1-s}. \tag{6}$$

The last example will consider anisotropic random walks, in which the X_{r_n} are drawn according to an asymmetric binary distribution with probabilities depending on n.

Example 4. Let $r_n = \max\{\lfloor \prod_{k=1}^{n}(1 + 1/(k+1))\rfloor, r_{n-1} + 1\}$ for $n \geq 2$, with $r_1 = 1$. Let $\mathbf{X}$ be a sequence of independent variables such that $\mathbb{P}(X_{r_n} = 1) = 1 - \mathbb{P}(X_{r_n} = -1) = p_n$, with p_n solution of $(p_n)^s + (1 - p_n)^s = 1 + 1/(n+1)$, while all other X_n are deterministic. By construction, the Dirichlet series associated with X_n is $l_n = \Lambda(X_n, s) = 1 + 1/(n+1)$ which converges to 1. Theorem 1 yields extensivity of Tsallis entropy.

Further, Examples 1 become Tsallis-extensive by introducing the respective delays $R = \{\lfloor 2^{n(1-s)}n!, \ n > 0\rfloor\}$ and $R = \{\lfloor 2^{n(n+1)(1-s)/2}\rfloor\}$ and applying Theorem 1.

Note that large classes of Tsallis-extensive DRWs can be built from Theorem 1, a construction that was the main aim of the paper.

References

1. Ciuperca, G., Girardin, V., Lhote, L.: Computation and estimation of generalized entropy rates for denumerable markov chains. IEEE Trans. Inf. Theory **57**(7), 4026–4034 (2011). https://doi.org/10.1109/TIT.2011.2133710
2. Cover, T.M., Thomas, J.A.: Elements of Information Theory. Wiley, 2 edn. (2006). https://onlinelibrary.wiley.com/doi/book/10.1002/047174882X
3. Girardin, V.: On the different extensions of the ergodic theorem of information theory. In: Baeza-Yates, R., Glaz, J., Gzyl, H., Hüsler, J., Palacios, J.L. (eds.) Recent Advances in Applied Probability, pp. 163–179. Springer, US (2005). https://doi.org/10.1007/0-387-23394-6_7
4. Girardin, V., Regnault, P.: Linear (h, φ)-entropies for quasi-power sequences with a focus on the logarithm of Taneja entropy. Phys. Sci. Forum **5**(1), 9 (2022). https://doi.org/10.3390/psf2022005009
5. Hanel, R., Thurner, S.: When do generalized entropies apply? How phase space volume determines entropy. Europhys. Lett. **96**(50003) (2011). https://doi.org/10.1209/0295-5075/96/50003
6. Hanel, R., Thurner, S.: Generalized (c, d)-entropy and aging random walks. Entropy **15**(12), 5324–5337 (2013). https://doi.org/10.3390/e15125324, https://www.mdpi.com/1099-4300/15/12/5324
7. Marsh, J.A., Fuentes, M.A., Moyano, L.G., Tsallis, C.: Influence of global correlations on central limit theorems and entropic extensivity. Phys. A **372**(2), 183–202 (2006). https://doi.org/10.1016/j.physa.2006.08.009
8. Regnault, P.: Différents problèmes liés à l'estimation de l'entropie de Shannon d'une loi, d'un processus de Markov (2011). https://www.theses.fr/2011CAEN2042
9. Thurner, S., Klimek, P., Hanel, R.: Introduction to the Theory of Complex Systems. Oxford University Press (2018). https://doi.org/10.1093/oso/9780198821939.001.0001
10. Tsallis, C.: Introduction to Nonextensive Statistical Mechanics. Springer (2009). https://link.springer.com/book/10.1007/978-0-387-85359-8
11. Vallée, B.: Dynamical sources in information theory: Fundamental intervals and word prefixes. Algorithmica **29**(1), 262–306 (2001). https://doi.org/10.1007/BF02679622

Empirical Likelihood with Censored Data

Mohamed Boukeloua[1] and Amor Keziou[2(✉)]

[1] Ecole Nationale Polytechnique de Constantine, Laboratoire LAMASD,
Université Les Frères Mentouri, Constantine, Algérie
[2] Laboratoire de Mathématiques de Reims, UMR 9008, Université de Reims
Champagne-Ardenne, Reims, France
amor.keziou@univ-reims.fr

Abstract. In this paper, we consider semiparametric models defined by moment constraints, with unknown parameter, for right censored data. We derive estimates, confidence regions and tests for the parameter of interest, by means of minimizing empirical divergences between the considered models and the Kaplan-Meier empirical measure. This approach leads to a new natural adaptation of the empirical likelihood method to the present context of right censored data. The asymptotic properties of the proposed estimates and tests are studied, including consistency and asymptotic distributions. Simulation results are given, illustrating the performance of the proposed estimates and confidence regions.

Keywords: Survival analysis · Confidence regions · Tests · Censored data · Moment condition models · Minimum divergence · Duality

1 Introduction

Let X be a nonnegative real random variable with cumulative distribution function F_X. We consider semiparametric statistical models defined by moment condition equations, of the form

$$\mathbb{E}\left[g(X,\theta)\right] = 0, \tag{1}$$

where $\theta \in \Theta \subset \mathbb{R}^d$ is the parameter of interest, $g(\cdot,\cdot) := (g_1(\cdot,\cdot),\ldots,g_\ell(\cdot,\cdot))^\top \in \mathbb{R}^\ell$ is some known $\mathbb{R}^\ell$-valued function defined on $\mathbb{R} \times \Theta$, $\ell \geq d$, and $\mathbb{E}[\cdot]$ is used to denote the mathematical expectation. For complete observations (i.e., for observations of X without censoring), these models have been widely studied in statistics and econometrics literature. We quote for example [22] who used the empirical likelihood (EL) method, see [19], to define estimates for θ and to construct confidence regions and tests on the parameter. [9] introduced the generalized method of moments (GMM). As an alternative to the GMM estimates, [26] introduced a class of generalized empirical likelihood (GEL) estimates. The properties of these estimates have been studied in [16], who have compared them with the GMM approach. In many practical situations, a censorship phenomenon may prevent the complete observation of the variable of interest X. There exist

F. Nielsen and F. Barbaresco (Eds.): GSI 2023, LNCS 14071, pp. 125–135, 2023.
https://doi.org/10.1007/978-3-031-38271-0_13

various kinds of censorship, but we focus, in the present paper, on the right censoring one. In this case, instead of observing X, we have at disposal a sample $(Z_1, \Delta_1), \ldots, (Z_n, \Delta_n)$ of independent copies of the pair $(Z := \min(X, R), \Delta)$, where R is a nonnegative censoring variable and $\Delta := \mathbb{1}_{\{Z=X\}}$ is the indicator of censorship, taking the value 1 if $Z = X$ and 0 otherwise. To deal with model (1) in this situation, several adaptations of the EL method have been introduced in the literature for the right censoring context. We cite the EL method of [27] and the weighted EL method (WEL) introduced by [23]. The first one studied model (1) with $g(x, \theta) = \xi(x) - \theta$, where ξ is some real valued function. Remarking that $\mathbb{E}\left[\frac{\xi(Z_i)\Delta_i}{S_R(Z_i)}\right] = \theta$, where $S_R(\cdot)$ is the survival function of R, they defined, by analogy with the case of complete data, the "estimated" empirical likelihood profile

$$L(\theta) := \max_{p_1, \ldots, p_n \in [0,1]} \prod_{i=1}^{n} p_i \qquad (2)$$

subject to the constraints

$$\sum_{i=1}^{n} p_i V_{ni} = \theta \quad \text{and} \quad \sum_{i=1}^{n} p_i = 1, \qquad (3)$$

where $V_{ni} := \frac{\xi(Z_i)\Delta_i}{S_R^{(n)}(Z_i)}$, $S_R^{(n)}(\cdot)$ being the Kaplan-Meier estimator of $S_R(\cdot)$. They solved this optimization problem and defined the log-likelihood ratio in the same way as that introduced by [17]. They also showed that the obtained log-likelihood ratio, multiplied by some estimated quantity, converges to a chi-square distribution with one degree of freedom. Therefore, they used this asymptotic distribution to construct confidence interval for θ. This approach has been widely adopted in the literature. [21] used it to estimate the mean residual life of X, taking $g(x, \theta) = (x - x_0 - \theta)\mathbb{1}_{\{x > x_0\}}$, where x_0 is a fixed point at which the mean residual life is evaluated. Independently, [23] considered model (1), with $g(x, \theta) = x - \theta$, and proposed weighted empirical likelihood ratio (WELR) confidence interval for θ. [23] estimated the lower and the upper bounds of this interval, similarly to the work of [18], by solving these two optimization problems

$$\widehat{X}_{L,n} := \min_{p_1, \ldots, p_m \in [0,1]} \sum_{i=1}^{m} p_i Z_i'$$

and

$$\widehat{X}_{U,n} := \max_{p_1, \ldots, p_m \in [0,1]} \sum_{i=1}^{m} p_i Z_i',$$

both subject to

$$\sum_{i=1}^{m} p_i = 1 \quad \text{and} \quad \prod_{i=1}^{m} (p_i/\widehat{p}_i)^{m\widehat{p}_i} \geq c,$$

where $(Z_i')_{1 \leq i \leq m}$ $(m \leq n)$ are the distinct values, in increasing order, of $(Z_i)_{1 \leq i \leq n}$ and $\widehat{p}_i := \mathbb{P}_n^{KM}(Z_i')$, $\mathbb{P}_n^{KM}(\cdot)$ being the "Kaplan-Meier empirical measure" described in Sect. 2 below. The constant c is calculated according to the level of the confidence interval, using the asymptotic distribution of the WELR, which is a scaled chi-square distribution, as shown in the same paper. A similar approach has been employed by [24] in quantile estimation and by [25] for two-sample semiparametric problem. Otherwise, based on the approach of [11,27] used some influence functions to give another adaptation of the EL method to the censoring context. The same device has been used by [10] to treat the case of the presence of covariables. [29] applied the adjusted empirical likelihood of [5] in the case of right censored data. Furthermore, other versions of the EL method for censored data have been employed in different problems, including the regression ones such as in [8], [1,28] and [20]. On the other hand, the theory of φ-divergences and duality, developed in [2] have been intensively used in inferential statistics, see [14] and [3] for parametric problems, and [15] for two-sample semiparametric density ratio models. The study of parametric models via φ-divergences has been extended to the case of censored data by [6]. [4] applied the theory of φ-divergences to the study of model (1), generalizing the EL approach. Moreover, they gave a new point of view of the EL method, showing that it is equivalent to minimizing the "modified" Kullback-Leibler divergence (KL_m), called also the likelihood-divergence (LD), between the model and the empirical measure of the data. In the present paper, we follow this point of view and we construct our estimators by minimizing divergences between the model and the Kaplan-Meier empirical measure. This leads to a new more natural adaptation of the EL approach to the right censoring context using the particular case of the likelihood divergence. We carry out an extensive simulation study to compare our approach with the EL method of [27] and the WEL method of [23]. The simulation results show that the proposed likelihood divergence-based method has generally better performance than the existing ones. Concerning our theoretical results, we establish weak consistency and asymptotic normality for the proposed estimates of the parameter θ. For that, we apply a central limit theorem (CLT) in the case of censored data, see [30], and a uniform strong law of large numbers (USLLN) in the same case. We build confidence regions for θ and perform tests on the model and the parameter θ. We provide the asymptotic distributions of the proposed test statistics, both under the null and the alternative hypotheses. The theoretical results we obtain have many applications such as those given in the following examples.

Example 1 *(Confidence intervals for the survival function, and the mean residual life). Taking $g(x, \theta) = \mathbb{1}_{\{x > x_0\}} - \theta$ in model (1), where x_0 is a fixed point, one can see that the true value of θ is the survival function of X evaluated at x_0. Therefore, the techniques we study can be used to construct confidence bound for the survival function. Taking $g(x, \theta) = (x - x_0 - \theta)\mathbb{1}_{\{x > x_0\}}$, one can obtain confidence bounds for the mean residual life.*

Example 2 *(Confidence intervals for the mean residual life). At a fixed point x_0, the mean residual life of X is defined by the conditional expectation $M(x_0) := \mathbb{E}[X - x_0 \mid X > x_0]$. It represents the expected value of the remaining lifetimes after x_0. The mean residual life exposes the survival characteristics of the phenomenon of interest, better than the popular hazard function (see e.g. [12]).*

The study of $M(x_0)$ can be considered in the framework of model (1) taking $g(x, \theta) = (x - x_0 - \theta)\,\mathbb{1}_{\{x > x_0\}}$. Therefore, we can construct, as in the previous example, confidence intervals for $M(x_0)$.

Example 3 *Consider the model (1) with $g(x, \theta) = (x - \theta, x^2 - h(\theta))^\top$, where $h(\cdot)$ is some known function. In this case, the true value of θ is the mean of X. These models include all the distributions for which the second order moment can be written as an explicit function of the first order one. Many usual distributions in survival analysis belong to this class of distributions, such as the exponential and the Rayleigh ones. Considering the semiparametric model (1) instead of a fully parametric one, leads to more robust estimates of the mean of X. Moreover, the divergence estimates we propose can be used to construct estimators of the distribution function of X.*

The rest is organized as follows. Section 2 describes the Kaplan-Meier empirical measure. In Sect. 3, we introduce the minimum divergence estimates. In Sect. 4, we provide the asymptotic properties of the proposed estimates and test statistics. Section 5 presents some simulation results. The proofs of our theoretical results, as well as many other simulation results, are available from the authors.

2   The Kaplan-Meier Empirical Measure

Let X be a nonnegative lifetime random variable with a continuous distribution function $F_X(\cdot)$. We suppose that X is right censored by a nonnegative random variable R, independent of X. The available observations consist of a sample $(Z_1, \Delta_1), \ldots, (Z_n, \Delta_n)$ of independent copies of the pair (Z, Δ), where $Z := \min(X, R)$ and $\Delta := \mathbb{1}_{\{X \leq R\}}$ ($\mathbb{1}_{\{.\}}$ denotes the indicator function). In all the sequel, for any real random variable V, $F_V(x) := P(V \leq x)$, $S_V(x) := 1 - F_V(x)$ and $T_V := \sup\{x \in \mathbb{R} \text{ such that } F_V(x) < 1\}$ denote, respectively, the distribution function, the survival function and the upper endpoint of the support of V. Furthermore, for any right continuous function $K : \mathbb{R} \to \mathbb{R}$, we set $K(x^-) := \lim_{\varepsilon \geq 0} K(x - \varepsilon)$ and $\Delta_K(x) := K(x) - K(x^-)$, whenever the limit exists. We assume that $T_X \leq T_R$, which ensures that we can observe X on the whole of its support. A popular estimator of F_X, in the context of right censoring, is the following product-limit estimator, introduced by [13],

$$F_n(x) := 1 - S_n(x) := 1 - \prod_{i \,|\, Z_i' \leq x} \left(1 - \frac{D(Z_i')}{U(Z_i')}\right), \tag{4}$$

where $(Z_i')_{1 \leq i \leq m}$ ($m \leq n$) are the distinct values, in increasing order, of $(Z_i)_{1 \leq i \leq n}$, $D(Z_i') := \sum_{j=1}^{n} \Delta_j \mathbb{1}_{\{Z_j = Z_i'\}}$ and $U(Z_i') := \sum_{j=1}^{n} \mathbb{1}_{\{Z_j \geq Z_i'\}}$. $S_n(x) := \prod_{i \,|\, Z_i' \leq x} \left(1 - \frac{D(Z_i')}{U(Z_i')}\right)$ is the Kaplan-Meier estimate of the survival function $S_X(\cdot)$ of X. The empirical measure of $F_n(.)$, denote it $\mathbb{P}_n^{KM}(\cdot)$, can be written as

$$\mathbb{P}_n^{KM}(\cdot) := \sum_{i=1}^{m} \Delta_{F_n}(Z_i')\, \delta_{Z_i'}(\cdot) = \sum_{i=1}^{n} \frac{\Delta_{F_n}(Z_i)}{D(Z_i)}\, \delta_{Z_i}(\cdot),$$

with the convention $\frac{0}{0} = 0$, $\delta_x(\cdot)$ being the Dirac measure which puts all the mass at the point x, for all x. Let $S_R^{(n)}(\cdot)$ be the "product-limit" estimator, of the survival function $S_R(\cdot)$ of the censorship variable R, given by $S_R^{(n)}(x) :=$ $\prod_{i \mid Z_i' \leq x} \left(1 - \frac{\overline{D}(Z_i')}{U(Z_i')}\right)$, where $\overline{D}(Z_i') := \sum_{j=1}^{n} (1 - \Delta_j) \mathbb{1}_{\{Z_j = Z_i'\}}$. Note that the above estimates $S_n(\cdot)$ and $S_R^{(n)}(\cdot)$ can be written as $S_n(x) = \prod_{j \mid Z_{(j)} \leq x} \left(\frac{n-j}{n-j+1}\right)^{\Delta_j}$ and $S_R^{(n)}(x) = \prod_{j \mid Z_{(j)} \leq x} \left(\frac{n-j}{n-j+1}\right)^{1-\Delta_j}$, where $Z_{(1)}, \ldots, Z_{(n)}$ are the order statistics of the observations $Z_1, \ldots, Z_n$. Using these two relations, one can show that,

$$\mathbb{P}_n^{KM}(\cdot) = \sum_{i=1}^{n} \frac{\Delta_i}{n\, S_R^{(n)}(Z_i^-)} \delta_{Z_i}(\cdot) =: \sum_{i=1}^{n} w_i\, \delta_{Z_i}(\cdot). \tag{5}$$

with the weights

$$w_i := \frac{\Delta_i}{n\, S_R^{(n)}(Z_i^-)}, \ \forall i = 1 \ldots, n. \tag{6}$$

3   Minimum Divergence Estimators

Denote by M the space of all signed finite measures (s.f.m.) on $(\mathbb{R}, \mathcal{B}(\mathbb{R}))$, and consider the statistical model $\mathcal{M} := \bigcup_{\theta \in \Theta} \mathcal{M}_\theta$, $\mathcal{M}_\theta$ being the set of all $Q \in M$ s.t. $\int_{\mathbb{R}} dQ(x) = 1$ and $\int_{\mathbb{R}} g(x, \theta)\, dQ(x) = 0$, $g := (g_1, \ldots, g_\ell)^\top \in \mathbb{R}^\ell$ being some specified $\mathbb{R}^\ell$-valued function of $x \in \mathbb{R}$ and vector parameter $\theta \in \Theta \subset \mathbb{R}^d$. We denote θ_T, if it exists, the true value of the parameter, i.e., the value such that (s.t.) $P_X \in \mathcal{M}_{\theta_T}$, where P_X is the probability distribution of the lifetime X. We will define estimates, confidence areas and test statistics for θ_T. For that, we will use some results on the theory of divergences and duality from [2] and [4]. Let φ be a convex function from $\mathbb{R}$ to $[0, +\infty]$ such that its domain $\mathrm{dom}_\varphi := \{x \in \mathbb{R} \text{ s.t. } \varphi(x) < \infty\}$ is an interval with endpoints $a_\varphi < 1 < b_\varphi$ (which may be bounded or not, open or not). We assume that $\varphi(1) = 0$ and that φ is closed. For any probability measure (p.m.) P on the measurable space $(\mathbb{R}, \mathcal{B}(\mathbb{R}))$ and for any s.f.m. $Q \in M$, the φ-divergence between Q and P, when Q is absolutely continuous with respect to (a.c.w.r.t.) P, is defined by $D_\varphi(Q, P) := \int_{\mathbb{R}} \varphi\left(\frac{dQ}{dP}(x)\right) dP(x)$, where $\frac{dQ}{dP}(\cdot)$ is the Radon-Nikodym derivative. When Q is not a.c.w.r.t. P, we set $D_\varphi(Q, P) := +\infty$. Recall that the divergence associated to the convex function $\varphi(x) = x \log x - x + 1$, is called Kullback-Leibler divergence (KL-divergence), and that the divergence associated to the convex function $\varphi(x) = -\log x + x - 1$, is called "modified" Kullback-Leibler divergence (KL_m), or Likelihood-divergence (LD). In all the sequel, for simplicity and convenience, we will use the notation (LD) for this particular divergence. Let θ be a given value in Θ. The "plug-in" estimate of $D_\varphi(\mathcal{M}_\theta, P_X)$ is $\widehat{D}_\varphi(\mathcal{M}_\theta, P_X) := \inf_{Q \in \mathcal{M}_\theta} D_\varphi(Q, \mathbb{P}_n^{KM})$. If the projection $Q_\theta^{(n)}$ of $\mathbb{P}_n^{KM}$ on $\mathcal{M}_\theta$ exists, then it is clear that $Q_\theta^{(n)}$ is a.c.w.r.t. $\mathbb{P}_n^{KM}$; this means that the support of $Q_\theta^{(n)}$ must

be included in the support of $\mathbb{P}_n^{KM}$, which is the set of the uncensored data $\{Z_i \text{ s.t. } i \in \{1, \ldots, n\} \text{ and } \Delta_i = 1\}$. Therefore, define the sets of discrete s.f.m
$$\mathcal{M}_\theta^{(n)} := \left\{ Q \text{ a.c.w.r.t. } \mathbb{P}_n^{KM}, \sum_{i=1}^n Q(Z_i) = 1 \text{ and } \sum_{i=1}^n Q(Z_i)g(Z_i, \theta) = 0 \right\},$$
then the estimate $\widehat{D}_\varphi(\mathcal{M}_\theta, P_X)$ can be written as

$$\widehat{D}_\varphi(\mathcal{M}_\theta, P_X) = \inf_{Q \in \mathcal{M}_\theta^{(n)}} D_\varphi(Q, \mathbb{P}_n^{KM}). \tag{7}$$

This constrained optimization problem can be transformed to unconstrained one, using the convex conjugate φ^* of φ. In fact, using Proposition 4.1 of [4], one can show that

$$\widehat{D}_\varphi(\mathcal{M}_\theta, P_X) = \sup_{t \in \mathbb{R}^{1+\ell}} \left\{ t_0 - \sum_{i=1}^n w_i \, \varphi^* \left(t_0 + \sum_{j=1}^\ell t_j g_j(Z_i, \theta) \right) \right\}. \tag{8}$$

Taking into account this result, we will redefine the estimate $\widehat{D}_\varphi(\mathcal{M}_\theta, P_X)$ as follows. Let $t := (t_0, t_1, \ldots, t_\ell)^\top \in \mathbb{R}^{1+\ell}$, $\overline{g} := (1_{\mathbb{R} \times \Theta}, g_1, \ldots, g_\ell)^\top$, $t^\top \overline{g}(x, \theta) := t_0 + \sum_{j=1}^\ell t_j g_j(x, \theta)$, $m(x, \theta, t) := \varphi^*(t^\top \overline{g}(x, \theta))$,

$$\Lambda_\theta^{(n)} := \{t \in \mathbb{R}^{1+\ell} \text{ s.t. } a_\varphi^* < t^\top \overline{g}(Z_i, \theta) < b_\varphi^*, \text{ for all } i = 1, \ldots, n \text{ with } \Delta_i = 1\},$$

and

$$\Lambda_\theta := \left\{ t \in \mathbb{R}^{1+\ell} \text{ s.t. } \int_{\mathbb{R}} |\varphi^*(t_0 + \sum_{j=1}^\ell t_j g_j(x, \theta))| \, dP_X(x) < \infty \right\}.$$

Denote, for any p.m. P on $(\mathbb{R}, \mathcal{B}(\mathbb{R}))$ and for any function f integrable with respect to P, $Pf := \int_{\mathbb{R}} f(x) \, dP(x)$. We redefine $\widehat{D}_\varphi(\mathcal{M}_\theta, P_X)$ as follows

$$\widehat{D}_\varphi(\mathcal{M}_\theta, P_X) := \sup_{t \in \Lambda_\theta^{(n)}} \{t_0 - \mathbb{P}_n^{KM} m(\theta, t)\}, \tag{9}$$

and we estimate $D_\varphi(\mathcal{M}, P_X)$ and θ_T, by analogy with the case of complete data (see [4]), by

$$\widehat{D}_\varphi(\mathcal{M}, P_X) := \inf_{\theta \in \Theta} \sup_{t \in \Lambda_\theta^{(n)}} \{t_0 - \mathbb{P}_n^{KM} m(\theta, t)\} \tag{10}$$

and

$$\widehat{\theta}_\varphi := \arg\inf_{\theta \in \Theta} \sup_{t \in \Lambda_\theta^{(n)}} \{t_0 - \mathbb{P}_n^{KM} m(\theta, t)\}. \tag{11}$$

For the particular case of the likelihood divergence, which corresponds to the EL approach in complete data (see [4] Remark 4.4), we have $\varphi(x) := -\log x + x - 1$ and $\varphi^*(x) := -\log(1-x)$. Hence, we obtain the following new version of the EL estimate of θ_T for the present context of censored data

$$\widehat{\theta}_{LD} := \arg\inf_{\theta \in \Theta} \sup_{t \in \Lambda_\theta^{(n)}} \left\{ t_0 + \sum_{i=1}^n w_i \log\left(1 - t^\top \overline{g}(Z_i, \theta)\right) \right\}. \tag{12}$$

4   Asymptotic Properties of the Proposed Estimates

In order to state our results, we need to define the advanced time transformation of a function $h : \mathbb{R} \to \mathbb{R}$ with respect to F_X, introduced by [7]. It is given by

$$\widetilde{h}(x) := \frac{1}{1 - F_X(x)} \int_x^{+\infty} h(u)\, dF_X(u), \ \forall x \in \mathbb{R}.$$

Assumption 1: (a) $P_X \in \mathcal{M}$ and $\theta_T \in \Theta$ is the unique solution in θ of $\mathbb{E}(g(X,\theta)) = 0$; (b) $\Theta \subset \mathbb{R}^d$ is compact; (c) $g(X,\theta)$ is continuous at each $\theta \in \Theta$ with probability one; (d) $\mathbb{E}(\sup_{\theta \in \Theta} \|g(X,\theta)\|^\alpha) < \infty$ for some $\alpha > 2$; (e) the matrix $\Omega := \mathbb{E}(g(X,\theta_T)g(X,\theta_T)^\top)$ is nonsingular; (f) the matrix $V_1 := (v_{i,j}^{(1)})_{1 \le i,j \le \ell}$, where $v_{i,j}^{(1)} := \int_{\mathbb{R}} (g_i(x,\theta_T) - \widetilde{g}_i(x,\theta_T))(g_j(x,\theta_T) - \widetilde{g}_j(x,\theta_T)) \frac{dF_X(x)}{S_R(x^-)}$, is well defined and nonsingular.

Theorem 1 *Under Assumption 1, with probability approaching one as $n \to \infty$, the estimate $\widehat{\theta}_\varphi$ exists, and converges to θ_T in probability. $\frac{1}{n} \sum_{i=1}^n w_i g(Z_i, \widehat{\theta}_\varphi) = O_P(1/\sqrt{n})$, $\widehat{t}(\widehat{\theta}_\varphi) := \arg\sup_{t \in \Lambda_{\widehat{\theta}_\varphi}^{(n)}} \left\{ t_0 - \mathbb{P}_n^{KM} m(\widehat{\theta}_\varphi, t) \right\}$ exists and belongs to $\mathrm{int}(\Lambda_{\widehat{\theta}_\varphi}^{(n)})$ with probability approaching one as $n \to \infty$, and $\widehat{t}(\widehat{\theta}_\varphi) = O_P(1/\sqrt{n})$.*

Additional assumptions are needed to establish the asymptotic normality. Consider the matrices $G := \mathbb{E}(\partial g(X,\theta_T)/\partial\theta), \Sigma := (G^\top \Omega^{-1} G)^{-1}, H := \Sigma G^\top \Omega^{-1}, P := \Omega^{-1} - \Omega^{-1} G \Sigma G^\top \Omega^{-1}$. Denote also $\underline{\widehat{t}}(\widehat{\theta}_\varphi) := (\widehat{t}_1, \ldots, \widehat{t}_\ell)^\top$, where $\widehat{t}_0, \widehat{t}_1, \ldots, \widehat{t}_\ell$ are the components of the vector $\widehat{t}(\widehat{\theta}_\varphi)$. **Assumption 2 :** (a) $\theta_T \in \mathrm{int}(\Theta)$; (b) with probability one, $g(X,\theta)$ is continuously differentiable in a neighborhood N_{θ_T} of θ_T, and $\mathbb{E}(\sup_{\theta \in N_{\theta_T}} \|\partial g(X,\theta)/\partial\theta\|) < \infty$; (c) $\mathrm{rank}(G) = d =: \dim(\Theta)$.

Theorem 2 *Under Assumptions 1 and 2,*

1. *$\sqrt{n}(\widehat{\theta}_\varphi - \theta_T, \underline{\widehat{t}}(\widehat{\theta}_\varphi)^\top)^\top$ converges in distribution to a centered normal random vector with covariance matrix $V := \begin{pmatrix} HV_1H^\top & HV_1P \\ PV_1H^\top & PV_1P \end{pmatrix}$.*

2. *The statistic $2n\widehat{D}_\varphi(\mathcal{M}, P_X)$ converges in distribution to $Y^\top Y$, where Y is a centered normal random vector with covariance matrix $(\Omega^{1/2})^\top PV_1P\Omega^{1/2}$, $\Omega =: \Omega^{1/2}(\Omega^{1/2})^\top$ being the Cholesky decomposition of Ω.*

At a fixed $\theta \in \Theta$, we will show that $\widehat{D}_\varphi(\mathcal{M}_\theta, P_X)$ converges in probability to $D_\varphi(\mathcal{M}_\theta, P_X)$ and we will give the limiting distribution of $\widehat{D}_\varphi(\mathcal{M}_\theta, P_X)$ both when $P_X \in \mathcal{M}_\theta$ and when $P_X \notin \mathcal{M}_\theta$. **Assumption 3 :** (a) $P_X \in \mathcal{M}_\theta$ and

θ is the unique solution of $\mathbb{E}(g(X, \theta)) = 0$; (b) $\mathbb{E}(\|g(X, \theta)\|^{\alpha}) < \infty$ for some $\alpha > 2$; (c) the matrix $\Omega := \mathbb{E}(g(X, \theta)g(X, \theta)^{\top})$ is nonsingular; (d) the matrix $V_1 := (v_{i,j}^{(1)})_{1 \leq i,j \leq \ell}$, where $v_{i,j}^{(1)} := \int_{\mathbb{R}} (g_i(x, \theta) - \widetilde{g}_i(x, \theta))(g_j(x, \theta) - \widetilde{g}_j(x, \theta)) \frac{dF_X(x)}{S_R(x^-)}$, is well defined and nonsingular.

Theorem 3 *Under Assumption 3,*

1. $\widehat{t}(\theta) := \arg\sup_{t \in \Lambda_\theta^{(n)}} \{ t_0 - \mathbb{P}_n^{KM} m(\theta, t) \}$ *exists and belongs to* $int(\Lambda_\theta^{(n)})$ *with probability approaching one as* $n \to \infty$*, and* $\widehat{t}(\theta) = O_P(1/\sqrt{n})$;

2. *The statistic* $2n\widehat{D}_\varphi(\mathcal{M}_\theta, P_X)$ *converges in distribution to* $Y^{\top}Y$*, where* Y *is a centered normal random vector with covariance matrix* $(\Omega^{1/2})^{-1}V_1$ $((\Omega^{1/2})^{-1})^{\top}$.

Remark 1 (Confidence region for the parameter). *Let* $\alpha \in]0, 1[$ *be a fixed level, according to part 2 of this Theorem, the set* $\{\theta \in \Theta$ *s.t.* $2n\widehat{D}_\varphi(\mathcal{M}_\theta, P_X) \leq q_{(1-\alpha)}\}$ *is an asymptotic confidence region for* θ_T*, where* $q_{(1-\alpha)}$ *is the* $(1 - \alpha)$*-quantile of the distribution of* $Y^{\top}Y$*, with* Y *a centered normal random vector with covariance matrix* $(\widehat{\Omega}^{1/2})^{-1}\widehat{V}_1((\widehat{\Omega}^{1/2})^{-1})^{\top}$*, where* $\widehat{\Omega}$ *and* $\widehat{V}_1$ *are the empirical counterparts of* Ω *and* V_1 *respectively. With an appropriate choice of the function* $g(x, \theta)$*, one can construct confidence intervals for the mean, the survival function and the mean residual life of* X*. A simulation study, available from the authors, shows that when we use the likelihood divergence (the modified Kullback-Leibler divergence), these confidence intervals outperform those based on the EL methods of [27], [21] and [23].*

5 Simulation Results

Consider the model $\mathcal{M} := \bigcup_{\theta \in \Theta} \mathcal{M}_\theta$, where $\Theta :=]0, \infty[$ and $g(x, \theta) := (x - \theta, x^2 - 2\theta^2)^{\top}$. The nonnegative random variable X is distributed as $\mathcal{E}(a)$, which belongs to this model with $\theta_T = \mathbb{E}(X) = 1/a$. We compare the estimates $\widehat{\theta}_{LD}$ with the EL estimator based on the method of [27] (which we denote by $\widehat{\theta}_{EL}$) and the plug-in estimator of $\mathbb{E}(X)$, $\widetilde{\theta} := \int_{\mathbb{R}} x \, d\mathbb{P}_n^{KM}(x)$. The variable of censoring R is distributed as $\mathcal{E}(b)$. We take different values of the parameters a and b to get different rates of censoring. For the estimation of θ_T, we generate samples of size $n = 100$ of the latent variables and we calculate the estimates $\widehat{\theta}_{LD}$, $\widehat{\theta}_{EL}$ and $\widetilde{\theta}$ corresponding to each sample. We compute the mean squared error (MSE) for each estimate, based on 1000 replications. We obtain the following results (Table 1).

Table 1. The obtained results for the estimation of the parameter in the exponential model.

a	4	7	3	2
b	1	3	2	3
Rate of censoring	20%	30%	40%	60%
$\theta_T = 1/a$	0.25	0.1429	0.3333	0.5
$\widehat{\theta}_{LD}$	**0.2398**	**0.1307**	**0.3165**	**0.4829**
$\widehat{\theta}_{EL}$	0.2391	0.1302	0.3118	0.4531
$\widetilde{\theta}$	0.2425	0.1343	0.2965	0.3445
$n \times MSE(\widehat{\theta}_{LD})$	**0.0428**	**0.0394**	**0.0578**	**0.0697**
$n \times MSE(\widehat{\theta}_{EL})$	0.0432	0.0410	0.0679	0.2277
$n \times MSE(\widetilde{\theta})$	0.0922	0.0485	0.4446	3.8423

From these results, one can see that our proposed estimate $\widehat{\theta}_{LD}$ outperforms both estimates $\widehat{\theta}_{EL}$ and $\widetilde{\theta}$.

References

1. Bouadoumou, M., Zhao, Y., Lu, Y.: Jackknife empirical likelihood for the accelerated failure time model with censored data. Comm. Statist. Simulation Comput. **44**(7), 1818–1832 (2015). https://doi.org/10.1080/03610918.2013.833234
2. Broniatowski, M., Keziou, A.: Minimization of ϕ-divergences on sets of signed measures. Studia Sci. Math. Hungar. **43**(4), 403–442 (2006). https://doi.org/10.1556/SScMath.43.2006.4.2
3. Broniatowski, M., Keziou, A.: Parametric estimation and tests through divergences and the duality technique. J. Multivariate Anal. **100**(1), 16–36 (2009). https://doi.org/10.1016/j.jmva.2008.03.011
4. Broniatowski, M., Keziou, A.: Divergences and duality for estimation and test under moment condition models. J. Statist. Plann. Inference **142**(9), 2554–2573 (2012). https://doi.org/10.1016/j.jspi.2012.03.013
5. Chen, J., Variyath, A.M., Abraham, B.: Adjusted empirical likelihood and its properties. J. Comput. Graph. Statist. **17**(2), 426–443 (2008). https://doi.org/10.1198/106186008X321068
6. Cherfi, M.: Dual divergences estimation for censored survival data. J. Statist. Plann. Inference **142**(7), 1746–1756 (2012). https://doi.org/10.1016/j.jspi.2012.02.052
7. Efron, B., Johnstone, I.M.: Fisher's information in terms of the hazard rate. Ann. Statist. **18**(1), 38–62 (1990). https://doi.org/10.1214/aos/1176347492
8. Fang, K.T., Li, G., Lu, X., Qin, H.: An empirical likelihood method for semiparametric linear regression with right censored data. Comput. Math. Methods Med. pp. Art. ID 469373, 9 (2013). https://doi.org/10.1155/2013/469373
9. Hansen, L.P.: Large sample properties of generalized method of moments estimators. Econometrica **50**(4), 1029–1054 (1982). https://doi.org/10.2307/1912775

10. He, S., Liang, W.: Empirical likelihood for right censored data with covariables. Sci. China Math. **57**(6), 1275–1286 (2014). https://doi.org/10.1007/s11425-014-4808-0

11. He, S., Liang, W., Shen, J., Yang, G.: Empirical likelihood for right censored lifetime data. J. Amer. Statist. Assoc. **111**(514), 646–655 (2016). https://doi.org/10.1080/01621459.2015.1024058

12. Jeong, J.-H.: Statistical Inference on Residual Life. SBH, Springer, New York (2014). https://doi.org/10.1007/978-1-4939-0005-3

13. Kaplan, E.L., Meier, P.: Nonparametric estimation from incomplete observations. J. Amer. Statist. Assoc. 53, 457–481 (1958), http://links.jstor.org/sici?sici=0162-1459(195806)53:282<457:NEFIO>2.0.CO;2-Z&origin=MSN

14. Keziou, A.: Dual representation of ϕ-divergences and applications. C. R. Math. Acad. Sci. Paris **336**(10), 857–862 (2003). https://doi.org/10.1016/S1631-073X(03)00215-2

15. Keziou, A., Leoni-Aubin, S.: On empirical likelihood for semiparametric two-sample density ratio models. J. Statist. Plann. Inference **138**(4), 915–928 (2008). https://doi.org/10.1016/j.jspi.2007.02.009

16. Newey, W.K., Smith, R.J.: Higher order properties of GMM and generalized empirical likelihood estimators. Econometrica **72**(1), 219–255 (2004). https://doi.org/10.1111/j.1468-0262.2004.00482.x

17. Owen, A.: Empirical likelihood ratio confidence regions. Ann. Statist. **18**(1), 90–120 (1990). https://doi.org/10.1214/aos/1176347494

18. Owen, A.B.: Empirical likelihood ratio confidence intervals for a single functional. Biometrika **75**(2), 237–249 (1988). https://doi.org/10.1093/biomet/75.2.237

19. Owen, A.B.: Empirical Likelihood. Chapman and Hall/CRC (2001)

20. Pan, X.R., Zhou, M.: Empirical likelihood ratio in terms of cumulative hazard function for censored data. J. Multivariate Anal. **80**(1), 166–188 (2002). https://doi.org/10.1006/jmva.2000.1977

21. Qin, G., Zhao, Y.: Empirical likelihood inference for the mean residual life under random censorship. Statist. Probab. Lett. **77**(5), 549–557 (2007). https://doi.org/10.1016/j.spl.2006.09.018

22. Qin, J., Lawless, J.: Empirical likelihood and general estimating equations. Ann. Statist. **22**(1), 300–325 (1994). https://doi.org/10.1214/aos/1176325370

23. Ren, J.J.: Weighted empirical likelihood ratio confidence intervals for the mean with censored data. Ann. Inst. Statist. Math. **53**(3), 498–516 (2001). https://doi.org/10.1023/A:1014612911961

24. Ren, J.J.: Smoothed weighted empirical likelihood ratio confidence intervals for quantiles. Bernoulli **14**(3), 725–748 (2008). https://doi.org/10.3150/08-BEJ129

25. Ren, J.J.: Weighted empirical likelihood in some two-sample semiparametric models with various types of censored data. Ann. Statist. **36**(1), 147–166 (2008). https://doi.org/10.1214/009053607000000695

26. Smith, R.J.: Alternative semi-parametric likelihood approaches to generalized method of moments estimation. Econ. J. **107**(441), 503–519 (1997)

27. Wang, Q.H., Jing, B.Y.: Empirical likelihood for a class of functionals of survival distribution with censored data. Ann. Inst. Statist. Math. **53**(3), 517–527 (2001). https://doi.org/10.1023/A:1014617112870

28. Wu, T.T., Li, G., Tang, C.: Empirical likelihood for censored linear regression and variable selection. Scand. J. Stat. **42**(3), 798–812 (2015). https://doi.org/10.1111/sjos.12137

29. Zheng, J., Shen, J., He, S.: Adjusted empirical likelihood for right censored lifetime data. Statist. Papers **55**(3), 827–839 (2014). https://doi.org/10.1007/s00362-013-0529-7

30. Zhou, M.: Empirical likelihood method in survival analysis. CRC Press, Boca Raton, FL, Chapman & Hall/CRC Biostatistics Series (2016)

Aggregated Tests Based on Supremal Divergence Estimators for Non-regular Statistical Models

Jean-Patrick Baudry[1,3]([✉]), Michel Broniatowski[1], and Cyril Thommeret[1,2]

[1] Sorbonne Université and Université Paris Cité, CNRS, Laboratoire de Probabilités,
Statistique et Modélisation, 75005 Paris, France
`jean-patrick.baudry@sorbonne-universite.fr`
[2] Safran Group, Paris, France
[3] 4 place Jussieu, 75005 Paris, France

Abstract. A methodology is proposed to build statistical test procedures pertaining to models with incomplete information; the lack of information corresponds to a nuisance parameter in the description of the model. The supremal approach based on the dual representation of CASM divergences (or f−divergences) is fruitful; it leads to M-estimators with simple and standard limit distribution, and it is versatile with respect to the choice of the divergence. Duality approaches to divergence-based optimisation are widely considered in statistics, data analysis and machine learning: indeed, they avoid any smoothing or grouping technique which would be necessary for a more direct divergence minimisation approach for the same problem. We are interested in a widely considered but still open problem which consists in testing the number of components in a parametric mixture. Although common, this is still a challenging problem since the corresponding model is non-regular particularly because of the true parameter lying on the boundary of the parameter space. This range of problems has been considered by many authors who tried to derive the asymptotic distribution of some statistic under boundary conditions. The present approach based on supremal divergence M-estimators makes the true parameter an interior point of the parameter space, providing a simple solution for a difficult question. To build a composite test, we aggregate simple tests.

Keywords: Non-regular models · Dual form of f-divergences · Statistical test aggregation · Number of components in mixture models

1 Dual Representation of the φ-Divergences and Tests

We consider CASM divergences (see [15] for definitions and properties):

$$D_\varphi(Q, P) = \begin{cases} \int \varphi(\frac{dQ}{dP})dP & \text{if } Q << P \\ +\infty & \text{otherwise} \end{cases}$$

where Q and P are probability measures on the same probability space. Extensions to divergences between probability measures and signed measures can be found in

© The Author(s), under exclusive license to Springer Nature Switzerland AG 2023
F. Nielsen and F. Barbaresco (Eds.): GSI 2023, LNCS 14071, pp. 136–144, 2023.
https://doi.org/10.1007/978-3-031-38271-0_14

[22]. Dual formulations of divergences can be found in [7,16]. Another interpretation of these formulations can be found in [9, Section 4.6]. They are widely considered in statistics, data analysis and machine learning (see e.g. [4,20]).

As in [1], let $\mathcal{F}$ be some class of $\mathcal{B}$-measurable (borelian) real valued functions and let $\mathcal{M}_{\mathcal{F}} = \{P \in \mathcal{M} : \int |f| dP < \infty, \forall f \in \mathcal{F}\}$ where $\mathcal{M}$ is the space of probability measures. Let any $P^* \in \mathcal{M}$, which shall be the underlying true unknown probability law in a statistical context in the following sections. Assume that φ is differentiable and strictly convex. Then, for all $P \in \mathcal{M}_{\mathcal{F}}$ such that $D_\varphi(P, P^*)$ is finite and $\varphi'(dP/dP^*)$ belongs to $\mathcal{F}$, D_φ admits the dual representation (see Theorem 4.4 in [6]):

$$D_\varphi(P, P^*) = \sup_{f \in \mathcal{F}} \int f dP - \int \varphi^\#(f) dP^*, \tag{1}$$

where $\varphi^\#(x) = \sup_{t \in \mathbb{R}} tx - \varphi(t)$ is the Fenchel-Legendre convex conjugate. Moreover, the supremum is uniquely attained at $f = \varphi'(dP/dP^*)$.

This result can be used in two directions. First, a statistical model, e.g. a parametrical model $\{P_\theta : \theta \in \Theta\}$ with P_θ is absolutely continuous with respect to some dominating measure μ for any θ, naturally induces a family $\mathcal{F} = \{\varphi'(p_\theta/p_{\theta'}) : \theta, \theta' \in \Theta\}$. This is the main framework of this paper.

Conversely, a class of functions $\mathcal{F}$ defines the distribution pairs P and Q that can be compared, which are these such that $\varphi'(dP/dQ) \in \mathcal{F}$. Furthermore it induces a divergence D_φ on these pairs. A typical example is the logistic model.

The KLm divergence is defined by the generator $\varphi : x \in \mathbb{R} \mapsto -logx + x - 1$ and leads to the maximum likelihood estimator for both forms of estimation for the supremal estimator, once of which is defined bellow (see Remark 3.2 in [7]).

We consider in this paper the problem of testing the number of components in a mixture model. This question has been considered by various authors. [2, 10,12,14,17] have considered likelihood ratio tests and showed some difficulties with those due to the fact that the likelihood ratio statistic is unbounded with respect to n. [17] prove that its distribution is driven by a $\log \log n$ term in a specific simple Gaussian mixture model. The test statistic needs to be calibrated in accordance with this result. But first, as stated by [17], the convergence to the limit distribution is extremely slow, making this result unpractical. And second, it seems very difficult to derive the corresponding term for a different model, and even more so for a general situation.

Our approach to this problem is suggested by the dual representation of the divergence. For the KLm divergence, it amounts to considering the maximum likelihood estimator itself as a test statistic instead of the usual maximum value of the likelihood function. This leads to a well-defined limiting distribution for the test statistic under the null. This holds for a class of estimators obtained by substituting KLm by any regular divergence. This approach also eliminates the curse of irregularity encountered by many authors for the problem of testing the number of components in a mixture.

Since we are interested in composite hypotheses, there is no justification in this context that the likelihood ratio test would be the best (in terms of uniform power) as is usually considered (e.g. [8,18]) and [13] showed what difficulties likelihood ratio tests can encounter in this context.

[8] considered tests based on an estimation of the minimum divergence between the true distribution and the null model. In we make use of the unicity of the optimiser of the dual representation of the divergence in (1) and of the supremal divergence estimator introduced by [7]. An immediate practical advantage of this choice as compared to estimating the minimum divergence is that one less optimisation is needed. Moreover [23] showed that this estimator is robust for several choices of the divergence.

Our procedure for composite hypotheses consists in the aggregation of simple tests in the spirit of [11]. [5] used a similar aggregation procedure for testing between two distributions under noisy data and obtained some control of the resulting test power.

2 Notation and Hypotheses

Let $\{f_1(\,.\,;\theta_1) : \theta_1 \in \Theta_1\}$, $\Theta_1 \subset \mathbb{R}^p$, and $\{f_2(\,.\,;\theta_2) : \theta_2 \in \Theta_2\}$, $\Theta_2 \subset \mathbb{R}^q$, be probability density families with respect to a σ-finite measure λ on $(\mathcal{X}, \mathcal{B})$. For some fixed open interval $]a, b[\ni 0$, let $\Theta \subset]a, b[\times\Theta_1 \times \Theta_2$, and

$$g_{\pi,\theta} = (1 - \pi)f_1(\,.\,;\theta_1) + \pi f_2(\,.\,;\theta_2)$$

for any $(\pi, \theta) \in \Theta$ with $\theta = (\theta_1, \theta_2)$.

Assume that $x_1, \ldots, x_n \in \mathbb{R}$ have been observed and they are modelled as a realisation of the i.i.d. sample $X_1, \ldots, X_n$ which distribution $\mathbb{P}^* := g_{\pi^*,\theta^*}.\lambda$ is known up to the parameters $(\pi^*, \theta^*) \in \Theta$. Our aim is to test the hypothesis $H_0 : \pi^* = 0$.

Assume that $g_{\pi,\theta} = g_{\pi^*,\theta^*} \Rightarrow \pi = \pi^*, \theta_1 = \theta_1^*$ and, if $\pi^* \neq 0$, $\theta_2 = \theta_2^*$.

Let g be a probability density with respect to λ such that $Supp(g) \subset Supp(g_{\pi,\theta})$ for any $(\pi, \theta) \in \Theta$ such that

$$\forall (\pi, \theta) \in \Theta, \int \left| \varphi'(\frac{g}{g_{\pi,\theta}}) \right| g d\lambda < \infty.$$

Let us define for any $(\pi, \theta) \in \Theta$,

$$m_{\pi,\theta} : x \in \mathcal{X} \mapsto \int \varphi'\left(\frac{g}{g_{\pi,\theta}}\right) g d\lambda - \varphi^\#\left(\frac{g}{g_{\pi,\theta}}\right)(x)$$

and assume that $(\pi, \theta) \mapsto m_{\pi,\theta}(x)$ is continuous for any $x \in \mathcal{X}$. Let us also assume that

$$\forall (\tilde{\pi}, \tilde{\theta}) \in \Theta, \exists r_0 > 0 / \forall r \leq r_0, \ P^* \Big| \sup_{d((\tilde{\pi},\tilde{\theta}),(\pi,\theta))<r} m_{\pi,\theta} \Big| < \infty$$

where $d(\cdot, \cdot)$ denotes the Euclidean distance and where, as usual, the operator-type notation $\mathbb{P}^*Y$ denotes the expectation—with respect to the probability measure $\mathbb{P}^*$—of the random variable Y.

Theorem 1. *For any* $(\pi^*, \theta^*) \in \Theta$

$$D_\varphi(g.\lambda, g_{\pi^*,\theta^*}.\lambda) = \sup_{(\pi,\theta)\in\Theta} P^* m_{\pi,\theta},$$

which we call the supremal form *of the divergence. Moreover attainment holds uniquely at* $(\pi, \theta) = \overline{(\pi^*, \theta^*)}$.

Definition 1. *Let* $\mathbb{P}_n$ *denote the empirical measure pertaining to the sample* $X_1, \ldots, X_n$. *Define*

$$(\hat{\pi}, \hat{\theta}) := \arg\max_{(\pi,\theta)} \mathbb{P}_n m_{\pi,\theta}$$

the supremal estimator *of* (π^*, θ^*).

The existence of $(\hat{\pi}, \hat{\theta})$ can be guaranteed by assuming that Θ is compact. When uniqueness does not hold, consider any maximizer. This class of estimators has been introduced in [7], under the name dual φ-divergence estimators.

3   Consistency of the Supremal Divergence Estimator

Let us first state the consistency of the supremal divergence estimator of the proportion and the parameters of the existing component, when the non-existing component parameters are fixed, uniformly over the latter.

Here and below, by abuse of notation, we let $\varphi'\left(\frac{g}{g_{\pi,\theta}}\right)$ stand for $x \mapsto \varphi'\left(\frac{g(x)}{g_{\pi,\theta}(x)}\right)$, and so on.

Remark that, for $\pi^* = 0$ and any $\theta_1^* \in \Theta_1$ and $\theta_2 \in \Theta_2$, we can unambiguously write m_{π^*,θ_1^*} for $m_{\pi^*,\theta_1^*,\theta_2}$ since the parameter θ_2 is not involved in the expression of $m_{0,\theta_1^*,\theta_2}$.

Theorem 2. *Assume that* $\pi^* = 0$ *and let for any* $\theta_2 \in \Theta_2$, $(\hat{\pi}(\theta_2), \hat{\theta}_1(\theta_2)) \in$ $]a, b[\times\Theta_1$ *such that*

$$\inf_{\theta_2\in\Theta_2} \mathbb{P}_n m_{\hat{\pi}(\theta_2),\hat{\theta}_1(\theta_2),\theta_2} \geq \mathbb{P}_n m_{\pi^*,\theta_1^*} - o_{P^*}(1). \tag{2}$$

Then

$$\sup_{\theta_2\in\Theta_2} d\big((\hat{\pi}(\theta_2), \hat{\theta}_1(\theta_2)), (0, \theta_1^*)\big) \xrightarrow[n\to\infty]{P^*} 0.$$

The convergence holds a.s. in the particular case of (2) *when, a.s.,*

$$\forall \theta_2 \in \Theta_2, (\hat{\pi}(\theta_2), \hat{\theta}_1(\theta_2)) \in \operatorname*{argmax}_{(\pi,\theta_1)\in]a,b[\times\Theta_1} \mathbb{P}_n m_{\pi,\theta_1,\theta_2}. \tag{3}$$

4   Asymptotic Distribution of the Supremal Divergence Estimator

Under H_0 ($\pi^* = 0$), the joint asymptotic distribution of $(\hat{\pi}(\theta_2), \hat{\pi}(\theta_2'))$ is provided by the following theorem. The interior of Θ will be denoted by $\overset{\circ}{\Theta}$.

Theorem 3. *Let $\theta_2 \in \Theta_2$ and $\theta_2' \in \Theta_2$ such that $(\pi^*, \theta_1^*, \theta_2) \in \overset{\circ}{\Theta}$ and $(\pi^*, \theta_1^*, \theta_2') \in \overset{\circ}{\Theta}$. Write*

$$\Theta(\theta_2) = \{(\pi, \theta_1) \in] - \infty, 1[\times \Theta_1 : (\pi, \theta_1, \theta_2) \in \Theta\}$$
$$\Theta(\theta_2') = \{(\pi, \theta_1) \in] - \infty, 1[\times \Theta_1 : (\pi, \theta_1, \theta_2') \in \Theta\},$$

and let $(\hat{\pi}, \hat{\theta}_1)$ and $(\hat{\pi}', \hat{\theta}_1')$ be such that

$$(\hat{\pi}, \hat{\theta}_1) \in \underset{(\pi, \theta_1) \in \Theta(\theta_2)}{\mathrm{argmax}} \; P_n m_{\pi, \theta_1, \theta_2}$$

$$(\hat{\pi}', \hat{\theta}_1') \in \underset{(\pi, \theta_1) \in \Theta(\theta_2')}{\mathrm{argmax}} \; P_n m_{\pi, \theta_1, \theta_2'}.$$

Assume that $\pi^* = 0$.
Moreover, assume that :

– *$(\pi, \theta_1) \in \Theta(\theta_2) \mapsto m_{\pi, \theta_1, \theta_2}(x)$ (resp. $(\pi, \theta_1) \in \Theta(\theta_2') \mapsto m_{\pi, \theta_1, \theta_2'}(x)$) is differentiable λ-a.e. with derivative $\psi_{\pi, \theta_1} = \left(\begin{smallmatrix} \frac{\partial}{\partial \pi} m_{\pi, \theta_1, \theta_2} \\ \frac{\partial}{\partial \theta_1} m_{\pi, \theta_1, \theta_2} \end{smallmatrix} \right)_{|(\pi, \theta_1)}$ (resp. $\psi_{\pi, \theta_1}' = \left(\begin{smallmatrix} \frac{\partial}{\partial \pi} m_{\pi, \theta_1, \theta_2'} \\ \frac{\partial}{\partial \theta_1} m_{\pi, \theta_1, \theta_2'} \end{smallmatrix} \right)_{|(\pi, \theta_1)}$) such that $P^* \psi_{\pi^*, \theta_1^*} = 0$ (resp. $P^* \psi_{\pi^*, \theta_1^*}' = 0$).*

– *$(\pi, \theta_1) \in \Theta(\theta_2) \mapsto P^* \psi_{\pi, \theta_1}$ (resp. $(\pi, \theta_1) \in \Theta(\theta_2') \mapsto P^* \psi_{\pi, \theta_1}'$) is differentiable at π^*, θ_1^* with invertible derivative matrix $H = D(P^* \psi)_{|\left(\begin{smallmatrix} \pi^* \\ \theta_1^* \end{smallmatrix} \right)}$ (resp. $H' = D(P^* \psi')_{|\left(\begin{smallmatrix} \pi^* \\ \theta_1^* \end{smallmatrix} \right)}$).*

– *$\{\psi_{\pi, \theta_1} : (\pi, \theta_1) \in \Theta(\theta_2)\}$ and $\{\psi_{\pi, \theta_1}' : (\pi, \theta_1) \in \Theta(\theta_2')\}$ are P^*-Donsker.*

– *$\int (\psi_{\hat{\pi}, \hat{\theta}_1}(x) - \psi_{\pi^*, \theta_1^*}(x))^2 dP^*(x) \xrightarrow{P^*} 0$ and $\int (\psi_{\hat{\pi}', \hat{\theta}_1'}'(x) - \psi_{\pi^*, \theta_1^*}'(x))^2 dP^*(x) \xrightarrow{P^*} 0$.*

Assume that $H = D(P^* \psi)_{|\left(\begin{smallmatrix} \pi^* \\ \theta_1^* \end{smallmatrix} \right)} = P^* D^2(h)_{|\left(\begin{smallmatrix} \pi^* \\ \theta_1^* \end{smallmatrix} \right)}$ (resp. $H' = D(P^* \psi')_{|\left(\begin{smallmatrix} \pi^* \\ \theta_1^* \end{smallmatrix} \right)} = P^* D^2(h')_{|\left(\begin{smallmatrix} \pi^* \\ \theta_1^* \end{smallmatrix} \right)}$) with $P^* |D^2(h)_{|\left(\begin{smallmatrix} \pi^* \\ \theta_1^* \end{smallmatrix} \right)}| < \infty$ (resp. $P^* |D^2(h')_{|\left(\begin{smallmatrix} \pi^* \\ \theta_1^* \end{smallmatrix} \right)}| < \infty$) where $h : (\pi, \theta_1) \in \Theta(\theta_2) \mapsto m_{\pi, \theta_1, \theta_2}(x)$ (resp. $h' : (\pi, \theta_1) \in \Theta(\theta_2') \mapsto m_{\pi, \theta_1, \theta_2'}(x)$).

Then with a_n (resp. a_n') being the $(1,1)$-entry of the matrix $H_n^{-1} \cdot \left(P_n \psi_{\hat{\pi}, \hat{\theta}_1} \psi_{\hat{\pi}, \hat{\theta}_1}^T \right) \cdot H_n^{-1}$ (resp. of $H_n'^{-1} \cdot \left(P_n \psi_{\hat{\pi}, \hat{\theta}_1}' \psi_{\hat{\pi}, \hat{\theta}_1}'^T \right) \cdot H_n'^{-1}$)—where H_n (resp. H_n') denotes the Hessian matrix of $(\pi, \theta_1) \mapsto P_n m_{\pi, \theta_1, \theta_2}$ (resp. of $(\pi, \theta_1) \mapsto P_n m_{\pi, \theta_1, \theta_2'}$) at the point $(\hat{\pi}, \hat{\theta}_1)$, which is supposed to be invertible with high probability—one gets

$$\begin{pmatrix} \sqrt{\frac{n}{a_n}}(\hat{\pi} - \pi^*) \\ \sqrt{\frac{n}{a_n'}}(\hat{\pi}' - \pi^*) \end{pmatrix} \xrightarrow{\mathcal{L}} \mathcal{N}(0, U)$$

with

$$U = \begin{pmatrix} 1 & \frac{b}{\sqrt{aa'}} \\ \frac{b}{\sqrt{aa'}} & 1 \end{pmatrix} \tag{4}$$

where b is the $(1,1)$-entry of the matrix $H^{-1} \cdot \left(\mathbb{P}\psi_{\pi^*, \theta_1^*} \psi_{\pi^*, \theta_1^*}^T \right) \cdot H'^{-1}$, and a (resp. a') the $(1,1)$-entry of the matrix $H^{-1} \cdot \left(\mathbb{P}\psi_{\pi^*, \theta_1^*} \psi_{\pi^*, \theta_1^*}^T \right) \cdot H^{-1}$ (resp. of $H'^{-1} \cdot \left(\mathbb{P}\psi'_{\pi^*, \theta_1^*} \psi'^T_{\pi^*, \theta_1^*} \right) \cdot H'^{-1}$).

This result naturally generalises to k-tuples. The marginal result for θ actually also holds when $\pi^* > 0$ and $\theta_2 = \theta_2^*$, which is useful to control the power of the test procedure to be defined.

Let us consider as a test statistic $T_n = \sup_{\theta_2} \sqrt{\frac{n}{a_n}} \hat{\pi}$ and let us reject H_0 when T_n takes large values. It seems sensible to reduce $\hat{\pi}$ for each value of θ_2 so that, under H_0, it is asymptotically distributed as a $\mathcal{N}(0,1)$ and that the (reduced) values of $\hat{\pi}$ for different values of θ_2 can be compared. In practice, the asymptotic variance has to be estimated hence the substitution of $\hat{\pi}$, $\hat{\theta}_1$, and P_n for π^*, θ_1^*, and P^* in $H^{-1} P^* \psi_{\pi^*, \theta_1^*} \psi_{\pi^*, \theta_1^*}^T H^{-1}$. This choice is justified in [19].

The Bonferoni aggregation rule is not sensible here since the tests for different values of θ_2 are obviously not independent so that such a procedure would lead to a conservative test. Hence the need in Theorem 3 for the joint asymptotic distribution to take the dependence between $\hat{\pi}$ for different values of θ_2. This leads to the study of the asymptotic distribution of T_n which should be the distribution of $\sup W$ where W is a Gaussian process which covariance structure is given by Theorem 3. This will be proved in the forthcoming section.

5 Asymptotic Distribution of the Supremum of Supremal Divergence Estimators

H_0 is assumed to hold in this section.

It is stated that the asymptotic distribution of T_n is that of the supremum of a Gaussian process with the covariance $\frac{b}{\sqrt{aa'}}$, as in (4).

Then it is stated that the distribution of the latter can be approximated by maximising the Gaussian process with the covariance $\frac{b_n}{\sqrt{a_n a'_n}}$, where a_n, a'_n, and b_n are estimations of the corresponding quantities, on a finite grid of values for θ_2.

Let X be the centred Gaussian process over Θ_2 with

$$\forall \theta_2, \theta'_2 \in \Theta_2, r(\theta_2, \theta'_2) = \mathrm{Cov}(X_{\theta_2}, X_{\theta'_2}) = \frac{b}{\sqrt{aa'}}$$

where a and b are defined in Theorem 3.

Theorem 4. *Under general regularity conditions pertaining to the class of derivatives of m (Glivenko-Cantelli classes), we have*

$$\sqrt{\frac{n}{a_n}} (\hat{\pi}_n - \pi^*) \xrightarrow{\mathcal{L}} X.$$

This results from [21] when $dim(\Theta_2) = 1$ and [24] when $dim(\Theta_2) > 1$.

Theorem 5. *Under the same general regularity conditions as above, we have*

$$T_n \xrightarrow{\mathcal{L}} \sup_{\theta_2 \in \Theta_2} X(\theta_2).$$

The proof of the last result when $dim(\Theta_2) = 1$ makes use of the fact that $\theta_2 \mapsto \hat{\pi}(\theta_2)$ is cadlag ([21]). This is a reasonable assumption, which holds in the examples which we considered. We are eager for counter-examples! When $dim(\Theta_2) > 1$, the result holds also by [3].

Let now X^n be the centred Gaussian process over Θ_2 with

$$\forall \theta_2, \theta_2' \in \Theta_2, \operatorname{Cov}(X^n_{\theta_2}, X^n_{\theta_2'}) = \frac{b_n}{\sqrt{a_n a_n'}}$$

where a_n, a_n' are defined in Theorem 3 and b_n is defined analogously.

Theorem 6. *Let, for any $\delta > 0$, Θ_2^δ be a finite set such that $\forall \theta_2 \in \Theta_2, \exists \tilde{\theta}_2 \in \Theta_2^\delta / \|\theta_2 - \tilde{\theta}_2\| \leq \delta$. Then*

$$M_n^\delta = \sup_{\theta_2 \in \Theta_2^\delta} X^n_{\theta_2} \xrightarrow[\substack{n \to \infty \\ \delta \to 0}]{\mathcal{L}} M = \sup_{\theta_2 \in \Theta_2} X_{\theta_2}.$$

6   Algorithm

Our algorithm for testing that the data was sampled from a single-component mixture ($H_0 : \pi^* = 0$) against a two-component mixture ($H_1 : \pi^* > 0$) is presented in Algorithm 1.

In this algorithm, $\hat{\pi}(\theta_2)$ is defined in (2) and (3). It depends on g. This Theorems hold as long as g fulfils $Supp(g) \subset Supp(g_{\pi,\theta})$ for any $(\pi, \theta) \in \Theta$. However it has to be chosen with care. The constants in the asymptotic distribution in Theorem 3 depend on it. Moreover [23] argue that the choice of g can influence the robustness properties of the procedure.

The choice of φ is also obviously crucial (see also [23] for the induced robustness properties).

The choice of φ and g are important practical questions which are work in progress.

As already stated, the supremal estimator for the modified Kullback-Leibler divergence $\varphi : x \in \mathbb{R}^{+*} \mapsto -\log x + x - 1$ is the usual maximum likelihood estimator. In this instance the estimator does not depend on g.

Algorithm 1: Test of H_0: one component vs H_1: two components

---

Input : φ, $\{f_1(\,.\,;\theta_1) : \theta_1 \in \Theta_1\}$, $\{f_2(\,.\,;\theta_2) : \theta_2 \in \Theta_2\}$, n, K, Θ_2^δ, $p \in [0,1]$

1. let $t = \sup_{\theta_2 \in \Theta_2} \sqrt{\frac{n}{a_n(\theta_2)}} \hat{\pi}(\theta_2)$
2. for $k \in \{1, \ldots, K\}$
   (a) sample $(X_t)_{t \in \Theta_2^\delta} \sim \mathcal{N}\big(0, (\frac{b_n(t,t')}{\sqrt{a_n(t)a_n(t')}})_{t,t' \in \Theta_2^\delta}\big)$
   (b) let $\tilde{t}_k = \max_{t \in \Theta_2^\delta} x_t$
3. if $t \geq$ empirical_quantile$((\tilde{t}_k)_{k \in \{1,\ldots,K\}}, 1 - p)$ reject H_0 else don't reject H_0 vs H_1.

Acknowledgements. The authors gratefully acknowledge the reviewers for their valuable comments and suggestions which helped improving the article.

References

1. Al Mohamad, D.: Towards a better understanding of the dual representation of phi divergences. Stat. Pap. **59**(3), 1205–1253 (2016). https://doi.org/10.1007/s00362-016-0812-5
2. Bickel, P.J., Chernoff, H.: Asymptotic distribution of the likelihood ratio statistic in a prototypical non regular problem. In: Statistics and Probability: A Raghu Raj Bahadur Festschrift, pp. 83–96 (1993)
3. Bickel, P.J., Wichura, M.J.: Convergence criteria for multiparameter stochastic processes and some applications. Ann. Math. Stat. **42**(5), 1656–1670 (1971)
4. Birrell, J., Dupuis, P., Katsoulakis, M.A., Pantazis, Y., Rey-Bellet, L.: (f, γ)-divergences: interpolating between f-divergences and integral probability metrics. J. Mach. Learn. Res. **23**(1), 1816–1885 (2022)
5. Broniatowski, M., Jurečková, J., Moses, A.K., Miranda, E.: Composite tests under corrupted data. Entropy **21**(1), 63 (2019)
6. Broniatowski, M., Keziou, A.: Minimization of φ-divergences on sets of signed measures. Studia Scientiarum Mathematicarum Hungarica **43**(4), 403–442 (2006)
7. Broniatowski, M., Keziou, A.: Parametric estimation and tests through divergences and the duality technique. J. Multivar. Anal. **100**(1), 16–36 (2009)
8. Broniatowski, M., Miranda, E., Stummer, W.: Testing the number and the nature of the components in a mixture distribution. In: Nielsen, F., Barbaresco, F. (eds.) GSI 2019. LNCS, vol. 11712, pp. 309–318. Springer, Cham (2019). https://doi.org/10.1007/978-3-030-26980-7_32
9. Broniatowski, M., Stummer, W.: Some universal insights on divergences for statistics, machine learning and artificial intelligence. In: Geometric Structures of Information, pp. 149–211 (2019)
10. Feng, Z., McCulloch, C.E.: Statistical inference using maximum likelihood estimation and the generalized likelihood ratio when the true parameter is on the boundary of the parameter space. Stat. Probabil. Lett. **13**(4), 325–332 (1992)
11. Garel, B.: Asymptotic theory of the likelihood ratio test for the identification of a mixture. J. Stat. Plan. Infer. **131**(2), 271–296 (2005)
12. Ghosh, J.K., Sen, P.K.: On the asymptotic performance of the log likelihood ratio statistic for the mixture model and related results. Technical report, North Carolina State University. Department of Statistics (1984)

13. Hall, P., Stewart, M.: Theoretical analysis of power in a two-component normal mixture model. J. Stat. Plan. Infer. **134**(1), 158–179 (2005)
14. Hartigan, J.A.: Statistical theory in clustering. J. Classif. **2**(1), 63–76 (1985)
15. Liese, F., Vajda, I.: Convex statistical distances, vol. 95. Teubner (1987)
16. Liese, F., Vajda, I.: On divergences and informations in statistics and information theory. IEEE Trans. Inf. Theory **52**(10), 4394–4412 (2006)
17. Liu, X., Shao, Y.: Asymptotics for the likelihood ratio test in a two-component normal mixture model. J. Stat. Plan. Infer. **123**(1), 61–81 (2004)
18. Lo, Y., Mendell, N.R., Rubin, D.B.: Testing the number of components in a normal mixture. Biometrika **88**(3), 767–778 (2001)
19. McLachlan, G., Peel, D.: Finite Mixture Models. Wiley, New York (2000)
20. Nguyen, X., Wainwright, M.J., Jordan, M.I.: Estimating divergence functionals and the likelihood ratio by convex risk minimization. IEEE Trans. Inf. Theory **56**(11), 5847–5861 (2010)
21. Pollard, D.: Convergence of Stochastic Processes. Springer, Heidelberg (2012). https://doi.org/10.1007/978-1-4612-5254-2
22. Rüschendorf, L.: On the minimum discrimination information theorem. In: Statistical Decisions, pp. 263–283 (1984)
23. Toma, A., Broniatowski, M.: Dual divergence estimators and tests: robustness results. J. Multivar. Anal. **102**(1), 20–36 (2011)
24. van der Vaart, A., Wellner, J.A.: Weak Convergence and Empirical Processes. Springer, Heidelberg (1996). https://doi.org/10.1007/978-1-4757-2545-2

Computational Information Geometry
and Divergences

Quasi-arithmetic Centers, Quasi-arithmetic Mixtures, and the Jensen-Shannon ∇-Divergences

Frank Nielsen$^{(\boxtimes)}$ (iD)

Sony Computer Science Laboratories Inc., Tokyo, Japan
Frank.Nielsen@acm.org

Abstract. We first explain how the information geometry of Bregman manifolds brings a natural generalization of scalar quasi-arithmetic means that we term quasi-arithmetic centers. We study the invariance and equivariance properties of quasi-arithmetic centers from the viewpoint of the Fenchel-Young canonical divergences. Second, we consider statistical quasi-arithmetic mixtures and define generalizations of the Jensen-Shannon divergence according to geodesics induced by affine connections.

Keywords: Legendre-type function · quasi-arithmetic means · co-monotonicity · information geometry · statistical mixtures · Jensen-Shannon divergence

1 Introduction

Let $\Delta_{n-1} = \{(w_1, \ldots, w_n) : w_i \geq 0, \sum_i w_i = 1\} \subset \mathbb{R}^d$ denotes the closed $(n-1)$-dimensional standard simplex sitting in $\mathbb{R}^n$, ∂ be the set boundary operator, and $\Delta_{n-1}^\circ = \Delta_{n-1} \backslash \partial \Delta_{n-1}$ the open standard simplex. Weighted quasi-arithmetic means [12] (QAMs) generalize the ordinary weighted arithmetic mean $A(x_1, \ldots, x_n; w) = \sum_i w_i x_i$ as follows:

Definition 1 (Weighted quasi-arithmetic mean (1930's)). *Let $f : I \subset \mathbb{R} \to \mathbb{R}$ be a strictly monotone and differentiable real-valued function. The weighted quasi-arithmetic mean (QAM) $M_f(x_1, \ldots, x_n; w)$ between n scalars $x_1, \ldots, x_n \in I \subset \mathbb{R}$ with respect to a normalized weight vector $w \in \Delta_{n-1}$, is defined by*

$$M_f(x_1, \ldots, x_n; w) := f^{-1}\left(\sum_{i=1}^n w_i f(x_i)\right).$$

Let us write for short $M_f(x_1, \ldots, x_n) := M_f(x_1, \ldots, x_n; \frac{1}{n}, \ldots, \frac{1}{n})$, and $M_{f,\alpha}(x, y) := M_f(x, y; \alpha, 1 - \alpha)$ for $\alpha \in [0, 1]$, the weighted bivariate QAM. A QAM satisfies the in-betweenness property:

$$\min\{x_1, \ldots, x_n\} \leq M_f(x_1, \ldots, x_n; w) \leq \max\{x_1, \ldots, x_n\},$$

© The Author(s), under exclusive license to Springer Nature Switzerland AG 2023
F. Nielsen and F. Barbaresco (Eds.): GSI 2023, LNCS 14071, pp. 147–156, 2023.
https://doi.org/10.1007/978-3-031-38271-0_15

and we have [16] $M_g(x, y) = M_f(x, y)$ if and only if $g(t) = \lambda f(t) + c$ for $\lambda \in \mathbb{R} \backslash \{0\}$ and $c \in \mathbb{R}$. The power means $M_p(x, y) := M_{f_p}(x, y)$ are obtained for the following continuous family of QAM generators indexed by $p \in \mathbb{R}$:

$$f_p(t) = \begin{cases} \frac{t^p - 1}{p}, & p \in \mathbb{R} \backslash \{0\}, \\ \log(t), & p = 0. \end{cases}, \quad f_p^{-1}(t) = \begin{cases} (1 + tp)^{\frac{1}{p}}, & p \in \mathbb{R} \backslash \{0\}, \\ \exp(t), & p = 0. \end{cases},$$

Special cases of the power means are the harmonic mean $(H = M_{-1})$, the geometric mean $(G = M_0)$, the arithmetic mean $(A = M_1)$, and the quadratic mean also called root mean square $(Q = M_2)$. A QAM is said positively homogeneous if and only if $M_f(\lambda x, \lambda y) = \lambda M_f(x, y)$ for all $\lambda > 0$. The power means M_p are the only positively homogeneous QAMs [12].

In Sect. 2, we define a generalization of quasi-arithmetic means called quasi-arithmetic centers (Definition 3) induced by a Legendre-type function. We show that the gradient maps of convex conjugate functions are co-monotone (Proposition 1). We then study their invariance and equivariance properties (Proposition 2). In Sect. 4, we define quasi-arithmetic mixtures (Definition 4), show their connections to geodesics, and define a generalization of the Jensen-Shannon divergence with respect to affine connections (Definition 5).

2   Quasi-arithmetic Centers and Information Geometry

2.1   Quasi-arithmetic Centers

To generalize scalar QAMs to other non-scalar types such as vectors or matrices, we face two difficulties:

1. we need to ensure that the generator $G : \mathbb{X} \to \mathbb{R}$ admits a global inverse[1] G^{-1}, and
2. we would like the smooth function G to bear a generalization of monotonicity of univariate functions.

We consider a well-behaved class $\mathcal{F}$ of non-scalar functions G (i.e., vector or matrix functions) which admits global inverse functions G^{-1} belonging to the same class $\mathcal{F}$: Namely, we consider the gradient maps of Legendre-type functions where Legendre-type functions are defined as follows:

Definition 2 (Legendre type function [24]). (Θ, F) *is of Legendre type if the function* $F : \Theta \subset \mathbb{X} \to \mathbb{R}$ *is strictly convex and differentiable with* $\Theta \neq \emptyset$ *an open convex set and*

$$\lim_{\lambda \to 0} \frac{d}{d\lambda} F(\lambda \theta + (1 - \lambda)\bar{\theta}) = -\infty, \quad \forall \theta \in \Theta, \forall \bar{\theta} \in \partial \Theta. \tag{1}$$

---

[1] The inverse function theorem [10,11] in multivariable calculus states only the local existence of an inverse continuously differentiable function G^{-1} for a multivariate function G provided that the Jacobian matrix of G is not singular.

Legendre-type functions $F(\Theta)$ admits a convex conjugate $F^*(\eta)$ of Legendre type via the Legendre transform (Theorem 1 [24]):

$$F^*(\eta) = \langle \nabla F^{-1}(\eta), \eta \rangle - F(\nabla F^{-1}(\eta)),$$

where $\langle \theta, \eta \rangle$ denotes the inner product in $\mathbb{X}$ (e.g., Euclidean inner product $\langle \theta, \eta \rangle = \theta^\top \eta$ for $\mathbb{X} = \mathbb{R}^d$, the Hilbert-Schmidt inner product $\langle A, B \rangle := \text{tr}(AB^\top)$ where $\text{tr}(\cdot)$ denotes the matrix trace for $\mathbb{X} = \text{Mat}_{d,d}(\mathbb{R})$, etc.), and $\eta \in H$ with H the image of the gradient map $\nabla F : \Theta \to H$. Moreover, we have $\nabla F^* = (\nabla F)^{-1}$ and $\nabla F = (\nabla F^*)^{-1}$, i.e., gradient maps of conjugate functions are reciprocal to each others.

The gradient of a strictly convex function of Legendre type exhibit a generalization of the notion of monotonicity of univariate functions: A function $G : \mathbb{X} \to \mathbb{R}$ is said strictly increasing co-monotone if

$$\forall \theta_1, \theta_2 \in \mathbb{X}, \theta_1 \neq \theta_2, \quad \langle \theta_1 - \theta_2, G(\theta_1) - G(\theta_2) \rangle > 0.$$

and strictly decreasing co-monotone if $-G$ is strictly increasing co-monotone.

Proposition 1 (Gradient co-monotonicity [25]). *The gradient functions* $\nabla F(\theta)$ *and* $\nabla F^*(\eta)$ *of the Legendre-type convex conjugates F and F^* in $\mathcal{F}$ are strictly increasing co-monotone functions.*

Proof. We have to prove that

$$\langle \theta_2 - \theta_1, \nabla F(\theta_2) - \nabla F(\theta_1) \rangle > 0, \quad \forall \theta_1 \neq \theta_2 \in \Theta \qquad (2)$$
$$\langle \eta_2 - \eta_1, \nabla F^*(\eta_2) - \nabla F^*(\eta_1) \rangle > 0, \quad \forall \eta_1 \neq \eta_2 \in H \qquad (3)$$

The inequalities follow by interpreting the terms of the left-hand-side of Eq. 2 and Eq. 3 as Jeffreys-symmetrization [17] of the dual Bregman divergences [9] B_F and B_{F^*}:

$$B_F(\theta_1 : \theta_2) = F(\theta_1) - F(\theta_2) - \langle \theta_1 - \theta_2, \nabla F(\theta_2) \rangle \geq 0,$$
$$B_{F^*}(\eta_1 : \eta_2) = F^*(\eta_1) - F^*(\eta_2) - \langle \eta_1 - \eta_2, \nabla F^*(\eta_2) \rangle \geq 0,$$

where the first equality holds if and only if $\theta_1 = \theta_2$ and the second inequality holds iff $\eta_1 = \eta_2$. Indeed, we have the following Jeffreys-symmetrization of the dual Bregman divergences:

$$B_F(\theta_1 : \theta_2) + B_F(\theta_2 : \theta_1) = \langle \theta_2 - \theta_1, \nabla F(\theta_2) - \nabla F(\theta_1) \rangle > 0, \quad \forall \theta_1 \neq \theta_2$$
$$B_{F^*}(\eta_1 : \eta_2) + B_{F^*}(\eta_2 : \eta_1) = \langle \eta_2 - \eta_1, \nabla F^*(\eta_2) - \nabla F^*(\eta_1) \rangle > 0, \quad \forall \eta_1 \neq \eta_2$$

$\square$

Definition 3 (Quasi-arithmetic centers, QACs)). *Let* $F : \Theta \to \mathbb{R}$ *be a strictly convex and smooth real-valued function of Legendre-type in $\mathcal{F}$.*

The weighted quasi-arithmetic average of $\theta_1, \ldots, \theta_n$ and $w \in \Delta_{n-1}$ is defined by the gradient map ∇F as follows:

$$M_{\nabla F}(\theta_1, \ldots, \theta_n; w) := \nabla F^{-1}\left(\sum_i w_i \nabla F(\theta_i)\right), \qquad (4)$$

$$= \nabla F^*\left(\sum_i w_i \nabla F(\theta_i)\right), \qquad (5)$$

where $\nabla F^* = (\nabla F)^{-1}$ is the gradient map of the Legendre transform F^* of F.

We recover the usual definition of scalar QAMs M_f (Definition 1) when $F(t) = \int_a^t f(u)\mathrm{d}u$ for a strictly increasing or strictly decreasing and continuous function f: $M_f = M_{F'}$ (with $f^{-1} = (F')^{-1}$). Notice that we only need to consider F to be strictly convex or strictly concave and smooth to define a multivariate QAM since $M_{\nabla F} = M_{-\nabla F}$.

Example 1 (Matrix example). Consider the strictly convex function [8] F : $\mathrm{Sym}_{++}(d) \to \mathbb{R}$ with $F(\theta) = -\log\det(\theta)$, where $\det(\cdot)$ denotes the matrix determinant. Function $F(\theta)$ is strictly convex and differentiable [8] on the domain of d-dimensional symmetric positive-definite matrices $\mathrm{Sym}_{++}(d)$ (open convex cone). We have

$$F(\theta) = -\log\det(\theta),$$
$$\nabla F(\theta) = -\theta^{-1} =: \eta(\theta),$$
$$\nabla F^{-1}(\eta) = -\eta^{-1} =: \theta(\eta)$$
$$F^*(\eta) = \langle\theta(\eta), \eta\rangle - F(\theta(\eta)) = -d - \log\det(-\eta),$$

where the dual parameter η belongs to the d-dimensional negative-definite matrix domain, and the inner matrix product is the Hilbert-Schmidt inner product $\langle A, B \rangle := \mathrm{tr}(AB^\top)$, where $\mathrm{tr}(\cdot)$ denotes the matrix trace. It follows that

$$M_{\nabla F}(\theta_1, \theta_2) = 2(\theta_1^{-1} + \theta_2^{-1})^{-1},$$

is the matrix harmonic mean [1] generalizing the scalar harmonic mean $H(a, b) = \frac{2ab}{a+b}$ for $a, b > 0$. Other examples of matrix means are reported in [7].

2.2   Quasi-arithmetic Barycenters and Dual Geodesics

A Bregman generator $F : \Theta \to \mathbb{R}$ induces a dually flat space [4]

$$(\Theta, g(\theta) = \nabla_\theta^2 F(\theta), \nabla, \nabla^*)$$

that we call a Bregman manifold (Hessian manifold with a global chart), where ∇ is the flat connection with Christoffel symbols $\Gamma_{ijk}(\theta) = 0$ and ∇^* is the dual connection with respect to g such that $\Gamma^{*ijk}(\eta) = 0$.

In a Bregman manifold, the primal geodesics $\gamma_\nabla(P, Q; t)$ are obtained as line segments in the θ-coordinate system (because the Christoffel symbols of the connection ∇ vanishes in the θ-coordinate system) while the dual geodesics $\gamma_{\nabla^*}(P, Q; t)$ are line segments in the η-coordinate system (because the Christoffel symbols of the dual connection ∇^* vanishes in the η-coordinate system). The dual geodesics define interpolation schemes $(PQ)^\nabla(t) = \gamma_\nabla(P, Q; t)$ and $(PQ)^{\nabla^*}(t) = \gamma_{\nabla^*}(P, Q; t)$ between input points P and Q with $P = \gamma_\nabla(P, Q; 0) = \gamma_{\nabla^*}(P, Q; 0)$ and $Q = \gamma_\nabla(P, Q; 1) = \gamma_{\nabla^*}(P, Q; 1)$ when t ranges in $[0, 1]$. We express the coordinates of the interpolated points on γ_∇ and γ_{∇^*} using quasi-arithmetic averages as follows:

$$(PQ)^\nabla(t) = \gamma_\nabla(P, Q; t) = \begin{bmatrix} M_{\mathrm{id}}(\theta(P), \theta(Q); 1 - t, t) \\ M_{\nabla F^*}(\eta(P), \eta(Q); 1 - t, t) \end{bmatrix}, \tag{6}$$

$$(PQ)^{\nabla^*}(t) = \gamma_{\nabla^*}(P, Q; t) = \begin{bmatrix} M_{\nabla F}(\theta(P), \theta(Q); 1 - t, t) \\ M_{\mathrm{id}}(\eta(P), \eta(Q); 1 - t, t) \end{bmatrix}, \tag{7}$$

where id denotes the identity mapping. See Fig. 1.

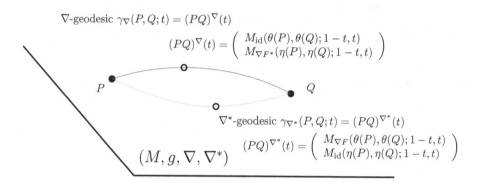

∇-geodesic $\gamma_\nabla(P, Q; t) = (PQ)^\nabla(t)$

$$(PQ)^\nabla(t) = \begin{pmatrix} M_{\mathrm{id}}(\theta(P), \theta(Q); 1 - t, t) \\ M_{\nabla F^*}(\eta(P), \eta(Q); 1 - t, t) \end{pmatrix}$$

P Q

∇^*-geodesic $\gamma_{\nabla^*}(P, Q; t) = (PQ)^{\nabla^*}(t)$

$$(PQ)^{\nabla^*}(t) = \begin{pmatrix} M_{\nabla F}(\theta(P), \theta(Q); 1 - t, t) \\ M_{\mathrm{id}}(\eta(P), \eta(Q); 1 - t, t) \end{pmatrix}$$

(M, g, ∇, ∇^*)

Fig. 1. The points on dual geodesics in a dually flat spaces have dual coordinates expressed with quasi-arithmetic averages.

Quasi-arithmetic centers were also used by a geodesic bisection algorithm to approximate the circumcenter of the minimum enclosing balls with respect to the canonical divergence in Bregman manifolds in [21], and for defining the Riemannian center of mass between two symmetric positive-definite matrices with respect to the trace metric in [15]. See also [22, 23].

3   Invariance and Equivariance Properties

A dually flat manifold [4] (M, g, ∇, ∇^*) has a canonical divergence [2] D_{∇, ∇^*} which can be expressed either as a primal Bregman divergence in the ∇-affine coordinate system θ (using the convex potential function $F(\theta)$) or as a dual Bregman divergence in the ∇^*-affine coordinate system η (using the convex conjugate potential function $F^*(\eta)$), or as dual Fenchel-Young divergences [18] using the mixed coordinate systems θ and η. The dually flat manifold (M, g, ∇, ∇^*) (a particular case of Hessian manifolds [26] which admit a global coordinate system) is thus characterized by $(\theta, F(\theta); \eta, F^*(\eta))$ which we shall denote by $(M, g, \nabla, \nabla^*) \leftarrow \mathrm{DFS}(\theta, F(\theta); \eta, F^*(\eta))$ (or in short $(M, g, \nabla, \nabla^*) \leftarrow (\Theta, F(\theta))$). However, the choices of parameters θ and η and potential functions $F(\theta)$ and $F^*(\eta)$ are *not unique* since they can be chosen up to affine reparameterizations and additive affine terms [4]: $(M, g, \nabla, \nabla^*) \leftarrow \mathrm{DFS}([\theta, F(\theta); \eta, F^*(\eta)])$ where $[\cdot]$ denotes the equivalence class that has been called purposely the affine Legendre invariance in [14]:

First, consider changing the potential function $F(\theta)$ by adding an affine term: $\bar{F}(\theta) = F(\theta) + \langle c, \theta \rangle + d$. We have $\nabla \bar{F}(\theta) = \nabla F(\theta) + c = \bar{\eta}$. Inverting $\nabla \bar{F}(x) = \nabla F(x) + c = y$, we get $\nabla \bar{F}^{-1}(y) = \nabla F(y - c)$. We check that $B_F(\theta_1 : \theta_2) = B_{\bar{F}}(\theta_1 : \theta_2) = D_{\nabla, \nabla^*}(P_1 : P_2)$ with $\theta(P_1) =: \theta_1$ and $\theta(P_2) =: \theta_2$. It is indeed well-known that Bregman divergences modulo affine terms coincide [5]. For the quasi-arithmetic averages $M_{\nabla \bar{F}}$ and $M_{\nabla F}$, we thus obtain the following invariance property:

$$M_{\nabla \bar{F}}(\theta_1, \ldots; \theta_n; w) = M_{\nabla F}(\theta_1, \ldots; \theta_n; w).$$

Second, consider an affine change of coordinates $\bar{\theta} = A\theta + b$ for $A \in \mathrm{GL}(d)$ and $b \in \mathbb{R}^d$, and define the potential function $\bar{F}(\bar{\theta})$ such that $\bar{F}(\bar{\theta}) = F(\theta)$. We have $\theta = A^{-1}(\bar{\theta} - b)$ and $\bar{F}(x) = F(A^{-1}(x - b))$. It follows that

$$\nabla \bar{F}(x) = (A^{-1})^{\top} \nabla F(A^{-1}(x - b)),$$

and we check that $B_{\bar{F}(\overline{\theta_1} : \overline{\theta_2})} = B_F(\theta_1 : \theta_2)$:

$$B_{\bar{F}(\bar{F}(\overline{\theta_1} : \overline{\theta_2}))} = \bar{F}(\overline{\theta_1}) - \bar{F}(\overline{\theta_2}) - \langle \overline{\theta_1} - \overline{\theta_2}, \nabla \bar{F}(\overline{\theta_2}) \rangle,$$

$$= F(\theta_1) - F(\theta_2) - (A(\theta_1 - \theta_2))^{\top}(A^{-1})^{\top} \nabla F(\theta_2),$$

$$= F(\theta_1) - F(\theta_2) - (\theta_1 - \theta_2)^{\top} \underbrace{A^{\top}(A^{-1})^{\top}}_{(A^{-1}A)^{\top} = I} \nabla F(\theta_2) = B_F(\theta_1 : \theta_2).$$

This highlights the invariance that $D_{\nabla, \nabla^*}(P_1 : P_2) = B_F(\theta_1 : \theta_2) = B_{\bar{F}(\overline{\theta_1} : \overline{\theta_2})}$, i.e., the canonical divergence does not change under a reparameterization of the ∇-affine coordinate system. For the induced quasi-arithmetic averages $M_{\nabla \bar{F}}$ and $M_{\nabla F}$, we have $\nabla \bar{F}(x) = (A^{-1})^{\top} \nabla F(A^{-1}(x - b)) = y$, we calculate

$$x = \nabla \bar{F}(x)^{-1}(y) = A \nabla \bar{F}^{-1}(((A^{-1})^{\top})^{-1} y) + b,$$

and we have

$$M_{\nabla \bar{F}}(\bar{\theta}_1, \ldots, \bar{\theta}_n; w) := \nabla \bar{F}^{-1}\left(\sum_i w_i \nabla \bar{F}(\bar{\theta}_i)\right),$$

$$= (\nabla \bar{F})^{-1}\left((A^{-1})^{\top} \sum_i w_i \nabla F(\theta_i)\right),$$

$$= A \nabla F^{-1}\left(\underbrace{((A^{-1})^{\top})^{-1}(A^{-1})^{\top}}_{=I} \sum_i w_i \nabla F(\theta_i)\right) + b,$$

$$M_{\nabla \bar{F}}(\bar{\theta}_1, \ldots, \bar{\theta}_n; w) = A M_{\nabla F}(\theta_1, \ldots, \theta_n; w) + b$$

More generally, we may define $\bar{F}(\bar{\theta}) = F(A\theta + b) + \langle c, \theta \rangle + d$ and get via Legendre transformation $\bar{F}^*(\bar{\eta}) = F^*(A^*\eta + b^*) + \langle c^*, \eta \rangle + d^*$ (with A^*, b^*, c^* and d^* expressed using A, b, c and d since these parameters are linked by the Legendre transformation).

Third, the canonical divergences should be considered relative divergences (and not absolute divergences), and defined according to a prescribed arbitrary "unit" $\lambda > 0$. Thus we can scale the canonical divergence by $\lambda > 0$, i.e., $D_{\lambda, \nabla, \nabla^*} := \lambda D_{\nabla, \nabla^*}$. We have $D_{\lambda, \nabla, \nabla^*}(P_1 : P_2) = \lambda B_F(\theta_1 : \theta_2) = \lambda B_{F^*}(\eta_2 : \eta_1)$, and $\lambda B_F(\theta_1 : \theta_2) = B_{\lambda F}(\theta_1 : \theta_2)$ (and $\nabla \lambda F = \lambda \nabla F$). We check the scale invariance of quasi-arithmetic averages: $M_{\lambda \nabla F} = M_{\nabla F}$.

Proposition 2 (Invariance and equivariance of QACs). *Let $F(\theta)$ be a function of Legendre type. Then $\bar{F}(\bar{\theta}) := \lambda(F(A\theta+b)+\langle c, \theta \rangle + d)$ for $A \in \mathrm{GL}(d)$, $b, c \in \mathbb{R}^d$, $d \in \mathbb{R}^d$ and $\lambda \in \mathbb{R}_{>0}$ is a Legendre-type function, and we have*

$$M_{\nabla \bar{F}} = A\, M_{\nabla F} + b.$$

This proposition generalizes the invariance property of scalar QAMs, and untangles the role of scale $\lambda > 0$ from the other invariance roles brought by the Legendre transformation.

Consider the Mahalanobis divergence Δ^2 (i.e., the squared Mahalanobis distance Δ) as a Bregman divergence obtained for the quadratic form generator $F_Q(\theta) = \frac{1}{2}\theta^\top Q\theta + c\theta + \kappa$ for a symmetric positive-definite $d \times d$ matrix Q, $c \in \mathbb{R}^d$ and $\kappa \in \mathbb{R}$. We have:

$$\Delta^2(\theta_1, \theta_2) = B_{F_Q}(\theta_1 : \theta_2) = \frac{1}{2}(\theta_2 - \theta_1)^\top Q\,(\theta_2 - \theta_1).$$

When $Q = I$, the identity matrix, the Mahalanobis divergence coincides with the Euclidean divergence[2] (i.e., the squared Euclidean distance). The Legendre convex conjugate is

$$F^*(\eta) = \frac{1}{2}\eta^\top Q^{-1}\eta = F_{Q^{-1}}(\eta),$$

and we have $\eta = \nabla F_Q(\theta) = Q\theta$ and $\theta = \nabla F_Q^*(\eta) = Q^{-1}\eta$. Thus we get the following dual quasi-arithmetic averages:

$$M_{\nabla F_Q}(\theta_1, \ldots, \theta_n; w) = Q^{-1}\left(\sum_{i=1}^{n} w_i Q\theta_i\right) = \sum_{i=1}^{n} w_i\theta_i = M_{\mathrm{id}}(\theta_1, \ldots, \theta_n; w),$$

$$M_{\nabla F_Q^*}(\eta_1, \ldots, \eta_n; w) = Q\left(\sum_{i=1}^{n} w_i Q^{-1}\eta_i\right) = M_{\mathrm{id}}(\eta_1, \ldots, \eta_n; w).$$

The dual quasi-arithmetic centers $M_{\nabla F_Q}$ and $M_{\nabla F_Q^*}$ induced by a Mahalanobis Bregman generator F_Q coincide since $M_{\nabla F_Q} = M_{\nabla F_Q^*} = M_{\mathrm{id}}$. This

---

[2] The squared Euclidean/Mahalanobis divergence are not metric distances since they fail the triangle inequality.

means geometrically that the left-sided and right-sided centroids of the underlying canonical divergences match. The average $M_{\nabla F_Q}(\theta_1, \ldots, \theta_n; w)$ expresses the centroid $C = \bar{C}_R = \bar{C}_L$ in the θ-coordinate system $(\theta(C) = \underline{\theta})$ and the average $M_{\nabla F_Q^*}(\eta_1, \ldots, \eta_n; w)$ expresses the same centroid in the η-coordinate system $(\eta(C) = \underline{\eta})$. In that case of self-dual flat Euclidean geometry, there is an affine transformation relating the θ- and η-coordinate systems: $\eta = Q\theta$ and $\theta = Q^{-1}\eta$. As we shall see this is because the underlying geometry is self-dual Euclidean flat space $(M, g_{\text{Euclidean}}, \nabla_{\text{Euclidean}}, \nabla^*_{\text{Euclidean}} = \nabla_{\text{Euclidean}})$ and that both dual connections coincide with the Euclidean connection (i.e., the Levi-Civita connection of the Euclidean metric). In this particular case, the dual coordinate systems are just related by affine transformations.

4   Quasi-arithmetic Mixtures and Jensen-Shannon-type Divergences

Consider a quasi-arithmetic mean M_f and n probability distributions $P_1, \ldots, P_n$ all dominated by a measure μ, and denote by $p_1 = \frac{dP_1}{d\mu}, \ldots, p_n = \frac{dP_n}{d\mu}$ their Radon-Nikodym derivatives. Let us define *statistical M_f-mixtures* of $p_1, \ldots, p_n$:

Definition 4. *The M_f-mixture of n densities $p_1, \ldots, p_n$ weighted by $w \in \Delta_n^\circ$ is defined by*

$$(p_1, \ldots, p_n; w)^{M_f}(x) := \frac{M_f(p_1(x), \ldots, p_n(x); w)}{\int M_f(p_1(x), \ldots, p_n(x); w) d\mu(x)}.$$

The quasi-arithmetic mixture (QAMIX) $(p_1, \ldots, p_n; w)^{M_f}$ generalizes the ordinary statistical mixture $\sum_{i=1}^d w_i p_i(x)$ when $f(t) = t$ and $M_f = A$ is the arithmetic mean. A statistical M_f-mixture can be interpreted as the M_f-integration of its weighted component densities, the densities p_i. The power mixtures $(p_1, \ldots, p_n; w)^{M_p}(x)$ (including the ordinary and geometric mixtures) are called α-mixtures in [3] with $\alpha(p) = 1 - 2p$ (or equivalently $p = \frac{1-\alpha}{2}$). A nice characterization of the α-mixtures is that these mixtures are the *density centroids* of the weighted mixture components with respect to the α-divergences [3] (proven by calculus of variation):

$$(p_1, \ldots, p_n; w)^{M_\alpha} = \arg\min_p \sum_i w_i D_\alpha(p_i, p),$$

where D_α denotes the α-divergences [4,20]. See also the entropic means defined according to f-divergences [6]. M_f-mixtures can also been used to define a generalization of the Jensen-Shannon divergence [17] between densities p and q as follows:

$$D_{\text{JS}}^{M_f}(p, q) := \frac{1}{2}\left(D_{\text{KL}}(p : (pq)^{M_f}) + D_{\text{KL}}(q : (pq)^{M_f})\right) \geq 0, \tag{8}$$

where $D_{\mathrm{KL}}(p:q) = \int p(x) \log \frac{p(x)}{q(x)} d\mu(x)$ is the Kullback-Leibler divergence, and $(pq)^{M_f} := (p, q; \frac{1}{2}, \frac{1}{2})^{M_f}$. The ordinary JSD is recovered when $f(t) = t$ and $M_f = A$:

$$D_{\mathrm{JS}}(p, q) = \frac{1}{2}\left(D_{\mathrm{KL}}\left(p : \frac{p+q}{2}\right) + D_{\mathrm{KL}}\left(q : \frac{p+q}{2}\right)\right).$$

In general, we may consider quasi-arithmetic paths between densities on the space $\mathcal{P}$ of probability density functions with a common support all dominated by a reference measure. On $\mathcal{P}$, we can build a parametric statistical model called a M_f-mixture family of order n as follows:

$$\mathcal{F}^{M_f}_{p_0, p_1, \ldots, p_n} := \left\{(p_0, p_1, \ldots, p_n; (\theta, 1))^{M_f} : \theta \in \Delta_n^\circ\right\}.$$

In particular, power q-paths have been investigated in [13] with applications in annealing importance sampling and other Monte Carlo methods.

To conclude, let us give a geometric definition of a generalization of the Jensen-Shannon divergence on $\mathcal{P}$ according to an arbitrary affine connection [4,27] ∇:

Definition 5 (Affine connection-based ∇-Jensen-Shannon divergence). *Let ∇ be an affine connection on the space of densities $\mathcal{P}$, and $\gamma_\nabla(p, q; t)$ the geodesic linking density $p = \gamma_\nabla(p, q; 0)$ to density $q = \gamma_\nabla(p, q; 1)$. Then the ∇-Jensen-Shannon divergence is defined by:*

$$D_\nabla^{\mathrm{JS}}(p, q) := \frac{1}{2}\left(D_{\mathrm{KL}}\left(p : \gamma_\nabla\left(p, q; \frac{1}{2}\right)\right) + D_{\mathrm{KL}}\left(q : \gamma_\nabla\left(p, q; \frac{1}{2}\right)\right)\right). \tag{9}$$

When $\nabla = \nabla^m$ is chosen as the mixture connection [4], we end up with the ordinary Jensen-Shannon divergence since $\gamma_{\nabla^m}(p, q; \frac{1}{2}) = \frac{p+q}{2}$. When $\nabla = \nabla^e$, the exponential connection, we get the geometric Jensen-Shannon divergence [17] since $\gamma_{\nabla^e}(p, q; \frac{1}{2}) = (pq)^G$ is a statistical geometric mixture. We may consider the α-connections [4] ∇^α of parametric or non-parametric statistical models, and skew the geometric Jensen-Shannon divergence to define the β-skewed ∇^α-JSD:

$$D_{\nabla^\alpha, \beta}^{\mathrm{JS}}(p, q) = \beta\, D_{\mathrm{KL}}(p : \gamma_{\nabla^\alpha}(p, q; \beta)) + (1 - \beta)\, D_{\mathrm{KL}}(q : \gamma_{\nabla^\alpha}(p, q; \beta)). \tag{10}$$

A longer technical report of this work is available [19].

References

1. Alić, M., Mond, B., Pečarić, J., Volenec, V.: The arithmetic-geometric-harmonic-mean and related matrix inequalities. Linear Algebra Appl. **264**, 55–62 (1997)
2. Amari, S.i.: Differential-geometrical methods in statistics. Lecture Notes on Statistics 28, 1 (1985). https://doi.org/10.1007/978-1-4612-5056-2
3. Amari, S.i.: Integration of stochastic models by minimizing α-divergence. Neural Comput. **19**(10), 2780–2796 (2007)
4. Amari, S.: Information Geometry and Its Applications. AMS, vol. 194. Springer, Tokyo (2016). https://doi.org/10.1007/978-4-431-55978-8

5. Banerjee, A., Merugu, S., Dhillon, I.S., Ghosh, J., Lafferty, J.: Clustering with Bregman divergences. J. Mach. Learn. Res. **6**(10) (2005)
6. Ben-Tal, A., Charnes, A., Teboulle, M.: Entropic means. J. Math. Anal. Appl. **139**(2), 537–551 (1989)
7. Bhatia, R., Gaubert, S., Jain, T.: Matrix versions of the Hellinger distance. Lett. Math. Phys. **109**(8), 1777–1804 (2019). https://doi.org/10.1007/s11005-019-01156-0
8. Boyd, S., Boyd, S.P., Vandenberghe, L.: Convex optimization. Cambridge University Press (2004)
9. Bregman, L.M.: The relaxation method of finding the common point of convex sets and its application to the solution of problems in convex programming. USSR Comput. Math. Math. Phys. **7**(3), 200–217 (1967)
10. Clarke, F.: On the inverse function theorem. Pac. J. Math. **64**(1), 97–102 (1976)
11. Dontchev, A.L., Rockafellar, R.T.: Implicit Functions and Solution Mappings. SMM, Springer, New York (2009). https://doi.org/10.1007/978-0-387-87821-8
12. Hardy, G.H., Littlewood, J.E., Pólya, G., Pólya, G.: Inequalities. Cambridge University Press (1952)
13. Masrani, V., et al.: q-paths: Generalizing the geometric annealing path using power means. In: Uncertainty in Artificial Intelligence, pp. 1938–1947. PMLR (2021)
14. Nakajima, N., Ohmoto, T.: The dually flat structure for singular models. Inform. Geometry **4**(1), 31–64 (2021). https://doi.org/10.1007/s41884-021-00044-8
15. Nakamura, Y.: Algorithms associated with arithmetic, geometric and harmonic means and integrable systems. J. Comput. Appl. Math. **131**(1–2), 161–174 (2001)
16. Niculescu, C.P., Persson, L.-E.: Convex Functions and Their Applications. CBM, Springer, New York (2006). https://doi.org/10.1007/0-387-31077-0
17. Nielsen, F.: On the Jensen-Shannon symmetrization of distances relying on abstract means. Entropy **21**(5), 485 (2019)
18. Nielsen, F.: Statistical divergences between densities of truncated exponential families with nested supports: duo bregman and duo jensen divergences. Entropy **24**(3), 421 (2022)
19. Nielsen, F.: Beyond scalar quasi-arithmetic means: Quasi-arithmetic averages and quasi-arithmetic mixtures in information geometry (2023)
20. Nielsen, F., Nock, R., Amari, S.i.: On clustering histograms with k-means by using mixed α-divergences. Entropy **16**(6), 3273–3301 (2014)
21. Nock, R., Nielsen, F.: Fitting the smallest enclosing bregman ball. In: Gama, J., Camacho, R., Brazdil, P.B., Jorge, A.M., Torgo, L. (eds.) ECML 2005. LNCS (LNAI), vol. 3720, pp. 649–656. Springer, Heidelberg (2005). https://doi.org/10.1007/11564096_65
22. Ohara, A.: Geodesics for dual connections and means on symmetric cones. Integr. Eqn. Oper. Theory **50**, 537–548 (2004)
23. Pálfia, M.: Classification of affine matrix means. arXiv preprint arXiv:1208.5603 (2012)
24. Rockafellar, R.T.: Conjugates and Legendre transforms of convex functions. Can. J. Math. **19**, 200–205 (1967)
25. Rockafellar, R.T.: Convex analysis, vol. 11. Princeton University Press (1997)
26. Shima, H., Yagi, K.: Geometry of Hessian manifolds. Differential Geom. Appl. **7**(3), 277–290 (1997)
27. Zhang, J.: Nonparametric information geometry: from divergence function to referential-representational biduality on statistical manifolds. Entropy **15**(12), 5384–5418 (2013)

Geometry of Parametric Binary Choice Models

Hisatoshi Tanaka[✉]

School of Political Science and Economics, Waseda University, 1-6-1 Nishiwaseda,
Shinjuku, Tokyo 169-8050, Japan
hstnk@waseda.jp

Abstract. In this study, we consider parametric binary choice models from the perspective of information geometry. The set of models is a dually flat manifold with dual connections, which are naturally derived from the Fisher information metric. Under the dual connections, the canonical divergence and the Kullback–Leibler (KL) divergence of the binary choice model coincide if and only if the model is a logit. The results are applied to a logit estimation with linear constraints.

Keywords: Binary Choice Models · Discrete Choice Models · Logit · Multinomial Logit · Single-Index Models

1 Introduction

Information geometry has been applied to econometric models such as the standard linear model, Poisson regression, Wald tests, the ARMA model, and many other examples [3,4,8,10]. In the present study, we apply the method to a standard binary choice model. Let x be an $\mathbb{R}^d$-valued random vector. Let $y \in \{0,1\}$ be a binary outcome such that

$$y = \begin{cases} 1 & \text{if } y^* \geq 0 \\ 0 & \text{if } y^* < 0 \end{cases}, \tag{1}$$

where $\theta \in \mathbb{R}^d$, $y^* = x \cdot \theta - \epsilon$, $\epsilon \perp\!\!\!\perp x$, and $E[\epsilon] = 0$. The choice probability conditioned on x is given by

$$\mathbf{P}\{y = 1 \mid x\} = \mathbf{P}\{\epsilon \leq x \cdot \theta \mid x\} = F(x \cdot \theta), \tag{2}$$

where the distribution F of ϵ is known to a statistician. Let p_θ be the density of the binary response model given by

$$p_\theta(y, x) = F(x \cdot \theta)^y (1 - F(x \cdot \theta))^{1-y} p_X(x), \quad (y, x) \in \{0,1\} \times \mathbb{R}^d, \tag{3}$$

where p_X denotes the marginal density of x.

The model is commonly used in social sciences to describe the choices made by decision-makers between two alternatives. These alternatives may represent

F. Nielsen and F. Barbaresco (Eds.): GSI 2023, LNCS 14071, pp. 157–166, 2023.
https://doi.org/10.1007/978-3-031-38271-0_16

school, labor supply, marital status, or transportation choices. See [9,11] for a list of empirical applications in the social sciences.

The model is referred to as *probit* when F is the standard normal distribution, that is,

$$F(u) = \int_{-\infty}^{u} \frac{1}{\sqrt{2\pi}} \exp\left(-\frac{s^2}{2}\right) ds,$$

and *logit* when F is the standard logistic distribution, that is,

$$F(u) = \frac{\exp u}{1 + \exp u}. \tag{4}$$

The logit model is particularly popular due to its closed-form choice probability $F(x \cdot \theta)$, which is easily interpretable [11]. We aim to show that among parametric binary response models, the logit model exhibits good geometric properties owing to its 'conditional' exponentiality.

The remainder of this paper is organized as follows. In Sect. 2, the geometry of the binary choice model is formulated. In Sect. 3, we introduce the canonical divergence and the Kullback–Leibler (KL) divergence. In particular, we demonstrate that the logit is a unique model, the canonical divergence of which is equal to the KL divergence. In Sect. 4, we consider the logit model with linear constraints. In Sect. 5, we summarize our conclusions.

2 Geometry of the Binary Choice Models

Assume that $F : \mathbb{R} \to [0,1]$ is a smooth distribution function of ϵ. Let $\Theta \subset \mathbb{R}^d$ be an open set of parameters θ. Given p_X, the set of models

$$\mathcal{P} = \{p_\theta \mid \theta \in \Theta\} \tag{5}$$

is considered as a d-dimensional C^∞ manifold with a canonical coordinate system $\theta \mapsto p_\theta$.

The tangent space of $\mathcal{P}$ at p_θ is simply denoted as $T_\theta \mathcal{P}$ and is given by

$$T_\theta \mathcal{P} = \mathrm{Span}\left\{(\partial_1)_\theta, \cdots, (\partial_d)_\theta\right\},$$

where $\partial_i = \frac{\partial}{\partial \theta_i}$ for $i = 1, \cdots, d$. The score of the model is

$$\frac{\partial}{\partial \theta} \log p_\theta(y,x) = \frac{y - F(x \cdot \theta)}{F(x \cdot \theta)(1 - F(x \cdot \theta))} f(x \cdot \theta) x \tag{6}$$

and the Fisher information matrix is

$$G(\theta) = E\left(\frac{\partial}{\partial \theta} \log p_\theta\right)\left(\frac{\partial}{\partial \theta} \log p_\theta\right)^\top = E\left[r(x \cdot \theta) x x^\top\right],$$

where

$$r(u) = \frac{f(u)^2}{F(u)(1 - F(u))}.$$

In the following, we assume that the integral $E\left[r(x\cdot\theta)xx^\top\right]$ is finite for every $\theta\in\Theta$. The assumption is trivially satisfied if f is continuous and positive everywhere on $\mathbb{R}$ and if x has a bounded support.

The Fisher information metric g is introduced on $T_\theta\mathcal{P}$ by

$$g_\theta(X,Y)=E\left[X\left(\int_0^{x\cdot\theta}\sqrt{r(u)}du\right)Y\left(\int_0^{x\cdot\theta}\sqrt{r(u)}du\right)\right]$$

for $X,Y\in T_\theta\mathcal{P}$. In particular, the (i,j) component of g at θ is

$$g_{ij}(\theta)=g_\theta(\partial_i,\partial_j)=E\left[r(x\cdot\theta)x_ix_j\right].$$

Hence, the Levi–Civita connection ∇ of $(\mathcal{P},g)$ is given by the connection coefficients

$$\Gamma_{ij,k}(\theta)=\frac{1}{2}\left[(\partial_i)_\theta g_{kj}(\theta)+(\partial_j)_\theta g_{ik}(\theta)-(\partial_k)_\theta g_{ij}(\theta)\right]$$
$$=\frac{1}{2}E\left[r'(x\cdot\theta)x_ix_jx_k\right]$$

for $1\le i,j,k\le n$. The coefficients show symmetry on (i,j,k). In particular, $\Gamma_{ij,k}(\theta)=\Gamma_{ji,k}(\theta)$, which implies that $(\mathcal{P},g,\nabla)$ is a tortion-free manifold. The symmetry of the connection is caused by the single-index structure of the model. $G(\theta)$ depends on θ only through the linear index $x\cdot\theta$.

Due to the symmetry of the Levi–Civita connection, the α-connection $\nabla^{(\alpha)}$ is defined naturally by

$$\Gamma_{ij,k}^{(\alpha)}(\theta)=\frac{1-\alpha}{2}\Gamma_{ij,k}(\theta) \tag{7}$$

for each $\alpha\in\mathbb{R}$. A pair $(\nabla^{(\alpha)},\nabla^{(-\alpha)})$ provides the dual connections of $(\mathcal{P},g)$ such that

$$Xg_\theta(Y,Z)=g_\theta(\nabla_X^{(\alpha)}Y,Z)+g_\theta(Y,\nabla_X^{(-\alpha)}Z)$$

for every $X,Y,Z\in\mathcal{X}(\mathcal{P})$, where $\mathcal{X}(\mathcal{P})$ is the family of smooth vector fields on $\mathcal{P}$.

Theorem 1. $(\mathcal{P},g,\nabla^{(+1)},\nabla^{(-1)})$ *is a dually flat space with dual affine coordinates* (θ,η); θ *is the* $\nabla^{(+1)}$-*affine coordinate, and* $\eta=(\eta_1,\cdots,\eta_d)$ *given by*

$$\eta_j=E\left[\left(\int_0^{x\cdot\theta}r(u)du\right)x_j\right] \tag{8}$$

for $1\le j\le d$ *is the* $\nabla^{(-1)}$-*affine coordinate.*

Proof. For $\alpha = 1$, $\Gamma_{ij,k}^{(+1)} \equiv 0$ holds for all i, j, and k. Moreover, given that

$$g_{ij}(\theta) = \partial_i \partial_j \psi(\theta)$$

holds with potential $\psi : \Theta \to \mathbb{R}$ defined by

$$\psi(\theta) = E\left[\int_0^{x \cdot \theta} \left(\int_0^v r(u)du \right) dv \right],$$

the dual-affine coordinates are obtained as

$$\eta_j = \partial_j \psi(\theta) = E\left[\left(\int_0^{x \cdot \theta} r(u)du \right) x_j \right]$$

for $1 \le j \le d$. □

For convenience, we denote the inverse function of

$$\partial \psi : \Theta \to \mathbb{R}^d, \theta \mapsto \eta = (\partial_1 \psi(\theta), \cdots, \partial_d \psi(\theta))$$

by $(\partial \psi)^{-1}$. Because the Hessian $\partial^2 \psi(\theta)$ is equal to the Fisher information matrix $G(\theta)$ and is therefore positive definite, $\partial \psi : \Theta \to \partial \psi(\Theta)$ is invertible at any $\eta \in \partial \psi(\Theta)$.

The dual potential $\varphi(\eta)$ is given by

$$\varphi(\eta) = \max_\theta \ \eta \cdot \theta - \psi(\theta) = \eta \cdot (\partial \psi)^{-1}(\eta) - \psi((\partial \psi)^{-1}(\eta)), \tag{9}$$

which is the Legendre transformation of $\psi(\theta)$. Let $\partial^i = \frac{\partial}{\partial \eta_i}$ for $1 \le i \le d$, then $\theta^i = \partial^i \varphi(\eta)$ holds.

Corollary 1. *The $\nabla^{(\pm 1)}$-geodesic path connecting $p, q \in \mathcal{P}$ is given by $t \in [0, 1] \to p_{\theta_t^{(\pm 1)}} \in \mathcal{P}$, where*

$$\theta_t^{(+1)} = (1 - t)\theta_p + t\theta_q \tag{10}$$

and

$$\theta_t^{(-1)} = (\partial \psi)^{-1}((1 - t)\eta_p + t\eta_q) \tag{11}$$

for $0 \le t \le 1$.

The $\nabla^{(-1)}$-geodesic is a solution to the ordinary differential equation,

$$\dot{\theta}_t = G(\theta_t)^{-1}(\eta_q - \eta_p), \ \theta_0 = \theta_p.$$

To see this, let $\eta_t^{(-1)} = \partial \psi(\theta_t^{(-1)}) = (1 - t)\eta_p + t\eta_q$. Then,

$$\frac{d}{dt} \eta_t^{(-1)} = \partial^2 \psi(\theta_t^{(-1)}) \frac{d}{dt} \theta_t^{(-1)} = \eta_q - \eta_p,$$

where $G(\theta_t^{(-1)}) = \partial^2 \psi(\theta_t^{(-1)})$.

3   The Logit Model

For a dually flat manifold with dual affine coordinates (θ, η) and dual potentials (ψ, φ), the *canonical divergence* (or *U-divergence with* $U = \psi$) is defined as

$$D(p\|q) = \varphi(\eta_p) + \psi(\theta_q) - \eta_p \cdot \theta_q \tag{12}$$

[2,6,7]. For the binary response model, the divergence is given as

$$D(p\|q) = [\eta_p \cdot \theta_p - \psi(\theta_p)] + \psi(\theta_q) - \eta_p \cdot \theta_q$$
$$= E\left[\int_{x\cdot\theta_p}^{x\cdot\theta_q} \left(\int_0^v r(u)du\right) dv\right] - E\left[\left(\int_0^{x\cdot\theta_p} r(u)du\right) x \cdot (\theta_q - \theta_p)\right] \tag{13}$$

for each p and q in $\mathcal{P}$, because $\varphi(\eta_p) = \eta_p \cdot \theta_p - \psi(\theta_p)$ and

$$\psi(\theta_q) - \psi(\theta_p) = E\left[\int_{x\cdot\theta_p}^{x\cdot\theta_q} \left(\int_0^v r(u)du\right) dv\right].$$

The following results are standard.

Theorem 2. *Let p, q, r be in $\mathcal{P}$. Let $\theta^{(+1)}$ be the $\nabla^{(+1)}$-geodesic path connecting p and q, and let $\theta^{(-1)}$ be the $\nabla^{(-1)}$-geodesic path connecting q and r. If $\theta^{(+1)}$ and $\theta^{(-1)}$ are orthogonal at the intersection q in the sense that*

$$g_q\left(\left(\frac{d}{dt}\right)_q \theta_t^{(+1)}, \left(\frac{d}{dt}\right)_q \theta_q^{(-1)}\right) = 0,$$

then we have

$$D(p\|r) = D(p\|q) + D(q\|r). \tag{14}$$

Corollary 2. *The Pythagorean formula (14) holds if $(\eta_p - \eta_q) \cdot (\theta_q - \theta_r) = 0$.*

An alternative for the divergence on $\mathcal{P}$ is the KL divergence

$$KL(p\|q) = E_p\left[\log \frac{p(y,x)}{q(y,x)}\right].$$

In the case of the binary response model, the KL divergence is

$$KL(p\|q) = E_p\left[\log \frac{yF(x\cdot\theta_p) + (1-y)(1-F(x\cdot\theta_p))}{yF(x\cdot\theta_q) + (1-y)(1-F(x\cdot\theta_q))}\right]$$
$$= E\left[F(x\cdot\theta_p)\log\left(\frac{F(x\cdot\theta_p)}{F(x\cdot\theta_q)}\right)\right]$$
$$+E\left[(1-F(x\cdot\theta_p))\log\left(\frac{1-F(x\cdot\theta_p)}{1-F(x\cdot\theta_q)}\right)\right] \tag{15}$$

because $E_p[y|x] = F(x\cdot\theta_p)$. The canonical divergence (13) and the KL divergence (15) generally do not coincide. As shown below, in a special case where F is a logistic distribution, they are equivalent.

Theorem 3. $D = KL$ *holds for arbitrary* p_X *if and only if* F *is a logistic distribution; that is,*

$$F(u) = \frac{\exp(\beta u)}{1 + \exp(\beta u)}, \tag{16}$$

where $\beta > 0$.

Proof. If F is a logistic distribution, then $\beta F(1 - F) = f$. This equation is substituted on the right-hand side of (13) to obtain $D = KL$.

Now, we assume that $D \equiv KL$ holds for an arbitrary p_X. Because

$$(\partial_\theta)_p (\partial_\theta)_q D(p\|q) = -E\left[\frac{f(x \cdot \theta_p)^2}{F(x \cdot \theta_p)(1 - F(x \cdot \theta_p))} x x^\top\right]$$

and

$$(\partial_\theta)_p (\partial_\theta)_q KL(p\|q) = -E\left[\frac{f(x \cdot \theta_p) f(x \cdot \theta_q)}{F(x \cdot \theta_q)(1 - F(x \cdot \theta_q))} x x^\top\right],$$

$D(p\|q) \equiv KL(p\|q)$ implies that

$$\frac{f(x \cdot \theta_p)^2}{F(x \cdot \theta_p)(1 - F(x \cdot \theta_p))} \equiv \frac{f(x \cdot \theta_p) f(x \cdot \theta_q)}{F(x \cdot \theta_q)(1 - F(x \cdot \theta_q))}$$

for arbitrary p and q. By the principle of the separation of variables, this is possible only if there exists a positive constant β such that

$$\frac{f(u)}{F(u)(1 - F(u))} \equiv \beta.$$

Therefore, F is the logistic distribution.   □

In the case of $\beta = 1$, the results in the previous section are largely simplified. The Fisher information metric is given by

$$g_{ij}(\theta) = E\left[f(x \cdot \theta) x_i x_j\right]$$

for $1 \le i, j \le d$. The $\nabla^{(-1)}$-affine coordinate η is expressed as

$$\eta_j = E\left[F(x \cdot \theta) x_j\right]$$

for $1 \le j \le d$. The potential is given as

$$\psi(\theta) = E\left[\log\left(1 + \exp(x \cdot \theta)\right)\right]. \tag{17}$$

The canonical divergence is expressed as follows:

$$D(p\|q) = E\left[\log\left(\frac{1 + \exp(x \cdot \theta_q)}{1 + \exp(x \cdot \theta_p)}\right)\right] - E\left[\frac{\exp(x \cdot \theta_p)}{1 + \exp(x \cdot \theta_p)} x\right] \cdot (\theta_q - \theta_p), \tag{18}$$

which is equal to $KL(p\|q)$.

The logit model exhibits geometrically desirable properties because of the explicit integrability of F. We say that a statistical model $\mathcal{P} = \{p_\theta \mid \theta \in \Theta\}$ is an exponential family if it is expressed as

$$p(z, \theta) = \exp\left[C(z) + \sum_{i=1}^{d} \theta^i \beta_i(z) - \psi(\theta)\right]. \tag{19}$$

It is widely known that the (curved) exponential family possesses desirable properties such as higher-order efficiency of the maximum likelihood estimation [1,5]. Although the logit model is not truly exponential, the conditional density $p_\theta(y|x)$ is still written as

$$p_\theta(y|x) = \exp\left((x \cdot \theta)\delta_1(y) + \delta_0(y) - \psi(\theta|x)\right), \tag{20}$$

where

$$\delta_i(y) = \begin{cases} 1 & \text{if} \quad y = i \\ 0 & \text{if} \quad y \neq i \end{cases},$$

and

$$\psi(\theta|x) = \log\left(1 + \exp(x \cdot \theta)\right).$$

Conditioned by x, the model (20) belongs to an exponential family with potential $\psi(\theta|x)$. Notably, $\psi(\theta) = E\left[\psi(\theta|x)\right]$.

The marginal density p_X does not appear in the score of the model (6). Hence, p_X plays a minor role in the estimation of θ. The statistical properties of the model are primarily determined by $p_\theta(y|x)$, and the following result is obtained.

Theorem 4. *Assume that the density of z conditioned on w with respect to some positive measure $\nu(dz)$ is given by*

$$q_\theta(z|w) = \exp\left(\theta \cdot \beta(z|w) - \psi(\theta|w)\right), \tag{21}$$

where $\beta(z|w)$ is an $\mathbb{R}^d$-valued function of (z, w), and

$$\psi(\theta|w) := \log \int \exp\left(\theta \cdot \beta(z|w)\right) dz. \tag{22}$$

Then, the KL divergence of $\mathcal{Q} = \{q_\theta | \theta \in \Theta\}$ is equivalent to the canonical divergence D of $\mathcal{Q}$ with potential $\psi(\theta) = E\left[\psi(\theta \mid w)\right]$.

We can generalize Theorem 3 to cover the multinomial discrete choice model. Let $\{1, \cdots, k\}$ be the set of choices. Assume that the choice probability conditioned on x is now given by

$$\mathbf{P}\{y = i \mid x\} = F(x \cdot \theta_i)$$

for $1 \leq i \leq k$, where F is a smooth distribution function and $\theta = \begin{bmatrix} \theta_1 \cdots \theta_k \end{bmatrix} \in (\mathbb{R}^d)^k$ with $\theta_i = (\theta_i^1, \cdots, \theta_i^d) \in \mathbb{R}^d$. Let p_X be the marginal density of x and let $\Theta \subset (\mathbb{R}^d)^k$ be the set of parameters. Then, the multinomial choice model $\{\rho_\theta \mid \theta \in \Theta\}$ is given by

$$\rho_\theta(y, x) = \sum_{i=1}^{k} \delta_i(y) F(x \cdot \theta_i) p_X(x). \qquad (23)$$

In particular, when F is the standard logit distribution, the model becomes the multinomial logit model with the choice probability

$$\rho_\theta(y = i \mid x) = \frac{\exp(x \cdot \theta_i)}{\sum_{j=1}^{k} \exp(x \cdot \theta_j)} \qquad (24)$$

for $1 \leq i \leq k$. The model is a conditional exponential family because it is expressed as

$$\rho_\theta(y \mid x) = \exp\left[\sum_{i=1}^{k} \delta_i(y) x \cdot \theta_i - \psi(\theta \mid x) \right]$$

with conditional potential $\psi(\theta \mid x) = \log \sum_{j=1}^{k} \exp(x \cdot \theta_j)$. Hence, the set of models $\{\rho_\theta \mid \theta \in \Theta\}$ is a dually flat space with dual affine coordinates (θ, η) and potential $\psi(\theta) = E\left[\log \sum_{j=1}^{k} \exp(x \cdot \theta_j) \right]$, where $\eta = \begin{bmatrix} \eta_1 \cdots \eta_k \end{bmatrix} \in (\mathbb{R}^d)^k$, $\eta_i = (\eta_{i,1}, \cdots, \eta_{i,d}) \in \mathbb{R}^d$, and

$$\eta_{i,l} = E\left[\frac{\exp(x \cdot \theta_i)}{\sum_{j=1}^{k} \exp(x \cdot \theta_j)} x_l \right]$$

for $1 \leq i \leq k$ and $1 \leq l \leq d$. Furthermore, the canonical divergence D is equivalent to the KL divergence as a result of Theorem 4.

4   Linearly Constraint Logistic Regression

In this section, we applied the results of the previous section to the logit model with linear constraints. In empirical applications, we typically aim to estimate θ under the linear constraint hypothesis, $H_0 : H^\top \theta = c$, where $H = \begin{bmatrix} h_1 \cdots h_m \end{bmatrix}$ is an $d \times m$ matrix with rank$(H) = m < d$, and $c = (c_1, \cdots, c_m) \in \mathbb{R}^m$. Let $\mathcal{P}_\mathcal{H} = \{p_\theta \in \mathcal{P} \mid \theta \in \mathcal{H}\}$, where $\mathcal{H} = \{\theta \in \Theta \mid H^\top \theta = c\}$. Suppose that the *true* model p does not belong to $\mathcal{P}(\mathcal{H})$. Then, the KL projection $\Pi : \mathcal{P} \to \mathcal{P}(\mathcal{H})$ is given by

$$\Pi p = \arg\min \ KL(p \| q) \quad \text{subject to} \quad q \in \mathcal{P}(\mathcal{H}). \qquad (25)$$

In the following, we assumed that F is a standard logistic distribution.

Theorem 5. $q = \Pi p$ *if and only if* $\eta_q - \eta_p \in \text{Image}(H)$.

Proof. Let $\mathcal{L}(\theta, \lambda) = KL(p\|p_\theta) - \sum_{i=1}^{m} \lambda^i(h_i \cdot \theta - c_i)$ be the Lagrangian corresponding to (25) with Lagrange multipliers $\lambda = (\lambda^1, \cdots, \lambda^m)$. As $\theta \mapsto KL(p\|p_\theta)$ is convex, a necessary and sufficient condition for minimization is given by

$$\frac{\partial}{\partial\theta}KL(p\|p_\theta) = \sum_{i=1}^{m} \lambda^i h_i \in \text{Image}(H).$$

As $KL = D$, in contrast,

$$\frac{\partial}{\partial\theta}KL(p\|p_\theta) = \frac{\partial}{\partial\theta}\left[\varphi(\eta_p) + \psi(\theta) - \eta_p \cdot \theta\right] = \eta_q - \eta_p.$$

$\square$

The condition $\eta_q - \eta_p \in \text{Image}(H)$ is satisfied if $\lambda^1, \cdots, \lambda^m \in \mathbb{R}$ such that $\eta_q - \eta_p = \sum_{i=1}^{m} \lambda^i h_i$ exists. Hence, the conditions given in the theorem are written as

$$\begin{cases} \eta - \sum_{i=1}^{m} \lambda^i h_i = \eta_p \\ H^\top \theta = c \\ \eta = \partial\psi(\theta) \end{cases},$$

which are $2d + m$ equations with $2d + m$ variables (θ, η, λ). The solution is given by $\theta = (\partial\psi)^{-1}(\eta_p + H\lambda)$, where λ solves $H^\top(\partial\psi)^{-1}(\eta_p + H\lambda) = c$. The solution is approximated well by $\lambda = (H^\top G(\theta_p)^{-1}H)^{-1}(c - H^\top\theta_p)$ if θ_p locates sufficiently close to $\mathcal{H}$ and might be recursively updated by the standard Newton-Raphson method.

5 Conclusions

In this study, we have investigated the geometry of parametric binary response models. The model was established as a dually flat space, where the canonical coefficient parameter θ acts as an affine coordinate. The dual flat property introduces a canonical divergence into the model. The divergence is equivalent to the KL divergence if and only if the model is a logit. As an application example, the KL projection of the logit model onto an affine linear subspace was geometrically characterized.

The dual flatness of the binary response model is caused by the single-index structure of the model, which depends on the parameter θ only through the linear index $x \cdot \theta$, making the Levi–Civita connection coefficients $\Gamma_{ij,k}$ symmetrical on (i, j, k). Therefore, the results of this study can be extended to a more general class of single-index models, including nonlinear regressions, truncated regressions, and ordered discrete response models.

References

1. Amari, S.: Differential geometry of curved exponential families-curvature and information loss. Ann. Stat. **10**(2), 357–385 (1982)
2. Amari, S., Nagaoka, H.: Methods of Information Geometry. Oxford University Press, Tokyo (2000)
3. Andrews, I., Mikusheva, A.: A geometric approach to nonlinear econometric models. Econometrica **84**(3), 1249–1264 (2016)
4. Critchley, F., Marriott, P., Salmon, M.: On the differential geometry of the Wald test with nonlinear restrictions. Econometrica **64**(5), 1213–1222 (1996)
5. Eguchi, S.: Second order efficiency of minimum contrast estimators in a curved exponential family. Ann. Stat. **11**(3), 793–803 (1983)
6. Eguchi, S., Komori, O.: Minimum Divergence Methods in Statistical Machine Learning. From an Information Geometric Viewpoint. Springer, Tokyo (2022). https://doi.org/10.1007/978-4-431-56922-0
7. Eguchi, S., Komori, O., Ohara, A.: Duality of maximum entropy and minimum divergence. Entropy **16**(7), 3552–3572 (2014)
8. Kemp, G.C.R.: Invariance and the Wald test. J. Econ. **104**(2), 209–217 (2001)
9. Lee, M.J.: Micro-Econometrics: Methods of Moments and Limited Dependent Variables, 2nd edn. Springer, New York (2010). https://doi.org/10.1007/b60971
10. Marriott, P., Salmon, M.: An introduction to differential geometry in econometrics. In: Applications of Differential Geometry to Econometrics, pp. 7–63. Cambridge University Press, Cambridge (2000)
11. Train, K.E.: Discrete Choice Methods with Simulations. Cambridge University Press, New York (2003)

A q-Analogue of the Family of Poincaré Distributions on the Upper Half Plane

Koichi Tojo[1]($\boxtimes$)(iD) and Taro Yoshino[2]

[1] RIKEN Center for Advanced Intelligence Project, Nihonbashi 1-chome Mitsui Building, 15th floor, 1-4-1 Nihonbashi, Chuo-ku, Tokyo 103-0027, Japan
koichi.tojo@riken.jp
[2] Graduate School of Mathematical Science, The University of Tokyo, 3-8-1 Komaba, Meguro-ku, Tokyo 153-8914, Japan
yoshino@ms.u-tokyo.ac.jp

Abstract. The authors suggested a family of Poincaré distributions on the upper half plane, which is essentially the same as a family of hyperboloid distributions on the two dimensional hyperbolic space. This family has an explicit form of normalizing constant and is $SL(2, \mathbb{R})$-invariant. In this paper, as a q-analogue of Poincaré distributions, we propose a q-exponential family on the upper half plane with an explicit form of normalizing constant and show that it is also $SL(2, \mathbb{R})$-invariant.

Keywords: q-exponential family · upper half plane · Poincaré distribution

1 Introduction

In the field of Tsallis statistics, q-exponential families play an essential role. In fact, the family of q-Gaussian distributions (including Cauchy distributions) is an important family of distributions on $\mathbb{R}$ and is widely used. It is a q-analogue of the family of Gaussian distributions and has the following good properties:

(i) normalizing constant has an explicit form,
(ii) it is scaling and translation invariant as a family ($\mathrm{Aff}(\mathbb{R})$-invariant).

The property (i) enables us to perform practical inferences easily. The other property (ii) is closely related to transformation models, which have been studied well, for example, in [6,7] and are useful for Bayesian learning ([11]).

On the other hand, the upper half plane has a good exponential family called Poincaré distributions ([15]), which is essentially the same as a family of hyperboloid distributions ([5,10]) on the 2-dimensional hyperbolic space, and as a family of Gauss densities on Poincaré unit disk ([3]) based on Souriau's symplectic model of statistical mechanics (see [2,12] for example).

In this paper, we introduce a q-analogue of the family of Poincaré distributions on the upper half plane $\mathcal{H}$. This family also has the following good properties:

(i) normalizing constant has an explicit form,
(ii) it is $SL(2, \mathbb{R})$-invariant as a family.

F. Nielsen and F. Barbaresco (Eds.): GSI 2023, LNCS 14071, pp. 167–175, 2023.
https://doi.org/10.1007/978-3-031-38271-0_17

2   Preliminary

In this section, we recall the definition of a q-exponential family (Definition 2), the notion of G-invariance of a family of distributions (Definition 4) and the definition of a relatively G-invariant measure (Definition 6). Let X be a locally compact Hausdorff space and $\mathcal{R}(X)$ the set of all Radon measures on X.

Definition 1 (q**-logarithmic function,** q**-exponential function** [17]**, see also** [1, **Sect. 2.1**], [13, **Sect. 2.2**]**for more details**). *For* $q \in \mathbb{R}$, *we define the* q-logarithmic function $\ln_q \colon \mathbb{R}_{>0} \to \mathbb{R}$ *by*

$$\ln_q(x) := \int_1^x \frac{1}{t^q} dt = \begin{cases} \frac{1}{1-q}(x^{1-q} - 1) & (q \neq 1), \\ \log x & (q = 1). \end{cases}$$

We denote the image of $\ln_q$ *by* I_q. *Namely,*

$$I_q = \{x \in \mathbb{R} \mid 1 + (1 - q)x > 0\}.$$

We define the q-exponential function $\exp_q \colon I_q \to \mathbb{R}_{>0}$ *as the inverse function of* $\ln_q$, *That is,*

$$\exp_q(x) := \begin{cases} (1 + (1 - q)x)^{\frac{1}{1-q}} & (q \neq 1), \\ \exp x & (q = 1). \end{cases}$$

Definition 2 (q**-exponential family,** [1, **Sect. 2.2**],[13, **Definition 2.1**]). *A subset* $\mathcal{P} \subset \mathcal{R}(X)$ *consisting of probability measures is a* q-exponential family *on* X *if there exists a triple* (μ, V, T) *such that*

(i) $\mu \in \mathcal{R}(X)$,
(ii) V *is a finite dimensional vector space over* $\mathbb{R}$,
(iii) $T : X \to V$ *is a continuous map,*
(iv) *for any* $\nu \in \mathcal{P}$, *there exists* $\theta \in V^\vee$ *and* $\varphi(\theta) \in \mathbb{R}$ *such that*
  (a) $-\langle \theta, T(x) \rangle - \varphi(\theta) \in I_q$ *for any* $x \in X$ *and*
  (b) $d\nu(x) = \exp_q(-\langle \theta, T(x) \rangle - \varphi(\theta))d\mu(x)$,
  where $V^\vee$ *denotes the dual space of* V *and* $\langle \cdot, \cdot \rangle : V^\vee \times V \to \mathbb{R}$ *denotes the natural pairing.*

We call the triple (μ, V, T) *a* realization *of* $\mathcal{P}$. *(*[4] *call it* representation *for* $q = 1$.)

Remark 1. In the case where $q = 1$, a q-exponential family on X is nothing but an exponential family on X.

From now on, let G be a locally compact group, and X a locally compact Hausdorff space equipped with a continuous G-action.

Definition 3. *The group* G *naturally acts on* $\mathcal{R}(X)$ *by pushforward as follows:*

$$G \times \mathcal{R}(X) \to \mathcal{R}(X), \ (g, \mu) \mapsto g \cdot \mu.$$
$$(g \cdot \mu)(B) := \mu(g^{-1}B) \ \text{ for a Borel measurable set } B \subset X.$$

Definition 4. *A subset $\mathcal{P} \subset \mathcal{R}(X)$ is said to be G-invariant if*

$$g \cdot p \in \mathcal{P} \text{ for any } p \in \mathcal{P}, g \in G.$$

Definition 5 ([9, Sect. 2.3]). *For $\mu, \nu \in \mathcal{R}(X)$, we say μ and ν are strongly equivalent if there exists a continuous function $f \colon X \to \mathbb{R}_{>0}$ satisfying $\nu(B) = \int_B f d\mu$ for any Borel measurable subset B of X. This relation is an equivalence relation on $\mathcal{R}(X)$.*

Definition 6 ([9, Sect. 2.6], [8, (7.1.1) Definition 1]). *A measure $\mu \in \mathcal{R}(X)$ is said to be relatively G-invariant if μ and $g \cdot \mu$ are strongly equivalent and there exists a continuous group homomorphism $\chi \colon G \to \mathbb{R}_{>0}$ satisfying*

$$d(g \cdot \mu)(x) = \chi(g)^{-1} d\mu(x) \quad (g \in G, x \in X).$$

We say μ is G-invariant if $\chi \equiv 1$.

3 Construction of a q-Exponential Family

Let X be a locally compact Hausdorff space. In this section, we write down a method to construct q-exponential families on X. Concretely, first, we choose a triple (μ, V, T), and construct a q-exponential family on X with the realization (μ, V, T). We have two such ways. Here, μ is a Radon measure on X, V is a finite dimensional vector space over $\mathbb{R}$, and $T \colon X \to V$ is a continuous map.

Definition 7 (construction of a q-exponential family). *We put*

$$\Theta_q := \{\theta \in V^\vee \mid \text{there exists } \psi_\theta \in \mathbb{R} \text{ satisfying the conditions (i) and (ii)}\},$$

(i) $- \langle \theta, T(x) \rangle - \psi_\theta \in I_q$ for any $x \in X$,
(ii) $\int_{x \in X} \exp_q(- \langle \theta, T(x) \rangle - \psi_\theta) d\mu(x) = 1$,

and define

$$d\nu_\theta(x) := \exp_q(- \langle \theta, T(x) \rangle - \psi_\theta) d\mu(x) \in \mathcal{R}(X) \quad (\theta \in \Theta_q).$$

Then, the obtained family $\mathcal{P}_q := \{\nu_\theta\}_{\theta \in \Theta_q}$ is a q-exponential family on X with a realization (μ, V, T). We call Θ_q the parameter space of $\mathcal{P}_q$.

Definition 8 (another construction of a q-exponential family). *We put*

$$\tilde{\Theta}_q := \{\theta \in V^\vee \mid - \langle \theta, T(x) \rangle \in I_q \text{ for any } x \in X\},$$

$$d\tilde{\eta}_\theta(x) := \exp_q(- \langle \theta, T(x) \rangle) d\mu(x) \text{ for } \theta \in \tilde{\Theta}_q,$$

$$\Theta_q^o := \left\{ \theta \in \tilde{\Theta}_q \,\middle|\, \int_X d\tilde{\eta}_\theta < \infty \right\},$$

$$\eta_\theta := c_\theta \tilde{\eta}_\theta \in \mathcal{R}(X), \quad c_\theta := \left(\int_X d\tilde{\eta}_\theta \right)^{-1} \quad (\theta \in \Theta_q^o).$$

Then, the obtained family $\mathcal{P}_q^o := \{\eta_\theta \mid \theta \in \Theta_q^o\}$ is a q-exponential family on X. We call Θ_q^o the parameter space of $\mathcal{P}_q^o$.

The assertion that the obtained family $\mathcal{P}_q^o$ is a q-exponential family on X in the sense of Definition 2 with a realization (μ, V, T) follows from the following:

Proposition 1. $\mathcal{P}_q^o \subset \mathcal{P}_q$.

Proof. Take any $\theta \in \Theta_q^o$. Then there exists $c_\theta > 0$ such that $\eta_\theta = c_\theta \tilde{\eta}_\theta$. We put

$$\ell_\theta := \ln_q c_\theta \in I_q, \quad \theta' := (1 + (1 - q)\ell_\theta)\theta \in V^\vee.$$

By applying Note 1 below for $\alpha := -\langle \theta, T(x) \rangle$ and $\beta := \ell_\theta$, we have

$$d\eta_\theta(x) = c_\theta d\tilde{\eta}_\theta(x) = \exp_q(-\langle \theta', T(x) \rangle + \ell_\theta)d\mu(x).$$

Therefore, we have $\eta_\theta \in \mathcal{P}_q$.

The following Note 1 can be verified by a direct calculation easily.

Note 1. For $\alpha, \beta \in I_q$, the following properties hold.

(i) $(1 + (1 - q)\beta)\alpha + \beta \in I_q$,
(ii) $(\exp_q \alpha)(\exp_q \beta) = \exp_q((1 + (1 - q)\beta)\alpha + \beta)$.

Remark 2. Sometimes $\mathcal{P}_q = \mathcal{P}_q^o$ holds. In fact, the example given in Sect. 4 is the case where $\mathcal{P}_q = \mathcal{P}_q^o$. However, we do not give its proof in this article. On the other hand, there is an example where $\mathcal{P}_q^o \subsetneq \mathcal{P}_q$ (see Example 1 below).

Example 1. We put $X := \mathbb{R}$ and $q := 2$. We take a triple (μ, V, T) as follows: μ is the Lebesgue measure on X, $V := \mathbb{R}$, $T: X \to V$, $x \mapsto 1 + x^2$. Then we have

$$\mathcal{P}_q^o = \left\{ \frac{1}{\pi} \frac{\lambda}{\lambda^2 + x^2} dx \right\}_{\lambda > 1},$$

$$\mathcal{P}_q = \left\{ \frac{1}{\pi} \frac{\lambda}{\lambda^2 + x^2} dx \right\}_{\lambda > 0},$$

where dx denotes the Lebesgue measure on $\mathbb{R}$. Especially, we have $\mathcal{P}_q^o \subsetneq \mathcal{P}_q$.

4   A q-Exponential Family on the Upper Half Plane

We give an example of Definition 8 on the upper half plane.

We consider the upper half plane $X := \mathcal{H} = \{z \in \mathbb{C} \mid \operatorname{Im} z > 0\}$ with the Poincaré metric. Let μ be the Riemannian measure on X, that is, $d\mu(z) = y^{-2}dxdy$. We put $V := \operatorname{Sym}(2, \mathbb{R})$ and consider the following continuous map:

$$T: X \to V, \ z := x + \sqrt{-1}y \mapsto \frac{1}{y} \begin{pmatrix} x^2 + y^2 & x \\ x & 1 \end{pmatrix}.$$

We identify $V^\vee = \operatorname{Sym}(2, \mathbb{R})^\vee$ with $\operatorname{Sym}(2, \mathbb{R})$ by the following inner product:

$$\operatorname{Sym}(2, \mathbb{R}) \times \operatorname{Sym}(2, \mathbb{R}) \to \mathbb{R}, \quad (X, Y) \mapsto \operatorname{Tr}(XY).$$

We denote by $\operatorname{Sym}^+(2, \mathbb{R})$ the set of all positive definite symmetric matrices of size two.

Theorem 1. *Let $q \in \mathbb{R}$. The obtained q-exponential family $\mathcal{P}_q^o$ on X from the triple (μ, V, T) above by Definition 8 is given as follows: In the case where $q < 1$ or $2 \leq q$, $\mathcal{P}_q^o = \emptyset$, and in the case where $q \in [1, 2)$,*

$$\mathcal{P}_q^o = \left\{ c_\theta \exp_q \left(-\frac{a(x^2 + y^2) + 2bx + c}{y} \right) \frac{dxdy}{y^2} \right\}_{\theta := \begin{pmatrix} a & b \\ b & c \end{pmatrix} \in \mathrm{Sym}^+(2,\mathbb{R})},$$

where $c_\theta = \dfrac{(2-q)D}{\pi(\exp_q(-2D))^{2-q}}, \quad D := \sqrt{ac - b^2}.$

This theorem follows from Lemmas 1, 2, Proposition 2, and the calculation of normalizing constant in Sect. 5. Since the case of $q = 1$ is known in [15, Example 1], we assume $q \neq 1$.

Lemmas 1, 2 below follows from easy calculation.

Lemma 1. *We have*

$$\tilde{\Theta}_q = \begin{cases} \left\{ \begin{pmatrix} a & b \\ b & c \end{pmatrix} \middle| (a > 0, ac - b^2 \geq 0) \text{ or } (a = b = 0, c \geq 0) \right\} & (q > 1), \\ \left\{ \begin{pmatrix} a & b \\ b & c \end{pmatrix} \middle| (a < 0, ac - b^2 \geq 0) \text{ or } (a = b = 0, c \leq 0) \right\} & (q < 1). \end{cases}$$

Lemma 2. *In the case of $q > 1$,* $\begin{pmatrix} 0 & 0 \\ 0 & c \end{pmatrix} \notin \Theta_q^o$ *for $c \geq 0$.*

Proposition 2. *In the case of $q < 1$, the obtained family $\mathcal{P}_q^o$ from the triple (μ, V, T) is empty.*

Proof. Take any $\theta := (a, b, c) \in \tilde{\Theta}_q$. Then, we have $(a < 0$ and $ac - b^2 \geq 0)$ or $(a = b = 0$ and $c \leq 0)$. In the both cases, $\langle \theta, T(z) \rangle \leq ay$ for any $z \in \mathcal{H}$ holds, which implies

$$\exp_q(-\langle \theta, T(z) \rangle) \geq \exp_q(-ay).$$

Since the right hand side depends only on y, the function $\exp_q(-\langle \theta, T(z) \rangle)$ is not integrable.

Now, let us consider the $SL(2, \mathbb{R})$-invariance of the family $\mathcal{P}_q^o$. To see the natural $SL(2, \mathbb{R})$-action on $\mathcal{R}(X)$, let us recall that the Lie group $SL(2, \mathbb{R})$ acts on the upper half plane $\mathcal{H}$ as a fractional transformation. Namely,

$$g \cdot z := \frac{az + b}{cz + d}, \quad \text{where } g = \begin{pmatrix} a & b \\ c & d \end{pmatrix} \in SL(2, \mathbb{R}) \text{ and } z \in \mathcal{H}.$$

Theorem 2. *The obtained q-exponential family $\mathcal{P}_q^o$ on the upper half plane in Theorem 1 is $SL(2, \mathbb{R})$-invariant, and the $SL(2, \mathbb{R})$-action on the parameter space $\mathrm{Sym}^+(2, \mathbb{R})$ is given as follows:*

$$SL(2, \mathbb{R}) \times \mathrm{Sym}^+(2, \mathbb{R}) \to \mathrm{Sym}^+(2, \mathbb{R}), (g, S) \mapsto {}^t g^{-1} S g^{-1}.$$

Proof. The upper half plane with $SL(2,\mathbb{R})$-action can be identified with the homogeneous space $G/H = SL(2,\mathbb{R})/SO(2)$ by the G-equivariant map

$$\mathcal{H} \to G/H, \quad z \mapsto \alpha(z)H, \text{ where } \alpha(z) := \begin{pmatrix} \sqrt{y} & \frac{x}{\sqrt{y}} \\ 0 & \frac{1}{\sqrt{y}} \end{pmatrix} \in G \quad (z = x + \sqrt{-1}y),$$

which satisfies $g \cdot \alpha(z) = \alpha(g \cdot z)$ for $g \in G$, $z \in \mathcal{H}$. For $\theta = \begin{pmatrix} a & b \\ b & c \end{pmatrix} \in \mathrm{Sym}^+(2,\mathbb{R})$,

$$d\eta_\theta(z) = c_\theta \exp_q \left(-\frac{a(x^2 + y^2) + 2bx + c}{y} \right) d\mu(z)$$
$$= c_\theta \exp_q(-\mathrm{Tr}(\theta\alpha(z)^t\alpha(z)))d\mu(z).$$

Let $g \in G$, $\theta \in \Theta_q^o$ and put $\theta' := {}^tg^{-1}\theta g^{-1} \in \Theta_q^o$. Since the Riemannian measure μ on $\mathcal{H}$ is G-invariant, we have

$$d(g \cdot \eta_\theta)(z) \propto \exp_q \left(-\mathrm{Tr}\left(\theta g^{-1}\alpha(z)^t(g^{-1}\alpha(z))\right) \right) d(g \cdot \mu)(z)$$
$$= \exp_q(-\mathrm{Tr}({}^tg^{-1}\theta g^{-1}\alpha(z)^t\alpha(z)))d\mu(z) = d\tilde{\eta}_{\theta'}(z)$$

Since the pushforward of a probability measure is also a probability measure, we get $g \cdot \eta_\theta = \eta_{\theta'}$. Theorem 2 was proved.

5   Calculation of the Normalizing Constant

In this section, we verify that the family $\mathcal{P}_q^o$ has the explicit form as in Theorem 1 by calculating the normalizing constant. From Lemmas 1, 2 in Sect. 4, it is enough to consider the following:

Setting 3. *Let* $q > 1$, $\theta = \begin{pmatrix} a & b \\ b & c \end{pmatrix} \in \tilde{\Theta}_q$ *with* $a > 0$, $ac - b^2 \geq 0$ *and* $D := \sqrt{\det\theta} = \sqrt{ac - b^2}$. *We take* $r > 0$ *such that* $q = 1 + \frac{2}{r}$.

The normalizing constant of the family follows from the following:

Proposition 3. *Under Setting 3,*

$$\int_0^\infty \int_{-\infty}^\infty \exp_q \left(-\frac{a(x^2 + y^2) + 2bx + c}{y} \right) \frac{dxdy}{y^2}$$
$$= \begin{cases} \frac{\pi}{(2-q)D}(\exp_q(-2D))^{2-q} & (q < 2 \text{ and } ac - b^2 > 0), \\ \infty & (q \geq 2 \text{ or } ac - b^2 = 0). \end{cases}$$

To prove proposition above, we prepare a definition and lemmas.

Definition 9 (Wallis's integral). *For* $r \in \mathbb{R}$, *we call* W_r *below Wallis's integral.*

$$W_r := \int_0^{\frac{\pi}{2}} \cos^r \theta d\theta \in (0, \infty].$$

Lemma 3. *(i)* $W_r < \infty \iff r > -1$,

(ii) $\int_{-\infty}^{\infty} (x^2 + 1)^{-r} dx = 2W_{2(r-1)}$,

(iii) $\int_{-\infty}^{\infty} (ax^2 + b)^{-r} dx = \frac{2}{\sqrt{a}} b^{\frac{1}{2}-r} W_{2(r-1)}$ $(a, b > 0)$,

(iv) $\int_{0}^{\infty} (a(x^2 + x^{-2}) + 1)^{-r} dx = \frac{1}{\sqrt{a}} (2a + 1)^{\frac{1}{2}-r} W_{2(r-1)}$ $(a > 0, r \in \mathbb{R})$,

(v) $W_r = \frac{1}{2} B(\frac{1}{2}, \frac{r+1}{2})$ $(r > -1)$,

(vi) $W_r W_{r-1} = \frac{\pi}{2r}$ $(r > 0)$.

Proof. Since (i), (ii), (iii) and (v) are easy, we prove only (iv) and (vi).

(iv): Put $I := \int_{0}^{\infty} (a(x^2 + x^{-2}) + 1)^{-r} dx$ and $x = t^{-1}$, then we get

$$I = \int_{0}^{\infty} (a(t^2 + t^{-2}) + 1)^{-r} t^{-2} dt.$$

Therefore

$$I = \frac{1}{2} \int_{0}^{\infty} (1 + x^{-2})(a(x^2 + x^{-2}) + 1)^{-r} dx.$$

Put $u = x - x^{-1}$, then $I = \frac{1}{2} \int_{-\infty}^{\infty} (au^2 + 2a + 1)^{-r} du$. Thus (iv) follows from (iii).

(vi): From (v), we have

$$W_r W_{r-1} = \frac{1}{4} B\left(\frac{1}{2}, \frac{r+1}{2}\right) B\left(\frac{1}{2}, \frac{r}{2}\right) = \frac{1}{4} \frac{\Gamma(\frac{1}{2})\Gamma(\frac{r+1}{2})}{\Gamma(\frac{r}{2}+1)} \frac{\Gamma(\frac{1}{2})\Gamma(\frac{r}{2})}{\Gamma(\frac{r+1}{2})} = \frac{\pi}{2r}.$$

Here, we used $\Gamma(\frac{1}{2}) = \sqrt{\pi}$ and $\Gamma(s+1) = s\Gamma(s)$ for $s > 0$.

Lemma 4. *Under Setting 3,*

$$\int_{-\infty}^{\infty} \exp_q\left(-\frac{a(x^2 + y^2) + 2bx + c}{y}\right) dx = \sqrt{\frac{2ry}{a}} \left(\frac{2}{r}\left(ay + \frac{D^2}{ay}\right) + 1\right)^{\frac{1-r}{2}} W_{r-2}.$$

Proof. Since we have

$$\exp_q\left(-\frac{a(x^2 + y^2) + 2bx + c}{y}\right) = \exp_q\left(-\frac{a}{y}\left(x + \frac{b}{a}\right)^2 - \left(ay + \frac{D^2}{ay}\right)\right)$$

$$= \left(\left((q-1)\frac{a}{y}\right)\left(x + \frac{b}{a}\right)^2 + \left((q-1)\left(ay + \frac{D^2}{ay}\right) + 1\right)\right)^{-\frac{r}{2}}.$$

By applying Lemma 3 (iii), we get the desired equality.

Proof (of Proposition 3). By Tonelli's theorem and Lemma 4,

$$\int_{0}^{\infty} \int_{-\infty}^{\infty} \exp_q\left(-\frac{a(x^2 + y^2) + 2bx + c}{y}\right) \frac{dx\,dy}{y^2} \tag{1}$$

$$= W_{r-2} \sqrt{\frac{2r}{a}} \int_{0}^{\infty} \left(\frac{2}{r}\left(ay + \frac{D^2}{ay}\right) + 1\right)^{\frac{1-r}{2}} y^{-\frac{3}{2}} dy. \tag{2}$$

(Case1): $q \geq 2$.
Since we have $q \geq 2 \iff r \leq 1$, by Lemma 3 (i), this integral does not converge.
(Case2): $ac - b^2 = 0$.
We may assume $r > 1$. Since we have $D = 0$,

$$\int_0^\infty \left(\frac{2}{r}\left(ay + \frac{D^2}{ay} \right) + 1 \right)^{\frac{1-r}{2}} y^{-\frac{3}{2}} dy \geq \int_0^1 \left(\frac{2}{r}a + 1 \right)^{\frac{1-r}{2}} y^{-\frac{3}{2}} dy = \infty.$$

As a result, the integral (2) does not converge.
(Case3): Otherwise. Put $s := \sqrt{\frac{D}{ay}}$, and by using Lemma 3 (iv), we have

$$(2) = 2W_{r-2}\sqrt{\frac{2r}{D}} \int_0^\infty \left(\frac{2D}{r}(s^2 + s^{-2}) + 1 \right)^{\frac{1-r}{2}} ds = W_{r-2}W_{r-3}\frac{2r}{D}\left(\frac{4}{r}D + 1 \right)^{1-\frac{r}{2}}.$$

Here, $\left(\frac{4}{r}D + 1 \right)^{1-\frac{r}{2}} = (\exp_q(-2D))^{2-q}$ and from Lemma 3 (vi), $W_{r-2}W_{r-3}\frac{2r}{D} = \frac{\pi}{(2-q)D}$.

Acknowledgements. The first author is supported by Special Postdoctoral Researcher Program at RIKEN.

References

1. Amari, S., Ohara, A.: Geometry of q-exponential family of probability distributions. Entropy **13**, 1170–1185 (2011)
2. Barbaresco, F., Gay-Balmaz, F.: Lie group cohomology and (multi)symplectic integrators: new geometric tools for Lie group machine learning based on Souriau Geometric statistical mechanics. Entropy **22**(5), 498 (2022)
3. Barbaresco, F.: Symplectic theory of heat and information geometry. In: Handbook of Statistics, chap. 4, vol. 46, pp. 107–143. Elsevier (2022)
4. Barndorff-Nielsen, O.E.: Information and exponential families in statistical theory. Wiley Series in Probability and Statistics, John Wiley & Sons Ltd., Chichester (1978)
5. Barndorff-Nielsen, O.E.: Hyperbolic distributions and distribution on hyperbolae. Scand. J. Stat. **8**, 151–157 (1978)
6. Barndorff-Nielsen, O.E., Blæsild, P., Eriksen, P.S.: Decomposition and invariance of measures, and statistical transformation models. Springer-Verlag Lecture Note in Statistics (1989)
7. Barndorff-Nielsen, O.E., Blæsild, P., Jensen, J.L., Jørgensen, B.: Exponential transformation models. Proc. Roy. Soc. Lond. Ser. A **379**(1776), 41–65 (1982)
8. Bourbaki, N.: Integration (translated by Sterlin K. Beberian). Springer, Heidelberg (2004)
9. Folland, G.B.: A Course in Abstract Harmonic Analysis. CRC Press, Boca Raton (1995)
10. Jensen, J.L.: On the hyperboloid distribution. Scand. J. Statist. **8**, 193–206 (1981)
11. Kiral, E.M., Möllenhoff, T., Khan, M.E.: The lie-group bayesian learning rule. In: Proceedings of 26th International Conference on Artificial Intelligence and Statistics (AISTATS), vol. 206. PMLR (2023)

12. Marle, C.-M.: On Gibbs states of mechanical systems with symmetries. J. Geom. Symm. Phys. **57**, 45–85 (2020)
13. Matsuzoe, H., Ohara, A.: Geometry of q-exponential families. Recent Progress in Differential Geometry and its Related Fields (2011)
14. Tojo, K., Yoshino, T.: Harmonic exponential families on homogeneous spaces. Inf. Geo. **4**, 215–243 (2021)
15. Tojo, K., Yoshino, T.: An exponential family on the upper half plane and its conjugate prior. In: Barbaresco, F., Nielsen, F. (eds.) SPIGL 2020. SPMS, vol. 361, pp. 84–95. Springer, Cham (2021). https://doi.org/10.1007/978-3-030-77957-3_4
16. Tojo, K., Yoshino, T.: A method to construct exponential families by representation theory. Inf. Geo. **5**, 493–510 (2022)
17. Tsallis, C.: What are the numbers that experiments provide? Quimica Nova **17**, 468–471 (1994)

On the f-Divergences Between Hyperboloid and Poincaré Distributions

Frank Nielsen[1]([⊠]) [iD] and Kazuki Okamura[2] [iD]

[1] Sony Computer Science Laboratories Inc., Tokyo, Japan
Frank.Nielsen@acm.org
[2] Department of Mathematics, Faculty of Science,
Shizuoka University, Shizuoka, Japan
okamura.kazuki@shizuoka.ac.jp

Abstract. Hyperbolic geometry has become popular in machine learning due to its capacity to embed discrete hierarchical graph structures with low distortions into continuous spaces for further downstream processing. It is thus becoming important to consider statistical models and inference methods for data sets grounded in hyperbolic spaces. In this work, we study the statistical f-divergences between two kinds of hyperbolic distributions: The Poincaré distributions and the related hyperboloid distributions. By exhibiting maximal invariants of group actions, we show how these f-divergences can be expressed as functions of canonical terms.

Keywords: exponential family · group action · maximal invariant · Csiszár's f-divergence · hyperbolic distributions

1 Introduction

Hyperbolic geometry[1] [2] is very well suited for embedding tree graphs with low distortions [20] as hyperbolic Delaunay subgraphs of embedded tree nodes. So a recent trend in machine learning and data science is to embed *discrete* hierarchical graphs into *continuous* spaces with low distortions for further downstream processing. There exists many models of hyperbolic geometry [2] like the Poincaré disk or upper-half plane conformal models, the Klein non-conformal disk model, the Beltrami hemisphere model, the Minkowski or Lorentz hyperboloid model, etc. We can transform one model of hyperbolic geometry to another model by a bijective mapping yielding a corresponding isometric embedding [11]. As a byproduct of the low-distortion hyperbolic embeddings of hierarchical graphs, many embedded data sets are nowadays available in hyperbolic model spaces, and those data sets need to be further processed. Thus it is important to build *statistical models* and *inference methods* for these hyperbolic data

---

[1] Hyperbolic geometry has constant negative curvature and the volume of hyperbolic balls increases exponentially with respect to their radii rather than polynomially as in Euclidean spaces.

© The Author(s), under exclusive license to Springer Nature Switzerland AG 2023
F. Nielsen and F. Barbaresco (Eds.): GSI 2023, LNCS 14071, pp. 176–185, 2023.
https://doi.org/10.1007/978-3-031-38271-0_18

sets using probability distributions with support hyperbolic model spaces, and to consider statistical mixtures in those spaces for modeling arbitrary smooth densities.

Let us quickly review some of the various families of probability distributions defined in hyperbolic models as follows: One of the very first proposed family of such "hyperbolic distributions" was proposed in 1981 [16] and are nowadays commonly called the *hyperboloid distributions* in reference to their support: The hyperboloid distributions are defined on the Minkowski upper sheet hyperboloid by analogy to the von-Mises Fisher distributions [3] which are defined on the sphere. Another work by Barbaresco [4] defined the so-called Souriau-Gibbs distributions (2019) in the Poincaré disk (Eq. 57 of [4], a natural exponential family) with its Fisher information metric coinciding with the Poincaré hyperbolic Riemannian metric (the Poincaré unit disk is a homogeneous space where the Lie group $SU(1,1)$ acts transitively).

In this paper, we focus on Ali-Silvey-Csiszár's f-divergences between hyperbolic distributions [1,14]. In Sect. 2, we prove using Eaton's method of group action maximal invariants [15,19] that all f-divergences (including the Kullback-Leibler divergence) between Poincaré distributions [21] can be expressed canonically as functions of three terms (Proposition 1 and Theorem 1). Then, we deal with the hyperboloid distributions in dimension 2 in §3. We also consider q-deformed family of these distributions [23]. We exhibit a correspondence in §4 between the upper-half plane and the Minkowski hyperboloid 2D sheet. The f-divergences between the hyperboloid distributions are in spirit very geometric because it exhibits a beautiful and clear maximal invariant which has connections with the side-angle-side congruence criteria for triangles in hyperbolic geometry. This paper summarizes the preprint [18] with some proofs omitted: We refer the reader to the preprint for more details and other topics than f-divergences.

2 The Poincaré Distributions

Tojo and Yoshino [21–23] described a versatile method to build exponential families of distributions on homogeneous spaces which are invariant under the action of a Lie group G generalizing the construction in [13]. They exemplify their so-called "G/H-method" on the upper-half plane $\mathbb{H} := \{(x,y) \in \mathbb{R}^2 : y > 0\}$ by constructing an exponential family with probability density functions invariant under the action of Lie group $G = SL(2, \mathbb{R})$, the set of invertible matrices with unit determinant. We call these distributions the Poincaré distributions, since their sample space $\mathcal{X} = G/H \simeq \mathbb{H}$, and we study this set of distributions as an exponential family [8]: The probability density function (pdf) of a Poincaré distribution [21] expressed using a 3D vector parameter $\theta = (a, b, c) \in \mathbb{R}^3$ is given by

$$p_\theta(x,y) := \frac{\sqrt{ac - b^2}\exp(2\sqrt{ac - b^2})}{\pi} \exp\left(-\frac{a(x^2 + y^2) + 2bx + c}{y}\right)\frac{1}{y^2}, \quad (1)$$

where θ belongs to the parameter space

$$\Theta := \{(a,b,c) \in \mathbb{R}^3 : a > 0, c > 0, \ ac - b^2 > 0\}.$$

The set Θ forms an open 3D convex cone. Thus the Poincaré distribution family has a 3D parameter cone space and the sample space is the hyperbolic upper plane. We can also use a matrix form to express the pdf. Indeed, we can naturally identify Θ with the set of real symmetric positive-definite matrices $\mathrm{Sym}^+(2, \mathbb{R})$ by the mapping $(a, b, c) \mapsto \begin{bmatrix} a & b \\ b & c \end{bmatrix}$. Hereafter, we denote the determinant of θ by $|\theta| := ac - b^2 > 0$ and the trace of θ by $\mathrm{tr}(\theta) = a + c$ for $\theta = (a, b, c) \simeq \begin{bmatrix} a & b \\ b & c \end{bmatrix}$.

The f-divergence [1,14] induced by a convex generator $f : (0, \infty) \to \mathbb{R}$ between two pdfs $p(x, y)$ and $q(x, y)$ defined on the support $\mathbb{H}$ is defined by

$$D_f[p : q] := \int_{\mathbb{H}} p(x, y) f\left(\frac{q(x, y)}{p(x, y)}\right) \, \mathrm{d}x \, \mathrm{d}y. \tag{2}$$

Since $D_f[p : q] \geq f(1)$, we consider convex generators $f(u)$ such that $f(1) = 0$. Moreover, in order to satisfy the law of the indiscernibles (i.e., $D_f[p : q] = 0$ iff $p(x, y) = q(x, y)$), we require f to be strictly convex at 1. The class of f-divergences includes the total variation distance ($f(u) = |u - 1|$), the Kullback-Leibler divergence ($f(u) = -\log(u)$, and its two common symmetrizations, namely, the Jeffreys divergence and the Jensen-Shannon divergence), the squared Hellinger divergence, the Pearson and Neyman sided χ^2-divergences, the α-divergences, etc.

We state the notion of maximal invariant by following [15]: Let G be a group acting on a set X. We denote the group action by $(g, x) \mapsto gx$.

Definition 1 (Maximal invariant). *We say that a map φ from X to a set Y is maximal invariant if it is invariant, specifically, $\varphi(gx) = \varphi(x)$ for every $g \in G$ and $x \in X$, and furthermore, whenever $\varphi(x_1) = \varphi(x_2)$ there exists $g \in G$ such that $x_2 = gx_1$.*

It can be shown that every invariant map is a function of a maximal invariant [15]. Specifically, if a map ψ from X to a set Z is invariant, then, there exists a unique map Φ from $\varphi(X)$ to Z such that $\Phi \circ \varphi = \psi$.

These invariant/maximal invariant concepts can be understood using group orbits: For each $x \in X$, we may consider its orbit $O_x := \{gx \in X : g \in G\}$. A map is invariant when it is constant on orbits and maximal invariant when orbits have distinct map values.

We denote by $A^\top e$ the transpose of a square matrix A and $A^{-\top}$ the transpose of the inverse matrix A^{-1} of a regular matrix A. It holds that $A^{-\top} = (A^\top)^{-1}$.

Let $\mathrm{SL}(2, \mathbb{R})$ be the group of 2×2 real matrices with unit determinant.

Proposition 1. *Define a group action of $\mathrm{SL}(2, \mathbb{R})$ to $\mathrm{Sym}^+(2, \mathbb{R})^2$ by*

$$(g, (\theta, \theta')) \mapsto (g^{-\top}\theta g^{-1}, g^{-\top}\theta' g^{-1}). \tag{3}$$

Define a map $S : \mathrm{Sym}^+(2, \mathbb{R})^2 \to (\mathbb{R}_{>0})^2 \times \mathbb{R}$ by

$$S(\theta, \theta') := \left(|\theta|, |\theta'|, \mathrm{tr}(\theta'\theta^{-1})\right). \tag{4}$$

Then, the map S is maximal invariant of the group action.

Proof. Observe that S is invariant with respect to the group action: $S(\theta, \theta') = S(g.\theta, g.\theta')$. Assume that $S\left(\theta^{(1)}, \theta^{(2)}\right) = S\left(\widetilde{\theta^{(1)}}, \widetilde{\theta^{(2)}}\right)$. We see that there exists $g_{\theta^{(1)}} \in \mathrm{SL}(2, \mathbb{R})$ such that $g_{\theta^{(1)}}.\theta^{(1)} = g_{\theta^{(1)}}^{-\top}\theta^{(1)}g_{\theta^{(1)}}^{-1} = \sqrt{|\theta^{(1)}|}I_2$, where I_2 denotes the 2×2 identity matrix. Then, $\theta^{(1)} = \sqrt{|\theta^{(1)}|}g_{\theta^{(1)}}^{\top}g_{\theta^{(1)}}$. Let $\theta^{(3)} := g_{\theta^{(1)}}.\theta^{(2)} = g_{\theta^{(1)}}^{-\top}\theta^{(2)}g_{\theta^{(1)}}^{-1}$. Then $\mathrm{tr}\left(\theta^{(3)}\right) = \mathrm{tr}\left(\theta^{(2)}g_{\theta^{(1)}}^{-1}g_{\theta^{(1)}}^{-\top}\right) = \sqrt{|\theta^{(1)}|}\,\mathrm{tr}\left(\theta^{(2)}(\theta^{(1)})^{-1}\right)$. We define $g_{\widetilde{\theta^{(1)}}}$ and $\widetilde{\theta^{(3)}}$ in the same manner. Then, $\mathrm{tr}\left(\theta^{(3)}\right) = \mathrm{tr}\left(\widetilde{\theta^{(3)}}\right)$ and $|\theta^{(3)}| = \left|\widetilde{\theta^{(3)}}\right|$. Hence the set of eigenvalues of $\theta^{(3)}$ and $\widetilde{\theta^{(3)}}$ are identical with each other. By this and $\theta^{(3)}, \widetilde{\theta^{(3)}} \in \mathrm{Sym}(2, \mathbb{R})$, there exists $h \in \mathrm{SO}(2)$ such that $h.\theta^{(3)} = \widetilde{\theta^{(3)}}$. Hence $(hg_{\theta^{(1)}}).\theta^{(2)} = g_{\widetilde{\theta^{(1)}}}\widetilde{\theta^{(2)}}$. We also see that

$$(hg_{\theta^{(1)}}).\theta^{(1)} = g_{\theta^{(1)}}.\theta^{(1)} = \sqrt{|\theta^{(1)}|}\,I_2 = \sqrt{|\widetilde{\theta^{(1)}}|}\,I_2 = g_{\widetilde{\theta^{(1)}}}.\widetilde{\theta^{(1)}}.$$

Thus we have $\left(\widetilde{\theta^{(1)}}, \widetilde{\theta^{(2)}}\right) = (g_{\theta^{(1)}}^{-1}hg_{\theta^{(1)}}).(\theta^{(1)}, \theta^{(2)})$.

Remark 1 (This is pointed by an anonymous referee.). We can consider an extension of Proposition 1 to a case of higher degree of matrices. Let $n \geq 2$ and assume that $\theta, \theta' \in \mathrm{Sym}(n, \mathbb{R})$. Let $P_{\theta,\theta'}(t) := |(1-t)\theta + t\theta'|$ for $t \in \mathbb{R}$. where $|A|$ denotes the determinant of a square matrix A. This is a polynomial in t with degree n. Assume that $P_{\theta_1,\theta_1'} = P_{\theta_2,\theta_2'}$ for $\theta_1, \theta_1', \theta_2, \theta_2' \in \mathrm{Sym}(n, \mathbb{R})$. We can factor θ_i as $\theta_i = L_i^{\top}L_i$ for some L_i, $i = 1, 2$. Let I_n be the identity matrix of degree n. Then, $P_{\theta_i,\theta_i'}(t) = |\theta_i||I_n + t(L_i^{-\top}\theta_i'L_i^{-1} - I_n)|$, $i = 1, 2$. Since $L_i^{-\top}\theta_i'L_i^{-1} \in \mathrm{Sym}(n, \mathbb{R})$, the set of eigenvalues of $L_1^{-\top}\theta_1'L_1^{-1}$ and $L_2^{-\top}\theta_2'L_2^{-1}$ is identical with each other. Hence there exists an orthogonal matrix Q such that $L_2^{-\top}\theta_2'L_2^{-1} = Q^{\top}L_1^{-\top}\theta_1'L_1^{-1}Q$. Let $G := L_1^{-1}QL_2$. Then, $\theta_2 = G^{\top}\theta_1G$ and $\theta_2' = G^{\top}\theta_1'G$. We finally remark that $P_{\theta_1,\theta_1'} = P_{\theta_2,\theta_2'}$ holds if and only if $P_{\theta_1,\theta_1'}(t) = P_{\theta_2,\theta_2'}(t)$ for $n+1$ different values of t.

If $n = 2$, then,

$$P_{\theta,\theta'}(t) = (1-t)^2|\theta|^2 + t^2|\theta'|^2 + t(1-t)|\theta|\,\mathrm{tr}(\theta'\theta^{-1}). \tag{5}$$

Hence the arguments above give an alternative proof of Proposition 1.

Proposition 2 (Invariance of f-divergences under group action).

$$D_f\left[p_\theta : p_{\theta'}\right] = D_f\left[p_{g^{-\top}\theta g^{-1}} : p_{g^{-\top}\theta'g^{-1}}\right].$$

For $g \in \mathrm{SL}(2, \mathbb{R})$, we denote the pushforward measure of a measure ν on $\mathbb{H}$ by the map $z \mapsto g.z$ on $\mathbb{H}$ by $\nu \circ g^{-1}$.

The latter part of the following proof utilizes the method used in the proof of [21, Proposition 1].

Proof. We first see that for $g \in \mathrm{SL}(2, \mathbb{R})$,

$$D_f [p_\theta : p_{\theta'}] = D_f [p_\theta \circ g^{-1} : p_{\theta'} \circ g^{-1}]. \tag{6}$$

Let $\mu(\mathrm{d}x\mathrm{d}y) := \mathrm{d}x\mathrm{d}y/y^2$. Then it is well-known that μ is invariant with respect to the action of $\mathrm{SL}(2, \mathbb{R})$ on $\mathbb{H}$, that is, $\mu = \mu \circ g^{-1}$ for $g \in \mathrm{SL}(2, \mathbb{R})$.
Define a map $\varphi : \Theta \times \mathbb{H} \to \mathbb{R}_{>0}$ by

$$\varphi(\theta, x + yi) := \frac{a(x^2 + y^2) + 2bx + c}{y}, \quad \theta = \begin{bmatrix} a & b \\ b & c \end{bmatrix}.$$

Then, $\varphi(\theta, z) = \varphi(g.\theta, g.z)$ for $g \in \mathrm{SL}(2, \mathbb{R})$.
Since

$$p_\theta(x, y)\mathrm{d}x\mathrm{d}y = \frac{\sqrt{|\theta|}\exp(2\sqrt{|\theta|})}{\pi} \exp(-\varphi(\theta, x + yi))\mu(\mathrm{d}x\mathrm{d}y),$$

we have $p_\theta \circ g^{-1} = p_{g.\theta}$. Hence,

$$D_f [p_\theta \circ g^{-1} : p_{\theta'} \circ g^{-1}] = D_f [p_{g.\theta} : p_{g.\theta'}]. \tag{7}$$

The assertion follows from (6) and (7).

By Propositions 1 and 2, we get

Theorem 1. *Every f-divergence between two Poincaré distributions p_θ and $p_{\theta'}$ is a function of $\left(|\theta|, |\theta'|, \mathrm{tr}\left(\theta'\theta^{-1}\right)\right)$ and invariant with respect to the $\mathrm{SL}(2, \mathbb{R})$-action.*

We obtained exact formulae for the Kullback-Leibler divergence, the squared Hellinger divergence, and the Neyman chi-squared divergence.

Proposition 3. *We have the following results for two Poincaré distributions p_θ and $p_{\theta'}$.*

(i) (Kullback-Leibler divergence) Let $f(u) = -\log u$. Then,

$$D_f [p_\theta : p_{\theta'}] = \frac{1}{2} \log \frac{|\theta|}{|\theta'|} + 2 \left(\sqrt{|\theta|} - \sqrt{|\theta'|} \right) + \left(\frac{1}{2} + \sqrt{|\theta|} \right) (\mathrm{tr}(\theta'\theta^{-1}) - 2). \tag{8}$$

(ii) (squared Hellinger divergence) Let $f(u) = (\sqrt{u} - 1)^2/2$. Then,

$$D_f[p_\theta : p_{\theta'}] = 1 - \frac{2|\theta|^{1/4}|\theta'|^{1/4} \exp\left(|\theta|^{1/2} + |\theta'|^{1/2}\right)}{|\theta + \theta'|^{1/2} \exp\left(|\theta + \theta'|^{1/2}\right)}. \tag{9}$$

(iii) (Neyman chi-squared divergence) Let $f(u) := (u-1)^2$. Assume that $2\theta' - \theta \in \Theta$. Then,

$$D_f[p_\theta : p_{\theta'}] = \frac{|\theta'| \exp(4|\theta'|^{1/2})}{|\theta|^{1/2}|2\theta' - \theta|^{1/2} \exp\left(2(|\theta|^{1/2} + |2\theta' - \theta|^{1/2})\right)} - 1. \tag{10}$$

We remark that $|\theta + \theta'|$ and $|2\theta' - \theta|$ can be expressed by using $|\theta|, |\theta'|$, and $\operatorname{tr}(\theta'\theta^{-1})$. Indeed, we have

$$|\theta + \theta'| = |\theta| + |\theta'| + |\theta|\operatorname{tr}(\theta'\theta^{-1}),$$
$$|2\theta' - \theta| = 4|\theta'| + |\theta| - 2|\theta|\operatorname{tr}(\theta'\theta^{-1}).$$

Thus the KLD between two Poincaré distributions is asymmetric in general. The situation is completely different from the Cauchy distribution whose f-divergences are always symmetric [19, 24].

Recently, Tojo and Yoshino [23] introduced a notion of deformed exponential family associated with their G/H method in representation theory. As an example of it, they considered a family of *deformed Poincaré distributions* with index $q > 1$. For $x \in I_q := \{x \in \mathbb{R} : (1 - q)x + 1 > 0\}$, let

$$\exp_q(x) := ((1 - q)x + 1)^{1/(1-q)}, \quad x \in I_q.$$

For $q \in [1, 2)$, let a q-deformed Poincaré distribution be the distribution

$$p_\theta(x, y) := c_q(\sqrt{|\theta|})\exp_q\left(-\frac{a(x^2 + y^2) + 2bx + c}{y}\right)\frac{1}{y^2}, \tag{11}$$

where $\theta \in \Theta$ and $c_q(x) := \dfrac{(2 - q)x}{\pi(\exp_q(-2x))^{2-q}}$. In this case, Proposition 2 holds for q-deformed Poincaré distributions, so we also obtain that

Theorem 2. *Let* $q \in [1, 2)$. *Every* f-*divergence between two* q-*deformed Poincaré distributions* p_θ *and* $p_{\theta'}$ *is a function of* $\left(|\theta|, |\theta'|, \operatorname{tr}\left(\theta'\theta^{-1}\right)\right)$.

We can show this by Theorem 4 below and the correspondence principle in §4.

3   The Two-Dimensional Hyperboloid Distributions

We first give the definition of the Lobachevskii space (in reference to Minkowski hyperboloid model of hyperbolic geometry also called the Lorentz model) and the parameter space of the hyperboloid distribution. We focus on the bidimensional case $d = 2$. Let

$$\mathbb{L}^2 := \left\{(x_0, x_1, x_2) \in \mathbb{R}^3 : x_0 = \sqrt{1 + x_1^2 + x_2^2}\right\},$$

and

$$\Theta_{\mathbb{L}^2} := \left\{(\theta_0, \theta_1, \theta_2) \in \mathbb{R}^3 : \theta_0 > \sqrt{\theta_1^2 + \theta_2^2}\right\}.$$

Let the Minkowski inner product [12] be

$$[(x_0, x_1, x_2), (y_0, y_1, y_2)] := x_0 y_0 - x_1 y_1 - x_2 y_2.$$

We have $\mathbb{L}^2 = \{x \in \mathbb{R}^3 : [x, x] = 1\}$.

Now we define the *hyperboloid distribution* by following [5,7,9]. Hereafter, for ease of notation, we let $|\theta| := [\theta, \theta]^{1/2}$, $\theta \in \Theta_{\mathbb{L}^2}$. For $\theta \in \Theta_{\mathbb{L}^2}$, we define a probability measure P_θ on $\mathbb{L}^d \simeq \mathbb{R}^d$ by

$$P_\theta(\mathrm{d}x_1\mathrm{d}x_2) := c_2(|\theta|)\exp(-[\theta, \widetilde{x}])\mu(\mathrm{d}x_1\mathrm{d}x_2), \tag{12}$$

where we let $c_2(t) := \frac{t\exp(t)}{2(2\pi)^{1/2}}$, $t > 0$, $\widetilde{x} := \left(\sqrt{1 + x_1^2 + x_2^2}, x_1, x_2\right)$, and $\mu(\mathrm{d}x_1\mathrm{d}x_2) := \frac{1}{\sqrt{1+x_1^2+x_2^2}}\mathrm{d}x_1\mathrm{d}x_2$.

Remark 2. The 1D hyperboloid distribution was first introduced in statistics in 1977 [6] to model the log-size distributions of particles from aeolian sand deposits, but the 3D hyperboloid distribution was later found already studied in statistical physics in 1911 [17]. The 2D hyperboloid distribution was investigated in 1981 [10].

Now we consider group actions on the space of parameters $\Theta_{\mathbb{L}^2}$. Let the indefinite special orthogonal group be

$$\mathrm{SO}(1,2) := \left\{A \in \mathrm{SL}(3, \mathbb{R}) : [Ax, Ay] = [x, y] \; \forall x, y \in \mathbb{R}^3\right\},$$

and $\mathrm{SO}_0(1,2) := \left\{A \in \mathrm{SO}(1,2) : A(\mathbb{L}^2) = \mathbb{L}^2\right\}$.

An action of $\mathrm{SO}_0(1,2)$ to $(\Theta_{\mathbb{L}^2})^2$ is defined by

$$\mathrm{SO}_0(1,2) \times (\Theta_{\mathbb{L}^2})^2 \ni (A, (\theta, \theta')) \mapsto (A\theta, A\theta') \in (\Theta_{\mathbb{L}^2})^2.$$

Proposition 4. $(\theta, \theta') \mapsto ([\theta, \theta], [\theta', \theta'], [\theta, \theta'])$ *is maximal invariant for the action of* $\mathrm{SO}_0(1,2)$ *to* $(\Theta_{\mathbb{L}^2})^2$.

In the following proof, all vectors are column vectors.

Proof. It is clear that the map is invariant with respect to the group action. Assume that

$$\left([\theta^{(1)}, \theta^{(1)}], [\theta^{(2)}, \theta^{(2)}], [\theta^{(1)}, \theta^{(2)}]\right) = \left(\left[\widetilde{\theta^{(1)}}, \widetilde{\theta^{(1)}}\right], \left[\widetilde{\theta^{(2)}}, \widetilde{\theta^{(2)}}\right], \left[\widetilde{\theta^{(1)}}, \widetilde{\theta^{(2)}}\right]\right).$$

Let $\psi_i := \frac{\theta^{(i)}}{|\theta^{(i)}|}$, $\widetilde{\psi}_i := \frac{\widetilde{\theta^{(i)}}}{|\widetilde{\theta^{(i)}}|}$, $i = 1, 2$. Then, $[\psi_1, \psi_2] = \left[\widetilde{\psi}_1, \widetilde{\psi}_2\right]$.

We first consider the case that $\psi_1 = \widetilde{\psi}_1 = (1, 0, 0)^\top$. Let $\psi_i = (x_{i0}, x_{i1}, x_{i2})^\top$, $\widetilde{\psi}_i = (\widetilde{x}_{i0}, \widetilde{x}_{i1}, \widetilde{x}_{i2})^\top$, $i = 1, 2$. Then, $x_{20} = \widetilde{x}_{20} > 0$, $x_{21}^2 + x_{22}^2 = \widetilde{x}_{21}^2 + \widetilde{x}_{22}^2$ and hence there exists a special orthogonal matrix P such that $P(x_{21}, x_{22})^\top = (\widetilde{x}_{21}, \widetilde{x}_{22})^\top$. Let $A := \begin{pmatrix} 1 & 0 \\ 0 & P \end{pmatrix}$. Then, $A \in \mathrm{SO}_0(1,2)$, $A\psi_1 = (1, 0, 0)^\top = \widetilde{\psi}_1$ and $A\psi_2 = \widetilde{\psi}_2$.

We second consider the general case. Since the action of $\mathrm{SO}_0(1,2)$ to $\mathbb{L}^2$ defined by $(A, \psi) \mapsto A\psi$ is transitive, there exist $A, B \in \mathrm{SO}_0(1,2)$ such that $A\psi_1 = B\widetilde{\psi}_1 = (1, 0, 0)^\top$. Thus this case is attributed to the first case.

We regard μ as a probability measure on $\mathbb{L}^2$. We recall that $[A\theta, A\widetilde{x}] = [\theta, \widetilde{x}]$ for $A \in SO_0(1, 2)$. We remark that μ is an $SO(1, 2)$-invariant Borel measure [16] on $\mathbb{L}^2$. Now we have that

Theorem 3. *Every f-divergence between p_θ and $p_{\theta'}$ is invariant with respect to the action of $SO_0(1, 2)$, and is a function of the triplet $([\theta, \theta], [\theta', \theta'], [\theta, \theta'])$, i.e., the pairwise Minkowski inner products of θ and θ'.*

There is a clear geometric interpretation of this fact: The side-angle-side theorem for triangles in Euclidean geometry states that if two sides and the included angle of one triangle are equal to two sides and the included angle of another triangle, then the triangles are congruent. This is also true for the hyperbolic geometry and it corresponds to Proposition 4 above. Every f-divergence is determined by the triangle formed by a pair of the parameters (θ, θ') when f is fixed.

Proposition 5. *We have the following results for two hyperboloid distributions p_θ and $p_{\theta'}$.*

(i) (Kullback-Leibler divergence) Let $f(u) = -\log u$. Then,

$$D_f[p_\theta : p_{\theta'}] = \log\left(\frac{|\theta|}{|\theta'|}\right) - |\theta'| + \frac{[\theta, \theta']}{[\theta, \theta]} + \frac{[\theta, \theta']}{|\theta|} - 1. \tag{13}$$

(ii) (squared Hellinger divergence) Let $f(u) = (\sqrt{u} - 1)^2/2$. Then,

$$D_f[p_\theta : p_{\theta'}] = 1 - \frac{2|\theta|^{1/2}|\theta'|^{1/2}\exp(|\theta|/2 + |\theta'|/2)}{|\theta + \theta'|\exp(|\theta + \theta'|/2)}. \tag{14}$$

(iii) (Neyman chi-squared divergence) Let $f(u) := (u-1)^2$. Assume that $2\theta' - \theta \in \Theta_{\mathbb{L}^2}$. Then,

$$D_f[p_\theta : p_{\theta'}] = \frac{|\theta'|^2\exp(2|\theta'|)}{|\theta||2\theta' - \theta|\exp(|\theta| + |2\theta' - \theta|)} - 1. \tag{15}$$

Now we consider deformations of the hyperboloid distribution. For $q \in [1, 2)$, we let a q-deformed hyperboloid distribution be the distribution

$$p_\theta(x_1, x_2) := c_q(|\theta|)\exp_q\left(-[\theta, \widetilde{x}]\right)\frac{1}{\sqrt{1 + x_1^2 + x_2^2}}, \tag{16}$$

where $c_q(z) := \dfrac{(2 - q)z}{2\pi(\exp_q(-z))^{2-q}}$.

In the same manner as in the derivation of Theorem 3, we obtain that

Theorem 4 (Canonical terms of the f-divergences between deformed hyperboloid distributions). *Let $q \in [1, 2)$. Then, every f-divergence between q-deformed hyperboloid distributions p_θ and $p_{\theta'}$ is invariant with respect to the action of $SO_0(1, 2)$, and is a function of the triplet $([\theta, \theta], [\theta', \theta'], [\theta, \theta'])$.*

4 Correspondence Principle

It is well-known that there exists a correspondence between the 2D Lobachevskii space $\mathbb{L} = \mathbb{L}^2$ (hyperboloid model) and the Poincaré upper-half plane $\mathbb{H}$.

Proposition 6 (Correspondence between the parameter spaces). *For* $\theta = (a, b, c) \in \Theta_{\mathbb{H}} := \{(a, b, c) : a > 0, c > 0, ac > b^2\}$, *let* $\theta_{\mathbb{L}} := (a+c, a-c, 2b) \in \Theta_{\mathbb{L}}$. *We denote the f-divergence on $\mathbb{L}$ and $\mathbb{H}$ by $D_f^{\mathbb{L}}[\cdot : \cdot]$ and $D_f^{\mathbb{H}}[\cdot : \cdot]$ respectively. Then,*

(i) For $\theta, \theta' \in \Theta_{\mathbb{H}}$,

$$|\theta_{\mathbb{L}}|^2 = [\theta_{\mathbb{L}}, \theta_{\mathbb{L}}] = 4|\theta|, \ |\theta_{\mathbb{L}}'|^2 = [\theta_{\mathbb{L}}', \theta_{\mathbb{L}}'] = 4|\theta'|, \ [\theta_{\mathbb{L}}, \theta_{\mathbb{L}}'] = 2|\theta|\mathrm{tr}(\theta'\theta^{-1}). \quad (17)$$

(ii) For every f and $\theta, \theta' \in \mathbb{H}$,

$$D_f^{\mathbb{L}}\left[p_{\theta_{\mathbb{L}}} : p_{\theta_{\mathbb{L}}'}\right] = D_f^{\mathbb{H}}\left[p_\theta : p_{\theta'}\right]. \quad (18)$$

For (i), at its first glance, there seems to be an inconsistency in notation. However, $|\theta|$ is the Minkowski norm for $\theta \in \theta_{\mathbb{L}}$, and, $|\theta|$ is the determinant for $\theta \in \Theta_v$, so the notation is consistent in each setting. By this assertion, it suffices to compute the f-divergences between the hyperboloid distributions on $\mathbb{L}$.

Let $\mu_{\mathbb{H}}(\mathrm{d}x\mathrm{d}y) := \dfrac{\mathrm{d}x\mathrm{d}y}{y^2}$ and $\mu_{\mathbb{L}}(\mathrm{d}x\mathrm{d}y) := \dfrac{\mathrm{d}x\mathrm{d}y}{\sqrt{1 + x^2 + y^2}}$. By the change of variable

$$\mathbb{H} \ni (x, y) \mapsto (X, Y) = \left(\frac{1 - x^2 - y^2}{2y}, \frac{x}{y}\right) \in \mathbb{R}^2,$$

by recalling the correspondence between the parameters in Eq. (17), it holds that $y^2 p_\theta(x, y) = \sqrt{1 + X^2 + Y^2} p_{\theta_{\mathbb{L}}}(X, Y)$, and $\mu_{\mathbb{H}}(\mathrm{d}x\mathrm{d}y) = \mu_{\mathbb{L}}(\mathrm{d}X\mathrm{d}Y)$.

Acknowledgements. The authors are grateful to two anonymous referees for valuable comments. It is worth of special mention that Remark 1 is suggested by one referee and the proof of Proposition 2 is suggested by the other referee.

References

1. Ali, S.M., Silvey, S.D.: A general class of coefficients of divergence of one distribution from another. J. Roy. Stat. Soc.: Ser. B (Methodol.) **28**(1), 131–142 (1966)
2. Anderson, J.W.: Hyperbolic geometry. Springer Science & Business Media (2006)
3. Banerjee, A., Dhillon, I.S., Ghosh, J., Sra, S., Ridgeway, G.: Clustering on the Unit Hypersphere using von Mises-Fisher Distributions. J. Mach. Learn. Res. **6**(9) (2005)
4. Barbaresco, F.: Lie group machine learning and gibbs density on Poincaré unit disk from souriau lie groups thermodynamics and su(1,1) coadjoint orbits. In: Nielsen, F., Barbaresco, F. (eds.) GSI 2019. LNCS, vol. 11712, pp. 157–170. Springer, Cham (2019). https://doi.org/10.1007/978-3-030-26980-7_17
5. Barndorff-Nielsen, O.: Hyperbolic distributions and distributions on hyperbolae. Scand. J. Stat. **5**, 151–157 (1978)

6. Barndorff-Nielsen, O.: Exponentially decreasing distributions for the logarithm of particle size. Proc. Royal Society of London. A. Math. Phys. Sci. **353**(1674), 401–419 (1977)

7. Barndorff-Nielsen, O.: The hyperbolic distribution in statistical physics. Scandinavian J. Stat. **9**(1), 43–46 (1982)

8. Barndorff-Nielsen, O.: Information and exponential families: in statistical theory. John Wiley & Sons (2014)

9. Barndorff-Nielsen, O., Blaesild, P.: Hyperbolic distributions and ramifications: Contributions to theory and application. In: Statistical distributions in scientific work, pp. 19–44. Springer (1981). https://doi.org/10.1007/978-94-009-8549-0_2

10. Blæsild, P.: The two-dimensional hyperbolic distribution and related distributions, with an application to Johannsen's bean data. Biometrika **68**(1), 251–263 (1981)

11. Cannon, J.W., Floyd, W.J., Kenyon, R., Parry, W.R., et al.: Hyperbolic geometry. Flavors of geometry **31**(59–115), 2 (1997)

12. Cho, H., DeMeo, B., Peng, J., Berger, B.: Large-margin classification in hyperbolic space. In: The 22nd International Conference on Artificial Intelligence and Statistics, pp. 1832–1840. PMLR (2019)

13. Cohen, T., Welling, M.: Harmonic exponential families on manifolds. In: International Conference on Machine Learning, pp. 1757–1765. PMLR (2015)

14. Csiszár, I.: Eine informationstheoretische ungleichung und ihre anwendung auf beweis der ergodizitaet von markoffschen ketten. Magyer Tud. Akad. Mat. Kutato Int. Koezl. 8, 85–108 (1964)

15. Eaton, M.L.: Group invariance applications in statistics. Hayward, CA: Institute of Mathematical Statistics; Alexandria, VA: American Statistical Association (1989)

16. Jensen, J.L.: On the hyperboloid distribution. Scandinavian J. Stat.**8**(4), 193–206 (1981)

17. Jüttner, F.: Das Maxwellsche Gesetz der Geschwindigkeitsverteilung in der Relativtheorie. Ann. Phys. **339**(5), 856–882 (1911)

18. Nielsen, F., Okamura, K.: Information measures and geometry of the hyperbolic exponential families of Poincaré and hyperboloid distributions. arXiv preprint arXiv:2205.13984 (2022)

19. Nielsen, F., Okamura, K.: On f-divergences between Cauchy distributions. IEEE Trans. Inf. Theory **69**(5), 3150–3171 (2023)

20. Sarkar, R.: Low distortion delaunay embedding of trees in hyperbolic plane. In: van Kreveld, M., Speckmann, B. (eds.) GD 2011. LNCS, vol. 7034, pp. 355–366. Springer, Heidelberg (2012). https://doi.org/10.1007/978-3-642-25878-7_34

21. Tojo, K., Yoshino, T.: An exponential family on the upper half plane and its conjugate prior. In: Barbaresco, F., Nielsen, F. (eds.) SPIGL 2020. SPMS, vol. 361, pp. 84–95. Springer, Cham (2021). https://doi.org/10.1007/978-3-030-77957-3_4

22. Tojo, K., Yoshino, T.: Harmonic exponential families on homogeneous spaces. Inform. Geometry 4(1), 215–243 (2021)

23. Tojo, K., Yoshino, T.: Classification problem of invariant q-exponential families on homogeneous spaces. In: Mathematical optimization and statistical theories using geometric methods OCAMI Reports, Vol. 8, pp. 85–104. Osaka Central Advanced Mathematical Institute (OCAMI) (2022). https://doi.org/10.24544/ocu.20221208-007

24. Verdú, S.: The Cauchy Distribution in Information Theory. entropy **25**(2), 346 (2023)

λ-Deformed Evidence Lower Bound (λ-ELBO) Using Rényi and Tsallis Divergence

Kaiming Cheng[1] and Jun Zhang[1,2(✉)]

[1] Department of Statistics, University of Michigan, Ann Arbor, MI 48109, USA
[2] Department of Psychology, University of Michigan, Ann Arbor, MI 48109, USA
{kaimingc,junz}@umich.edu

Abstract. We investigate evidence lower bound (ELBO) with generalized/deformed entropy and generalized/deformed divergence, in place of Shannon entropy and KL divergence in the standard framework. Two equivalent forms of deformed ELBO have been proposed, suitable for either Tsallis or Rényi deformation that have been unified in the recent framework of λ-deformation (Wong and Zhang, 2022, IEEE Trans Inform Theory). The decomposition formulae are developed for λ-deformed ELBO, or λ-ELBO in short, now for real-valued λ (with λ = 0 reducing to the standard case). The meaning of the deformation factor λ in the λ-deformed ELBO and its performance for variational autoencoder (VAE) are investigated. Naturally emerging from our formulation is a deformation homotopy probability distribution function that extrapolates encoder distribution and the latent prior. Results show that λ values around 0.5 generally achieve better performance in image reconstruction for generative models.

Keywords: evidence lower bound · λ-deformation · Tsallis and Rényi divergence · variational autoencoder · deformation homotopy

1 Introduction

The concept of evidence lower bound (ELBO) was first proposed in [9], which introduced an efficient inference and learning framework where prior distribution and likelihood are explicitly modeled, in the presence of continuous latent variables to overcome intractable posterior distributions. ELBO has its name because it provides a lower bound for the log-likelihood of the observed data. Since optimization of lower bound amounts to a minimization of divergences [9,12], ELBO has founded its use in many applications such as variational autoencoder (VAE) and Bayesian neural networks.

One research direction extending the standard ELBO formulation is through the use of q-logarithmic function [22], defined as $\log_q(x) = \frac{x^{1-q}-1}{1-q}$ for real number q, instead of the log function ($q = 1$) in defining Shannon entropy and KL

© The Author(s), under exclusive license to Springer Nature Switzerland AG 2023
F. Nielsen and F. Barbaresco (Eds.): GSI 2023, LNCS 14071, pp. 186–196, 2023.
https://doi.org/10.1007/978-3-031-38271-0_19

divergence. Introduced by Tsallis in 1988 in the context of generalizing KL divergence, this scheme parallels Rényi's generalization of Shannon entropy in 1961. Making use of the associated Rényi divergence, tighter ELBO bounds (Rényi bound) was investigated [12]. Later, Tsallis divergence was likewise adopted for ELBO, and a q-deformed lower bound was obtained in [10] and [19]. That Rényi and Tsallis entropies and divergences are both monotonically related, see e.g. [15], actually reflects a deformation of an underlying dually flat Hessian geometry [24–26]. So Rényi bound and q-deformed lower bound represent generalizations to the original ELBO framework that are still analytically trackable.

Other studies have extended the ELBO framework to the more general f-divergence [8], used K-sample generalized evidence lower bound [3], or applied the ideas of Tsallis deformation to model predictive control (MPC) algorithm [23]. Numerical experiments on different tasks showed that proposed new ELBOs are tighter and much closer to the estimated true log-likelihood [19], and that the models trained with these bounds outperformed the baseline comparisons [12,19]. Put in this backdrop, our current work will reexamine deformation to the ELBO by the Tsallis/Rényi deformation by developing corresponding formulae, and provide a systematic interpretation of the effect of deformation parameter on the generalized bounds.

The rest of our paper is organized as follows. After a brief review of the vanilla VAE framework (Sect. 2.1) and Tsallis and Rényi deformation (Sect. 2.2), we define the λ-deformed ELBO for both the Tsallis and Rényi scenario (Sect. 2.3) and then provide the corresponding decomposition formulae (Sect. 2.4). Numerical results are reported in Sect. 3. Section 4 provides the conclusion with a discussion of the λ-deformation in relation to robustness (Sect. 4.1) and to other divergence functions (Sect. 4.2). The overarching goal of our paper is to extend the above suite of formulae with KL divergence to more generalized cases with Rényi and Tsallis divergence, unified under λ-deformation framework of Wong and Zhang (2022).

2 λ-ELBO

2.1   A Review on Vanilla VAE

Let us first have a look at the basic framework of VAE, which is an application of ELBO to the encoder/decoder architecture in computer vision [9]. Assume a dataset $X = \{x^{(i)}\}_{i=1}^N$ consists of N i.i.d. samples generated by a probability distribution $p(x)$ with a latent variable z. The decoder distribution for reconstructing images x is denoted as $p_\theta(x|z)$, and the encoder distribution is denoted as $q_\phi(z|x)$, with intractable original posterior distribution $p(z|x)$. The symbols θ and ϕ represent learnable parameters for the decoder p and the encoder q, respectively. Without confusion, we also use $p(z)$ to denote the prior distribution of z.

The log-likelihood of the data x can be decomposed as

$$\log p(x) = E_{q_\phi(z|x)}[\log p(x)] \tag{1}$$

$$= E_{q_\phi(z|x)}\left[\log \frac{p_\theta(x|z)p(z)q_\phi(z|x)}{p(z|x)q_\phi(z|x)}\right] \tag{2}$$

$$= E_{q_\phi(z|x)}[\log p_\theta(x|z)] - D[q_\phi(z|x)||p(z)] + D[q_\phi(z|x)||p(z|x)]. \tag{3}$$

Here $E_{q_\phi(z|x)}$ denotes expectation over random variable $z \sim q_\phi(z|x)$. $D[q||p]$ denotes the KL divergence between two distributions q and p. Then the following identity

$$\log p(x) - D[q_\phi(z|x)||p(z|x)] = E_{q_\phi(z|x)|}[\log p_\theta(x|z)] - D[q_\phi(z|x)||p(z)] \tag{4}$$

motivates the definition of *evidence lower bound* (ELBO) as the right hand-side of the above, such that maximizing ELBO is equivalent to minimizing the KL divergence between $q_\phi(\cdot|x)$ and $p(\cdot|x)$ since $\log p(x)$ is a constant.

Definition 1. *ELBO (Evidence Lower Bound), denoted as L, is*

$$L(x) = \log p(x) - D[q_\phi(z|x)||p(z|x)] \tag{5}$$

$$= E_{q_\phi(z|x)}[\log p_\theta(x|z)] - D[q_\phi(z|x)||p(z)]. \tag{6}$$

The second line is a rearrangement of terms; it combines reconstruction $-E_{q_\phi(z|x)}[\log p_\theta(x|z)]$ and variational approximation loss $D[q_\phi(z|x)||p(z)]$. For VAE, the importance of this decomposition lies in its use in training [16,18,20].

2.2   Tsallis and Rényi Deformation

Let us recall Rényi and Tsallis divergence. In the following definitions, P and Q are two probability distributions with respect to a reference measure μ such that $P \ll \mu$ and $Q \ll \mu$. We also denote $p = \frac{dP}{d\mu}$ and $q = \frac{dQ}{d\mu}$.

Definition 2. *The Tsallis divergence from Q to P is*

$$D_\lambda^T[q||p] = -E_q\left[\log_\lambda \frac{p}{q}\right] = -\int q \cdot \frac{(\frac{p}{q})^\lambda - 1}{\lambda}d\mu = \frac{\int q^{1-\lambda}p^\lambda d\mu - 1}{-\lambda} \tag{7}$$

Definition 3. *The Rényi divergence from Q to P is*

$$D_\lambda^R[q||p] = \frac{1}{-\lambda}\log E_q\left[\left(\frac{p}{q}\right)^\lambda\right] = \frac{\log \int q^{1-\lambda}p^\lambda d\mu}{-\lambda} \tag{8}$$

Lemma 1. *The relationship between Tsallis and Rényi divergence is*

$$D_\lambda^T[q||p] = \frac{e^{(-\lambda \cdot D_\lambda^R[q||p])} - 1}{-\lambda} \iff D_\lambda^R[q||p] = \frac{\log[-\lambda \cdot D_\lambda^T[q||p] + 1]}{-\lambda}. \tag{9}$$

Denote a pair of mutually inverse monotonic increasing functions

$$\kappa_\lambda(t) = \frac{\log[\lambda t + 1]}{\lambda} \iff \gamma_\lambda(t) = \frac{e^{\lambda t} - 1}{\lambda}. \tag{10}$$

Then the above relationship can be written using the scaling functions $\kappa_\lambda, \gamma_\lambda$:

$$D_\lambda^T[q||p] = \gamma_{-\lambda}(D_\lambda^R[q||p]) \iff D_\lambda^R[q||p] = \kappa_{-\lambda}(D_\lambda^T[q||p]).$$

This parallels the relationship between Tsallis entropy H_λ^T and Rényi entropy H_λ^R:

$$H_\lambda^T[q||p] = \gamma_\lambda(H_\lambda^R[q||p]) \iff H_\lambda^R[q||p] = \kappa_\lambda(H_\lambda^T[q||p]).$$

2.3 λ-ELBO Defined

In this subsection we develop λ-deformed ELBO formulae under Tsallis and Rényi deformation, which we write as $L_\lambda^T(x)$ and $L_\lambda^R(x)$. We call these λ-deformed ELBO, or λ-ELBO for short. Throughout the rest of the paper, we consider $\lambda \in \mathbf{R}, \lambda \neq 0$; the vanilla ELBO will be recovered as a special case when taking the limit of $\lambda \to 0$.

Definition 4. *The Tsallis ELBO is defined as [19]*

$$L_\lambda^T(x) = \frac{1}{\lambda} E_{q_\phi(z|x)} \left[\left(\frac{p(x,z)}{q_\phi(z|x)} \right)^\lambda - 1 \right] = E_q \left[\log_\lambda \left(\frac{p(x,z)}{q_\phi(z|x)} \right) \right] \tag{11}$$

Definition 5. *The Rényi ELBO is defined as [12]*

$$L_\lambda^R(x) = \frac{1}{\lambda} \log E_{q_\phi(z|x)} \left[\left(\frac{p(x,z)}{q_\phi(z|x)} \right)^\lambda \right] \tag{12}$$

One immediately finds that

Proposition 1. *The relationship between Tsallis ELBO and Rényi ELBO is*

$$L_\lambda^R(x) = \frac{1}{\lambda} \log \left(\lambda L_\lambda^T(x) + 1 \right) = \kappa_\lambda(L_\lambda^T(x)) , \tag{13}$$

and

$$L_\lambda^T(x) = \frac{1}{\lambda} \left(e^{\lambda L_\lambda^R(x)} - 1 \right) = \gamma_\lambda(L_\lambda^R(x)). \tag{14}$$

This shows that Tsallis ELBO and Rényi ELBO are intrinsically equivalent from an optimization perspective.

2.4 Decomposition Formulae

We now show (proof omitted) that the deformed ELBOs are related to the deformed divergence via the following decomposition formulae.

Theorem 1. *The λ-deformed ELBO and λ-deformed divergence in either Tsallis or Rényi form satisfy the generalized decomposition equation*

$$\log_\lambda p(x) = L_\lambda^T + D_\lambda^T[q_\phi(z|x)||p(z|x)] \cdot p(x)^\lambda \,, \tag{15}$$

$$\log p(x) = L_\lambda^R + D_\lambda^R[q_\phi(z|x)||p(z|x)] \,. \tag{16}$$

Theorem 2. *The decomposition of λ-deformed ELBO, in terms of reconstruction minus divergence-to-prior, is:*

(i) The case of Tsallis ELBO

$$L_\lambda^T(x) = E_{u_\lambda(z|x)}[\log_\lambda(p(x|z))] \cdot C_\lambda(x) - D_\lambda^T[q_\phi(z|x)||p(z)]; \tag{17}$$

(ii) The case of Rényi ELBO

$$L_\lambda^R(x) = \kappa_\lambda(E_{u_\lambda(z|x)}(\log_\lambda(p(x|z)))) - D_\lambda^R[q_\phi(z|x)||p(z)]. \tag{18}$$

In both (i) and (ii), an extrapolating function u_λ is defined by

$$u_\lambda(z|x) = \frac{q_\phi(z|x)^{1-\lambda}p(z)^\lambda}{C_\lambda(x)} = \frac{q_\phi(z|x)^{1-\lambda}p(z)^\lambda}{\int q_\phi(z|x)^{1-\lambda}p(z)^\lambda dz}. \tag{19}$$

with normalization $C_\lambda(x)$ given by

$$C_\lambda(x) = \int q_\phi(z|x)^{1-\lambda}p(z)^\lambda dz.$$

Note that
$$u_0(z|x) = q_\phi(z|x), \quad u_1(z|x) = p(z),$$

and $\int u_\lambda(z|x)dz = 1$ for every λ, meaning that this is a probability distribution function — it extrapolates the encoder distribution $q_\phi(z|x)$ and the latent prior $p(z)$ with the deformation parameter λ as extrapolating parameter. In information-theoretic language, it is actually an e-geodesic connecting $p(\cdot)$ and $q_\phi(\cdot|x)$. We call it *deformation homotopy function*.

3 Numerical Simulation

3.1 Monte Carlo Approximation of the Rényi Bound

In the previous Section, it is shown (Lemma 1) that the optimization of Tsallis and Rényi lower bound are equivalent. So in simulation studies below, we adopt Rényi lower bound as the loss function since it is more numerically effective than that of Tsallis. In order to optimize the λ-deformed ELBO, we need to find a way to approximate the bound when $\lambda \neq 0$. Ref [12] proposed a method to calculate the Rényi bound using K independent samples $z_k \sim q_\phi(z|x)$ for $k = 1, ..., K$ using

$$\hat{L}_{\lambda,K}(x) = \frac{1}{\lambda} \log \frac{1}{K} \sum_{k=1}^{K} \left[\left(\frac{p(z_k, x)}{q_\phi(z_k|x)} \right)^\lambda \right]. \tag{20}$$

This is the numerical estimator for

$$L_\lambda^R(x) = \frac{1}{\lambda} \log E_{q_\phi(z|x)} \left[\left(\frac{p(x, z)}{q_\phi(z|x)} \right)^\lambda \right]. \tag{21}$$

Note that although this is a biased estimator for L_R, the bias will approach zero as $K \to \infty$ [12]. In order to calculate $\hat{L}_{\lambda,K}(x)$, in each iteration we calculate $p(x|z_k)$, $p(z_k)$ and $q_\phi(z_k|x)$ in turn. If the Gaussian assumption holds for the prior p, for the encoder $q(\cdot|x)$, and for the decoder $p(\cdot|z)$, then the likelihood can be explicitly calculated with input x and sampled z_k.

3.2   Deformation Homotopy $u_\lambda(z|x)$ under Gaussian assumption

Consider two multivariate Gaussian distributions p and q with location and scale parameter μ_p, Σ_p and μ_q, Σ_q. Suppose that they have the same dimension d as determined by the dimension of latent variable z, and assume that the marginal distributions are independent. Then the deformed homotopy $u_\lambda(z|x)$ will still be a multivariate Gaussian distribution with mean vector and variance matrix given by,

$$\mu_u = [\lambda \Sigma_q \cdot \mu_p + (1 - \lambda)\Sigma_p \cdot \mu_q] \oslash [\lambda \Sigma_q \cdot \mathbf{1}_{d \times 1} + (1 - \lambda)\Sigma_p \cdot \mathbf{1}_{d \times 1}], \tag{22}$$

$$\Sigma_u = [\Sigma_p \otimes \Sigma_q] \oslash [\lambda \Sigma_q + (1 - \lambda)\Sigma_p]. \tag{23}$$

with $\otimes$ and $\oslash$ as Hadamard product and division.

3.3   Experimental Results

Here we perform numerical experiments on λ-deformed ELBO using the VAE architecture. The results will be compared for different λ values from $-\infty$ to $+\infty$ with a main attention to the range $(0, 1)$. Some classical variants of VAE including IWAE and VR-max are included as special cases of λ-deformed VAE, and are well approximated using MC sampled ELBO.

We used MNIST as the dataset for our numerical experiments. The encoder and decoder are designed using convolutional layers, deconvolutional layers, max-pooling layers, Relu layers and batchnorm layers. The latent space is in dimension of 50, which defines two random vectors representing the mean and variance of the distribution (of the same dimension). These two vectors are output by the encoder using two separate linear layers. The approximated posterior is assumed as a standard Gaussian distribution, with a diagonal covariance matrix. The output model is a Bernoulli distribution for each pixel. We trained each model for 500 epochs using Adam optimizer with a batch size of 128 and a learning rate of 5e-4. The complete Python codes for are available here https://github. com/kaimingCheng/lambda_deformed_VAE.

In testing we adopted the metric the negative log-likelihood (NLL) $-\log p(x) \approx -\hat{L}_{0,5000}(x)$, the same metric as reported in the previous works [2,16]. It was shown that with a large K the IWAE estimator $\hat{L}_K$ will be very close to the log-likelihood $\ln p(x)$, which will serve as a good indicator of the model performance. And also we list the negative log-likelihood (NLL) values by category for the negative log-likelihood of testing samples.

Figure 1a shows that the best λ values for the reconstruction and generating are around 0.5 (typically 0.4 to 0.6). The model will become less capable either when $\lambda \to \infty$ or $\lambda \to -\infty$, but it is more steady when going to positive infinity. Some special cases are IWAE ($\lambda = 1.0$) and VR-max ($\lambda = +\infty$), which are both better than vanilla VAE ($\lambda = 0.0$). Figure 1b presents the negative log likelihood (NLL) values across different categories of each λ. As expected, the better model is always uniformly better across all categories of inputs, meaning that the generative/reconstruction abilities are also uniformly better.

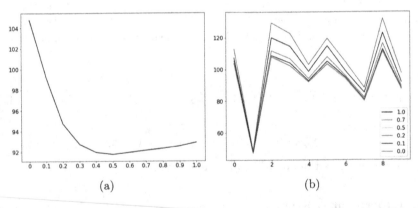

(a)   (b)

Fig. 1. a). The negative log-likelihoods (Y-axis) is plotted against λ values. (b). The negative log-likelihoods (Y-axis) plotted against the ten categories of digits (X-axis), for different λ values.

Fig. 2. From top left to bottom right: the reconstructed samples of the 25 samples on the right using $\lambda \in \{1.0, 0.7, 0.5, 0.2, 0.1, 0.0\}$. Each λ value corresponds to one 5×5 panel, running from top left ($\lambda = 1.0$) to bottom right ($\lambda = 0.0$).

Fig. 3. From top to bottom: randomly generated samples, with each row corresponding to a fixed λ value taken from $\lambda \in \{1.0, 0.7, 0.5, 0.2, 0.1, 0.0\}$.

4 Conclusions and Discussions

Synthesizing previous generalizations of ELBO based on Rényi and Tsallis divergences, this paper provided a suite of formulae for decomposing the λ-deformed evidence lower bound (ELBO) as a generalized reconstruction loss minus deformed divergence. We discovered a deformation homotopy function $u_\lambda(z|x)$, which is a family of distribution lying in-between (i.e., extrapolating) the encoding distribution and the prior for latent variable. Our newly developed formulae revealed that the deformed reconstruction loss is to be calculated as expectation with respect to the $u_\lambda(z|x)$ distribution rather than with respect to the encoder distribution q_ϕ under both Rényi and Tsallis cases. We also performed simulations on variational autoencoder (VAE) using different λ values. The results show that the λ-VAE achieves the best results when λ is set around 0.5, and the negative log likelihood (NLL) are uniformly better under each category in MNIST dataset.

4.1   Robustness Vs. Flexibility

To explain why some λ values in the deformation homotopy distribution $u_\lambda(z|x)$ lead to better results, we note that a $\lambda \to 1$ value close to 1 will make the deformed prior distribution $u_\lambda(z|x)$ more close to a fixed standard normal as we assumed, while $\lambda \to 0$ will make it close to the encoder $q_\phi(z|x)$, which is variant in training. When taking expectation with $u_1(z|x) \equiv p(z)$, then the training process will be robust because $u_1(z|x)$ now is independent of x. On the other hand, the standard VAE framework, the reconstruction loss is calculated using $u_0(z|x) \equiv q_\phi(z|x)$, which is more flexible. A trade-off between these two extreme cases of robustness vs flexibility will lead to a best λ value lying in between 0.4 to 0.6. It is consistent with the strategy of sacrificing efficiency to achieve robustness in parameter estimation — algorithms with robustness property handle outliers and additive noise well with improved accuracy in statistical inference, hypothesis testing and optimization problems [5,11,17,21]. Using power divergence (called β-divergence) to achieve robustness was studied in great length by S. Eguchi, e.g. [7,13,14].

4.2   Related Divergences

The Tsallis-Rényi type divergence is equivalent in forms (apart from a scaling factor) to the alpha divergences [4]

$$D_\alpha[q||p] = \frac{1}{\alpha(1-\alpha)}\left[1 - \int q^\alpha p^{1-\alpha} d\mu\right]$$

and power divergences [1,6,13]

$$D_\lambda[q||p] = \frac{1}{\lambda(\lambda+1)}\left[\int q^{\lambda+1}p^{-\lambda}d\mu - 1\right].$$

Ref. [5] showed that the alpha and beta divergences can be generated from each other, and from another one called gamma divergences [7], and that alpha divergences are flexible and efficient for nonnegative matrix and tensor factorizations, while the beta and gamma divergences are applied for robust PCA, clustering and parameter estimation. So our findings of best performance for $\lambda \simeq 0.5$ is consistent with these earlier approaches to robustness in machine learning. Future research would focus on analytic properties of deformation homotopy function, which is revealed by our current approach.

References

1. Basu, A., Harris, I.R., Hjort, N.L., Jones, M.: Robust and efficient estimation by minimising a density power divergence. Biometrika **85**(3), 549–559 (1998)
2. Burda, Y., Grosse, R., Salakhutdinov, R.: Importance weighted autoencoders (2015). https://doi.org/10.48550/ARXIV.1509.00519

3. Chen, L., Tao, C., Zhang, R., Henao, R., Duke, L.C.: Variational inference and model selection with generalized evidence bounds. In: International Conference on Machine Learning, pp. 893–902. PMLR (2018)
4. Chernoff, H.: A measure of asymptotic efficiency for tests of a hypothesis based on the sum of observations. The Annals of Mathematical Statistics, pp. 493–507 (1952)
5. Cichocki, A., Amari, S.I.: Families of alpha-beta-and gamma-divergences: Flexible and robust measures of similarities. Entropy **12**(6), 1532–1568 (2010)
6. Cressie, N., Read, T.R.: Multinomial goodness-of-fit tests. J. Roy. Stat. Soc.: Ser. B (Methodol.) **46**(3), 440–464 (1984)
7. Fujisawa, H., Eguchi, S.: Robust parameter estimation with a small bias against heavy contamination. J. Multivar. Anal. **99**(9), 2053–2081 (2008)
8. Gimenez, J.R., Zou, J.: A unified f-divergence framework generalizing vae and gan (2022). https://doi.org/10.48550/ARXIV.2205.05214
9. Kingma, D.P., Welling, M.: Auto-encoding variational bayes (2013). https://doi.org/10.48550/ARXIV.1312.6114
10. Kobayashis, T.: q-VAE for disentangled representation learning and latent dynamical systems. IEEE Robotics Autom. Lett. **5**(4), 5669–5676 (2020). https://doi.org/10.1109/lra.2020.3010206
11. Lafferty, J.: Additive models, boosting, and inference for generalized divergences. In: Proceedings of the Twelfth Annual Conference on Computational Learning Theory, pp. 125–133 (1999)
12. Li, Y., Turner, R.E.: Rényi divergence variational inference (2016). https://doi.org/10.48550/ARXIV.1602.02311
13. Mihoko, M., Eguchi, S.: Robust blind source separation by beta divergence. Neural Comput. **14**(8), 1859–1886 (2002)
14. Mollah, M.N.H., Sultana, N., Minami, M., Eguchi, S.: Robust extraction of local structures by the minimum β-divergence method. Neural Netw. **23**(2), 226–238 (2010)
15. Nielsen, F., Nock, R.: On rényi and tsallis entropies and divergences for exponential families. arXiv: 1105.3259 (2011)
16. Prokhorov, V., Shareghi, E., Li, Y., Pilehvar, M.T., Collier, N.: On the importance of the Kullback-Leibler divergence term in variational autoencoders for text generation. In: Proceedings of the 3rd Workshop on Neural Generation and Translation, pp. 118–127. Association for Computational Linguistics, Hong Kong (Nov 2019). https://doi.org/10.18653/v1/D19-5612
17. Regli, J.B., Silva, R.: Alpha-beta divergence for variational inference (2018)
18. Sajid, N., Faccio, F., Da Costa, L., Parr, T., Schmidhuber, J., Friston, K.: Bayesian brains and the rényi divergence (2021). https://doi.org/10.48550/ARXIV.2107.05438
19. Sârbu, S., Malagò, L.: Variational autoencoders trained with q-deformed lower bounds (2019)
20. Sârbu, S., Volpi, R., Peşte, A., Malagò, L.: Learning in variational autoencoders with kullback-leibler and renyi integral bounds (2018). https://doi.org/10.48550/ARXIV.1807.01889
21. Taneja, I.J.: New developments in generalized information measures. In: Advances in Imaging and Electron Physics, vol. 91, pp. 37–135. Elsevier (1995)
22. Tsallis, C.: What are the numbers that experiments provide. Quim. Nova **17**, 468–471 (1994)
23. Wang, Z., et al.: Variational inference mpc using tsallis divergence (2021). https://doi.org/10.48550/ARXIV.2104.00241

24. Wong, T.K.L., Zhang, J.: Tsallis and rényi deformations linked via a new λ-duality. IEEE Trans. Inf. Theory **68**(8), 5353–5373 (2022). https://doi.org/10.1109/TIT. 2022.3159385
25. Zhang, J., Wong, T.K.L.: λ-Deformed probability families with subtractive and divisive normalizations, vol. 45, pp. 187–215 (Jan 2021). https://doi.org/10.1016/bs.host.2021.06.003
26. Zhang, J., Wong, T.K.L.: λ-deformation: A canonical framework for statistical manifolds of constant curvature. Entropy **24**(2) (2022). https://doi.org/10.3390/e24020193

Statistical Manifolds and Hessian Information Geometry

On the Tangent Bundles of Statistical Manifolds

Barbara Opozda[⊠]

Faculty of Mathematics and Computer Science, Jagiellonian University, ul.
Łojasiewicza 6, 30-348 Cracow, Poland
barbara.opozda@im.uj.edu.pl

Abstract. Curvature properties of the Sasakian metric on the tangent
bundle of statistical manifolds are discussed.

Keywords: statistical structure · curvature · Sasakian metric

1 Introduction

One of the most natural structures on the tangent bundle of a Riemannian manifold is the Sasakian metric introduced in [11] and then studied and generalized in various ways, e.g. [1,3,6,12]. In particular, having a statistical structure on a manifold $\mathcal{M}$, that is, a pair (g, ∇), where g is a Riemannian metric tensor field and ∇ is a torsion-free connection, for which ∇g is a symmetric cubic form, one produces a metric on the tangent bundle $T\mathcal{M}$ depending on g and ∇. Although this structure is not, in general, the one introduced by Sasaki, it is called Sasakian.

An important case is when a statistical structure is Hessian. In this case the structure on the tangent bundle is Kähler. Its holomorphic sectional curvature and other properties are discussed in [13]. If a statistical structure is not Hessian, controlling the curvature on the tangent bundle is more complicated. There are various curvatures on statistical manifolds, see [8,9]. The curvatures appear in the formulas for the curvature tensors on the tangent bundle. In this paper we present and discuss some of such formulas and their consequences. We propose some results in relation with a classical theorem of Sasaki. In particular, we have

Theorem 1. *If the Sasakian metric on the tangent bundle of a statistical manifold is flat, then the statistical structure is Hessian, its Riemannian metric is flat and its difference tensor is parallel relative to the statistical connection. The converse is also true.*

The paper does not contain proofs of the results. The proofs are provided in [10].

© The Author(s), under exclusive license to Springer Nature Switzerland AG 2023
F. Nielsen and F. Barbaresco (Eds.): GSI 2023, LNCS 14071, pp. 199–206, 2023.
https://doi.org/10.1007/978-3-031-38271-0_20

2   Curvatures on Statistical Manifolds

In this section we collect some facts dealing with the geometry of statistical structures. The facts are used to formulate and prove the results of this paper. All details for this section can be found, for instance, in [8] and [9].

A statistical structure on a manifold $\mathcal{M}$ is a pair (g, ∇), where g is a metric tensor field and ∇ is a torsion-free connection (called a statistical connection for g) such that $(\nabla_X g)(Y, Z) = (\nabla_Y g)(X, Z)$ for every X, Y, Z. In this paper $X, Y, Z, ...$ stand for tangent vectors or vector fields (local) on $\mathcal{M}$, depending on the context. The following notation is adopted: $\nabla T(X, X_1, ..., X_k) = (\nabla_X T)(X_1, ..., X_k)$ for any tensor field of type (l, k). Thus, for a statistical structure (g, ∇) the cubic form ∇g is totally symmetric.

Let (g, ∇) be a statistical structure on $\mathcal{M}$. The difference tensor of this structure is given by the formula $K(X, Y) := K_X Y := \nabla_X Y - \nabla^g_X Y$, where ∇^g stands for the Levi-Civita connection of g. The difference tensor is symmetric and symmetric relative to g. The following relation holds

$$-2g(K(X, Y), Z) = \nabla g(X, Y, Z)$$

for every X, Y, Z. The dual connection ∇^* for ∇ is given by

$$Xg(Y, Z) = g(\nabla_X Y, Z) + g(X, \nabla^*_X Z). \tag{1}$$

If ∇ is a statistical connection for g, then so is the connection ∇^*. For the dual structure (g, ∇^*) the difference tensor is equal to $-K$.

A statistical structure (g, ∇) is called trivial, if $\nabla = \nabla^g$, i.e. if $K = 0$.

The curvature tensor for ∇ (resp. ∇^*, ∇^g) will be denoted by R (resp. R^*, R^g). The two curvature tensors R and R^* are related by the formula

$$g(R(X, Y)Z, W) = -g(R^*(X, Y)W, Z). \tag{2}$$

The following conditions are equivalent:
1) $R = R^*$,
2) the $(1, 3)$-tensor field $\nabla^g K$ is symmetric,
3) $g(R(X, Y)Z, W)$ is skew-symmetric relative to Z, W, for every X, Y, Z, W.

Statistical structures satisfying the second condition were called conjugate symmetric in [4]. The above conditions are well-known in affine differential geometry, where they characterize equiaffine spheres.

Except for the curvature tensors R and R^* we have the curvature tensor $[K, K]$ given by

$$[K, K](X, Y)Z = [K_X, K_Y]Z = K_X K_Y Z - K_Y K_X Z. \tag{3}$$

This curvature tensor has all algebraic properties needed to define a sectional curvature. In particular, $g([K_X, K_Y]Z, W) = -g([K_X, K_Y]W, Z)$. Therefore, we can define the K-sectional curvature k^K as follows

$$k^K(\Pi) = g([K, K](X, Y)Y, X), \tag{4}$$

for a vector plane $\Pi \subset T_p\mathcal{M}$ and its orthonormal basis X, Y.

There exist a few nice formulas relating various curvatures on a statistical manifold. Recall those, which are important for formulating and proving the formulas from the last section of this paper. Firstly, we have

$$R(X,Y) = R^g(X,Y) + (\nabla^g_X K)_Y - (\nabla^g_Y K)_X + [K_X, K_Y], \qquad (5)$$

$$R(X,Y) = R^g(X,Y) + (\nabla_X K)_Y - (\nabla_Y K)_X - [K_X, K_Y], \qquad (6)$$

Writing (5) for the dual structure (g, ∇^*) and adding both equalities we get

$$R(X,Y) + R^*(X,Y) = 2R^g(X,Y) + 2[K_X, K_Y]. \qquad (7)$$

The tensor field

$$\mathcal{R}(X,Y) = \frac{1}{2}(R(X,Y) + R^*(X,Y)) \qquad (8)$$

has the same algebraic properties as the Riemannian curvature tensor. In particular, $g(\mathcal{R}(X,Y)Z, W) = -g(\mathcal{R}(X,Y)W, Z)$. Therefore, we can define a sectional curvature of (g, ∇) by means of $\mathcal{R}$, which, in general, is not possible using separately R or R^*. If a statistical structure is conjugate symmetric, then $\mathcal{R} = R$ and we can define the sectional curvature just for ∇. Of course, we still have the sectional curvature k^g determined by g and ∇^g.

For each above curvature tensor we can define the Ricci tensor in the standard manner. Denote by Ric^g, Ric, Ric^*, $\mathcal{R}ic$, Ric^K the Ricci tensors for the curvature tensors R^g, R, R^*, $\mathcal{R}$ and $[K, K]$ respectively. In particular, we have

$$\mathcal{R}ic = \frac{\mathrm{Ric} + \mathrm{Ric}^*}{2}. \qquad (9)$$

Having the Ricci tensors, one can define scalar curvatures as the traces relative to g of the corresponding Ricci tensors. In particular, we have the scalar curvature ρ for ∇, that is,

$$\rho = \mathrm{tr}\,_g \mathrm{Ric}. \qquad (10)$$

Similarly we can define the scalar curvature ρ^* for ∇^*, but for any statistical structure $\rho = \rho^*$ (by (2)).

We have the 1-form τ on $\mathcal{M}$ given by $\tau(X) = \mathrm{tr}\, K_X$. It is called the first Koszul form (or the Czebyszev form – in affine differential geomtry, e.g. [5]). If ν_g is the volume form determined by g, then $\nabla_X \nu_g = -\tau(X)\nu_g$. We now have

$$d\tau(X,Y) = \mathrm{tr}\, R(X,Y). \qquad (11)$$

Hence, the Ricci tensor Ric of ∇ is symmetric if and only if τ is closed. The same holds for Ric*. In particular, the Ricci tensor Ric is symmetric if and only if the $(0,2)$- tensor $\nabla\tau$ is symmetric (equivalently $\nabla^g\tau$ is symmetric). The second Koszul form is the $(0,2)$-tensor field $\nabla\tau$. If a statistical structure is conjugate symmetric, then its Ricci tensor Ric is automatically symmetric. The Ricci tensor is symmetric if and only if Ric $=$ Ric*.

If $\tau = 0$, the statistical structure is called traceless (or apolar). Such structures play a fundamental role in the classical affine differential geometry, see [5,7].

The form τ appears also in subsequent curvature formulas. For instance, we have

$$\mathrm{Ric}^K(X,Y) = -\mathrm{tr}\,(K_X K_Y) + \tau(K(X,Y)), \tag{12}$$

$$\mathrm{Ric}\,(X,Y) = \mathrm{Ric}^g(X,Y) + \mathrm{Ric}^K(X,Y) + (\mathrm{div}\,^{\nabla^g} K)(X,Y) - \nabla^g \tau(X,Y), \tag{13}$$

where $(\mathrm{div}\,^{\nabla^g} K)(X,Y) = \mathrm{tr}\,\{T_p\mathcal{M} \ni U \to (\nabla^g_U K)(X,Y) \in T_p\mathcal{M}\}$.

Recall also the following relation between the scalar curvature ρ and the metric scalar curvature ρ^g. The formula is known in affine differential geometry as the affine theorema egregium.

$$\rho^g = \rho + \|K\|^2 - \|\tau\|^2. \tag{14}$$

For later use we introduce a tensor field r of type $(1,3)$ by the formula

$$2g(W, r(X,Y,Z)) = g(R(W,X)Y, Z). \tag{15}$$

A statistical structure (g, ∇) is Hessian if and only if the statistical connection ∇ is flat, [13]. The structure is automatically conjugate symmetric and, by (5), $R^g = -[K,K]$. Conversely, using again (5), one sees that if a statistical structure is conjugate symmetric and $R^g = -[K,K]$, then the structure is Hessian. Note that each Hessian manifold can be locally realized on an equiaffine sphere, which, as a manifold, is a locally strongly convex hypersurface in an affine space $\mathbf{R}^{n+1}$, where $n = \dim \mathcal{M}$. In general, the theory of equiaffine (or, in another terminology, relative) hypersurfaces in $\mathbf{R}^{n+1}$ provides a lot of examples of statistical structures satisfying many remarkable properties. But, in general, not all statistical structures can be realized (even locally) on affine hypersurfaces. In particular, a necessary condition for the local realizability is the projective flatness of the dual connection. This condition is very restrictive. For the theory of affine hypersurfaces we refer to [5] or [7].

3   The Sasaki Metric Determined by A statistical Structure

Let $\mathcal{M}$ be an n-dimensional manifold. Consider the tangent bundle $\pi : T\mathcal{M} \to \mathcal{M}$. Denote by $TT\mathcal{M}$ the tangent bundle of $T\mathcal{M}$.

Take the differential $\pi_* : TT\mathcal{M} \to T\mathcal{M}$. For each point $\xi \in T\mathcal{M}$ the space $\mathcal{V}_\xi = \ker\,(\pi_*)_\xi$ is called the vertical vertical space at ξ. The space $\mathcal{V}_\xi$ can be easily identified with $T_{\pi(\xi)}\mathcal{M}$. Namely, let $p = \pi(\xi)$ and $X \in T_p\mathcal{M}$. We define the vertical lift X^v_ξ of X to ξ by saying that the curve $t \to \xi + tX$ lying in the affine space $T_p\mathcal{M}$ is an integral curve of the vector X^v_ξ. We shall write it as follows: $X^v_\xi = [t \to \xi + tX]$. It is clear that $\mathcal{V}_\xi = \{X^v_\xi : \quad X \in T_p\mathcal{M}\}$ and the assignment $T_p\mathcal{M} \ni X \to X_\xi^v \in \mathcal{V}_\xi$ is a linear isomorphism. The vertical bundle $\mathcal{V} = \bigcup_{\xi \in T\mathcal{M}}$ is a smooth n-dimensional vector subbundle of $TT\mathcal{M}$, so it is a distribution in $TT\mathcal{M}$. The distribution is clearly integrable.

Let ∇ be a connection on $\mathcal{M}$. The connection determines the horizontal distribution $\mathcal{H} = \bigcup_{\xi \in T\mathcal{M}} \mathcal{H}_\xi$ in $TT\mathcal{M}$ complementary to $\mathcal{V}$, that is, $\mathcal{H}_\xi \oplus \mathcal{V}_\xi = T_\xi T\mathcal{M}$ for each $\xi \in T\mathcal{M}$. We now briefly explain how to obtain the horizontal distribution. For details we refer to [1]. The horizontal lift X_ξ^h of a vector $X \in T_p\mathcal{M}$ can be obtained as follows. Take a geodesic (relative to ∇) $\gamma(t)$ determined by X. By using the parallel transport of the vector ξ along γ relative to ∇ we obtain a curve $t \to \xi_t$ in $T\mathcal{M}$, where $\xi_0 = \xi$. We set $X_\xi^h = [t \to \xi_t]$. We see that $(\pi_*)_\xi(X_\xi^h) = [t \to \pi(\xi_t)] = [t \to \gamma(t)] = X$. We now set $\mathcal{H}_\xi = \{X_\xi^h; \ X \in T_p\mathcal{M}\}$. The assignment $X \in T_p\mathcal{M} \to X_\xi^h \in \mathcal{H}_\xi$ is a linear isomorphism. If X is a smooth vector field on $\mathcal{M}$, then X^v and X^h will stand for the vertical and horizontal lifts to $T\mathcal{M}$. The vertical and horizontal lifts are smooth vector fields on $T\mathcal{M}$. Of course, X can be a locally defined vector field, say on $\mathcal{U} \subset \mathcal{M}$, and then the lifts are defined on $T\mathcal{M}_{|\mathcal{U}}$.

For a smooth function f on $\mathcal{M}$ we have the function $F = f \circ \pi$ on $T\mathcal{M}$. It is clear that

$$X^v F = 0, \qquad (X^h F)_\xi = (Xf)_{\pi(\xi)}. \tag{16}$$

We have the following formulas for the Lie bracket of vertical and horizontal lifts of vector fields on $\mathcal{M}$, e.g. [1].

Lemma 1. *If $X, Y \in \mathcal{X}(\mathcal{M})$, then we have*

$$[X^v, Y^v] = 0, \qquad [X^h, Y^v] = (\nabla_X Y)^v, \qquad [X^h, Y^h]_\xi = [X, Y]_\xi^h - (R(X,Y)\xi)_\xi^v, \tag{17}$$

where R is the curvature tensor of ∇.

The almost complex J structure on the manifold $T\mathcal{M}$ is given as follows:

$$JX^v = X^h, \qquad JX^h = -X^v \quad for \quad X \in \mathcal{X}(\mathcal{M}). \tag{18}$$

Using Lemma 1 and computing the Nijenjuis product $[J, J]$ one proves, [1].

Proposition 1. *The almost complex structure J is complex if and only if the connection ∇ is without torsion and its curvature tensor R vanishes.*

Assume now that we have a statistical structure (g, ∇) on $\mathcal{M}$. The metric tensor field g can be lifted to $T\mathcal{M}$ in various ways. It is natural to use a combination of the ideas proposed in [11] and [1]. In the literature concerning the lifts of structures to tangent bundles a few other types of lifts are studied.

We consider the following metric tensor field $\tilde{g}$ on TM:

$$\tilde{g}(X^v, Y^v) = g(X, Y) = \tilde{g}(X^h, Y^h), \qquad \tilde{g}(X^v, Y^h) = 0. \tag{19}$$

The metric tensor $\tilde{g}$ is called the Sasaki metric, although Sasaki used in his construction the Levi-Civita connection ∇^g.

The metric tensor $\tilde{g}$ is almost Hermitian relative to J. One also sees that the almost symplectic form ω of this almost Hermitian structure is symplectic. To prove that $d\omega = 0$ it is sufficient to use (1) and the first Bianchi identity for the dual connection ∇^*, [12].

4   Curvatures of the Sasaki Metric

Using properties of the curvatures mentioned in Sect. 2 and making suitable computations, see [10], one obtains formulas for the curvatures of $\tilde{g}$.

Proposition 2. *The curvature tensor $R_{\tilde{\xi}}^{\tilde{g}}$ of $\tilde{g}$ is given by the following formulas:*

$$
\begin{aligned}
R^{\tilde{g}}(X^v, Y^v)Z^v = &\ ([K_Y, K_X]Z)^v \\
&+ \{r(K(X,Z), \xi, Y) - r(K(Y,Z), \xi, X)\}^h,
\end{aligned}
\tag{20}
$$

$$
\begin{aligned}
R^{\tilde{g}}(X^v, Y^v)Z^h = &\ \{K(r(Z, \xi, Y), X) - K(r(Z, \xi, X), Y)\}^v \\
&+ \{R^g(X,Y)Z + r(r(Z, \xi, Y), \xi, X) - r(r(Z, \xi, X), \xi, Y)\}^h,
\end{aligned}
\tag{21}
$$

$$
\begin{aligned}
R^{\tilde{g}}(X^v, Y^h)Z^h = &\ \{-\tfrac{1}{2}R(Y,Z)X + [K_X, K_Y]Z + \nabla K(Y, X, Z) \\
&\quad - \tfrac{1}{2}R(Y, r(Z, \xi, X))\xi\}^v \\
&+ \{-\tfrac{1}{2}K(X, R(Y,Z)\xi) + r(K(Y,Z), \xi, X) + (\nabla_Y r)(Z, \xi, X) \\
&\quad - K(Y, r(Z, \xi, X)) - r(Y, \xi, K(X,Z))\}^h,
\end{aligned}
\tag{22}
$$

$$
\begin{aligned}
R^{\tilde{g}}(X^v, Y^h)Z^v = &\ \{K(r(Y, \xi, Z), X) + \tfrac{1}{2}R(Y, K(X,Z))\}^v \\
&+ \{[K_Y, K_X]Z + r(r(Y, \xi, Z), \xi, X) \\
&\quad - r(Y, X, Z) - (\nabla K)(Y, X, Z)\}^h,
\end{aligned}
\tag{23}
$$

$$
\begin{aligned}
R^{\tilde{g}}(X^h, Y^h)Z^h = &\ (R^g(X,Y)Z)^h \\
&+ \tfrac{1}{2}\{r(X, \xi, R(Y,Z)\xi) - r(Y, \xi, R(X,Z)\xi) - 2r(Z, \xi, R(X,Y)\xi)\}^h \\
&\tfrac{1}{2}\{(\nabla_Z R)(X,Y) - R(Y, K(X,Z))\xi + R(X, K(Y,Z))\xi \\
&\quad + K(X, R(Y,Z)\xi) - K(Y, R(X,Z)\xi) - 2K(Z, R(X,Y)\xi)\}^v
\end{aligned}
\tag{24}
$$

$$
\begin{aligned}
R^{\tilde{g}}(X^h, Y^h)Z^v = &\ (R^g(X,Y)Z)^v \\
&+ \tfrac{1}{2}\{R(X, r(Y, \xi, Z))\xi - R(Y, r(X, \xi, Z))\xi\}^v \\
&+ \{(\nabla_Y r)(X, \xi, Z) - (\nabla_X r)(Y, \xi, Z) \\
&\quad + K(X, r(Y, \xi, Z)) - K(Y, r(X, \xi, Z) + K(R(X,Y)\xi, Z) \\
&\quad + r(X, \xi, K(Y,Z))\}^h
\end{aligned}
\tag{25}
$$

In what follows $e_1, \ldots, e_n$ will stand for an orthonormal basis of $T_p\mathcal{M}$, where $p = \pi(\xi)$.

Proposition 3. *For any statistical structure and the induced Riemannian structure in the tangent bundle we have*

$$
\begin{aligned}
\mathrm{Ric}^{\tilde{g}}(Y^v, Z^v) = &\ \tfrac{1}{2}(\mathrm{Ric}\,(Y,Z) - \mathrm{Ric}^*(Y,Z)) + \nabla\tau(Y,Z) \\
&+ \tfrac{1}{4}\sum_{ij=1}^n g(R(e_i, e_j)\xi, Y)g(R(e_i, e_j)\xi, Z) \\
= &\ -(\mathrm{div}^{\nabla^{\tilde{g}}} K)(Y,Z) - \tau(K(Y,Z)) \\
&+ \tfrac{1}{4}\sum_{ij=1}^n g(R(e_i, e_j)\xi, Y)g(R(e_i, e_j)\xi, Z)
\end{aligned}
\tag{26}
$$

$$
\begin{aligned}
\mathrm{Ric}^{\tilde{g}}(Y^h, Z^h) = &\ \mathcal{R}ic(Y,Z) + \tfrac{1}{2}\nabla\tau(Y,Z) + \tfrac{1}{2}\nabla\tau(Z,Y) \\
&- \tfrac{1}{2}\sum_{i=1}^n g(R(e_i, Y)\xi, R(e_i, Z)\xi).
\end{aligned}
\tag{27}
$$

As a consequence we obtain

Corollary 1. *For any statistical structure we have the following formula for the scalar curvature $\rho^{\tilde{g}}$ of the Sasakian metric on the tangent bundle*

$$\rho_{\xi}^{\tilde{g}} = \rho + 2\mathrm{tr}_g \nabla \tau - \frac{1}{4} \sum_{ij=1}^{n} \|R(e_i, e_j)\xi\|^2. \tag{28}$$

Denote by $k^{\tilde{g}}(X^a \wedge Y^b)$, where $a, b = v$ or h, the sectional curvature of $\tilde{g}$ by $span\{X^a, Y^b\}$. Let X^a, Y^b be a pair of unit vectors in $T_\xi T\mathcal{M}$. By Proposition 2 we have

Proposition 4. *For any statistical structure we have*

$$k^{\tilde{g}}(X^v \wedge Y^v) = -k^K(X \wedge Y), \tag{29}$$

$$k^{\tilde{g}}(X^h \wedge Y^h) = k^g(X \wedge Y) - \frac{3}{4}\|R(X,Y)\xi\|^2, \tag{30}$$

$$k^{\tilde{g}}(X^v \wedge Y^h) = k^K(X \wedge Y) + g(\nabla K(Y,X,Y),X) + \|r(Y,\xi,X)\|^2. \tag{31}$$

For a Hessian structure we have

$$k^{\tilde{g}}(X^v \wedge Y^v) = k^g(X \wedge Y), \tag{32}$$

$$k^{\tilde{g}}(X^h \wedge Y^h) = k^g(X \wedge Y), \tag{33}$$

$$k^{\tilde{g}}(X^v \wedge Y^h) = -k^g(X \wedge Y) + g(\nabla K(Y,X,Y),X). \tag{34}$$

Recall a classical theorem of Sasaki, [3,11].

Theorem 2. *Let g be a metric tensor on $\mathcal{M}$ and $\tilde{g}$ be the Sasaki metric on $T\mathcal{M}$ determined by g and ∇^g. Then $\tilde{g}$ is flat if and only if g is flat.*

In [2] a few generalizations of this theorem were proved. For instance,

Theorem 3. *Let g be a metric tensor on $\mathcal{M}$ and $\tilde{g}$ the Sasaki metric on $T\mathcal{M}$ determined by g and ∇^g. If the sectional curvature of $\tilde{g}$ is bounded on $T\mathcal{M}$, then g is flat on $\mathcal{M}$.*
The scalar curvature of $\tilde{g}$ is constant on $T\mathcal{M}$ if and only if g is flat on $\mathcal{M}$.

Using the formulas presented in this section we can formulate the following generalizations of the above theorems.

Theorem 4. *Let (g, ∇) be a statistical structure on $\mathcal{M}$ and $\tilde{g}$ be the Sasakian metric on $T\mathcal{M}$ determined by g and ∇. If ∇ is not flat, then the sectional curvature of $\tilde{g}$ on $T\mathcal{M}$ is unbounded from above and below.*
If ∇ is not flat, then the scalar curvature of $\tilde{g}$ is unbounded from below on $T\mathcal{M}$.

Note that in the case of the Sasakian metrics determined by statistical structures, the flatness of a statistical connection (i.e. for Hessian structures) does not imply the flatness of the Sasakian metric. Namely, we have

Theorem 5. *If the Sasakian metric on the tangent bundle of a statistical manifold is flat, then the statistical structure is Hessian, its Riemannian metric is flat and its difference tensor is parallel relative to the statistical connection. The converse is also true.*

The Hessian structure from the last theorem, although it satisfies many conditions, does not have to be trivial. Namely, consider the following example

Example 1. Let $\mathcal{M} = \{x = (x_1, ..., x_n) \in \mathbf{R}^{n+1} : \quad x_i > 0 \; \forall i = 1, ..., n\}$. Let g be the standard flat metric tensor field on $\mathcal{M}$ and let $e_1, ..., e_n$ be the canonical basis in $\mathbf{R}^{n+1}$. Define a symmetric $(1, 2)$-tensor field K on $\mathcal{M}$ as follows

$$
\begin{aligned}
K(e_i, e_j) &= 0 \quad for \quad i \neq j, \\
K(e_i, e_i) &= \lambda_i e_i \quad for \quad i = 1, ..., n,
\end{aligned}
\tag{35}
$$

where $\lambda_i(x) = -x_i^{-1}$. One sees that K is symmetric relative to g and $[K, K] = 0$. When we define $\nabla = \nabla^g + K$, then the statistical structure (g, ∇) is non-trivial, Hessian, g is flat and $\nabla K = 0$.

References

1. Dombrowski, P.: On the geometry of the tangent bundle. J. Reine Angew. Math. **210**, 73–88 (1962)
2. Gudmundsson, S., Kappos, E.: On the geometry of tangent bundles. Expo. Math. **20**, 1–41 (2002)
3. Kowalski, O.: Curvature of the induced Riemannian metric on the tangent bundle of a Riemannian manifold. J. Reine Angew. Math. **250**, 124–129 (1971)
4. Lauritzen, S.L.: Statistical manifolds. Siff. Geom. Stat. Infer. **10**, 163–216 (1987)
5. Li, A-M., Simon, U., Zhao, G.: Global Affine Differential Geometry of Hypersurfaces, Walter de Gruyter (1993)
6. Matsuzoe, H., Inoguchi, J.: Statistical structures on tangent bundles. Appl. Sci. **5**, 55–75 (2003)
7. Nomizu, K., Sasaki, T.: Affine Differential Geometry. Cambridge University Press, Cambridge (1994)
8. Opozda, B.: Bochner's technique for statistical structures. Ann. Glob. Anal. Geom. **48**, 357–395 (2015)
9. Opozda, B.: A sectional curvature for statistical structures. Linear Algebra Appl. **497**, 134–161 (2016)
10. Opozda, B.: Curvature properties of the Sasaki metric of statistical manifolds (2019)
11. Sasaki, S.: On the differential geometry of tangent bundles of Riemannian manifolds. Tôhoku Math. J. **10**, 338–354 (1958)
12. Satoh, H.: Almost Hermitian structures on tangent byndles. arXiv 1908.10824 (2019)
13. Shima, H.: The Geometry Hessian Structures. World Scientific Publishing, Singapore (2007)

Geometric Properties of Beta Distributions

Prosper Rosaire Mama Assandje◉ and Joseph Dongho$^{(\boxtimes)}$◉

University of Maroua, 814, Maroua, Cameroon
josephdongho@yahoo.fr
http://www.fs.univ.maroua.cm

Abstract. The aim of this work is to prove that the Amari manifold of beta distributions of the first kind distribution have dual potential, dual coordinate pairs and his corresponding gradient system is linearizable and Hamiltonian.

Keywords: Hamiltonian system · statistical manifold · gradient system

1 Introduction

In 1993, Nakamura's work [7], pointed out that certain gradient flows on Gaussian and multinomial distributions are completely integrable Hamiltonian systems. In the same year, Fujiwara's work [6] propose a prove of a theorem giving a method of studying the complete integrability of gradient systems for some even dimensional statistical manifold with a potential function. This work is focuses on the study of gradient systems defined on

$$S = \left\{ p_\theta : [0;1] \longrightarrow [0;1]; \int_0^1 p_\theta(x)dx = 1 \right\}, \ \theta = (a,b) \in \mathbb{R}_+^* \times \mathbb{R}_+^*;$$

where

$$p_\theta(x) = \frac{1}{B(a,b)} x^{a-1}(1-x)^{b-1} \mathbb{1}_{[0,1]}(x),$$

is the Beta distribution law of the first kind. It is also obvious that this family have a potential function see [8]. Therefore, according to Amari's Theorem 3.4 in [1], it have a pair of dual coordinates $\{\theta, \eta\}$. More explicitly, as in [8],

$$p_\theta(x) = \exp\left[\theta_1 \log(x) + \theta_2 \log(1-x) - \Phi(\theta)\right]$$

with $\theta = (\theta_1, \theta_2)$, $\theta_1 = a - 1, \theta_2 = b - 1$ and $\Phi(\theta) = \log(B(\theta_1 + 1, \theta_2 + 1))$. Therefore, according to Ovidiu [8], Φ is its related potential function. We can then conclude that our model have

$$(\theta_i, \eta_i = \partial_{\theta_i}\Phi)$$

Supported by UFD-SF-UMa.

as dual coordinate pair. In this work we show that (θ, η) satisfying the following biorthogonal property:

$$g(\partial_{\theta_i}, \partial_{\eta_j}) = \delta_i^j.$$

With g the Fisher information metric. Where the natural basis of the tangent space T_pS at a point $p \in S$ is

$$\{\partial_{\theta_i} = \frac{\partial}{\partial\theta_i}\}$$

with respect to the coordinate system θ, and

$$\{\partial_{\eta_i} = \frac{\partial}{\partial\eta_i}\}$$

with respect to the associated dual coordinate system η. Furthermore, we deduce from the Legendre equation the following expression of the dual potential function

$$\Psi = \theta_i\eta_i - \Phi.$$

It should be noted that the dual potential function is known as entropy in Koszul's work in [4]. We can then conclude that the gradient system of our model (S, g) is linearizable and it is equivalent to

$$\dot{\eta}_i = -\eta_i.$$

We show that the dual potential function can be put in the following form:

$$\Psi(\eta) = \mathbb{E}\left[-\log q(X, \theta)\right]$$

where q_θ is the density at maximum entropy and it is defined by:

$$q(x, \theta) = \frac{e^{-(\theta_1 \log x + \theta_2 \log(1-x))}}{\int_0^1 e^{-(\theta_1 \log x + \theta_2 \log(1-x))} dx}.$$

Using q_θ, we show that Hamiltonian function's which is associate to our gradient system is

$$\mathcal{H} = \frac{\mathbb{E}[-\log(1-x)]}{\mathbb{E}[-\log(x)]}.$$

2   Riemannian Structure on Set of Beta Distributions

According to Ovidiu's [8], the Beta law of the first kind distribution given by:

$$p_\theta(x) = \frac{1}{B(a,b)} x^{a-1}(1-x)^{b-1} \mathbb{1}_{[0,1]}(x), \tag{1}$$

with

$$(a, b) \in \mathbb{R}_+^* \times \mathbb{R}_+^*, \ B(a,b) = \int_0^1 x^{a-1}.(1-x)^{b-1} dx, \ \int_0^1 p_\theta(x) dx = 1.$$

In this section we present the geometric structure that we will study came from the work J.L.Koszul's work recalled by Barbaresco [2–4]. According to Ovidiu [8], the manifold defined from exponential families and admit a potential function. Let us state the following proposition

Proposition 1. *The probability density*

$$p_\theta(x) = \frac{1}{B(\theta_1 + 1, \theta_2 + 1)} x^{\theta_1}(1 - x)^{\theta_2} \mathbb{1}_{[0,1]}(x)$$

is of exponential family, with

$$\theta = (\theta_1, \theta_2) \in]-1; +\infty[\times]-1; +\infty[, \ B(\theta_1 + 1, \theta_2 + 1) = \int_0^1 x^{\theta_1}.(1 - x)^{\theta_2} dx.$$

Proof. The density function (1) can be rewritten as follows:

$$p_\theta(x) = \exp\left[-\log(x) - \log(1 - x) + a\log(x) + b\log(1 - x) - \log(B(a, b))\right].$$

Which imply that

$$\log p_\theta(x) = -\log(x) - \log(1 - x) + a\log(x) + b\log(1 - x) - \log(B(a, b)).$$

According to Ovidiu Calin [8], (1) is of exponential family. We write,

$$\log p_\theta(x) = (a - 1)\log(x) + (b - 1)\log(1 - x) - \log(B(a, b)).$$

And denote,

$$\theta_1 = a - 1, \ \theta_2 = b - 1; \ \theta = (\theta_1, \theta_2) \in]-1; +\infty[\times]-1; +\infty[$$

a new parametrization. The Beta law of the first kind distribution (1) become

$$p_\theta(x) = \frac{1}{B(\theta_1 + 1, \theta_2 + 1)} x^{\theta_1}(1 - x)^{\theta_2} \mathbb{1}_{[0,1]}(x) \tag{2}$$

and

$$B(\theta_1 + 1, \theta_2 + 1) = \int_0^1 x^{\theta_1}.(1 - x)^{\theta_2} dx.$$

Then

$$\log p_\theta(x) = \theta_1 \log(x) + \theta_2 \log(1 - x) - \log(B(\theta_1 + 1, \theta_2 + 1)).$$

According to Ovidiu Calin [8], (2) is also of exponential family. Since

$$l(x, \theta) = \log p_\theta(x) = \theta_1 f_1(x) + \theta_2 f_2(x) - \Phi(\theta), \tag{3}$$

with

$$\Phi(\theta) = \log(B(\theta_1 + 1, \theta_2 + 1)) = \log \int_0^1 x^{\theta_1}.(1 - x)^{\theta_2} dx. \tag{4}$$

$$f_1(x) = \log(x), \ f_2(x) = \log(1 - x).$$

$\square$

2.1   Existence of a Pair of Dual Coordinates

Let

$$S = \left\{ p_\theta(x) = \frac{1}{B(\theta_1 + 1, \theta_2 + 1)} x^{\theta_1}(1 - x)^{\theta_2}, \begin{array}{l} \theta_1, \theta_2 \in]-1; +\infty[\\ x \in [0; 1] \\ B(\theta_1 + 1, \theta_2 + 1) = \\ \int_0^1 x^{\theta_1} \cdot (1 - x)^{\theta_2} dx \\ \int_0^1 p_\theta(x) dx = 1 \end{array} \right\}$$

be the statistical manifold. The function Φ defined by (4) is called potential function. As this family admits a potential function then it follow from Amari's Theorem 3.5 and 3.4 that there exists functions Φ and Ψ such that:

$$\eta_i = \partial_i \Phi(\theta), \; \theta_i = \partial^i \Psi(\eta). \tag{5}$$

are dual coordinates of S. It is well known that Φ and Ψ satisfy the following Legendre equation

$$\Psi(\eta) = \theta_i \eta_i - \Phi(\theta). \tag{6}$$

Using (4) and (5) we have

$$\eta_1 = \frac{\int_0^1 \log(x) e^{(\theta_1 \log x + \theta_2 \log(1-x))} dx}{\int_0^1 e^{(\theta_1 \log x + \theta_2 \log(1-x))} dx}, \; \eta_2 = \frac{\int_0^1 \log(1 - x) e^{(\theta_1 \log x + \theta_2 \log(1-x))} dx}{\int_0^1 e^{(\theta_1 \log x + \theta_2 \log(1-x))} dx}. \tag{7}$$

We define,

$$g_{ij} = \langle \partial_{\theta_i}, \partial_{\theta_j} \rangle_g := g\left(\partial_{\theta_i}, \partial_{\theta_j}\right). \tag{8}$$

Then we can use the Amari's [1], l−representation associated respectively to coordinate θ and η; and define the following scalar production on T_θ where T_θ is the l−representation of the tangent space with respect to the coordinate θ,

$$g_{ij} = \mathbb{E}\left[\partial_{\theta_i} l(\theta, x) \partial_{\theta_j} l(\theta, x)\right] = -\mathbb{E}\left[\partial_{\theta_i} \partial_{\theta_j} l(\theta, x)\right]. \tag{9}$$

Using (8), (9) and (3) we have:

$$g\left(\partial_{\theta_1} l(x, \theta), \partial_{\eta_1} l(x, \theta)\right) = \mathbb{E}\left[\partial_{\eta_1}\left(\frac{\int_0^1 \log(x) e^{(\theta_1 \log x + \theta_2 \log(1-x))} dx}{\int_0^1 e^{(\theta_1 \log x + \theta_2 \log(1-x))} dx}\right)\right] = \mathbb{E}\left[1\right] = 1.$$

$$g\left(\partial_{\theta_1} l(x, \theta), \partial_{\eta_2} l(x, \theta)\right) = \mathbb{E}\left[\partial_{\eta_2}\left(\frac{\int_0^1 \log(x) e^{(\theta_1 \log x + \theta_2 \log(1-x))} dx}{\int_0^1 e^{(\theta_1 \log x + \theta_2 \log(1-x))} dx}\right)\right] = 0.$$

$$g\left(\partial_{\theta_2} l(x, \theta), \partial_{\eta_2} l(x, \theta)\right) = \mathbb{E}\left[\partial_{\eta_2}\left(\frac{\int_0^1 \log(1 - x) e^{(\theta_1 \log x + \theta_2 \log(1-x))} dx}{\int_0^1 e^{(\theta_1 \log x + \theta_2 \log(1-x))} dx}\right)\right] = 1.$$

Then the two bases $\{\partial_{\theta_i}\}_{i=1}^2$ and $\{\partial_{\eta_j}\}_{j=1}^2$ are said to be biorthogonal. Therefore,

$$\Psi(\eta) = \frac{\int_0^1 \theta_1 \log(x) e^{(\theta_1 \log x + \theta_2 \log(1-x))} dx}{\int_0^1 e^{(\theta_1 \log x + \theta_2 \log(1-x))} dx} + \frac{\int_0^1 \theta_2 \log(1-x) e^{(\theta_1 \log x + \theta_2 \log(1-x))} dx}{\int_0^1 e^{(\theta_1 \log x + \theta_2 \log(1-x))} dx}$$
$$- \log \int_0^1 e^{(\theta_1 \log x + \theta_2 \log(1-x))} dx$$

represent the dual potential of the potential Φ. In the following section we show that this dual potential is an entropy.

2.2   Potential Function and Shannon Entropy

The potential function is given by:

$$\Phi(\theta) = -\log \frac{1}{\int_0^1 e^{(\theta_1 \log x + \theta_2 \log(1-x))} dx} = -\log \int_0^1 e^{-(\theta_1 \log x + \theta_2 \log(1-x))} dx.$$

(10)

Let us state the following Theorem which will allow us to prove that the dual potential an entropy.

Theorem 1. *Let S be the statistical manifold. Then, the associate dual potential function Ψ, solution of Legendre equation (6) is*

$$\Psi(\eta) = \mathbb{E}\left[-\log q(X, \theta)\right]$$

(11)

with

$$q(x, \theta) = \frac{e^{-(\theta_1 \log x + \theta_2 \log(1-x))}}{\int_0^1 e^{-(\theta_1 \log x + \theta_2 \log(1-x))} dx}.$$

Proof. Using (10) we have

$$\left.\begin{array}{l} \eta_1 = -\dfrac{\int_0^1 -\log(x) e^{-(\theta_1 \log x + \theta_2 \log(1-x))} dx}{\int_0^1 e^{-(\theta_1 \log x + \theta_2 \log(1-x))} dx}; \\[3mm] \eta_2 = -\dfrac{\int_0^1 -\log(1-x) e^{-(\theta_1 \log x + \theta_2 \log(1-x))} dx}{\int_0^1 e^{-(\theta_1 \log x + \theta_2 \log(1-x))} dx}. \end{array}\right\}$$

(12)

According to Amari's [1] the entropy function is given by the relation (6). Therefore

$$\left.\begin{array}{l} \Psi(\eta) = \log \int_0^1 e^{-(\theta_1 \log x + \theta_2 \log(1-x))} dx - \dfrac{\int_0^1 -\theta_1 \log(x) e^{-(\theta_1 \log x + \theta_2 \log(1-x))} dx}{\int_0^1 e^{-(\theta_1 \log x + \theta_2 \log(1-x))} dx} \\[3mm] \qquad\quad - \dfrac{\int_0^1 -\theta_2 \log(1-x) e^{-(\theta_1 \log x + \theta_2 \log(1-x))} dx}{\int_0^1 e^{-(\theta_1 \log x + \theta_2 \log(1-x))} dx} \end{array}\right\}$$

(13)

so,

$$q(x, \theta) = \frac{-e^{(\theta_1 \log x + \theta_2 \log(1-x))}}{\int_0^1 e^{-(\theta_1 \log x + \theta_2 \log(1-x))} dx}$$

(14)

is the maximum entropy density. By applying the one $1-$form of Koszul

$$\eta = \alpha = d\Phi(\theta)$$

and Koszul's normalisation contraint as in [4], We have the following constraints:

$$\left.\begin{array}{l} \eta_1 = -\int_0^1 \dfrac{-\log(x) e^{-(\theta_1 \log x + \theta_2 \log(1-x))}}{\int_0^1 e^{-(\theta_1 \log x + \theta_2 \log(1-x))} dx} dx; \\[3mm] \eta_2 = -\int_0^1 \dfrac{-\log(1-x) e^{-(\theta_1 \log x + \theta_2 \log(1-x))}}{\int_0^1 e^{-(\theta_1 \log x + \theta_2 \log(1-x))} dx} dx \end{array}\right\}$$

(15)

and

$$\int_0^1 q(x, \theta) dx = \int_0^1 \frac{e^{-(\theta_1 \log x + \theta_2 \log(1-x))}}{\int_0^1 e^{-(\theta_1 \log x + \theta_2 \log(1-x))} dx} dx = 1.$$

We denote

$$\Psi(\eta) = \log \int_0^1 e^{-(\theta_1 \log x + \theta_2 \log(1-x))} dx - \int_0^1 -\theta_1 \log(x) \frac{e^{-(\theta_1 \log x + \theta_2 \log(1-x))}}{\int_0^1 e^{-(\theta_1 \log x + \theta_2 \log(1-x))} dx} dx \left.\right\}$$
$$- \int_0^1 -\theta_2 \log(1-x) \frac{e^{-(\theta_1 \log x + \theta_2 \log(1-x))}}{\int_0^1 e^{-(\theta_1 \log x + \theta_2 \log(1-x))} dx} dx. \qquad (16)$$

An we obtain

$$\Psi(\eta) = \log \int_0^1 e^{-(\theta_1 \log x + \theta_2 \log(1-x))} dx \left.\right\}$$
$$- \int_0^1 (-\theta_1 \log x - \theta_2 \log(1-x)) \frac{e^{-(\theta_1 \log x + \theta_2 \log(1-x))}}{\int_0^1 e^{-(\theta_1 \log x + \theta_2 \log(1-x))} dx} dx. \qquad (17)$$

The relation (17) becomes:

$$\Psi(\eta) = \log \int_0^1 e^{-(\theta_1 \log x + \theta_2 \log(1-x))} dx \left.\right\}$$
$$- \int_0^1 \left(\log e^{-(\theta_1 \log x + \theta_2 \log(1-x))}\right) \cdot \frac{e^{-(\theta_1 \log x + \theta_2 \log(1-x))}}{\int_0^1 e^{-(\theta_1 \log x + \theta_2 \log(1-x))} dx} dx. \qquad (18)$$

So we can write

$$\Psi(\eta) = - \left[- \log \int_0^1 e^{-(\theta_1 \log x + \theta_2 \log(1-x))} dx. \left(\int_0^1 \frac{e^{-\theta_1 \log x - \theta_2 \log(1-x)}}{\int_0^1 e^{-(\theta_1 \log x + \theta_2 \log(1-x))} dx} dx \right) \right.$$
$$\left. + \int_0^1 \log e^{-(\theta_1 \log x + \theta_2 \log(1-x))} \cdot \frac{e^{-(\theta_1 \log x + \theta_2 \log(1-x))}}{\int_0^1 e^{-(\theta_1 \log x + \theta_2 \log(1-x))} dx} dx \right] \left.\right\}$$
$$\qquad (19)$$

and obtain the following relation

$$\Psi(\eta) = - \left[\int_0^1 \frac{e^{-\theta_1 \log x - \theta_2 \log(1-x)}}{\int_0^1 e^{-(\theta_1 \log x + \theta_2 \log(1-x))} dx} \cdot \log \left(\frac{e^{-\theta_1 \log x - \theta_2 \log(1-x)}}{\int_0^1 e^{-(\theta_1 \log x + \theta_2 \log(1-x))} dx} \right) dx \right]$$
$$\qquad (20)$$

the relation (20) becomes

$$\Psi(\eta) = \int_0^1 (-\log q_\theta(x)) . q_\theta(x) dx. \qquad (21)$$

Therefore:

$$\Psi(\eta) = \mathbb{E}\left[-\log q(X, \theta) \right].$$

$\square$

Therefore it follow from this Theorem and Koszul's work that (14) is the density function at maximum entropy.

2.3 Associated Information Metric

To make the link with the Koszul's work, recalled by Barbaresco [2,5] we note the Fisher metric by $I(\theta) = [g_{ij}]$ given by Amari [1] is a Hessian metric, which is written as the Hessian of a potential function given by:

$$[g_{ij}] = \begin{pmatrix} -\frac{\partial^2 \Phi(\theta)}{\partial \theta_1 \partial \theta_1} & -\frac{\partial^2 \Phi(\theta)}{\partial \theta_1 \partial \theta_2} \\ -\frac{\partial^2 \Phi(\theta)}{\partial \theta_2 \partial \theta_1} & -\frac{\partial^2 \Phi(\theta)}{\partial \theta_2 \partial \theta_2} \end{pmatrix}.$$

Thus we have the Koszul's metric associated with the model on the manifold S:

$$[g_{ij}] = \begin{bmatrix} A_1(\theta) & A_2(\theta) \\ A_2(\theta) & A_3(\theta) \end{bmatrix}$$

with

$$A_1(\theta) = -\frac{\int_0^1 \log^2(x)e^{(\theta_1 \log x + \theta_2 \log(1-x))}dx}{\int_0^1 e^{(\theta_1 \log x + \theta_2 \log(1-x))}dx} + \frac{\left(\int_0^1 \log(x)e^{(\theta_1 \log x + \theta_2 \log(1-x))}dx\right)^2}{\left(\int_0^1 e^{(\theta_1 \log x + \theta_2 \log(1-x))}dx\right)^2};$$

$$A_2(\theta) = -\frac{\int_0^1 \log(x)\log(1-x)e^{(\theta_1 \log x + \theta_2 \log(1-x))}dx}{\int_0^1 e^{(\theta_1 \log x + \theta_2 \log(1-x))}dx} + \frac{\left(\int_0^1 \log(x)e^{(\theta_1 \log x + \theta_2 \log(1-x))}dx\right)\left(\int_0^1 \log(1-x)e^{(\theta_1 \log x + \theta_2 \log(1-x))}dx\right)}{\left(\int_0^1 e^{(\theta_1 \log x + \theta_2 \log(1-x))}dx\right)^2};$$

$$A_3(\theta) = -\frac{\int_0^1 \log^2(1-x)e^{(\theta_1 \log x + \theta_2 \log(1-x))}dx}{\int_0^1 e^{(\theta_1 \log x + \theta_2 \log(1-x))}dx} + \frac{\left(\int_0^1 \log(1-x)e^{(\theta_1 \log x + \theta_2 \log(1-x))}dx\right)^2}{\left(\int_0^1 e^{(\theta_1 \log x + \theta_2 \log(1-x))}dx\right)^2}.$$

2.4   Associated Gradient System

The objective in this part is to build the gradient system defined on S.
The inverse of the matrix:

$$[g_{ij}]^{-1} = \begin{bmatrix} B_1(\theta) & B_2(\theta) \\ B_2(\theta) & B_3(\theta) \end{bmatrix}$$

with

$$B_1(\theta) = -\frac{\left(\int_0^1 \log^2(1-x)e^{(\theta_1 \log x + \theta_2 \log(1-x))}dx\right)\left(\int_0^1 e^{(\theta_1 \log x + \theta_2 \log(1-x))}dx\right)}{m(\theta)\left(\int_0^1 e^{(\theta_1 \log x + \theta_2 \log(1-x))}dx\right)^2}$$
$$+\frac{\left(\int_0^1 \log(1-x)e^{(\theta_1 \log x + \theta_2 \log)}dx\right)^2}{m(\theta)\left(\int_0^1 e^{(\theta_1 \log x + \theta_2 \log(1-x))}dx\right)^2};$$

$$B_2(\theta) = \frac{\left(\int_0^1 \log(x)\log(1-x)e^{(\theta_1 \log x + \theta_2 \log(1-x))}dx\right)\left(\int_0^1 e^{(\theta_1 \log x + \theta_2 \log(1-x))}dx\right)}{m(\theta)\left(\int_0^1 e^{(\theta_1 \log x + \theta_2 \log(1-x))}dx\right)^2}$$
$$-\frac{\left(\int_0^1 \log(x)e^{(\theta_1 \log x + \theta_2 \log(1-x))}dx\right)\left(\int_0^1 \log(1-x)e^{(\theta_1 \log x + \theta_2 \log(1-x))}dx\right)}{m(\theta)\left(\int_0^1 e^{(\theta_1 \log x + \theta_2 \log(1-x))}dx\right)^2};$$

$$B_3(\theta) = -\frac{\left(\int_0^1 \log^2(x)e^{(\theta_1 \log x + \theta_2 \log(1-x))}dx\right)\left(\int_0^1 e^{(\theta_1 \log x + \theta_2 \log(1-x))}dx\right)}{m(\theta)\left(\int_0^1 e^{(\theta_1 \log x + \theta_2 \log(1-x))}dx\right)^2}$$
$$-\frac{\left(\int_0^1 \log(x)e^{(\theta_1 \log x + \theta_2 \log(1-x))}dx\right)^2}{m(\theta)\left(\int_0^1 e^{(\theta_1 \log x + \theta_2 \log(1-x))}dx\right)^2}.$$

where

$$m(\theta) = -\frac{\left(\int_0^1 \log^2(x)e^{(\theta_1 \log x+\theta_2 \log(1-x))}dx\right)\left(\int_0^1 \log^2(1-x)e^{(\theta_1 \log x+\theta_2 \log(1-x))}dx\right)}{\left(\int_0^1 e^{(\theta_1 \log x+\theta_2 \log(1-x))}dx\right)^4} \times$$

$$\left(\int_0^1 e^{(\theta_1 \log x+\theta_2 \log(1-x))}dx\right)^2 - \frac{\left(\int_0^1 \log^2(x)e^{(\theta_1 \log x+\theta_2 \log(1-x))}dx\right)}{\left(\int_0^1 e^{(\theta_1 \log x+\theta_2 \log(1-x))}dx\right)^4} \times$$

$$\left(\int_0^1 \log(1-x)e^{(\theta_1 \log x+\theta_2 \log(1-x))}dx\right)^2 \left(\int_0^1 e^{(\theta_1 \log x+\theta_2 \log(1-x))}dx\right)$$

$$-\frac{\left(\int_0^1 \log(x)e^{(\theta_1 \log x+\theta_2 \log(1-x))}dx\right)^2\left(\int_0^1 \log(1-x)e^{(\theta_1 \log x+\theta_2 \log(1-x))}dx\right)}{\left(\int_0^1 e^{(\theta_1 \log x+\theta_2 \log(1-x))}dx\right)^4} \times$$

$$\left(\int_0^1 e^{(\theta_1 \log x+\theta_2 \log(1-x))}dx\right) + \frac{\left(\int_0^1 \log(x)e^{(\theta_1 \log x+\theta_2 \log(1-x))}dx\right)^2}{\left(\int_0^1 e^{(\theta_1 \log x+\theta_2 \log(1-x))}dx\right)^4} \times$$

$$\left(\int_0^1 \log(1-x)e^{(\theta_1 \log x+\theta_2 \log(1-x))}dx\right)^2$$

$$-\frac{\left(\left(\int_0^1 \log(x)\log(1-x)e^{(\theta_1 \log x+\theta_2 \log(1-x))}dx\right)\left(\int_0^1 e^{(\theta_1 \log x+\theta_2 \log(1-x))}dx\right)\right)}{\left(\int_0^1 e^{(\theta_1 \log x+\theta_2 \log(1-x))}dx\right)^4}$$

$$-\frac{\left(\left(\int_0^1 \log(x)e^{(\theta_1 \log x+\theta_2 \log(1-x))}dx\right)\left(\int_0^1 \log(1-x)e^{(\theta_1 \log x+\theta_2 \log(1-x))}dx\right)\right)^2}{\left(\int_0^1 e^{(\theta_1 \log x+\theta_2 \log(1-x))}dx\right)^4}.$$

So, the gradient system will therefore be written as follows:

$$\dot{\theta}_1 = \frac{\left(\int_0^1 \log^2(1-x)e^{(\theta_1 \log x+\theta_2 \log(1-x))}dx\right)\left(\int_0^1 \log(x)e^{(\theta_1 \log x+\theta_2 \log(1-x))}dx\right)}{m(\theta)\left(\int_0^1 e^{(\theta_1 \log x+\theta_2 \log(1-x))}dx\right)^3} \times$$

$$\left(\int_0^1 e^{(\theta_1 \log x+\theta_2 \log(1-x))}dx\right) - \frac{\left(\int_0^1 \log(x)e^{(\theta_1 \log x+\theta_2 \log(1-x))}dx\right)^3}{m(\theta)\left(\int_0^1 e^{(\theta_1 \log x+\theta_2 \log(1-x))}dx\right)^3}$$

$$-\frac{\left(\int_0^1 \log(1-x)e^{(\theta_1 \log x+\theta_2 \log(1-x))}dx\right)\left(\int_0^1 \log(x)\log(1-x)e^{(\theta_1 \log x+\theta_2 \log(1-x))}dx\right)}{m(\theta)\left(\int_0^1 e^{(\theta_1 \log x+\theta_2 \log(1-x))}dx\right)^3} \times$$

$$\left(\int_0^1 e^{(\theta_1 \log x+\theta_2 \log(1-x))}dx\right) - \frac{\left(\int_0^1 \log(x)e^{(\theta_1 \log x+\theta_2 \log(1-x))}dx\right)}{m(\theta)\left(\int_0^1 e^{(\theta_1 \log x+\theta_2 \log(1-x))}dx\right)^3} \times$$

$$\left(\int_0^1 \log(1-x)e^{(\theta_1 \log x+\theta_2 \log(1-x))}dx\right)^2$$

$$\dot{\theta}_2 = -\frac{\left(\int_0^1 \log(x)e^{(\theta_1 \log x+\theta_2 \log(1-x))}dx\right)\left(\int_0^1 \log(x)\log(1-x)e^{(\theta_1 \log x+\theta_2 \log(1-x))}dx\right)}{m(\theta)\left(\int_0^1 e^{(\theta_1 \log x+\theta_2 \log(1-x))}dx\right)^3} \times$$

$$\left(\int_0^1 e^{(\theta_1 \log x+\theta_2 \log(1-x))}dx\right) + \frac{\left(\int_0^1 \log(x)e^{(\theta_1 \log x+\theta_2 \log(1-x))}dx\right)^2}{m(\theta)\left(\int_0^1 e^{(\theta_1 \log x+\theta_2 \log(1-x))}dx\right)^3} \times$$

$$\left(\int_0^1 \log(1-x)e^{(\theta_1 \log x+\theta_2 \log(1-x))}dx\right) + \frac{\left(\int_0^1 \log(1-x)e^{(\theta_1 \log x+\theta_2 \log(1-x))}dx\right)}{m(\theta)\left(\int_0^1 e^{(\theta_1 \log x+\theta_2 \log(1-x))}dx\right)^3} \times$$

$$\left(\int_0^1 \log^2(1-x)e^{(\theta_1 \log x+\theta_2 \log(1-x))}dx\right)\left(\int_0^1 e^{(\theta_1 \log x+\theta_2 \log(1-x))}dx\right)$$

$$-\frac{\left(\int_0^1 \log(1-x)e^{(\theta_1 \log x+\theta_2 \log(1-x))}dx\right)^3}{m(\theta)\left(\int_0^1 e^{(\theta_1 \log x+\theta_2 \log(1-x))}dx\right)^3}.$$

(22)

We determine the linear system associated with this system gradient. So, we calculate: $\dot{\vec{\eta}} = [g_{ij}]\dot{\vec{\theta}}$ and obtain:

$$\begin{pmatrix}\dot{\eta}_1 \\ \dot{\eta}_2\end{pmatrix} = \begin{bmatrix}A_1(\theta) & A_2(\theta) \\ A_2(\theta) & A_3(\theta)\end{bmatrix} \cdot \begin{pmatrix}\eta_1 \\ \eta_2\end{pmatrix}.$$

And then

$$\begin{pmatrix}\dot{\eta}_1 \\ \dot{\eta}_2\end{pmatrix} = \begin{pmatrix}-\frac{\int_0^1 e^{(\theta_1 \log x+\theta_2 \log(1-x))}\log x\, dx}{\int_0^1 e^{(\theta_1 \log x+\theta_2 \log(1-x))}dx} \\ -\frac{\int_0^1 e^{(\theta_1 \log x+\theta_2 \log(1-x))}\log(1-x)\, dx}{\int_0^1 e^{(\theta_1 \log x+\theta_2 \log(1-x))}dx}\end{pmatrix}.$$

According the relation (7), we conclude that:

$$\begin{pmatrix} \dot{\eta}_1 \\ \dot{\eta}_2 \end{pmatrix} = \begin{pmatrix} -\eta_1 \\ -\eta_2 \end{pmatrix}.$$

So, the linear system associated with our gradient system is

$$\dot{\eta} = -\eta.$$

We have the following Theorem.

Theorem 2. *The Hamiltonian of the gradient system (22) for the beta family of the first species manifold is given by:*

$$\mathcal{H} = \frac{\int_0^1 \log(1-x)e^{(\theta_1 \log(x)+\theta_2 \log(1-x))}dx}{\int_0^1 \log(x)e^{(\theta_1 \log x+\theta_2 \log(1-x))}dx}$$

and

$$\mathcal{H} = \frac{\mathbb{E}[-\log(1-x)]}{\mathbb{E}[-\log(x)]},$$

which is a constant of motion of the dynamical system (22).

Proof. According to Fujiwara's Theorem [6] we have

$$\mathcal{H} = \frac{\eta_2}{\eta_1}. \tag{23}$$

Using (7), we obtain:

$$\mathcal{H} = \frac{\int_0^1 \log(1-x)e^{(\theta_1 \log(x)+\theta_2 \log(1-x))}dx}{\int_0^1 \log(x)e^{(\theta_1 \log x+\theta_2 \log(1-x))}dx}.$$

Using (10) we have the expression (12). By consider (14) the density at maximum entropy we have the relation (15). Therefore (23) become:

$$\mathcal{H} = \frac{\mathbb{E}[-\log(1-x)]}{\mathbb{E}[-\log(x)]}.$$

By a direct calculation we obtain

$$\frac{d\mathcal{H}}{dt} = \frac{\partial \mathcal{H}}{\partial \theta_1}\frac{d\theta_1}{dt} + \frac{\partial \mathcal{H}}{\partial \theta_2}\frac{d\theta_2}{dt}. \tag{24}$$

with

$$\left.\begin{aligned} \frac{\partial \mathcal{H}}{\partial \theta_1} &= -\frac{\int_0^1 \log(1-x)\log(x)e^{(\theta_1 \log x+\theta_2 \log(1-x))}dx}{\int_0^1 \log(x)e^{(\theta_1 \log x+\theta_2 \log(1-x))}dx} \\ &+ \frac{\left(\int_0^1 \log(1-x)e^{(\theta_1 \log x+\theta_2 \log(1-x))}dx\right)\left(\int_0^1 \log^2(x)e^{(\theta_1 \log x+\theta_2 \log(1-x))}dx\right)}{\left(\int_0^1 \log(x)e^{(\theta_1 \log x+\theta_2 \log(1-x))}dx\right)^2} \end{aligned}\right\} \tag{25}$$

and

$$\frac{\partial \mathcal{H}}{\partial \theta_2} = -\frac{\int_0^1 \log^2(1-x)e^{(\theta_1 \log x + \theta_2 \log(1-x))}dx}{\int_0^1 \log(x)e^{(\theta_1 \log x + \theta_2 \log(1-x))}dx} \\ + \frac{\left(\int_0^1 \log(1-x)e^{(\theta_1 \log x + \theta_2 \log(1-x))}dx\right)\left(\int_0^1 \log(x) \log(1-x)e^{(\theta_1 \log x + \theta_2 \log(1-x))}dx\right)}{\left(\int_0^1 \log(x)e^{(\theta_1 \log x + \theta_2 \log(1-x))}dx\right)^2}. \Bigg\} \quad (26)$$

Using (25) and (26) in (24) and the gradient system (22) we have:

$$\frac{d\mathcal{H}}{dt} = 0.$$

$\mathcal{H}$ is a constant of motion of the dynamical system (22).   □

3   Conclusion

The beta distribution of the first kind present the geometric structure that we will study based on the work J.L.Koszul recalled by Barbaresco [2].

Acknowledgements. I gratefully acknowledge all my discussions with members of ERAG of the University of Maroua. Thanks are due to Dr. Kemajou Theophile for fruitful discussions.

References

1. Amari, S.: Differential-Geometrical Methods in Statistics. 2nd Printing, vol. 28. Springer Science, Tokyo (2012). https://doi.org/10.1007/978-1-4612-5056-2
2. Barbaresco, F.: Jean-Louis Koszul and the elementary structures of information geometry. In: Nielsen, F. (ed.) Geometric Structures of Information. Signals and Communication Technology, pp. 333–392. Springer, Cham (2019). https://doi.org/10.1007/978-3-030-02520-5-12
3. Barbaresco, F.: Les densités de probabilité distinguées et l'équation d'Alexis Clairaut: regards croisés de Maurice Fréchet et de Jean-Louis Koszul. GRETSI (2017). http://gretsi.fr/colloque2017/myGretsi/programme.php
4. Barbaresco, F.: Science géométrique de l'Information: Géométrie des matrices de covariance, espace métrique de Fréchet et domaines bornés homogénes de Siegel. C mh **2**(1), 1–188 (2011)
5. Barbaresco, F.: La géometrie de l'information des systemes dynamiques de Jean-Louis Koszul et Jean-Marie Souriau: thermodynamique des groupes de lie et métrique de Fisher-Koszul-Souriau. Thales (1), 1–7 (1940–1942)
6. Fujiwara, A.: Dynamical systems on statistical models (state of art and perspectives of studies on nonliear integrable systems). RIMS Kôkyuroku **822**, 32–42 (1993). http://hdl.handle.net/2433/83219
7. Nakamura, Y.: Completely integrable gradient systems on the manifolds of Gaussian and multinomial distributions. Jpn. J. Ind. Appl. Math. **10**(2), 179–189 (1993). https://doi.org/10.1007/BF03167571
8. Calin, O., Udriște, C.: Geometric Modeling in Probability and Statistics, vol. 121. Springer, Cham (2014). https://doi.org/10.1007/978-3-319-07779-6

KV Cohomology Group of Some KV Structures on $\mathbb{R}^2$

Mopeng Herguey[1,3]($\boxtimes$) and Joseph Dongho[2]

[1] Université de Palar, Faculté de Sciences, Palar, Chad
mopenpeter@gmail.com
[2] University of Maroua, Faculty of Sciences, Maroua, Cameroon
[3] University of Douala, Faculty of Sciences, Douala, Cameroon
http://www.univ-douala.cm, http://www.univ-maroua.cm

Abstract. The main concern of this paper is to prove that the vector space $\mathbb{R}^2$ have non trivial KV structures and some of them have non trivail KV cohomology. We propose the explicit computation of one of them.

Keywords: KV algebra · KV cohomology · Jacobi element

1 Introduction

In other to consolidate the following Gertenhaber assumption, **every restricted deformation theory generates its proper cohomology theory**, [5], Boyom complete the Nijenhuis work and define the complex of KV cohomology in [3]. Since that period it style hard to find paper in which KV structure and their KV cohology are constructed on a given vector space. The aims of this paper is to prove that $\mathbb{R}^2$ admit non trivial KV structure and that some of KV structure on $\mathbb{R}^2$ have non trivial KV cohomology.

2 On the Space of KV Structures on $\mathbb{R}^2$

Note (e_1, e_2) the canonical basis of $\mathbb{R}^2$; (e_1^*, e_2^*) its dual basis and $Sol((\mathbb{R}^2, KV)$ the set of KV-structures on $\mathbb{R}^2$. For any element μ of $Sol((\mathbb{R}^2, KV)$, there exist two bilinear forms Γ_i on $\mathbb{R}^2$; $i = 1, 2$ such that $\mu = (\Gamma_1, \Gamma_2)$. Each such $\Gamma_i : \mathbb{R}^2 \longrightarrow \mathbb{R}$, $i = \{1, 2\}$, is entirely determine by a 2×2 matrix. Let us consider the following such representations.

$$\Gamma_1 = \begin{pmatrix} \Gamma_{11}^1 & \Gamma_{12}^1 \\ \Gamma_{21}^1 & \Gamma_{22}^1 \end{pmatrix}, \Gamma_2 = \begin{pmatrix} \Gamma_{11}^2 & \Gamma_{12}^2 \\ \Gamma_{21}^2 & \Gamma_{22}^2 \end{pmatrix}$$

By a straightforward computation, we have the following result.

Supported by UFD-SF-UMa-2023.

Lemma 1. $\mu = (\Gamma_1, \Gamma_2)$ define a KV structure on $\mathbb{R}^2$ if and only if $\Gamma_{11}^1, \Gamma_{22}^1, \Gamma_{11}^2, \Gamma_{22}^2$ satisfy the following relation.

$$\begin{cases} \Gamma_{21}^1(\Gamma_{12}^2 - \Gamma_{11}^1 - \Gamma_{21}^2) + \Gamma_{12}^1(\Gamma_{11}^1 - \Gamma_{21}^2) + \Gamma_{22}^1\Gamma_{11}^2 = 0 \\ \Gamma_{22}^1(2\Gamma_{12}^2 - \Gamma_{11}^1 - \Gamma_{21}^2) + \Gamma_{12}^1(\Gamma_{12}^1 - \Gamma_{22}^2) = 0 \\ \Gamma_{11}^2(\Gamma_{12}^1 - 2\Gamma_{21}^1 + \Gamma_{22}^2) + \Gamma_{21}^2(\Gamma_{11}^1 - \Gamma_{21}^2) = 0 \\ \Gamma_{12}^2(\Gamma_{12}^1 - \Gamma_{21}^1 + \Gamma_{22}^2) + \Gamma_{21}^2(\Gamma_{12}^1 - \Gamma_{22}^2) - \Gamma_{11}^2\Gamma_{22}^1 = 0 \end{cases} \qquad (1)$$

We denote, $Sol^0(\mathbb{R}, KV)$ the set solution $\mu = (\Gamma^1, \Gamma^2)$ of equation (1) such that Γ^1, Γ^2 are not symmetric and $\Gamma_{ij}^k \neq 0$ for all $k, i, j \in \{1, 2\}$.

From relation (1) and the definition of the KV structures, we can easily prove the following theorem.

Theorem 1. The set $Sol((\mathbb{R}^2, KV)$ of KV structures on $\mathbb{R}^2$ is;

$$\begin{aligned} Sol((\mathbb{R}^2, KV) \simeq &\ (\mathbb{R}e_2^* \otimes e_1^* + \mathbb{R}^*e_2^* \otimes e_2^*) \times (\mathbb{R}^*e_2^* \otimes e_2^*)\cup \\ &\ (\mathbb{R}e_2^* \otimes e_1^* + \mathbb{R}^*e_1^* \otimes e_2^* + \mathbb{R}^*e_2^* \otimes e_2^*) \times (\mathbb{R}e_1^* \otimes e_2^*)\cup \\ &\ (\mathbb{R}e_1^* \otimes e_2^* + \mathbb{R}e_2^* \otimes e_2^*) \times (\mathbb{R}e_2^* \otimes e_1^* + \mathbb{R}e_1^* \otimes e_2^* + \mathbb{R}e_2^* \otimes e_2^*)\cup \\ &\ \mathbb{R}^*e_1^* \otimes e_1^* \times (\mathbb{R}^*e_1^* \otimes e_1^* + \mathbb{R}e_1^* \otimes e_2^*)\cup \\ &\ \mathbb{R}^*e_1^* \otimes e_1^* \times (\mathbb{R}^*e_1^* \otimes e_1^* + \mathbb{R}^*e_1^* \otimes e_1^* + \mathbb{R}^*e_1^* \otimes e_2^*)\cup \\ &\ (\mathbb{R}^*e_1^* \otimes e_1^* + \mathbb{R}e_2^* \otimes e_1^*) \times (\mathbb{R}^*e_1^* \otimes e_1^* + \mathbb{R}^*e_2^* \otimes e_1^*)\cup \\ &\ \mathbb{R}e_1^* \otimes e_2^* \cup \mathbb{R}e_1^* \otimes e_1^* \cup (\mathbb{R}e_1^* \otimes e_1^* + \mathbb{R}e_2^* \otimes e_2^*) \times (\mathbb{R}e_1^* \otimes e_1^* + \mathbb{R}e_2^* \otimes e_2^*)\cup \\ &\ Sol^0(\mathbb{R}, KV) \end{aligned} \qquad (2)$$

and $Sol^0(\mathbb{R}, KV) \neq \emptyset$.

To illustrate this theorem we will now study the set $\Gamma_{abc} = \{\mu_{abc} = (ae_2^* \otimes e_1^* + be_2^* \otimes e_2^*)e_1 + (ce_2^* \otimes e_2^*)e_2, a, b, c \in \mathbb{R}^*\}$. This is the main objective of the following lemma.

Lemma 2. $\Gamma_{abc} = \{\mu_{abc}; a, b, c\} \subset Sol(\mathbb{R}, KV)$

Proof. For any $a, b, c \in \mathbb{R}$, we denote

$$\mu_{abc} = (\alpha e_2^* \otimes e_1^* + \beta e_2^* \otimes e_2^*)e_1 + c(e_2^* \otimes e_2^*)e_2. \qquad (3)$$

More explicitly, we have:

$\mu_{abc}(u, v) = (au_2v_1 + bu_2v_2)e_1 + cu_2v_2e_2.$

For all $u, v, w \in \mathbb{R}$, such that $u = u_1e_1 + u_2e_2, v = v_1e_1 + v_2e_2$ and $w = w_1e_1 + w_2e_2$. Denote

(KV1) $(u, v, w) = (uv)w - u(vw)$
(KV2) $(v, u, w) = (vu)w - v(uw)$
(KV3) $(u, v, w) - (v, u, w)$

According to the definition of μ_{abc}, we have: Firstly

$$uv = (au_2v_1 + bu_2v_2, cu_2v_2) = (X, Y)$$

then
$$(uv)w = (aYw_1 + bYw_2, cYw_2)$$
$$= (acu_2v_2w_1 + bcu_2v_2w_2, c^2u_2v_2w_2)$$

Secondly $vw = (av_2w_1 + bv_2w_2, cv_2w_2) = (T, Z)$.

$$u(vw) = (au_2T + bu_2Z, cu_2Z)$$
$$= (au_2(av_2w_1 + bv_2w_2) + bu_2(cv_2w_2), c^2u_2v_2w_2)$$
$$= (a^2u_2v_2w_1 + acu_2v_2w_2 + bcu_2v_2w_2, c^2u_2v_2w_2)$$

And then
$$(u, v, w) = (au_2v_2(cw_1 - bw_2) - a^2u_2v_2w_1, 0). \tag{4}$$

In other hand, we have: Firstly $vu = (av_2u_1 + bv_2u_2, cv_2u_2) = (X, Y)$ and

$$(vu)w = (aYw_1 + bYw_2, cYw_2)$$
$$= (acv_2u_2w_1 + bcv_2u_2w_2, c^2v_2u_2w_2)$$

Secondly $uw = (au_2w_1 + bu_2w_2, cu_2w_2) = (T, Z)$ and

$$v(uw) = (av_2T + bv_2Z, cv_2Z)$$
$$= (a^2v_2u_2w_1 + abv_2u_2w_2) + bcv_2u_2w_2, c^2v_2u_2w_2)$$

And then
$$(v, u, w) = (av_2u_2(cw_1 - bw_2) - a^2v_2u_2w_1, 0) \tag{5}$$

The relations (4) and (5) allows us to complete the proof.   □

In what follows, μ will designate μ_{abc} for a given $(a, b, c) \in \mathbb{R}^* \times \mathbb{R}^* \times \mathbb{R}^*$ of non zero real numbers.

3   On the Koszul-Vulberg Cohomology of μ

Let $q > 1$ be a positive integer. Let $C^q(\mathbb{R}^2)$ be the vector space of all q-linear maps from $\mathbb{R}^2$ to $\mathbb{R}^2$. Recall that $C^q(\mathbb{R}^2)$ is a $\mathbb{R}^2$-KV-module with respect to the following two actions of $\mathbb{R}^2$: For all $a, a_j \in \mathbb{R}^2; j = 1, ..., q$ and $\varphi \in C^q(\mathbb{R}^2, \mathbb{R}^2)$:

$$(a\varphi)(a_1, ..., a_q) = \mu(a, (\varphi(a_1, ..., a_q))) - \sum_{j=1}^q \varphi(a_1, ..., \mu(a, a_j), ..., a_q)$$
$$(\varphi a)(a_1...a_q) = \mu((\varphi(a_1...a_q)), a) \tag{6}$$

For any $\rho = 1, ..., q$ define $e_\rho(a) : C^q(\mathbb{R}^2) \to C^{q-1}(\mathbb{R}^2)$, by;

$$(e_\rho(a)\varphi)(a_1, ..., a_{q-1}) = \varphi(a_1, ..., a_{\rho-1}, a, a_\rho, ...a_{q-1})$$

The KV coboundary operator of order q, associated to μ, is the additive application $\delta^q : C^q(\mathbb{R}^2) \to C^{q+1}(\mathbb{R}^2)$ such that for any $\varphi \in C^q(\mathbb{R}^2)$ and $(a_1, ..., a_{q+1}) \in (\mathbb{R}^2)^{q+1}$, $\delta^q(\varphi) \in C^{q+1}(\mathbb{R}^2)$ is given by the following formula.

$$(\delta^q\varphi)(a_1, ..., a_{q+1}) = \sum_{1 \le j \le q+1} (-1)^j \{(a_j\varphi)(a_1, ..., \hat{a}_j, ..., a_{q+1}) + (e_q(a_j)(\varphi a_{q+1})(a_1, ..., \hat{a}_j, ..., a_{\hat{q}+1})\} \tag{7}$$

It is prove in [3] that $\delta^2 = 0$. But to become a complex, he complete $C(\mathbb{R}^2) = \bigoplus_{q \geq 1} C^q(\mathbb{R}^2)$ by $J(A) = \{\xi \in A; (a, b, \xi) = 0, \forall a, b \in A\}$; named the set of Jacobi elements of A. We recall that

$$\left.\begin{aligned}
&\delta^0 : J(A) \to C^1(A, A) : \xi \mapsto \delta^0(\xi) : u \mapsto -u\xi + \xi u.\\
&\delta^1(\varphi)(u, v) = -u\varphi(v) + \varphi(uv) - \varphi(u)v\\
&\delta^2\varphi(u, v, w) = v\varphi(u, w) - u\varphi(v, w) + \varphi(v, uw) - \varphi(u, vw) +\\
&\quad \varphi(uv, w) - \varphi(vu, w) + \varphi(u, v)w - \varphi(v, u)w
\end{aligned}\right\} \quad (8)$$

From these expressions, we have:

$$\begin{aligned}
\delta^1(\delta^0(\xi))(u, v) &= -u\delta^0(\xi)(v) + \delta^0(\xi)(uv) - \delta^0(\xi)(u)v\\
&= -u(-v\xi + \xi v) + (-uv\xi + \xi uv) - (-u\xi + \xi u)v\\
&= uv\xi - u\xi v - uv\xi + \xi uv + u\xi v - \xi uv\\
&= uv\xi - uv\xi + u\xi - u\xi + \xi uv - \xi uv\\
&= 0
\end{aligned}$$

Therefore,

$$\delta^1 \circ \delta^0 = 0 \quad (9)$$

In other hand, for all $f \in C^1(\mathbb{R}^2)$, we have;

$$\begin{aligned}
&\delta^2(\delta^1(f))(a, b, c)\\
&= -a(\delta^1)(b, c) + (\delta^1 f)(ab, c) + (\delta^1 f)(b, ac) - (\delta^1 f)(b, a)c\\
&\quad + b(\delta^1 f)(a, c) - (\delta^1)(ba, c) - (\delta^1 f)(a, bc) + (\delta^1 f)(a, b)c\\
&= -a\{-bf(c) + f(bc) - f(b)c\} + \{-(ab)f(c) + f((ab)c)\\
&\quad - f(ab)c\} + \{-bf(ac) + f(b(ac)) - f(b)(ac)\}\\
&\quad + -\{-(bf(a))c + f(ba)c - (f(b)a)c\} + b\{-af(c) + f(ac)\\
&\quad - f(a)c\} - \{-(ba)f(c) + f((ba)c) - f(ba)c\}\\
&\quad + -\{-af(bc) + f(a(bc)) - f(a)(bc)\} + \{-af(b) + f(ab) - f(a)b\}c\\
&= +a(bf(c)) - af(bc) + a(f(b)c) - (ab)f(c) + f((ab)c)\\
&\quad - f(ab)c - bf(ac) + f(b(ac)) - f(b)(ac)\\
&\quad + (bf(a))c - f(ba)c + (f(b)a)c - b(af(c)) + bf(ac)\\
&\quad - b(f(a)c) + (ba)f(c) - f((ba)c) + f(ba)c\\
&\quad + af(bc) - f(a(bc)) + f(a)(bc) - (af(b))c + f(ab)c - (f(a)b)c\\
&= [+a(bf(c)) - (ab)f(c) - b(af(c)) + (ba)f(c)]\\
&\quad + [+a(f(b)c) - f(b)(ac) + (f(b)a)c - (af(b))c]\\
&\quad + [+f(a)(bc) - (f(a)b)c - b(f(a)c) + (bf(a))c]\\
&\quad + [+f((ab)c) + f(b(ac)) - f(a(bc))f((ba)c)]\\
&\quad + [+f(ab)c - f(ab)c] + [+af(bc) - af(bc)]\\
&\quad + [+f(ba)c - f(ba)c] + [+bf(ac) - bf(ac)]
\end{aligned}$$

It is obvious that the last line of above expression is zero. More explicitly, for all $f \in C^1(\mathbb{R}^2), a, b, c \in \mathbb{R}^2$, we have:

$$\begin{aligned}
\delta^2(\delta^1 f)(a, b, c) &= [(a, b, f(c)) - (b, a, f(c))]\\
&\quad + [(a, f(b), c) - (f(b), a, c)] + [(f(a), b, c) - (b, f(a), c)]\\
&\quad + [f[(b, a, c) - (a, b, c)]] + [f(ab - ab)]c + a[f(bc - bc)]\\
&\quad + [f(ba - ba)]c + b[f(ac - ac)] = 0
\end{aligned} \quad (10)$$

It follow from (10) that

$$\delta^2 \circ \delta^1 = 0 \tag{11}$$

Therefore the following modules are well defined. $H^0_{KV}(\mu), H^1_{KV}(\mu), H^2_{KV}(\mu)$ The next section is devoted to their computation.

4   KV Cohomology of the KV Structure μ on $\mathbb{R}^2$.

Let $\mu \in Sol((\mathbb{R}^2, KV)$ defined by; $\mu(u,v) = (ayx' + byy', cyy')$ with $a \neq 0, b \neq 0, c \neq 0$.

4.1   Computation of $H^0_{KV}(\mu)$

In this subsection, we will prove the following proposition

Proposition 1. *The first KV cohomology of the KV structure on $\mathbb{R}^2$ defined by $\mu = (ae_1^* \otimes e_1^* + be_2^* \otimes e_2^*)e_1 + c(e_2^* \otimes e_2^*)e_2$ is:*

$$H^0_{KV}(\mu) \approx \mathbb{R}(-be_1 + (c-a)e_2) \tag{12}$$

Proof. The application

$$\delta^0 : C^0(\mathbb{R}^2) = J_\mu(\mathbb{R}^2) \to C^1(\mathbb{R}^2)$$
$$\xi \qquad\qquad\qquad \mapsto \delta^0(\xi)$$

is defined by $\delta^0(\xi)(u) = -u\xi + \xi u, \forall \xi \in J_\mu(\mathbb{R}^2)$ and $u \in \mathbb{R}^2$ but, the jacobian of $\mathbb{R}^2$ is:

$$J_\mu(\mathbb{R}^2) = \{\xi \in \mathbb{R}^2, Ass_\mu(u,v,\xi) = 0, \forall u,v \in \mathbb{R}^2\}.$$

Let $u = (x,y)$, $v = (x',y')$ and $\xi = (x'',y'')$ be three vectors of $\mathbb{R}^2$. We have:

$$Ass_\mu(u,v,\xi) = (a(c-a)yy'x'' - abyy'y'', 0)$$

So: If $a = 0$ then $J_\mu(\mathbb{R}^2) = \mathbb{R}^2$. If not,

$$Ass_\mu(u,v,\xi) = 0 \iff Ass_\mu(e_i, e_j, \xi) = 0, 1 \leq i,j, \leq 2$$
$$\iff (c-c)x'' - by'' = 0$$

which is the vectorial line directed by $-be_1 + (c-a)e_2$. Therefore,

$$J_\mu(\mathbb{R}^2) \simeq \begin{cases} \mathbb{R}(-be_1 + (c-a)e_2) & if \quad a \neq 0 \\ \mathbb{R}^2 & if \quad a = 0 \end{cases} \tag{13}$$

and then,

$$Ker\delta^0 = \{\xi \in \mathbb{R}^2, \delta^0(\xi)(u) = 0, u \in \mathbb{R}\}$$
$$\simeq \mathbb{R}(-ae_1 + (c-a)e_2)$$

We conclude that

$$H^0_{KV}(\mu) \simeq \mathbb{R}(-be_1 + (c-a)e_2) \tag{14}$$

$\square$

4.2   Computation of $H^1_{KV}(\mu)$

In this subsection we will essentially prove the following proposition

Proposition 2. *The second KV cohomology of the KV structure on $\mathbb{R}^2$ defined by $\mu = (ae_1^* \otimes e_1^* + be_2^* \otimes e_2^*)e_1 + c(e_2^* \otimes e_2^*)e_2$ is:*

$$
H^1_{KV}(\mu) \approx
\begin{cases}
\mathbb{R}\begin{pmatrix} 1 & \frac{b}{a-c} \\ 0 & 0 \end{pmatrix} & if \quad a \neq c \\[3mm]
\mathbb{R}\begin{pmatrix} 0 & 1 \\ 1 & 0 \end{pmatrix} & if \quad a = c
\end{cases}
\tag{15}
$$

Proof. It ensure from the definition of

$$
\begin{aligned}
\delta^1 : C^1(\mathbb{R}^2) &\to C^2(\mathbb{R}^2) \\
f_1 &\mapsto \delta^1 f_1,
\end{aligned}
$$

that; for all u, v elements of $\mathbb{R}^2$, we have:

$$
\delta^1 f_1(u, v) = -u f_1(v) + f_1(uv) - f_1(u)v
$$

where f_1 is represented by its matrix $A = \begin{pmatrix} \alpha & \beta \\ \gamma & \lambda \end{pmatrix}$

Let $u = (x, y)$ and $v = (x', y')$ be two vectors of $\mathbb{R}^2$. One has the following expression of $\delta^1 f_1(u, v)$:
$\delta^1 f_1(u,v) = ((-a\gamma)xx' + (-b\gamma)xy' + (-b\gamma - a\lambda)x'y + (b\alpha + (c-a)\beta - 2b\lambda)yy',$
$(-c\gamma)xy' + (a-c)\gamma x'y + (b\gamma - c\lambda)yy').$
Then

$$
\delta^1 f_1(u, v) = 0 \iff \delta^1 f_1(e_i, e_j) = 0, \forall i, j = 1, 2.
$$

In other words, α, β, γ and λ are solution of the following equation.

$$
\gamma = 0, \quad \lambda = 0, \quad b\alpha + (c-a)\beta = 0
$$

One distinguishes two cases.

(i) If $c \neq a$, the we obtain:

$$
\gamma = 0, \quad \lambda = 0, \quad \beta = \frac{b}{a-c}\alpha
$$

$$
f_1 = \begin{pmatrix} \alpha & \frac{b}{a-c}\alpha \\ 0 & 0 \end{pmatrix}
$$

From where we deduce that

$$
Ker\delta^1 = \left\langle \begin{pmatrix} 1 & \frac{b}{a-c} \\ 0 & 0 \end{pmatrix} \right\rangle_{\mathbb{R}}
$$

(ii) If $c = a$, then we obtain:

$$\gamma = 0, \quad \lambda = 0, \quad \alpha = 0$$

So,

$$f_1 = \begin{pmatrix} 0 & \beta \\ 0 & 0 \end{pmatrix}$$

Therefore,

$$Ker\delta^1 = \langle \begin{pmatrix} 0 & 1 \\ 0 & 0 \end{pmatrix} \rangle_{\mathbb{R}}$$

Let g be a linear map from $\mathbb{R}^2$ to $\mathbb{R}^2$ defined by his matrix $B = \begin{pmatrix} u_{11} & u_{12} \\ u_{21} & u_{22} \end{pmatrix}$.
It follows from the definition that $g \in Im\delta_\mu^0$ if and only if there exist $\xi \in J_\mu(\mathbb{R}^2)$ such that $\delta_\mu^0(\xi) = g$.

$$g \in Im\delta^0 \Leftrightarrow \delta^0\xi(u) = g(u), \forall u \in \mathbb{R}^2$$
$$\Leftrightarrow -u\xi + \xi u = g(u)$$
$$\Leftrightarrow g(u) = 0$$

Therefore, $Im\delta^0 = \{0\}$ and

$$H_{KV}^1(\mu) \approx Ker\delta^1,$$

$\square$

4.3   Computation of $H_{KV}^2(\mu)$

The main objective of this section is the prove the following proposition

Proposition 3. *Let* $\mu = (ae_1^* \otimes e_1^* + be_2^* \otimes e_2^*)e_1 + c(e_2^* \otimes e_2^*)e_2$ *be a KV structure on* $\mathbb{R}^2$ *with* $(a, b, c) \neq (0, 0, 0)$, *then:*

(i) If $c \neq 2a$, *then the third KV cohomology of* μ *on* $\mathbb{R}^2$ *is:*

$$H_{KV}^2(\mu) \approx \mathbb{R}[(e_1^* \otimes e_1^* + \frac{b}{c}e_1^* \otimes e_2^*)e_1 + (\frac{c-a}{c}e_2^* \otimes e_1^* + e_1^* \otimes e_2^*)e_2] \oplus \mathbb{R}(e_1^* \otimes e_1^*)e_1$$

(ii) If $c = 2a$, *then the third KV cohomology of* μ *on* $\mathbb{R}^2$ *is*

$$H_{KV}^2(\mu) \approx \mathbb{R}[(-3e_1^* \otimes e_1^* + \frac{b}{a}e_1^* \otimes e_2^*)e_1 + (-\frac{2a}{b}e_1^* \otimes e_1^* + e_2^* \otimes e_1^*)e_2]$$

Proof. According to its definition, the application

$$\delta^2 \colon C^2(\mathbb{R}^2) \to C^3(\mathbb{R}^2)$$
$$f_2 \qquad \mapsto \delta^2 f_2$$

is defined by, for all u, v, w elements of $\mathbb{R}^2$, we have:

$$\delta^2 f_2(u, v, w) = vf_2(u, w) - uf_2(v, w) + f_2(v, uw) - f_2(u, vw) + f_2(uv, w) - f_2(vu, w)$$
$$+ f_2(u, v)w - f_2(v, u)w$$

Let

$$\left(\begin{pmatrix} e & f \\ g & h \end{pmatrix}, \begin{pmatrix} i & j \\ k & l \end{pmatrix}\right)$$

denote the matrix representation of f_2 with respect to the basis $\{e_1, e_2\}$ of $\mathbb{R}^2$. Let $u = (x, y)$, $v = (x', y')$ and $w = (x'', y'')$ be three elements of $\mathbb{R}^2$. We have the following expression of $\delta^2 f_2(u, v, w)$:

$$\delta^2 f_2(u, v, w) = \begin{pmatrix} ((-ae + bi + aj - ak)yx'x'' \\ -(-ae + bi + aj - ak)xy'x'' + (-2a + c)iyx'y'' - (-2a + c)ixy'y'', \\ (-be - cf + 2bj - ak)yx'x'' - (-be - cf + 2bj - ak)xy'x'' \\ +(-bi + (-a+c)j - ck)yx'y'' - (-bi + (-a+c)j - ck)xy'y'') \end{pmatrix}.$$

Then

$$\delta^2 f_2(u, v, w) = 0 \iff \delta^2 f_2(e_i, e_j, e_k) = 0, 1 \le i, j, k \le 2$$

$$\iff \begin{cases} -ae + bi + aj - ak = 0 \\ (-2a + c)i = 0 \\ -be - cf + 2bj - ak = 0 \\ -bi + (-a+c)j - ck = 0 \end{cases}$$

We distinguish the case.

(i) $c \ne 2a$.

In this case, an element f_2 of $C^2(\mathbb{R}^2)$ is in $ker\delta^2$ if and only if

$$f_2 \in \left\{ \begin{array}{l} [(j - k)(e_1^* \otimes e_1^*) + j\frac{b}{c}(e_1^* \otimes e_2^*) + g(e_2^* \otimes e_1^*) + h(e_2^* \otimes e_2^*)]e_1 \\ +[j(e_1^* \otimes e_2^*) + j\frac{c-a}{c}(e_2^* \otimes e_1^*) + l(e_2^* \otimes e_2^*)]e_2; \\ j, k, h \in \mathbb{R}, c \ne 2a \end{array} \right\}$$

Therefore,

$Ker\delta^2 \approx \mathbb{R}[(e_1^* \otimes e_1^*) + \frac{b}{c}(e_1^* \otimes e_2^*)]e_1 \oplus \mathbb{R}[(e_1^* \otimes e_2^*) + \frac{c-a}{c}(e_2^* \otimes e_1^*)]e_2 \oplus$
$\mathbb{R}[-(e_1^* \otimes e_1^*)]e_1 \oplus \mathbb{R}[(e_2^* \otimes e_1^*)]e_1 \oplus \mathbb{R}[(e_2^* \otimes e_2^*)]e_1 \oplus \mathbb{R}[(e_2^* \otimes e_2^*)]e_2$.

Let

$$g = (u_{11}e_1^* \otimes e_1^* + u_{12}e_1^* \otimes e_2^* + u_{21}e_2^* \otimes e_1^* + u_{22}e_2^* \otimes e_2^*)e_1 + \\ (v_{11}e_1^* \otimes e_1^* + v_{12}e_1^* \otimes e_2^* + v_{21}e_2^* \otimes e_1^* + v_{22}e_2^* \otimes e_2^*)e_2 \;;$$

be an element of $C^2(\mathbb{R}^2)$; g is in $Im\delta^1$ if and only if there exist $f_1 \in C(\mathbb{R}^2)$ such that; for all $u, v \in \mathbb{R}^2$, we have $\delta^1(f_1)(u, v) = g(u, v)$. In other words, f_1 is solution of a linear equation system having g as right hand side and of which the matrix representation is:

$$\begin{pmatrix} b & c-a & 0 & -2b & : u_{22} \\ 0 & 0 & -a & 0 & : u_{11} \\ 0 & 0 & 0 & 0 & : bu_{11} - au_{12} \\ 0 & 0 & -b & -a & : u_{21} \\ 0 & 0 & 0 & 0 & : cu_{11} - av_{12} \\ 0 & 0 & 0 & 0 & : (a - c)u_{11} + av_{21} \\ 0 & 0 & 0 & -(a+c) & : u_{21} + v_{22} \\ 0 & 0 & 0 & 0 & : v_{11} \end{pmatrix}$$

What allows us to write;

$$g \in Im\delta^1 \approx ker\delta^2 \cap \left\{ \begin{array}{l} (u_{11}e_1^* \otimes e_1^* + \frac{b}{a}u_{11}e_1^* \otimes e_2^* + u_{21}e_2^* \otimes e_1^* + u_{22}e_2^* \otimes e_2^*)e_1 + \\ (\frac{c}{a}u_{11}e_1^* \otimes e_2^* + \frac{a-c}{c}u_{11}e_2^* \otimes e_1^* + v_{22}e_2^* \otimes e_2^*)e_2; \\ u_{11}, u_{21}, u_{22}, v_{22} \in \mathbb{R}; c \neq 2a, c^2 - a^2 \neq 0 \quad bc \neq -a^2 \end{array} \right\}$$

Therefore, the space of 1-coboundary is

$$Im\delta^1 \approx \left\{ \begin{array}{l} (u_{21}e_1^* \otimes e_1^* + u_{22}e_2^* \otimes e_2^*)e_1 + (v_{22}e_2^* \otimes e_2^*)e_2; \\ u_{21}, u_{22}, v_{22} \in \mathbb{R}; c \neq 2a, c^2 - a^2 \neq 0 \quad bc \neq -a^2 \end{array} \right\}$$

The result follows from the fact that

$$H_{KV}^2(\mu) = \{\overline{x}, x \in Ker\delta^2\}$$
$$= \{x + Im\delta^1, x \in Ker\delta^2\}$$

(ii) $c = 2a$.

In this case, an element f_2 of $C^2(\mathbb{R}^2)$ is in $ker\delta^2$ if and only if

$$f_2 \in \left\{ \begin{array}{l} [(2j - 3k)(e_1^* \otimes e_1^*) + \frac{b}{a}k(e_1^* \otimes e_2^*) + g(e_2^* \otimes e_1^*) + h(e_2^* \otimes e_2^*)]e_1 \\ + [(\frac{a}{b}j - 2\frac{a}{b}k)(e_1^* \otimes e_1^*) + j(e_1^* \otimes e_2^*) + k(e_2^* \otimes e_1^*) + l(e_2^* \otimes e_2^*)]e_2 \end{array} \right\}$$

Therefore,

$$Ker\delta^2 \approx \mathbb{R}[2(e_1^* \otimes e_1^*)e_1 + \frac{a}{b}(e_1^* \otimes e_1^*) + (e_1^* \otimes e_2^*)e_2] \oplus \mathbb{R}[-3(e_1^* \otimes e_1^*) + \frac{b}{a}(e_1^* \otimes e_2^*)e_1 + \frac{-a2}{b}(e_1^* \otimes e_1^*) + (e_2^* \otimes e_1^*)e_2] \oplus \mathbb{R}[(e_2^* \otimes e_1^*)]e_1 \oplus \mathbb{R}[(e_2^* \otimes e_2^*)]e_1 \mathbb{R}[(e_2^* \otimes e_2^*)e_2]$$

By a straightforward computation, we prove that

$$Im\delta^1 \approx \left\{ \begin{array}{l} (u_{11}e_1^* \otimes e_1^* + \frac{b}{a}u_{11}e_1^* \otimes e_2^* + u_{21}e_2^* \otimes e_1^* + u_{22}e_2^* \otimes e_2^*)e_1 + \\ (2u_{11}e_1^* \otimes e_2^* + u_{11}e_2^* \otimes e_1^* + v_{22}e_2^* \otimes e_2^*); \\ u_{11}, u_{22}, u_{21}, v_{22} \in \mathbb{R}; 2b \neq -a \end{array} \right\}$$

Therefore, we deduce the result from the fact that,

$$H_{KV}^2(\mu) = \{\overline{x}, x \in Ker\delta^2\}$$
$$= \{x + Im\delta^1, x \in Ker\delta^2\}$$

$\square$

Acknowledgements. We kindly thank the Mama Assandje Rosaire Prospere and Dr. Tsimi Armand for their comments.

References

1. Boyom, M, N., Wolak, R.: Local structure of Koszul-Vinberg and of Lie algebroids. Bulletin des sciences mathématiques **128**(6), 467–479 (2004) https://doi.org/10.1016/j.bulsci.2004.02.007
2. Boyom, M.N.: KV-cohomology of Koszul-Vinberg algebroids and Poisson Manifolds. Internat. J. Mathemat. **16**(09), 1033–1061 (2005). https://doi.org/10.1142/S0129167X0500320X
3. Boyom, M. N.: The cohomology of Koszul-Vinberg algebras. Pacific J. Mathem. **225**(1), 119–153 (2006) https://doi.org/10.2140/pjm.2006.225.119
4. Boyom, M.N.: Cohomology of Koszul-Vinberg algebroide and Poisson manifolds I. Banach Center Publicat. **54**(1), 99–110 (2000). https://doi.org/10.4064/bc54-0-7
5. Gerstenhaber, M.: On deformation of rings and algebras. Anals Mathem., 59–103 (1964) https://doi.org/10.2307/1970484

Alpha-parallel Priors on a One-Sided Truncated Exponential Family

Masaki Yoshioka[1]([✉])[ID] and Fuyuhiko Tanaka[1,2][ID]

[1] Graduate School of Engineering Science, Osaka University, Osaka, Japan
yoshioka@sigmath.es.osaka-u.ac.jp, ftanaka.celas@osaka-u.ac.jp
[2] Center for Education in Liberal Arts and Sciences, Osaka University, Osaka, Japan

Abstract. In conventional information geometry, the deep relationship between differential geometrical structures such as the Fisher metric and α-connections and statistical theory has been investigated for statistical models satisfying regularity conditions. However, the study of information geometry on non-regular statistical models has not been fully investigated. A one-sided truncated exponential family (oTEF) is a typical example. In this study, we define the Riemannian metric on the oTEF model not in a formal way but in the way compatible with the asymptotic properties of MLE in statistical theory. Then, we define alpha-parallel priors and show that the one-parallel prior exists on the oTEF model.

Keywords: truncated exponential family · information geometry · noninformative priors · alpha-parallel prior

1 Introduction

Information geometry is the study of statistical models by differential geometry. From the standpoint of geometry, a statistical model consisting of a collection of parameterized probability distributions can be regarded as a manifold. Then, when the statistical model satisfies certain regularity conditions, Chentsov's theorem leads to a natural differential geometrical structure [8]. The natural differential geometrical structure consists of the Riemannian metric defined by the Fisher information matrix and a one-parameter family of affine connections. They are called the *Fisher metric* and *α-connections* respectively. Information geometry of regular statistical models has been studied for a long time, and a deep relationship between the above geometrical structures and the statistical properties of statistical models has been revealed [4,5].

However, for non-regular statistical models, their geometric properties have not been fully investigated. One reason is the inapplicability of Chentsov's theorem to non-regular models, which prevents the geometrical structure from being determined naturally. For example, while the Fisher information matrix on a regular statistical model has two equivalent definitions, these definitions do not agree with each other on a statistical model where the support of a probability density function depends on the parameter. In prior research, Amari [3] provides

© The Author(s), under exclusive license to Springer Nature Switzerland AG 2023
F. Nielsen and F. Barbaresco (Eds.): GSI 2023, LNCS 14071, pp. 226–235, 2023.
https://doi.org/10.1007/978-3-031-38271-0_23

an idea that the Finsler geometry is familiar to non-regular models, especially for the location family.

In the present study, we take the one-sided truncated exponential family (oTEF) [7] as one of the most important non-regular statistical models, and discuss desirable geometrical structures. This family covers many practical examples like Pareto distributions, truncated normal distributions, and two-parameter exponential distributions. Statistical properties of the oTEF model, in particular, those related to point estimation theory, have been investigated by several authors [2, 7]. Recently, in information geometry, Yoshioka and Tanaka [13] shows the existence of 1-parallel prior on the oTEF.

This paper aims to define a family of appropriate affine connections for the oTEF. In the quest for suitable affine connections, we emphasize that an exponential family possesses α-parallel priors for all values of α. Using formally defined α connections, the existence of only a 1-connection is observed. As a substitute for the α-connections, we introduce a 1-parameter family of affine connections as β-connections. Subsequently, we show that β-connections are equiaffine for all β values. This property is congruent with the case of the exponential family.

We introduce a one-sided truncated exponential family (oTEF) in Sect. 2. Then, Sect. 3 defines a Riemannian metric on the oTEF. In Sect. 4, we define β-connections on the oTEF and prove there exist β-parallel priors for all β. Section 4 also reviews α-parallel priors in regular cases.

2   One-Sided Truncated Exponential Family

First, we introduce the one-sided truncated exponential family as follows:

Definition 1 (One-sided Truncated Exponential Family [7]). *Let the parameter space Θ be an open subset of $\mathbb{R}^n$ and $I = (I_1, I_2)$ be an open interval. Let $\mathcal{P} = \{P_{\theta,\gamma} : \theta \in \Theta, \gamma \in I\}$ be a parametrized family of probability distributions. When each probability distribution $P_{\theta,\gamma}$ has its probability density function*

$$p(x; \theta, \gamma) = \exp\left\{ \sum_{i=1}^{n} \theta^i F_i(x) + C(x) - \psi(\theta, \gamma) \right\} \cdot \mathbb{1}_{[\gamma, I_2)}(x) \quad (x \in I), \quad (1)$$

$C \in C(I)$, $F_i \in C^{\infty}(I) (i = 1, \ldots, n)$, $\psi \in C^{\infty}(\Theta \times I)$ with respect to the Lebesgue measure, we call $\mathcal{P}$ as a one-sided truncated exponential family (oTEF).

We have two kinds of parameters, θ is called the *natural parameter* and γ is called the *truncation parameter*. Since $\theta \in \mathbb{R}^n$, $\gamma \in \mathbb{R}$, the oTEF $\mathcal{P}$ is an $(n + 1)$-dimensional statistical model. We often use the abbreviated notation: $\partial_i = \partial/\partial \theta^i (i = 1, \ldots, n)$, $\partial_\gamma = \partial/\partial \gamma$ below.

Clearly, the oTEF model is a non-regular statistical model, which does not satisfy regularity conditions. Indeed, the support of the probability density function $p(x; \theta, \gamma)$ as a function of x is written as the interval $[\gamma, I_2]$, which depends on the parameter γ.

On the other hand, when the truncation parameter γ is prescribed, considering the n-dimensional submodel $\mathcal{E}_\gamma = \{P_{\theta,\gamma} : \theta \in \Theta\}$, the oTEF is an exponential family.

3 Definition of Riemannian Structure on the oTEF Model

In this section, let us define the Riemannian metric on the oTEF model.

Yoshioka and Tanaka [13] provide a Riemannian metric for the oTEF based on the asymptotic variance of MLEs. The asymptotic behavior of MLEs on the oTEF is discussed by Akahira [1].

Definition 2 (Riemannian metric on the oTEF model [13]). *Let $\mathcal{P}$ denote the oTEF model with the probability density (1). We define the Riemannian metric g on the oTEF model $\mathcal{P}$ as*

$$g_{ij} = \boldsymbol{E}[\partial_i l(X; \theta, \gamma) \partial_j l(X; \theta, \gamma)],$$
$$g_{i\gamma} = 0,$$
$$g_{\gamma\gamma} = \{-\partial_\gamma \psi(\theta, \gamma)\}^2$$

for $i, j = 1, \ldots, n$, where $l(x; \theta, \gamma) = \log p(x; \theta, \gamma)$.

This Riemannian metric is different from the negative Hessian form $-\boldsymbol{E}[\partial_a \partial_b \log p]$, because the expectation $\boldsymbol{E}[\partial_\gamma \log p]$ does not vanish. As shown by Li *et al.* [10], the negative Hessian form is sometimes not positive definite on the Pareto model (See Example 1 in Sect. 4). On the Pareto model, we obtain

$$(-\boldsymbol{E}[\partial_a \partial_b \log p])_{a,b} = \begin{pmatrix} \frac{1}{\theta^2} & -\frac{1}{\gamma} \\ -\frac{1}{\gamma} & \frac{\theta}{\gamma^2} \end{pmatrix},$$

which is not positive definite except $0 < \theta < 1$.

4 α-parallel Priors on the oTEF Model

Next, we consider affine connections on the non-regular model, a oTEF. Our purpose is to provide a suitable extension of α-connections in regular models for the oTEF.

To establish our approach, let us return to the geometric property of an exponential family, one of the typical regular models. We focus on the fact that the exponential family is famous as statistically equiaffine for α-connections [12]. This property confirms the exponential family has α-parallel prior for all $\alpha \in \mathbb{R}$.

This section provides an extension of the α-connections, referred to as the β-connection. Furthermore, we will show that the β-connection is equiaffine for any β.

4.1   Equiaffine Structure on Regular Models

Before the discussion of equiaffine connections on the oTEF, we briefly review equiaffine connections and α-parallel priors in regular cases. Please see Takeuchi and Amari [12] for more details.

In Bayesian statistics, for a given statistical model $\mathcal{P}$, we need a probability distribution over the model parameter space, which is called a *prior distribution*, or simply a *prior*. We often denote a prior density function as π. ($\pi(\xi) \geq 0$ and $\int_{\Xi} \pi(\xi)d\xi = 1$.) If we have certain information on the parameter in advance, then the prior should reflect this, and such a prior is often called a *subjective prior*. If not, we adopt a certain criterion and use a prior obtained through the criterion. Such priors are called *noninformative priors*.

A volume element on an n-dimensional model manifold corresponds to a prior density function over the parameter space $(\xi \in \Xi \subset \mathbb{R}^n)$ in a one-to-one manner. For a prior $\pi(\xi)$, its corresponding volume element ω is an n-form (differential form of degree n) and is written as

$$\omega = \pi(\xi)d\xi^1 \wedge \cdots \wedge d\xi^n$$

in the local coordinate system. We identify a prior density π with a volume element ω below.

To define α-parallel priors, we introduce a geometric property of affine connections.

Definition 3 (equiaffine). *Let $\mathcal{P}$ be an n-dimensional manifold with an affine connection induced by a covariant derivative ∇.*

An affine connection ∇ is equiaffine if there exists a volume element ω such that

$$\nabla \omega = 0$$

holds everywhere in $\mathcal{P}$. Furthermore, such a volume element ω is said to be a parallel volume element with respect to ∇.

For an affine connection ∇, a necessary and sufficient condition to be equiaffine is described by its curvature. The following proposition holds for a manifold with an affine connection ∇. Let $R_{ijk}{}^l$ be the components of the Riemannian curvature tensor [5] of ∇, defined as

$$R_{ijk}{}^l = \partial_i \Gamma_{jk}{}^l - \partial_j \Gamma_{ik}{}^l + \Gamma_{im}{}^l \Gamma_{jk}{}^m - \Gamma_{jm}{}^l \Gamma_{ik}{}^m, \quad (i,j,k,l = 1, \ldots, n)$$

where $\Gamma_{jk}{}^l$ denotes the connection coefficients of ∇.

Proposition 1 (Nomizu and Sasaki [11]). *The following conditions are equivalent:*

- *∇ is equiaffine,*
- *$R_{ijk}{}^k = 0$,*

where $R_{ijk}{}^k = \sum_{k=1}^n R_{ijk}{}^k$.

Note that the Levi-Civita connection is always equiaffine [11].

Returning to the statistical model, we define the α-parallel prior distribution. Let $\mathcal{P}$ be an n-dimensional regular statistical manifold.

α-parallel priors are a family of priors that generalizes the Jeffreys prior from the standpoint of information geometry. The definition is as follows:

Definition 4. *Let $\overset{(\alpha)}{\nabla}$ denote a covariant derivative operator with respect to an α-connection. When the α-connection is equiaffine, a parallel volume element $\omega^{(\alpha)}$ exists.*

Then, the prior π corresponding to the volume element $\omega^{(\alpha)}$ is called an α-parallel prior.

In regular statistical models, the Jeffreys prior is the 0-parallel prior and necessarily exists. However, there is no guarantee that an α-parallel prior exists when $\alpha \neq 0$. In other words, a geometrical structure on a model manifold determines the existence of such priors.

Takeuchi and Amari [12] give a necessary and sufficient condition for α-parallel priors to exist.

Proposition 2 (Takeuchi and Amari [12]). *For a model manifold $\mathcal{P}$, if*

$$\partial_i T_{jk}{}^k - \partial_j T_{ik}{}^k = 0 \quad (i, j = 1, \ldots, n), \tag{2}$$

then the α-parallel prior exists for any $\alpha \in \mathbb{R}$. Otherwise, only the 0-parallel prior exists.

If a statistical model satisfies the condition (2), the model is called *statistically equiaffine*. An exponential family is one of the statistically equiaffine models [12], which has α-parallel priors for all $\alpha \in \mathbb{R}$. We will construct a family of affine connections based on this fact.

4.2 α-parallel Priors on a oTEF

Let us return to the discussion of affine connections in the oTEF.

First, we define α-connections formally in the same way as regular models [5].

Definition 5 (α-connection). *For every $\alpha \in \mathbb{R}$, we define an α-connection on the oTEF model with the coefficients*

$$\overset{(\alpha)}{\Gamma}_{ab,c} (\theta, \gamma) = E\left[\partial_a \partial_b l\, \partial_c l\right] + \frac{1-\alpha}{2} E\left[\partial_a l\, \partial_b l\, \partial_c l\right] \quad (a, b, c = 1, \ldots, n, \gamma),$$

where $l = \log p(X; \theta, \gamma)$.

We write the covariant derivative with respect to the α-connection as $\overset{(\alpha)}{\nabla}$.

Unlike the regular cases, the equation

$$\overset{(\alpha)}{\Gamma}_{ab,c} (\theta, \gamma) = \overset{g}{\Gamma}_{ab,c} - \frac{\alpha}{2} E\left[\partial_a l \partial_b l \partial_c l\right]$$

does not hold, where $\overset{g}{\Gamma}_{ab,c}$ denotes the connection coefficients of the Levi-Civita connection. Then, the α-connections do not include the Levi-Civita connection in the o*TEF*. For example, in the Pareto model, it holds that

$$\overset{(\alpha)}{\Gamma}_{\gamma\gamma,1}(\theta,\gamma) = 0,$$

$$\overset{g}{\Gamma}_{\gamma\gamma,1}(\theta,\gamma) = -\frac{\theta}{\gamma^2}.$$

Let us consider the o*TEF* model with the α-connection. Then we obtain the following theorem.

Theorem 1 (Yoshioka and Tanaka [13]**).** *Let $\mathcal{P}$ denote the o*TEF* model endowed with the Riemannian metric in Definition 2 and the α-connections in Definition 5. Then, an α-parallel volume element exists when $\alpha = 1$. Let $\pi^{(1)}$ denote the density of the 1-parallel volume element. It is written as*

$$\pi^{(1)}(\theta,\gamma) \propto -\partial_\gamma\psi(\theta,\gamma).$$

Proof. Fix any $\alpha \in \mathbb{R}$ arbitrarily.
We will check the condition in Proposition 1,

$$\overset{(\alpha)}{R}_{abc}{}^c = 0.$$

By calculation of the α-connection coefficients, we have

$$\overset{(\alpha)}{R}_{ija}{}^a = 0,$$

$$\overset{(\alpha)}{R}_{i\gamma a}{}^a = \frac{1-\alpha}{2}\left\{\partial_i\partial_\gamma\log\left(\det(g_{kl})\right) + (n+1)\partial_i\partial_\gamma\psi\right\}$$

for $i,j = 1,\ldots,n$, where $\psi = \psi(\theta,\gamma)$.
Thus, $\overset{(\alpha)}{R}_{abc}{}^c = 0$ and there exists an α-parallel prior when $\alpha = 1$.
According to Takeuchi and Amari [12], the 1-parallel prior $\pi^{(1)}$ satisfies

$$\partial_a\log\pi^{(1)} = \overset{(1)}{\Gamma}_{ab}{}^b$$
$$= \partial_a\log(-\partial_\gamma\psi)$$

for $a = 1,\ldots,n,\gamma$.
Therefore, the 1-parallel prior for $\mathcal{P}$ is given as

$$\pi^{(1)}(\theta,\gamma) \propto -\partial_\gamma\psi.$$

The above 1-parallel prior coincides with a certain noninformative prior called the *(Bernardo's) reference prior* on a non-regular statistical model [9]. It is obtained through an information-theoretic argument. We apply their argument to the o*TEF* model setting γ as the parameter of interest and θ as the nuisance parameter. Then, we obtain $\pi_{\mathrm{Ghosal}}(\theta,\gamma) \propto -\partial_\gamma\psi$, which agrees with the 1-parallel prior in Theorem 1.

4.3 β-parallel Priors on a oTEF

The above α-connections are not equiaffine for almost all α. This result differs from the case of the exponential family. For example, in a family of left-truncated exponential distributions [1],

$$\overset{(\alpha)}{R}_{1\gamma a}{}^{a} = -\frac{(1-\alpha)(n+1)}{2}$$

holds for all α since $\psi = -\theta\gamma - \log\theta$. This model has only 1-parallel prior.

However, Theorem 1 provides an equiaffine connection, the 1-connection $\overset{(1)}{\nabla}$. We now construct another extension of the regular α-connections based on the two equiaffine connections: the 1-connection and the Levi-Civita connection.

The following lemma is elemental for the extension.

Lemma 1. *Let $\mathcal{P}$ be an n-dimensional manifold, and let ∇^0, ∇^1 be equiaffine connections on $\mathcal{P}$. For each $\beta \in \mathbb{R}$, we define an affine connection $\overset{(\beta)}{\nabla}$ on $\mathcal{P}$ by*

$$\overset{(\beta)}{\nabla} = (1-\beta)\nabla^1 + \beta\nabla^0.$$

Then, for each $\beta \in \mathbb{R}$, $\overset{(\beta)}{\nabla}$ is also an equiaffine connection.

Proof. Let $\theta = (\theta^1, \ldots, \theta^n) \in \Theta$ be a coordinate system on $\mathcal{P}$. In the θ-coordinates, we write the connection coefficients of ∇^0, ∇^1 as $\left\{\overset{0}{\Gamma}_{ij}{}^{k}\right\}, \left\{\overset{1}{\Gamma}_{ij}{}^{k}\right\}$, respectively. Also, we denote the curvature tensor of $\overset{(\beta)}{\nabla}$ by $\overset{(\beta)}{R}$.

Then, the connection coefficients of $\overset{(\beta)}{\nabla}$ are given by

$$\overset{(\beta)}{\Gamma}_{ij}{}^{k} = (1-\beta)\overset{0}{\Gamma}_{ij}{}^{k} + \beta\overset{1}{\Gamma}_{ij}{}^{k} \quad (i,j,k=1,\ldots,n).$$

We need to verify the condition in Proposition 1:

$$\overset{(\beta)}{R}_{ijk}{}^{k} = 0 \quad (i,j=1,\ldots,n).$$

Calculating the left-hand side, we get

$$\overset{(\beta)}{R}_{ijk}{}^{k} = (1-\beta)\left(\partial_i \overset{0}{\Gamma}_{jk}{}^{k} - \partial_j \overset{0}{\Gamma}_{ik}{}^{k}\right) + \beta\left(\partial_i \overset{1}{\Gamma}_{jk}{}^{k} - \partial_j \overset{1}{\Gamma}_{ik}{}^{k}\right).$$

Here, since ∇^0, ∇^1 are equiaffine connections, we have

$$\partial_i \overset{0}{\Gamma}_{jk}{}^{k} - \partial_j \overset{0}{\Gamma}_{ik}{}^{k} = 0,$$

$$\partial_i \overset{1}{\Gamma}_{jk}{}^{k} - \partial_j \overset{1}{\Gamma}_{ik}{}^{k} = 0$$

for all $i,j = 1,\ldots,n$.

Therefore, in the affine connection $\overset{(\beta)}{\nabla}$, we have

$$\overset{(\beta)}{R}_{ijk}{}^{k} = 0 \quad (i,j=1,\ldots,n).$$

Based on the above lemma, we define a family of affine connections that connect the 1-connection and the Levi-Civita connection via a parameter β.

Definition 6 (β-**connection**). *For every* $\beta \in \mathbb{R}$, *we define* β-*connections on the oTEF model with the coefficients*

$$\overset{(\beta)}{\Gamma}_{ab,c}(\theta, \gamma) = \beta \boldsymbol{E}\left[\partial_a \partial_b l \, \partial_c l\right] + (1 - \beta)\Gamma^g_{ab,c} \quad (a, b, c = 1, \ldots, n, \gamma),$$

where Γ^g *denotes the Levi-Civita connection.*

In regular models, the β-connections coincide with the α-connections when $\beta = \alpha$. They represent a new extension of Amari's geometric structure.

Theorem 2. *Let* $\mathcal{P}$ *denote the oTEF model endowed with the Riemannian metric in Definition 2 and* β-*connections in Definition 6. Then, for every* β, *the* β-*parallel volume element exists. Let* $\pi^{(\beta)}$ *denote the density of the* β-*parallel volume element. It is written as*

$$\pi^{(\beta)} \propto \{\det (g_{ij})\}^{\frac{1-\beta}{2}} \left(-\partial_\gamma \psi\right).$$

Proof. Fix any $\beta \in \mathbb{R}$ arbitrarily.

Lemma 1 and Definition 6 confirm that the β-connection is equiaffine. The representation of β-parallel priors is given as follows.

The sum of the β-connection coefficients satisfies that

$$\overset{(\beta)}{\Gamma}_{ia}{}^a = \frac{1 - \beta}{2}\partial_i \log (\det(g_{jk})) + \partial_i \log (-\partial_\gamma \psi),$$

$$\overset{(\beta)}{\Gamma}_{\gamma a}{}^a = \frac{1 - \beta}{2}\partial_\gamma \log (\det(g_{jk})) + \partial_\gamma \log (-\partial_\gamma \psi).$$

Therefore, from Proposition 1 by Takeuchi and Amari [12], we obtain:

$$\pi^{(\beta)} \propto \{\det (g_{ij})\}^{\frac{1-\beta}{2}} \left(-\partial_\gamma \psi\right).$$

When $\beta = 0$, $\pi^{(\beta)}$ agrees with the Jeffreys prior and when $\beta = 1$, it agrees with the 1-parallel prior in Theorem 1.

Example 1 (Pareto model). A family of Pareto distributions [6], having a density function

$$p(x, \theta, \gamma) = \frac{\theta \gamma^\theta}{x^{\theta+1}}\mathbb{1}_{[\gamma, \infty)} \quad (x \in (0, \infty))$$

with parameter $\theta, \gamma \in (0, \infty)$, is one of the famous oTEFs.

Since $\psi(\theta, \gamma) = -\log \theta - \theta \log \gamma$, the Riemannian metric of the Pareto model is represented as

$$\begin{pmatrix} \frac{1}{\theta^2} & 0 \\ 0 & \left(\frac{\theta}{\gamma}\right)^2 \end{pmatrix}.$$

Then, the β-parallel priors for $\beta \in \mathbb{R}$ is

$$\pi^{(\beta)} \propto \frac{\theta^\beta}{\gamma}.$$

5   Concluding Remarks

In the present study, we proposed a new extension of α-connections for a oTEF, referred to as β-connections, based on the fact that exponential families are statistically equiaffine. As shown in Sect. 4, the β-connections in the oTEF have β-parallel priors for all β. Then, the β-connections have the same geometric property as α-connections in exponential families.

On the other hand, in the statistical inference of regular models, α-connections appear in the higher-order asymptotic behavior of estimators. It is interesting to consider how β-connections work on statistical inference on the oTEF. Also, to reveal the relationship with Amari's Finsler structure on location family [3] is a topic of interest.

Acknowledgements. This work was supported by JSPS KAKENHI Grant Number 19K11860 and JST SPRING Grant Number JPMJSP2138. This work was also supported by the Research Institute for Mathematical Sciences, an International Joint Usage/Research Center located in Kyoto University.

References

1. Akahira, M.: Statistical Estimation for Truncated Exponential Families. SS, Springer, Singapore (2017). https://doi.org/10.1007/978-981-10-5296-5
2. Akahira, M.: Maximum likelihood estimation for a one-sided truncated family of distributions. Japanese J. Statist. Data Sci. **4**(1), 317–344 (2020). https://doi.org/10.1007/s42081-020-00098-5
3. Amari, S.: Finsler geometry of non-regular statistical models. RIMS Kokyuroku (in Japanese) **538**, 81–95 (1984)
4. Amari, S.: Differential-Geometrical Methods in Statistics. LNS, vol. 28. Springer-Verlag, Berlin (1985). https://doi.org/10.1007/978-1-4612-5056-2
5. Amari, S., Nagaoka, H.: Methods of Information Geometry, Translations of Mathematical Monographs, vol. 191. American Mathematical Society, Providence (2000). https://doi.org/10.1090/mmono/191
6. Arnold, B.C.: Pareto Distributions, Statistical Distributions in Scientific Work, vol. 5. International Co-operative Publishing House, Burtonsville (1983)
7. Bar-Lev, S.K.: Large sample properties of the mle and mcle for the natural parameter of a truncated exponential family. Ann. Inst. Stat. Math. **36**(2), 217–222 (1984). https://doi.org/10.1007/BF02481966
8. Chentsov, N.N.: Statistical Decision Rules and Optimal Inference, Translations of Mathematical Monographs, vol. 53. American Mathematical Society, Providence (1982). https://doi.org/10.1090/mmono/053
9. Ghosal, S.: Reference priors in multiparameter nonregular cases. TEST **6**(1), 159–186 (1997). https://doi.org/10.1007/BF02564432
10. Li, M., Sun, H., Peng, L.: Fisher-Rao geometry and Jeffreys prior for Pareto distribution. Commun. Statist. - Theory Methods **51**(6), 1895–1910 (2022). https://doi.org/10.1080/03610926.2020.1771593
11. Nomizu, K., Sasaki, T.: Affine Differential Geometry, Cambridge Tracts in Mathematics, vol. 111. Cambridge University Press, Cambridge (1994)

12. Takeuchi, J., Amari, S.: Alpha-parallel prior and its properties. IEEE Trans. Inform. Theory **51**(3), 1011–1023 (2005). https://doi.org/10.1109/TIT.2004. 842703

13. Yoshioka, M., Tanaka, F.: Information-geometric approach for a one-sided truncated exponential family. Entropy **25**(5), 769 (2023). https://doi.org/10.3390/ e25050769

Conformal Submersion with Horizontal Distribution and Geodesics

K. S. Subrahamanian Moosath[1(✉)] and T. V. Mahesh[2]

[1] Department of Mathematics, Indian Institute of Space Science and Technology, Thiruvananthapuram 695547, Kerala, India
smoosath@iist.ac.in
[2] Department of Mathematics, Mahatma Gandhi College, Thiruvananthapuram 695004, Kerala, India

Abstract. In this paper, we compare geodesics for conformal submersion with horizontal distribution. Then, proved a condition for the completeness of statistical connection for a conformal submersion with horizontal distribution.

Keywords: Conformal Submersion · Fundamental equation · Geodesics

1 Introduction

Riemannian submersion is a special tool in differential geometry and it has got application in different areas such as Kaluza-Klein theory, Yang-Mills theory, supergravity and superstring theories, statistical machine learning processes, medical imaging, theory of robotics and the statistical analysis on manifolds. O'Neill [6] defined a Riemannian submersion and obtained the fundamental equations of Riemannian submersion for Riemannian manifolds. In [7], O'Neill compare the geodesics of $\mathbf{M}$ and $\mathbf{B}$ for a semi-Riemannian submersion $\pi : \mathbf{M} \longrightarrow \mathbf{B}$. Abe and Hasegawa [2] defined an affine submersion with horizontal distribution which is a dual notion of affine immersion and obtained the fundamental equations. They compare the geodesics of $\mathbf{M}$ and $\mathbf{B}$ for an affine submersion with horizontal distribution $\pi : \mathbf{M} \longrightarrow \mathbf{B}$.

Conformal submersion and the fundamental equations of conformal submersion were also studied by many researchers, see [4,8] for example. Horizontally conformal submersion is a generalization of the Riemannian submersion. Horizontally conformal submersion is a special horizontally conformal map which got introduced independently by Fuglede [3] and Ishihara [5]. Their study focuses on the conformality relation between metrics on the Riemannian manifolds and Levi-Civita connections. The present authors defined conformal submersion with horizontal distribution and studied statistical manifold structure obtained using conformal submersion with horizontal distribution [9,10]. In this paper, we compare the geodesics for conformal submersion with horizontal distribution and obtained some interesting results. In Sect. 2, relevant basic concepts are given.

F. Nielsen and F. Barbaresco (Eds.): GSI 2023, LNCS 14071, pp. 236–243, 2023.
https://doi.org/10.1007/978-3-031-38271-0_24

In Sect. 3, we obtained some fundamental equations of conformal submersion with horizontal distribution. In Sect. 4, we proved a necessary and sufficient condition for $\pi \circ \sigma$ to be a geodesic of $\mathbf{B}$ when σ is a geodesic of $\mathbf{M}$ for $\pi : (\mathbf{M}, \nabla, g_m) \to (\mathbf{B}, \nabla^*, g_b)$ a conformal submersion with horizontal distribution. Completeness of statistical connection with respect to conformal submersion with horizontal distribution is given in Sect. 5. Throughout this paper, all the objects are assumed to be smooth.

2 Preliminaries

In this section, the concepts like submersion with horizontal distribution and affine submersion with horizontal distribution are given.

Let $\mathbf{M}$ and $\mathbf{B}$ be Riemannian manifolds with dimension n and m respectively with $n > m$. An onto map $\pi : \mathbf{M} \longrightarrow \mathbf{B}$ is called a submersion if $\pi_{*p} : T_p\mathbf{M} \longrightarrow T_{\pi(p)}\mathbf{B}$ is onto for all $p \in \mathbf{M}$. For a submersion $\pi : \mathbf{M} \longrightarrow \mathbf{B}$, $\pi^{-1}(b)$ is a submanifold of $\mathbf{M}$ of dimension $(n - m)$ for each $b \in \mathbf{B}$. These submanifolds $\pi^{-1}(b)$ are called fibers. Set $\mathcal{V}(\mathbf{M})_p = Ker(\pi_{*p})$ for each $p \in \mathbf{M}$.

Definition 1. *A submersion $\pi : \mathbf{M} \longrightarrow \mathbf{B}$ is called a submersion with horizontal distribution if there is a smooth distribution $p \longrightarrow \mathcal{H}(M)_p$ such that*

$$T_p\mathbf{M} = \mathcal{V}(\mathbf{M})_p \bigoplus \mathcal{H}(M)_p. \tag{1}$$

We call $\mathcal{V}(\mathbf{M})_p$ $(\mathcal{H}(M)_p)$ the vertical (horizontal) subspace of $T_p\mathbf{M}$. $\mathcal{H}$ and $\mathcal{V}$ denote the projections of the tangent space of $\mathbf{M}$ onto the horizontal and vertical subspaces, respectively.

Note 1. Let $\pi : \mathbf{M} \longrightarrow \mathbf{B}$ be a submersion with horizontal distribution $\mathcal{H}(M)$. Then, $\pi_* |_{\mathcal{H}(M)_p} : \mathcal{H}(M)_p \longrightarrow T_{\pi(p)}\mathbf{B}$ is an isomorphism for each $p \in \mathbf{M}$.

Definition 2. *A vector field Y on $\mathbf{M}$ is said to be projectable if there exists a vector field Y_* on $\mathbf{B}$ such that $\pi_*(Y_p) = Y_{*\pi(p)}$ for each $p \in \mathbf{M}$, that is Y and Y_* are π- related. A vector field X on $\mathbf{M}$ is said to be basic if it is projectable and horizontal. Every vector field X on $\mathbf{B}$ has a unique smooth horizontal lift, denoted by $\tilde{X}$, to $\mathbf{M}$.*

Definition 3. *Let ∇ and ∇^* be affine connections on $\mathbf{M}$ and $\mathbf{B}$ respectively. $\pi : (\mathbf{M}, \nabla) \longrightarrow (\mathbf{B}, \nabla^*)$ is said to be an affine submersion with horizontal distribution if $\pi : \mathbf{M} \longrightarrow \mathbf{B}$ is a submersion with horizontal distribution and satisfies $\mathcal{H}(\nabla_{\tilde{X}}\tilde{Y}) = (\nabla^*_X Y)\tilde{\,}$, for vector fields X, Y in $\mathcal{X}(\mathbf{B})$, where $\mathcal{X}(\mathbf{B})$ denotes the set of all vector fields on $\mathbf{B}$.*

Note 2. Abe and Hasegawa [2] proved that the connection ∇ on $\mathbf{M}$ induces a connection ∇' on $\mathbf{B}$ when $\pi : \mathbf{M} \longrightarrow \mathbf{B}$ is a submersion with horizontal distribution and $\mathcal{H}(\nabla_{\tilde{X}}\tilde{Y})$ is projectable for all vector fields X and Y on $\mathbf{B}$.

A connection $\mathcal{V}\nabla\mathcal{V}$ on the subbundle $\mathcal{V}(\mathbf{M})$ is defined by $(\mathcal{V}\nabla\mathcal{V})_E V = \mathcal{V}(\nabla_E V)$ for any vertical vector field V and any vector field E on $\mathbf{M}$. For each $b \in \mathbf{B}$, $\mathcal{V}\nabla\mathcal{V}$ induces a unique connection $\hat{\nabla}^b$ on the fiber $\pi^{-1}(b)$. Abe and Hasegawa [2] proved that if ∇ is torsion free, then $\hat{\nabla}^b$ and ∇' are also torsion free.

3   Conformal Submersion with Horizontal Distribution

In this section, we consider the conformal submersion with horizontal distribution which is a generalization of the affine submersion with horizontal distribution [9]. Here, we obtained some fundamental equations of conformal submersion with horizontal distribution for Riemannian manifolds.

Definition 4 [9]. *Let* $\pi : (\mathbf{M}, g_m) \longrightarrow (\mathbf{B}, g_b)$ *be a conformal submersion and let* ∇ *and* ∇^* *be affine connections on* $\mathbf{M}$ *and* $\mathbf{B}$, *respectively. Then,* $\pi : (\mathbf{M}, \nabla) \longrightarrow (\mathbf{B}, \nabla^*)$ *is said to be a conformal submersion with horizontal distribution* $\mathcal{H}(\mathbf{M}) = \mathcal{V}(\mathbf{M})^\perp$ *if*

$$\mathcal{H}(\nabla_{\tilde{X}}\tilde{Y}) = (\widetilde{\nabla^*_X Y}) + \tilde{X}(\phi)\tilde{Y} + \tilde{Y}(\phi)\tilde{X} - \mathcal{H}(grad_\pi\phi)g_m(\tilde{X}, \tilde{Y}),$$

for some $\phi \in C^\infty(\mathbf{M})$ *and for all* $X, Y \in \mathcal{X}(\mathbf{B})$.

Note 3. If ϕ is constant, it turns out to be an affine submersion with horizontal distribution.

As in the case of affine submersion with horizontal distribution we have,

Lemma 1. *Let* $\pi : (\mathbf{M}, \nabla) \longrightarrow (\mathbf{B}, \nabla^*)$ *be conformal submersion with horizontal distribution, then*

$$\mathcal{H}(Tor(\nabla)(\tilde{X}, \tilde{Y})) = (\widetilde{Tor(\nabla^*)(X, Y)}). \tag{2}$$
$$\mathcal{V}(Tor(\nabla)(V, W)) = (Tor(\hat{\nabla})(V, W)). \tag{3}$$

Proof. Proof follows immediately form the definition of the conformal submersion with horizontal distribution.

Corollary 1. *If* ∇ *is torsion-free, then* ∇^* *and* $\hat{\nabla}$ *are also torsion-free.*

Fundamental tensors T and A for a conformal submersion with horizontal distribution $\pi : (\mathbf{M}, \nabla) \longrightarrow (\mathbf{B}, \nabla^*)$ are defined for E and F in $\mathcal{X}(\mathbf{M})$ by

$$T_E F = \mathcal{H}\nabla_{\mathcal{V}E}(\mathcal{V}F) + \mathcal{V}\nabla_{\mathcal{V}E}(\mathcal{H}F).$$

and

$$A_E F = \mathcal{V}\nabla_{\mathcal{H}E}(\mathcal{H}F) + \mathcal{H}\nabla_{\mathcal{H}E}(\mathcal{V}F).$$

Note that these are $(1, 2)$-tensors.

We have the fundamental equations correspond to the conformal submersion with horizontal distribution. Let R be the curvature tensor of $(\mathbf{M}, \nabla)$ defined by

$$R(E, F)G = \nabla_{[X,Y]}G - \nabla_E\nabla_F G + \nabla_F\nabla_E G,$$

for E, F and G in $\mathcal{X}(\mathbf{M})$. Similarly, we denote the curvature tensor of ∇^* (respectively $\hat{\nabla}$) by R^* (respectively $\hat{R}$). Define $(1, 3)$-tensors R^{P_1, P_2, P_3} for conformal submersion with horizontal distribution by

$$R^{P_1, P_2, P_3}(E, F)G = P_3\nabla_{[P_1 E, P_2 F]}P_3 G - P_3\nabla_{P_1 E}(P_3\nabla_{P_2 F}P_3 G)$$
$$+ P_3\nabla_{P_2 F}(P_3\nabla_{P_1 E}P_3 G),$$

where $P_i = \mathcal{H}$ or $\mathcal{V}$ $(i = 1, 2, 3)$ and E, F, G are in $\mathcal{X}(\mathbf{M})$. Then, the following fundamental equations for conformal submersion with horizontal distribution can be obtained.

Theorem 1. *Let X, Y, Z be horizontal and U, V, W vertical vector fields in $\mathbf{M}$. Then,*

$$\mathcal{V}R(U, V)W = R^{\mathcal{V}\mathcal{V}\mathcal{V}}(U, V)W + T_V T_U W - T_U T_V W.$$

$$\mathcal{H}R(U, V)W = \mathcal{H}(\nabla_V T)_U W - \mathcal{H}(\nabla_U T)_V W - T_{Tor(\nabla)(U,V)} W.$$

$$\mathcal{V}R(U, V)X = \mathcal{H}(\nabla_V T)_U X - \mathcal{V}(\nabla_U T)_V X - T_{Tor(\nabla)(U,V)} X.$$

$$\mathcal{H}R(U, V)X = R^{\mathcal{V}\mathcal{V}\mathcal{H}}(U, V)X + T_V T_U X - T_U T_V X.$$

$$\mathcal{V}R(U, X)V = R^{\mathcal{V}\mathcal{H}\mathcal{V}}(U, X)V - T_U A_X V - A_X T_U V.$$

$$\mathcal{H}R(U, X)V = \mathcal{H}(\nabla_X T)_U V - \mathcal{H}(\nabla_U A)_X V - A_{A_X U} V + T_{T_U X} V$$
$$\qquad -T_{Tor(\nabla)(U,X)} V - A_{Tor(\nabla)(U,X)} V.$$

$$\mathcal{V}R(U, X)Y = \mathcal{V}(\nabla_X T)_U Y - \mathcal{V}(\nabla_U A)_X Y - A_{A_X U} Y + T_{T_U X} Y$$
$$\qquad -T_{Tor(\nabla)(U,X)} Y - A_{Tor(\nabla)(U,X)} Y.$$

$$\mathcal{H}R(U, X)Y = R^{\mathcal{V}\mathcal{H}\mathcal{H}}(U, X)Y - T_V A_X Y + A_X T_U Y.$$

$$\mathcal{V}R(X, Y)U = R^{\mathcal{H}\mathcal{H}\mathcal{V}}(X, Y)U + A_Y A_X U - A_X A_Y U.$$

$$\mathcal{H}R(X, Y)U = \mathcal{H}(\nabla_Y A)_X U - \mathcal{H}(\nabla_X A)_Y U + T_{A_X Y} U - T_{A_Y X} U$$
$$\qquad -T_{Tor(\nabla)(X,Y)} U - A_{Tor(\nabla)(X,Y)} U.$$

$$\mathcal{V}R(X, Y)Z = \mathcal{V}(\nabla_Y A)_X Z - \mathcal{V}(\nabla_X A)_Y Z + T_{A_X Y} Z - T_{A_Y X} Z$$
$$\qquad -T_{Tor(\nabla)(X,Y)} Z - A_{Tor(\nabla)(X,Y)} Z.$$

$$\mathcal{H}R(X, Y)Z = R^{\mathcal{H}\mathcal{H}\mathcal{H}}(X, Y)Z + A_Y A_X Z - A_X A_Y Z.$$

Proof. Proof follows directly from the definition of conformal submersion with horizontal distribution.

4   Geodesics

In this section, for a conformal submersion with horizontal distribution we prove a necessary and sufficient condition for $\pi \circ \sigma$ to be a geodesic of $\mathbf{B}$ when σ is a geodesic of $\mathbf{M}$. Let $\mathbf{M}, \mathbf{B}$ be Riemannian manifolds and $\pi : \mathbf{M} \to \mathbf{B}$ be a submersion. Let E be a vector field on a curve σ in $\mathbf{M}$ and the horizontal part $\mathcal{H}(E)$ and the vertical part $\mathcal{V}(E)$ of E be denoted by H and V, respectively. $\pi \circ \sigma$ is a curve in $\mathbf{B}$ and E_* denote the vector field $\pi_*(E) = \pi_*(H)$ on the curve $\pi \circ \sigma$ in $\mathbf{B}$. E'_* denote the covariant derivative of E_* and is a vector field on $\pi \circ \sigma$. The horizontal lift to σ of E'_* is denoted by $\tilde{E}'_*$. In [7], O'Neil compared the geodesics for semi-Riemannian submersion and Abe and Hasegawa [2] have done it for affine submersion with horizontal distribution.

Let $\pi : (\mathbf{M}, \nabla, g_m) \to (\mathbf{B}, \nabla^*, g_b)$ be a conformal submersion with horizontal distribution $\mathcal{H}(\mathbf{M})$. Throughout this section we assume ∇ is torsion free.

A curve σ is a geodesic if and only if $\mathcal{H}(\sigma'') = 0$ and $\mathcal{V}(\sigma'') = 0$, where σ'' is the covariant derivative of σ'. So, first we obtain the equations for $\mathcal{H}(E')$ and $\mathcal{V}(E')$ for a vector field E on the curve σ in $\mathbf{M}$ for a conformal submersion with horizontal distribution.

Theorem 2. *Let* $\pi : (\mathbf{M}, \nabla, g_m) \to (\mathbf{B}, \nabla^*, g_b)$ *be a conformal submersion with horizontal distribution, and let* $E = H + V$ *be a vector field on curve* σ *in* $\mathbf{M}$. *Then, we have*

$$\pi_*(\mathcal{H}(E')) = E'_* + \pi_*(A_X U + A_X V + T_U V) - e^{2\phi}\pi_*(grad_\pi\phi)g_b(\pi_*X, \pi_*H)$$
$$+X(\phi)\pi_*H + H(\phi)\pi_*X,$$
$$\mathcal{V}(E') = A_X H + T_U H + \mathcal{V}(V'),$$

where $X = \mathcal{H}(\sigma')$ *and* $U = \mathcal{V}(\sigma')$.

Proof. Consider a neighborhood of an arbitrary point $\sigma(t)$ of the curve σ in $\mathbf{M}$. By choosing the base fields $W_1, ..., W_n$, where $n = dim\mathbf{B}$, near $\pi(\sigma(t))$ on $\mathbf{B}$ and an appropriate vertical base field near $\sigma(t)$, we can derive

$$(E'_*)_t = \sum_i r^{i'}(t)(W_i)_{\pi(\sigma(t))} + \sum_{i,k} r^i(t)s^k(t)(\nabla^*_{W_k}W_i)_{\pi(\sigma(t))}, \qquad (4)$$

$$\pi_*(\mathcal{H}(E')_t) = \sum_i r^{i'}(t)(W_i)_{\pi(\sigma(t))} + \sum_{i,k} r^i(t)s^k(t)\pi_*(\mathcal{H}(\nabla_{\tilde{W}_k}\tilde{W}_i))_{\pi(\sigma(t))}$$
$$+\pi_*((A_H U) + (A_X V) + (T_U V))_{\pi(\sigma(t))}, \qquad (5)$$

where $\tilde{W}_i$ be the horizontal lift of W_i, for $i = 1, 2, ...n$ and $r^i(t)$ (respectively $s^k(t)$) be the coefficients of H (respectively of X) in the representation using the base fields $\tilde{W}_i$ restricted to σ.

Since π is a conformal submersion with horizontal distribution

$$\pi_*(\mathcal{H}(\nabla_H X)) = \nabla^*_{\pi_*(H)}\pi_*X + X(\phi)\pi_*H + H(\phi)\pi_*X$$
$$-e^{2\phi}\pi_*(grad_\pi\phi)g_b(\pi_*X, \pi_*H).$$

Hence

$$\pi_*(\mathcal{H}(E')) = E'_* + \pi_*(A_X U + A_X V + T_U V) - e^{2\phi}\pi_*(grad_\pi\phi)g_b(\pi_*X, \pi_*H)$$
$$+X(\phi)\pi_*H + H(\phi)\pi_*X.$$

Similarly we can prove $\mathcal{V}(E') = A_X H + T_U H + \mathcal{V}(V')$.

For σ'' we have

Corollary 2. *Let σ be a curve in* **M** *with $X = \mathcal{H}(\sigma')$ and $U = \mathcal{V}(\sigma')$. Then,*

$$\pi_*(\mathcal{H}(\sigma'')) = \sigma''_* + \pi_*(2A_X U + T_U U) - e^{2\phi}\pi_*(grad_\pi\phi)g_b(\pi_*X, \pi_*X)$$
$$+2X(\phi)\pi_*X, \tag{6}$$
$$\mathcal{V}(\sigma'') = A_X X + T_U X + \mathcal{V}(U'), \tag{7}$$

where σ''_* denotes the covariant derivative of $(\pi \circ \sigma)'$.

Now for a conformal submersion with horizontal distribution we prove a necessary and sufficient condition for $\pi \circ \sigma$ to become a geodesic of **B** when σ is a geodesic of **M**.

Theorem 3. *Let $\pi : (\mathbf{M}, \nabla, g_m) \to (\mathbf{B}, \nabla^*, g_b)$ be a conformal submersion with horizontal distribution. If σ is a geodesic of* **M**, *then $\pi \circ \sigma$ is a geodesic of* **B** *if and only if*

$$\pi_*(2A_X U + T_U U)) + 2X(\phi)\pi_*X = \pi_*(grad_\pi\phi) \parallel X \parallel^2,$$

where $X = \mathcal{H}(\sigma')$ and $U = \mathcal{V}(\sigma')$ and $\parallel X \parallel^2 = g_m(X, X)$.

Proof. Since σ is a geodesic on **M** from the Eq. (6)

$$\sigma''_* = \pi_*(grad_\pi\phi) \parallel X \parallel^2 -\pi_*(2A_X U + T_U U) - 2d\phi(X)\pi_*X.$$

Hence, $\pi \circ \sigma$ is a geodesic on **B** if and only if

$$\pi_*(2A_X U + T_U U)) + 2X(\phi)\pi_*X = \pi_*(grad_\pi\phi) \parallel X \parallel^2 .$$

Remark 1. If σ is a horizontal geodesic (that is, σ is a geodesic with $\mathcal{V}(\sigma') = 0$), then $\pi \circ \sigma$ is a geodesic if and only if $2X(\phi)\pi_*X = \pi_*(grad_\pi\phi) \parallel X \parallel^2$.

5   Completeness of Statistical Connections

Completeness of connection on statistical manifolds is an interesting area of study in information geometry. In [11], Nagouchi started the study of the completeness of the statistical connection on a certain type of statistical manifolds (In his study, he referred this kind of statistical manifolds as "special" statistical manifolds). In [12], Barbara Opozda obtained some results on completeness of statistical connection. In this section, we give a condition for completeness of statistical connection for conformal submersion with horizontal distribution.

For a torsion-free affine connection ∇ and a semi-Riemannian metric g_m, we say $(\mathbf{M}, \nabla, g_m)$ is a statistical manifold if ∇g_m is a symmetric $(0,3)$- tensor. For statistical manifolds the dual connections are also torsion-free [1].

Definition 5. *Let* $(\mathbf{M}, \nabla, g_m)$ *and* $(\mathbf{B}, \nabla^*, g_b)$ *be statistical manifolds,* $\pi : \mathbf{M} \to \mathbf{B}$ *be a conformal submersion with horizontal distribution and* α *be a smooth curve in* $\mathbf{B}$. *Let* α' *be the tangent vector field of* α *and* $(\tilde{\alpha}')$ *be its horizontal lift. Define the horizontal lift of the smooth curve* α *as the integral curve* σ *on* $\mathbf{M}$ *of* $(\tilde{\alpha}')$.

Now, we have

Proposition 1. *Let* $(\mathbf{M}, \nabla, g_m)$ *and* $(\mathbf{B}, \nabla^*, g_b)$ *be statistical manifolds,* $\pi :$ $\mathbf{M} \to \mathbf{B}$ *be a conformal submersion with horizontal distribution such that* $A_Z Z = 0$ *for all horizontal vector fields* Z. *Then, every horizontal lift of a geodesic of* $\mathbf{B}$ *is a geodesic of* $\mathbf{M}$ *if and only if* $2X(\phi)\pi_* X = \pi_*(grad_\pi \phi) \parallel X \parallel^2$, *where* X *is the horizontal part of the tangent vector field of the horizontal lift of the geodesic on* $\mathbf{B}$.

Proof. Let α be a geodesic on $\mathbf{B}$, σ be the horizontal lift of α. Then, we have $\pi \circ \sigma = \alpha$ and $\sigma'(t) = (\tilde{\alpha}'(t))$. Let $X = \mathcal{H}(\sigma'(t))$ and $U = \mathcal{V}(\sigma'(t))$, clearly $X = (\tilde{\alpha}'(t))$ and $U = 0$. Then, from the Eqs. (6) and (7)

$$\pi_*(\mathcal{H}(\sigma'')) = \alpha'' - \pi_*(grad_\pi \phi) \parallel X \parallel^2 + 2X(\phi)\pi_* X.$$
$$\mathcal{V}(\sigma'') = A_X X.$$

Since α is a geodesic and $A_X X = 0$ we have, $\sigma'' = 0$ if and only if $2X(\phi)\pi_* X = \pi_*(grad_\pi \phi) \parallel X \parallel^2$. That is, every horizontal lift of a geodesic of $\mathbf{B}$ is a geodesic of $\mathbf{M}$ if and only if $2X(\phi)\pi_* X = \pi_*(grad_\pi \phi) \parallel X \parallel^2$.

Theorem 4. *Let* $(\mathbf{M}, \nabla, g_m)$ *and* $(\mathbf{B}, \nabla^*, g_b)$ *be statistical manifolds,* $\pi : \mathbf{M} \to$ $\mathbf{B}$ *be a conformal submersion with horizontal distribution such that* $A_Z Z = 0$ *for all horizontal vector fields* Z. *Then,* ∇^* *is geodesically complete if* ∇ *is geodesically complete and* $2X(\phi)\pi_* X = \pi_*(grad_\pi \phi) \parallel X \parallel^2$, *where* X *is the horizontal part of the tangent vector field of the horizontal lift of the geodesic on* $\mathbf{B}$.

Proof. Let α be a geodesic of $\mathbf{B}$ and $\tilde{\alpha}$ be its horizontal lift to $\mathbf{M}$, by Proposition (1) $\tilde{\alpha}$ is a geodesic on $\mathbf{M}$. Since ∇ is geodesically complete, $\tilde{\alpha}$ can be defined on the entire real line. Then, the projected curve of the extension of $\tilde{\alpha}$ is a geodesic and is the extension of α, that is ∇^* is geodesically complete.

Acknowledgements. During this work the second named author was supported by the Doctoral Research Fellowship from the Indian Institute of Space Science and Technology (IIST), Department of Space, Government of India.

References

1. Lauritzen, S.L.: Statistical manifolds. Differ. Geom. Stat. Infer. **10**, 163–216 (1987)
2. Abe, N., Hasegawa, K.: An affine submersion with horizontal distribution and its applications. Differ. Geom. Appl. **14**, 235–250 (2001)

3. Fuglede, B.: Harmonic morphisms between Riemannian manifolds. Ann. de l'institut Fourier **28**, 107–144 (1978)
4. Gudmundsson, S.: On the geometry of harmonic morphisms. Math. Proc. Camb. Philos. Soc. **108**, 461–466 (1990)
5. Ishihara, T.: A mapping of Riemannian manifolds which preserves harmonic functions. J. Math Kyoto University **19**, 215–229 (1979)
6. O'Neill, B.: The fundamental equations of a submersion. Mich. Math. J. **13**, 459–469 (1966)
7. O'Neill, B.: Submersions and geodesics. Duke Math. J. **34**, 363–373 (1967)
8. Ornea, L., Romani, G.: The fundamental equations of conformal submersions. Beitr. Algebra Geom. **34**, 233–243 (1993)
9. Mahesh, T.V., Subrahamanian Moosath, K.S.: Harmonicity of conformally-projectively equivalent statistical manifolds and conformal statistical submersions. In: Nielsen, F., Barbaresco, F. (eds.) GSI 2021. LNCS, vol. 12829, pp. 397–404. Springer, Cham (2021). https://doi.org/10.1007/978-3-030-80209-7_44
10. Mahesh, T.V., Subrahamanian Moosath, K.S.: Affine and conformal submersions with horizontal distribution and statistical manifolds. Balkan J. Geom. Appl. **26**, 34–45 (2021)
11. Noguchi, M.: Geometry of statistical manifolds. Differ. Geom. Appl. **2**, 197–222 (1992)
12. Opozda, B.: Completeness in affine and statistical geometry. Ann. Glob. Anal. Geom. **59**(3), 367–383 (2021). https://doi.org/10.1007/s10455-021-09752-x

Statistics, Information and Topology

Higher Information from Families of Measures

Tom Mainiero[(⊠)]

St. Joseph's University, Patchogue, NY 11772, USA
mainiero@physics.utexas.edu

Abstract. We define the notion of a measure family: a pre-cosheaf of finite measures over a finite set; every joint measure on a product of finite sets has an associated measure family. To each measure family there is an associated index, or "Euler characteristic", related to the Tsallis deformation of mutual information. This index is further categorified by a (weighted) simplicial complex whose topology retains information about the correlations between various subsystems.

1 Introduction

Questions relating to the independence of random variables have a deep relationship to questions of topology and geometry: given the data of a multipartite, a.k.a. "joint" measure, there is an emergent "space" that encodes the relationship between various subsystems. Topological invariants of this emergent space capture non-trivial correlations between different subsystems: this includes numerical invariants—such as the Euler characteristic—which roughly indicate *how much* information is shared characteristic, as well as "higher" invariants—such as cohomology—that capture *what* information is shared. In [11] these ideas were explored, using a language engineered for an audience interested in the purely quantum regime, i.e., "non-commutative" measure theory. This note provides a sketch of the categorical underpinnings of these ideas in the opposite "classical" or "commutative" extreme, focusing on finite atomic measure spaces for brevity. Some of these underpinnings are partly outlined in the recorded talks [12,13].

The majority of this note is dedicated to formalizing the working parts that underlie the "commutative diagram" in Fig. 1. The word "space" is taken to mean a (semi-)simplicial measure or a (weighted semi-)simplicial set, and the grayed out mystery box indicates a suspected "weighted" version of cohomology that may provide a novel measure of shared information. The classical picture that is presented here is unified with the quantum picture of [11] using the language of von Neumann algebras.[1] A reader wishing to learn how this fits into a larger picture should consult [11] and the talks [12,13]. The upcoming paper [9] is a related spin-off of the categorical and W*-algebraic underpinnings of some ideas discussed here.

---

[1] See [15] for a precise categorical equivalence between commutative W*-algebras and (localizable) measurable spaces.

© The Author(s), under exclusive license to Springer Nature Switzerland AG 2023
F. Nielsen and F. Barbaresco (Eds.): GSI 2023, LNCS 14071, pp. 247–257, 2023.
https://doi.org/10.1007/978-3-031-38271-0_25

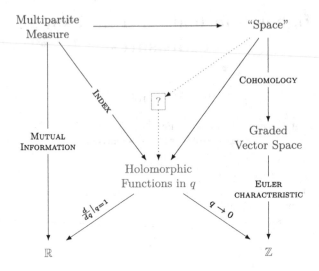

Fig. 1. A "commutative diagram" summarizing the big picture behind the definitions and results stated in this note.

Our categorical perspective of measures on finite sets has close ties to the work of Baez, Fritz, and Leinster [2,3], and the quantum mechanical generalization is related to the work of Parzygnat [14]. The homotopical or homological perspective has strong relations to the work of Baudot and Bennequin [4]; Vigneaux [18]; Sergeant-Perthuis [16]; and Drummond-Cole, Park, and Terilla [7,8]. Ideas around the index (Sect. 4.3) bear relation to the work of Lang, Baudot, Quax, and Forré [10].

2   Preliminaries

In this note, a (finite) *measure* μ consists of the data of a finite set Ω_μ and a function $\mathrm{Subset}(\Omega_\mu) \to \mathbb{R}_{\geq 0}$ that evaluates to zero on $\emptyset$ and satisfies the *additivity condition*: the value on a subset U reduces to the sum of its evaluation on points of U. In a mild abuse of notation, we use μ to denote the function $\mathrm{Subset}(\Omega_\mu) \to \mathbb{R}_{\geq 0}$. We allow for measures to be identically zero on a set, and also allow for the *empty measure*: the unique measure on the empty set.[2]

If μ and ν are measures with $\Omega_\mu = \Omega_\nu = \Omega$ we write $\mu \leq \nu$ if $\mu(U) \leq \nu(U)$ for all $U \subseteq \Omega$. Given a measure μ, and a function between sets $\underline{f} \colon \Omega_\mu \to \Gamma$, the *pushforward measure* $\underline{f}_*\mu$ is the measure with set $\Omega_{\underline{f}_*\mu} := \Gamma$ and $(\underline{f}_*\mu)(U) := \mu[\underline{f}^{-1}(U)]$ for any $U \subseteq \Gamma$. When $\underline{f}_*\mu = \nu$ we call $\underline{f}$ *measure-preserving*.

---

[2] The empty measure corresponds to the zero expectation value on the zero algebra.

3   The Category of Finite Measures

Definition 1. Meas *is the category with objects given by finite measures, and a morphism* $f\colon \mu \to \nu$ *defined by an underlying function on sets* $\underline{f}\colon \Omega_\mu \to \Omega_\nu$ *such that* $\underline{f}_*\mu \leq \nu$.

Remark 1. Meas utilizes the relation $\leq$ to define a larger class of morphisms than the similarly named category in the work of Baez-Fritz-Leinster [2,3], who define morphisms as measure-preserving functions. Nevertheless, isomorphisms are measure-preserving bijections (Lemma 2); so the notion of isomorphism coincides with that of Baez-Fritz-Leinster.

Lemma 1. Meas *has:*

1. *A symmetric monoidal structure induced by the product of underlying sets: Let* μ *and* ν *be measures, then* $\mu \otimes \nu$ *is the product measure on* $\Omega_\mu \times \Omega_\nu$.
2. *Coproduct* $\boxplus$ *induced by the disjoint union* $\coprod$ *of sets: for any measures* μ *and* ν, $\Omega_{\mu \boxplus \nu} := \Omega_\mu \coprod \Omega_\nu$ *and* $(\mu \boxplus \nu)(U) := \mu(U \cap \Omega_\mu) + \nu(U \cap \Omega_\nu)$ *for any* $U \subseteq \Omega_\mu \coprod \Omega_\nu$.

Proof. Verifying that $\otimes$ provides a symmetric monoidal structure is straightforward. To see that $\boxplus$ is a coproduct, note that the inclusion map $\iota_\mu\colon \Omega_\mu \to \Omega_\mu \coprod \Omega_\nu$ defines a valid morphism $\iota_\mu\colon \mu \to \mu \boxplus \nu$ as $(\iota_\mu)_*\mu = \mu \boxplus 0_{\Omega_\nu} \leq \mu \boxplus \nu$. Similarly, $\iota_\nu\colon \Omega_\nu \to \Omega_\mu \coprod \Omega_\nu$ defines a morphism $\iota_\nu\colon \nu \to \mu \boxplus \nu$. The universal property follows in part by the fact that $\coprod$ is a coproduct for sets. $\square$

Remark 2. The operation $\otimes$ is a categorical product in Meas, but it is *not* a categorical product in the quantum-classical enlargement of Meas.

Remark 3. The fact that $\boxplus$ is a coproduct relies on the presence of non-probability measures and maps that are not measure preserving.[3]

3.1   The Rig of Isomorphism Classes of Measures

The following lemma is straightforward.

Lemma 2. $f\colon \mu \to \nu$ *is an isomorphism if and only if* $\underline{f}$ *is a bijection and* $\underline{f}_*\mu = \nu$.

Remark 4. One can generalize the class of morphisms in Meas to the class of stochastic maps that manifest algebraically as completely positive contractions on $*$-algebras of $\mathbb{C}$-valued random variables. Even in this situation, isomorphisms would still be measure-preserving bijections.

---

[3] If one works with probability measures and measure-preserving maps, $\boxplus$ instead manifests as an operadic structure which encapsulates the ability to take convex linear combinations of probability measures; this is the approach taken by [6].

The collection of isomorphism classes [Meas] of Meas is a set.[4] Moreover, letting $[\mu]$ denote the isomorphism class of an object μ, we can define binary operations $+$ and $\cdot$ by: $[\mu] + [\nu] := [\mu \boxplus \nu]$ and $[\mu] \cdot [\nu] := [\mu \otimes \nu]$ for any pair of objects μ and ν. These equip [Meas] with the structure of a commutative *rig*: a commutative ring dropping the condition that there are additive inverses (a ring without "negatives"). The empty measure $\emptyset$ provides an additive $(+)$ unit, and the unit measure on the one-point set provides a multiplicative $(\cdot)$ unit.

The following theorem is a take on the observations of Baez-Fritz-Leinster in [2].

Theorem 1. *Let $\mathcal{O}(\mathbb{C})$ denote the ring of holomorphic functions on $\mathbb{C}$. There is a unital homomorphism* $\dim \colon [\mathrm{Meas}] \to \mathcal{O}(\mathbb{C})$, *defined on any $[\mu]$ by:*

$$\dim[\mu] \colon \mathbb{C} \longrightarrow \mathbb{C}$$

$$q \longmapsto \sum_{\omega \in \Omega_\mu} \mu(\{\omega\})^q =: \dim_q[\mu],$$

where, for any $\lambda \geq 0$, $\lambda^0 := \lim_{q \to 0} \lambda^q$: i.e., $\lambda^0 = 1$ if $\lambda > 0$ and $0^0 := 0$.

Remark 5. There is a reflective full subcategory of Meas generated by "faithful" measures: measures μ such that $\mu(\{\omega\}) > 0$ for all $\omega \in \Omega_\mu$. The homomorphism dim is an isomorphism on this full subcategory (see [2]).

Remark 6. An extension of Theorem 1 to finite-dimensional quantum-classical systems appears in the constructions of [11, Sect. 8.4.1].

Remark 7. The parameter q in $\dim_q$ has several potential interpretations:

1. As a character of a continuous complex irreducible representation of the multiplicative group $\mathbb{R}_{>0}$: every such representation is of the form $m_q \colon \mathbb{R}_{>0} \to \mathrm{Aut}(\mathbb{C})$ for some $q \in \mathbb{C}$ such that $m_q(\lambda)z = \lambda^q z$.
2. As a (negative) inverse temperature: $\dim_{-\beta}[\mu]$ is the partition function $\sum_{\omega \in \Omega_\mu} e^{-\beta E(\omega)}$, associated to the classical system with state space Ω_μ and energy function $E \colon \Omega_\mu \to \mathbb{R}$ given by $E(\omega) := \log[\mu(\{\omega\})]$.
3. As the parameter defining a q-norm for an L^q space.

A detailed justification for the first and third interpretation is left for future work. The second interpretation is is also discussed in [2].

4   Measure Families

Definition 2. *A measure family μ is the data of a finite set P_μ and a functor*

$$\mathrm{Subset}(P_\mu) \longrightarrow \mathrm{Meas},$$

---

[4] It is easy to write down a natural bijection of [Meas] with $\coprod_{n=0}^{\infty} (\mathbb{R}_{\geq 0})^{\times n}$, taking $\mathbb{R}^{\times 0} := \{\star\}$ to correspond to the empty measure. This observation can be used to equip [Meas] with a topology as in [3].

where Subset(P_μ) *is the category with objects given by subsets of* P_μ, *and a unique morphism* $T \to V$ *if and only if* $T \subseteq V$. *Analogous to the situation for measures, we abuse notation and denote the functor by* μ.

Given a function between finite sets $\underline{f} \colon P_\mu \to Q$, we can define the push-forward of a measure family μ as the measure family $\underline{f}_* \mu \colon$ Subset(Q) $\to$ Meas defined by

$$(\underline{f}_* \mu)(T) := \mu[\underline{f}^{-1}(T)]$$

for every $T \subseteq Q$. Given two measure families μ and ν with $P = P_\mu = P_\nu$, we say $\mu \leq \nu$ if and only if $\Omega_{\mu(T)} = \Omega_{\nu(T)}$ and $\mu(T) \leq \nu(T)$ for every $T \subseteq P$.

Definition 3. *The category of measure families* MeasFam *is the category with objects given by measure families, and a morphism* $f \colon \mu \to \nu$ *defined by a function* $\underline{f} \colon P_\mu \to P_\nu$ *such that* $\underline{f}_* \mu \leq \nu$.

There are various versions of "lifts" of the monoidal operations $\boxplus$ and $\otimes$ on Meas to monoidal operations on MeasFam; the following versions will be useful.

Definition 4. *Let* μ *and* ν *be measure families, then* $\mu \boxplus \nu$ *and* $\mu \otimes \nu$ *are measure families with* $P_{\mu \boxplus \nu}$ *and* $P_{\mu \otimes \nu}$ *both defined as the disjoint union* $P_\mu \coprod P_\nu$. *On a subset* $T \subseteq P_\mu \coprod P_\nu$ *we define* $(\mu \boxplus \nu)(T) := \mu(T \cap P_\mu) \boxplus \nu(T \cap P_\nu)$, *and* $(\mu \otimes \nu)(T) := \mu(T \cap P_\mu) \otimes \nu(T \cap P_\nu)$. *The definitions on inclusions follow from the obvious induced morphisms.*

Definition 5. *Let* μ *be a measure family, and* $T \subseteq P_\mu$; *then* $\mu|_T \colon$ Subset(T) $\to$ Meas *denotes the obvious restriction. We say* μ *is a 2-measure if* $\mu(\emptyset)$ *is the empty measure and there is an isomorphism* $\mu \xrightarrow{\sim} \boxplus_{p \in P} \mu|_{\{p\}}$.

2-measures are measure families where all global data is given by gluing together local data.[5] This is a categorified notion of the additivity condition for a measure.

4.1   2-Measures from Measures

Let μ be a measure, then there is a measure family $\mathsf{R}^\mu \colon$ Subset(Ω_μ) $\to$ Meas given by the restriction of μ to subsets of P. On objects, it acts in the following way: for $T \subseteq \Omega_\mu$ nonempty, $\mathsf{R}^\mu T := \mu|_T$, where $\mu|_T$ the restriction of μ to subsets of T; to the empty set we assign the empty measure. To every inclusion $T \subseteq V$, it assigns the morphism $\mathsf{R}^\mu T \to \mathsf{R}^\mu V$ whose underlying map is the inclusion map $T \hookrightarrow V$. The additivity condition on a measure requires that for any subset $T \subseteq P$ the identity map $T \to T$ induces an isomorphism of measures: $\mathsf{R}^\mu T \xrightarrow{\sim} \boxplus_{t \in T} \mathsf{R}^\mu(\{t\})$. As a result, R^μ is a 2-measure. Conversely, any 2-measure that reduces to a coproduct of measure families on one point sets defines a measure on P_μ.

---

[5] In some sense a 2-measure is an "acyclic cosheaf" of measures.

4.2   Measure Families from Multipartite Measures

Definition 6. *A multipartite measure* μ *over a finite subset* P *is a collection of sets* $\{\Omega_p\}_{p\in P}$, *and a measure* $\mu_P \colon \mathsf{Subset}(\prod_{p\in P}\Omega_p) \to \mathbb{R}_{\geq 0}$.

Given a multipartite measure μ over P, define Ω_T as $\prod_{t\in T}\Omega_t$ if $T\neq\emptyset$, and the one-point set $\{\star\}$ if $T=\emptyset$. Let $\underline{p}_T\colon \Omega_P \to \Omega_T$ denote the projection map if $T\neq\emptyset$ and the map to the point otherwise. Then to each subset $T\subseteq P$, we can assign a *reduced* (or "marginal") measure $\mu_T := (\underline{p}_T)_*\mu_P$.

The data of the reduced measures collects into a functor $\mathsf{Subset}(P)^{\mathrm{op}} \to$ Meas that takes a subset T of P to μ_T, and takes an inclusion $T\subseteq V$ to the morphism $\mu_V \to \mu_T$ provided by the projection $\Omega_V \to \Omega_T$. Because this functor is contravariant, it is *not* a measure family; however, we can make it one by composing with the functor $(-)_P^c\colon \mathsf{Subset}(\Omega)^{\mathrm{op}} \to \mathsf{Subset}(\Omega)$ that takes a subset of P to its complement. The result is a measure family

$$\mathsf{A}^\mu \colon \mathsf{Subset}(P) \longrightarrow \mathsf{Meas},$$

which acts on objects by taking T to μ_{T^c}. Factorizability questions about μ are equivalent to factorizability questions about A^μ: e.g. $\mu_P = \bigotimes_{p\in P}\mu_p$, if and only if there is an isomorphism $\mathsf{A}^\mu \xrightarrow{\sim} \bigotimes_{p\in P}\mathsf{A}^\mu|_{\{p\}}$.

4.3   The Index

Definition 7. *Let* μ *be a measure family. The index of* μ *is defined as the holomorphic function*

$$\mathfrak{X}[\mu] := \sum_{k=0}^{|P_\mu|}(-1)^k \dim\left[\boxplus_{|T|=k}\mu(T^c)\right].$$

The evaluation of $\mathfrak{X}[\mu]$ *at* $q\in\mathbb{C}$ *is denoted as* $\mathfrak{X}_q[\mu]$.

The complement in the definition is for convenience;[6] without it, the definition would be the same up to the overall sign $(-1)^{|P_\mu|}$. Theorem 1 and manipulations of the inclusion-exclusion relation defining the index lead to the following results.

Theorem 2. *The index only depends on isomorphism classes of measure families; moreover,* $\mathfrak{X}[\mu\otimes\nu] = \mathfrak{X}[\mu]\mathfrak{X}[\nu]$ *for any measure families* μ *and* ν.

Proposition 1. *If* μ *and* ν *measure families with* P_μ *and* P_ν *non-empty, then* $\mathfrak{X}_q[\mu\boxplus\nu] = 0$. *In particular,* $\mathfrak{X}$ *vanishes on any 2-measure* μ *with* $|P_\mu|\geq 2$.

According to Proposition 1, a non-vanishing index indicates that there is an obstruction to an "additive" ($\boxplus$) descent of data. For a multipartite measure μ, we are more interested in an obstruction to a "*multiplicative*" ($\otimes$) descent of

---

[6] One can define an index with respect to any cover of P_μ; but our primary interest will be the cover that is the complement of the finest partition of P_μ.

data, i.e., a failure to factorize. As the discussion below indicates, this can be detected by looking at the derivative of $q \mapsto \mathfrak{X}_q[A^\mu]$ at $q = 1$ (where $\mathfrak{X}_1[A^\mu] = 0$).

Tsallis Mutual Information and the $q \to 1$ Limit. For a multipartite measure μ we have:

$$\mathfrak{X}_q[A^\mu] = \sum_{\emptyset \subseteq T \subseteq P} (-1)^{|T|} \left(\sum_{\omega \in \Omega_T} \mu_T(\{\omega\})^q \right).$$

If μ is a multipartite probability measure ($\mu_P(P) = 1$), then a bit of manipulation demonstrates that $I_q[\mu] := \frac{1}{q-1} \mathfrak{X}_q[A^\mu]$ can be rewritten as:

$$I_q[\mu] = \sum_{\emptyset \neq T \subseteq P} (-1)^{|T|-1} S_q^{\mathrm{Ts}}(\mu_T),$$

where $S_q^{\mathrm{Ts}}(\mu) := \frac{1}{q-1}[1 - \sum_{\omega \in \Omega} \mu(\{\omega\})^q]$ is the Tsallis deformation of mutual information. Multipartite mutual information is recovered in the limit that $q \to 1$. If μ is a multipartite measure on a one-element set, then $I_q[\mu]$ is simply the Tsallis entropy. This observation can be combined with multiplicativity of the index (Theorem 2), to demonstrate that the multipartite mutual information of a multipartite measure on P must vanish if the measure factorizes with respect to any partition of P: see [11, Sect. 8.5].

The $q \to 0$ Limit. $\dim_0 \mu$, which is $\lim_{q \to 0} \dim_q \mu$ by definition, is an integer counting the number of points of Ω_μ with non-vanishing measure. Thus, for μ any measure family, $\mathfrak{X}_0[\mu]$ is an integer. This is a hint that $\mathfrak{X}_0[\mu]$ is related to the Euler characteristic of a topological space.

4.4   (Semi-)Simplicial Objects

By viewing a measure family μ as a pre-cosheaf and applying Čech techniques with respect to a cover of P_μ, we can construct an (augmented)[7] semi-simplicial object in Meas: an (augmented) *semi-simplicial measure*. In this note, we specialize to the "complementary cover" $\{\{p\}^c\}_{p \in P_\mu}$ of P_μ and choose a total order on P.[8] Using the fact that the intersection of complements is the complement of a union, the non-trivial part of the resulting augmented semi-simplicial measure can be summarized by a diagram in Meas of the form:

$$\mu(\emptyset^c) \longleftarrow \underset{|T|=1}{\boxplus} \mu(T^c) \overset{\longleftarrow}{\underset{\longleftarrow}{\longleftarrow}} \cdots \underset{n-1 \text{ arrows}}{\vdots} \underset{|T|=n-1}{\boxplus} \mu(T^c) \underset{n \text{ arrows}}{\vdots} \mu(P_\mu^c) \quad , \quad (1)$$

Degree -1 Degree 0 Degree $n-2$ Degree $n-1$

---

[7] Augmented in this context means there is an additional degree -1 component and a single map from the degree 0 component to the degree -1 component.

[8] All interesting quantities are equivariant under change of total order.

where: $n = |P|$, degree $\geq n$ components are taken to be the empty measure, and the arrows satisfy the face-map relations of an augmented semi-simplicial object. The holomorphic function $-\mathfrak{X}[\mu]$, where $\mathfrak{X}[\mu]$ is the index of Definition 7, can be thought of as the graded dimension or "Euler characteristic" of (1).

Remark 8. In the special case that $\mu = \mathsf{A}^\mu$ for a multipartite measure μ over P, the complements disappear: for any $T \subseteq P$, we have $\mathsf{A}^\mu(T^c) = \mu_T$ (the reduced measure on $\prod_{t \in T} \Omega_t$). With this specialization, diagram (1) becomes:

$$
\underbrace{\mu_\emptyset}_{\text{Degree } -1} \leftarrow \underbrace{\boxplus_{|T|=1} \mu_T}_{\text{Degree } 0} \overset{\leftarrow}{\underset{\leftarrow}{\leftarrow}} \cdots \; \vdots \; \underbrace{\overset{\leftarrow}{\underset{\leftarrow}{\leftarrow}} \boxplus_{|T|=|P|-1} \mu_T}_{\text{Degree } |P|-2} \; \vdots \; \underbrace{\overset{\leftarrow}{\underset{\leftarrow}{\leftarrow}} \mu_P}_{\text{Degree } |P|-1} \; , \tag{2}
$$

where, as before, we ignore the empty measures in degree $\geq |P|$ components. The augmentation map into the degree -1 component has underlying map given by the unique map $\coprod_{p \in P} \Omega_p \to \{\star\} = \Omega_\emptyset$, and the remaining face maps are induced by projection maps composed with inclusions into disjoint unions. For instance, if $P = \{1, 2\}$, the two face maps out of the degree 1 component have underlying maps given by the following compositions (letting $i \in \{1, 2\}$):

$$
\Omega_1 \times \Omega_2 \xrightarrow{\text{project}} \Omega_i \xhookrightarrow{\text{include}} \Omega_1 \coprod \Omega_2.
$$

Remark 9. One can also apply Čech techniques to produce (augmented) simplicial objects rather than (augmented) *semi*-simplicial objects (see, [5] and [1, Sect. 25.1–25.5]). Simplicial objects include additional "degeneracies" and extend the diagram (1) infinitely far to the right with possibly non-empty measures. The invariants of the underlying measure family that are discussed in this note—Euler characteristics, indices, and cohomology—can be recovered by passing through either version: yielding results that are equivalent, or canonically isomorphic. This note focuses on the semi-simplicial version for pedagogical reasons and immediate connections to the computational underpinnings of [11].

(Semi-)Simplicial Sets. From the semi-simplicial measure (1), one can derive an (augmented) semi-simplicial set: a slight generalization of a simplicial complex. Indeed, there is a functor:

$$
\mathsf{S} \colon \mathsf{FinMeas} \longrightarrow \mathsf{FinSet}
$$

that assigns to a measure μ, its support:

$$
\mathsf{S}\mu := \{\omega \in \Omega_\mu \colon \mu(\omega) \neq 0\},
$$

and assigns to a morphism $f \colon \mu \to \nu$, the morphism $\mathsf{S}f \colon \mathsf{S}\mu \to \mathsf{S}\nu$ whose underlying function on sets is the restriction $\underline{f}|_{\mathsf{S}\mu}$: a valid assignment as $\underline{f}(\mathsf{S}\mu) \subseteq \mathsf{S}\nu$ due to the condition $\underline{f}_*\mu \leq \nu$. Applying S to our semi-simplicial measure, we obtain an augmented semi-simplicial set $\Delta[\mu]$ whose non-trivial part is summarized by the following diagram in FinSet (with $n = |P_\mu|$):

$$\overbrace{\mathsf{S}\mu(\emptyset^c)}^{\text{Degree } -1} \longleftarrow \overbrace{\coprod_{|T|=1} \mathsf{S}\mu(T^c)}^{\text{Degree } 0} \overset{\longleftarrow}{\longleftarrow} \cdots \overset{\longleftarrow}{\underset{\longleftarrow}{:}} \overbrace{\coprod_{|T|=n-1} \mathsf{S}\mu(T^c)}^{\text{Degree } n-2} \overset{\longleftarrow}{\underset{\longleftarrow}{:}} \overbrace{\mathsf{S}\mu(P_\mu^c)}^{\text{Degree } n-1} .$$

The Euler characteristic of $\Delta[\mu]$, denoted $\chi(\Delta[\mu])$, is the negative of the $q = 0$ evaluation (or $q \to 0$ limit) of the index:

$$\chi(\Delta[\mu]) := \sum_{k=-1}^{|P_\mu|-1} (-1)^k \left(\sum_{|T|=k+1} |\mathsf{S}\mu(T^c)| \right) = -\mathfrak{X}_0[\mu].$$

Geometric Realizations. Specialize to $\mu = \mathtt{A}^\mu$, and define $\Delta_\mu := \Delta[\mathtt{A}^\mu]$, which is summarized by the diagram:

$$\mathsf{S}\mu_\emptyset \longleftarrow \coprod_{|T|=1} \mathsf{S}\mu_T \overset{\longleftarrow}{\longleftarrow} \cdots \overset{\longleftarrow}{\underset{\longleftarrow}{:}} \coprod_{|T|=|P|-1} \mathsf{S}\mu_T \overset{\longleftarrow}{\underset{\longleftarrow}{:}} \mathsf{S}\mu_P.$$

Let Δ'_μ denote the semi-simplicial set obtained by removing the augmentation map[9] into $\mathsf{S}\mu_\emptyset$. To construct the geometric realization $|\Delta'_\mu|$, observe that 0-simplices are given by points $\omega \in \coprod_{p \in P} \Omega_p$ such that $\mu_{\{p\}}(\{\omega\}) \neq 0$ for every $p \in P$. Higher k-simplices are given by collections of 0-simplices with non-vanishing measure as computed with respect to the reduced measure $\boxplus_{|T|=k+1} \mu_T$. The geometric realization is simple: first identify $P = P_{\mathtt{A}^\mu}$ with the set $\{1, \cdots, n\}$, then for each $(\omega_1, \cdots, \omega_n) \in \prod_{i=1}^{n} \Omega_i$ with $\mu(\{(\omega_1, \cdots, \omega_n)\}) \neq 0$, draw an $(n-1)$-simplex with vertices $\omega_1, \omega_2, \ldots, \omega_n$.

When μ is a bipartite measure on the set $P = \{1,2\}$, the geometric realization $|\Delta'_\mu|$ might look familiar: it is a bipartite (directed) graph whose vertices are colored by points in P. A bit of experimentation demonstrates that the connectivity of this graph is closely related to the correlations between "subsystems" 1 and 2.

Cohomology. If we apply the functor[10] $\mathrm{Hom}_{\mathsf{FinSet}}(-, \mathbb{C}) \colon \mathsf{FinSet}^{\mathrm{op}} \to \mathsf{FinVect}_{\mathbb{C}}$ to Δ_μ, we obtain a cosimplicial vector space; this can can be turned into a cochain complex by taking alternating sums of face maps. For $\mu = \mathtt{A}^\mu$, this complex looks like (letting $\mathbb{C}[-]$ be shorthand for $\mathrm{Hom}_{\mathsf{FinSet}}(-, \mathbb{C})$):

$$\overbrace{\mathbb{C}}^{\text{Degree -1}} \longrightarrow \overbrace{\prod_{|T|=1} \mathbb{C}[\mathsf{S}\mu_T]}^{\text{Degree } 0} \longrightarrow \overbrace{\prod_{|T|=2} \mathbb{C}[\mathsf{S}\mu_T]}^{\text{Degree } 1} \longrightarrow \cdots \longrightarrow \overbrace{\mathbb{C}[\mathsf{S}\mu_P]}^{\text{Degree } |P| - 1} \longrightarrow 0 \longrightarrow \cdots$$

---

[9] If μ_P does not vanish everywhere, then $|\mathsf{S}\mu_\emptyset| = |\{\star\}| = 1$. Consequently, one can show that $\mathfrak{X}_0[\mathtt{A}^\mu] = 1 - \chi(\Delta'_\mu)$.

[10] This is a specialization of a functor from (localizable) measurable spaces to the Banach space underlying the W^*-algebra of essentially bounded measurable functions.

The cohomology $H^\bullet$ of this cochain complex is the *reduced* simplicial cohomology of the geometric realization $|\Delta'_\mu|$ with coefficients in $\mathbb{C}$. Representatives of H^k can be interpreted as assignments of $\mathbb{C}$-valued random variables to all subsystems (subsets of $P_{\lambda\mu}$) of size $k+1$, such that these assignments have linear correlations that do not reduce to correlations on subsystems of size k. Random variables on the subsystem of size 0, coming from the degree -1 component, are the constant random variables. For a bipartite measure on $P = \{1, 2\}$, a representative of a non-zero element in H^0 is a pair (r_1, r_2) of random variables such that (for $i \in \{1, 2\}$): r_i is a random variable on Ω_i that is not almost everywhere (a.e.) equal to zero, r_i is non-constant, and $r_1 \otimes 1 - 1 \otimes r_2$ is a.e. equal to zero with respect to the measure on $\Omega_1 \times \Omega_2$. See [11, Sect. 7.4] and [11, Sect. 6.5] for interpretations of the quantum mechanical analogs of H^0 and higher H^k; these interpretations can be translated into precise statements for the classical context of this note.[11]

Remark 10. As a source of new invariants of multipartite measures, one might also study the cohomology *ring* of $|\Delta_\mu|$ with coefficients in a commutative ring R: a graded R-algebra. In fact, the story in this section can be souped-up to use part of the monoidal Dold-Kan correspondence in order to produce a differential graded R-algebra from a measure family.

5   Some Future Directions

The reduced measures associated to a multipartite measure supply a "weight" to the simplices of its associated semi-simplicial set. Thus, in some sense, the index is a weighted Euler characteristic. It is natural to suspect that there is a weighted version of cohomology categorifying the index for all values of q: the mystery box of Fig. 1. Moreover, the existence of a canonically associated semi-simplicial set to a multipartite measure may open the door to measures of shared information using combinatorial invariants, such as the Stanley-Reisner ring (the "face ring of a simplicial complex" in [17]).

References

1. Stacks Project. https://stacks.math.columbia.edu. Accessed 16 May 2023
2. Baez, J., Fritz, T., Leinster, T.: Entropy as a functor (2011). https://ncatlab.org/johnbaez/revision/Entropy+as+a+functor/55
3. Baez, J., Fritz, T., Leinster, T.: A characterization of entropy in terms of information loss. Entropy **13**(11), 1945–1957 (2011)
4. Baudot, P., Bennequin, D.: The homological nature of entropy. Entropy **17**, 3253–3318 (2015). https://doi.org/10.3390/e17053253
5. Bennequin, D., Peltre, O., Sergeant-Perthuis, G., Vigneaux, J.P.: Extra-fine sheaves and interaction decompositions (2020). arXiv:2009.12646

---

[11] The classical analog of the GNS and the commutant complexes of [11] both reduce to the alternating sum of face maps complex that is used in this note.

6. Bradley, T.D.: Entropy as a topological operad derivation. Entropy **23**(9), 1195 (2021)
7. Drummond-Cole, G., Park, J.S., Terilla, J.: Homotopy probability theory I (2015). arXiv:1302.3684
8. Drummond-Cole, G., Park, J.S., Terilla, J.: Homotopy probability theory II (2015). arXiv:1302.5385
9. Geiko, R., Mainiero, T., Moore, G.: A categorical triality: Matrix product factors, positive maps and von Neumann bimodules (2022)
10. Lang, L., Baudot, P., Quax, R., Forré, P.: Information decomposition diagrams applied beyond shannon entropy: a generalization of Hu's theorem (2022). arXiv:2202.09393
11. Mainiero, T.: Homological tools for the quantum mechanic (2019). arXiv: 1901.02011
12. Mainiero, T.: The secret topological life of shared information (2020). https://youtu.be/XgbZSwRlAjU, String-Math
13. Mainiero, T.: Higher entropy. In: Symposium on Categorical Semantics of Entropy at CUNY (2022). https://youtu.be/6cFDviX0hUs,
14. Parzygnat, A.J.: A functorial characterization of von Neumann entropy (2020). arXiv:2009.07125
15. Pavlov, D.: Gelfand-type duality for commutative von Neumann algebras. J. Pure Appl. Algebra **226**(4), 106884 (2022). arXiv:2005.05284
16. Sergeant-Perthuis, G.: Intersection property, interaction decomposition, regionalized optimization and applications (2021). https://doi.org/10.13140/RG.2.2.19278. 38729
17. Stanley, R.P.: Combinatorics and Commutative Algebra, Progress in Mathematics, vol. 41, 2nd edn. Birkhäuser Boston Inc., Boston (1996)
18. Vigneaux, J.: The structure of information: from probability to homology (2017). arXiv: 1709.07807

A Categorical Approach to Statistical Mechanics

Grégoire Sergeant-Perthuis[✉]

Inria Paris, OURAGAN, Paris, France
gregoireserper@gmail.com

Abstract. 'Rigorous' Statistical Mechanics is centered on the mathematical study of statistical systems. In this article, we show that important concepts in this field have a natural expression in terms of category theory. We show that statistical systems are particular representations of partially ordered sets (posets) and express their phases as invariants of these representations. This work opens the way to the use of homological algebra to compute phases of statistical systems. In particular we compute the invariants of projective poset representations. We remark that in this formalism finite-size systems are allowed to have several phases.

Keywords: Applied category theory · Statistical mechanics

1 Introduction

Statistical physics is a framework that focuses on the probabilistic description of complex systems: a collection of interacting 'particles' or components of a whole in most generality [19]. Its main feature is to introduce an energy function H for the system, that associates to any of its configurations a real value; the probability p of a configuration is given in terms of the Boltzmann distribution $(p = e^{-\beta H} / \int dx e^{-\beta H(x)})$, which associates high energy configurations to unlikely events. We will call 'statistical system' the complex system provided with the Boltzmann distribution. Statistical physics serves as a rich framework for probabilistic modeling 'in general'. It has several names depending on the community [15]; for example, it is called 'energy-based modeling' in machine learning, the probabilistic model is called a Gibbs Random Field in graphical modeling and is a particular case of statistical system. It is widely used in engineering, two examples are: computational structural biology (computational statistical physics: Hamiltonian Monte Carlo) [5,16], robotics (reinforcement learning: Markov Chains, Markov Decision Processes) [26].

Applied Category Theory: New foundations, based on topology and geometry were proposed for probability theory, information theory, deep learning [2,6,8,12,25]. More generally they fall in a field of research that has recently emerged, Applied Category Theory, which focuses on applying these principles to engineering [1,6–8,18,25].

F. Nielsen and F. Barbaresco (Eds.): GSI 2023, LNCS 14071, pp. 258–267, 2023.
https://doi.org/10.1007/978-3-031-38271-0_26

Contribution: In this paper, we propose to build a bridge between geometry and (rigorous) statistical physics by showing that the standard formal definition of a statistical system, a 'specification' (Definition 1.23 [10]), can be identified with a particular representation of a partially ordered set (poset).

Representations of posets have a precise geometric interpretation [4,13,23,24, 27], and there is a rich literature coming from algebra, geometry, and topology to study them. In particular, we hope that this work opens the way to the systematic algebraic study of statistical systems.

Characterizing the phases (Gibbs measures) of a statistical system is a central subject of statistical mechanics. As one step in this direction, we show that phases of statistical systems are geometric invariants of these representations and compute them for 'projective' poset representation. In particular in this setting system of 'finite size' can have multiple phases, which is not possible in the current formalism for phase transition.

2 Structure of the Paper and Contribution

We start by recalling some important notions of (rigorous) statistical mechanics [10] (specifications, Gibbs measures). We then show that the standard concept that encodes the statistical system, which is the notion of 'specifications', can be extended into a poset representation; we also show that the Gibbs measures of a statistical system are geometric invariants of this representation. From there, we propose a novel categorical formulation for statistical mechanics; we define 'generalized specifications' and 'generalized Gibbs measures' for those specifications. We give a first result on characterizing generalized Gibbs measures for a certain class of generalized specifications (projective poset representations). Such specifications were characterized in [21] and relate to independent random variables. On the other hand, injective representations were characterized in [20] and their relationship with the marginal extension problem is studied in [3]; a unifying perspective for Hilbert spaces can be found in [22]. In the formalism we propose statistical systems of 'finite' size' can have multiple phases.

3 Background: Rigorous Statistical Mechanics

We will follow the presentation of Georgii's reference book *Gibbs Measures and Phase Transitions* [10].

Definition 1 (Markov Kernel). A Markov kernel π from the measurable space $(E, \mathscr{E})$ to the measurable space $(E_1, \mathscr{E}_1)$ is a function $\pi : \mathscr{E}_1 \times E \to [0, 1]$ such that

1. $\forall \omega \in E$, $\pi(.|\omega)$ is a measure on E_1
2. $\forall A \in \mathscr{E}_1$, $\pi(A|.)$ is a measurable map from E to $\mathbb{R}$
3. $\forall \omega \in E$, $\pi(E_1|\omega) = 1$, i.e. $\pi(.|\omega)$ is a probability measure.

We will denote a Markov kernel π from $(E, \mathscr{E})$ to $(E_1, \mathscr{E}_1)$ as $\pi : (E, \mathscr{E}) \to (E_1, \mathscr{E}_1)$. We denote $[(E, \mathscr{E}), (E_1, \mathscr{E}_1)]_K$, the set of kernels from $(E, \mathscr{E})$ to $(E_1, \mathscr{E}_1)$; if there is no ambiguity on the σ-algebras the spaces are provided with, we will simply denote it as $[E, E_1]_K$. We denote $\mathbb{P}(E)$, the space of probability distributions over E; it is a measurable space for the smallest σ-algebra that makes the evaluation maps, on measurable sets of E, measurable.

Markov kernels can be composed as follows. Let $\pi : E \to E_1$, $\pi_1 : E_1 \to E_2$ be two Markov kernels, then the composition $\pi_1 \circ \pi : E \to E_2$ is the following Markov kernel: for any $A \in \mathscr{E}_2$ and $\omega \in E$,

$$\pi_1 \circ \pi(A|\omega) = \int \pi_1(A|\omega_1)\pi(d\omega_1|\omega) \tag{3.1}$$

A measurable map $f : (E, \mathscr{E}) \to (E_1, \mathscr{E}_1)$ between two measurable spaces can be extended into the following Markov kernel: for any $A \in \mathscr{E}_1$ and $\omega \in E$,

$$\pi_f(A|\omega) = 1[f(\omega) \in A] \tag{3.2}$$

To avoid having too many notations, we will denote π_f as f and the context will specify if f refers to the measurable map or its extension; for example, a composition $\pi \circ f$ between a Markov kernel π and f necessarily means that, here, f refers to π_f. We will also denote $\pi_1 \circ \pi$ as $\pi_1\pi$.

A probability measures $p \in \mathbb{P}(E)$ can also be identified with the following Markov kernel π_p from $*$, the measurable space with one element, to $(E, \mathscr{E})$; for any $A \in \mathscr{E}$, $\pi_p(A|*) = p(A)$. Similarly we identify p and π_p.

Remark 1. Measurable spaces and measurable maps form a category. Giry [11] and Lawvere [14] are the first to have remarked that measurable spaces and Markov kernels also form a category; the latter category is 'dual' to the first, it is its Kleisli category.

Definition 2 (Proper Kernel, Sect. 1.1. [10]). Let $\mathscr{E}_1 \subseteq \mathscr{E}$ be two σ-algebras of a set E, a kernel $\pi \in [(E, \mathscr{E}_1), (E, \mathscr{E})]_K$ is proper if and only if, for any $A \in \mathscr{E}$, $B \in \mathscr{E}_1$,

$$\pi(A \cap B|.) = \pi(A|.)1_B \tag{3.3}$$

Let us set the notations. I is the set of components of a complex system. $(E_i, \mathscr{E}_i, i \in I)$ is a collection of measurable spaces, with each E_i being the space of configuration (state space) of the component $i \in I$. $(E, \mathscr{E})$ denotes the state space of the system; it is the product $E := \prod_{i \in I} E_i$ with the product σ-algebra. For a sub-collection of components $a \subseteq I$, we denote $E_a := \prod_{i \in a} E_i$ the associated state space, and $\mathscr{E}_a$ the associated product σ-algebra. $i^a : E \to E_a$ is the projection that sends a configuration $\omega := (\omega_i, i \in I)$ to the configuration of the sub-collection a, $(\omega_i, i \in a)$. Finally, let us denote $\mathscr{P}_f(I)$ the set of finite subsets of I.

Definition 3 (Specification, Adaptation of Def. 1.23 [10]). A specification γ with state space $(E, \mathscr{E})$ is a collection $(\gamma_a, a \in \mathscr{P}_f(I))$ of proper Markov kernels such that for any $a \in \mathscr{P}_f(I)$, $\gamma_a \in [(E_{\overline{a}}, \mathscr{E}_{\overline{a}}), (E, \mathscr{E})]_K$ and which satisfies that for any $b \subseteq a$, i.e. $\overline{a} \subseteq \overline{b}$ and any $A \in \mathscr{E}$,

$$\gamma_b \circ i^{\overline{b}} \circ \gamma_a(A|.) = \gamma_a(A|.) \tag{3.4}$$

Remark 2. In Definition (1.23) [10], E is the product over the same measurable space X over I and I is a countably infinite set.

In the standard definition of specification (Definition 3) the Markov kernels encode border conditions for experiments that only involve finite numbers of components $(a \in \mathscr{P}_f(I))$. It is what formally encodes the statistical system. The next definition defines the 'phases' of the system.

Definition 4 (Gibbs measures, Def. 1.23 [10]). Let γ be a specification with state space E; the set of probability measures,

$$\mathscr{G}(\gamma) := \{p \in \mathbb{P}(E) : \quad \mathbb{E}_p(A|\mathscr{E}_{\overline{a}}) = \gamma_a(A|.) \ p \text{ a.s.}\} \tag{3.5}$$

is the set of Gibbs measures of γ.

One of the central problems of (rigorous) statistical mechanics is to understand the relationship between a specification γ (statistical system) and its set of Gibbs measures $\mathscr{G}(\gamma)$ (its phases).

4   Statistical Systems as Poset Representations

Theorem 1. *Let γ be a specification with state space E. For any $a, b \in \mathscr{P}_f(I)$ such that $b \subseteq a$, there is a unique Markov kernel $F_b^a : E_{\overline{a}} \to E_{\overline{b}}$ such that the following diagram commutes,*

$$
\begin{array}{ccc}
E_{\overline{a}} & \xrightarrow{\ \gamma_a\ } & E \\
{\scriptstyle F_b^a}\Big\downarrow & {\scriptstyle \gamma_b}\ \nearrow & \\
E_{\overline{b}} & &
\end{array}
\tag{4.1}
$$

i.e. such that $\gamma_b \circ F_b^a = \gamma_a$. Furthermore for any collection $a, b, c \in \mathscr{P}_f(I)$ with $a \subseteq b \subseteq c$,

$$F_c^b \circ F_b^a = F_c^a \tag{4.2}$$

Proof. Let $b \subseteq a$, let F_b^a satisfy the commutative diagram 4.1, then,

$$i^{\overline{b}}\gamma_b F_b^a = i^{\overline{b}}\gamma_a \tag{4.3}$$

therefore,

$$F_b^a = i^{\overline{b}}\gamma_a \tag{4.4}$$

For any $a, b \in \mathscr{A}$ such that $b \subseteq a$ let $F_b^a = i^{\overline{b}}\gamma_a$ then for $c \subseteq b \subseteq a$,

$$F_c^b F_b^a = i^{\overline{c}}\gamma_b i^{\overline{b}}\gamma_a \tag{4.5}$$

Equation 3.4 can be rewritten as, for $b \subseteq a$,

$$\gamma_b i^{\overline{b}}\gamma_a = \gamma_a \tag{4.6}$$

therefore,

$$F_c^b F_b^a = F_c^a \tag{4.7}$$

Definition 5 (Partially ordered set). A partially ordered set (poset), $(\mathscr{A}, \leq)$, is a set $\mathscr{A}$ provided with a binary relation $\leq: \mathscr{A} \times \mathscr{A} \to \{0, 1\}$, such that

1. reflexive: $\forall a \in \mathscr{A}, a \leq a$
2. transitive: if $c \leq b$ and $b \leq a$ then $c \leq a$
3. antisymmetric: if $b \leq a$ and $a \leq b$ then $a = b$

$(\mathscr{P}_f(I), \subseteq)$ is a poset for the inclusion relation; $(\mathscr{P}_f(I), \supseteq)$ is also a poset for the reversed inclusion relation: $b \supseteq a \iff a \subseteq b$. The convention is to denote $(\mathscr{P}_f(I), \supseteq)$ as $\mathscr{P}_f(I)^{op}$ because what relates $(\mathscr{P}_f(I), \supseteq)$ to $\mathscr{P}_f(I)$ is the fact that the order is 'opposed'; the same convention holds for any poset; $\mathscr{A}^{op}$ is the set $\mathscr{A}$ with opposed order.

We call a representation of the poset $\mathscr{A}$: a collection of 'spaces' $(G(a), a \in \mathscr{A})$ and a collection of 'maps' $(G_a^b : G(b) \to G(a); b, a \in \mathscr{A}, b \leq a)$ which satisfies for any $c \leq b \leq a$, $G_c^b \circ G_b^a = G_c^a$. We keep the notion of 'poset representation' a bit vague for now (we do not say what 'spaces' or 'maps' are); we keep this notion vague at this stage but in the next section will make this notion formal by introducing the concept of category and of functor.

Theorem 1 implies that a specification γ can be promoted to a representation F of $(\mathscr{P}_f(I), \supseteq)$.

Theorem 2. *Let γ be a specification with state space E, let $p \in \mathscr{G}(\gamma)$. For any $a \in \mathscr{A}$ let $p_a := i^{\overline{a}} \circ p$; it is the marginal distribution on $E_{\overline{a}}$ of p. Then, for any $a, b \in \mathscr{P}_f(I)$ such that $b \subseteq a$,*

$$F_b^a \circ p_a = p_b \tag{4.8}$$

Proof. For any $a, b \in \mathscr{P}_f(I)$ such that $b \subseteq a$,

$$\gamma_b \circ (F_b^a p_a) = p \tag{4.9}$$

therefore, $F_b^a p_a = i^{\overline{b}}p$ and

$$F_b^a p_a = p_b \tag{4.10}$$

5   Categorical Formulation

A category 'plays the role of' collections of spaces and of continuous maps between those spaces; spaces are called 'objects', and continuous maps are called 'morphisms' (see I.1 [17] for a formal definition). Categories are usually denoted in bold, e.g. **C**. A classical example of a category is the category of sets, **Set**, that has as objects sets and as morphisms maps. We will denote **Mes** the category that has as objects measurable space and as morphisms measurable maps (Sect. 1 [11]); we will denote **Kern** the category that has as object measurable spaces and as morphisms Markov kernels (the Kleisli category of the monad **Mes**).

Definition 6 (Functor, I.3 [17]). A functor $F : \mathbf{C} \to \mathbf{C}_1$ is a 'generalized function':

1. that sends an object O of the source category $\mathbf{C}$ to an object $F(O)$ of the target category $\mathbf{C}_1$
2. that sends a morphism $f : O \to O_1$ of $\mathbf{C}$ to a morphism $F(f) : F(O) \to F(O_1)$ of $\mathbf{C}_1$
3. that respects the composition of two morphisms in $\mathbf{C}$: for two morphisms of $\mathbf{C}$, $f : O \to O_1$ and $f_1 : O_1 \to O_2$, then $F(f_1 \circ f) = F(f_1) \circ F(f)$.

A poset, $(\mathscr{A}, \leq)$, can be seen as a category, **A**, with at most one morphism between two objects: the objects of **A** are the elements of $\mathscr{A}$ and for any two elements $b, a \in \mathscr{A}$ there is one morphism $b \to a$ when $b \leq a$. From now we will drop the bold notation for the category **A** and denote it simply as $\mathscr{A}$. A functor G from a poset $\mathscr{A}$ to a category $\mathbf{C}$ is precisely a collection of maps G_a^b for $b, a \in \mathscr{A}$ such that $b \leq a$, which satisfy $G_a^b \circ G_b^c = G_a^c$ for any three elements $c \leq b \leq a$. A functor from $\mathscr{A}$ to some target category is what we will call a representation of the poset $\mathscr{A}$; in general, the target category is the category of vector spaces or modules [23]. For this article, the target category will be **Mes** and **Kern**.

Consider a functor $G : \mathscr{A} \to \mathbf{Set}$ from a poset $\mathscr{A}$ to the category of sets. A collection $(x_a, a \in \mathscr{A})$ is called a section of G if for any $b \leq a$, $G_a^b(x_b) = x_a$; the set of sections of a poset representation is called the limit of G (III.4 [17]) and denoted $\lim G$. It is an 'invariant' of G that can be computed using homological algebra when the target category of G is enriched with some algebraic structure (Chap. 13 [9]).

In Theorem 1 we showed that we can associate to a specification γ, a functor from $(\mathscr{P}_f(I), \supseteq)$ to **Kern**. The convention is to call a functor with source $\mathscr{A}^{op}$, a *presheaf*. In Theorem 2 we showed that Gibbs measures of γ are 'sections' of F. We will denote this set of 'sections' as $[*, F]_{K, \mathscr{A}}$; more precisely for a functor $F : \mathscr{A} \to \mathbf{Kern}$,

$$[*, F]_{K, \mathscr{A}} := \{(p_a \in \mathbb{P}(F(a)), a \in \mathscr{A}) | \quad \forall b \leq a, F_a^b p_b = p_a\} \tag{5.1}$$

Introducing the presheaf F and $[*, F]_{K,\mathscr{A}}$ is our way to emphasize that the compatible measures $p \in [*, F]_{K,\mathscr{A}}$ don't have to be measures over the whole space E. There is no need to define statistical systems 'globally', one can also define them 'locally'.

Let us now introduce the more general, categorical setting we propose for statistical systems.

Definition 7 (Generalized Specification, $\mathscr{A}$-Specifications). Let $\mathscr{A}$ be a poset, a generalized specification over $\mathscr{A}$, or simply $\mathscr{A}$-specification, is a couple (G, F) of a presheaf and a functor where $G : \mathscr{A}^{op} \to \textbf{Mes}$ and $F : \mathscr{A} \to \textbf{Kern}$ are such that for any $a, b \in \mathscr{A}$ with $b \leq a$,

$$G_b^a F_a^b = \text{id} \tag{5.2}$$

In the previous definition G encodes the collection of projections $i_b^a : E_a \to E_b$ for $b \subseteq a$; it is in some way the 'skeleton' of the spaces of observables of the statistical system. It is a key ingredient for the generalization of (rigorous) statistical mechanics to a categorical framework.

Definition 8 (Gibbs measures for $\mathscr{A}$-specifications). Let $\gamma = (G, F)$ be an $\mathscr{A}$-specification, we call the Gibbs measures of γ the sections of F,

$$\mathscr{G}_g(\gamma) := [*, F]_{K,\mathscr{A}} \tag{5.3}$$

Proposition 1 (Category of $\mathscr{A}$-Specification). *Let us call objects all $\mathscr{A}$-specifications. Call morphisms between two $\mathscr{A}$-specifications, (G, F) and (G_1, F_1) couples of natural transformations (ϕ, ψ) with*

1. *$\phi : G \to G_1$ a morphism of presheaves in* **Mes**
2. *$\psi : F \to F_1$ a morphism of functors in* **Kern**

that satisfy for any $a \in \mathscr{A}$,

$$\phi_a \psi_a = \text{id} \tag{5.4}$$

Then this collection of objects and morphisms forms a category. We will call this category the category of $\mathscr{A}$-specifications and denote it as $\textbf{Sp}(\mathscr{A})$.

6 Gibbs Measure of Projective Specifications

Let E be a measurable space; we denote $L^\infty(E)$ the set of bounded, real-valued, measurable functions over E. L^∞ defines a presheaf from **Mes** and **Kern** to the category of vector spaces **Vect**. For the definition and characterization of projective presheaf over a poset see [4, 13, 27].

Definition 9 (Projective $\mathscr{A}$-specifications). An $\mathscr{A}$-specification (G, F) is called projective when $L^\infty \circ F$ is a projective presheaf (in **Vect**). In other words, there is a collection of presheaves $(S_a, a \in \mathscr{A})$ such that, $L^\infty \circ F \cong \bigoplus_{a \in \mathscr{A}} S_a$ where for any $b \geq a$, $S_a(b)$ is a constant vector space denoted S_a and $S_{a\,c}^{\,b} = $ id for any $a \leq c \leq b$ and $S_a(b) = 0$ if $b \not\geq a$. The collection of presheaves $(S_a, a \in \mathscr{A})$ is called the decomposition of (G, F).

For any poset $\mathscr{A}$, symmetrizing the order defines the following equivalence relation,

$$\forall a, b \in \mathscr{A}, \ a \sim b \iff a \leq b \text{ or } b \leq a \tag{6.1}$$

The equivalence classes of this equivalence relation are the connected components of $\mathscr{A}$ that we will denote as $\mathscr{C}(\mathscr{A})$. To each element of $a \in \mathscr{A}$ one can associate its connected component $\mathscr{C}(a)$. If each connected component has a minimum element, in other words, if for any $C \in \mathscr{C}(\mathscr{A})$, and any $b \in C$, there is $c \in C$ such that, $c \leq b$, then we shall denote, $\mathscr{C}_*(\mathscr{A})$ as the collection of these minimum elements; if not $\mathscr{C}_*(\mathscr{A}) = \emptyset$.

To conclude this article let us characterize Gibbs measures of projective $\mathscr{A}$-specifications.

Theorem 3. *Let γ be a projective $\mathscr{A}$-specification. If at least one of the connected components of $\mathscr{A}$ does not have a minimum element, i.e. when,*

$$\mathscr{C}_*(\mathscr{A}) = \emptyset \tag{6.2}$$

then,

$$\mathscr{G}_g(\gamma) = \emptyset \tag{6.3}$$

if not,

$$\mathscr{G}_g(\gamma) = \prod_{a \in \mathscr{C}_*(\mathscr{A})} \mathbb{P}(\gamma(a)) \tag{6.4}$$

Proof. Let us denote $L^\infty G$ as i and, $L^\infty F$ as π; (i, π) is decomposable, let $(S_a, a \in \mathscr{A})$ be its decomposition. For any $a, b \in \mathscr{A}$ such that $b \leq a$ and $\mu \in \mathscr{G}(\gamma)$, let us denote $L^\infty \mu$ as ν.

$$\nu_b \pi_b^a \left(\sum_{c \leq a} S_c(a)(v) \right) = \nu_a \left(\sum_{c \leq a} S_c(a)(v) \right) \tag{6.5}$$

therefore,

$$\nu_b \left(\sum_{c \leq b} S_{cb}^a(v) \right) = \nu_a i_a^b \left(\sum_{c \leq b} S_{cb}^a(v) \right) = \nu_a \left(\sum_{c \leq a} S_c(a)(v) \right) \tag{6.6}$$

and so,

$$\nu_a\Big(\sum_{\substack{c \leq a \\ c \nleq b}} S_c(a)\Big) = 0 \qquad (6.7)$$

Therefore for any $a \notin \mathscr{C}_*(\mathscr{A})$, $\nu_a|_{S_a(a)} = 0$. Furthermore,

$$\mathrm{colim}\, i \cong \bigoplus_{a \in \mathscr{A}} S_a(a). \qquad (6.8)$$

ν is uniquely determined by $(\nu_a|_{S_a(a)}, a \in \mathscr{A})$; if there is a connected component $C \in \mathscr{C}(\mathscr{A})$ that does not have a minimal element, for any $a \in C$,

$$\nu_a|_{S_a} = 0 \qquad (6.9)$$

Therefore for any $a \in C$, $\nu_a = 0$; this is contradictory with the fact that $\mu_a \in \mathbb{P}(\gamma(a))$ and so,

$$\mathscr{G}_g(\gamma) = \emptyset \qquad (6.10)$$

When $\mathscr{C}_*(\mathscr{A})$ is non empty for any functor, H, from $\mathscr{A}$ to **Set**,

$$\lim H \cong \prod_{a \in \mathscr{C}_*(\mathscr{A})} H(a) \qquad (6.11)$$

therefore,

$$\mathscr{G}_g(\gamma) = \prod_{a \in \mathscr{C}_*(\mathscr{A})} \mathbb{P}(\gamma(a)) \qquad (6.12)$$

Acknowledgements. We express our gratitude to the reviewers for their insightful feedback and diligent contributions in enhancing the quality of our manuscript.

References

1. Baudot, P., Bennequin, D.: The homological nature of entropy. Entropy **17**(5), 3253–3318 (2015)
2. Belfiore, J.C., Bennequin, D.: Topos and stacks of deep neural networks. ArXiv (2021)
3. Bennequin, D., Peltre, O., Sergeant-Perthuis, G., Vigneaux, J.P.: Extra-fine sheaves and interaction decompositions, September 2020. https://arxiv.org/abs/2009.12646
4. Brown, A., Draganov, O.: Computing minimal injective resolutions of sheaves on finite posets (2021). https://doi.org/10.48550/ARXIV.2112.02609. https://arxiv.org/abs/2112.02609
5. Chipot, C., Pohorille, A. (eds.): Free Energy Calculations: Theory and Applications in Chemistry and Biology. No. 86 in Springer Series in Chemical Physics. Springer, New York (2007), oCLC: ocm79447449
6. Fritz, T.: A synthetic approach to Markov kernels, conditional independence and theorems on sufficient statistics. Adv. Math. (2020)

7. Fritz, T., Perrone, P.: A probability monad as the colimit of spaces of finite samples. Theory Appl. Categories (2019)
8. Fritz, T., Spivak, D.I.: Internal probability valuations. In: Workshop Categorical Probability and Statistics (2020)
9. Gallier, J., Quaintance, J.: Homology, Cohomology, and Sheaf Cohomology for Algebraic Topology, Algebraic Geometry, and Differential Geometry. World Scientific (2022)
10. Georgii, H.O.: GIBBS Measures and Phase Transitions. De Gruyter, New York (2011). https://doi.org/10.1515/9783110250329
11. Giry, M.: A categorical approach to probability theory. In: Banaschewski, B. (ed.) Categorical Aspects of Topology and Analysis. LNM, vol. 915, pp. 68–85. Springer, Heidelberg (1982). https://doi.org/10.1007/BFb0092872
12. Gromov, M.: In a search for a structure, part 1: On entropy (2013)
13. Hu, C.S.: A brief note for sheaf structures on posets (2020). https://doi.org/10.48550/ARXIV.2010.09651
14. Lawvere, E.W.: The category of probabilistic mappings (1962). https://ncatlab.org/nlab/files/lawvereprobability1962.pdf. Link to draft
15. Lecun, Y., Chopra, S., Hadsell, R., Ranzato, M., Huang, F.: A Tutorial on Energy-Based Learning. MIT Press (2006)
16. Lelièvre, T., Rousset, M., Stoltz, G.: Free Energy Computations. Imperial College Press (2010). https://doi.org/10.1142/p579. https://www.worldscientific.com/doi/abs/10.1142/p579
17. Mac Lane, S.: Categories for the Working Mathematician. GTM, vol. 5. Springer, New York (1978). https://doi.org/10.1007/978-1-4757-4721-8
18. Marcolli, M.: Gamma spaces and information. J. Geometry Phys. (2019)
19. Ruelle, D.: Statistical Mechanics. Imperial College Press (1999). https://doi.org/10.1142/4090. https://www.worldscientific.com/doi/abs/10.1142/4090
20. Sergeant-Perthuis, G.: Intersection property and interaction decomposition, April 2019. https://arxiv.org/abs/1904.09017
21. Sergeant-Perthuis, G.: Interaction decomposition for presheaves, August 2020. https://arxiv.org/abs/2008.09029
22. Sergeant-Perthuis, G.: Interaction decomposition for Hilbert spaces (2021). https://arxiv.org/abs/2107.06444
23. Simson, D.: Linear representations of partially ordered sets and vector space categories (1993)
24. Spiegel, E., O'Donnell, C.J.: Incidence Algebras. CRC Press (1997)
25. Vigneaux, J.P.: Information structures and their cohomology. Theory Appl. Categories **35**(38), 1476–1529 (2020)
26. Wiering, M., van Otterlo, M. (eds.): Reinforcement Learning: State-of-the-Art, Adaptation, Learning, and Optimization, vol. 12. Springer, Heidelberg (2012). https://doi.org/10.1007/978-3-642-27645-3. https://link.springer.com/10.1007/978-3-642-27645-3
27. Yanagawa, K.: Sheaves on finite posets and modules over normal semi-group rings. J. Pure Appl. Algebra **161**(3), 341–366 (2001). https://doi.org/10.1016/S0022-4049(00)00095-5. https://www.sciencedirect.com/science/article/pii/S0022404900000955

Categorical Information Geometry

Paolo Perrone[(⊠)](ID)

University of Oxford, Oxford, UK
paoloperrone@cs.ox.ac.uk
http://www.paoloperrone.org

Abstract. Information geometry is the study of interactions between random variables by means of metric, divergences, and their geometry. Categorical probability has a similar aim, but uses algebraic structures, primarily monoidal categories, for that purpose. As recent work shows, we can unify the two approaches by means of enriched category theory into a single formalism, and recover important information-theoretic quantities and results, such as entropy and data processing inequalities.

Keywords: Category Theory · Markov Categories · Graphical Models · Divergences · Information Geometry

1 Metrics and Divergences on Monoidal Categories

A *monoidal category* [15, Sect. VII.1] is an algebraic structure used to describe processes that can be composed both sequentially and in parallel. Before introducing their metric enrichment, we sketch their fundamental aspects and their graphical representation. See the reference above for the full definition and for the details.

1.1 Monoidal Categories and Their Graphical Calculus

First of all, a category C consists of *objects*, which we can view as spaces of possible states, or alphabets, and which we denote by capital letters such as X, Y, A, B. We also have *morphisms* or *arrows* between them. A morphism $f : A \to B$ can be seen as a process or a channel with input from A and output in B. Graphically, we represent objects as wires and morphisms as boxes, to be read from left to right.

$$X \longrightarrow \boxed{f} \longrightarrow Y$$

Morphisms can be composed sequentially, with their composition represented as follows,

$$X \longrightarrow \boxed{f} \overset{Y}{\longrightarrow} \boxed{g} \longrightarrow Z$$

Supported by Sam Staton's ERC grant "BLaSt – A Better Language for Statistics".

F. Nielsen and F. Barbaresco (Eds.): GSI 2023, LNCS 14071, pp. 268–277, 2023.
https://doi.org/10.1007/978-3-031-38271-0_27

and having a *category* means that the composition is associative and unital. A relevant example is the category FinStoch of finite sets, which we view as finite alphabets, and stochastic matrices between them, which we view as noisy channels. A stochastic matrix $f : X \to Y$ is a matrix of entries $f(y|x) \in [0,1]$, which we can view as transition probabilities, such that $\sum_y f(y|x) = 1$ for each $x \in X$.

A *monoidal structure* on C is what allows us to have morphisms with several inputs and outputs, represented as follows.

$$X \quad \boxed{g} \quad Y$$
$$A \qquad\qquad B$$

This is accomplished by forming, for each two object X and A, a new object, which we call $X \otimes A$. This assignment is moreover *functorial*, in the sense of a two-variable functor $\otimes : C \times C \to C$, meaning that we also multiply *morphisms*. Given $f : X \to Y$ and $g : A \to B$ we get a morphism $f \otimes g : X \otimes A \to Y \otimes B$, which we represent as follows,

and which we can interpret as executing f and g independently and in parallel.

We also have morphisms with *no* inputs or outputs. For example, a *state* or *source* is a morphism with no inputs.

This is accomplished by means of a distinguished object I, called the *unit*, with the property that $X \otimes I \cong I \otimes X \cong X$, so that it behaves similarly to a neutral element for the tensor product. In FinStoch, I is the one-element set: stochastic matrices $I \to X$ are simply probability measures on X.

A monoidal category is then a category C equipped with a distinguished object I, called the *unit*, and a product functor $\otimes : C \times C \to C$ which is associative and unital up to particular isomorphisms. This makes the structure analogous to a monoid, hence the name. A monoidal category is *symmetric* whenever there is a particular involutive isomorphism

$$X \diagdown \qquad \diagup Y$$
$$Y \diagup \qquad \diagdown X$$

for each pair of objects X, Y, analogously to commutative monoids. For the details, see once again [15, Sect. VII.1].

Let's now equip these structures with metrics and divergences.

1.2   Metrics and Divergences, and a Fundamental Principle

Definition 1. *A* divergence *or* statistical distance *on a set X is a function*

$$X \times X \xrightarrow{\ D\ } [0, \infty]$$
$$(x, y) \longmapsto D(x \parallel y)$$

such that $D(x \parallel x) = 0$.
We call the pair (X, D) a divergence space.
We call the divergence D strict *if $D(x \parallel y) = 0$ implies $x = y$.*

Every metric is a strict divergence which is moreover finite, symmetric, and satisfies a triangle inequality. The Kullback-Leibler divergence (see the next section) is an example of a non-metric divergence.

Definition 2. *A* divergence *on a monoidal category* C *amounts to*

– *For each pair of objects X and Y, a divergence $D_{X,Y}$ on the set of morphisms $X \to Y$, or more briefly just D;*

such that

– *The composition of morphisms in the following form*

$$X \underset{f'}{\overset{f}{\rightrightarrows}} Y \underset{g'}{\overset{g}{\rightrightarrows}} Z$$

satisfies the following inequality,

$$D(g \circ f \parallel g' \circ f') \leq D(f \parallel f') + D(g \parallel g'); \qquad (1)$$

– *The tensor product of morphisms in the following form*

$$X \otimes A \underset{f' \otimes h'}{\overset{f \otimes h}{\rightrightarrows}} Y \otimes B$$

satisfies the following inequality,

$$D\big((f \otimes h) \parallel (f' \otimes h')\big) \leq D(f \parallel f') + D(h \parallel h'). \qquad (2)$$

We can interpret this definition in terms of the following *fundamental principle of categorical information geometry*: *We can bound the distance between complex configurations in terms of their simpler components.*

For example, the distance or divergence between the two systems depicted below is bounded by $D(p, p') + D(f, f') + D(g, g')$. More generally, for any string diagrams of any configuration, the distance or divergence between the resulting constructions will always be bounded by the divergence between the basic building blocks.

An important consequence of this principle, which can be obtained by setting, for example, $p = p'$ and $f = f'$ but not $g = g'$ in the example above, is that *adding the same block to both sides, in any sequential or parallel direction, cannot increase the distance or divergence.* This is a wide generalization of Shannon's data processing inequalities, which say that the divergence between two sources cannot increase by processing them in the same way. (See [18, Sect. 2.1])

In the next two sections we are going to see two main examples of monoidal categories with divergences: Markov categories, in particular FinStoch, and categories of couplings.

2   Markov Categories and Divergences

Markov categories are particular monoidal categories with a structure that makes them very well suited for modeling probabilistic processes.[1] They were defined in their current form in [5], building up on previous work (see Sect. 2.1).

A Markov category is a symmetric monoidal category where each object X is equipped with two particular maps called "copy" and "discard", and represented as follows.

$$X \multimap \overset{X}{\underset{X}{\bigg\langle}} \qquad\qquad X \longrightarrow \bullet$$

Note that the copy map has output $X \otimes X$ and the discard map has output I, i.e. "no output". These maps have to satisfy some properties (commutative comonoid axioms) which ensure that the interpretation as "copy" and "discard" maps is indeed consistent. See [5] as well as [18] for more details on this.

Example 1 (The category FinStoch). We can construct a category of finite alphabets and noisy channels, called FinStoch, as follows.

– Its objects are finite sets, which we denote by X, Y, Z, etc.
– A morphism $X \to Y$ is a *stochastic matrix*, i.e. a matrix of nonnegative entries with columns indexed by the elements of X, and rows indexed by the elements of Y,

$$X \times Y \overset{f}{\longrightarrow} [0, 1]$$

$$(x, y) \longmapsto f(y|x)$$

---

[1] Despite the name, Markov categories are not only suited to model Markov processes, but arbitrary stochastic processes. Indeed, arbitrary joint distributions can be formed, and the *Markov property* states that the stochastic dependencies between the variables are faithfully represented by a particular graph. If the graph is (equivalent to) a single chain, we have a Markov process. In general, the graph is more complex. In this respect, Markov categories are similar to, but more general than, Markov random fields. See [6] for more details on this.

such that each column sums to one,

$$\sum_{y \in Y} f(y|x) = 1 \qquad \text{for every } x \in X.$$

We can interpret $f(y|x)$ as a conditional or transition probability from state $x \in X$ to state $y \in Y$, or we can interpret f as a family of probability measures f_x over Y indexed by the elements of X.

FinStoch is a Markov category with the copy maps $X \to X \otimes X$ given by mapping x to (x, x) deterministically for each $x \in X$, and discard maps given by the unique stochastic matrix $X \to 1$, i.e. a row matrix of entries 1.

2.1 A Brief History of the Idea

The first known study of some aspects of probability theory via categorical methods is due to Lawvere [12], where he defined the category FinStoch outlined above, as well as its generalization to arbitrary measurable spaces. Some of those ideas reappeared in the work of Giry [9] in terms of monads. Lawvere was also the first to see the potential of enriched category theory in metric geometry [13], although it seems that he never used these ideas to study probability theory.

The same categories of probabilistic mappings were defined independently by Chentsov [1], and used to set the stage for the (differential) geometry of probability distributions [2]. Interestingly, Chentsov's work involves *categories* of probabilistic mappings as well as their *geometry*, but he never merged the two approaches into a geometric enrichment of the category of kernels (most likely because at that time, enriched category theory was still in its infancy).[2] The influence of Chentsov on the present work is therefore two-fold, and the main challenge of this work is integrating his two approaches, geometric and categorical, into one unified formalism.

Markov categories, and the more general GS or CD categories, first appeared in [8] in the context of graph rewriting. Similar structures reappeared independently in the work of Golubtsov [10], and were applied for the first time to probability, statistics and information theory. The idea of using "copy" and "discard" maps to study probability came independently to several other authors, most likely initially unaware of each other's work, such as Fong [4], Cho and Jacobs [3], and Fritz [5]. (Here we follow the conventions and terminology of [5].)

Finally, the idea to use both category theory and geometry to study the properties of entropy was inspired by the work of Gromov [11]. This work has a similar philosophy, but follows a different approach.

For more information on the history of these ideas, we refer the reader to [5, Introduction], to [7, Remark 2.2], and to [18, Introduction].

---

[2] The *geometry of the category of Markov kernels* studied by Chentsov in [2, Sects. 4 and 6] is not *metric geometry*, it is a study of invariants in the sense of Klein's Erlangen Program. More related to the present work are, rather, the *invariant information characteristics* of Sect. 8 of [2]. Much of classical information geometry, and hence indirectly this work, is built upon those notions.

2.2 Divergences on Markov Categories

As Markov categories are monoidal categories, one can enrich them in divergences according to Definition 2 (see also [18, Sect. 2]).

Here are two important examples of divergences that we can put on FinStoch.

Example 2 (The Kullback-Leibler divergence). Let X and Y be finite sets, and let $f, g : X \to Y$ be stochastic matrices. The *relative entropy* or *Kullback-Leibler divergence* between f and g is given by

$$D_{KL}(f \parallel g) := \max_{x \in X} \sum_{y \in Y} f(y|x) \log \frac{f(y|x)}{g(y|x)},$$

with the convention that $0 \log(0/0) = 0$ and $p \log(p/0) = \infty$ for $p \neq 0$.

Example 3 (The total variation distance). Let X and Y be finite sets, and let $f, g : X \to Y$ be stochastic matrices. The *total variation distance* between f and g is given by

$$D_T(f \parallel g) := \max_{x \in X} \frac{1}{2} \sum_{y \in Y} |f(y|x) - g(y|x)|.$$

See [18] for why these examples satisfy the conditions of Definition 2. This in particular implies that all these quantities satisfy a very general version of the data processing inequality, see the reference above for more information.

Remark 1. It is well known that the KL divergence and the total variation distance, as well as Rényi's α-divergences, are special cases of f-*divergences* [16]. It is still an open question whether all f-divergences give an enrichment on FinStoch. However, Tsallis' q-divergences do not [18, Sect. 2.3.4].

3 Categories of Couplings and Divergences

Besides Markov categories, another example of divergence-enriched categories relevant for the purposes of information theory are *categories of couplings*. The category FinCoup has

- As objects, finite probability spaces, i.e. pairs (X, p) where X is a finite set and p is a probability distribution on it;
- As morphisms $(X, p) \to (Y, q)$, *couplings* of p and q, i.e. probability measures s on $X \otimes Y$ which have p and q as their respective marginals;
- The identity $(X, p) \to (X, p)$ is given by the pushforward of p along the diagonal map $X \to X \otimes X$;
- The composition of couplings is given by the *conditional product*: for $s : (X, p) \to (Y, q)$ and $y : (Y, q) \to (Z, r)$

$$(t \circ s)(x, z) := \sum_y \frac{s(x, y) \, t(y, z)}{q(y)},$$

where the sum is taken over the $y \in Y$ such that $q(y) > 0$.

More information about this category, and its generalization to the continuous case, can be found in [17].

The two choices of divergence outlined in the previous section also work for the category FinCoup.

Example 4 (Kullback-Leibler divergence). Let (X, p) and (Y, q) be finite probability spaces, and let s and t be couplings of p and q. The Kullback-Leibler divergence

$$D_{KL}(s \parallel t) := \sum_{x,y} s(x, y) \log \frac{s(x, y)}{t(x, y)}$$

can be extended to a divergence on the whole of FinCoup, i.e. the conditions of Definition 2 are satisfied.

Example 5 (Total variation distance). Let (X, p) and (Y, q) be finite probability spaces, and let s and t be couplings of p and q. The total variation distance

$$D_T(s, t) := \frac{1}{2} \sum_{x,y} |s(x, y) - t(x, y)|$$

can be extended to a divergence on the whole of FinCoup, i.e. the conditions of Definition 2 are satisfied.

The category FinCoup is moreover an *enriched dagger category*. A coupling $(X, p) \to (Y, q)$ can also be seen as a coupling $(Y, q) \to (X, p)$, and this choice does not have any effect on the metrics or divergences. This property is analogous to, but independent from, the symmetry of the distance in a metric space.

Categories of couplings and Markov categories are tightly related, for more information see [5, Definition 13.7 and Proposition 13.8]. Further links between the two structures will be established in future work.

4   Recovering Information-Theoretic Quantities

One of the most interesting features of categorical information geometry is that *basic information-theoretic quantities can be recovered from categorical prime principles*. These include Shannon's entropy and mutual information for discrete sources. Here are some examples, more details can be found in [18].

4.1   Measures of Randomness

Markov categories come equipped with a notion of *deterministic morphisms*, [5, Definition 10.1]. Let's review here the version for sources.

Definition 3. *A source p on X in a Markov category is called* deterministic *if and only if copying its output has the same effect as running it twice independently:*

$$\tag{3}$$

Let's try to interpret this notion. First of all, if p is a source which outputs deterministically a single element x of X, then both sides output the ordered pair (x, x), and hence they are equal. Instead, if p is random, the left-hand side will have perfectly correlated output, while the right-hand side will display identically distributed, but independent outputs.

In FinStoch Eq. (3) reduces to

$$p(x) = p(x)^2$$

for all $x \in X$, so that deterministic sources are precisely those probability distributions p whose entries are only zero and one, i.e. the "Dirac deltas". It is then natural to define as our measure of randomness the discrepancy between the two sides of Eq. (3).

Definition 4. *Let* C *be a Markov category with divergence D. The* entropy *of a source p is the quantity*

$$H(p) := D\big(\text{copy} \circ p \,\|\, (p \otimes p)\big), \tag{4}$$

i.e. the divergence between the two sides of (3). (Note that the order matters.)

Example 6. In FinStoch, equipped with the KL divergence, our notion of entropy recovers exactly Shannon's entropy:

$$\begin{aligned}
H_{KL}(p) &= D_{KL}\big(\text{copy} \circ p \,\|\, (p \otimes p)\big) \\
&= \sum_{x,x' \in X} p(x)\, \delta_{x,x'} \log \frac{p(x)\, \delta_{x,x'}}{p(x)\, p(x')} \\
&= -\sum_{x \in X} p(x) \log p(x).
\end{aligned}$$

Example 7. FinStoch, equipped with the total variation distance, our notion of entropy gives the Gini-Simpson index [14], used for example in ecology to quantify diversity:

$$\begin{aligned}
H_T &= \frac{1}{2} \sum_{x,x' \in X} \big| p(x)\, \delta_{x,x'} - p(x)\, p(x')\big| \\
&= \frac{1}{2} \sum_{x \in X} p(x) \left(1 - p(x) + \sum_{x' \neq x} p(x') \right) \\
&= 1 - \sum_{x \in X} p(x)^2.
\end{aligned}$$

Rényi's α-entropies can also be obtained in this way (see [18, Sect. 4.2.2]), while it is still unclear whether Tsallis' q-entropies can be obtained in this way for $q \neq 2$ (see [18, Question 4.4]).

The fundamental principle of Sect. 1 implies a data processing inequality for entropy generalizing the traditional one. See [18, Sect. 4] for more details.

4.2   Measures of Stochastic Interaction

Just as for determinism, Markov categories are equipped with a notion of stochastic and conditional independence [5, Definition 12.12 and Lemma 12.11]. For sources it reads as follows.

Definition 5. *A joint source* h *on* $X \otimes Y$ *in a Markov category displays* independence *between* X *and* Y *if and only if*

$$\langle h \mid \begin{matrix} X \\ Y \end{matrix} = \begin{matrix} \langle h \mid \begin{matrix} X \\ \bullet \end{matrix} \\ \langle h \mid \begin{matrix} \bullet \\ Y \end{matrix} \end{matrix} \tag{5}$$

For discrete probability measures, this is exactly the condition

$$p(x, y) = p(x)\, p(y),$$

i.e. that p is the product of its marginals. It is a natural procedure in information theory to quantify the stochastic dependence of the variables X and Y by taking the divergence between both sides of the equation.

Definition 6. *Let* C *be a Markov category with a divergence* D. *The* mutual information *displayed by a joint source* h *on* $X \otimes Y$ *is the divergence between the two sides of Eq.* (5),

$$\mathrm{I}_D(h) := D\big(h \parallel (h_X \otimes h_Y)\big).$$

Note that the order of the arguments of D *matters.*

In FinStoch, with the KL divergence, one recovers exactly Shannon's mutual information. This is well known fact in information theory, and through our formalism, it acquires categorical significance. Using other notion of divergences one can obtain other analogues of mutual information, such as a total variation-based one. Moreover, once again the fundamental principle of Sect. 1 implies a data processing inequality for mutual information generalizing the traditional one. See [18, Sect. 3] for more on this.

Acknowledgements. The author would like to thank Tobias Fritz, Tomáš Gonda and Sam Staton for the helpful discussions and feedback, the anonymous reviewers for their constructive comments, and Swaraj Dash for the help with translating from Russian.

References

1. Chentsov, N.N.: The categories of mathematical statistics. Dokl. Akad. Nauk SSSR **164**, 511–514 (1965)
2. Chentsov, N.N.: Statistical decision rules and optimal inference. Nauka (1972)

3. Cho, K., Jacobs, B.: Disintegration and Bayesian inversion via string diagrams. Math. Struct. Comput. Sci. **29**, 938–971 (2019). https://doi.org/10.1017/S0960129518000488

4. Fong, B.: Causal theories: a categorical perspective on Bayesian networks. Master's thesis, University of Oxford (2012). arXiv:1301.6201

5. Fritz, T.: A synthetic approach to Markov kernels, conditional independence and theorems on sufficient statistics. Adv. Math. **370**, 107239 (2020). arXiv:1908.07021

6. Fritz, T., Klingler, A.: The d-separation criterion in categorical probability (2022). arXiv:2207.05740

7. Fritz, T., Liang, W.: Free GS-monoidal category and free Markov categories (2022). arXiv:2204.02284

8. Gadducci, F.: On the algebraic approach to concurrent term rewriting. Ph.D. thesis, University of Pisa (1996)

9. Giry, M.: A categorical approach to probability theory. In: Banaschewski, B. (ed.) Categorical Aspects of Topology and Analysis. LNM, vol. 915, pp. 68–85. Springer, Heidelberg (1982). https://doi.org/10.1007/BFb0092872

10. Golubtsov, P.V.: Axiomatic description of categories of information transformers. Problemy Peredachi Informatsii **35**(3), 80–98 (1999)

11. Gromov, M.: In search for a structure, Part 1: on entropy (2013). https://www.ihes.fr/ gromov/wp-content/uploads/2018/08/structre-serch-entropy-july5-2012.pdf

12. Lawvere, F.W.: The category of probabilistic mappings (1962). Unpublished notes

13. Lawvere, W.: Metric spaces, generalized logic and closed categories. Rendiconti del seminario matematico e fisico di Milano **43** (1973). http://www.tac.mta.ca/tac/reprints/articles/1/tr1abs.html

14. Leinster, T.: Entropy and Diversity. Cambridge University Press (2021)

15. Mac Lane, S.: Categories for the Working Mathematician. Graduate Texts in Mathematics, vol. 5, 2nd edn. Springer, New York (1998). https://doi.org/10.1007/978-1-4757-4721-8

16. Morozova, E., Chentsov, N.N.: Natural geometry on families of probability laws. Itogi Nauki i Tekhniki. Sovremennye Problemy Matematiki. Fundamental'nye Napravleniya **83**, 133–265 (1991)

17. Perrone, P.: Lifting couplings in Wasserstein spaces (2021). arXiv:2110.06591

18. Perrone, P.: Markov categories and entropy (2022). arXiv:2212.11719

Categorical Magnitude and Entropy

Stephanie Chen[1] and Juan Pablo Vigneaux[2(✉)]

[1] California Institute of Technology, Pasadena, CA 91125, USA
schen7@caltech.edu
[2] Department of Mathematics, California Institute of Technology,
Pasadena, CA 91125, USA
vigneaux@caltech.edu

Abstract. Given any finite set equipped with a probability measure, one may compute its Shannon entropy or information content. The entropy becomes the logarithm of the cardinality of the set when the uniform probability is used. Leinster introduced a notion of Euler characteristic for certain finite categories, also known as magnitude, that can be seen as a categorical generalization of cardinality. This paper aims to connect the two ideas by considering the extension of Shannon entropy to finite categories endowed with probability, in such a way that the magnitude is recovered when a certain choice of "uniform" probability is made.

Keywords: Entropy · Magnitude · Categories · Information measure · Topology

1 Introduction

Given a finite set X endowed with a probability measure p, its Shannon entropy [1] is given by

$$H(p) = -\sum_{x \in X} p(x) \ln p(x). \tag{1}$$

In particular, taking the uniform probability $u : x \mapsto 1/|X|$ yields $H(u) = \ln |X|$. We may thus view Shannon entropy as a *probabilistic* generalization of cardinality. A *categorical* generalization of cardinality may be found in the Euler characteristic or magnitude of finite ordinary categories [2], defined as follows.

Let $\mathbf{A}$ be a finite category. The zeta function $\zeta : \mathrm{Ob}(\mathbf{A}) \times \mathrm{Ob}(\mathbf{A}) \to \mathbb{Q}$ is given by $\zeta(x, y) = |\mathrm{Hom}(x, y)|$, the cardinality of the hom-set, for any $x, y \in \mathrm{Ob}(\mathbf{A})$. A weighting on $\mathbf{A}$ is a function $k^\bullet : \mathrm{Ob}(\mathbf{A}) \to \mathbb{Q}$ such that

$$\sum_{b \in \mathrm{Ob}(\mathbf{A})} \zeta(a, b) k^b = 1 \tag{2}$$

for all $a \in \mathrm{Ob}(\mathbf{A})$. Similarly, a coweighting on $\mathbf{A}$ is a function $k_\bullet : \mathrm{Ob}(\mathbf{A}) \to \mathbb{Q}$ such that

$$\sum_{b \in \mathrm{Ob}(\mathbf{A})} \zeta(b, a) k_b = 1 \tag{3}$$

SC acknowledges the support of Marcella Bonsall through her SURF fellowship.

F. Nielsen and F. Barbaresco (Eds.): GSI 2023, LNCS 14071, pp. 278–287, 2023.
https://doi.org/10.1007/978-3-031-38271-0_28

for all $a \in \mathrm{Ob}(\mathbf{A})$. Equivalently, one may view a coweighting on $\mathbf{A}$ as a weighting on $\mathbf{A}^{\mathrm{op}}$. If $\mathbf{A}$ admits both a weighting and a coweighting, then

$$\sum_{a \in \mathrm{Ob}(\mathbf{A})} k^a = \sum_{a \in \mathrm{Ob}(\mathbf{A})} k_a; \tag{4}$$

in this case, the *magnitude* of $\mathbf{A}$, denoted $\chi(\mathbf{A})$, is defined as the common value of both sums.

Magnitude enjoys algebraic properties reminiscent of cardinality, such as $\chi(\mathbf{A} \coprod \mathbf{B}) = \chi(\mathbf{A}) + \chi(\mathbf{B})$ and $\chi(\mathbf{A} \times \mathbf{B}) = \chi(\mathbf{A})\chi(\mathbf{B})$. Moreover, when $\mathbf{A}$ is a discrete category (i.e. it only has identity arrows), $\chi(\mathbf{A}) = |\mathrm{Ob}(\mathbf{A})|$. Hence magnitude may be regarded as a categorical generalization of cardinality.

We ask if there is an extension of Shannon entropy from finite sets to finite categories that gives a probabilistic generalization of the magnitude; in particular, we want this extension to give us the logarithm of the magnitude under some "uniform" choice of probabilities and to coincide with Shannon entropy when specialized to discrete categories.

The rest of this paper is organized as follows. In Sect. 2, we introduce the category **ProbFinCat** whose objects are categorical probabilistic triples $(\mathbf{A}, p, \phi)$ and whose morphisms are probability-preserving functors. In Sect. 3, we define a function $\mathcal{H}$ of categorical probabilistic triples that shares analogous properties to those used by Shannon [1] to characterize the entropy (1). This function $\mathcal{H}$ allows us to recover the set-theoretical Shannon entropy and the categorical magnitude for particular choices of p or ϕ. In Sect. 4, we discuss the possibility of characterizing the "information loss" given by $\mathcal{H}$ in the spirit of [3].

2 Probabilistic Categories

Definition 1. *A categorical probabilistic triple* $(\mathbf{A}, p, \phi)$ *consists of*

1. *a finite category* $\mathbf{A}$,
2. *a probability* p *on* $\mathrm{Ob}(\mathbf{A})$, *and*
3. *a function* $\phi : \mathrm{Ob}(\mathbf{A}) \times \mathrm{Ob}(\mathbf{A}) \to [0, \infty)$ *such that* $\phi(a, a) > 0$ *for all objects* a *of* $\mathbf{A}$, *and* $\phi(b, b') = 0$ *whenever there is no arrow from* b *to* b' *in* $\mathbf{A}$.

The definition gives a lot of flexibility for ϕ, provided it reflects the incidence relations in the category. It might be the ζ function introduced above. Alternatively, it might be a measure of similarity between two objects, see next section. Finally, it might be a transition kernel, in which case for every $a \in \mathrm{Ob}(\mathbf{A})$, the function $\phi(\cdot, a)$ is a probability mass function on the objects b such that an arrow $a \to b$ exists in $\mathbf{A}$; we treat this case in more detail in Sect. 4.

Remark 1. Given a categorical probabilistic triple $(\mathbf{A}, p, \phi)$, set $N = |\mathrm{Ob}(\mathbf{A})|$ and enumerate the objects of $\mathbf{A}$, in order to introduce a matrix Z_ϕ of size $N \times N$ whose (i, j)-component $(Z_\phi)_{ij}$ is $\phi(a_i, a_j)$. A linear system $\vec{f} = Z_\phi \vec{g}$ expresses each $f(a_i)$ as $\sum_{a_j : a_i \to a_j} \phi(a_i, a_j) g(a_j)$. In certain cases the matrix Z_ϕ can be inverted to express $\vec{g}$ as a function of $\vec{f}$. For instance, when $\mathbf{A}$ is a poset and $\phi = \zeta$, the matrix $Z = Z_\zeta$ is invertible and its inverse is known as the Möbius function; it was introduced by Rota in [4], as a generalization of the number-theoretic Möbius function. Similarly, if ϕ is a probabilistic transition kernel and Z_ϕ is invertible, this process might be seen as an inversion of a system of conditional expectations.

We define a category **ProbFinCat** of probabilistic (finite) categories whose objects are categorical probabilistic triples. A morphism $F : (\mathbf{A}, p, \phi) \to (\mathbf{B}, q, \theta)$ in **ProbFinCat** is given by a functor $F : \mathbf{A} \to \mathbf{B}$ such that for all $b \in \mathrm{Ob}(\mathbf{B})$,

$$q(b) = F_*p(b) = \sum_{a \in F^{-1}(b)} p(a). \tag{5}$$

and for all $b, b' \in \mathrm{Ob}(\mathbf{B})$,

$$\theta(b, b') = F_*\phi(b, b') = \begin{cases} \dfrac{\sum_{a' \in F^{-1}(b')} p(a') \sum_{a \in F^{-1}(b)} \phi(a, a')}{F_*p(b')} & F_*p(b') > 0 \\ 1 & b = b', F_*p(b') = 0 \\ 0 & b \neq b', F_*p(b') = 0 \end{cases} \tag{6}$$

Remark that (5) corresponds to the push-forward of probabilities under the function induced by F on objects. In turn, when $b \mapsto \phi(b, a)$ is a probability of transition, (6) gives a probability of a transition $\theta(b, b')$ from b' to b in $\mathbf{B}$ as a weighted average of all the transitions from preimages of b' to preimages of b.

Lemma 1 shows that the function $F_*\phi$ defined by (6) is compatible with our definition of a categorical probabilistic triple. Lemma 2 establishes the functoriality of (5) and (6).

Lemma 1. *Let $(\mathbf{A}, p, \phi)$ be a categorical probabilistic triple, $\mathbf{B}$ a finite category, and. $F : \mathbf{A} \to \mathbf{B}$ be a functor. Then $F_*\phi(b, b) > 0$ for all $b \in \mathrm{Ob}(\mathbf{B})$, and $F_*\phi(b, b') = 0$ whenever $\mathrm{Hom}(b, b') = \emptyset$.*

Proof. Let b, b' be objects of $\mathbf{B}$ and suppose that $F_*p(b) > 0$ (otherwise $F_*\phi(b, b) > 0$ and $F_*\phi(b, b') = 0$ by definition).

To prove the first claim, remark that

$$F_*\phi(b, b) \geq \sum_{a \in F^{-1}(b)} p(a)\phi(a, a) \geq \min_{a \in F^{-1}(b)} \phi(a, a)F_*p(b) > 0. \tag{7}$$

If $\mathrm{Hom}(b, b') = \emptyset$ then $\mathrm{Hom}(a, a') = \emptyset$ for any $a \in F^{-1}(b)$ and $a' \in F^{-1}(b')$; it follows that $\phi(a, a') = 0$ by Definition 1. Then it is clear from (6) that $F_*\theta(b, b')$ vanishes.

Lemma 2. *Let $(\mathbf{A}, p, \phi) \xrightarrow{F} (\mathbf{B}, q, \theta) \xrightarrow{G} (\mathbf{C}, r, \psi)$ be a diagram in **ProbFinCat**. Then $(G \circ F)_*p = G_*(F_*p) = G_*q$ and $(G \circ F)_*\phi = G_*(F_*\phi) = G_*\theta$.*

Proof. For any $c \in \mathbf{C}$,

$$(G \circ F)_* p(c) = \sum_{a \in (G \circ F)^{-1}(c)} p(a) = \sum_{b \in G^{-1}(c)} \sum_{a \in F^{-1}(b)} p(a) = \sum_{b \in G^{-1}(c)} F_*(b).$$

Similarly, for any $c, c' \in \mathrm{Ob}(\mathbf{C})$,

$$(G \circ F)_* \phi(c, c') = \frac{1}{G_*(F_*p)(c)} \sum_{b' \in G^{-1}(c')} \sum_{a' \in F^{-1}(b')} p(a') \sum_{b \in G^{-1}(c)} \sum_{a \in F^{-1}(b)} \phi(a, a')$$

$$= \frac{1}{G_*q(c)} \sum_{b' \in G^{-1}(c)} q(b') \sum_{b \in G^{-1}(c)} \left(\sum_{a' \in F^{-1}(b')} \frac{p(a')}{F_*p(b')} \sum_{a \in F^{-1}(b)} \phi(a, a') \right)$$

$$= G_*(F_*\phi)(c, c') = G_*\theta(c, c').$$

We also consider probability preserving products and weighted sums as follows. For any two categorical probabilistic triples $(\mathbf{A}, p, \phi), (\mathbf{B}, q, \theta)$, we define their probability preserving product $(\mathbf{A}, p, \phi) \otimes (\mathbf{B}, q, \theta)$ to be the triple $(\mathbf{A} \times \mathbf{B}, p \otimes q, \phi \otimes \theta)$ where for any $\langle a, b \rangle, \langle a', b' \rangle \in \mathrm{Ob}(\mathbf{A} \times \mathbf{B})$,

$$(p \otimes q)(\langle a, b \rangle) = p(a)q(b), \quad \text{and} \quad (\theta \otimes \phi)(\langle a, b \rangle, \langle a', b' \rangle) = \theta(a, a')\phi(b, b'). \quad (8)$$

Given any $\lambda \in [0, 1]$, we define the weighted sum $(\mathbf{A}, p, \phi) \oplus_\lambda (\mathbf{B}, q, \theta)$ by $(\mathbf{A} \coprod \mathbf{B}, p \oplus_\lambda q, \phi \oplus \theta)$ where

$$(p \oplus_\lambda q)(x) = \begin{cases} \lambda p(x) & x \in \mathrm{Ob}(\mathbf{A}) \\ (1 - \lambda)q(x) & x \in \mathrm{Ob}(\mathbf{B}) \end{cases} \quad (9)$$

and

$$(\phi \oplus \theta)(x, y) = \begin{cases} \phi(x, y) & x, y \in \mathrm{Ob}(\mathbf{A}) \\ \theta(x, y) & x, y \in \mathrm{Ob}(\mathbf{B}) \ . \\ 0 & \text{otherwise} \end{cases} \quad (10)$$

Given morphisms $f_i : (\mathbf{A}_i, p_i \phi_i) \to (\mathbf{B}_i, q_i, \theta_i)$, for $i = 1, 2$, there is a unique morphism

$$\lambda f_1 \oplus (1 - \lambda) f_2 : (\mathbf{A}_1, p_1 \phi_1) \oplus_\lambda (\mathbf{A}_2, p_2 \phi_2) \to (\mathbf{B}_1, q_1, \theta_1) \oplus_\lambda (\mathbf{B}_2, q_2, \theta_2) \quad (11)$$

that restricts to f_1 on $\mathbf{A}_1$ and to f_2 on $\mathbf{A}_2$.

Finally, we introduce a notion of continuity for functions defined on categorical probabilistic triples. Let $\{(\mathbf{A}_k, p_k, \phi_k)\}_{k \in \mathbb{N}}$ be a sequence of categorical probabilistic triples; we say that $\{(\mathbf{A}_k, p_k, \phi_k)\}_k$ converges to an object $(\mathbf{A}, p, \phi)$ of **ProbFinCat** if $\mathbf{A}_k = \mathbf{A}$ for sufficiently large k and $\{p_k\}_k, \{\phi_k\}_k$ converge pointwise as sequences of functions to p and ϕ respectively. A function $G : \mathrm{Ob}(\mathbf{ProbFinCat}) \to \mathbb{R}$ is continuous if for any convergent sequence $\{(\mathbf{A}_k, p_k, \phi_k)\}_{k \in \mathbb{N}} \to (\mathbf{A}, p, \phi)$, the sequence $\{G(\mathbf{A}_k, p_k, \phi_k)\}_{k \in \mathbb{N}}$ converges to $G(\mathbf{A}, p, \phi)$.

Similarly, a sequence of morphisms $f_n : (\mathbf{A}_n, p_n, \theta_n) \to (\mathbf{B}_n, q_n, \psi_n)$ coverges to $f : (\mathbf{A}, p, \theta) \to (\mathbf{B}, q, \psi)$ if $\mathbf{A}_n = \mathbf{A}$, $\mathbf{B}_n = \mathbf{B}$ and $f_n = f$ (as functors) for n big enough, and $p_n \to p$ and $\theta_n \to \theta$ converge pointwise. Functors from **ProbFinCat** to a topological (semi)group—seen as a one-point category—are continuous if they are sequentially continuous. One obtains in this way a generalization of the notions of continuity discussed in [3, p. 4] to our setting, which is compatible with the faithful embedding of the category **FinProb** of finite probability spaces considered there into **ProbFinCat** which maps a finite set equipped with a probability (A, p) to the triple $(\mathbf{A}, p, \delta)$, where $\mathbf{A}$ is the discrete category with object set A and δ is the Kronecker delta function, given by $\delta(x, x) = 1$ for all $x \in \mathrm{Ob}\, \mathbf{A}$, and $\delta(x, y) = 0$ whenever $x \neq y$.

3   Categorical Entropy

Let Z be a general matrix in $\mathrm{Mat}_{N \times N}([0, \infty))$ with strictly positive diagonal entries and p a probability on a finite set of cardinality N. In the context of [5, Ch. 6], which discusses measures of ecological diversity, Z corresponds to a species similarity matrix and p to the relative abundance of a population of N different species. The species diversity of order 1 is given by

$$^1D^Z(p) = \prod_{i=1}^{N} \frac{1}{(Zp)_i^{p_i}} \tag{12}$$

and its logarithm is a generalization of Shannon entropy, which is recovered when Z is the identity matrix.

Inspired by this, we consider the following entropic functional:

$$\mathcal{H}(\mathbf{A}, p, \phi) := - \sum_{a \in \mathrm{Ob}\, \mathbf{A}} p(a) \ln \left(\sum_{b \in \mathrm{Ob}\, \mathbf{A}} p(b)\phi(a, b) \right) = - \sum_{i=1}^{N} p_i \ln((Z_\phi p)_i). \tag{13}$$

As we required $\phi(a, a) > 0$ for any $a \in \mathrm{Ob}(\mathbf{A})$, we have that $\sum_{b \in \mathrm{Ob}\, \mathbf{A}} p(b)\phi(a, b) \geq p(a)\phi(a, a) > 0$ whenever $p(a) \neq 0$. In order to preserve continuity (see Proposition 5 below) while making $\mathcal{H}$ well defined, we take the convention that

$$p(a) \ln \left(\sum_{b \in \mathrm{Ob}(\mathbf{A})} p(b)\phi(a, b) \right) = 0$$

whenever $p(a) = 0$.

The rest of this section establishes certain properties of the functional $\mathcal{H}$. These properties only depend on the operations $\oplus$ and $\otimes$, as well as the topology on the resulting semiring (a.k.a. *rig*) of categorical probabilistic triples. The morphisms in **ProbFinCat** only appear in the next section, in connection with an algebraic characterization of $\mathcal{H}$.

The functional $\mathcal{H}$ generalizes Shannon entropy in the following sense.

Proposition 1. *If* $\phi = \delta$, *the Kronecker delta, then* $\mathcal{H}(\mathbf{A}, p, \phi) = H(p) = -\sum p_i \ln p_i$.

We now consider properties of $\mathcal{H}$ that are analogous to those used to characterize Shannon entropy, see [6].

Proposition 2. $\mathcal{H}((\mathbf{A}, p, \phi) \otimes (\mathbf{B}, q, \theta)) = \mathcal{H}(\mathbf{A}, p, \phi) + \mathcal{H}(\mathbf{B}, q, \theta)$.

Proof. For simplicity, we write $\mathbf{A}_0 := \mathrm{Ob}(\mathbf{A})$, etc. Recall that $(\mathbf{A} \times \mathbf{B})_0 = \mathbf{A}_0 \times \mathbf{B}_0$. By a direct computation,

$$\mathcal{H}((\mathbf{A}, p, \phi) \otimes (\mathbf{B}, q, \theta))$$

$$= - \sum_{(a,b) \in \mathbf{A}_0 \times \mathbf{B}_0} p(a)q(b) \ln \left(\sum_{(a',b') \in \mathbf{A}_0 \times \mathbf{B}_0} p(a')q(b')\phi(a,a')\theta(b,b') \right)$$

$$= - \sum_{(a,b) \in \mathbf{A}_0 \times \mathbf{B}_0} p(a)q(b) \ln \left(\left(\sum_{a' \in \mathbf{A}_0} p(a')\phi(a,a') \right) \left(\sum_{b' \in \mathbf{B}_0} q(b')\theta(b,b') \right) \right),$$

from which the result follows. More concisely, one might remark that $Z_{\phi \otimes \theta}$ equals $Z_\phi \otimes Z_\theta$, the Kronecker product of matrices, and that $(Z_\phi \otimes Z_\theta)(p \otimes q) = Z_\phi p \otimes Z_\theta q$.

Proposition 3. *Given any* $\lambda \in [0,1]$, *let* $(\mathbf{2}, \Lambda, \delta)$ *be the categorical probability triple where* $\mathbf{2}$ *denotes the discrete category with exactly two objects,* $\mathrm{Ob}(\mathbf{2}) = \{x_1, x_2\}$, Λ *is such that* $\Lambda(x_1) = \lambda, \Lambda(x_2) = 1 - \lambda$, *and* δ *denotes the Kronecker delta. Then,*

$$\mathcal{H}((\mathbf{A}, p, \phi) \oplus_\lambda (\mathbf{B}, q, \theta)) = \lambda \mathcal{H}(\mathbf{A}, p, \phi) + (1 - \lambda)\mathcal{H}(\mathbf{B}, q, \theta) + \mathcal{H}(\mathbf{2}, \Lambda, \delta).$$

Proof. By a direct computation: $\mathcal{H}((\mathbf{A}, p, \phi) \oplus_\lambda (\mathbf{B}, q, \theta))$ equals

$$- \sum_{a \in \mathrm{Ob}(\mathbf{A})} \lambda p(a) \ln \left(\sum_{b \in \mathrm{Ob}(\mathbf{A})} \lambda p(b)\phi(a,b) \right)$$

$$- \sum_{x \in \mathrm{Ob}(\mathbf{B})} (1 - \lambda)q(x) \ln \left(\sum_{y \in \mathrm{Ob}(\mathbf{B})} (1 - \lambda)q(y)\theta(x,y) \right)$$

from which the result follows, because $\mathcal{H}(\mathbf{2}, \Lambda, \delta) = -\lambda \ln(\lambda) - (1 - \lambda) \ln(1 - \lambda)$. Similar to the product case, one might remark that $Z_{\phi \oplus \theta} = Z_\phi \oplus Z_\theta$ and $(Z_\phi \oplus Z_\theta)(p \oplus_\lambda q) = Z_\phi(\lambda p) \oplus Z_\theta((1 - \lambda)q)$.

More generally, given a finite collection $(\mathbf{A}_1, p_1, \phi_1), \ldots, (\mathbf{A}_m, p_m, \phi_m)$ of categorical probabilistic triples and a probability vector $(\lambda_1, \ldots, \lambda_m)$, we can define

$$\bigoplus_{i=1}^m \lambda_i (\mathbf{A}_i, p_i, \phi_i) = \left(\coprod_{i=1}^m \mathbf{A}_i, \bigoplus_{i=1}^m \lambda_i p_i, \bigoplus_{i=1}^m \phi_i \right)$$

where $(\bigoplus_{i=1}^{m} \lambda_i p_i)(x) = \lambda_i p_i(x)$ for $x \in \mathrm{Ob}(\mathbf{A}_i)$, and

$$\left(\bigoplus_{i=1}^{m} \phi_i\right)(x,y) = \begin{cases} \phi_i(x,y) & \text{if } x, y \in \mathrm{Ob}(\mathbf{A}_i) \\ 0 & \text{otherwise} \end{cases}. \tag{14}$$

Proposition 4. *Take notation as in the preceding paragraph. Let* $(\mathbf{m}, \Lambda, \delta)$ *be the categorical probability triple where* $\mathbf{m}$ *is the finite discrete category of cardinality* m*, with objects* $\{x_1, \ldots, x_m\}$*,* $\Lambda(x_i) = \lambda_i$*, and* δ *is again the Kronecker delta. Then,*

$$\mathcal{H}\left(\bigoplus_{i=1}^{m} \lambda_i(\mathbf{A}_i, p_i, \phi_i)\right) = \sum_{i=1}^{m} \lambda_i \mathcal{H}(\mathbf{A}_i, p_i, \phi_i) + \mathcal{H}(\mathbf{m}, \Lambda, \delta).$$

Proof. Similar to the proof of Proposition 3, observe that $Z_{\bigoplus_{i=1}^{m} \phi_i} = \bigoplus_{i=1}^{m} Z_{\phi_i}$ and $(\bigoplus_{i=1}^{m} Z_{\phi_i})(\bigoplus_{i=1}^{m} \lambda_i p_i) = \bigoplus_{i=1}^{m} Z_{\phi_i}(\lambda_i p_i)$.

Proposition 5. *The entropy functional* $\mathcal{H}$ *is continuous.*

Proof. Let $\{(\mathbf{A}_k, p_k, \phi_k)\}_{k \in \mathbb{N}}$ be a sequence of categorical probabilistic triples converging to the triple $(\mathbf{A}_\infty, p_\infty, \phi_\infty)$. It follows that, for any $a \in \mathrm{Ob}\,\mathbf{A}$, the sequence $f_k(a) := \sum_{b \in \mathrm{Ob}(\mathbf{A})} p_k(b)\phi_k(a,b)$ converges pointwise to f_∞ as $k \to \infty$. Since we may rewrite

$$\mathcal{H}(\mathbf{A}_k, p_k, \phi_k) = - \sum_{a \in \mathrm{Ob}(\mathbf{A})} p_k(a) \ln(f_k(a))$$

for $k = 0, 1, 2, 3, \ldots, \infty$, it suffices to show that $\{p_k(a) \ln(f_k(a))\}_{k \in \mathbb{N}}$ converges to $p_\infty(a) \ln(f_\infty(a))$ for each $a \in \mathrm{Ob}(\mathbf{A})$.

Fix $a \in \mathrm{Ob}(\mathbf{A})$. Assume first that $p_\infty(a) > 0$. Then for sufficiently large k, $p_k(a) > 0$, and hence $f_k(a) \geq p_k(a)\phi_k(a,a) > 0$ and $f_\infty(a) \geq p_\infty(a)\phi_\infty(a,a) > 0$. By the continuity of $\ln(x)$, we then have that $\ln(f_k(a))$ converges to $\ln(f_\infty(a))$, hence $\lim_{k \to \infty} p_k(a) \ln(f_k(a)) = p_\infty(a) \ln(f_\infty(a))$.

Assume now that $p_\infty(a) = 0$. If $f_\infty(a) > 0$, then $f_k(a) > 0$ for sufficiently large k, so we may again use the continuity of $\ln(x)$. If $f_\infty(a) = 0$, pick $1 > \epsilon, \epsilon_0 > 0$ so that there exists $N \in \mathbb{N}$ such that $f_k(a) < \epsilon$ and $\phi_k(a,a) > \epsilon_0$ for all $k > N$. Then,

$$1 > \epsilon > f_k(a) \geq p_k(a)\phi_k(a,a) \geq p_k(a)\epsilon_0.$$

We deduce from this that

$$0 \geq p_k(a) \ln(f_k(a)) \geq p_k(a) \ln(p_k(a)) + p_k(a) \ln(\epsilon_0).$$

Using that $\lim_{x \to 0^+} x \ln(x) = 0$, we conclude that $\lim_{k \to \infty} p_k(a) \ln(f_k(a)) = 0 = p_\infty(a) \ln(f_\infty(a))$. We thus have that $\mathcal{H}$ is continuous.

We take a brief detour to recall a definition of magnitude for a matrix.

Definition 2. *Let M be a matrix in $Mat_{n \times n}(\mathbb{R})$. Denote by $1_n \in Mat_{n \times 1}(\mathbb{R})$ the column vector of ones. A matricial weighting of M is a (column) vector $w \in Mat_{n \times 1}(\mathbb{R})$ such that $Mw = 1_n$ Similarly, a matricial coweighting of M is a vector $\bar{w} \in Mat_{n \times 1}(\mathbb{R})$ such that $\bar{w}^T M = 1_n$.*

If M has both a weighting w and a coweighting $\bar{w}$, then we say that M has magnitude, with the magnitude of M given by

$$\|M\| := \sum_{i=1}^{n} w_i = \sum_{i=1}^{n} \bar{w}_i. \tag{15}$$

In fact, it follows easily from the definitions that if M has both a weighting and a coweighting the sums in (15) are equal.

Proposition 6. *Let $(\mathbf{A}, p, \phi)$ be a probabilistic category. Suppose Z_ϕ has magnitude (i.e. has a weighting and a coweighting). If Z_ϕ has a nonnegative weighting w, then $u = w/\|Z_\phi\|$ is a probability distribution and $\mathcal{H}(\mathbf{A}, u, \phi) = \ln \|Z_\phi\|$.*

Proof. Remark that if the weighting is nonnegative then necessarily $\|Z_\phi\| > 0$. Additionally,

$$\mathcal{H}(\mathbf{A}, u, \phi) = -\sum_{i} \frac{w_i}{\|Z_\phi\|} \ln \left(\frac{1}{\|Z_\phi\|} \right). \tag{16}$$

Specialized to the case $\phi = \zeta$, Proposition 6 tells us that if $\mathbf{A}$ has magnitude, and the category (equivalently: the matrix Z representing ζ) has a positive weighting w, then $u = w/\chi(\mathbf{A})$ satisfies $\mathcal{H}(\mathbf{A}, u, \zeta) = \ln \chi(\mathbf{A})$. Therefore the categorical entropy generalizes both Shannon entropy and the logarithm of the categorical magnitude, as we wanted.

Remark 2. A known result on the maximization of diversity [7, Thm. 2] can be restated as follows: the supremum of $\mathcal{H}(\mathbf{A}, p, \phi)$ over all probability distributions p on $Ob(\mathbf{A})$ equals $\ln(\max_B \|Z_B\|)$, where the maximum is taken over all subsets B of $\{1, ..., n\}$ such that the submatrix $Z_B := ((Z_\phi)_{i,j})_{i,j \in B}$ of Z_ϕ has a nonnegative weighting.

Remark 3. One might generalize the definitions, allowing p to be a *signed probability*: a function $p : Ob(\mathbf{A}) \to \mathbb{R}$ such that $\sum_{a \in Ob(\mathbf{A})} p(a) = 1$. In this case, we define $\mathcal{H}(\mathbf{A}, p, \phi)$ as $-\sum_{i=1}^{N} p_i \ln |(Z_\phi p)_i|$. Then the previous proposition generalizes as follows: if $\|Z_\phi\| \neq 0$ and w is any weighting, the vector $u = w/\|Z_\phi\|$ is a signed probability and $\mathcal{H}(\mathbf{A}, u, \phi) = \ln |\|Z_\phi\||$.

4   Towards a Characterization of the Categorical Entropy

Let $\mathbf{R}_+$ be the additive semigroup of non-negative real numbers seen as a one-object category, that is, $Ob(\mathbf{R}_+) = \{*\}$ and $Hom(*, *) = [0, \infty)$, with $+$ as composition of arrows. A functor $F : \mathbf{ProbFinCat} \to \mathbf{R}_+$ is:

– **convex-linear** if for all morphisms f, g and all scalars $\lambda \in [0,1]$,

$$F(\lambda f \oplus (1-\lambda)q) = \lambda F(f) + (1-\lambda)F(g).$$

– **continuous** if $F(f_n) \to F(f)$ whenever f_n is a sequence of morphisms converging to f (see Sect. 2).

Via the embedding described at the end of Sect. 2, we obtain similar definitions for functors $F : \mathbf{FinProb} \to \mathbf{R}_+$, which correspond to those introduced by Baez, Fritz, and Leinster in [3, p. 4]. They showed there that if a functor $F : \mathbf{FinProb} \to \mathbf{R}_+$ is convex linear and continuous, then there exists a constant $c \geq 0$ such that each arrow $f : (A,p) \to (B,q)$ is mapped to $F(f) = c(H(p) - H(q))$. This is an algebraic characterization of the *information loss* $H(p) - H(q)$ given by Shannon entropy.

In our setting, the "loss" functor $L : \mathbf{ProbFinCat} \to \mathbf{R}_+$ that maps $f : (\mathbf{A}, p, \theta) \to (\mathbf{B}, q, \psi)$ to $L(f) = \mathcal{H}(\mathbf{A}, p, \theta) - \mathcal{H}(\mathbf{B}, q, \psi)$ is continuous and convex linear: this can be easily seen from Propositions 4 and 5 above. However, are all the convex-linear continuous functors $G : \mathbf{ProbFinCat} \to \mathbf{R}_+$ positive multiples of L?

Baez, Fritz, and Leinster's proof uses crucially that $\mathbf{FinProb}$ has a terminal object. $\mathbf{ProbFinCat}$ has no terminal object: even a category with a unique object $*$ accepts an arbitrary value of $\theta(*, *)$. To fix this, we introduce the full subcategory $\mathbf{TransFinCat}$ of $\mathbf{ProbFinCat}$ given by triples $(\mathbf{A}, p, \phi)$ such that ϕ is a transition kernel; this means that for all $a \in \mathrm{Ob}(\mathbf{A})$, the function $b \mapsto \phi(b, a)$ is a probability measure. It is easy to verify that if $F : (\mathbf{A}, p, \phi) \to (\mathbf{B}, q, \theta)$ is a morphism in $\mathbf{ProbFinCat}$ and ϕ is a transition kernel, then θ is a transition kernel too.

Note that the embedding of $\mathbf{FinProb}$ into $\mathbf{ProbFinCat}$ described previously induces a fully faithful embedding of $\mathbf{FinProb}$ into $\mathbf{TransFinCat}$; moreover, when $\mathbf{A}$ is a discrete category, the only choice of transition kernel is given by the Kronecker delta δ.

If p represents the probability of each object at an initial time, we can regard $\hat{p} : \mathrm{Ob}(\mathbf{A}) \to [0,1]$, $a \mapsto \sum_{b \in \mathrm{Ob}(\mathbf{A})} \phi(a,b)p(b)$ as the probability of each object after one transition has happened. Quite remarkably, there is a compatibility between the push-forward of probabilities (5) and the push-forward of transition kernels (6) in the following sense: if $F : (\mathbf{A}, p, \phi) \to (\mathbf{B}, q, \theta)$ is a morphism in $\mathbf{TransFinCat}$, then $\hat{q} = F_* \hat{p}$, because

$$\hat{q}(b) := \sum_{b' \in \mathrm{Ob}(\mathbf{B})} \theta(b, b')q(b') \tag{17}$$

$$= \sum_{b' \in \mathrm{Ob}(\mathbf{B})} \left(\frac{\sum_{a' \in F^{-1}(b')} p(a') \sum_{a \in F^{-1}(b)} \phi(a, a')}{q(b')} \right) q(b') \tag{18}$$

$$= \sum_{a \in F^{-1}(b)} \sum_{a' \in \mathrm{Ob}(\mathbf{A})} \phi(a, a')p(a') = \sum_{a \in F^{-1}(b)} \hat{p}(a). \tag{19}$$

In turn, the entropy (13) can be rewritten as

$$\mathcal{H}(\mathbf{A}, p, \phi) = - \sum_{a \in \text{Ob } \mathbf{A}} p(a) \ln \hat{p}(a) = H(p) - D(\hat{p}|p), \qquad (20)$$

where $H(p)$ is Shannon entropy (1), and $D(\hat{p}|p) = \sum_{a \in \text{Ob } \mathbf{A}} p(a) \ln(\hat{p}(a)/p(a))$ is the Kullback-Leibler divergence

Let us return to the characterization problem described above. Let G : $\mathbf{TransFinCat} \to \mathbf{R}_+$ be a functor that is convex linear, continuous, and possibly subject to other constraints. Let $\top$ be the triple (one-point discrete category, trivial probability, trivial transition kernel), which is the terminal object of $\mathbf{TransFinCat}$. Denote by $I_G(\mathbf{A}, p, \phi)$ the image under G of the unique morphism $(\mathbf{A}, p, \phi) \to \top$. The functoriality of G implies that, for any morphism $f : (\mathbf{A}, p, \phi) \to (\mathbf{B}, q, \theta)$, one has $G(f) + I_G(\mathbf{B}, q, \theta) = I_G(\mathbf{A}, p, \phi)$. In particular, $G(\text{id}_{(\mathbf{A}, p, \phi)}) = 0$ for any triple $(\mathbf{A}, p, \phi)$ and $I_G(\top) = 0$. Moreover, if f is invertible, then $G(f) + G(f^{-1}) = 0$, so $G(f) = 0$ and I_G is invariant under isomorphisms. In turn, convex linearity implies that

$$I_G \left(\bigoplus_{i=1}^{m} \lambda_i (\mathbf{A}_i, p_i, \phi_i) \right) = I_G(\mathbf{m}, (\lambda_1, ..., \lambda_m), \delta) + \sum_{i=1}^{m} \lambda_i I_G(\mathbf{A}_i, p_i, \phi_i), \qquad (21)$$

for any vector of probabilities $(\lambda_1, ..., \lambda_m)$ and any $(\mathbf{A}_i, p_i, \phi_i) \in$ Ob($\mathbf{TransFinCat}$), $i = 1, .., m$. This is a system of functional equations reminiscent of those used by Shannon, Khinchin, Fadeev, etc. to characterize the entropy [1,6], see in particular [3, Thm. 6]. It is not clear if a similar result holds for $\mathcal{H}$ in our categorical setting. A fundamental difference is that every finite probability space (A, p) can be expressed nontrivially in many ways as convex combinations $\bigoplus_{i=1}^{m} \lambda_i (A_i, p_i)$ in $\mathbf{FinProb}$, but this is not true for categorical probabilistic triples. Remark also that the function $I_G(\mathbf{A}, p, \phi) = - \sum_{a \in \text{Ob}(\mathbf{A})} p(a) \ln(\sum_{b \in \mathbf{A}} \phi(b, a) p(b))$ defines a continuous, convex-linear functor G. Hence G needs to satisfy some additional properties in order to recover L up to a positive multiple.

References

1. Shannon, C.E.: A mathematical theory of communication. Bell Syst. Tech. J. **27**, 379–423 (1948)
2. Leinster, T.: The Euler characteristic of a category. Documenta Mathematica **13**, 21–49 (2008)
3. Leinster, T., Baez, J.C., Fritz, T.: A characterization of entropy in terms of information loss. Entropy **13**, 1945–1957 (2011)
4. Rota, G.-C.: On the foundations of combinatorial theory I. Theory of Möbius functions. Z. Wahrscheinlichkeitstheorie und Verw. Gebiete **2**, 340–368 (1964)
5. Leinster, T.: Entropy and Diversity: The Axiomatic Approach. Cambridge University Press (2021)
6. Csiszár, I.: Axiomatic characterizations of information measures. Entropy **10**(3), 261–273 (2008)
7. Leinster, T., Meckes, M.W.: Maximizing diversity in biology and beyond. Entropy **18**(3), 88 (2016)

Information Theory and Statistics

A Historical Perspective
on Schützenberger-Pinsker Inequalities

Olivier Rioul[(✉)]

LTCI, Télécom Paris, Institut Polytechnique de Paris, Palaiseau, France
olivier.rioul@telecom-paris.fr
https://perso.telecom-paristech.fr/rioul/

Abstract. This paper presents a tutorial overview of so-called Pinsker inequalities which establish a precise relationship between information and statistics, and whose use have become ubiquitous in many information theoretic applications. According to Stigler's law of eponymy, no scientific discovery is named after its original discoverer. Pinsker's inequality is no exception: Years before the publication of Pinsker's book in 1960, the French medical doctor, geneticist, epidemiologist, and mathematician Marcel-Paul (Marco) Schützenberger, in his 1953 doctoral thesis, not only proved what is now called Pinsker's inequality (with the optimal constant that Pinsker himself did not establish) but also the optimal second-order improvement, more than a decade before Kullback's derivation of the same inequality. We review Schützenberger and Pinsker contributions as well as those of Volkonskii & Rozanov, Sakaguchi, McKean, Csiszár, Kullback, Kemperman, Vajda, Bretagnolle & Huber, Krafft & Schmitz, Toussaint, Reid & Williamson, Gilardoni, as well as the optimal derivation of Fedotov, Harremoës, & Topsøe.

Keywords: Pinsker inequality · Total variation · Kullback-Leibler divergence · Statistical Distance · Mutual Information · Data processing inequality

1 Introduction

How far is one probability distribution from another? This question finds many different answers in information geometry, statistics, coding and information theory, cryptography, game theory, learning theory, and even biology or social sciences. The common viewpoint is to define a "distance" $\Delta(p, q)$ between probability distributions p and q, which should at least satisfy the basic property that it is *nonnegative* and *vanishes only when the two probability distributions coincide*: $p = q$ in the given statistical manifold [1].

Strictly speaking, distances $\Delta(p, q)$ should also satisfy the two usual requirements of *symmetry* $\Delta(p, q) = \Delta(q, p)$ and *triangle inequality* $\Delta(p, q) + \Delta(q, r) \geq \Delta(p, r)$. In this case the probability distribution space becomes a metric space.

F. Nielsen and F. Barbaresco (Eds.): GSI 2023, LNCS 14071, pp. 291–306, 2023.
https://doi.org/10.1007/978-3-031-38271-0_29

Examples include the Lévy-Prokhorov and the Fortet-Mourier (a.k.a. "Wasser-stein" or Kantorovich-Rubinstein) distances (which metrize the weak convergence or convergence in distribution), the (stronger) Kolmogorov-Smirnov distance (which metrizes the uniform convergence in distribution), the Radon distance (which metrizes the strong convergence), the Jeffreys (a.k.a. Hellinger[1]) distance, and many others[2].

In this paper, we focus on the *total variation distance*, which is one of the strongest among the preceding examples. Arguably, it is also the simplest—as a L^1-norm distance—and the most frequently used in applications, particularly those related to Bayesian inference.

In many information theoretic applications, however, other types of "distances," that do not necessarily satisfy the triangle inequality, are often preferred. Such "distances" are called *divergences* $D(p,q)$. They may not even satisfy the symmetry property: In general, $D(p,q)$ is the divergence *of q from p*, and not "between p and q"[3]. Examples include the Rényi α-divergence, the Bhattacharyya divergence (a variation of the Jeffreys (Hellinger) distance), Lin's "Jensen-Shannon" divergence, the triangular divergence, Pearson's χ^2 divergence, the "Cauchy-Schwarz" divergence, the (more general) Sundaresan divergence, the Itakura-Saito divergence, and many more.

In this paper, we focus on the *Kullback-Leibler divergence*[4], historically the most popular type of divergence which has become ubiquitous in information theory. Two of the reasons of its popularity are its relation to Shannon's entropy (the Kullback-Leibler divergence is also known as the *relative entropy*); and the fact that it tensorizes nicely for products of probability distributions, expressed in terms of the sum of the individual divergences[5] (which give rise to useful chain rule properties).

---

[1] What is generally known as the "Hellinger distance" was in fact introduced by Jeffreys in 1946. The Hellinger integral (1909) is just a general method of integration that can be used to define the Jeffreys distance. The Jeffreys ("Hellinger") distance should not be confused with the "Jeffreys divergence", which was studied by Kullback as a symmetrized Kullback-Leibler divergence (see below).

[2] Some stronger types of convergence can also be metrized, but by distances between *random variables* rather than between distributions. For example, the Ky Fan distance metrizes the convergence in probability.

[3] Evidently, such divergences can always be symmetrized by considering $(D(p,q) + D(q,p))/2$ instead of $D(p,q)$.

[4] Two fairly general classes of divergences are Rényi's f-divergences and the Bregman divergences. Some (square root of) f-divergences also yield genuine distances, like the Jeffreys (Hellinger) distance or the square root of the Jensen-Shannon divergence. It was recently shown that the Kullback-Leibler divergence is the only divergence that is both a f-divergence and a Bregman divergence [13].

[5] Incidentally, this tensorization property implies that the corresponding divergence is unbounded, while, by contrast, most of the above examples of distances (like the total variation distance) are bounded and can always be normalized to assume values between 0 and 1.

A Pinsker-type inequality can be thought of as a general inequality of the form

$$D \geq \varphi(\Delta) \tag{1}$$

relating divergence $D = D(p, q)$ to distance $\Delta = \Delta(p, q)$ and holding for any probability distributions p and q. Here $\varphi(x)$ should assume positive values for $x > 0$ with $\varphi(0) = 0$ in accordance with the property that both $D(p, q)$ and $\Delta(p, q)$ vanish only when $p = q$. Typically φ is also increasing, differentiable, and often convex. Any such Pinsker inequality implies that the topology induced by D is finer[6] than that induced by Δ. Many Pinsker-type inequalities have been established, notably between f-divergences.

In this paper, we present historical considerations of the classical *Pinsker inequality* where D is the Kullback-Leibler divergence and Δ is the total variation distance. This inequality is by far the most renowned inequality of its kind, and finds many applications, e.g., in statistics, information theory, and computer science. Many considerations in this paper, however, equally apply to other types of distances and divergences.

2   Preliminaries

Notations. We assume that all considered probability distributions over a given measurable space $(\Omega, \mathcal{A})$ admit a σ-finite *dominating measure* μ, with respect to which they are absolutely continuous. This can always be assumed when considering finitely many distributions. For example, p and q admit $\mu = (p+q)/2$ as a dominating measure since $p \ll \mu$ and $q \ll \mu$. By the the Radon-Nikodym theorem, they admit *densities* with respect to μ, which we again denote by p and q, respectively. Thus for any event[7] $A \in \mathcal{A}$, $p(A) = \int_A p \, d\mu = \int_A p(x) \, d\mu(x)$, and similarly for q. Two distributions p, q are equal if $p(A) = q(A)$ for all $A \in \mathcal{A}$, that is, $p = q$ μ-a.e. in terms of densities.

If μ is a counting measure, then p is a discrete probability distribution with $\int_A p \, d\mu = \sum_{x \in A} p(x)$; if μ is a Lebesgue measure, then p is a continuous probability distribution with $\int_A p \, d\mu = \int_A p(x) \, dx$. We also consider the important case where p and q are binary (Bernoulli) distributions with parameters again denoted p and q, respectively. Thus for $p \sim \mathcal{B}(p)$ we have $p(x) = p$ or $1 - p$. This ambiguity in notation should be easily resolved from the context.

Distance. The *total variation distance* $\Delta(p, q)$ can be defined in two different ways. The simplest is to set

$$\Delta(p, q) \triangleq \frac{1}{2} \int |p - q| \, d\mu, \tag{2}$$

---

[6] If, in addition, a *reverse* Pinsker inequality $\Delta \geq \psi(D)$ holds, then the associated topologies are equivalent.

[7] This is an overload in notations and one should not confuse $p(\{x\})$ with $p(x)$.

that is, half the $L^1(\mu)$-norm of the difference of densities. It is important to note that this definition does *not* depend on the choice of the dominating measure μ. Indeed, if $\mu \ll \mu'$, with density $\frac{d\mu}{d\mu'} = f$, then the densities w.r.t. μ' become $p' = pf$ and $q' = qf$ so that $\int |p' - q'| \, d\mu' = \int |p - q| \, d\mu$.

That Δ is a distance (metric) is obvious from this definition. Since $\int (p - q) \, d\mu = 0$, we can also write $\Delta(p, q) = \int (p - q)^+ \, d\mu = \int (p - q)^- \, d\mu$ (positive and negative parts) or $\Delta(p, q) = \int p \vee q \, d\mu - 1 = 1 - \int p \wedge q \, d\mu$ in terms of the maximum and minimum. The normalization factor $1/2$ ensures that $0 \leq \Delta(p, q) \leq 1$, with maximum value $\Delta(p, q) = 1 - \int p \wedge q \, d\mu = 1$ if and only if $p \wedge q = 0$ μ-a.e., that is, p and q have "non-overlapping" supports. Note that the total variation distance between *binary* distributions $\mathcal{B}(p)$ and $\mathcal{B}(q)$ is simply

$$\delta(p, q) = |p - q|. \tag{3}$$

The alternate definition of the total variation distance is to proceed from the discrete case to the general case as follows. One can define

$$\Delta(p, q) \triangleq \frac{1}{2} \sup \sum_i |p(A_i) - q(A_i)|, \tag{4}$$

where the supremum is taken all *partitions* of Ω into a countable number of (disjoint) $A_i \in \mathcal{A}$. When $\Omega \subset \mathbb{R}$, this supremum can simply be taken over partitions of *intervals* A_i, and (apart from the factor $1/2$) this exactly corresponds to the usual notion of *total variation* of the corresponding cumulative distribution f of the signed measure $p - q$. This is a well-known measure of the one-dimensional arclength of the curve $y = f(x)$, introduced by Jordan in the 19th century, and justifies the name "total variation" given to Δ.

That the two definitions (2) and (4) coincide can easily be seen as follows. First, by the triangular inequality, the sum $\sum_i |p(A_i) - q(A_i)|$ in (4) can only increase by subpartitioning, hence (4) can be seen as a limit for finer and finer partitions. Second, consider the subpartition $A_i^+ = A_i \cap A^+$, $A_i^- = A_i \cap A^-$, where, say, $A^+ = \{p > q\}$ and $A^- = \{p \leq q\}$. Then the corresponding sum already equals $\sum_i (p-q)(A_i^+) + (q-p)(A_i^-) = (p-q)(\sum_i A_i^+) + (q-p)(\sum_i A_i^-) = (p-q)(A^+) + (q-p)(A^-) = \int (p-q)^+ + (p-q)^- \, d\mu = \int |p - q| \, d\mu$.

As a side result, the supremum in (4) is attained for binary partitions $\{A^+, A^-\}$ of the form $\{A, A^C\}$, so that $\Delta(p, q) = \frac{1}{2} \sup (|p(A) - q(A)| + |p(A^C) - q(A^C)|)$, that is,

$$\Delta(p, q) = \sup_A |p(A) - q(A)| \tag{5}$$

(without the $1/2$ factor). This important property ensures that *a sufficiently small value of $\Delta(p, q)$ implies that no statistical test can effectively distinguish between the two distributions p and q*. In fact, given some observation X following either p (null hypothesis H_0) or q (alternate hypothesis H_1), such a statistical test takes the form "is $X \in A$?" (then accept H_0, otherwise reject it). Then since $|p(X \in A) - q(X \in A)| \leq \Delta$ is small, type-I or type-II errors have total probability $p(X \notin A) + q(X \in A) \geq 1 - \Delta$. Thus in this sense the two hypotheses p and q

are Δ-undistinguishable. For the case of independent observations we are faced with the evaluation of the total variation distance for products of distributions. In this situation, Pinsker's inequality is particularly useful since it relates it to the Kullback-Leibler divergence which nicely tensorizes, thus allowing a simple evaluation.

Divergence. The Kullback-Leibler divergence [19], also known as statistical divergence, or simply divergence, can similarly be defined in two different ways. One can define

$$D(p\|q) \triangleq \int p \log \frac{p}{q} \, d\mu, \tag{6}$$

where since $x \log x \geq -(\log e)/e$, the negative part of the integral is finite[8]. Therefore, this integral is always meaningful and can be finite, or infinite $= +\infty$. Again note that this definition does *not* depend on the choice of the dominating measure μ. Indeed, if $\mu \ll \mu'$, with density $\frac{d\mu}{d\mu'} = f$, then the densities w.r.t. μ' become $p' = pf$ and $q' = qf$ so that $\int p' \log \frac{p'}{q'} \, d\mu' = \int p \log \frac{p}{q} \, d\mu$.

By Jensen's inequality applied to the convex function $x \log x$, $D(p\|q)$ is non-negative and vanishes if only if the two distributions p and q coincide. For products of distributions $p = \bigotimes_i p_i$, $q = \bigotimes_i q_i$, it is easy to establish the useful tensorization property $D(p\|q) = \sum_i D(p_i\|q_i)$. The divergence between binary distributions $\mathcal{B}(p)$ and $\mathcal{B}(q)$ is simply

$$d(p\|q) = p \log \frac{p}{q} + (1 - p) \log \frac{1 - p}{1 - q}. \tag{7}$$

The double bar notation '$\|$' (instead of a comma) is universally used but may look exotic. Kullback and Leibler did not originate this notation in their seminal paper [19]. They rather used $I(1 : 2)$ for alternatives p_1, p_2 with a semi colon to indicate non commutativity. Later the notation $I(P \mid Q)$ was used but this collides with the notation '$\mid$' for conditional distributions. The first occurence of the double bar notation I could find was by Rényi in the form $I(P\|Q)$ in the same paper that introduced Rényi entropies and divergences [27]. This notation was soon adopted by researchers of the Hungarian school of information theory, notably Csiszár (see, e.g., [5–7]).

The alternate definition of divergence is again to proceed from the discrete case to the general case as follows. One can define

$$D(p\|q) \triangleq \sup \sum_i p(A_i) \log \frac{p(A_i)}{q(A_i)} \tag{8}$$

where the supremum is again taken all *partitions* of Ω into a countable number of (disjoint) $A_i \in \mathcal{A}$. By the *log-sum inequality*, the sum $\sum_i p(A_i) \log \frac{p(A_i)}{q(A_i)}$ in (8) can only increase by subpartitioning, hence (8) can be seen as a limit for finer and finer partitions. Also, when $\Omega \subset \mathbb{R}$ or R^d, this supremum can simply be taken

---

[8] The logarithm (log) is considered throughout this paper in *any* base.

over partitions of *intervals* A_i (this is the content of Dobrushin's theorem [24, § 2]). That the two definitions (6) and (8) coincide (in particular when (8) is finite, which implies $p \ll q$) is the content of a theorem by Gel'fand & Yaglom [10] and Perez [23].

Statistical Distance and Mutual Information. How does some observation Y affect the probability distribution of some random variable X? This can be measured as the distance or divergence of X from X given Y, averaged over the observation Y. Using the total variation distance, one obtains the notion of *statistical distance* between the two random variables:

$$\Delta(X;Y) = \mathbb{E}_y \, \Delta(p_{X|y}, p_X) = \Delta(p_{XY}, p_X \otimes p_Y), \tag{9}$$

and using the statistical divergence, one obtains the celebrated *mutual information*[9]:

$$I(X;Y) = \mathbb{E}_y \, D(p_{X|y} \| p_X) = D(p_{XY}, \| p_X \otimes p_Y) \tag{10}$$

introduced by Fano [8], based on Shannon's works. From these definitions, it follows that any Pinsker inequality (1) can also be interpreted as an inequality relating statistical distance $\Delta = \Delta(X;Y)$ to mutual information $I = I(X;Y)$:

$$I \geq \varphi(\Delta) \tag{11}$$

for any two random variables X and Y, with the same φ as in (1). In particular, in terms of sequences of random variables, $I(X_n;Y_n) \to 0$ implies $\Delta(X_n;Y_n) \to 0$, a fact first proved by Pinsker [24, §2.3].

Binary Reduction of Pinsker's Inequality. A straightforward observation, that greatly simplifies the derivation of Pinsker inequalities, follows from the alternative definitions (4) and (8). We have seen that the supremum in (4) is attained for binary partitions of the form $\{A, A^C\}$. On the other hand, the supremum in (8) is obviously greater then that for such binary partitions. Therefore, any Pinsker inequality (1) is equivalent to the inequality expressed in term of binary distributions (3), (7):

$$d \geq \varphi(\delta) \tag{12}$$

relating binary divergence $d = d(p\|q)$ to binary distance $\delta = |p - q|$ and holding for any parameters $p, q \in [0, 1]$. Thus, the binary case, which writes

$$p \log \frac{p}{q} + (1 - p) \log \frac{1 - p}{1 - q} \geq \varphi(|p - q|) \tag{13}$$

is equivalent to the general case, but is naturally easier to prove. This binary reduction principle was first used by Csiszár [6] but as a consequence of a more general *data processing inequality* for any transition probability kernel (whose full generality is not needed here).

---

[9] Here, the semicolon ";" is often used to separate the variables. The comma "," rather denotes joint variables and has higher precedence than ";" as in $I(X;Y,Z)$ which denotes the mutual information between X and (Y, Z).

Comparison of Pinsker Inequalities. The following is sometimes useful to compare two different Pinsker inequalities (1) of the form $D \geq \varphi_1(\Delta)$ and $D \geq \varphi_2(\Delta)$ where both φ_1 and φ_2 are nonnegative differentiable functions such that $\varphi_1(0) = \varphi_2(0) = 0$. By comparison of derivatives, $\varphi_1' \geq \varphi_2'$ implies that $D \geq \varphi_1(\Delta) \geq \varphi_2(\Delta)$. This comparison principle can be stated as follows: *lower derivative φ' implies weaker Pinsker inequality.*

3 Pinsker and Other Authors in the 1960s

It is generally said that Pinsker, in his 1960 book [24], proved the classical Pinsker inequality in the form

$$D \geq c \cdot \log e \cdot \Delta^2 \tag{14}$$

with a suboptimal constant c, and that the optimal (maximal) constant $c = 2$ was later found independently by Kullback [20], Csiszár [6] and Kemperman [16], hence the alternative name Kullback-Csiszár-Kemperman inequality.

In fact, Pinsker did not explicitly state Pinsker's inequality in this form, not even in the general form (1) for some other function φ. First of all, he only investigated mutual information vs. statistical distance with $p = p_{X,Y}$ and $q = p_X \otimes p_Y$—yet his results do easily carry over to the general case of arbitrary distributions p and q. More important, he actually showed two separate inequalities[10] $\Delta \leq \int p |\log \frac{p}{q}| \, d\mu \leq D + 10\sqrt{D}$ with a quite involved proof for the second inequality[11] [24, pp. 14–15]. As noticed by Verdú [34], since one can always assume $\Delta \leq D + 10\sqrt{D} \leq 1$ (otherwise the inequality is vacuous), then two Pinsker inequalities imply $\Delta^2 \leq (D + 10\sqrt{D})^2 = D(D + 20\sqrt{D}) + 100D \leq 102D$ which indeed gives (14) with the suboptimal constant $c = \frac{1}{102}$. But this was nowhere mentioned in Pinsker's book [24].

The *first explicit occurrence of a Pinsker inequality of the general form* (1) occurs even *before* the publication of Pinsker's book, by Volkonskii and Rozanov [35, Eq. (V)] in 1959. They gave a simple proof of the following inequality:

$$D \geq 2 \log e \cdot \Delta - \log(1 + 2\Delta). \tag{15}$$

It is easily checked, from the second-order Taylor expansion of $\varphi(x) = 2 \log e \cdot x - \log(1 + 2x)$, that this inequality is strictly weaker than the classical Pinsker inequality (14) with the optimal constant $c = 2$, although both are asymptotically optimal near $D = \Delta = 0$.

The *first explicit occurrence of a Pinsker inequality of the classical form* (14) appeared as an exercise in Sakaguchi's 1964 book [28, pp. 32–33]. He proved $D \geq H^2 \log e \geq \Delta^2 \log e$ where H is the Hellinger distance, which gives (14) with the suboptimal constant $c = 1$. Unfortunately, Sakaguchi's book remained unpublished.

---

[10] In nats (natural units), that is, when the logarithm is taken to base e.

[11] Decades later, Barron [2, Cor. p. 339] proved this second inequality (with the better constant $\sqrt{2}$ instead of 10) as an easy consequence of Pinsker's inequality itself with the optimal constant $c = 2$.

The *first published occurence of a Pinsker inequality of the classical form* (14) was by McKean [22, § 9a)] in 1966, who was motivated by a problem in physics related to Boltzmann's H-theorem. He proved (14) with the suboptimal constant $c = \frac{1}{e}$ (worse than Sakaguchi's) under the (unnecessary) assumption that q is Gaussian.

The *first mention of the classical Pinsker inequality* (14) *with the optimal constant c = 2* was by Csiszár [5], in a 1966 manuscript received just one month after McKean's. In his 1966 paper, however, Csiszár only proved (14) with the suboptimal constant $c = \frac{1}{4}$ [5, Eq. 13], which is worse than McKean's. But he also acknowledged the preceding result of Sakaguchi (with the better constant $c = 1$) and stated (without proof) that the best constant is $c = 2$. He also mentioned the possible generalization to f-divergences. On this occasion he credited Pinsker for having found an inequality of the type (14) (which as we have seen was only implicit).

The *first published proof of the classical Pinsker inequality* (14) *with the optimal constant c = 2* was again by Csiszár one year later [6, Thm. 4.1] using binary reduction. His proof can be written as a one-line proof as follows:

$$d(p\|q) = \underbrace{d(p\|p)}_{=0} + \int_p^q \frac{\partial d(p\|r)}{\partial r}\, dr = \int_p^q \frac{r - p}{r(1 - r)}\, dr \geq 4 \int_p^q (r - p)\, dr = 2(p - q)^2,$$
(16)

where we used natural logarithms and the inequality $r(1 - r) \leq \frac{1}{4}$ for $r \in [0, 1]$. That $c = 2$ is not improvable follows from the expansion $d(p\|q) = 2(p - q)^2 + o((p - q)^2)$, which also shows that this inequality (like the Volkonskii-Rozanov inequality (15)) is asymptotically optimal near $D = \Delta = 0$.

In a note added in proof, however, Csiszár mentions an earlier independent derivation of Kullback, published in the same year 1967 in [20], with an improved inequality of the form $D \geq 2 \log e \cdot \Delta^2 + \frac{4}{3} \log e \cdot \Delta^4$. In his correspondance, Kullback acknowledged the preceding result of Volkonskii and Rozanov. Unfortunately, as Vajda noticed [33] in 1970, the constant $\frac{4}{3}$ is wrong and should be corrected as $\frac{4}{9}$ [21] (see explanation in the next section).

Finally, in an 1968 Canadian symposium presentation [15]—later published as a journal paper [16] in 1969, Kemperman, apparently unaware of the 1967 papers by Csiszár and Kullback, again derived the classical Pinsker inequality with optimal constant $c = 2$. His ad-hoc proof (repeated in the renowned textbook [32]) is based in the inequality $\frac{4+2x}{3}(x \log x - x + 1) \geq (x - 1)^2$, which is much less satisfying than the one-line proof (16).

To acknowledge all the above contributions, it is perhaps permissible to rename Pinsker's inequality as the Volkonskii-Rozanov-Sakaguchi-McKean-Csiszár-Kullback-Kemperman inequality. However, this would unfairly obliterate the pioneer contribution of Schützenberger, as we now show.

4   Schützenberger's Contribution (1953)

Seven years before the publication of Pinsker's book, the French medical doctor, geneticist, epidemiologist, and mathematician Marcel-Paul (Marco) Schützenberger, in his 1953 doctoral thesis [29] (see Fig. 3), proved:

$$D \geq 2\log e \cdot \Delta^2 + \frac{4}{9}\log e \cdot \Delta^4 \tag{17}$$

Not only does this contain the classical Pinsker inequality (14) with the optimal constant $c = 2$, but also the second-order improvement, with the (correct) optimal constant $\frac{4}{9}$ for the second-order term, seventeen years before Kullback! Admittedly, Schützenberger only considered the binary case, but due to the binary reduction principle, this does not entail any loss of generality.

Dans le cas dichotomique, on a l'inégalité suivante qui semble nouvelle. Ecrivons :

$D = p\ (\theta_0) - p\ (\theta_i) = q(\theta_i) - q\ (\theta_0)$

$$W \geq 2D^2 \tfrac{4}{9}D^4$$

Posons en effet 2 p (θ_0) = 1-x et 2 p(θ_i) = 1-y après avoir choisi p de telle sorte que x soit positif.

On peut développer W en série de puissance de x et de y :

2 W = (1-x) Log(1+x)/(1-y) + (1+x) Log (1+x)(1+y).

On trouve :

$W = \sum_{i=1}^{\infty} (4\ i^4 - 2i\ )-1\ (x^{2i}-2ixy^{2i-1} + (2\ i-1)y^{2i})$

Tous les termes sont positifs car le polynome $t^{2i}-2it+2i-1$ a un unique extremum pour t = 1 et prend en ce point la valeur 0.

Bien plus :

$x^{2i} - 2ixy^{2i-1} + (2i-1)y^{2i} = 4\ D^2\,(x^{2i-2} + 2x^{2i-3}\,y + 3x^{2i-4}y^2 + \ldots$

$\ldots + (2\ i-1)y^{2i-2}\ )$

Par conséquent W est plus grand que la somme des deux premiers termes de son développement qui sont :

4 D²/2 et 4 D²/12 (x⁴+2x y+ 3 y²) et la valeur de ce dernier polynome étant supérieure pour D fixe à D²/3 on trouve bien le résultat.

Fig. 1. Left: Pinsker before Pinsker: In Schützenberger's notation, W is for Wald's information, which is Kullback-Leibler divergence, and $D = p - q$. There is a typo at the end: minimizing $x^2 + 2xy + 3y^2$ for fixed $2D = y - x$ is said to give $\frac{D^2}{3}$ instead of the correct $\frac{4D^2}{3}$. Right: Marcel-Paul (Marco) Schützenberger at his first marriage, in London, Aug. 30th, 1948.

In fact, leaving aside the use of binary reduction, Kullback's derivation [20] is just a mention of Schützenberger's inequality with the wrong constant $\frac{4}{3}$ instead of $\frac{4}{9}$. However, Vajda [33] asserts that the wrong constant comes from Schützenberger's manuscript itself, and that it was corrected in 1969 by Krafft [17]. In fact, Krafft does not refer to Schützenberger's thesis but rather to a 1966 paper by Kambo and Kotz [14] which contains a verbatim copy of Schützenberger's derivation (with the wrong constant and without citing the initial reference). While the correct constant $\frac{4}{9}$ does appear in the publicly available manuscript of Schützenberger (Fig. 3), it is apparent from the zooming in of Fig. 2 that the denominator was in fact carefully corrected by hand from a "3" to a "9". It is likely that the correction in Schützenberger's manuscript was made after 1970, when the error was discovered.

(a) Denominator in the fraction 4/9, zoomed in.  (b) Digits from the same manuscript.

Fig. 2. Schützenberger's correction from "3" to "9": the correction clearly follows the shape of a "3" in the original manuscript.

Nevertheless, Schützenberger's derivation is correct and gives the best constants 2 and 4/9 in (17) as an easy consequence of his identity

$$d = \sum_{k\geq1} \frac{x^{2k} - 2kxy^{2k-1} + (2k-1)y^{2k}}{2k(2k-1)} = 2\delta^2 \sum_{k\geq1} \frac{x^{2k-2} + 2x^{2k-3}y + \cdots + (2k-1)y^{2k-2}}{k(2k-1)}$$

(18)

where $x = 1 - 2p$ and $y = 1 - 2q$ (see Fig. 1). In 1969, Krafft and Schmitz [18] extended Schützenberger's derivation by one additional term in $\frac{2}{9}\log e \cdot \Delta^6$, which was converted into a Pinsker inequality in 1975 by Toussaint [31]. But, in fact, the constant $\frac{2}{9}$ is not optimal; the optimal constant $\frac{32}{135}$ was found in 2001 by Topsøe [30]. Topsøe also derived the optimal constant for the additional term $\frac{7072}{42525}\log e \cdot \Delta^8$, whose proof is given in [9]. It is quite remarkable that all of such derivations are crucially based on the original Schützenberger's identity (18).

5   More Recent Improvements (1970s to 2000s)

So far, all derived Schützenberger-Pinsker inequalities are only useful when D and Δ are small, and become uninteresting as D or Δ increases. For example, the classical Pinsker inequality (14) with optimal constant $c = 2$ become vacuous as soon as $D > 2\log e$ (since $\Delta \leq 1$). Any improved Pinsker inequality of the form (1) should be such that $\varphi(1) = +\infty$ because $\Delta(p,q) = 1$ (non overlapping supports) implies $D(p\|q) = +\infty$.

The first Pinsker inequality of this kind is due to Vajda in his 1970 paper [33]. He explicitly stated the problem of finding the optimal Pinsker inequality and proved

$$D \geq \log \frac{1+\Delta}{1-\Delta} - 2\log e \cdot \frac{\Delta}{1+\Delta}.$$

(19)

where the lower bound becomes infinite as Δ approaches 1, as it should. This inequality is asymptotically optimal near $D = \Delta = 0$ since $\log\frac{1+\Delta}{1-\Delta} - 2\log e \cdot \frac{\Delta}{1+\Delta} = 2\log e \cdot \Delta^2 + o(\Delta^2)$.

In a 1978 French seminar, Bretagnolle and Huber [3,4] derived yet another Pinsker inequality similar to Vajda's (where the lower bound becomes infinite for $\Delta = 1$) but with a simpler expression:

$$D \geq \log \frac{1}{1-\Delta^2}.$$

(20)

By the comparison principle, for natural logarithms and $0 < \Delta < 1$, $\frac{d}{d\Delta} \log \frac{1}{1-\Delta^2} = \frac{2\Delta}{1-\Delta^2} < \frac{4\Delta}{(1+\Delta)(1-\Delta^2)} = \frac{d}{d\Delta}\left(\log \frac{1+\Delta}{1-\Delta} - \frac{2\Delta}{1+\Delta}\right)$ always, since $1 + \Delta < 2$. Therefore, the Bretagnolle-Huber inequality (20) is strictly *weaker* than Vajda's inequality (19). Moreover, it is not asymptotically optimal near $D = \Delta = 0$ since $\log \frac{1}{1-\Delta^2} \sim \log e \cdot \Delta^2$ is worse than the asymptotically optimal $2 \log e \cdot \Delta^2$. However, a nice property of the Bretagnolle-Huber inequality is that it can be inverted in closed form. In fact the authors expressed it as[12] $\Delta \leq \sqrt{1 - \exp(-D)}$.

The Bretagnolle-Huber inequality was popularized by Tsybakov in his 2009 book on nonparametric estimation [32, Eq. (2.25)], but with a different form $\Delta \leq 1 - \frac{1}{2}\exp(-D)$, or $D \geq \log \frac{1}{2(1-\Delta)}$, which is strictly *weaker* than the original, since $1 - \Delta^2 = (1 - \Delta)(1 + \Delta) < 2(1 - \Delta)$ for $0 < \Delta < 1$.

Today and to my knowledge, the best known explicit Pinsker inequality of this kind is

$$D \geq \log \frac{1}{1 - \Delta} - (1 - \Delta) \log(1 + \Delta). \tag{21}$$

derived by Gilardoni in 2008 [11] (see also [12]). Gilardoni's proof is based on considerations on symmetrized f-divergences. A simple proof is as follows:

Proof. One can always assume that $\delta = p - q > 0$, where $\delta \leq p \leq 1$ and $0 \leq q \leq 1 - \delta$. Then $d(p\|q) = (q+\delta) \log \frac{q+\delta}{q} + (1-q-\delta) \log \frac{1-q-\delta}{1-q} = \left[-q \log \frac{q+\delta}{q} - (1-q-\delta) \log \frac{1-q}{1-q-\delta}\right] + (2q + \delta) \log \frac{q+\delta}{q}$. Since $q + (1-q-\delta) = 1-\delta$ and $-\log$ is convex, the first term inside brackets is $\geq -(1 - \delta) \log(\frac{q+\delta}{1-\delta} + \frac{1-q}{1-\delta}) = (1 - \delta) \log \frac{1-\delta}{1+\delta}$. The second term writes $\delta \frac{(2+x) \log(1+x)}{x}$ where $x = \frac{\delta}{q}$. Now $(2 + x) \log(1 + x)$ is convex for $x \geq 0$ and vanishes for $x = 0$, hence the slope $\frac{(2+x) \log(1+x)}{x}$ is minimal for minimal x, that is, for maximal $q = 1 - \delta$. Therefore, the second term is $\geq (2 - 2\delta + \delta) \log \frac{1}{1-\delta} = (2 - \delta) \log \frac{1}{1-\delta}$. Summing the two lower bounds gives the inequality.   □

Note that Gilardoni's inequality adds the term $\Delta \log(1 + \Delta)$ to the Bretagnolle-Huber lower bound. In fact it uniformly improves Vajda's inequality [11]. In particular, it is also asymptotically optimal near $D = \Delta = 0$, which can easily be checked directly: $\log \frac{1}{1-\Delta} - (1 - \Delta) \log(1 + \Delta) = 2 \log e \cdot \Delta^2 + o(\Delta^2)$. Also by the comparison principle, for natural logarithms and $\Delta > 0$, $\frac{d}{d\Delta}\left(\log \frac{1}{1-\Delta} - (1 - \Delta) \log(1 + \Delta)\right) = \Delta \frac{3-\Delta}{1-\Delta^2} + \log(1 + \Delta) < 3\Delta + \Delta = 4\Delta = \frac{d}{d\Delta}(2\Delta^2)$ as soon as $\Delta \geq 3\Delta^2$, i.e., $\Delta \leq \frac{1}{3}$. Therefore, Gilardoni's inequality (21) is strictly weaker than the classical Pinsker inequality *at least* for $0 < \Delta < 1/3$ (in fact for $0 < \Delta < 0.569\ldots$). For Δ close to 1, however, Gilardoni's inequality is better (see below).

---

[12] Here the exponential is relative to the base considered, e.g., $\Delta \leq \sqrt{1 - e^{-D}}$ when D is expressed in nats (with natural logarithms) and $\Delta \leq \sqrt{1 - 2^{-D}}$ when D is expressed in bits (with logarithms to base 2).

6   The Optimal Pinsker Inequality

The problem of finding the *optimal* Pinsker inequality (best possible lower bound in (1)) was opened by Vajda [33] in 1970. It was found in 2003 in *implicit* form, using the Legendre-Fenchel transformation, by Fedotov, Harremoës, and Topsøe in [9], as a curve parametrized by hyperbolic trigonometric functions. We give the following equivalent but simpler parametrization with the following proof that is arguably simpler as it only relies of the well-known Lagrange multiplier method.

Theorem 1 (Optimal Pinsker Inequality). *The optimal Pinsker inequality* $D \geq \varphi^*(\Delta)$ *is given in parametric form as*

$$\begin{cases} \Delta & = \lambda(1-q)q \\ D & = \log(1 - \lambda q) + \lambda q(1 + \lambda(1-q)) \log e \end{cases} \tag{22}$$

where $\lambda \geq 0$ *is the parameter and* $q = q(\lambda) \triangleq \frac{1}{\lambda} - \frac{1}{e^{\lambda}-1} \in [0, \frac{1}{2}]$..

Proof. Using binary reduction, $d(p\|q) = p \log \frac{p}{q} + (1-p) \log \frac{1-p}{1-q}$ is to be minimized under the linear constraint $p - q = \delta \in [-1, 1]$. It is well known that divergence $d(p\|q)$ is strictly convex in (p, q). Given that the objective function is convex and the constraint is linear, the solution can be given by the Lagrange multiplier method. The Lagrangian is $L(p, q) = d(p\|q) - \lambda(p-q)$ and the solution is obtained as global minimum of L, which by convexity is obtained by setting the gradient w.r.t. p and q to zero. Assuming *nats* (natural logarithms), this gives

$$\begin{cases} \frac{\partial L}{\partial p} = \log \frac{p}{q} - \log \frac{1-p}{1-q} - \lambda = 0 \\ \frac{\partial L}{\partial q} = -\frac{p}{q} + \frac{1-p}{1-q} + \lambda = 0 \end{cases} \text{ or } \begin{cases} \lambda = \frac{p}{q} - \frac{1-p}{1-q} \\ e^{\lambda} = \frac{p}{q} \Big/ \frac{1-p}{1-q} \end{cases}. \tag{23}$$

Therefore, $\frac{p}{q} = \lambda + \frac{1-p}{1-q} = e^{\lambda} \frac{1-p}{1-q}$, and we have $\frac{1-p}{1-q} = \frac{\lambda}{e^{\lambda}-1}$ and $\frac{p}{q} = \frac{\lambda e^{\lambda}}{e^{\lambda}-1}$. Solving for q, then for p, one obtains $1 = 1 - p + p = (1-q)\frac{\lambda}{e^{\lambda}-1} + q\frac{\lambda e^{\lambda}}{e^{\lambda}-1}$, which gives $q = q(\lambda) = \frac{1}{\lambda} - \frac{1}{e^{\lambda}-1}$ as announced above and $p = q\lambda(1 + \frac{1}{e^{\lambda}-1}) = q\lambda(1 + \frac{1}{\lambda} - q) = q(1 + \lambda(1-q))$. Therefore, we obtain the desired parametrization $\delta = p - q = \lambda(1-q)q$ and $d(p\|q) = \log \frac{1-p}{1-q} + p\lambda = \log(1 - \lambda q) + \lambda q(1 + \lambda(1-q))$. Finally, observe that the transformation $(p, q) \mapsto (1-p, 1-q)$ leaves $d = d(p\|q)$ unchanged but changes $\delta \mapsto -\delta$. In the parametrization, this changes $\lambda \mapsto -\lambda$ and $q(\lambda) \mapsto q(-\lambda) = 1 - q(\lambda)$. Accordingly, this change of parametrization changes $(\delta, d) \mapsto (-\delta, d)$ as can be easily checked. Therefore, the resulting optimal φ^* is even. Restricting to $\delta = |p - q| = p - q \geq 0$ amounts to $p \geq q \iff \lambda \geq 0 \iff q \in [0, 1/2]$.   □

In 2009, Reid and Williamson [25, 26], using a particularly lengthy proof mixing learning theory, 0-1 Bayesian risks, and integral representations of f-divergences, claimed the following "explicit form" of the *optimal* Pinsker inequality: $D \geq \min_{|\beta| \leq 1 - \Delta} \frac{1 + \Delta - \beta}{2} \log \frac{1 + \Delta - \beta}{1 - \Delta - \beta} + \frac{1 - \Delta + \beta}{2} \log \frac{1 - \Delta + \beta}{1 + \Delta - \beta}$. This formula, however, is just a tautological definition of the optimal Pinsker lower bound: Indeed, by binary reduction, $d(p\|q) = p \log \frac{p}{q} + (1-p) \log \frac{1-p}{1-q}$ is to be minimized under the constraint $\delta = p - q$, hence $\delta \leq p \leq 1$ and $q \leq 1 - \delta$. Letting $\beta = 1 - p - q$, this amounts to minimizing over β in the interval $[\delta - 1, 1 - \delta]$ for fixed $\delta = p - q$. Since $p = \frac{1+\delta-\beta}{2}$ and $q = p - v = \frac{1-\delta-\beta}{2}$, this minimization boils down to the above expression for the lower bound.

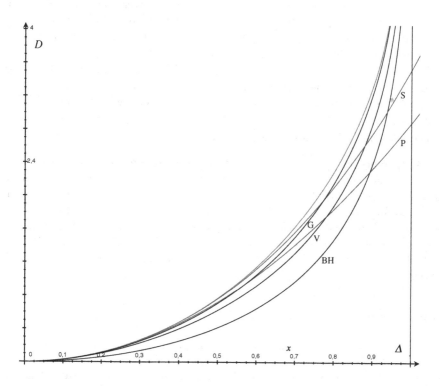

Fig. 3. Pinsker lower bounds of divergence D vs. total variation Δ. Red: Optimal (Theorem 1). Blue: Pinsker (P, Eq. 14 with $c = 2$) with optimal constant and Schützenberger (S, Eq. 17). Black: Bretagnolle-Huber (BH, Eq. 20), Vajda (V, Eq. 19) and Gilardoni (G, Eq. 21). (Color figure online)

7 Conclusion

Figure 3 illustrates the main Pinsker inequalities seen in this paper. As a temporary conclusion, from the implicit form using the exact parametrization of Theorem 1, it is likely that the optimal Pinsker inequality cannot be written as a *closed-form* expression with standard operations and functions. Also, the problem of finding an explicit Pinsker inequality which *uniformly* improve all the preceding ones (in particular, the classical Pinsker inequality with optimal constant and Gilardoni's inequality) is still open.

Interestingly, asymptotic optimality near the two extremes $(V = D = 0$ as $\lambda \to 0$ or $V = 1$, $D = +\infty$ as $\lambda \to \infty)$ can easily be obtained from the parametrization of Theorem 1:

- As $\lambda \to 0$, by Taylor expansion one obtains $q = \frac{1}{2} - \frac{\lambda}{12} + o(\lambda)$, $\Delta = \frac{\lambda}{4} + o(\lambda)$, and (in nats) $D = \frac{\lambda^2}{8} + o(\lambda^2)$. Thus, one recovers that $D \sim 2\Delta^2$ near $D = \Delta = 0$. In particular, the classical Pinsker inequality (with optimal constant) and its improvements, as well as Vajda's and Gilardoni's inequality, are asymptotically optimal near $D = \Delta = 0$.

- As $\lambda \to +\infty$, $q = \frac{1}{\lambda} + o(\frac{1}{\lambda})$, $\exp d = \frac{\lambda}{e^{\lambda}-1}e^{\lambda+o(1)} \sim \lambda \sim \frac{1}{1-\Delta}$. Thus it follows that $\exp D \sim \frac{1}{1-\Delta}$ near $\Delta = 1$ and $D = +\infty$. Vajda's and the Bretagnolle-Huber inequalities are such that $\exp D \sim \frac{c}{1-\Delta}$ there, with suboptimal constants $c = \frac{2}{e} = 0.7357\ldots < 1$ and $c = \frac{1}{2} < 1$, respectively. Only Gilardoni's inequality is optimal in this region with $c = 1$.

As a perspective, one may envision that the exact parametrization of Theorem 1 can be exploited to find new explicit bounds. Indeed, since $\lambda = \varphi^{*\prime}(\Delta)$ in the parametrization of Theorem 1, from the comparison principle, any inequality of the form $\varphi'(\Delta) \leq \lambda = \varphi^{*\prime}(\Delta)$ is equivalent to a corresponding Pinsker inequality (1) associated to φ. For example, since $4\Delta = 4\lambda(1 - q)q \leq \lambda$ always in the parametrization, one recovers the classical Pinsker inequality (14) with optimal constant $c = 2$. Thus, the search of new Pinsker inequality amounts to solving the inequality in $\lambda > 0$: $\varphi'\big(\lambda\big(1 - q(\lambda)\big)q(\lambda)\big) \leq \lambda$ for φ.

References

1. Amari, S.: Information Geometry and Its Applications. AMS, vol. 194. Springer, Tokyo (2016). https://doi.org/10.1007/978-4-431-55978-8
2. Barron, A.R.: Entropy and the central limit theorem. Ann. Probab. **14**(1), 336–342 (1986)
3. Bretagnolle, J., Huber, C.: Estimation des densités : risque minimax. In: Séminaire de probabilités (Strasbourg 1976/77), vol. 12, pp. 342–363. Springer (1978)
4. Bretagnolle, J., Huber, C.: Estimation des densités?: risque minimax. Zeitschrift für Wahrscheinlichkeitstheorie und Verwandte Gebiete **47**, 119–137 (1979)
5. Csiszár, I.: A note on Jensen's inequality. Stud. Sci. Math. Hung. **1**, 185–188 (1966)
6. Csiszár, I.: Information-type measures of difference of probability distributions and indirect observations. Stud. Sci. Math. Hung. **2**, 299–318 (1967)
7. Csiszár, I., Körner, J.: Information Theory. Coding Theorems for Discrete Memoryless Systems. Cambridge University Press, 2nd edn. (2011) (1st edn 1981)
8. Fano, R.M.: Class notes for course 6.574: Transmission of Information. MIT, Cambridge (1952)
9. Fedotov, A.A., Harremoës, P., Topsøe, F.: Refinements of Pinsker's inequality. IEEE Trans. Inf. Theory **49**(6), 1491–1498 (2003)
10. Gel'fand, S.I., Yaglom, A.M.: О вычислении количества информации о случайной функции, содержащейся в другой такой функции (calculation of the amount of information about a random function contained in another such function). Usp. Mat. Nauk. **12**(1), 3–52 (1959)

11. Gilardoni, G.L.: An improvement on Vajda's inequality. In: In and Out of Equilibrium 2, Progress in Probability, vol. 60, pp. 299–304. Birkhäuser (Nov 2008)
12. Gilardoni, G.L.: On Pinsker's and Vajda's type inequalities for Csiszár's f-divergences. IEEE Trans. Inf. Theory **56**(11), 5377–5386 (2010)
13. Jiao, J., Courtade, T.A., No, A., Venkat, K., Weissman, T.: Information measures: The curious case of the binary alphabet. IEEE Trans. Inf. Theory **60**(12), 7616–7626 (2014)
14. Kambo, N.S., Kotz, S.: On exponential bounds for binomial probabilities. Ann. Inst. Statist. Math. **18**, 277–287 (1966)
15. Kemperman, J.H.B.: On the optimum rate of transmitting information. In: Behara, M., Krickeberg, K., Wolfowitz, J. (eds.) Probability and Information Theory. LNM, vol. 89, pp. 126–169. Springer, Heidelberg (1969). https://doi.org/10.1007/BFb0079123
16. Kemperman, J.H.B.: On the optimum rate of transmitting information. Ann. Math. Stat. **40**(6), 2156–2177 (1969)
17. Krafft, O.: A note on exponential bounds for binomial probabilities. Ann. Institut für Mathematische Statistik **21**, 219–220 (1969)
18. Krafft, O., Schmitz, N.: A note on Hoeffding's inequality. J. Am. Stat. Assoc. **64**(327), 907–912 (1969)
19. Kullback, S., Leibler, R.A.: On information and sufficiency. Ann. Math. Stat. **22**(1), 79–86 (1951)
20. Kullback, S.: A lower bound for discrimination information in terms of variation. IEEE Trans. Inform. Theory **13**, 126–127 (1967)
21. Kullback, S.: Correction to "A lower bound for discrimination information in terms of variation". IEEE Trans. Informat. Theory **16**, 652 (1970)
22. McKean, H.P., Jr.: Speed of approach to equilibrium for Kac's caricature of a Maxwellian gas. Arch. Rational Mech. Anal. **21**, 343–367 (1966)
23. Perez, A.: Information theory with an abstract alphabet generalized forms of McMillan's limit theorem for the case of discrete and continuous times. Theory Probabil. Applicat. **4**(1), 99–102 (1959)
24. Pinsker, M.S.: Информация и информационная устойчивость случайных величин и процессов (Information and Information Stability of Random Variables and Processes). Izv. Akad. Nauk (1960) English translation Holden-Day, San Francisco (1964)
25. Reid, M.D., Williamson, R.C.: Generalised Pinsker inequalities. In: 22nd Annual Conference on Learning Theory (COLT 2009), Montreal, Canada, 18–21 June (2009)
26. Reid, M.D., Williamson, R.C.: Information, divergence and risk for binary experiments. J. Mach. Learn. Res. **12**, 731–817 (2011)
27. Rényi, A.: Az információelmélet néhány alapvető kérdése (some basic questions in information theory). Magyar Tud. Akad. Mat. Fiz. Oszt. Közl **10**, 251–282 (1960) (In Hungarian)
28. Sakaguchi, M.: Information Theory and Decision Making. unpublished. George Washington University, Whasington D.C. (June 1964)
29. Schützenberger, M.P.: Contribution aux applications statistiques de la théorie de l'information, Institut de statistique de l'Université de Paris (1954) Thèse de doctorat, vol. 3(1–2) (1953)
30. Topsøe, F.: Bounds for entropy and divergence for distributions over a two-element set. J. Inequalities Pure Appli. Mathemat. **2**(2, Art 25), 1–13 (2001)
31. Toussaint, G.T.: Sharper lower bounds for discrimination information in terms of variation. IEEE Trans. Inf. Theory **21**(1), 99–100 (1975)

32. Tsybakov, A.B.: Introduction to Nonparametric Estimation. Springer Series in Statistics. Springer (2009). https://doi.org/10.1007/b13794
33. Vajda, I.: Note on discrimination information and variation. IEEE Trans. Inf. Theory **16**, 771–773 (1970)
34. Verdú, S.: Total variation distance and the distribution of relative information. In: 2014 Information Theory and Applications Workshop (ITA), San Diego, CA, USA, 9–14 Feb (2014)
35. Volkonskii, V.A., Rozanov, Y.A.: Some limit theorems for random functions. I (English translation from Russian). Theory Probabil. Applicat. **IV(2)**, 178–197 (1959)

On Fisher Information Matrix, Array Manifold Geometry and Time Delay Estimation

Franck Florin$^{(\boxtimes)}$ (iD)

Thales, Paris, France
franck.florin@fr.thalesgroup.com

Abstract. The Fisher information matrix is used to evaluate the minimum variance-covariance of unbiased parameter estimation. It is also used, in natural gradient descent algorithms, to improve learning of modern neural networks. We investigate the Fisher information matrix related to the reception of a signal wave on a sensor array. The signal belongs to a parametric family. The objective of the receiver is to estimate the time of arrival, the direction of arrival and the other parameters describing the signal. Based on the parametric model, Fisher information matrix, array manifold and time delay variances are calculated. With an appropriate choice of parameters, the Fisher matrix is block diagonal and easily invertible. It is possible to estimate the direction of arrival on an array of sensors and the time of arrival whatever the signal parameters are. However, some signal characteristics may have an influence on the asymptotic estimation of the time delay. We give examples with a simple parametric family from the literature.

Keywords: Fisher Information Matrix · Array Manifold · Time Delay Estimation

1 Introduction

Time delay estimation (TDE) has been an active area of signal processing research and development for a large number of applications, including telecommunications, passive and active sonar, radar, electronic warfare, music, speech analysis, fault diagnosis and medical imaging. Many contributions involve techniques for estimating the time delays of a signal in receiving channels [1–3]. Some publications consider parametric expressions of the signal [4, 5], when others express the signal as wide-band, random and stationary [6, 7].

When an array of sensors receives an incident wave, the different time delays on each sensor give some information about the source direction. Array of sensors are used for signal detection and for the estimation of source direction parameters, as elevation and azimuth bearings. Manikas *and al.* Have developed models of arrays using the application of differential geometry in complex space [8–11]. They use the models to evaluate the resolution and the detection capabilities of the array [8].

A common statistical method to evaluate the performance of an estimator is to compute bias and variance [12]. For unbiased estimators, the minimal variance is given by

© The Author(s), under exclusive license to Springer Nature Switzerland AG 2023
F. Nielsen and F. Barbaresco (Eds.): GSI 2023, LNCS 14071, pp. 307–317, 2023.
https://doi.org/10.1007/978-3-031-38271-0_30

the Cramer-Rao lower bound (CRLB) [12, 13]. The Fisher information matrix is used to compute the CRLB.

In the domain of geometric science of information and Riemannian geometry, a statistical distance between two populations can defined by integrating infinitesimal element lengths along the geodesics, and the Fisher information matrix can be used to evaluate the Rao distance between two populations [13].

In modern deep learning, the natural gradient or Riemannian gradient method is more effective for learning [14]. It needs the inversion of the Fisher information matrix. When the Fisher matrix is quasi-diagonal, the inversion is made easier and learning speed is improved [15].

Considering the estimation of array parameters and time delays, Friedlander and Kopp showed in separate works that the Fisher matrix associated with a Gaussian noise model can be computed and exhibits a block diagonal structure [16–18].

We look at the reception on a sensor array of an incoming wave with a finite energy. We examine the geometric structure of the statistical information in this situation, assuming that the receiver may not know much about the transmitted signal.

The objective is to understand if the form of the signal parametric family and the array manifold structure may have an influence on the accuracy of the estimation of the time of arrival of the signal. We limit the approach to the one source problem.

Taking into account the model and results of [16–18], we extend the model to an unknown deterministic signal. The parametric family of the signal is explained and characterised. The Fisher information matrix is computed in order to examine the interdependencies of the various parameters.

Section 2 gives a description of the measurements acquired by the receiver and the general expression of the Fisher information matrix associated with the parameter estimation.

Section 3 details the hypotheses leading to a block diagonal structure of the Fisher information matrix.

Section 4 gives some examples of the CRLB calculations. These examples show the influence of the signal parameters on the CRLB for TDE.

2   Fisher Matrix General Expression

2.1   Receiving a Plane Wave on an Array of Sensors

A signal propagates from a transmitter to a receiver. The transmitted signal $s(t)$ is a function of time t and is known from the transmitter. For the receiver, the knowledge about the signal is not perfect. The power of transmission and the time of transmission are unknown, which leads to uncertainties about signal attenuation and time delay. In addition, we assume that, for the receiver, the signal belongs to a parametric family $\{s_\theta(t)\theta \in \mathbb{R}^P\}$. The receiver objective is to recover the signal parameters, expressed in its own time frame, and to estimate the time τ at which the signal is received. As an example, a sonar receiving marine mammal whistles or vocalisations may detect the signal and estimate the type of mammal. Estimating τ is a detection problem. Estimating the vector θ is a classification or a regression problem.

At the receiver level, the signal is time delayed and attenuated. We assume there is no Doppler shift due to sources, medium, or receiver speed, and no phase shift due to reflection. In addition, the noise, which represents some uncertainty due to the imperfect knowledge of the measurement conditions as sensors perturbations or uncertainties, and variations in environmental conditions, is assumed additive and Gaussian noise. Furthermore, the noise does not depend on the signal itself, as it would be the case in the presence of reverberation phenomena.

Thus, the observation at the receiver level is expressed as a function of time $x(t)$, which takes the form:

$$x(t) = \alpha \cdot s_\theta(t - \tau) + n(t) \tag{1}$$

where:

- α is an attenuation coefficient, time independent,
- $n(t)$ is the noise, function of time,
- τ is the time delay characterising reception at a reference location, time independent as no Doppler effect is assumed.

When the signal is received by several sensors at different positions $\vec{r_k}$, the passive observations can be addressed with a multichannel perspective. The simplest and most common situation, as illustrated on Fig. 1, assumes that, at long range from the source, the signal propagates as a plane wave and that it is received on an antenna array made of K omnidirectional sensors. The wave plane is described by a normal unit vector $\vec{u}$.

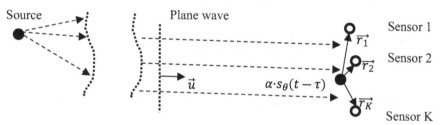

Fig. 1. Time delayed transmission received on an antenna array.

Each sensor k is described by its position $\vec{r_k}$ (3D vector) and its gain coefficient a_k. As the signal is received on the K different sensors, the observation is expressed on each sensor k in the same way as previously:

$$x_k(t) = a_k \cdot \alpha \cdot s_\theta(t - \tau - \tau_k) + n_k(t), k = 1, \ldots K \tag{2}$$

where, for each sensor k:

- a_k is a gain coefficient, time independent.
- $n_k(t)$ is the noise, function of time.
- τ_k is the time delay at sensor k, time independent:

$$\tau_k = \frac{1}{c}\vec{u}, \vec{r_k} \tag{3}$$

- c is the speed of the waves (the travelling speed of the signal).

- $\vec{u}, \vec{r_k}$ is the 3D scalar product between the vectors $\vec{u}$ and $\vec{r_k}$.

We assume that:

- The plane wave direction $\vec{u} = \vec{u}(\eta)$ is parametrized by a vector η representing the knowledge available about the source direction (typically the η components are bearing and elevation).
- The τ_k are functions of η and of the sensors positions $\vec{r_k}$.
- The a_k are related to the omnidirectional sensors characteristics and do not depend on η.
- The signal $s_\theta(t)$ does not depend on η, because the wave direction at reception does not affect the transmitted signal.

The question raised in the following is related to the dependencies between the performances of the estimations of the wave direction parameter η, the time delay τ, the attenuation α and the signal unknown characteristics θ.

2.2 Expression of the Array Manifold

The Fourier transform offers a way to write the Eqs. (2) in the spectral domain in such a way that they are expressed as a set of complex column vector relations:

$$X(\nu) = \alpha \cdot S_\theta(\nu) \cdot e^{-2i\pi\nu\tau} \cdot A_\eta(\nu) + N(\nu), \qquad (4a)$$

where, for each frequency ν in the observation bandwidth $\mathcal{B}$ ($\nu \in \mathcal{B}$):

- The bold quantities are vectors.
- $X(\nu)$ is the $K \times 1$ observation vector.
- $\alpha \cdot S_\theta(\nu) \cdot e^{+i\nu\tau}$ is the Fourier transform of the unknown signal with attenuation α and time delay τ.
- $N(\nu)$ is the $K \times 1$ noise vector.
- $A_\eta(\nu)$ is the $K \times 1$ array manifold vector, representing the sensors gains and phase shifts referred to the array phase center and expressed as:

$$A_\eta(\nu) = \begin{pmatrix} a_1 e^{-2i\pi\nu\tau_1(\eta)} \\ \vdots \\ a_K e^{-2i\pi\nu\tau_K(\eta)} \end{pmatrix} \qquad (4b)$$

- The knowledge about the signal does not suppose any prior statistical distribution of the signal parameters themselves. The parameters θ are deterministic. They are unknown but deterministic.

2.3 Prior Knowledge About the Noise

The noise is a stationary centred complex circular Gaussian variable, independent from one sensor to the other and from one frequency to the other. The noise covariance matrix is known and we can write:

$$E\left(N(\nu)N(\nu)^H\right) = \gamma_0(\nu) \cdot Id_K \qquad (5a)$$

$$E\left(N(v)N(v)^T\right) = 0 \cdot Id_K \tag{5b}$$

where, for each frequency v in the observation bandwidth $\mathcal{B}$:

- Id_K is the $K \times K$ identity matrix.
- $\gamma_0(v)$ is the noise spectral density at frequency v.
- Upper H denotes the transpose conjugate.
- Upper T denotes the transpose.
- $E()$ is the statistical expectation.

The spectral densities are supposed all the same on the different sensors. This condition imposes that the noise spectral densities can be deduced from one sensor to another by a homothetic transformation. If a receiver produces a noise power γ_0, for instance lower than the others, the output of the sensor can be multiplied by a factor to correct it. This is possible because the noise spectral densities are known.

2.4 Prior Knowledge About the Signal

In the general case, the signal $S_\theta(v)$ belongs to a set $\mathcal{H} = \left\{S_\theta(v)\theta \in \mathbb{R}^P, v \in \mathcal{B}\right\}$. This set contains all the signals that the receiver expects to receive. For the receiver, the signal is only considered for the frequency bandwidth $\mathcal{B}$ accepted by the sensors. As the observation bandwidth $\mathcal{B}$ is always limited, the received signal is only an approximation of the real transmitted signal. For instance, if the transmitted signal is a radar pulse with a finite duration, its theoretical bandwidth is infinite. So the observation through a limited bandwidth can only approximate the exact transmitted signal.

In practice, the set $\mathcal{H}$ varies with the application. As an example, for some fault diagnostic applications, the parametric family may be a Gaussian-modulated linear group delay model [19]. The set $\mathcal{H}$ must reflect the dependence with the frequency v.

2.5 Law of Probability and Log-Likelihood

Because of the modelling hypotheses, the observations are:

$$\mathcal{X} = \{X(v)|v \in \mathcal{B}\} \tag{6a}$$

The parametric law of probability of the observations is expressed as:

$$p(\mathcal{X}/\alpha, \tau, \theta, \eta) = \prod_v \frac{1}{\pi^K \gamma_0(v)^K} \exp\left(-\frac{\|X(v) - M(v, \alpha, \tau, \theta, \eta)\|^2}{\gamma_0(v)}\right) \tag{6b}$$

where, for each frequency v in $\mathcal{B}$, the observation mean is:

$$M(v, \alpha, \tau, \theta, \eta) = \alpha \cdot S_\theta(v) \cdot e^{-2i\pi v\tau} \cdot A_\eta(v) \tag{6c}$$

The log-likelihood is:

$$l(\mathcal{X}/\alpha, \tau, \theta, \eta) = -\sum_v K \ln \pi - \sum_v K \ln(\gamma_0(v)) - \sum_v \frac{1}{\gamma_0(v)} X(v) - M(v, \alpha, \tau, \theta, \eta)^2 \tag{7}$$

Thus maximising $l(\mathcal{X}/\alpha, \tau, \theta, \eta)$ as a function of $\alpha, \tau, \theta, \eta$ is the same as minimising:

$$J(\alpha, \tau, \theta, \eta) = \sum_\nu \frac{1}{\gamma_0(\nu)} \|X(\nu) - M(\nu, \alpha, \tau, \theta, \eta)\|^2 \tag{8}$$

2.6 Fischer Matrix Expression

By rearrangement of the parameters in one vector $\zeta^T = (\eta^T, \alpha, \tau, \theta^T)$, the Fisher matrix F of the parameters estimation problem is the matrix whose generic term at line l and column j is given by:

$$F_{lk} = E\left(\frac{\partial l(\mathcal{X}/\zeta)}{\partial \zeta_l} \cdot \frac{\partial l(\mathcal{X}/\zeta)}{\partial \zeta_j}\right) \tag{9a}$$

Considering the complex nature of the vectors, the generic term of the Fisher matrix can be expressed as:

$$F_{lj} = \sum_\nu \frac{2}{\gamma_0(\nu)} Re\left\{\frac{\partial(M(\nu,\alpha,\tau,\theta,\eta))}{\partial \zeta_l}^H \frac{\partial(M(\nu,\alpha,\tau,\theta,\eta))}{\partial \zeta_j}\right\} \tag{9b}$$

3 Fisher Matrix Structure

3.1 Separation of Array Manifold and Signal Parameters

Equation (9b) can be rewritten with ζ_l among α, τ, θ on one side and ζ_j among η on the other side. For example, with θ_l and η_j the generic term F_{lj} becomes:

$$F_{\theta_l \eta_j} = \sum_\nu \frac{2}{\gamma_0(\nu)} Re\left\{\frac{\partial\left(\alpha \cdot S_\theta(\nu) \cdot e^{-2i\pi\nu\tau}\right)}{\partial \theta_l} \cdot \alpha \cdot S_\theta(\nu) \cdot e^{-2i\pi\nu\tau} \cdot A_\eta(\nu)^H \frac{\partial(A_\eta(\nu))}{\partial \eta_j}\right\} \tag{10}$$

where $\overline{Z}$ is the complex conjugate of Z.

In addition, from Eqs. (3) and (4b), for any η_j, we get:

$$A_\eta(\nu)^H \frac{\partial(A_\eta(\nu))}{\partial \eta_j} = -2i\pi \frac{\nu}{c} \sum_{k=1}^K \alpha_k^2 \frac{\partial \vec{u}(\eta)}{\partial \eta_j}, \vec{r_k} = -2i\pi \frac{\nu}{c} \frac{\partial \vec{u}(\eta)}{\partial \eta_j}, \sum_{k=1}^K \alpha_k^2 \vec{r_k} \tag{11}$$

When the sensors positions are referenced to the phase centre of the array, we have, by definition of the phase centre:

$$\sum_{k=1}^K \alpha_k^2 \vec{r_k} = \vec{0} \tag{12}$$

And so, using (11) and (12), we get:

$$A_\eta(\nu)^H \frac{\partial(A_\eta(\nu))}{\partial \eta_j} = 0 \tag{13}$$

Equation (13) induces a block structure of the Fisher matrix with:

$$\begin{cases} F_{\theta_l \eta_j} = 0 \\ F_{\alpha \eta_j} = 0 \\ F_{\tau \eta_j} = 0 \end{cases} \tag{14}$$

These equations show that the Fisher information matrix F has a block diagonal structure as:

$$F = \begin{pmatrix} F_\eta & 0 \\ 0 & F_S \end{pmatrix} \tag{15}$$

- $F_\eta = \left(F_{\eta_l \eta_j}\right)$ is a block describing the estimation of the array manifold parameters.
- F_S is a block describing the estimation of the received signal parameters α, τ, θ.

We can compute the block related to the array manifold:

$$\begin{aligned} F_{\eta_l \eta_j} &= \sum_\nu \frac{2\alpha^2 \|S_\theta(\nu)\|^2}{\gamma_o(\nu)} Re \left\{ \frac{\partial (A_\eta(\nu))}{\partial \eta_l}^H \frac{\partial (A_\eta(\nu))}{\partial \eta_j} \right\} \\ &= \sum_\nu 8\pi^2 \frac{\nu^2}{c^2} \frac{\alpha^2 \|S_\theta(\nu)\|^2}{\gamma_o(\nu)} \sum_{k=1}^K a_k^2 \langle \frac{\partial \vec{u}(\eta)}{\partial \eta_l}, \vec{r_k} \rangle \langle \frac{\partial \vec{u}(\eta)}{\partial \eta_j}, \vec{r_k} \rangle \end{aligned} \tag{16}$$

The separation of the two blocks and the expression of F_η show that the estimation of the array manifold parameters depends on the signal-to-noise ratio (SNR) and on the array manifold structure (Eq. (16)).

3.2 Separating the Signal Fisher Matrix Block F_S into Two Blocks

The signal $S_\theta(\nu)$ is a complex number, which can be written with module and phase quantities. With the hypothesis that the parameters can be separated into two vectors, one for the parameters describing the module (ϕ) and one for the parameters describing the phase (φ), we can write:

$$\theta^T = \left(\phi^T, \varphi^T\right) \tag{17a}$$

$$S_\theta(\nu) = \rho(\nu, \phi) \cdot e^{i\psi(\nu, \varphi)} \tag{17b}$$

From the Paley-Wiener second theorem, a necessary and sufficient condition for $S_\theta(\nu)$ to be the Fourier transform of a causal signal with finite energy is that [20]:

$$\int_0^{+\infty} \frac{|\ln \rho(\nu, \phi)|}{1 + \nu^2} d\nu < \infty \tag{18}$$

In addition, if the signal is phase-minimum, its phase $\psi(\nu, \varphi)$ can be deduced from its module $\rho(\nu, \phi)$ [20, 21]. In such a case, the phase parameters φ are related to the module parameters ϕ by a functional relation. However, in the general case, we may estimate separately the parameters describing module and phase.

As ϕ and φ are separated parameters, the generic terms for the module and phase parameters show a block matrix structure and become:

$$F_{\phi_l\varphi_j} = \sum_\nu \frac{2\|A_\eta(\nu)\|^2(\alpha\cdot\rho(\nu,\phi))}{\gamma_0(\nu)} Re\left\{\frac{\partial(\alpha\cdot\rho(\nu,\phi))}{\partial\phi_l} \quad i \quad \frac{\partial(\psi(\nu,\varphi)-2\pi\nu\tau)}{\partial\varphi_j}\right\} = 0 \qquad (19a)$$

$$F_{\phi_l\phi_j} = \sum_\nu \frac{2\|A_\eta(\nu)\|^2}{\gamma_0(\nu)} \frac{\partial(\alpha\cdot\rho(\nu,\phi))}{\partial\phi_l} \frac{\partial(\alpha\cdot\rho(\nu,\phi))}{\partial\phi_j} \qquad (19b)$$

$$F_{\varphi_l\varphi_j} = \sum_\nu \frac{2\|A_\eta(\nu)\|^2}{\gamma_0(\nu)} |\alpha\cdot\rho(\nu,\phi)|^2 \frac{\partial(\psi(\nu,\varphi)-2\pi\nu\tau)}{\partial\varphi_l} \frac{\partial(\psi(\nu,\varphi)-2\pi\nu\tau)}{\partial\varphi_j} \qquad (19c)$$

These equations show that the Fisher information matrix block regarding the signal parameters has also a block diagonal structure:

$$F_S = \begin{pmatrix} F_{\alpha\phi} & 0 \\ 0 & F_{\tau\varphi} \end{pmatrix} \qquad (20)$$

The first block $F_{\alpha\phi}$ is related to the parameters describing the signal module and the block $F_{\tau\varphi}$ is related to the parameters describing the signal phase.

4 Examples

4.1 Simplifying Hypothesis

In the various examples, we consider that most of the signal energy is concentrated in the bandwidth $\left[\nu_0 - \frac{B}{2}, \nu_0 + \frac{B}{2}\right] \subset B$. We define the SNR at frequency ν:

$$\gamma_{SNR}(\nu)\underline{\underline{def}}\, 2\frac{\|A_\eta(\nu)\|^2}{\gamma_0(\nu)} |\alpha\cdot\rho(\nu)|^2 \qquad (21a)$$

And the total SNR:

$$\gamma \underline{\underline{def}} \sum_\nu \gamma_{SNR}(\nu) \qquad (21b)$$

As the Fisher matrix is block diagonal (Eq. (20)), we only look at the block $F_{\tau\varphi}$. As in [19], we assume a polynomial dependence of the phase with the frequency.

4.2 TDE with no Phase Parameter

Assuming there is no phase parameter φ means that the phase is exactly known (except for the time delay τ):

$$\psi(\nu, \varphi) = \psi_0(\nu) \qquad (22)$$

This hypothesis leads to:

$$F_{\tau\tau} = \sum_\nu \frac{8\pi^2\nu^2}{\gamma_0(\nu)} \sum_{k=1}^K \alpha_k^2 |\alpha\cdot\rho_0(\nu)|^2 = \sum_\nu (2\pi\nu)^2\gamma_{SNR}(\nu) \cong \gamma 4\pi^2\left(\nu_0^2 + \frac{B^2}{12}\right) \qquad (23)$$

By definition, the CRLB on the TDE is given by the first diagonal coefficient of the inverse of the Fisher matrix. Due to the diagonal structure of the Fisher matrix, with (22) and (23), we get:

$$\sigma_\tau^2 = \frac{1}{F_{\tau\tau}} \cong \frac{1}{\gamma} \cdot \frac{1}{4\pi^2\left(\nu_0^2+\frac{B^2}{12}\right)} \qquad (24)$$

4.3 TDE with One Phase Parameter

The phase is assumed constant within the bandwidth (except for the time delay τ):

$$\boldsymbol{\varphi}^T = (\varphi_1) \tag{25a}$$

$$\psi(\nu, \boldsymbol{\varphi}) = \varphi_1 \tag{25b}$$

With these values, we get the following expressions:

$$F_{\tau\tau} = \sum_\nu (2\pi\nu)^2 \gamma_{SNR}(\nu) \underset{=}{\text{def}} \gamma \, \omega_\tau^2 \cong \gamma 4\pi^2 \left(\nu_0^2 + \frac{B^2}{12}\right) \tag{26a}$$

$$F_{\tau\varphi_1} = \sum_\nu (2\pi\nu)\gamma_{SNR}(\nu) \underset{=}{\text{def}} \gamma \, \omega_1 \cong \gamma 2\pi(\nu_0) \tag{26b}$$

$$F_{\varphi_1\varphi_1} = \sum_\nu \gamma_{SNR}(\nu) \underset{=}{\text{def}} \gamma \tag{26c}$$

$$\boldsymbol{F}_{\tau\varphi} = \begin{pmatrix} F_{\tau\tau} & F_{\tau\varphi_1} \\ F_{\tau\varphi_1} & F_{\varphi_1\varphi_1} \end{pmatrix} \underset{=}{\text{def}} \gamma \begin{pmatrix} \omega_\tau^2 & \omega_1 \\ \omega_1 & 1 \end{pmatrix} \tag{27}$$

The first diagonal term of the inverse matrix of $\boldsymbol{F}_{\tau\varphi}$ is:

$$\sigma_\tau^2 = \frac{1}{\gamma} \cdot \frac{1}{\omega_\tau^2 - \omega_1^2} \cong \frac{1}{\gamma} \cdot \frac{3}{\pi^2 B^2} \tag{28}$$

4.4 TDE with Two Phase Parameters

We consider the Gaussian-modulated linear group delay model as expressed in [19]:

$$\boldsymbol{\varphi}^T = (\varphi_1, \varphi_2) \tag{29a}$$

$$\psi(\nu, \boldsymbol{\varphi}) = \varphi_1 + 4\pi^2 \varphi_2 \nu^2 \tag{29b}$$

With these values, in addition to Eqs. (26), we get the following expressions:

$$F_{\tau\varphi_2} = \sum_\nu (2\pi\nu)^3 \gamma_{SNR}(\nu) \underset{=}{\text{def}} \gamma \omega_2^3 \cong \gamma 8\pi^3 \left(\nu_0^3 + \nu_0 \frac{B^2}{4}\right) \tag{30a}$$

$$F_{\varphi_1\varphi_2} = \sum_\nu (2\pi\nu)^2 \gamma_{SNR}(\nu) = F_{\tau\tau} \underset{=}{\text{def}} \gamma \omega_\tau^2 \cong \gamma 4\pi^2 \left(\nu_0^2 + \frac{B^2}{12}\right) \tag{30b}$$

$$F_{\varphi_2\varphi_2} = \sum_\nu (2\pi\nu)^4 \gamma_{SNR}(\nu) \underset{=}{\text{def}} \gamma \omega_3^4 \cong \gamma 16\pi^4 \left(\nu_0^4 + \frac{1}{2}\nu_0^2 B^2 + \frac{B^4}{80}\right) \tag{30c}$$

$$\boldsymbol{F}_{\tau\varphi} = \begin{pmatrix} F_{\tau\tau} & F_{\tau\varphi_1} & F_{\tau\varphi_2} \\ F_{\tau\varphi_1} & F_{\varphi_1\varphi_1} & F_{\varphi_1\varphi_2} \\ F_{\tau\varphi_2} & F_{\varphi_1\varphi_2} & F_{\varphi_2\varphi_2} \end{pmatrix} \underset{=}{\text{def}} \gamma \begin{pmatrix} \omega_\tau^2 & \omega_1 & \omega_2^3 \\ \omega_1 & 1 & \omega_\tau^2 \\ \omega_2^3 & \omega_\tau^2 & \omega_3^4 \end{pmatrix} \tag{31}$$

The first diagonal term of the inverse matrix of $\boldsymbol{F}_{\tau\varphi}$ becomes:

$$\sigma_\tau^2 = \frac{1}{\gamma} \cdot \frac{\omega_3^4 - \omega_\tau^4}{\omega_\tau^2 \omega_3^4 - \omega_\tau^6 - \omega_1^2 \omega_3^4 + 2\omega_1 \omega_2^3 \omega_\tau^2 - \omega_2^6} \cong \frac{1}{\gamma} \cdot \frac{3(60\nu_0^2 + B^2)}{\pi^2 B^4} \tag{32}$$

5 Conclusion

The prior knowledge about the signal has a significant influence on the TDE. However, TDE is only affected by the parameters characterising the phase of the signal in the spectral domain. The estimation of the parameters describing the source direction and the signal amplitude in the spectral domain does not affect the TDE performance. Future work could investigate the impact of the parametric description fitting with various impulse signals (e.g. sonar signals, marine mammal whistles or vocalisations), and estimate some Rao distance between different signals.

References

1. Carter, G.: Time Delay estimation for passive sonar signal processing. IEEE Trans. Acoust. Speech Signal Process. ASSP **29**(3), 463–470 (1981)
2. Scarbrough, K., Ahmed, N., Carter, G.: On the simulation of a class of time delay estimation algorithms. IEEE Trans. Acoust. Speech Signal Process. ASSP **29**(3), 534–540 (1981)
3. Friedlander, B.: An efficient parametric technique for doppler-delay estimation. IEEE Trans. Signal Process. **60**(8), 3953–3963 (2012)
4. Miller, L., Lee, J.: Error analysis of time delay estimation using a finite integration time correlator. IEEE Trans. Acoust. Speech Signal Process. ASSP **29**(3), 490–496 (1981)
5. Pallas, M.-A., Jourdain, G.: Active high resolution time delay estimation for large BT signals. IEEE Trans. Signal Process. **39**(4), 781–787 (1991)
6. Azaria, M., Hertz, D.: Time delay estimation by generalized cross correlation methods. IEEE Trans. Acoust. Speech Signal Process. ASSP **32**(2), 280–285 (1984)
7. Knapp, C., Carter, G.: The generalized correlation method for estimation of time delay. IEEE Trans. Acoust. Speech Signal Process. ASSP **24**(4), 280–285 (1976)
8. Manikas, A., Karimi, H., Dacos, I.: Study of the detection and resolution capabilities of a one-dimensional array of sensors by using differential geometry. IEE Proc.-Radar Sonar Navig. **141**(2), 83–92 (1994)
9. Manikas, A., Proukakis, C.: Modeling and estimation of ambiguities in linear arrays. IEEE Trans. Signal Process. **46**(8), 2166–2179 (1998)
10. Manikas, A., Sleiman, A.: Manifold Studies of nonlinear antenna array geometries. IEEE Trans. Signal Process. **49**(3), 497–506 (2001)
11. Sleiman, A., Manikas, A., Dacos, I.: The impact of sensor positioning on the array manifold. IEEE Trans. Ant. Propag. **51**(9), 2227–2237 (2003)
12. Borovkov, A.: Statistique Mathématique. Editions Mir, Moscou (1987)
13. Nielsen, F.: Cramér-Rao Lower Bound and Information Geometry. https://doi.org/10.48550/arXiv.1301.3578 (2013)
14. Amari, S., Karakida, R., Oizumi, M.: Fisher Information and Natural Gradient Learning of Random Deep Networks. https://doi.org/10.48550/arXiv.1808.07172
15. Kaul, P., Lall, B.: Projective fisher information for natural gradient descent. IEEE Trans. Artif. Intell. **4**(2), 304–314 (2023)
16. Friedlander, B.: On the cramer-rao bound for time delay and doppler estimation. IEEE Trans. Inf. Theory IT **30**(3), 575–580 (1984)
17. Kopp, L., Thubert, D.: Cramer-Rao bounds and array processing.Part One Formal. Traitement du Signal 3(3), 111–125 (1986)
18. Kopp, L., Thubert, D.: Cramer-Rao bounds and array processing. Part Two Appl. Traitement du Signal 4(1), 57–71 (1987)

19. He, Z., Tu, X., Bao, W., Hu, Y., Li, F.: Second-order transient-extracting transform with application to time-frequency filtering. IEEE Trans. Instrum. Meas. **69**(8), 5428–5437 (2020)
20. Oppenheim, A., Schafer, R.: Digital Signal Processing. Prentice Hall, Upper Saddle River (1975)
21. Roubine, E.: Distributions – Signal, 2nd edn. Eyrolles, Paris (1999)

Revisiting Lattice Tiling Decomposition and Dithered Quantisation

Fábio C. C. Meneghetti[1] , Henrique K. Miyamoto[2] , Sueli I. R. Costa[1(✉)] ,
and Max H. M. Costa[3]

[1] IMECC, University of Campinas, Campinas SP, Brazil
fabiom@ime.unicamp.br, sueli@unicamp.br
[2] L2S, CentraleSupélec, Université Paris-Saclay, Gif-sur-Yvette, France
henrique.miyamoto@centralesupelec.fr
[3] FEEC, University of Campinas, Campinas SP, Brazil
max@fee.unicamp.br

Abstract. A lattice tiling decomposition induces dual operations: quantisation and wrapping, which map the Euclidean space to the lattice and to one of its fundamental domains, respectively. Applying such decomposition to random variables over the Euclidean space produces quantised and wrapped random variables. In studying the characteristic function of those, we show a 'frequency domain' characterisation for deterministic quantisation, which is dual to the known 'frequency domain' characterisation of uniform wrapping. In a second part, we apply the tiling decomposition to describe dithered quantisation, which consists in adding noise during quantisation to improve its perceived quality. We propose a non-collaborative type of dithering and show that, in this case, a wrapped dither minimises the Kullback-Leibler divergence to the original distribution. Numerical experiments illustrate this result.

Keywords: Characteristic function · Dithering · Lattice · Quantisation · Wrapping

1 Introduction

Lattices are discrete sets of points in $\mathbb{R}^n$ formed by all integer linear combinations of independent vectors [5,7], and have applications that include information theory (e.g., coding, quantisation) [6,18] and cryptography [4]. Translations of a fundamental domain by lattice points tile the entire Euclidean space, and this construction induces two operations: quantisation and wrapping, which consist in decomposing a point into its lattice and fundamental domain components, respectively. Accordingly, a random variable defined on the Euclidean space can produce two new random variables by applying these operations: a quantised and a wrapped one. These notions are recalled in Sect. 2.

Supported by the Brazilian National Council for Scientific and Technological Development (CNPq) grants 141407/2020-4 and 32441/2021-2.

In this paper, we extend a previous work [14] by further investigating properties of this dual decomposition. Specifically, in Sect. 3, we study relations between
the characteristic function of the involved random variables, and characterise
deterministic quantisation in the 'frequency domain', in an analogous way to
the 'frequency domain' characterisation of uniform wrapping [9,18].

In Sect. 4, we use the language of lattice tiling decomposition to describe
dithered quantisation [8], a technique that consists in adding controlled random
noise (dither) to improve the quality of the quantised signals, e.g., images [9],
audio [11,12]. In commonly used subtractive dithering, encoder and decoder
have to collaborate: the former adds a random noise and the later subtracts the
same realisation of the noise. While this scheme provides analytical convenience
and has been extensively studied (see, e.g., [18]), we propose a non-collaborative
scheme that could be of practical interest: the decoder receives a quantised signal
and can only add random noise to improve its perceptual quality. Following [11,
12], we assess the perceptual quality by a similarity measure between original
and quantised random variables, and show that the dither that minimises the
Kullback-Leibler (KL) divergence between these distributions is precisely the
one that has the same distribution of the wrapped random variable.

2 Lattices and Tiling

A *lattice* Λ is a discrete additive subgroup of $\mathbb{R}^n$, which can be described as
the set of all integer linear combinations of linearly independent vectors $\beta = \{b_1, \ldots, b_k\}$, i.e., $\Lambda := \{x_1 b_1 + \cdots + x_k b_k : x_1, \ldots, x_k \in \mathbb{Z}\}$. The set β is called a
basis and the matrix B whose columns are the vectors of β is called a *generator
matrix* for this lattice. Another matrix B' is also a generator matrix for the same
lattice if, and only if, $B' = BU$, for U an integer matrix with $\det U = \pm 1$. We
consider here only *full-rank* lattices, that is, $k = n$. The determinant of a lattice
is independent of the basis and defined as $\det \Lambda := |\det B|$. The *dual lattice* of Λ
is defined as $\Lambda^* := \{\omega \in \mathbb{R}^n : \langle \omega, \lambda \rangle \in \mathbb{Z}, \forall \lambda \in \Lambda\}$. If B is a generator matrix
of Λ, then $B^{-\top}$ is a generator matrix of Λ^*.

We say that a measurable set $\mathcal{D} \subset \mathbb{R}^n$ *tiles* $\mathbb{R}^k$ by Λ, or that $\mathcal{D}$ is
a *fundamental domain* of Λ, if (1) $\bigcup_{\lambda \in \Lambda}(\lambda + \mathcal{D}) = \mathbb{R}^k$, and (2) $(\lambda + \mathcal{D}) \cap (\lambda' + \mathcal{D}) = \emptyset$, for $\lambda \neq \lambda'$. A pair $(\mathcal{D}, \Lambda)$ such that $\mathcal{D}$ tiles $\mathbb{R}^n$
by Λ is called a *tiling pair*. The *fundamental parallelotope* of a basis
$\mathcal{P}(\beta) := \{\alpha_1 b_1 + \cdots + \alpha_n b_n : \alpha_1, \ldots, \alpha_n \in [0,1)\}$ is a fundamental domain.
The set of points that are closer to the origin than to any other lat-
tice point is called the *Voronoi region* of a lattice and denoted $\mathcal{V}(\Lambda) := \{x \in \mathbb{R}^n : \|x\| \leq \|x - \lambda\|, \forall \lambda \in \Lambda\}$. It becomes a fundamental domain after
parts of its boundary are excluded in such a way ties are broken in a systematic
manner. Any fundamental domain has the same volume $\operatorname{vol} \mathcal{D} = \det \Lambda$.

Given a tiling pair $(\mathcal{D}, \Lambda)$, we have that $\mathbb{R}^n = \bigsqcup_{\lambda \in \Lambda}(\lambda + \mathcal{D})$, which means
that any $x \in \mathbb{R}^n$ can be written in a unique way as $x = y + \lambda$, with $y \in \mathcal{D}$ and $\lambda \in \Lambda$. This partitioning induces two important functions: wrapping

and quantisation [14]. *Wrapping*[1] is the map $\pi\colon \mathbb{R}^n \to \mathcal{D}$, $\pi(y + \lambda) = y$, while *quantisation* is defined as $\mathcal{Q}\colon \mathbb{R}^n \to \Lambda$, $\mathcal{Q}(y + \lambda) = \lambda$, for $y \in \mathcal{D}$ and $\lambda \in \Lambda$.

Given a continuous random variable X over $\mathbb{R}^n$, we can define the associated wrapped and quantised random variables as $X_\pi := \pi(X)$ and $X_\mathcal{Q} := \mathcal{Q}(X)$, respectively. Since $\pi + \mathcal{Q} = \mathrm{id}$, we have $X = X_\pi + X_\mathcal{Q}$, and the joint random variable $(X_\pi, X_\mathcal{Q})$ is defined over the tiling pair $(\mathcal{D}, \Lambda)$. If X admits a probability density function (pdf) p, then the wrapped and quantised densities can be respectively obtained as $p_\pi(y) = \sum_{\lambda \in \Lambda} p(y + \lambda)$, $y \in \mathcal{D}$, and $p_\mathcal{Q}(\lambda) = \int_\mathcal{D} p(y + \lambda)\mathrm{d}y$, $\lambda \in \Lambda$ [14]. The marginal random variables X_π and $X_\mathcal{Q}$ are not necessarily independent; if that is the case, i.e., $p(y + \lambda) = p_\pi(y)p_\mathcal{Q}(\lambda)$, we call X a *tiled* random variable. *Lattice staircase* is a straightforward way to construct tiled random variables: if U is uniform over $\mathcal{D}$, and L is any random variable over Λ, chosen independently of U, then $X := U + L$ is a tiled random variable.

3   Characteristic Function and Fourier Transform

Let X be a continuous random variable over $\mathbb{R}^n$ with pdf p, and $X_\mathcal{Q}$ and X_π the corresponding quantised and wrapped random variables. The *characteristic function*[2] of X is the function $\Phi_X\colon \mathbb{R}^n \to \mathbb{C}$ given by

$$\Phi_X(\omega) := \mathrm{E}_X\left[e^{-2\pi i\langle \omega, X\rangle}\right] = \int_{\mathbb{R}^n} p(x)e^{-2\pi i\langle \omega, x\rangle}\mathrm{d}x, \tag{1}$$

which is equal to the Fourier transform of the density p, i.e., $\Phi_X(\omega) = \widehat{p}(\omega)$. Analogously, the characteristic function of X_π is $\Phi_{X_\pi}\colon \Lambda^* \to \mathbb{C}$, defined as $\Phi_{X_\pi}(\omega) := \mathrm{E}_{X_\pi}\left[e^{-2\pi i\langle \omega, X_\pi\rangle}\right] = \int_\mathcal{D} p_\pi(y)e^{-2\pi i\langle \omega, y\rangle}\mathrm{d}y$, and the characteristic function of $X_\mathcal{Q}$ is $\Phi_{X_\mathcal{Q}}\colon \mathcal{D}^* \to \mathbb{C}$, for a given fundamental domain $\mathcal{D}^*$ of Λ^*, defined as $\Phi_{X_\mathcal{Q}}(z) = \mathrm{E}_{X_\mathcal{Q}}\left[e^{-2\pi i\langle z, X_\mathcal{Q}\rangle}\right] = \sum_{\lambda \in \Lambda} p_\mathcal{Q}(\lambda)e^{-2\pi i\langle z, \lambda\rangle}$.

By the generalization of the Fourier transform to locally compact Abelian (LCA) groups [16], it is possible to define the Fourier transform of the densities p_π and $p_\mathcal{Q}$. Since p_π is Λ-periodic, it can be decomposed in a Fourier series with Fourier coefficients $\widehat{p_\pi}\colon \Lambda^* \to \mathbb{C}$ given by [2,7]

$$\widehat{p_\pi}(\omega) = \frac{1}{\det \Lambda}\int_\mathcal{D} p_\pi(y)e^{-2\pi i\langle \omega, y\rangle}\mathrm{d}y. \tag{2}$$

On the other hand, $p_\mathcal{Q}$ has discrete support and can be seen as the Fourier coefficients of the Λ^*-periodic function $\widehat{p_\mathcal{Q}}\colon \mathcal{D}^* \to \mathbb{C}$, whose Fourier series is [2,7]

$$\widehat{p_\mathcal{Q}}(z) = \sum_{\lambda \in \Lambda} p_\mathcal{Q}(\lambda)e^{2\pi i\langle z, \lambda\rangle}. \tag{3}$$

---

[1]   Wrapping can be thought of as the canonical projection $\mathbb{R}^n \to \mathbb{R}^n/\Lambda$ to the quotient space (isomorphic to an n-torus), composed with a choice of representatives.

[2]   The classical definition of characteristic function in statistics books does not include the factor -2π in the exponent, e.g. [13, § 3.3]. Here, we follow [18, § 4.2] and define the characteristic function as in (1) to simplify relations. There are slightly different definitions for the Fourier transform too: the form we adopt (as in [18]) has the advantage of simplifying the expression for the inverse Fourier transform [10, § 6.2].

We remark that these definitions of Fourier transforms in lattices have been used in the context of lattice-based cryptography, e.g., [2]. With these definitions, the characteristic functions for the wrapped and quantised random variables can be written as $\Phi_{X_\pi}(\omega) = (\det \Lambda)\,\widehat{p_\pi}(\omega)$ and $\Phi_{X_Q}(z) = \widehat{p_Q}(-z)$, respectively. These concepts are illustrated in Fig. 1 for Gaussian distributions on $\mathbb{R}$.

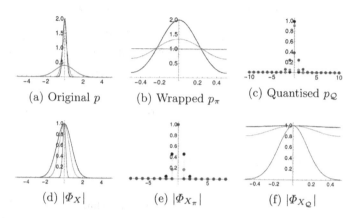

(a) Original p (b) Wrapped p_π (c) Quantised p_Q

(d) $|\Phi_X|$ (e) $|\Phi_{X_\pi}|$ (f) $|\Phi_{X_Q}|$

Fig. 1. Examples of zero-mean Gaussian distributions: densities (top) and characteristic functions (bottom), with $\Lambda = \Lambda^* = \mathbb{Z}$ and $\mathcal{D} = \mathcal{D}^* = [-\frac{1}{2}, \frac{1}{2}[$, for different variances: $\sigma^2 = 0.4$ (blue), $\sigma^2 = 0.9$ (orange) and $\sigma^2 = 1$ (green). (Color figure online)

Theorem 1 ([18, Eq. (4.21)]). $\Phi_{X_\pi} = \Phi_X|_{\Lambda^*}$

Proof. For any $\omega \in \Lambda^*$ and $\lambda \in \Lambda$, we have $e^{-2\pi i \langle \omega, \lambda \rangle} = 1$. Then

$$\Phi_{X_\pi}(\omega) = \int_{\mathcal{D}} p_\pi(y) e^{-2\pi i \langle \omega, y \rangle} dy = \sum_{\lambda \in \Lambda} \int_{\lambda + \mathcal{D}} p(y) e^{-2\pi i \langle \omega, y - \lambda \rangle} dy$$

$$= \int_{\mathbb{R}^n} p(x) e^{-2\pi i \langle \omega, x \rangle} dx = \Phi_X(\omega).$$

The previous result, that relates the characteristic functions of the wrapped and original random variables, can be used to readily prove the classical Poisson summation formula for probability density functions [7, Thm. 2.3],[17, p. 252]:

Theorem 2 (Poisson summation formula). *Let Λ be a lattice in $\mathbb{R}^n$ and X a random variable in $\mathbb{R}^n$ with density p. If p and Φ_X have fast decay of order $O(x^{-(n+\epsilon)})$ and $O(\omega^{-(n+\epsilon)})$, respectively, then*

$$\sum_{\lambda \in \Lambda} p(\lambda) = \frac{1}{\det \Lambda} \sum_{\omega \in \Lambda^*} \Phi_X(\omega). \tag{4}$$

Proof. Writing the Fourier series of the wrapped distribution p_π, we have $p_\pi(y) = \frac{1}{\det \Lambda} \sum_{\omega \in \Lambda^*} \Phi_{X_\pi}(\omega) e^{2\pi i \langle \omega, y \rangle}$. Evaluating it at $y = 0$ and using Theorem 1 yields the desired result.

A random variable X is said to be *modulo-uniform* if its wrapping X_π has uniform distribution over $\mathcal{D}$, i.e., $p_\pi \equiv 1/(\det \Lambda)$ [18, Def. 4.2.2]. Interestingly, modulo-uniformity has a 'frequency domain' characterisation [9,18]:

Theorem 3 ([9, **Thm. 1**],[18, **Lem. 4.2.4**]). *A random variable X is modulo-uniform if, and only if, $\Phi_{X_\pi}(\omega) = \delta_0(\omega)$, where δ_0 is the Dirac delta function of the origin, which takes value 1 when $\omega = 0$, and 0 otherwise.*

Proof. If $p_\pi \equiv 1/(\det \Lambda)$, then $\Phi_{X_\pi}(\omega) = \frac{1}{\det \Lambda} \int_\mathcal{D} e^{-2\pi i \langle \omega, y \rangle} \mathrm{d}y$, which is 1 for $\omega = 0$, and 0 for $\omega \neq 0$, thus $\Phi_{X_\pi} = \delta_0$. Now suppose that $\Phi_{X_\pi} = \delta_0$. Write the Fourier series $p_\pi(y) = \sum_{\omega \in \Lambda^*} \widehat{p_\pi}(\omega) e^{2\pi i \langle \omega, y \rangle} = \frac{1}{\det \Lambda} \sum_{\omega \in \Lambda^*} \delta_0(\omega) e^{2\pi i \langle \omega, y \rangle} = \frac{1}{\det \Lambda}$.

The quantised distribution has an analogous expression to Theorem 1:

$$\Phi_{X_\mathcal{Q}}(z) = \sum_{\lambda \in \Lambda} \left(\int_\mathcal{D} p(x + \lambda) \mathrm{d}x \right) e^{2\pi i \langle z, \lambda \rangle} = \int_{\mathbb{R}^n} p(x) e^{2\pi i \langle z, \mathcal{Q}(x) \rangle} \mathrm{d}x. \quad (5)$$

The fact that $e^{2\pi i \langle \omega, \mathcal{Q}(x) \rangle} = 1$ for $\omega \in \Lambda^*$ implies that $\Phi_{X_\mathcal{Q}}$ is a Λ^*-periodic function, in a similar behaviour to that of a wrapped distribution of the dual lattice. This similarity is emphasised by the following result, which is dual to Theorem 3.

Theorem 4. *A random variable X has deterministic quantisation $p_\mathcal{Q} = \delta_0$ if, and only if, $\Phi_{X_\mathcal{Q}}$ is constant.*

Proof. If $p_\mathcal{Q} = \delta_0$, then $\Phi_{X_\mathcal{Q}}(z) = \sum_{\lambda \in \Lambda} \delta_0(\lambda) e^{2\pi i \langle z, \lambda \rangle} = \delta_0(0) = 1$. On the other hand, if $\Phi_{X_\mathcal{Q}}(z)$ is constant, by evaluating at $z = 0$, we see that its value must necessarily be 1, and hence $\widehat{p_\mathcal{Q}} \equiv 1/(\det \Lambda^*)$. The inverse Fourier transform of $\widehat{p_\mathcal{Q}}$ is $p_\mathcal{Q}(\lambda) = \int_{\mathcal{D}^*} \widehat{p_\mathcal{Q}}(z) e^{-2\pi i \langle z, \lambda \rangle} \mathrm{d}z = \frac{1}{\det \Lambda^*} \int_{\mathcal{D}^*} e^{2\pi i \langle z, \lambda \rangle} \mathrm{d}z$. Since $p_\mathcal{Q}$ is a probability distribution and $p_\mathcal{Q}(0) = 1$, we conclude that $p_\mathcal{Q} = \delta_0$.

Remark 1. It should be noted that, while the previous results recover, in dimension one, classical results from signal processing literature, they are presented here in the more general form for multi-dimensional lattices. In particular, the 'frequency domain' characterisation in Theorem 4 is analogous to the Nyquist's criterion for inter-symbol interference from digital communications [10, § 11.3], in dimension one, with $\Lambda = \mathbb{Z}$. This correspondence been remarked in [9,18] for Theorem 3, which is somewhat analogous too, in the dual form.

4   Dithered Quantisation

4.1   Subtractive and Non-subtractive Dithering

Dithering is a technique that consists in adding noise in the quantisation process in order to improve the perceptual quality of the quantised signal [8], and has been studied also in the context of lattice quantisation [18, Ch. 4]. In *subtractive dithering*, a random variable $N \in \mathbb{R}^n$ called *dither* is added to the source signal $X \in \mathbb{R}^n$ before quantisation, and then subtracted after it, resulting in the reconstructed signal

$$\hat{X} = \mathcal{Q}(X + N) - N. \tag{6}$$

In this case, the quantisation error can be conveniently written as a wrapping: $\hat{X} - X = \mathcal{Q}(X + N) - (X + N) = -\pi(X + N)$.

The following result is a straightforward generalisation of [18, Lem. 4.2.1] to non-uniform dithering and shows that, without loss of generality, we can limit ourselves to dither defined over the fundamental domain $\mathcal{D}$.

Theorem 5. *Subtractive dithered quantisation with dither N over $\mathbb{R}^n$ is equivalent to subtractive dithered quantisation by its wrapping N_π.*

Proof. We show that the reconstruction is the same with dither N or its wrapping N_π. Denoting $N_\mathcal{Q} := \mathcal{Q}(N)$ and noting that $\mathcal{Q}(X + N_\mathcal{Q}) = \mathcal{Q}(X) + N_\mathcal{Q}$, we have $\mathcal{Q}(X + N_\pi) - N_\pi = \mathcal{Q}(X + N - N_\mathcal{Q}) - N_\pi = \mathcal{Q}(X + N) - N_\mathcal{Q} - N_\pi = \mathcal{Q}(X + N) - N$.

It can be shown that if N is modulo-uniform dither, then the quantisation error is independent of the source and is uniformly distributed over $\mathcal{D}$ [9, Thm. 4]. This explains the interest in modulo-uniform dither, even if other forms of dither have been considered in the early literature (e.g., sinusoidal, Gaussian) [3].

In another type of dithering, called *non-subtractive* [8], the dither random variable N is not subtracted after the quantisation. As compared to (6), in this case, the reconstructed signal is simply $\hat{X} = \mathcal{Q}(X + N)$.

4.2   Non-collaborative Dithering

The application of subtractive dithering schemes requires collaboration between encoder and decoder, as they have to agree on the dither random variable that is added and then subtracted during the quantisation—this can be done by generating pseudo-random variables with the same seed. We propose to consider a case in which such collaboration is not possible: the decoder receives a quantised signal and, nonetheless, wishes to improve its perceptual quality by adding random noise. In this case, the reconstructed signal is[3]

$$\hat{X} = \mathcal{Q}(X) + N = X_\mathcal{Q} + N. \tag{7}$$

We are interested in finding the best independent dither N to improve the reconstructed signal. Despite being a common distortion measure for quantised signals, the mean squared error (MSE) fails to quantify the perceptual quality of such signals, especially at low rates. Some works have proposed instead to compare the similarity between the original and quantised distributions as to

---

[3] A uniform 'dither' N of this form has been used to find discrete entropy upper bounds in Massey-type inequalities [15].

assess perceptual quality, particularly for audio coding [11,12]. Following these, we measure the distortion incurred by quantisation with the KL-divergence. The next result shows that the choice that minimises $D_{\mathrm{KL}}[p_X\|p_{X_\mathcal{Q}+N}]$ is precisely to let N have the same distribution as the wrapped random variable X_π. The minimising value $I(X_\pi, X_\mathcal{Q})$ achieved in this case has been studied in [14].

Theorem 6. *Let X be a random variable, $X_\mathcal{Q}$ its quantisation, and N a random variable over $\mathcal{D}$, independent of $X_\mathcal{Q}$. Then*

$$D_{\mathrm{KL}}[p_X\|p_{X_\mathcal{Q}+N}] \geq I(X_\pi; X_\mathcal{Q}), \tag{8}$$

with equality if, and only if, N is distributed according to p_π.

Proof. Denote p_X and p_N the pdf's of X and N, respectively; the pdf of $X_\mathcal{Q}+N$ is $p_{X_\mathcal{Q}+N}(x) = p_\mathcal{Q}(\mathcal{Q}(x))p_N(\pi(x))$. The result readily follows from the non-negativity of the Kullback-Leibler divergence:

$$D_{\mathrm{KL}}[p_X\|p_{X_\mathcal{Q}+N}] - I(X_\pi; X_\mathcal{Q}) = D_{\mathrm{KL}}[p_X\|p_{X_\mathcal{Q}+N}] - D_{\mathrm{KL}}[p_X\|p_{X_\mathcal{Q}+X_\pi}]$$
$$= \mathrm{E}_X\left[\log\frac{p_\pi(\pi(X))}{p_N(\pi(X))}\right] = \mathrm{E}_{X_\pi}\left[\log\frac{p_\pi(\pi(X))}{p_N(\pi(X))}\right] = D_{\mathrm{KL}}[p_\pi\|p_N] \geq 0,$$

with equality if, and only if, $p_\pi = p_N$ almost everywhere.

This result is illustrated with a simple experiment: we quantise samples from the set $\mathcal{A} := \{0, 1, \ldots, 255\}$ within k levels, using uniform, Gaussian[4] and wrapped dithered one-dimensional quantisers. First, we consider samples generated approximately according to a Gaussian and a Beta distribution; then we quantise the 8-bit greyscale 'Fishing Boat' image from the SIPI image database [1]. In practice, quantisation is done with a finite subset of the lattice and the quantised values are limited to some bounded range. We use a shifted

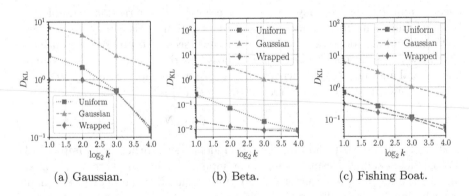

(a) Gaussian.   (b) Beta.   (c) Fishing Boat.

Fig. 2. Performance of different dithered quantisers.

---

[4] We note that the Gaussian dither is not confined to the fundamental region. Its variance was optimised by grid search.

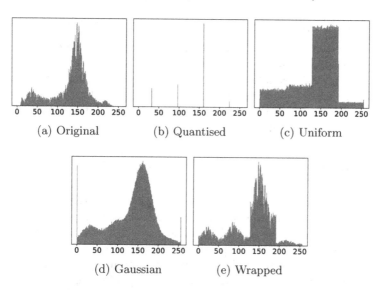

(a) Original   (b) Quantised   (c) Uniform

(d) Gaussian   (e) Wrapped

Fig. 3. Histograms of 'Fishing Boat' image before and after 2-bit quantisation.

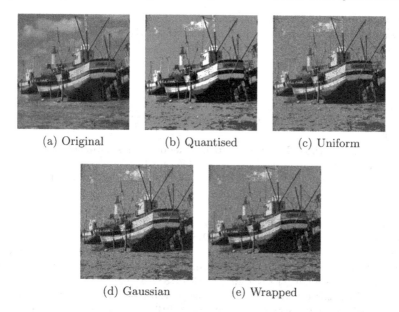

(a) Original   (b) Quantised   (c) Uniform

(d) Gaussian   (e) Wrapped

Fig. 4. 'Fishing Boat' image before and after 2-bit quantisation.

version of the $\left(\frac{255}{k}\right)\mathbb{Z}$ lattice (affine lattice). Shifting the lattice is equivalent to shifting the distribution, and this operation preserves Theorem 6.

We assess the KL-divergence from the original distribution to the quantised one for different number of levels k. To ensure this computation is possible and avoid numerical issues, the probability p_i of symbol $i \in \mathcal{A}$ is estimated from its absolute frequency n_i as $p_i = \frac{n_i+0.5}{\sum_{j=0}^{255}(n_i+0.5)}$, the Krichevsky-Trofimov estimator, which is a Bayesian estimation with Dirichlet prior with parameter $(0.5, \cdots, 0.5)$. We observe on Fig. 2 that, at extreme low rates, dithering with the wrapped distribution produces the smallest KL-divergence, suggesting improved perceptual quality. Figure 3 and 4 show the histograms and corresponding images before and after quantisation of the 'Fishing Boat' image with a two-bit quantiser. Gaussian dither uses $\sigma^2 = 500$, resulting from grid search.

5   Conclusion

In this work, we have first studied the characteristic functions of lattice tiling decomposition of random variables; then, we used this decomposition to describe dithered quantisation, showing that, in the non-collaborative setup, wrapped dither minimises the KL-divergence from the original distribution. Future directions include further studying the duality of wrapping and quantisation, particularly in connection to the flatness factor, and performing experiments with high-dimensional lattices, other types of signals and dithering schemes.

Acknowledgements. The authors thank the anonymous reviewers for their careful reading and valuable comments, which have improved the original manuscript.

References

1. The USC-SIPI image database, https://sipi.usc.edu/database/
2. Aharonov, D., Regev, O.: Lattice problems in NP ∩ CoNP. J. ACM **52**(5), 749–765 (2005)
3. Carbone, P., Petri, D.: Effect of additive dither on the resolution of ideal quantizers. IEEE Trans. Instrum. Meas. **43**(3), 389–396 (1994)
4. Chung, K.M., Dadush, D., Liu, F.H., Peikert, C.: On the lattice smoothing parameter problem. In: 2013 IEEE Conference on Computational Complexity. pp. 230–241. Stanford, USA (2013)
5. Conway, J.H., Sloane, N.J.A.: Sphere packings, lattices and groups. Springer, New York (1999). https://doi.org/10.1007/978-1-4757-6568-7
6. Costa, S.I.R., Oggier, F., Campello, A., Belfiore, J.-C., Viterbo, E.: Lattices Applied to Coding for Reliable and Secure Communications. SM, Springer, Cham (2017). https://doi.org/10.1007/978-3-319-67882-5
7. Ebeling, W.: Lattices and codes: a course partially based on lectures by Friedrich Hirzebruch, 3rd edn. Springer, Wiesbaden (2013). https://doi.org/10.1007/978-3-658-00360-9
8. Gray, R., Stockham, T.: Dithered quantizers. IEEE Trans. Inf. Theor. **39**(3), 805–812 (1993)
9. Kirac, A., Vaidyanathan, P.: Results on lattice vector quantization with dithering. IEEE Trans. Circuits Syst. II: Analog Digit. Signal Process **43**(12), 811–826 (1996)
10. Lapidoth, A.: A foundation in digital communication, 2nd edn. Cambridge Univ. Press, Cambridge (2017)

11. Li, M., Kleijn, W.B.: Quantization with constrained relative entropy and its application to audio coding. In: 127th Audio Engineering Society Convention. pp. 401–408. New York USA (2009)
12. Li, M., Klejsa, J., Kleijn, W.B.: Distribution preserving quantization with dithering and transformation. IEEE Signal Process. Lett. **17**(12), 1014–1017 (2010)
13. Mardia, K.V., Jupp, P.E.: Directional statistics. Wiley, Chichester (2000)
14. Meneghetti, F.C.C., Miyamoto, H.K., Costa, S.I.R.: Information properties of a random variable decomposition through lattices. Phys. Sci. Forum **5**(1), 1–9 (2022)
15. Rioul, O.: Variations on a theme by Massey. IEEE Trans. Inf. Theor. **68**(5), 2813–2828 (2022)
16. Rudin, W.: Fourier analysis on groups. Wiley, New York (1990)
17. Stein, E.M., Weiss, G.: Introduction to Fourier Analysis on Euclidean Spaces. Princenton Univ. Press, Princenton (1971)
18. Zamir, R.: Lattice coding for signals and networks: a structured coding approach to quantization, modulation and multiuser information theory. Cambridge Univ. Press, Cambridge (2014)

Geometric Reduction for Identity Testing of Reversible Markov Chains

Geoffrey Wolfer[1]([envelope])[iD] and Shun Watanabe[2][iD]

[1] RIKEN Center for AI Project, 1-4-1 Nihonbashi,
Chuo-ku, Tokyo 103-0027, Japan
geoffrey.wolfer@riken.jp
[2] Department of Computer and Information Sciences, Tokyo University of
Agriculture and Technology, 2-24-16 Nakamachi, Koganei-shi, Tokyo 184-8588, Japan
shunwata@cc.tuat.ac.jp

Abstract. We consider the problem of testing the identity of a reversible Markov chain against a reference from a single trajectory of observations. Employing the recently introduced notion of a lumping-congruent Markov embedding, we show that, at least in a mildly restricted setting, testing identity to a reversible chain reduces to testing to a symmetric chain over a larger state space and recover state-of-the-art sample complexity for the problem.

Keywords: Irreducible Markov chains · Information geometry · Identity testing · Markov embedding · Congruent embedding · Lumpability

1 Introduction

Uniformity testing is the flagship problem of the modern distribution testing [1] research program. From n independent observations sampled from an unknown distribution μ on a finite space $\mathcal{X}$, the goal is to distinguish between the two cases where μ is uniform and μ is ε-far from being uniform with respect to some notion of distance. The complexity of this problem in total variation is known to be [12] of the order[1] of $\Theta(\sqrt{|\mathcal{X}|}/\varepsilon^2)$, which compares favorably with the linear dependency in $|\mathcal{X}|$ required for estimating the distribution to precision ε [17]. Interestingly, the uniform distribution can be replaced by any arbitrary reference at same statistical cost. In fact, Goldreich [7] proved that the latter problem formally reduces to the former. Inspired by his approach, we seek and obtain a reduction result in the much less understood and more challenging Markovian setting.

---

[1] As is customary in the property testing literature, we respectively write $\Theta, \mathcal{O}$ and Ω for tight, upper and lower bounds, and the tilda notation suppresses lower-order logarithmic factors in any parameter.

© The Author(s), under exclusive license to Springer Nature Switzerland AG 2023
F. Nielsen and F. Barbaresco (Eds.): GSI 2023, LNCS 14071, pp. 328–337, 2023.
https://doi.org/10.1007/978-3-031-38271-0_32

Informal Markovian Problem Statement — The scientist is given the full description of a reference transition matrix $\overline{P}$ and a single Markov chain X_1^n sampled with respect to some unknown transition operator P and arbitrary initial distribution. For fixed proximity parameter $\varepsilon > 0$, the goal is to design an algorithm that distinguishes between the two cases $P = \overline{P}$ and $K(P, \overline{P}) > \varepsilon$, with high probability, where K is a contrast function[2] between stochastic matrices.

Related Work — Under the contrast function (1) described in Sect. 2, and the hypothesis that P and $\overline{P}$ are both irreducible and symmetric over a finite space $\mathcal{X}$, a first tester with sample complexity $\tilde{\mathcal{O}}(|\mathcal{X}|/\varepsilon + h)$, where h [4, Definition 3] is the hitting time of the reference chain, and a lower bound in $\Omega(|\mathcal{X}|/\varepsilon)$, were obtained in [4]. In [3], a graph partitioning algorithm delivers, under the same symmetry assumption, a testing procedure with sample complexity $\mathcal{O}(|\mathcal{X}|/\varepsilon^4)$, i.e. independent of hitting properties. More recently, [6] relaxed the symmetry requirement, replacing it with a more natural reversibility assumption. The algorithm therein has a sample complexity of $\mathcal{O}(1/(\overline{\pi}_\star \varepsilon^4))$, where $\overline{\pi}_\star$ is the minimum stationary probability of the reference $\overline{P}$, gracefully recovering [3] under symmetry. In parallel, [18] started and [2] complemented the research program of inspecting the problem under the infinity norm for matrices, and derived nearly minimax-optimal bounds.

Contribution — We show how to mostly recover [6] from [3] under additional assumptions (see Sect. 3), with a technique based on a geometry preserving embedding. We obtain a more economical proof than [6], which went through the process of re-deriving a graph partitioning algorithm for the reversible case. Furthermore, the impact of our approach, by its generality, stretches beyond the task at hand and is also applicable to related inference problems (see Remark 2).

2 Preliminaries

We let $\mathcal{X}, \mathcal{Y}$ be finite sets, and denote $\mathcal{P}(\mathcal{X})$ the set of all probability distributions over $\mathcal{X}$. All vectors are written as row vectors. For matrices A, B, $\rho(A)$ is the spectral radius of A, $A \circ B$ is the Hadamard product of A and B defined by $(A \circ B)(x, x') = A(x)B(x')$ and $A^{\circ 1/2}(x, x') = \sqrt{A(x, x')}$. For $n \in \mathbb{N}$, we use the compact notation $x_1^n = (x_1, \ldots, x_n)$. $\mathcal{W}(\mathcal{X})$ is the set of all row-stochastic matrices over the state space $\mathcal{X}$, and π is called a stationary distribution for $P \in \mathcal{W}(\mathcal{X})$ when $\pi P = \pi$.

Irreducibility and Reversibility — Let $(\mathcal{X}, \mathcal{D})$ be a digraph with vertex set $\mathcal{X}$ and edge-set $\mathcal{D} \subset \mathcal{X}^2$. When $(\mathcal{X}, \mathcal{D})$ is strongly connected, a Markov chain with connection graph $(\mathcal{X}, \mathcal{D})$ is said to be irreducible. We write $\mathcal{W}(\mathcal{X}, \mathcal{D})$ for the set

---

[2] General contrast functions under consideration satisfy identity of indiscernibles and non-negativity (e.g. proper metrics induced from matrix norms), and need not satisfy symmetry or the triangle inequality (e.g. information divergence rate between Markov processes).

of irreducible stochastic matrices over $(\mathcal{X}, \mathcal{D})$. When $P \in \mathcal{W}(\mathcal{X}, \mathcal{D})$, π is unique and we denote $\pi_\star \doteq \min_{x \in \mathcal{X}} \pi(x) > 0$ the minimum stationary probability. When P satisfies the detailed-balance equation $\pi(x)P(x, x') = \pi(x')P(x', x)$ for any $(x, x') \in \mathcal{D}$, we say that P is reversible.

Lumpability — In contradistinction with the distribution setting, merging symbols in a Markov chain may break the Markov property, resulting in a hidden Markov model. For $P \in \mathcal{W}(\mathcal{Y}, \mathcal{E})$ and a surjective map $\kappa \colon \mathcal{Y} \to \mathcal{X}$ merging elements of $\mathcal{Y}$ together, we say that P is κ-lumpable [10] when the output process still defines a Markov chain. Introducing $\mathcal{S}_x = \kappa^{-1}(\{x\})$ for the collection of symbols that merge into $x \in \mathcal{X}$, lumpability was characterized by [10, Theorem 6.3.2] as follows. P is κ-lumpable, when for any $x, x' \in \mathcal{X}$, and $y_1, y_2 \in \mathcal{S}_x$, it holds that

$$P(y_1, \mathcal{S}_{x'}) = P(y_2, \mathcal{S}_{x'}).$$

The lumped transition matrix $\kappa_\star P \in \mathcal{W}(\mathcal{X}, \kappa_2(\mathcal{E}))$, with connected edge set

$$\kappa_2(\mathcal{E}) \doteq \left\{ (x, x') \in \mathcal{X}^2 \colon \exists (y, y') \in \mathcal{E}, (\kappa(y), \kappa(y')) = (x, x') \right\},$$

is then given by

$$\kappa_\star P(x, x') = P(y, \mathcal{S}_{x'}), \text{ for some } y \in \mathcal{S}_x.$$

Contrast Function — We consider the following notion of discrepancy between two stochastic matrices $P, P' \in \mathcal{W}(\mathcal{X})$,

$$K(P, P') \doteq 1 - \rho \left(P^{\circ 1/2} \circ P'^{\circ 1/2} \right). \tag{1}$$

Although K made its first appearance in [4] in the context of Markov chain identity testing, its inception can be traced back to Kazakos [9]. K is directly related to the Rényi entropy of order $1/2$, and asymptotically connected to the Bhattacharyya/Hellinger distance between trajectories (see e.g. proof of Lemma 2). It is instructive to observe that K vanishes on chains that share an identical strongly connected component and does not satisfy the triangle inequality for reducible matrices, hence is not a proper metric on $\mathcal{W}(\mathcal{X})$ [4, p.10, footnote 13]. Some additional properties of K of possible interest are listed in [6, Section 7].

Reduction Approach for Identity Testing of Distributions — Problem reduction is ubiquitous in the property testing literature. Our work takes inspiration from [7], who introduced two so-called "stochastic filters" in order to show how in the distribution setting, identity testing was reducible to uniformity testing, thereby recovering the known complexity of $\mathcal{O}(\sqrt{|\mathcal{X}|}/\varepsilon^2)$ obtained more directly by [14]. Notable works also include [5], who reduced a collection of distribution testing problems to ℓ_2-identity testing.

3 The Restricted Identity Testing Problem

Let $\mathcal{V}_{\text{test}} \subset \mathcal{W}(\mathcal{X})$ be a class of stochastic matrices of interest, and let $\overline{P} \in \mathcal{V}_{\text{test}}$ be a fixed reference. The identity testing problem consists in determining, with high probability, from a single stream of observations $X_1^n = X_1, \ldots, X_n$ drawn according to a transition matrix $P \in \mathcal{V}_{\text{test}}$, whether

$$P \in \mathcal{H}_0 \doteq \left\{\overline{P}\right\}, \text{ or } P \in \mathcal{H}_1(\varepsilon) \doteq \left\{P \in \mathcal{V}_{\text{test}} \colon K(P, \overline{P}) > \varepsilon\right\}.$$

We note the presence of an exclusion region, and regard the problem as a Bayesian testing problem with a prior which is uniform over the two hypotheses classes $\mathcal{H}_0$ and $\mathcal{H}_1(\varepsilon)$ and vanishes on the exclusion region. Casting our problem in the minimax framework, the worst-case error probability $e_n(\phi, \varepsilon)$ of a given test $\phi \colon \mathcal{X}^n \to \{0, 1\}$ is defined as

$$2e_n(\phi, \varepsilon) \doteq \mathbb{P}_{X_1^n \sim \overline{\pi}, \overline{P}}\left(\phi(X_1^n) = 1\right) + \sup_{P \in \mathcal{H}_1(\varepsilon)} \mathbb{P}_{X_1^n \sim \pi, P}\left(\phi(X_1^n) = 0\right).$$

We subsequently define the minimax risk $\mathcal{R}_n(\varepsilon)$ as,

$$\mathcal{R}_n(\varepsilon) \doteq \min_{\phi \colon \mathcal{X}^n \to \{0, 1\}} e_n(\phi, \varepsilon),$$

where the minimum is taken over all —possibly randomized— testing procedures. For a confidence parameter δ, the sample complexity is

$$n_\star(\varepsilon, \delta) \doteq \min\left\{n \in \mathbb{N} \colon \mathcal{R}_n(\varepsilon) < \delta\right\}.$$

We briefly recall the assumptions made in [6]. For $(P, \overline{P}) \in (\mathcal{V}_{\text{test}}, \mathcal{H}_0)$,

(A.1) P and $\overline{P}$ are irreducible and reversible.
(A.2) P and $\overline{P}$ share the same[3] stationary distribution $\overline{\pi} = \pi$.

The following additional assumptions will make our approach readily applicable.

(B.1) $P, \overline{P}$ and share the same connection graph, $P, \overline{P} \in \mathcal{W}(\mathcal{X}, \mathcal{D})$.
(B.2) The common stationary distribution is rational, $\overline{\pi} \in \mathbb{Q}^{\mathcal{X}}$.

Remark 1. A sufficient condition for $\overline{\pi} \in \mathbb{Q}^{\mathcal{X}}$ is $\overline{P}(x, x') \in \mathbb{Q}$ for any $x, x' \in \mathcal{X}$.

Without loss of generality, we express $\overline{\pi} = \left(p_1, p_2, \ldots, p_{|\mathcal{X}|}\right)/\Delta$, for some $\Delta \in \mathbb{N}$, and $p \in \mathbb{N}^{|\mathcal{X}|}$ where $0 < p_1 \leq p_2 \leq \cdots \leq p_{|\mathcal{X}|} < \Delta$. We henceforth denote by $\mathcal{V}_{\text{test}}$ the subset of stochastic matrices that verify assumptions $(A.1), (A.2), (B.1)$ and $(B.2)$ with respect to the fixed $\overline{\pi} \in \mathcal{P}(\mathcal{X})$. Our below-stated theorem provides an upper bound on the sample complexity $n_\star(\varepsilon, \delta)$ in $\widetilde{\mathcal{O}}(1/(\overline{\pi}_\star \varepsilon))$.

---

[3] We note that [6] also slightly loosen the requirement of having a matching stationary distributions to being close in the sense where $\|\pi/\overline{\pi} - 1\|_\infty < \varepsilon$.

Theorem 1. Let $\varepsilon, \delta \in (0,1)$ and let $\overline{P} \in \mathcal{V}_{\text{test}} \subset \mathcal{W}(\mathcal{X}, \mathcal{D})$. There exists a randomized testing procedure $\phi \colon \mathcal{X}^n \to \{0,1\}$, with $n = \tilde{\mathcal{O}}(1/(\overline{\pi}_\star \varepsilon^4))$, such that the following holds. For any $P \in \mathcal{V}_{\text{test}}$ and X_1^n sampled according to P, ϕ distinguishes between the cases $P = \overline{P}$ and $K(P, \overline{P}) > \varepsilon$ with error probability less than δ.

Proof (sketch). Our strategy can be broken down into two steps. First, we employ a transformation on Markov chains, termed Markov embedding [20], in order to symmetrize both the reference chain (algebraically, by computing the new transition matrix) and the unknown chain (operationally, by simulating an embedded trajectory). Crucially, our transformation preserves the contrast between two chains and their embedded version (Lemma 2). Second, we invoke the known tester [3] for symmetric chains as a black box and report its output. The proof is deferred to Sect. 6.

Remark 2. Our reduction approach has applicability beyond recovery of the sample complexity of [6], for instance in the tolerant testing setting, where the two competing hypotheses are

$$K(P, \overline{P}) < \varepsilon/2 \text{ and } K(P, \overline{P}) > \varepsilon.$$

Even in the symmetric setting, this problem remains open. Our technique shows that future work can focus on solving the problem under a symmetry assumption, as we provide a natural extension to the reversible class.

4   Symmetrization of Reversible Markov Chains

Information geometry — Our construction and notation follow [11], who established a dually-flat structure

$$(\mathcal{W}(\mathcal{X}, \mathcal{D}), \mathfrak{g}, \nabla^{(e)}, \nabla^{(m)})$$

on the space of irreducible stochastic matrices, where $\mathfrak{g}$ is a Riemannian metric, and $\nabla^{(e)}, \nabla^{(m)}$ are dual affine (exponential and mixture) connections. Introducing a model $\mathcal{V} = \{P_\theta : \theta \in \Theta \subset \mathbb{R}^d\} \subset \mathcal{W}(\mathcal{X}, \mathcal{D})$, we write $P_\theta \in \mathcal{V}$ for the transition matrix at coordinates $\theta = (\theta^1, \ldots, \theta^d)$, and where d is the manifold dimension of $\mathcal{V}$. Using the shorthand $\partial_i \cdot \doteq \partial \cdot /\partial\theta^i$, the Fisher metric is expressed [11, (9)] in the chart induced basis $(\partial_i)_{i \in [d]}$ as

$$\mathfrak{g}_{ij}(\theta) = \sum_{(x,x') \in \mathcal{D}} \pi_\theta(x) P_\theta(x, x') \partial_i \log P_\theta(x, x') \partial_j \log P_\theta(x, x'), \text{ for } i, j \in [d]. \quad (2)$$

Following this formalism, it is possible to define mixture families (m-families) and exponential families (e-families) of stochastic matrices [8,11].

Example 1. The class $\mathcal{W}_{\text{rev}}(\mathcal{X}, \mathcal{D})$ of reversible Markov chains irreducible over a connection graph $(\mathcal{X}, \mathcal{D})$ forms both an e-family and an m-family of dimension

$$\dim \mathcal{W}_{\text{rev}}(\mathcal{X}, \mathcal{D}) = \frac{|\mathcal{D}| + |\ell(\mathcal{D})|}{2} - 1,$$

where $\ell(\mathcal{D}) \subset \mathcal{D}$ is the set of loops present in the connection graph [19, Theorem 3,5].

Embeddings — The operation converse to lumping is embedding into a larger space of symbols. In the distribution setting, Markov morphisms were introduced by Čencov [16] as the natural operations on distributions. In the Markovian setting, [20] proposed the following notion of an embedding for stochastic matrices.

Definition 1 (Markov embedding for Markov chains [20]). *We call Markov embedding, a map $\Lambda_\star \colon \mathcal{W}(\mathcal{X}, \mathcal{D}) \to \mathcal{W}(\mathcal{Y}, \mathcal{E}), P \mapsto \Lambda_\star P$, such that for any $(y, y') \in \mathcal{E}$,*

$$\Lambda_\star P(y, y') = P(\kappa(y), \kappa(y'))\Lambda(y, y'),$$

and where κ and Λ satisfy the following requirements

(i) $\kappa \colon \mathcal{Y} \to \mathcal{X}$ is a lumping function for which $\kappa_2(\mathcal{E}) = \mathcal{D}$.
(ii) Λ is a positive function over the edge set, $\Lambda \colon \mathcal{E} \to \mathbb{R}_+$.
(iii) Writing $\bigcup_{x \in \mathcal{X}} \mathcal{S}_x = \mathcal{Y}$ for the partition defined by κ, Λ is such that for any $y \in \mathcal{Y}$ and $x' \in \mathcal{X}$,

$$(\kappa(y), x') \in \mathcal{D} \implies (\Lambda(y, y'))_{y' \in \mathcal{S}_{x'}} \in \mathcal{P}(\mathcal{S}_{x'}).$$

The above embeddings are characterized as the linear maps over the space of lumpable matrices that satisfy a set of monotonicity requirements and are congruent with respect to the lumping operation [20, Theorem 3.1]. When for any $y, y' \in \mathcal{Y}$, it additionally holds that $\Lambda(y, y') = \Lambda(y')\delta\left[(\kappa(y), \kappa(y')) \in \mathcal{D}\right]$, the embedding $\Lambda_\star$ is called memoryless [20, Section 3.4.2] and is e/m-geodesic affine [20, Th. 3.2, Lemma 3.6], preserving both e-families and m-families of stochastic matrices.

Given $\overline{\pi}$ and Δ as defined in Sect. 3, from [20, Corollary 3.3], there exists a lumping function $\kappa \colon [\Delta] \to \mathcal{X}$, and a memoryless embedding $\sigma_\star^{\overline{\pi}} \colon \mathcal{W}(\mathcal{X}, \mathcal{D}) \to \mathcal{W}([\Delta], \mathcal{E})$ with $\mathcal{E} = \left\{(y, y') \in [\Delta]^2 \colon (\kappa(y), \kappa(y')) \in \mathcal{D}\right\}$, such that $\sigma_\star^{\overline{\pi}} P$ is symmetric. Furthermore, identifying $\mathcal{X} \cong \{1, 2, \dots, |\mathcal{X}|\}$, its existence is constructively given by

$$\kappa(j) = \operatorname*{arg\,min}_{1 \le i \le |\mathcal{X}|} \left\{ \sum_{k=1}^{i} p_k \ge j \right\}, \text{ with } \sigma^{\overline{\pi}}(j) = p_{\kappa(j)}^{-1}, \text{ for any } 1 \le j \le \Delta.$$

As a consequence, we obtain 1. and 2. below.

1. The expression of $\sigma_\star^{\overline{\pi}} P$ following algebraic manipulations in Definition 1.
2. A randomized algorithm to memorylessly simulate trajectories from $\sigma_\star^{\overline{\pi}} P$ out of trajectories from P (see [20, Section 3.1]). Namely, there exists a stochastic mapping $\Psi^{\overline{\pi}} \colon \mathcal{X} \to \Delta$ such that,

$$X_1, \dots, X_n \sim P \implies \Psi^{\overline{\pi}}(X_1^n) = \Psi^{\overline{\pi}}(X_1), \dots, \Psi^{\overline{\pi}}(X_n) \sim \sigma_\star^{\overline{\pi}} P.$$

5   Contrast Preservation

It was established in [20, Lemma 3.1] that similar to their distribution counterparts, Markov embeddings in Definition 1 preserve the Fisher information metric $\mathfrak{g}$ in (2), the affine connections $\nabla^{(e)}, \nabla^{(m)}$ and the informational (Kullback-Leibler) divergence between points. In this section, we show that memoryless embeddings, such as the symmetrizer $\sigma_*^{\overline{\pi}}$ introduced in Sect. 4, also preserve the contrast function K. Our proof will rely on first showing that the memoryless embeddings of [20, Section 3.4.2] induce natural Markov morphisms [15] from distributions over $\mathcal{X}^n$ to $\mathcal{Y}^n$.

Lemma 1. *Let a lumping function* $\kappa\colon \mathcal{Y} \to \mathcal{X}$, *and*

$$L_*\colon \mathcal{W}(\mathcal{X},\mathcal{D}) \to \mathcal{W}(\mathcal{Y},\mathcal{E})$$

be a κ-*congruent memoryless Markov embedding. For* $P \in \mathcal{W}(\mathcal{X},\mathcal{D})$, *let* $Q^n \in \mathcal{P}(\mathcal{X}^n)$ *(resp.* $\widetilde{Q}^n \in \mathcal{P}(\mathcal{Y}^n)$) *be the unique distribution over stationary paths of length* n *induced from* P *(resp.* L_*P). *Then there exists a Markov morphism* $M_*\colon \mathcal{P}(\mathcal{X}^n) \to \mathcal{P}(\mathcal{Y}^n)$ *such that* $M_*Q^n = \widetilde{Q}^n$.

Proof. Let $\kappa_n\colon \mathcal{Y}^n \to \mathcal{X}^n$ be the lumping function on blocks induced from κ,

$$\forall y_1^n \in \mathcal{Y}^n, \kappa_n(y_1^n) = (\kappa(y_t))_{1\le t\le n} \in \mathcal{X}^n,$$

and introduce

$$\mathcal{Y}^n = \bigcup_{x_1^n\in\mathcal{X}^n} \mathcal{S}_{x_1^n}, \text{ with } \mathcal{S}_{x_1^n} = \{y_1^n \in \mathcal{Y}^n\colon \kappa_n(y_1^n) = x_1^n\},$$

the partition associated to κ_n. For any realizable path $x_1^n, Q^n(x_1^n) > 0$, we define a distribution $M^{x_1^n} \in \mathcal{P}(\mathcal{Y}^n)$ concentrated on $\mathcal{S}_{x_1^n}$, and such that for any $y_1^n \in \mathcal{S}_{x_1^n}$, $M^{x_1^n}(y_1^n) = \prod_{t=1}^n L(y_t)$. Non-negativity of $M^{x_1^n}$ is immediate, and

$$\sum_{y_1^n\in\mathcal{Y}^n} M^{x_1^n}(y_1^n) = \sum_{y_1^n\in\mathcal{Y}^n\colon \kappa_n(y_1^n)=x_1^n} M^{x_1^n}(y_1^n) = \prod_{t=1}^n \left(\sum_{y_t\in\mathcal{S}_{x_t}} L(y_t)\right) = 1,$$

thus $M^{x_1^n}$ is well-defined. Furthermore, for $y_1^n \in \mathcal{Y}^n$, it holds that

$$\widetilde{Q}^n(y_1^n) = L_*\pi(y_1) \prod_{t=1}^{n-1} L_*P(y_t,y_{t+1}) \overset{(\spadesuit)}{=} \pi(\kappa(y_1))L(y_1) \prod_{t=1}^{n-1} P(\kappa(y_t),\kappa(y_{t+1}))L(y_{t+1})$$

$$= Q^n(\kappa(y_1),\ldots,\kappa(y_n)) \prod_{t=1}^n L(y_t) = Q^n(\kappa_n(y_1^n)) \prod_{t=1}^n L(y_t)$$

$$= \sum_{x_1^n\in\mathcal{X}^n} Q^n(\kappa_n(y_1^n))M^{x_1^n}(y_1^n) = M_*Q^n(y_1^n),$$

where ($\spadesuit$) stems from [20, Lemma 3.5], whence our claim holds.

Lemma 1 essentially states that the following diagram commutes

$$\begin{array}{ccc} \mathcal{W}(\mathcal{X},\mathcal{D}) & \overset{L_*}{\longrightarrow} & L_*\mathcal{W}(\mathcal{X},\mathcal{D}) \\ \downarrow & & \downarrow \\ \mathcal{Q}^n_{\mathcal{W}(\mathcal{X},\mathcal{D})} & \overset{M_*}{\longrightarrow} & \mathcal{Q}^n_{L_*\mathcal{W}(\mathcal{X},\mathcal{D})}, \end{array}$$

for the Markov morphism $M_\star$ induced by $L_\star$, and where we denoted $\mathcal{Q}^n_{\mathcal{W}(\mathcal{X},\mathcal{D})} \subset \mathcal{P}(\mathcal{X}^n)$ for the set of all distributions over paths of length n induced from the family $\mathcal{W}(\mathcal{X},\mathcal{D})$. As a consequence, we can unambiguously write $L_\star Q^n \in \mathcal{Q}^n_{L_\star \mathcal{W}(\mathcal{X},\mathcal{D})}$ for the distribution over stationary paths of length n that pertains to $L_\star P$.

Lemma 2. *Let $L_\star \colon \mathcal{W}(\mathcal{X},\mathcal{D}) \to \mathcal{W}(\mathcal{Y},\mathcal{E})$ be a memoryless embedding,*

$$K(L_\star P, L_\star \overline{P}) = K(P, \overline{P}).$$

Proof. We recall for two distributions $\mu, \nu \in \mathcal{P}(\mathcal{X})$ the definition of $R_{1/2}$ the Rényi entropy of order $1/2$,

$$R_{1/2}(\mu\|\nu) \doteq -2\log\left(\sum_{x \in \mathcal{X}} \sqrt{\mu(x)\nu(x)}\right),$$

and note that $R_{1/2}$ is closely related to the Hellinger distance between μ and ν. This definition extends to the notion of a divergence rate between stochastic processes $(X_t)_{t\in\mathbb{N}}, (X'_t)_{t\in\mathbb{N}}$ on $\mathcal{X}$ as follows

$$R_{1/2}\left((X_t)_{t\in\mathbb{N}}\|(X'_t)_{t\in\mathbb{N}}\right) = \lim_{n\to\infty} \frac{1}{n} R_{1/2}\left(X_1^n \| X_1'^n\right),$$

and in the irreducible time-homogeneous Markovian setting where $(X_t)_{t\in\mathbb{N}}$, $(X'_t)_{t\in\mathbb{N}}$ evolve according to transition matrices P and P', the above reduces [13] to

$$R_{1/2}\left((X_t)_{t\in\mathbb{N}}\|(X'_t)_{t\in\mathbb{N}}\right) = -2\log\rho(P^{\circ 1/2} \circ P'^{\circ 1/2}) = -2\log(1 - K(P, P')).$$

Reorganizing terms and plugging for the embedded stochastic matrices,

$$K(L_\star P, L_\star \overline{P}) = 1 - \exp\left(-\frac{1}{2}\lim_{n\to\infty}\frac{1}{n}R_{1/2}\left(L_\star Q^n \| L_\star \overline{Q}^n\right)\right),$$

where $L_\star \overline{Q}^n$ is the distribution over stationary paths of length n induced by the embedded $L_\star \overline{P}$. For any $n \in \mathbb{N}$, from Lemma 1 and information monotonicity of the Rényi divergence, $R_{1/2}\left(L_\star Q^n \| L_\star \overline{Q}^n\right) = R_{1/2}\left(Q^n \| \overline{Q}^n\right)$, hence our claim. $\quad\blacksquare$

6   Proof of Theorem 1

We assume that P and $\overline{P}$ are in $\mathcal{V}_{\text{test}}$. We reduce the problem as follows. We construct $\sigma_\star^{\overline{\pi}}$, the symmetrizer[4] defined in Sect. 4. We proceed to embed both the reference chain (using Definition 1) and the unknown trajectory (using the operational definition in [20, Section 3.1]). We invoke the tester of [3] as a black box, and report its answer.

---

[4] If we wish to test for the identity of multiple chains against the same reference, we only need to perform this step once.

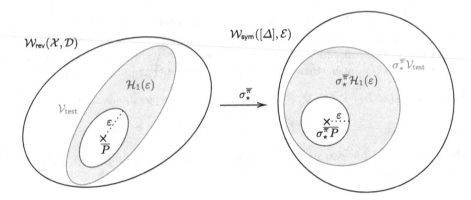

Fig. 1. Reduction of the testing problem by isometric embedding.

Completeness case. It is immediate that $P = \overline{P} \implies \sigma_\star^{\overline{\pi}} P = \sigma_\star^{\overline{\pi}} \overline{P}$.

Soundness case. From Lemma 2, $K(P, \overline{P}) > \varepsilon \implies K(\sigma_\star^{\overline{\pi}} P, \sigma_\star^{\overline{\pi}} \overline{P}) > \varepsilon$.

As a consequence of [3, Theorem 10], the sample complexity of testing is upper bounded by $\mathcal{O}(\Delta/\varepsilon^4)$. With $\overline{\pi}_\star = p_1/\Delta$ and treating p_1 as a small constant, we recover the known sample complexity.

Acknowledgements. GW is supported by the Special Postdoctoral Researcher Program (SPDR) of RIKEN and by the Japan Society for the Promotion of Science KAKENHI under Grant 23K13024. SW is supported in part by the Japan Society for the Promotion of Science KAKENHI under Grant 20H02144.

References

1. Canonne, C.L., et al.: Topics and techniques in distribution testing: A biased but representative sample. Found. Trends Commun. Inf. Theor. **19**(6), 1032–1198 (2022)
2. Chan, S.O., Ding, Q., Li, S.H.: Learning and testing irreducible Markov chains via the k-cover time. In: Algorithmic Learning Theory. pp. 458–480. PMLR (2021)
3. Cherapanamjeri, Y., Bartlett, P.L.: Testing symmetric Markov chains without hitting. In: Proceedings of the Thirty-Second Conference on Learning Theory. Proceedings of Machine Learning Research, vol. 99, pp. 758–785. PMLR (2019)
4. Daskalakis, C., Dikkala, N., Gravin, N.: Testing symmetric Markov chains from a single trajectory. In: Conference On Learning Theory. pp. 385–409. PMLR (2018)
5. Diakonikolas, I., Kane, D.M.: A new approach for testing properties of discrete distributions. In: 2016 IEEE 57th Annual Symposium on Foundations of Computer Science (FOCS). pp. 685–694. IEEE (2016)
6. Fried, S., Wolfer, G.: Identity testing of reversible Markov chains. In: Proceedings of The 25th International Conference on Artificial Intelligence and Statistics. Proceedings of Machine Learning Research, vol. 151, pp. 798–817. PMLR (2022)
7. Goldreich, O.: The uniform distribution is complete with respect to testing identity to a fixed distribution. In: Electron. Colloquium Comput. Complex. vol. 23, p. 15 (2016)

8. Hayashi, M., Watanabe, S.: Information geometry approach to parameter estimation in Markov chains. Ann. Stat. **44**(4), 1495–1535 (2016)
9. Kazakos, D.: The Bhattacharyya distance and detection between Markov chains. IEEE Trans. Inf. Theor. **24**(6), 747–754 (1978)
10. Kemeny, J.G., Snell, J.L.: Finite Markov chains: with a new appendix Generalization of a fundamental matrix. Springer (1983)
11. Nagaoka, H.: The exponential family of Markov chains and its information geometry. In: The proceedings of the Symposium on Information Theory and Its Applications. vol. 28(2), pp. 601–604 (2005)
12. Paninski, L.: A coincidence-based test for uniformity given very sparsely sampled discrete data. IEEE Trans. Inf. Theor. **54**(10), 4750–4755 (2008)
13. Rached, Z., Alajaji, F., Campbell, L.L.: Rényi's divergence and entropy rates for finite alphabet Markov sources. IEEE Trans. Inf. Theor. **47**(4), 1553–1561 (2001)
14. Valiant, G., Valiant, P.: An automatic inequality prover and instance optimal identity testing. SIAM J. Comput. **46**(1), 429–455 (2017)
15. Čencov, N.N.: Algebraic foundation of mathematical statistics. Series Stat. **9**(2), 267–276 (1978)
16. Čencov, N.N.: Statistical decision rules and optimal inference, transl. math. monographs, vol. 53. Amer. Math. Soc., Providence-RI (1981)
17. Waggoner, B.: l_p testing and learning of discrete distributions. In: Proceedings of the 2015 Conference on Innovations in Theoretical Computer Science. pp. 347–356 (2015)
18. Wolfer, G., Kontorovich, A.: Minimax testing of identity to a reference ergodic Markov chain. In: Proceedings of the Twenty Third International Conference on Artificial Intelligence and Statistics. vol. 108, pp. 191–201. PMLR (2020)
19. Wolfer, G., Watanabe, S.: Information geometry of reversible Markov chains. Inf. Geom. **4**(2), 393–433 (12 2021)
20. Wolfer, G., Watanabe, S.: Geometric aspects of data-processing of Markov chains (2022), arXiv:2203.04575

On the Entropy of Rectifiable and Stratified Measures

Juan Pablo Vigneaux[(⊠)]

Department of Mathematics, California Institute of Technology,
Pasadena, CA 91125, USA
vigneaux@caltech.edu

Abstract. We summarize some results of geometric measure theory concerning rectifiable sets and measures. Combined with the entropic chain rule for disintegrations (Vigneaux, 2021), they account for some properties of the entropy of rectifiable measures with respect to the Hausdorff measure first studied by (Koliander et al., 2016). Then we present some recent work on *stratified measures*, which are convex combinations of rectifiable measures. These generalize discrete-continuous mixtures and may have a singular continuous part. Their entropy obeys a chain rule, whose "conditional term" is an average of the entropies of the rectifiable measures involved. We state an asymptotic equipartition property (AEP) for stratified measures that shows concentration on strata of a few "typical dimensions" and that links the conditional term of the chain rule to the volume growth of typical sequences in each stratum.

Keywords: Entropy · stratified measures · rectifiable measures · chain rule · asymptotic equipartition property

1 Introduction

The starting point of our considerations is the asymptotic equipartition property:

Proposition 1 (AEP). *Let $(E_X, \mathfrak{B}, \mu)$ be a σ-finite measure space, and ρ a probability measure on $(E_X, \mathfrak{B})$ such that $\rho \ll \mu$. Suppose that the* entropy

$$H_\mu(\rho) := -\int_E \ln \frac{\mathrm{d}\rho}{\mathrm{d}\mu} \, \mathrm{d}\rho = \mathbb{E}_\rho\left(-\ln \frac{\mathrm{d}\rho}{\mathrm{d}\mu}\right) \tag{1}$$

is finite. For every $\delta > 0$, define the set of weakly δ-typical realizations

$$W_\delta^{(n)}(\rho; \mu) = \left\{ (x_1, ..., x_n) \in E_X^n \ : \ \left| -\frac{1}{n} \ln \prod_{i=1}^n \frac{\mathrm{d}\rho}{\mathrm{d}\mu}(x_i) - H_\mu(\rho) \right| < \delta \right\}. \tag{2}$$

Then, for every $\varepsilon > 0$, there exists $n_0 \in \mathbb{N}$ such that, for all $n \geq n_0$,

1. $\mathbb{P}\left(W_\delta^{(n)}(\rho; \mu)\right) > 1 - \varepsilon$ *and*

F. Nielsen and F. Barbaresco (Eds.): GSI 2023, LNCS 14071, pp. 338–346, 2023.
https://doi.org/10.1007/978-3-031-38271-0_33

2. $(1 - \varepsilon) \exp\{n(H_\mu(\rho) - \delta)\} \le \mu^{\otimes n}(W_\delta^{(n)}(\rho; \mu)) \le \exp\{n(H_\mu(\rho) + \delta)\}$.

The proof of this proposition only depends on the weak law of large numbers, which ensures the convergence in probability of $-\frac{1}{n} \sum_{i=1}^n \ln f(x_i)$ to its mean $H_\mu(\rho)$, see [11, Ch. 12] and [3, Ch. 8].

Shannon's discrete entropy and differential entropy are particular cases of (2): the former when μ is the counting measure on a discrete set equipped with the σ-algebra of all subsets; the latter when $E_X = \mathbb{R}^d$, $\mathfrak{B}$ is generated by the open sets, and μ is the Lebesgue measure. Up to a sign, the Kullback-Leibler divergence, also called *relative entropy*, is another particular case, that arises when μ is a probability measure.

One can imagine other examples, geometric in nature, that involve measures μ on $\mathbb{R}^d$ that are singular continuous. For instance, E_X could be a Riemannian manifold equipped with the Borel measure given by integration of its Riemannian volume form. The measure-theoretic nature of Proposition 1 makes the smoothness in this example irrelevant. It is more natural to work with geometric measure theory.

2 Some Elements of Geometric Measure Theory

Geometric measure theory "could be described as differential geometry, generalized through measure theory to deal with maps and surfaces that are not necessarily smooth, and applied to the calculus of variations" [9]. The place of smooth maps is taken by Lipschitz maps, which are differentiable almost everywhere [2, p. 46]. In turn, manifolds are replaced by rectifiable sets, and the natural notion of volume for such sets is the Hausdorff measure.

2.1 Hausdorff Measure and Dimension

To define the Hausdorff measure, recall first that the diameter of a subset S of the Euclidean space $(\mathbb{R}^d, \|\cdot\|_2)$ is $\mathrm{diam}(S) = \sup\{\, \|x - y\|_2 : x, y \in S \,\}$. For any $m \ge 0$ and any $A \subset \mathbb{R}^n$, define

$$\mathcal{H}^m(A) = \lim_{\delta \to 0} \inf_{\{S_i\}_{i \in I}} \sum_{i \in I} w_m \left(\frac{\mathrm{diam}(S_i)}{2} \right)^m, \tag{3}$$

where $w_m = \pi^{m/2} / \Gamma(m/2 + 1)$ is the volume of the unit ball $B(0, 1) \subset \mathbb{R}^m$, and the infimum is taken over all countable coverings $\{S_i\}_{i \in I}$ of A such that each set S_i has diameter at most δ. This is an outer measure and its restriction to the Borel σ-algebra is a measure i.e. a σ-additive $[0, \infty]$-valued set-function [2, Thm. 1.49]. Moreover, the measure $\mathcal{H}^d$ equals the standard Lebesgue measure $\mathcal{L}^d$ [2, Thm. 2.53] and $\mathcal{H}^0$ is the counting measure. More generally, when m is an integer between 0 and d, $\mathcal{H}^m$ gives a natural notion of m-dimensional volume. The 1- and 2-dimensional volumes coincide, respectively, with the classical notions of length and area, see Examples 1 and 2 below.

For a given set $A \subset \mathbb{R}^d$, the number $\mathcal{H}^m(A)$ is either 0 or ∞, except possibly for a single value of m, which is then called the Hausdorff dimension $\dim_H A$ of A. More precisely [2, Def. 2.51],

$$\dim_H A := \inf\{\, k \in [0, \infty) \,:\, \mathcal{H}^k(A) = 0 \,\} = \sup\{\, k \in [0, \infty) \,:\, \mathcal{H}^k(A) = \infty \,\}. \tag{4}$$

Finally, we remark here that the Hausdorff measure interacts very naturally with Lipschitz maps.

Lemma 1 ([2, Prop. 2.49]). *If $f : \mathbb{R}^d \to \mathbb{R}^{d'}$ is a Lipschitz function,*[1] *with Lipschitz constant $\mathrm{Lip}(f)$, then for all $m \geq 0$ and every subset E of $\mathbb{R}^d$,* $\mathcal{H}^m(f(E)) \leq \mathrm{Lip}(f)^k \mathcal{H}^m(E)$.

2.2   Rectifiable Sets

Definition 1 ([6, 3.2.14]). *A subset S of $\mathbb{R}^d$ is called m-rectifiable (for $m \leq d$) if it is the image of a bounded subset of $\mathbb{R}^m$ under a Lipschitz map and countably m-rectifiable if it is a countable union of m-rectifiable sets. The subset S is countably $(\mathcal{H}^m, m)$-rectifiable if there exist countable m-rectifiable set containing $\mathcal{H}^m$-almost all of S, this is, if there are bounded sets $A_k \subset \mathbb{R}^m$ and Lipschitz functions $f_k : A_k \to \mathbb{R}^d$, enumerated by $k \in \mathbb{N}$, such that $\mathcal{H}^m(S \backslash \bigcup_k f_k(A_k)) = 0$.*

By convention, $\mathbb{R}^0$ is a point, so that a countably 0-rectifiable set is simply a countable set. An example of countable m-rectifiable set is an m-dimensional C^1-submanifold E of $\mathbb{R}^d$, see [1, App. A]. A set that differs from an m-dimensional C^1-submanifold by a $\mathcal{H}^m$-null set is countably $(\mathcal{H}^m, m)$-rectifiable.

For every countably $(\mathcal{H}^m, m)$-rectifiable set $E \subset \mathbb{R}^d$, there exists (see [13, pp. 16-17] and the references therein) an $\mathcal{H}^m$-null set E_0, compact sets $(K_i)_{i \in \mathbb{N}}$ and injective Lipschitz functions $(f_i : K_i \to \mathbb{R}^d)_{i \in \mathbb{N}}$ such that the sets $f_i(K_i)$ are pairwise disjoint and

$$E \subset E_0 \cup \bigcup_{i \in \mathbb{N}} f_i(K_i). \tag{5}$$

It follows from Lemma 1 and the boundedness of the sets K_i that every $(\mathcal{H}^m, m)$-rectifiable set has σ-finite $\mathcal{H}^m$-measure. Because of monotonicity of $\mathcal{H}^m$, any subset of an $(\mathcal{H}^m, m)$-rectifiable set is $(\mathcal{H}^m, m)$-rectifiable. Also, the countable union of $(\mathcal{H}^m, m)$-rectifiable subsets is $(\mathcal{H}^m, m)$-rectifiable.

The area and coarea formulas may be used to compute integrals on an m-rectifiable set with respect to the $\mathcal{H}^m$-measure. As a preliminary, we define the area and coarea factors. Let V, W be finite-dimensional Hilbert spaces and

---

[1] A function $f : X \to Y$ between metric spaces (X, d_X) and (Y, d_Y) is called Lipschitz if there exists $C > 0$ such that

$$\forall x, x' \in X, \quad d_Y(f(x), f(x')) \leq C d_X(x, x').$$

The Lipschitz constant of f, denoted $\mathrm{Lip}(f)$, is the smallest C that satisfies this condition.

$L : V \to W$ be a linear map. Recall that the inner product gives explicit identifications $V \cong V^*$ and $W \cong W^*$ with their duals.

1. If $k := \dim V \le \dim V$, the k-dimensional Jacobian or area factor [2, Def. 2.68] is $J_k L = \sqrt{\det(L^* \circ L)}$, where $L^* : W^* \to V^*$ is the transpose of L.
2. If $\dim V \ge \dim W =: d$, then the d-dimensional coarea factor [2, Def. 2.92] is $C_d L = \sqrt{\det(L \circ L^*)}$.

Proposition 2 (Area formula, cf. [2, Thm. 2.71] and [6, Thm. 3.2.3]). *Let k, d be integers such that $k \le d$ and $f : \mathbb{R}^k \to \mathbb{R}^d$ a Lipschitz function. For any Lebesgue measurable subset E of $\mathbb{R}^k$ and $\mathcal{L}^k$-integrable function u, the function $y \mapsto \sum_{x \in E \cap f^{-1}(y)} u(x)$ on $\mathbb{R}^d$ is $\mathcal{H}^k$-measurable, and*

$$\int_E u(x) J_k f(x) \, \mathrm{d}\mathcal{L}^k(x) = \int_{\mathbb{R}^d} \sum_{x \in E \cap f^{-1}(y)} u(x) \, \mathrm{d}\mathcal{H}^k(y). \tag{6}$$

This implies in particular that, if f is injective on E, then $f(E)$ is $\mathcal{H}^k$-measurable and $\mathcal{H}^k(f(E)) = \int_E J_k f(x) \, \mathrm{d}\mathcal{L}^k(x)$.

Example 1 (Curves). Let $\varphi : [0,1] \to \mathbb{R}^d$, $t \mapsto (\varphi_1(t), ..., \varphi_d(t))$ be a C^1-curve; recall that a C^1-map defined on a compact set is Lipschitz. Then $d\varphi(t) = (\varphi_1'(t), ..., \varphi_d'(t))$, and $J_1 \varphi = \sqrt{(d\varphi)^*(d\varphi)} = \|d\varphi\|_2$. So we obtain the standard formula for the length of a curve,

$$\mathcal{H}^1(\varphi([0,1])) = \int_0^1 \|d\varphi(t)\|_2 \, \mathrm{d}t. \tag{7}$$

Example 2 (Surfaces). Let $\varphi = (\varphi_1, \varphi_2, \varphi_3) : V \subset \mathbb{R}^2 \to \mathbb{R}^3$ be a Lipschitz, differentiable, and injective map defining a surface. In this case, $d\varphi(u,v) = \big(\varphi_x(u,v) \; \varphi_y(u,v)\big)$ where $\varphi_x(u,v)$ is the column vector

$$(\partial_x \varphi_1(u,v), \partial_x \varphi_2(u,v), \partial_x \varphi_3(u,v))$$

and similarly for $\varphi_y(u,v)$. Then, if θ denotes the angle between $\varphi_x(u,v)$ and $\varphi_y(u,v)$,

$$J_2 \varphi = \sqrt{\det \begin{pmatrix} \|\varphi_x\|^2 & \varphi_x \bullet \varphi_y \\ \varphi_x \bullet \varphi_y & \|\varphi_y\|^2 \end{pmatrix}} = \|\varphi_x\| \, \|\varphi_y\| \sqrt{1 - \cos \theta} = \|\varphi_x \times \varphi_y\|. \tag{8}$$

Therefore,

$$\mathcal{H}^2(\varphi(V)) = \int_V \|\varphi_x \times \varphi_y(u,v)\| \, \mathrm{d}u \, \mathrm{d}v, \tag{9}$$

which is again the classical formula for the area of a parametric surface.

In many situations, it is useful to compute an integral over a countably $(\mathcal{H}^m, m)$-rectifiable set $E \subset \mathbb{R}^k$ as an iterated integral, first over the level sets $E \cap \{ x : f(x) = t \}$ of a Lipschitz function $f : \mathbb{R}^k \to \mathbb{R}^d$ with $d \le k$, and then over $t \in \mathbb{R}^d$. In particular, if $E = \mathbb{R}^k$ and f is a projection onto the space generated by some vectors of the canonical basis of $\mathbb{R}^k$, this procedure corresponds to Fubini's theorem. Its generalization to the rectifiable case is the coarea formula.

Proposition 3 (Coarea formula, [2, Thm. 2.93]). *Let* $f : \mathbb{R}^k \to \mathbb{R}^d$ *be a Lipschitz function,* E *a countably* $(\mathcal{H}^m, m)$-*rectifiable subset of* $\mathbb{R}^k$ *(with* $m \geq d$*) and* $g : \mathbb{R}^k \to [0, \infty]$ *a Borel function. Then, the set* $E \cap f^{-1}(t)$ *is countably* $(\mathcal{H}^{m-d}, m - d)$-*rectifiable and* $\mathcal{H}^{m-d}$-*measurable, the function* $t \mapsto \int_{E \cap f^{-1}(t)} g(y) \, d\mathcal{H}^{m-d}$ *is* $\mathcal{L}^d$-*measurable on* $\mathbb{R}^d$*, and*

$$\int_E g(x) C_d d^E f_x \, d\mathcal{H}^m(x) = \int_{\mathbb{R}^d} \left(\int_{f^{-1}(t)} g(y) \, d\mathcal{H}^{m-d}(y) \right) dt. \tag{10}$$

Here $d^E f_x$ is the *tangential differential* [2, Def. 2.89], the restriction of df_x to the approximate tangent space to E at x. The precise computation of this function is not essential here, but rather the fact that $C_k d^E f_x > 0$ $\mathcal{H}^m$-almost surely, hence $\mathcal{H}^m|_E$ has a $(f, \mathcal{L}^d)$-disintegration given by the measures $\{(C_k d^E f)^{-1} \mathcal{H}^{m-d}|_{E \cap f^{-1}(t)}\}_{t \in \mathbb{R}^k}$, which are well-defined $\mathcal{L}^d$-almost surely.[2]

Remark 1 (On disintegrations). Disintegrations are an even broader generalization of Fubini's theorem.

Let $T : (E, \mathfrak{B}) \to (E_T, \mathfrak{B}_T)$ be a measurable map and let ν and ξ be σ-finite measures on $(E, \mathfrak{B})$ and $(E_T, \mathfrak{B}_T)$ respectively. The measure ν has a (T, ξ)-disintegration $\{\nu_t\}_{t \in E_T}$ if

1. ν_t is a σ-finite measure on $\mathfrak{B}$ such that $\nu_t(T \neq t) = 0$ for ξ-almost every t;
2. for each measurable nonnegative function $f : E \to \mathbb{R}$, the map $t \mapsto \int_E f \, d\nu_t$ is measurable, and $\int_E f \, d\nu = \int_{E_T} \left(\int_E f(x) \, d\nu_t(x) \right) d\xi(t)$.

In case such a disintegration exists, any probability measure $\rho = r \cdot \nu$ has a $(T, T_*\rho)$-disintegration $\{\rho_t\}_{t \in E_t}$ such that each ρ_t is a probability measure with density $r / \int_E r \, d\nu_t$ w.r.t. ν_t, and the following chain rule holds [12, Prop. 3]:

$$H_\nu(\rho) = H_\xi(T_*\rho) + \int_{E_T} H_{\nu_t}(\rho_t) \, dT_*\rho(t). \tag{11}$$

3   Rectifiable Measures and Their Entropy

Let ρ be a locally finite measure and s a nonnegative real number. Marstrand proved that if the limiting density $\Theta_s(\rho, x) := \lim_{r \downarrow 0} \rho(B(x, r))/(w_s r^s)$ exists and is strictly positive and finite for ρ-almost every x, then s is an integer not greater than n. Later Preiss proved that such a measure is also s-rectifiable in the sense of the following definition. For details, see e.g. [5].

Definition 2 ([8, Def. 16.6]). *A Radon outer measure* ν *on* $\mathbb{R}^d$ *is called* m-*rectifiable if* $\nu \ll \mathcal{H}^m$ *and there exists a countably* $(\mathcal{H}^m, m)$-*rectifiable Borel set* E *such that* $\nu(\mathbb{R}^d \setminus E) = 0$.

---

[2] Given a measure μ on a σ-algebra $\mathfrak{B}$ and $B \in \mathfrak{B}$, $\mu|_B$ denotes the restricted measure $A \mapsto \mu|_B(A) := \mu(A \cap B)$.

The study of these measures from the viewpoint of information theory, particularly the properties of the entropy $H_{\mathcal{H}^m}(\rho)$ of an m-rectifiable probability measure ρ, was carried out relatively recently by Koliander, Pichler, Riegler, and Hlawatsch in [7]. We provide here an idiosyncratic summary of some of their results.

First, remark that in virtue of (5), an m-rectifiable measure ν is absolutely continuous with respect to the restricted measure $\mathcal{H}^m|_{E^*}$, where E^* is countably m-rectifiable and has the form $\bigcup_{i \in \mathbb{N}} f_i(K_i)$ with f_i injective and K_i Borel and bounded. (A refinement of this construction gives a similar set such that, additionally, the density of ρ is strictly positive [7, App. A].) Although the product of an $(\mathcal{H}^{m_1}, m_1)$-rectifiable set and an $(\mathcal{H}^{m_2}, m_2)$ rectifiable set is *not* $(\mathcal{H}^{m_1+m_2}, m_1 + m_2)$-rectifiable—see [6, 3.2.24]—the carriers behave better.

Lemma 2 (See [7, Lem. 27]and [13, Lem. 6]). *If S_i is a carrier of an m_i-rectifiable measure ν_i (for $i = 1, 2$), then $S_1 \times S_2$ is a carrier of $\nu_1 \otimes \nu_2$, of Hausdorff dimension $m_1 + m_2$. Additionally, the Hausdorff measure $\mathcal{H}^{m_1+m_2}|_{S_1 \times S_2}$ equals $\mathcal{H}^{m_1}|_{S_1} \otimes \mathcal{H}^{m_2}|_{S^2}$.*

Let ρ be an m-rectifiable measure, with carrier E (we drop the $*$ hereon). It holds that $\rho \ll \mathcal{H}^m|_E$ and $\mathcal{H}^m|_E$ is σ-finite. If moreover $H_{\mathcal{H}^m|_E}(\rho) < \infty$, Proposition 1 gives estimates for $(\mathcal{H}^m|_E)^{\otimes n}(\mathcal{W}_\delta^{(n)}(\rho; \mathcal{H}^m|_E))$. Lemma 2 tells us that E^n is mn-rectifiable and that $(\mathcal{H}^m|_E)^{\otimes n} = \mathcal{H}^{mn}|_{E^n}$, which is desirable because the Hausdorff dimension of E^n is mn and $\mathcal{H}^{mn}$ is the only nontrivial measure on it as well as on $W^{(n)}$, which as a subset of E^n is mn rectifiable too.

To apply the AEP we need to compute $H_{\mathcal{H}^m|_E}(\rho)$. In some cases, one can use the area formula (Proposition 2) to "change variables" and express $H_{\mathcal{H}^m|_E}(\cdot)$ in terms of the usual differential entropy. For instance, suppose A is a bounded Borel subset of $\mathbb{R}^k$ of nontrival $\mathcal{L}^k$-measure and f is an injective Lipschitz function on A. The set $f(A)$ is k-rectifiable. Moreover, if ρ is a probability measure such that $\rho \ll \mathcal{L}^k|_A$ with density r, then the area formula applied to $u = r/J_k f$ and $E = f^{-1}(B)$, for some Borel subset B of $\mathbb{R}^d$, shows that $f_*\rho \ll \mathcal{H}^k|_{f(A)}$ with density $(r/J_k f) \circ f^{-1}$, which is well-defined ($\mathcal{H}^k|_{f(A)}$)-almost surely. A simple computation yields

$$H_{\mathcal{H}^k|_{f(A)}}(f_*\rho) = H_{\mathcal{L}^k}(\rho) + \mathbb{E}_\rho \left(\ln J_k f \right). \tag{12}$$

There is a more general formula of this kind when A is a rectifiable subset of $\mathbb{R}^d$.

Finally, we deduce the chain rule for the entropy of rectifiable measures as a consequence of our general theorem for disintegrations (Remark 1). Let E be a countably $(\mathcal{H}^m, m)$-rectifiable subset of $\mathbb{R}^k$, $f : \mathbb{R}^k \to \mathbb{R}^d$ a Borel function (with $d \leq m$), and ρ a probability measure such that $\rho \ll \mathcal{H}^m|_E$. Because $\mathcal{H}^m|_E$ has an $(f, \mathcal{L}^d)$-disintegration $\{F^{-1}\mathcal{H}^{m-d}|_{E \cap f^{-1}(t)}\}_{t \in \mathbb{R}^k}$, with $F = C_d d^E f$, then $f_*\rho \ll \mathcal{L}^d$, and

$$H_{\mathcal{H}^m|_E}(\rho) = H_{\mathcal{L}^d}(f_*\rho) + \int_{\mathbb{R}^k} H_{F^{-1}\mathcal{H}^{m-d}|_{E \cap f^{-1}(t)}}(\rho_t) \, df_*\rho. \tag{13}$$

The probabilities ρ_t are described in Remark 1. If one insists in only using the Hausdorff measures as reference measures, one must rewrite the integrand in (13) using the chain rule for the Radon-Nikodym derivative:

$$H_{F^{-1}\mathcal{H}^{m-d}|_{E\cap f^{-1}(t)}}(\rho_t) = \mathbb{E}_{\rho_t}\left(-\ln\frac{\mathrm{d}\rho_t}{\mathrm{d}\mathcal{H}^{m-d}|_{E\cap f^{-1}(t)}}\frac{\mathrm{d}\mathcal{H}^{m-d}|_{E\cap f^{-1}(t)}}{\mathrm{d}F^{-1}\mathcal{H}^{m-d}|_{E\cap f^{-1}(t)}}\right)$$

$$= \mathbb{E}_{\rho_t}\left(-\ln\frac{\mathrm{d}\rho_t}{\mathrm{d}\mathcal{H}^{m-d}|_{E\cap f^{-1}(t)}}\right) - \mathbb{E}_{\rho_t}(\ln F).$$

One recovers in this way the formula (50) in [7].

4 Stratified Measures

Definition 3 (k-stratified measure). *A measure ν on $(\mathbb{R}^d, \mathcal{B}(\mathbb{R}^d))$ is k-stratified, for $k \in \mathbb{N}^*$, if there are integers $(m_i)_{i=1}^k$ such that $0 \leq m_1 < m_2 < \ldots < m_k \leq d$ and ν can be expressed as a sum $\sum_{i=1}^k \nu_i$, where each ν_i is a nonzero m_i-rectifiable measure.*

Thus 1-stratified measures are rectifiable measures. If ν is k-stratified for some k we simply say that ν is a *stratified measure*.

A fundamental nontrival example to bear in mind is a discrete-continuous mixture, which corresponds to $k = 2$, E_1 countable, and $E_2 = \mathbb{R}^d$. More generally, a stratified measure has a Lebesgue decomposition with a singular continuous part provided some m_i is strictly between 0 and d.

Let ρ be a probability measure that is stratified in the sense above. We can always put it in the *standard form* $\rho = \sum_{i=1}^k q_i\rho_i$, where each ρ_i is a rectifiable probability measure with carrier E_i of dimension m_i (so that $\rho_i = \rho_i|_{E_i}$), the carriers $(E_i)_{i=1}^k$ are disjoint, $0 \leq m_1 < \cdots < m_k \leq d$, and $(q_1, ..., q_k)$ is a probability vector *with strictly positive entries*. The carriers can be taken to be disjoint because if E has Hausdorff dimension m, then $\mathcal{H}^k(E) = 0$ for $k > m$, hence one can prove [13, Sec. IV-B] that $E_i \setminus (\bigcup_{j=1}^{i-1} E_j)$ is a carrier for ν_i, for $i = 2, ..., k$.

We can regard ρ as the law of a random variable X valued in $E_X := \bigcup_{i=1}^k E_i$ and the vector $(q_1, ..., q_k)$ as the law $\pi_*\rho$ of the discrete random variable Y induced by the projection π from E_X to $E_Y := \{1, ..., k\}$ that maps $x \in E_i$ to i. We denote by D the random variable $\dim_H E_Y$, with expectation $\mathbb{E}(D) = \sum_{i=1}^k m_i q_i$.

The measure ρ is absolutely continuous with respect to $\mu = \sum_{i=1}^k \mu_i$, where $\mu_i = \mathcal{H}^{m_i}|_{E_i}$, so it makes sense to consider the entropy $H_\mu(\rho)$; it has a concrete probabilistic meaning in the sense of Proposition 1. Moreover, one can prove that $\frac{\mathrm{d}\rho}{\mathrm{d}\mu} = \sum_{i=1}^m q_i\frac{\mathrm{d}\rho_i}{\mathrm{d}\mu_i}\mathbb{1}_{E_i}$ [13, Lem. 3] and therefore

$$H_\mu(\rho) = H(q_1, ..., q_n) + \sum_{i=1}^k q_i H_{\mu_i}(\rho_i) \tag{14}$$

holds [13, Lem. 4]. This formula also follows form the chain rule for general disintegrations (Remark 13), because $\{\rho_i\}_{i \in E_Y}$ is a $(\pi, \pi_*\rho)$-disintegration of ρ. The powers of ρ are also stratified. In fact, remark that

$$\rho^{\otimes n} = \sum_{\mathbf{y} = (y_1, \ldots, y_n) \in E_Y^n} q_1^{N(1;\mathbf{y})} \cdots q_k^{N(k;\mathbf{y})} \rho_{y_1} \otimes \cdots \otimes \rho_{y_n} \tag{15}$$

where $N(a; \mathbf{y})$ counts the appearances of the symbol $a \in E_Y$ in the word $\mathbf{y}$. Each measure $\rho_{\mathbf{y}} := \rho_{y_1} \otimes \cdots \otimes \rho_{y_n}$ is absolutely continuous with respect to $\mu_{\mathbf{y}} := \mu_{y_1} \otimes \cdots \otimes \mu_{y_n}$. It follows from Lemma 2 that for any $\mathbf{y} \in E_Y^n$, the *stratum* $\Sigma_{\mathbf{y}} := E_{y_1} \times \cdots \times E_{y_n}$ is also a carrier, of dimension $m(\mathbf{y}) := \sum_{j=1}^n \dim_H E_{y_j}$, and the product measure $\mu_{\mathbf{y}}$ equals $\mathcal{H}^{m(\mathbf{y})}|_{\Sigma_{\mathbf{y}}}$. Therefore each measure $\rho_{\mathbf{y}}$ is rectifiable. We can group together the $\rho_{\mathbf{y}}$ of the same dimension to put $\rho^{\otimes n}$ as in Definition 3.

By Proposition 1, one might approximate $\rho^{\otimes n}$ with an arbitrary level of accuracy by its restriction to the weakly typical realizations of ρ, provided n is big enough. In order to get additional control on the dimensions appearing in this approximation, we restrict it further, retaining only the strata that correspond to *strongly* typical realizations of the random variable Y.

Let us denote by Q the probability mass function (p.m.f) of $\pi_*\rho$. Recall that $\mathbf{y} \in E_Y^n$ induces a probability law $\tau_{\mathbf{y}}$ on E_Y, known as *empirical distribution*, given by $\tau_{\mathbf{y}}(\{a\}) = N(a; \mathbf{y})/n$. Csiszár and Körner [4, Ch. 2] define $\mathbf{y} \in E_Y^n$ to be *strongly* (Q, η)-*typical* if $\tau_{\mathbf{y}}$, with p.m.f P, is such that $\tau_{\mathbf{y}} \ll \pi_*\rho$ and, for all $a \in E_Y$, $|P(a) - Q(a)| < \eta$. We denote by $A_{\delta_n'}^{(n)}$ the set of these sequences when $\eta_n = n^{-1/2 + \xi}$. In virtue of the union bound and Hoeffding's inequality, $(\pi_*\rho)^{\otimes n}(A_{\delta_n'}^{(n)}) \geq 1 - \varepsilon_n$, where $\varepsilon_n = 2|E_Y|e^{-2n\eta_n^2}$; the choice of η_n ensures that $\varepsilon_n \to 0$ as $n \to \infty$. Moreover, the continuity of the discrete entropy in the total-variation distance implies that $A_{\delta_n'}^{(n)}$ is a subset of $W_{\delta_n'}^{(n)}$ with $\delta_n' = -|E_Y|\eta_n \ln \eta_n$, which explains our notation. See [13, Sec. III-D]

We introduce the set $T_{\delta, \delta_n'}^{(n)} = (\pi^{\times n})^{-1}(A_{\delta_n'}^{(n)}) \cap W_\delta^{(n)}(\rho)$ of *doubly typical sequences* in E_X^n, and call $T_{\delta, \delta_n'}^{(n)}(\mathbf{y}) = T_{\delta, \delta_n'}^{(n)} \cap (\pi^{\times n})^{-1}(\mathbf{y})$ a *doubly typical stratum* for any $\mathbf{y} \in A_{\delta_n'}^{(n)}$.

The main result of [13] is a refined version of the AEP for stratified measures that gives an interpretation for the conditional term in the chain rule (14).

Theorem 1 *(Setting introduced above).* For any $\varepsilon > 0$ there exists an $n_0 \in \mathbb{N}$ such that for any $n \geq n_0$ the restriction of ρ to $T_{\delta, \delta_n'}^{(n)}$, $\rho^{(n)} = \sum_{\mathbf{y} \in A_{\delta'}^{(n)}} \rho^{\otimes n}|_{T_{\delta, \delta'}^{(n)}(\mathbf{y})}$, satisfies $d_{TV}(\rho^{\otimes n}, \rho^{(n)}) < \varepsilon$. Moreover, the measure $\rho^{(n)}$ equals a sum of m-rectifiable measures for $m \in [n\mathbb{E}(D) - n^{1/2 + \xi}, n\mathbb{E}(D) + n^{1/2 + \xi}]$. The conditional entropy $H(X|Y) := \sum_{i=1}^k q_i H_{\mu_i}(\rho_i)$ quantifies the volume growth of most doubly typical fibers in the following sense:

1. For any $\mathbf{y} \in A_{\delta'}^{(n)}$, one has $n^{-1} \ln \mathcal{H}^{m(\mathbf{y})}(T_{\delta, \delta'}^{(n)}(\mathbf{y})) \leq H(X|Y) + (\delta + \delta_n')$.

2. *For any* $\varepsilon > 0$, *the set* $B_\varepsilon^{(n)}$ *of* $\mathbf{y} \in\subset A_{\delta'_n}^{(n)}$ *such that*

$$\frac{1}{n} \ln \mathcal{H}^{m(y)}(T_{\delta,\delta'}^{(n)}(\mathbf{y})) > H(X|Y) - \varepsilon + (\delta + \delta'_n),$$

satisfies

$$\limsup_{||(\delta,\delta'_n)||\to 0} \limsup_{n\to\infty} \frac{1}{n} \ln |B_\varepsilon^{(n)}| = H(Y) = \limsup_{||(\delta,\delta'_n)||\to 0} \limsup_{n\to\infty} \frac{1}{n} \ln |A_{\delta'_n}^{(n)}|.$$

This gives a geometric interpretation to the possibly noninteger dimension $\mathbb{E}(D) = \sum_{i=1}^{k} q_i m_i$, which under suitable hypotheses is the information dimension of ρ [13, Sec. V], thus answering an old question posed by Renyi in [10, p. 209].

References

1. Alberti, G., Bölcskei, H., De Lellis, C., Koliander, G., Riegler, E.: Lossless analog compression. IEEE Trans. Inf. Theory 65(11), 7480–7513 (2019)
2. Ambrosio, L., Fusco, N., Pallara, D.: Functions of Bounded Variation and Free Discontinuity Problems. Oxford Science Publications, Clarendon Press (2000)
3. Cover, T.M., Thomas, J.A.: Elements of Information Theory. A Wiley-Interscience Publication. Wiley (2006)
4. Csiszár, I., Körner, J.: Information Theory: Coding Theorems for Discrete Memoryless Systems. Probability and Mathematical Statistics. Academic Press (1981)
5. De Lellis, C.: Rectifiable Sets, Densities and Tangent Measures. Zurich Lectures in Advanced Mathematics. European Mathematical Society (2008)
6. Federer, H.: Geometric Measure Theory. Classics in Mathematics. Springer, Heidelberg (1969). https://doi.org/10.1007/978-3-642-62010-2
7. Koliander, G., Pichler, G., Riegler, E., Hlawatsch, F.: Entropy and source coding for integer-dimensional singular random variables. IEEE Trans. Inf. Theory 62(11), 6124–6154 (2016)
8. Mattila, P.: Geometry of Sets and Measures in Euclidean Spaces: Fractals and Rectifiability. Cambridge Studies in Advanced Mathematics. Cambridge University Press (1995)
9. Morgan, F.: Geometric Measure Theory: A Beginner's Guide. Elsevier Science (2008)
10. Rényi, A.: On the dimension and entropy of probability distributions. Acta Mathematica Academiae Scientiarum Hungarica 10(1), 193–215 (1959)
11. Vigneaux, J.P.: Topology of statistical systems: a cohomological approach to information theory. Ph.D. thesis, Université de Paris (2019)
12. Vigneaux, J.P.: Entropy under disintegrations. In: Nielsen, F., Barbaresco, F. (eds.) GSI 2021. LNCS, vol. 12829, pp. 340–349. Springer, Cham (2021). https://doi.org/10.1007/978-3-030-80209-7_38
13. Vigneaux, J.P.: Typicality for stratified measures. IEEE Trans. Inf. Theor. arXiv preprint arXiv:2212.10809 (2022, to appear)

Statistical Shape Analysis and more Non-Euclidean Statistics

Exploring Uniform Finite Sample Stickiness

Susanne Ulmer, Do Tran Van, and Stephan F. Huckemann[(✉)]

Felix-Bernstein-Institute for Mathematical Statistics in the Biosciences,
Georg-August-Universität at Göttingen, Göttingen, Germany
susanne.ulmer@stud.uni-goettingen.de, do.tranvan@uni-goettingen.de,
Stephan.Huckemann@mathematik.uni-goettingen.de

Abstract. It is well known, that Fréchet means on non-Euclidean spaces may exhibit nonstandard asymptotic rates depending on curvature. Even for distributions featuring standard asymptotic rates, there are non-Euclidean effects, altering finite sampling rates up to considerable sample sizes. These effects can be measured by the variance modulation function proposed by Pennec (2019). Among others, in view of statistical inference, it is important to bound this function on intervals of sampling sizes. In a first step into this direction, for the special case of a K-spider we give such an interval based only on folded moments and total probabilities of spider legs and illustrate the method by simulations.

1 Introduction to Stickiness and Finite Sample Stickiness

Data analysis has become an integral part of science due to the growing amount of data in almost every research field. This includes a plethora of data objects that do not take values in Euclidean spaces, but rather in a nonmanifold stratified space. For statistical analysis in such spaces, it is therefore necessary to develop probabilistic concepts. Fréchet (1948) was one of the first to generalize the concept of an expected value to a random variable X on an arbitrary metric space $(Q, \mathbf{d})$ as an minimizer of the expected squared distance:

$$\mu = \operatorname*{argmin}_{p \in Q} \mathbb{E}[d(X, p)^2],\tag{1}$$

nowadays called a *Fréchet mean* in his honor. Accordingly for a sample $X_1, \ldots, X_n \overset{\text{i.i.d.}}{\sim} X$, its Fréchet mean is given by

$$\mu_n = \operatorname*{argmin}_{p \in Q} \sum_{j=1}^{n} \mathbf{d}^2(X_j, p).\tag{2}$$

While on general space, these means can be empty or set valued, on *Hadamard spaces*, i.e. complete spaces of *global nonpositive curvature* (NPC), due to completeness, these means exists under very general conditions, and due to simple

F. Nielsen and F. Barbaresco (Eds.): GSI 2023, LNCS 14071, pp. 349–356, 2023.
https://doi.org/10.1007/978-3-031-38271-0_34

connectedness and nonpositive curvatures, they are unique, e.g. Sturm (2003), just as their Euclidean kin. They also share a law of strong numbers, i.e. that

$$\mu_n \overset{a.s.}{\to} \mu.$$

In contrast, however, their asymptotic distribution is often not normal, even worse, for some random variables their mean may be on a singular point stratum, and there may be a random sample size $N \in \mathbb{N}$ such that

$$\mu_n = \mu \text{ for all } n \geq N.$$

This phenomenon has been called *stickiness* by Hotz et al. (2013). It puts an end to statistical inference based on asymptotic fluctuation. While nonsticky means of random variables seem to feature the same asymptotic rate, as the Euclidean expected value, namely $1/\sqrt{n}$, it has been noted by Huckemann and Eltzner (2020) that for rather large sample sizes, the rates appear to be larger. This contribution is the first to systematically investigate this effect of *finite sample stickiness* on stratified spaces and we do this here for the model space of the K-spider introduced below. This effect is in some sense complementary to the effect of *finite sample smeariness*, where finite sample rates are smaller than $1/\sqrt{n}$. recently discovered by Hundrieser et al. (2020).

Definition 1. *With the above notation, assuming an existing Fréchet function $F(x) = \mathbb{E}[\mathbf{d}(X,x)^2] < \infty$ for all $x \in Q$, with existing Fréchet mean μ,*

$$\mathbf{m}_n = \frac{n\mathbb{E}\left[\mathbf{d}^2(\mu_n, \mu)\right]}{\mathbb{E}\left[\mathbf{d}^2(X, \mu)\right]}, \tag{3}$$

is the variance modulation *for sample size n (see Pennec (2019)), or simply* modulation.

If $(Q, \mathbf{d})$ is Euclidean, then $\mathbf{m}_n = 1$ for all $n \in \mathbb{N}$, *smeariness* governs the cases $\mathbf{m}_n \to \infty$, see Hotz and Huckemann (2015); Eltzner and Huckemann (2019); Eltzner (2022), finite sample smeariness the cases $1 < \mathbf{m}_n$, cf. Tran et al. (2021); Eltzner et al. (2021, 2023), stickiness the case that $\mu_n = \mu$ a.s. for $n > N$ with a finite random sample size N, see Hotz et al. (2013); Huckemann et al. (2015); Barden et al. (2013, 2018), and *finite sample stickiness* the case that

$$0 < \mathbf{m}_n < 1 \text{ for nonsticky } \mu.$$

Definition 2. *For a nonsticky mean, if there are integers $l \in \mathbb{N}_{\geq 2}$, and $N \in \mathbb{N}$ and $0 < \rho < 1$ such that*

$$0 < \mathbf{m}_n < 1 - \rho$$

for all $n \in \{N, N+1, ..., N^l\}$ then X is called finite sample sticky of level $\rho \in (0, 1)$, with scale l and basis N.

We note that Pennec (2019) has shown that finite sample stickiness affects all affine connection manifolds with constant negative sectional curvature. We conjecture that this is also the case for general Hadamard spaces.

A very prominent example of a nonmanifold Hadamard space is given by the *BHV tree spaces* introduced by Billera et al. (2001) modeling phylogenetic descendance trees. For a fixed number of species or taxa the BHV-space models all different tree topologies, where within each topology, lengths of internal edges reflect evolutionary mutation from unknown ancestors. For three taxa, there are three topologies featuring nonzero internal edges and a fourth one, the *star tree* featuring no internal edge. The corresponding BHV space thus carries the structure of a 3-spider as depicted in Fig. 1. For illustration of argument, in this contribution we consider K-spiders.

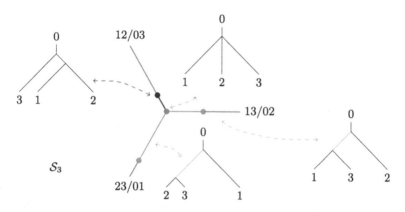

Fig. 1. Four different phylogenetic descendance trees for three taxa featuring one or none internal edge, modeled on the 3-spider $\mathcal{S}_3$.

2   A Model Space: The K-Spider

The following has been taken from Hotz et al. (2013).

Definition 3. *For $3 \le K \in \mathbb{N}$ the K-spider $\mathcal{S}_K$ is the space*

$$\mathcal{S}_K = [0, \infty) \times \{1, 2, ..., K\}/ \sim$$

where for $i, k \in \{1, 2, ..., K\}$ and $x \ge 0$, $(x, i) \sim (x, k)$ if $k = i$ or $x = 0$. The equivalence class of $(0, 1)$ is identified with the origin $\mathbf{0}$, so that $\mathcal{S}_K = \{\mathbf{0}\} \cup \bigcup_{k=1}^{K} L_k$ with the positive half-line $L_k := (0, \infty) \times \{k\}$ called the k-th leg. Further, for any $k \in \{1, 2, ..., K\}$ the map

$$F_k : \mathcal{S}_K \to \mathbb{R},$$

$$(x, i) \mapsto \begin{cases} x & if \quad i = k, \\ -x & else \end{cases}$$

is called the k-th folding map of $\mathcal{S}_K$ (Fig. 2).

The first two folded moments on the k-th leg of a random variable X are

$$m_k = \mathbb{E}[F_k(X)], \quad \sigma_k^2 = \mathbb{E}[(F_k(X) - m_k)^2]$$

in population version and the first folded moment in sample version for random variables $X_1, \ldots, X_n$ is

$$\eta_{n,k} = \frac{1}{n} \sum_{j=1}^{n} F_k(X_j).$$

We say the X is nondegenerate if $\mathbb{P}\{X \in L_k\} > 0$ for at least three different $k \in \{1, \ldots, K\}$.

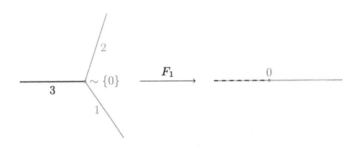

Fig. 2. The first folding map F_1 on the three-spider $\mathcal{S}_3$. The leg labelled 1 is mapped to the positive real line. The second and third leg are folded and then mapped to the negative half line.

Lemma 4. *For a sample of size n of a nondegenerate random variable, we have for every $1 \leq k \leq K$ that*

$$\eta_{n,k} > 0 \Leftrightarrow \mu_n \in L_k \ \Leftrightarrow \ F_k(\mu_n) > 0 \tag{4}$$

whereas

$$\eta_{n,k} = 0 \Rightarrow \mu_n = \mathbf{0} \in \mathcal{S}_K \Rightarrow \eta_{n,k} \leq 0.$$

In particular:

$$\eta_{n,k} \geq 0 \Leftrightarrow F_k(\mu_n) = \eta_{n,k} \tag{5}$$

and

$$\eta_{n,k} < 0 \Leftrightarrow \eta_{n,k} < F_k(\mu_n). \tag{6}$$

Proof. All but the last assertion are from Hotz et al. (2013, Lemma 3.3). The last is also in the proof of Huckemann and Eltzner (2024, Lemma 4.1.4), we prove it here for convenience.

Letting $h_{n,i} = \frac{1}{n} \sum_{X_j \in L_i} F_i(X_j)$ for $i \in \{1, \ldots, K\}$ we have

$$\eta_{n,k} = h_{n,k} - \sum_{\substack{i \neq k \\ 1 \leq i \leq K}} h_{n,i}. \tag{7}$$

If $\eta_{n,k} < 0$ then there must be $k \neq k' \in \{1, \ldots, K\}$ with $\mu_n \in L_{k'}$ and thus

$$
\begin{aligned}
0 < -F_k(\mu_n) = F_{k'}(\mu_n) &= \eta_{n,k'} = h_{n,k'} - h_{n,k} - \sum_{\substack{i \neq k, k' \\ 1 \leq i \leq K}} h_{n,i} \\
&= -\eta_{n,k} - 2 \sum_{\substack{i \neq k, k' \\ 1 \leq i \leq K}} h_{n,i} < -\eta_{n,k}
\end{aligned}
$$

as asserted, due to nondegeneracy. □

The case, $m_k < F_k(\mu)$ for all $k \in \{1, \ldots, K\}$ governs stickiness, cf. Hotz et al. (2013), and the case of $\eta_{n,k} < F_k(\mu_n) \leq 0 < F_k(\mu) = m_k$ for some $k \in \{1, \ldots, K\}$ governs finite sample stickiness, as detailed below. For this reason, we consider the following.

Corollary 5. *In case of $m_k > 0$, we have $|F_k(\mu_n) - m_k|^2 \leq |\eta_{n,k} - m_k|^2$ and $n\mathbb{E}[\mathbf{d}^2(\mu_n, \mu)] = n\mathbb{E}[|F_k(\mu_n) - m_k|^2] \leq \sigma_k^2$.*

Proof. If $\eta_{n,k} \geq 0$ we have by (5) that $F_k(\mu_n) = \eta_{n,k}$ so that $|F_k(\mu_n) - m_k|]$ $= |F_k(\mu_n) - m_k|$. If $\eta_{n,k} < 0$ we have by (6) that $\eta_{n,k} < F_k(\mu_n)$ and hence $|\eta_{n,k} - m_k| = m_k - \eta_{n,k} > m_k - F_k(\mu_n) = |m_k - F_k(\mu_n)|$, where the last equality is due to (4). □

3 Estimating Uniform Finite Sample Stickiness

Denote the standard-normal-cdf by $\Phi(x) = \frac{1}{\sqrt{2\pi}} \int_{-\infty}^{x} \exp(-t^2/2) dt$.

Theorem 6 (*Berry-Esseen, Esseen* (1945, *p. 42*), *Shevtsova* (2011)). *Let $Z_1, Z_2, \ldots, Z_n$ be iid random variables on $\mathbb{R}$ with mean zero and finite third moment $\mathbb{E}\left[|Z_1|^3\right] < \infty$. Denote by σ^2 the variance and by $\hat{F}_n(z)$ the cumulative distribution function of the random variable*

$$\frac{Z_1 + Z_2 + \ldots + Z_n}{\sqrt{n}\sigma}.$$

Then there is a finite positive constant $C_S \leq 0.4748$ such that

$$|\hat{F}_n(z) - \Phi(z)| \leq \frac{\mathbb{E}\left[|Z_1|^3\right]}{\sqrt{n}\sigma^3} C_S. \tag{8}$$

For all of the following let $X_1, \ldots, X_n \overset{\text{i.i.d.}}{\sim} X$ be nondegenerate random variables on $\mathcal{S}_K$ with Fréchet mean $\mu \in L_k$ for some $k \in \{1, \ldots, K\}$ and $\mathbb{E}\left[\mathbf{d}^3(X, \mathbf{0})\right] < \infty$. Let

$$p_n = \sum_{i=1}^{K} \Phi\left(\frac{\sqrt{n}m_i}{\sigma_i}\right) + \sum_{i=1}^{K} \frac{\mathbb{E}\left[|F_i(X) - m_i||^3\right]}{\sqrt{n}\sigma_i^3} C_S,$$

$$p_{n,k} = \Phi\left(\frac{\sqrt{n}m_k}{\sigma_k}\right) - \frac{\mathbb{E}\left[|F_k(X) - m_k||^3\right]}{\sqrt{n}\sigma_k^3} C_S.$$

For $i \in \{1, \ldots, K\}$ set

$$A_i = \{\eta_{n,i} = F_i(\mu_n)\}, \quad A = A_1 \cup \ldots \cup A_K.$$

Due nondegeneracy, (7) implies that the A_i $(i = 1, \ldots, K)$ are disjoint, see also (Hotz et al., 2013, Theorem 2.9).

Lemma 7. *For all $n \in \mathbb{N}$, $\mathbb{P}(A_k) \geq p_{n,k}$ and $\mathbb{P}(A) \leq p_n$.*

Proof. Fix $i \in \{1, ..., K\}$. By Lemma 4,

$$\mathbb{P}(A_i) = \mathbb{P}\{\eta_{n,i} \geq 0\} = \mathbb{P}\left\{\frac{\sqrt{n}(-\eta_{n,i} + m_i)}{\sigma_i} \leq \frac{\sqrt{n}m_i}{\sigma_i}\right\}.$$

Setting $Z_j = F_i(X_j) - m_i$ $(j = 1, \ldots, n)$ and $z = \frac{\sqrt{n}m_i}{\sigma_i}$ we can apply the Berry-Esseen theorem, Theorem 6, yielding for $i = k$

$$\mathbb{P}(A_k) \geq \Phi\left(\frac{\sqrt{n}m_k}{\sigma_k}\right) - \frac{\mathbb{E}\left[|F_k(X) - m_k||^3\right]}{\sqrt{n}\sigma_k^3} C_S = p_{n,k}$$

and

$$\mathbb{P}(A) \leq \sum_{i=1}^{K} \Phi\left(\frac{\sqrt{n}m_i}{\sigma_i}\right) + \sum_{i=1}^{K} \frac{\mathbb{E}\left[|F_i(X) - m_i|^3\right]}{\sqrt{n}\sigma_i^3} C_S = p_n.$$

Theorem 8. *A nondegenerate random variable on X on $\mathcal{S}_K$ with mean $\mu \in L_k$ and second folded moment σ_k^2, $k \in \{1, \ldots, K\}$, is finite sample sticky of level*

$$\rho = \min_{n \in \{N, N+1, \ldots, N^l\}} 1 - p_n - \frac{nm_k^2}{\sigma_k^2}(1 - p_{n,k})$$

with scale l and basis N, if there is $l \in \mathbb{N}_{\geq 2}$ such that $p_n < 1$ and $p_{n,k} \geq 0$, and if

$$p_n + \frac{nm_k^2}{\sigma_k^2}(1 - p_{n,k}) < 1$$

for all $n \in \{N, N+1, \ldots, N^l\}$.

Proof. By the law of total expectation, Corollary 5 and Lemma 7, exploiting that $A^c \subseteq A_k^c$, we obtain

$$\mathbf{m}_n = \mathbb{E}[\mathbf{m}_n] = \frac{n}{\sigma_k^2} \mathbb{E}\left[\mathbb{E}[\mathbf{d}^2(\mu_n, \mu) \mid A]\mathbb{P}(A) + \mathbb{E}[\mathbf{d}^2(\mu_n, \mu) \mid A^c]\mathbb{P}(A^c)\right]$$

$$= \frac{n}{\sigma_k^2} \mathbb{E}[\mathbf{d}^2(\mu_n, \mu)]\mathbb{P}(A) + \frac{nm_k^2}{\sigma_k^2}\mathbb{P}(A^c)$$

$$\leq p_n + \frac{nm_k^2}{\sigma_k^2}(1 - p_{n,k}).$$

4   Example and Simulations

Example 9. *For $t > 0$ let X_t be a random variable on $\mathcal{S}_K$ with probabilities*

$$\mathbb{P}\{X_t = (K - 1 + Kt, K)\} = \frac{1}{K} = \mathbb{P}\{X_t = (1, i)\}, \qquad k = i, ..., K - 1,$$

and first moments

$$m_i = \frac{4 - K(2 + t)}{K}, \qquad m_K = t.$$

Hence X_t is nonsticky with $\mu \in L_K$ for $t > 0$ and

$$\mathbb{E}[\mathbf{d}^3(X_t, 0)] = \frac{K - 1}{K} + \frac{(K - 1 + Kt)^3}{K} < \infty.$$

We can therefore apply Theorem 8. Figure 3 illustrates intervals of sample sizes displaying finite sample stickiness for the 3-spider and $t \in \{10^{-2}, 10^{-3}, 10^{-4}\}$. The explicit bound for the modulation derived by Theorem 8 is given as an orange dashed line.

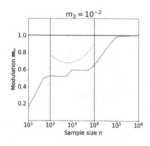 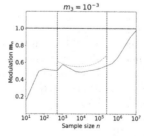 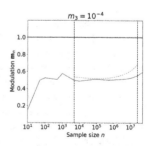

(a) *Finite sample stickiness of level $\rho = 0.1$ with scale 2 and base $N = 100$.*   (b) *Finite sample stickiness of level $\rho = 0.31$ with scale 2 and base $N = 500$.*   (c) *Finite sample stickiness of level $\rho = 0.31$ with scale 2 and base $N = 5000$.*

Fig. 3. Bounding the variance modulation function (blue) from above (orange) on various ranges of sampling sizes (vertical dashed lines). (Color figure online)

Acknowledgement. Acknowledging DFG HU 1575/7, DFG GK 2088, DFG SFB 1465 and the Niedersachsen Vorab of the Volkswagen Foundation. The work was done partially while the 2nd author was participating in the program of the Institute for Mathematical Sciences, National University of Singapore, in 2022.

References

Barden, D., Le, H., Owen, M.: Central limit theorems for Fréchet means in the space of phylogenetic trees. Electron. J. Probab. **18**(25), 1–25 (2013)

Barden, D., Le, H., Owen, M.: Limiting behaviour of Fréchet means in the space of phylogenetic trees. Ann. Inst. Stat. Math. **70**(1), 99–129 (2018)

Billera, L., Holmes, S., Vogtmann, K.: Geometry of the space of phylogenetic trees. Adv. Appl. Math. **27**(4), 733–767 (2001)

Eltzner, B.: Geometrical smeariness – a new phenomenon of Fréchet means. Bernoulli **28**(1), 239–254 (2022)

Eltzner, B., Hansen, P., Huckemann, S.F., Sommer, S.: Diffusion means in geometric spaces. Bernoulli (2023, to appear)

Eltzner, B., Huckemann, S.F.: A smeary central limit theorem for manifolds with application to high-dimensional spheres. Ann. Statist. **47**(6), 3360–3381 (2019)

Eltzner, B., Hundrieser, S., Huckemann, S.: Finite sample smeariness on spheres. In: Nielsen, F., Barbaresco, F. (eds.) GSI 2021. LNCS, vol. 12829, pp. 12–19. Springer, Cham (2021). https://doi.org/10.1007/978-3-030-80209-7_2

Esseen, C.-G.: Fourier analysis of distribution functions. A mathematical study of the Laplace-Gaussian law. Acta Math. **77**, 1–125 (1945)

Fréchet, M.: Les éléments aléatoires de nature quelconque dans un espace distancié. Ann. de l'Institut de Henri Poincaré **10**(4), 215–310 (1948)

Hotz, T., Huckemann, S.: Intrinsic means on the circle: uniqueness, locus and asymptotics. Ann. Inst. Stat. Math. **67**(1), 177–193 (2015)

Hotz, T., et al.: Sticky central limit theorems on open books. Ann. Appl. Probab. **23**(6), 2238–2258 (2013)

Huckemann, S., Mattingly, J.C., Miller, E., Nolen, J.: Sticky central limit theorems at isolated hyperbolic planar singularities. Electron. J. Probab. **20**(78), 1–34 (2015)

Huckemann, S.F., Eltzner, B.: Data analysis on nonstandard spaces. Comput. Stat., e1526 (2020)

Huckemann, S.F., Eltzner, B.: Foundations of Non-Euclidean Statistics. Chapman & Hall/CRC Press, London (2024). in preparation

Hundrieser, S., Eltzner, B., Huckemann, S.F.: Finite sample smeariness of Fréchet means and application to climate (2020). arXiv preprint arXiv:2005.02321

Pennec, X.: Curvature effects on the empirical mean in Riemannian and affine manifolds: a non-asymptotic high concentration expansion in the small-sample regime (2019). arXiv preprint arXiv:1906.07418

Shevtsova, I.: On the absolute constants in the Berry-Esseen type inequalities for identically distributed summands (2011)

Sturm, K.: Probability measures on metric spaces of nonpositive curvature. Contemp. Math. **338**, 357–390 (2003)

Tran, D., Eltzner, B., Huckemann, S.: Smeariness begets finite sample smeariness. In: Nielsen, F., Barbaresco, F. (eds.) GSI 2021. LNCS, vol. 12829, pp. 29–36. Springer, Cham (2021). https://doi.org/10.1007/978-3-030-80209-7_4

Types of Stickiness in BHV Phylogenetic Tree Spaces and Their Degree

Lars Lammers[1(✉)], Do Tran Van[1], Tom M. W. Nye[2],
and Stephan F. Huckemann[1]

[1] Felix-Bernstein-Institute for Mathematical Statistics in the Biosciences,
University of Göttingen, Goldschmidtstrasse 7, 37077 Göttingen, Germany
{lars.lammers,do.tranvan}@uni-goettingen.de,
stephan.huckemann@mathematik.uni-goettingen.de
[2] School of Mathematics, Statistics and Physics, Newcastle University,
Newcastle upon Tyne, UK
tom.nye@ncl.ac.uk

Abstract. It has been observed that the sample mean of certain probability distributions in Billera-Holmes-Vogtmann (BHV) phylogenetic spaces is confined to a lower-dimensional subspace for large enough sample size. This non-standard behavior has been called stickiness and poses difficulties in statistical applications when comparing samples of sticky distributions. We extend previous results on stickiness to show the equivalence of this sampling behavior to topological conditions in the special case of BHV spaces. Furthermore, we propose to alleviate statistical comparision of sticky distributions by including the directional derivatives of the Fréchet function: the degree of stickiness.

Keywords: Fréchet mean · Hadamard spaces · Wasserstein distance · statisical discrimination

1 Introduction

The Billera Holmes Vogtmann (BHV) spaces, first introduced in [4], are a class of metric spaces whose elements are rooted trees with labeled leaves. Classically, these describe potential evolutionary relations between species. However, there are many applications beyond the field of biology. For example in linguistics, the relationships within a language family might be represented by phylogenetic trees, e.g. [12].

Allowing for statistical analysis of samples of entire phylogenies, the BHV tree spaces have gained considerable attraction in recent years. They are particularly attractive from a mathematical point of view as they were shown to be Hadamard spaces [4, Lemma 4.1], i.e. complete metric spaces of global nonpositive curvature. This results in many convexity properties of the metric guaranteeing, e.g. unique (up to reparametrization) geodesics between any two points

---

Supported by DFG GK 2088 and DFG HU 1575/7.

F. Nielsen and F. Barbaresco (Eds.): GSI 2023, LNCS 14071, pp. 357–365, 2023.
https://doi.org/10.1007/978-3-031-38271-0_35

and the existence and uniqueness of Fréchet means (see (1) below) for distributions with finite first moment [13]. The Fréchet mean is a natural generalization of the expectation to any metric space (M, d) as the minimizer of the expected squared distance of a probability distribution $\mathbb{P} \in \mathcal{P}^1(M)$. Here $\mathcal{P}(M)$ denotes the family of all Borel probability distributions on M and

$$\mathcal{P}^1(M) = \left\{ \mathbb{P} \in \mathcal{P}(M) \mid \forall x \in M : \int_M d(x, y) \, d\mathbb{P}(x) < \infty \right\}.$$

For a distribution $\mathbb{P} \in \mathcal{P}^1(M)$, the Fréchet mean $b(\mathbb{P})$ is then the set of minimizers of the *Fréchet function*

$$F_\mathbb{P}(x) = \frac{1}{2} \int_M \left(d^2(x, y) - d^2(z, y) \right) d\mathbb{P}(y) \quad x \in M, \tag{1}$$

for arbitrary $z \in M$. In Hadamard spaces, while the Fréchet function is strictly convex, in certain spaces, sampling from some distributions leads to degenerate behavior of the sample Fréchet mean, where, after a finite random sample size, it is restricted to a lower dimensional subsets of the space. This phenomenon has been called stickiness and was studied for various spaces, including BHV spaces, see [2,3,8,9]. This absence of asymptotic residual variance or its reduction incapacitates or aggravates standard statistical methodology.

In [9], a topological notion of stickiness was proposed: given a certain topology on a set of probability spaces, a distribution *sticks* to $S \subset M$ if all distributions in a sufficiently small neighborhood have their Fréchet means in S. There, it was also shown for the so-called *kale* that *sample stickiness* is equivalent to *topological stickiness*, induced by equipping $\mathcal{P}^1(M)$ with the Wasserstein distance

$$W_1(\mathbb{P}, \mathbb{Q}) = \underset{\pi \in \Pi(\mathbb{P}, \mathbb{Q})}{\operatorname{argmin}} \int_{M \times M} d(x, y) \mathrm{d}\pi(x, y),$$

where $\Pi(\mathbb{P}, \mathbb{Q})$ denotes the set of all couplings of $\mathbb{P}, \mathbb{Q} \in \mathcal{P}^1(M)$.

In this paper, we provide this equivalence of both notions of stickiness for strata of BHV spaces with positive codimension by using directional derivatives of the Fréchet function. Furthermore, we propose using these directional derivatives as a tool to discriminate between sticky distributions whose means are indistinguishable.

2   The Billera-Holmes-Vogtmann Phylogenetic Tree Space

For $N \in \mathbb{N}$ and $N \geq 3$, the BHV tree space $\mathbb{T}_N$ represents rooted trees with N labelled leaves via positive lengths of *interior edges*. Here, an interior edge is a split of the set of leaves and the root with at least two elements in both parts. The set of splits of a tree determines its *topology*. Whenever new internal nodes appear or existing ones coalesce, the topology changes.

Definition 1. *Trees with common topology form a stratum* $\mathbb{S} \subset \mathbb{T}_N$. *We say the stratum is of codimension* $l \geq 0$ *if the topology features* $N - 2 - l$ *splits.*

The highest possible stratum dimension is $N - 2$, which happen in the topology of a binary, i.e. of a fully resolved tree.

Taking the Euclidean geometry within closed strata and gluing them together at their boundaries, [4] arrive at the separable Hadamard space $(\mathbb{T}_N, d)$). Thus, geodesics between two trees in a BHV tree space correspond to changing the length of splits present in both trees and the addition and removal of the other splits. Geodesics between two trees can be computed in polynomial time [11].

For $x \in \mathbb{T}_N$ let $B_\epsilon(x)$ be the open ball of radius $\epsilon > 0$ in $\mathbb{T}_N$. Then, in direct consequence of the construction, we have the following.

Lemma 1. *Let* $\mathbb{S} \subset \mathbb{T}_N$ *be any stratum* $l \geq 1$. *Then, for any* $x \in \mathbb{S}$, *there is* $\epsilon > 0$ *such that* $\overline{B_\epsilon(x)} \cap \mathbb{S}$ *is closed in* $\mathbb{S}$. *Furthermore, the topology of any* $y \in \overline{B_\epsilon(x)}$ *features all splits present in* x.

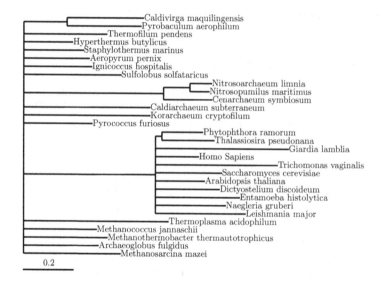

Fig. 1. Placing eukaryotes (homo sapiens) within the archaea: The Fréchet mean of a data set from [15] is a highly unresolved phylogenetic tree. Lengths of horizontal lines correspond to evolutionary distance, vertical lines to common nodes, the leftmost vertical stands for the common root.

3   Properties as Hadamard Spaces

In [4, Lemma 4.1], it was shown that the BHV tree spaces are Hadamard spaces, or spaces of global non-positive curvature, and as such enjoy many desirable properties. Notably, any two points in the space $\mathbb{T}_N$ are joined by a unqiue minimizing geodesic. For an extensive overview on Hadamard spaces, we refer to [5,6,13].

In a Hadamard space (M, d), a function $f : M \to \mathbb{R}$ is (strictly) convex if all compositions with geodesics are (strictly) convex. Convex sets are sets containing all geodesic segments between two points of the set. An example of a strictly convex function is the Fréchet function. Besides this convexity we will also require the following results.

Theorem 1 (Theorem 2.1.12 in [1]). *Let (M, d) be a Hadamard space and let $S \subset M$ be a closed convex set. Then the following statements hold true for the metric projection $\mathrm{P}_S : M \to S, x \mapsto \mathrm{argmin}_{y \in S} \, d(x, y)$.*

1. *The map $x \mapsto \mathrm{P}_S(x)$ is single-valued for all $x \in M$.*
2. *It holds for any $x, y \in M$ that $d\left(\mathrm{P}_S(y), \mathrm{P}_S(x)\right) \leq d(x, y)$.*

Remark 1. As the strata of BHV Spaces are locally Euclidean, the metric projection of a tree to an *adjacent* stratum of positive codimension corresponds to simply removing the redundant splits and keeping the splits featured in the topology of the stratum and their respective lengths.

Theorem 2 (Theorem 6.3 in [13]). *Let (M, d) be a separable Hadamard space. Then $d(b(\mathbb{P}), b(\mathbb{Q})) \leq W_1(\mathbb{P}, \mathbb{Q})$ for any two $\mathbb{P}, \mathbb{Q} \in \mathcal{P}^1(M)$.*

4   The Space of Directions

In a Hadamard space (M, d), it is possible to compute an angle between two (non-constant) geodesics γ, γ' starting at the same point $x \in M$. This angle is called the *Alexandrov angle* $\angle_x$ and can be computed as follows [6, Chapter 3]

$$\angle_x(\gamma, \gamma') = \lim_{t, t' \searrow 0} \arccos\left(\frac{t^2 + t'^2 - d(\gamma(t), \gamma'(t'))}{2t \cdot t'}\right).$$

Two geodesics have an equivalent direction at x if the Alexandrov angle between them is 0. The set of these equivalence classes is called the *space of directions* at x and is denoted by $\Sigma_x M$. Equipped with the Alexandrov angle, the space of directions becomes a spherical metric space itself. For an overview, see e.g. Chapter 9 in [6]. For points $x, y, z \in M$, we write $\mathrm{dir}_x(y)$ for the direction of the (unit speed) geodesic from x to y and we write $\angle_x(y, z) = \angle_x(\mathrm{dir}_x(y), \mathrm{dir}_x(z))$.

Let $\mathbb{S} \subset \mathbb{T}_N$ be a stratum with positive codimension $l \geq 1$ and set

$$(\Sigma_x \mathbb{T}_N)^\perp = \{\mathrm{dir}_x(z) | z \neq x, \ z \in \mathrm{P}_{\overline{\mathbb{S}}}^{-1}(\{x\})\},$$

$$(\Sigma_x \mathbb{T}_N)^\| = \{\mathrm{dir}_x(z) | z \neq x, \ z \in \mathbb{S}\}.$$

The following lemma is concerned with the structure of the space of directions in BHV tree spaces. It is inspired by the work of the tangent cone of orthant spaces in [3]. For two metric spaces $(M_1, d_1), (M_2, d_2)$, recall their *spherical join*

$$M_1 * M_2 = \left[0, \frac{\pi}{2}\right] \times M_1 \times M_2 / \sim \cong \left\{(\cos\theta\, p_1, \sin\theta\, p_2) : 0 \leq \theta \leq \frac{\pi}{2}, p_i \in M_i, i = 1, 2\right\}$$

with the metric

$$d\big((\theta, p_1, p_2), (\theta, p_1, p_2)\big) = \arccos\left(\cos\theta\cos\theta' d_1(p_1, p_1') + \sin\theta\sin\theta' d_2(p_2, p_2')\right).$$

In particular, this turns $M_1 * M_2$ into a sphere of dimension $n_1 + n_2 + 1$ if M_i is a sphere of dimension n_i, $i = 1, 2$.

Lemma 2. *Let* $\mathbb{S} \subset \mathbb{T}_N$ *be a stratum with positive codimension* $l \geq 1$ *and* $x \in \mathbb{S}$. *Then its space of directions can be given the structure of a spherical join*

$$\Sigma_x \mathbb{T}_N \cong (\Sigma_x \mathbb{S})^{\|} * (\Sigma_\mathbb{S} \mathbb{T}_N)^{\perp}, \quad (\Sigma_\mathbb{S} \mathbb{T}_N)^{\perp} \cong (\Sigma_x \mathbb{T}_N)^{\perp}$$

Proof. For sufficiently small $\epsilon > 0$, with $B_\epsilon(x)$ from Lemma 1 for any geodesic starting at x we have a one-to-one correspondence between its direction $\sigma \in \Sigma_x \mathbb{T}_N$ and a point $y_\sigma \in \partial B_\epsilon(x)$ with $d(x, y_\sigma) = \epsilon$. This gives rise to the angular part of the join

$$\theta : \Sigma_x \mathbb{T}_N \to [0, \pi/2], \sigma \mapsto \arcsin\left(\frac{d(y_\sigma, \mathbb{S})}{\epsilon}\right).$$

Furthermore, as remarked before Lemma 1, the topology of y features all splits of x and at most l additional splits. With the map $y \mapsto x_y : \partial \overline{B_\epsilon(x)} \to \overline{B_\epsilon(x)}$, adding to x all splits of y, not present in x, with their lengths from y, set

$$y^{\perp} := x_y, \quad y^{\|} := P_{\overline{\mathbb{S}}}(x_y),$$

where we identify the directions of x_y and x_y' at $x \in \mathbb{S}$ and $x' \in \mathbb{S}$, respectively, if their split lengths after removing those of x and x', respectively, agree. In conjunction with

$$\phi^{\perp}(\sigma) := \mathrm{dir}_x(y_\sigma^{\perp}), \quad \phi^{\|}(\sigma) := \mathrm{dir}_x(y_\sigma^{\|}),$$

thus obtain, with the second factor independent of the base point,

$$\Phi : \Sigma_x \mathbb{T}_N \to (\Sigma_x \mathbb{S})^{\|} * (\Sigma_x \mathbb{T}_N)^{\perp}, \quad \sigma \mapsto \left(\theta(\sigma), \phi^{\|}(\sigma), \phi^{\perp}(\sigma)\right).$$

Straightforward computation verifies that Φ is a bijection.

It remains to show that Φ is isometric. For notational simplicity, suppose $\epsilon = 1$. Let $\sigma_1, \sigma_2 \in \Sigma_x \mathbb{T}_N$ with $y_i := y_{\sigma_i}, i = 1, 2$ and $r_i^{\perp} := d(y_i, \mathbb{S}), r_i^{\|} := \sqrt{1 - (r_i^{\perp})^2}$. Exploiting $d^2(y_1, y_2) = d^2(y_1^{\|}, y_2^{\|}) + d^2(y_1^{\perp}, y_2^{\perp})$ by definition of the geodesic distance and that $\mathbb{T}_N$ is Euclidean in each stratum, we have indeed,

$$\cos(\angle_x(y_1, y_2)) = \frac{1 + 1 - d^2(y_1, y_2)}{2}$$

$$= \frac{(r_1^{\perp})^2 + (r_2^{\perp})^2 - d^2(y_1^{\perp}, y_2^{\perp})}{2} + \frac{(r_1^{\|})^2 + (r_2^{\|})^2 - d^2(y_1^{\|}, y_2^{\|})}{2}$$

$$= r_1^{\|} \cdot r_2^{\|} \cdot (\cos(\angle_x(y_1^{\|}, y_2^{\|})) + r_1^{\perp} \cdot r_2^{\perp} \cdot (\cos(\angle_x(y_1^{\perp}, y_2^{\perp}))$$

$$= \cos(\theta(\sigma_1)) \cos(\theta(\sigma_2)) \cos\left(\angle_x(\phi^{\|}(\sigma_1), \phi^{\|}(\sigma_2))\right)$$

$$\quad + \sin(\theta(\sigma_1)) \sin(\theta(\sigma_2)) \cos\left(\angle_x(\phi^{\perp}(\sigma_1), \phi^{\perp}(\sigma_2))\right).$$

In light of this fact, we shall henceforth abuse notation and will identify $(\Sigma_x \mathbb{T}_N)^{\perp} \cong (\Sigma_\mathbb{S} \mathbb{T}_N)^{\perp}$ and any $\sigma \in (\Sigma_\mathbb{S} \mathbb{T}_N)^{\perp}$ with natural embedding into $\Sigma_x M$ for all $x \in \mathbb{S}$.

5   Equivalent of Notions of Stickiness in BHV Spaces

For a sequence $(X_i)_{i \in \mathbb{N}}$ of i.i.d. random variables following a distribution $\mathbb{P}$, let $\mathbb{P}_n = \frac{1}{n} \sum_{i=1}^{n} X_i$ denote the empirical measure.

Definition 2 (Three Notions of Stickiness). *Let* $\mathbb{P} \in \mathcal{P}^1(\mathbb{T}_N)$ *be a probability distribution in a BHV space and* $\mathbb{S} \subset \mathbb{T}_N$ *be a stratum with codimension* $l \geq 1$. *Then on* $\mathbb{S}$, $\mathbb{P}$ *is called*

Wasserstein sticky *if there is* $\epsilon > 0$ *such that* $b(\mathbb{Q}) \in \mathbb{S}$ *for all* $\mathbb{Q} \in \mathcal{P}^1(\mathbb{T}_N)$ *with* $W_1(\mathbb{P}, \mathbb{Q}) < \epsilon$,
perturbation sticky *if for any* $\mathbb{Q} \in \mathcal{P}^1(\mathbb{T}_N)$ *there is* $t_\mathbb{Q} > 0$ *such that* $b((1 - t)\mathbb{P} + t\mathbb{Q}) \in \mathbb{S}$ *for all* $0 < t < t_\mathbb{Q}$,
sample sticky *if for any sequence of random variables* $(X_i)_{i \in \mathbb{N}} \overset{i.i.d.}{\sim} \mathbb{P}$ *there is a random* $N \in \mathbb{N}$ *such that* $b(\mathbb{P}_n) \in \mathbb{S}$ *for all* $n \geq N$ *a.s.*

Theorem 3 ([10]). *Let* (M, d) *be a Hadamard space,* $x \in M$, *and* $\gamma : [0, L] \to M$ *be a unit speed geodesic with direction* σ *at* $\gamma(0) = x$. *Then for any* $\mathbb{P} \in \mathcal{P}^1(M)$, *the directional derivative of the Fréchet function* $\nabla_\sigma F_\mathbb{P}(x) = \frac{d}{dt} F_\mathbb{P}(\gamma(t))|_{t=0}$ *exists, is well-defined and*

$$\nabla_\sigma F_\mathbb{P}(x) = - \int_M \cos(\angle_x(\sigma, \mathrm{dir}_x(z))) \cdot d(x, z) \, d\mathbb{P}(z).$$

In particular, it is

1. *Lipschitz continuous as a map* $\Sigma_x \mathbb{T}_N \to \mathbb{R}, \sigma \mapsto \nabla_\sigma F_\mathbb{P}(x)$, *and*
2. *1-Lipschitz continuous as a map* $\mathcal{P}^1(M) \to \mathbb{R}, \mathbb{P} \mapsto \nabla_\sigma F_\mathbb{P}(x)$.

The following result follows directly from Lemma 2 and Theorem 3.

Corollary 1. *Let* $\mathbb{S} \subset \mathbb{T}_N$ *be a stratum of positive codimension* $l \geq 1$, *let* $\mathbb{P} \in \mathcal{P}^1(\mathbb{T}_N)$ *and identify* $\sigma \in (\Sigma_\mathbb{S} \mathbb{T}_N)^{\perp}$ *for all* $x \in \mathbb{S}$ *across all spaces of directions in* $\mathbb{S}$. *Then* $x \mapsto \nabla_\sigma F_\mathbb{P}(x), \mathbb{S} \to \mathbb{R}$ *is constant.*

Assumption 1: For $X \sim \mathbb{P} \in \mathcal{P}^1(\mathbb{T}_N)$ with $b(\mathbb{P}) = \mu \in \mathbb{S}$ for a stratum $\mathbb{S} \subset \mathbb{T}_N$ assume that

$$\mathbb{P}\{\phi_\sigma(X) = 0\} < 1 \text{ for all } \sigma \in (\Sigma_\mathbb{S} \mathbb{T}_N)^{\perp},$$

where for $z \in \mathbb{T}_N$, $\phi_\sigma(z) = -d(z, \mu) \cos(\angle_\mu(\sigma, \mathrm{dir}_\mu(z)))$.

Theorem 4. *Let* $N \geq 4$ *and consider a stratum* $\mathbb{S} \subset \mathbb{T}_N$ *with positive codimension* $l \geq 1$. *Then the following statements are equivalent for a probability distribution* $\mathbb{P} \in \mathcal{P}^1(\mathbb{T}_N)$ *with* $\mu = b(\mathbb{P}) \in \mathbb{S}$.

1. $\mathbb{P}$ *is Wasserstein sticky on* $\mathbb{S}$.
2. $\mathbb{P}$ *is perturbation sticky on* $\mathbb{S}$.
3. $\mathbb{P}$ *is sample sticky on* $\mathbb{S}$ *and fulfills Assumption 1.*
4. *For any direction* $\sigma \in (\Sigma_{\mathbb{S}}\mathbb{T}_N)^{\perp}$, *we have that* $\nabla_{\sigma}F_{\mathbb{P}}(\mu) > 0$.

Proof. "1 $\implies$ 2" follows at once from $W_1(\mathbb{P}, (1-t)\mathbb{P} + t\mathbb{Q}) = tW_1(\mathbb{P}, \mathbb{Q})$ as verified by direct computation.

"2 $\implies$ 4": Let $\sigma \in (\Sigma_{\mathbb{S}}\mathbb{T}_N)^{\perp}$, $y \in \mathbb{T}_N$ such that $P_{\mathbb{S}}(y) = \mu$ and $\sigma = \mathrm{dir}_{\mu}(y)$, and let $\mathbb{Q}_t = (1-t)\mathbb{P} + t\delta_y$, for $0 \leq t \leq 1$. By hypothesis, we find $0 < t_y < 1$ such that $b(\mathbb{Q}_t) \in \mathbb{S}$ for all $t \leq t_y$. Since μ is the Fréchet mean of $\mathbb{P}$ and $P_{\mathbb{S}}(y) = \mu$, we have for any $x \in \mathbb{S} \setminus \{\mu\}$ that $F_{\mathbb{Q}_t}(x) > F_{\mathbb{Q}_t}(\mu)$. Thus, μ is be the Fréchet mean of $\mathbb{Q}_t$ for all $t \leq t_y$, and hence for all $t \leq t_y$,

$$0 \leq \nabla_{\sigma}F_{\mathbb{Q}_t}(\mu) = (1-t) \cdot \nabla_{\sigma}F_{\mathbb{P}}(\mu) - t \cdot d(\mu, y),$$

whence $\nabla_{\sigma}F_{\mathbb{P}}(\mu) \geq \frac{t_y}{1-t_y}d(\mu, y) > 0$.

"4 $\implies$ 1": Take arbitrary $\eta > 0$ such that $\overline{B_{\eta}(\mu)} \cap \mathbb{S}$ is closed in $\mathbb{S}$. As $(\Sigma_{\mathbb{S}}\mathbb{T}_N)^{\perp}$ is compact and the directional derivatives are continuous in directions (Theorem 3), there is a lower bound

$$0 < \zeta = \min_{\sigma \in (\Sigma_{\mathbb{S}}\mathbb{T}_N)^{\perp}} \nabla_{\sigma}F_{\mathbb{P}}(\mu).$$

Then, for any $\mathbb{Q} \in \mathcal{P}^1(\mathbb{T}_N)$ with $W_1(\mathbb{P}, \mathbb{Q}) < \epsilon := \min\{\zeta, \eta\}$, due to Theorem 2), it follows that $d(\mu, b(\mathbb{Q})) < \epsilon \leq \eta$. By Lemma 1, the topology of $\nu = b(\mathbb{Q})$ must feature all splits in the topology of μ.

It is left to show that $b(\mathbb{Q}) \notin \mathbb{S}$ cannot be. Otherwise, with $y = P_{\overline{\mathbb{S}}}(\nu)$, by Theorem 1, we have $d(\mu, y) \leq d(\mu, \nu) < \eta$ and hence, $y \in \mathbb{S}$. Furthermore, $\sigma = \mathrm{dir}_y(\nu) \in (\Sigma_{\mathbb{S}}\mathbb{T}_N)^{\perp}$ and thus by Theorem 1 and Corollary 1,

$$|\nabla_{\sigma}F_{\mathbb{P}}(\mu) - \nabla_{\sigma}F_{\mathbb{Q}}(y)| \leq W_1(\mathbb{P}, \mathbb{Q}) < \epsilon \leq \zeta.$$

Hence, $\nabla_{\sigma}F_{\mathbb{Q}}(y) > 0$, which implies, following σ, by strict convexity of the Fréchet function, that $F_{\mathbb{Q}}(y) < F_{\mathbb{Q}}(\nu)$, so that ν is not the Fréchet mean of $\mathbb{Q}$.

"3 $\implies$ 4": Since $b(\mathbb{P}) = \mu$, we have $\nabla_{\sigma}F_{\mathbb{P}}(\mu) \geq 0$ for all $\sigma \in (\Sigma_{\mathbb{S}}\mathbb{T}_N)^{\perp}$. We now show that $\nabla_{\sigma'}F_{\mathbb{P}}(\mu) = 0$ for some $\sigma' \in (\Sigma_{\mathbb{S}}\mathbb{T}_N)^{\perp}$ yields a contradiction. Indeed, then for $(X_i)_{i \in \mathbb{N}} \overset{i.i.d.}{\sim} \mathbb{P}$, there is $\mathbb{P}$-a.s. a random number $N \in \mathbb{N}$ such that $b(\mathbb{P}_n) \in \mathbb{S}$ for all $n \geq N$. Due to Theorem 3, we also have $\mathbb{P}$-a.s.,

$$\nabla_{\sigma'}F_{\mathbb{P}_n}(\mu) \geq 0 \quad \forall n \geq N. \tag{2}$$

Using the notation of Assumption 1, consider $S_n = \sum_{i=1}^{n} \phi_{\sigma'}(X_i)$, so that $\nabla_{\sigma'}F_{\mathbb{P}_n}(\mu) = S_n/n$. Recalling that we assumed that $\nabla_{\sigma'}F_{\mathbb{P}}(\mu) = \mathbb{E}(\phi(X_i)) = 0$, which implies that the random walk S_n is recursive (Theorem 5.4.8 in [7]), and hence (Exercise 5.4.1 in [7]) either

$$\mathbb{P}\{S_n = 0 : \text{ for all } n \in \mathbb{N}\} = 1 \text{ or } \mathbb{P}\left\{-\infty = \liminf_{n \in \mathbb{N}} S_n < \limsup_{n \in \mathbb{N}} S_n = \infty\right\} = 1.$$

The former violates Assumption 1, the latter contradicts (2), however. "1 and 4 $\implies$ 3": By [14, Theorem 6.9], $\mathbb{P}_n$ converges against $\mathbb{P}$ in W_1, hence $\mathbb{P}$ is sample sticky. As the directional derivatives of the Fréchet function for any $\sigma \in (\Sigma_{\mathbb{S}} \mathbb{T}_N)^{\perp}$ are non-zero by hypothesis, Assumption 1 holds.

6 Application: The Degrees of Stickiness

Definition 3. *Let* $\mathbb{P} \in \mathcal{P}^1(\mathbb{T}_N)$ *be a distribution that is Wasserstein sticky on a stratum* $\mathbb{S}$ *with positive codimension with* $\mu = b(\mathbb{P})$. *Then, we call* $D_{\sigma} F_{\mathbb{P}}(\mu)$ *the degree of stickiness of* $\mathbb{P}$ *in direction* σ.

We propose to use the degrees of stickiness as a way to discriminate between samples that are sticky on the same stratum. The following example illustrates such an application with two phylogenetic data sets X, Y from [15] with empirical distributions $\mathbb{P}^X$ and $\mathbb{P}^Y$, where each consists of 63 phylogenetic trees that were inferred from the same genetic data using two different methods. The resulting two Fréchet mean trees μ_X, μ_Y (after pruning very small splits) coincide in their topologies, as displayed in Fig. 1. We test the hypothesis

$$\mathcal{H}_0 : D_{\sigma} F_{\mathbb{P}^X}(\mu_X) = D_{\sigma} F_{\mathbb{P}^Y}(\mu_Y) \quad \forall \sigma \in \Sigma_{X,Y},$$

where we choose $\Sigma_{X,Y} \subset (\Sigma_{\mathbb{S}} \mathbb{T}_N)^{\perp}$ comprising only directions corresponding to a single split that is present in either X or Y and compatible with the topologies of μ_X and μ_Y. As there is a natural pairing, we performed a pairwise t-test for each of the directions and applied a Holm-correction, leading to a p-value of 0.0227. This endorses the observation in [15], that the two methods inferring phylogenetic trees differ significantly on this data set.

Acknowledgements. The 1st author gratefully acknowledges the DFG RTG 2088. The 2dn author gratefully acknowledges the DFG HU 1575/7. The 3rd author gratefully acknowledge the DFG CRC 1456. The work was done partially while the 2nd author was participating in the program of the Institute for Mathematical Sciences, National University of Singapore, in 2022.

References

1. Bacák, M.: Convex analysis and optimization in Hadamard spaces: De Gruyter (2014). https://doi.org/10.1515/9783110361629
2. Barden, D., Le, H., Owen, M.: Limiting behaviour of fréchet means in the space of phylogenetic trees (2014). https://doi.org/10.48550/ARXIV.1409.7602
3. Barden, D.M., Le, H.: The logarithm map, its limits and fréchet means in orthant spaces. Proc. London Mathem. Soc. **117** (2018)
4. Billera, L.J., Holmes, S.P., Vogtmann, K.: Geometry of the space of phylogenetic trees. Adv. Appl. Math. **27**(4), 733–767 (2001)
5. Bridson, M., Häfliger, A.: Metric Spaces of Non-Positive Curvature. Grundlehren der mathematischen Wissenschaften. Springer, Berlin Heidelberg (2011). https://doi.org/10.1007/978-3-662-12494-9

6. Burago, D., Burago, Y., Ivanov, S.: A Course in Metric Geometry. In: Crm Proceedings & Lecture Notes. American Mathematical Society (2001)
7. Durrett, R.: Probability: Theory and Examples. Cambridge Series in Statistical and Probabilistic Mathematics. Cambridge University Press (2019). https://books.google.de/books?id=b22MDwAAQBAJ
8. Hotz, T., et al.: Sticky central limit theorems on open books. Annals Appli. Probability **23**(6) (2013). https://doi.org/10.1214/12-aap899
9. Huckemann, S., Mattingly, J.C., Miller, E., Nolen, J.: Sticky central limit theorems at isolated hyperbolic planar singularities (2015)
10. Lammers, L., Van, D.T., Huckemann, S.F.: Types of stickiness, their degree and applications (2023), manuscript
11. Owen, M., Provan, J.S.: A fast algorithm for computing geodesic distances in tree space (2009). https://doi.org/10.48550/ARXIV.0907.3942
12. Shu, K., Ortegaray, A., Berwick, R., Marcolli, M.: Phylogenetics of indo-european language families via an algebro-geometric analysis of their syntactic structures (2019)
13. Sturm, K.T.: Probability measures on metric spaces of nonpositive curvature. Contemp. Math. 338 (2003). https://doi.org/10.1090/conm/338/06080
14. Villani, C.: Optimal Transport: Old and New. Grundlehren der mathematischen Wissenschaften. Springer, Berlin Heidelberg (2008). https://doi.org/10.1007/978-3-540-71050-9
15. Williams, T.A., Foster, P.G., Nye, T.M.W., Cox, C.J., Embley, T.M.: A congruent phylogenomic signal places eukaryotes within the archaea. Proc. Royal Soc. B: Biolog. Sci. **279**, 4870–4879 (2012)

Towards Quotient Barycentric Subspaces

Anna Calissano[✉][iD], Elodie Maignant[iD], and Xavier Pennec[iD]

Université Côte d'Azur and Inria, Epione Team, Sophia-Antipolis, Valbonne, France
{anna.calissano,elodie.maignant,xavier.pennec}@inria.fr

Abstract. Barycentric Subspaces have been defined in the context of manifolds using the notion of exponential barycenters. In this work, we extend the definition to quotient spaces which are not necessary manifolds. We define an alignment map and an horizontal logarithmic map to introduce Quotient Barycentric Subspaces (QBS). Due to the discrete group action and the quotient structure, the characterization of the subspaces and the estimation of the projection of a point onto the subspace is far from trivial. We propose two algorithms towards the estimation of the QBS and we discussed the results, underling the possible next steps for a robust estimation and their application to different data types.

Keywords: Discrete Group · Quotient Space · Barycentric Subspaces Analysis · Graph Space · Object Oriented Data Analysis

1 Introduction

Barycentric Subspace Analysis was introduced in [10]. Given a set of data points, it aims at estimating a set of subspaces of decreasing (or increasing) dimensions, which minimizes a loss function between the data and their projection onto the subspaces. In the specific context of manifold, the author defines barycentric subspaces using the Riemannian Exponential - referred as Exponential Barycentric Subspace (EBS). The derived EBS analysis is a promising dimensionality reduction technique for two main reasons. Firstly, it differs from tangent PCA [1,4] as it goes beyond a 1-dimensional subspace search, proposing an optimization over a flag of subspaces. Such property is useful in the context of complex data such as graphs, shapes or images, where dimensionality reduction plays a central role in terms of reducing the data complexity and interpreting the results [3,5,6]. Secondly, BSA can also be also defined as a "within data" statistics technique: the barycentric subspaces can be parametrized by data points. Such a choice allows to visualize and interpret the variability by looking at the data points which characterize the subspaces.

Complex data such as graphs or images are considered up to symmetries in many applications, such as node permutation for graphs or reflection for shapes. These type of data are usually embedded in quotient spaces obtained by applying a discrete group action, often resulting in a non manifold. In this geometric context, different dimensionality reduction techniques have been proposed [1,6]. To the best of our knowledge the majority of the techniques available are the equivalent of the estimation of a

---

A. Calissano—Authors equally contributed to this submission and they are alphabetically ordered.

F. Nielsen and F. Barbaresco (Eds.): GSI 2023, LNCS 14071, pp. 366–374, 2023.
https://doi.org/10.1007/978-3-031-38271-0_36

one-dimensional subspace at the time. Motivated by both high dimensional subspace search and interpretability through reference points, we propose a definition of Quotient Barycentric Subspaces (QBS). We underline where the complexity of the problem resides and we propose two algorithms for the estimation of QBS. We showcase the performance of the algorithm in a simple example of reflection action applied to $\mathbb{R}^2$. To the best of our knowledge, there are no works in the literature addressing the problem of barycentric analysis for quotient spaces which are not manifolds. Note that if the quotient space is a manifold, all the manifold statistic literature is applicable [8, 11], including exponential barycentric subspace analysis [10].

2   Quotient Barycentric Subspaces

Consider $X = \mathbb{R}^m$ a Euclidean space and $\mathcal{T}$ a discrete group acting on X. The orbit of a point $x \in X$ is the equivalence class $[x] = \{tx, t \in \mathcal{T}\}$. We are interested by the quotient space $X/\mathcal{T}$, made of the collection of these equivalence classes, when the action is not free. A group action is free if the only element of the group fixing a point is the identity [9]. If the action is not free, the resulting quotient space is not a manifold, but often only a stratified space. These spaces are very common, as non-free actions appear every time there are symmetries in the data.

Example 1. Consider $X = \mathbb{R}^2$ and the reflection action $\mathcal{R}$, giving the following equivalence relation: $(x_1, x_2) \sim (-x_1, -x_2)$. The resulting quotient space $X/\mathcal{R}$ is not a manifold - the point $(0, 0)$ is fixed by the whole group $\mathcal{R}$.

Example 2. Graph Space [2,7] is a quotient space used to study set of graphs with unlabelled nodes. It is obtained by applying permutation action to adjacency matrices. Consider a set of $n \times n$ undirected graphs represented as adjacency matrices $\{x_1, \ldots, x_k\}, x_i \in X = \mathbb{R}^{n \times n}$. The space of adjacency matrices X is equipped with a Frobenious norm. If we consider the nodes to not be labelled, we can represent the unlabelled graphs as equivalence classes of permuted graphs $[x] = \{px, p \in \mathcal{P}\}$, where $\mathcal{P}$ is the set of permutation matrices applied to the nodes, acting onto the matrices as $px = p^T xp$. Graph Space $X/\mathcal{P}$ is a discrete quotient space, equipped with a quotient metric $d_{X/\mathcal{P}}([x], [y]) = min_{p \in \mathcal{P}} d(x, py)$. As detailed in [3], Graph Space is not a manifold - as there are some permutations leaving the graph unchanged - making the extension of EBS not trivial.

2.1   Exponential Barycentric Subspaces

We first recall the definition of Exponential Barycentric Subspaces (EBS) introduced in [10]. Consider a Riemannian manifold $\mathcal{M}$ equipped with a Riemannian metric on each tangent space $T_x\mathcal{M}$ and a logarithmic map $log_x : \mathcal{M} \to T_x\mathcal{M}$.

Definition 1 (Exponential Barycentric Subspaces). *The Exponential Barycentric Subspaces (EBS) generated by the affinely independent reference points $(x_0, \ldots, x_k) \in \mathcal{M}^{k+1}$ is defined as*

$$EBS(x_0, \ldots, x_k) = \{x \in \mathcal{M}^*(x_0, \ldots, x_k) | \quad \exists w \in \mathbb{R}^{k+1} \setminus \{0\} : \sum_{i=0}^{k} w_i log_x(x_i) = 0\}.$$

As detailed in [10], the definition is only valid for points x that are outside the cut locus of the reference points. In other words, EBS are defined in the tangent space $T_x\mathcal{M}$, where x belongs to EBS. This gives an implicit definition.

2.2   Quotient Barycentric Subspaces

To extend EBS to Graph Space and more generally to quotient spaces, we need to define an analogous to the logarithmic map used in the definition.

Definition 2. *The Alignment Map* $a_x : X \to X$ *is defined as* $a_x(y) = t^*y$ *where* $t^* = \arg\min_{t \in T} d(x, ty)$ *is the group element which optimally aligns the two points.*

The Alignment Map at a point x is only defined on $X \setminus E([x])$, where $E([x])$ is the (null measure) equidistant set [12], which consists of points which are equidistant to at least two elements of the orbit $[x]$.

Given the alignment map, we can introduce an equivalent of the horizontal logarithmic map in the context of quotient space [6,9]:

Definition 3. *The Horizontal Logarithm* $log_x^H : X \setminus E(x) \to T_xX$ *is the logarithmic map of the total space X applied to the optimally aligned points:*

$$log_x^H(y) = log_x(a_x(y))$$

where $log_x : X \to T_xX$ *is the logarithmic map of the total space.*

As X is Euclidean, we can explicitly write the logarithmic map as $log_x^H y = log_x(a_x(t)) = a_x(y) - x$. Notice that due to the alignment map not being defined on the whole space, log_x^H does not descend to a proper logarithm on the quotient space.

We can now extend the definition to the quotient space.

Definition 4. *Consider a set of reference orbits* $\mathcal{X} = \{[x_1], \ldots, [x_k]\} \in X/\mathcal{T}$. *An orbit* $[y] \in X/\mathcal{T}$ *belongs to the Quotient Barycentric Subspace* $[y] \in QBS(\mathcal{X})$ *if* $\exists w_i \in \mathbb{R}$ *such that* $\sum_{i=1}^{k} w_i = 1$ *and*

$$\sum_{i=1}^{k} w_i log_y^H(x_i) = 0.$$

The above definition is implicit and corresponds to solving the following system:

$$\begin{cases} \sum_{i=1}^{k} w_i a_y(x_i) = y, \\ a_y(x_i) = t_i x_i, \quad t_i = \arg\min_{t \in \mathcal{T}} d(t x_i, y). \end{cases}$$

More explicitly, $[y]$ belongs to $QBS(\mathcal{X})$ if and only if its representative y can be written as a combination of some representatives of $\{[x_1], \ldots, [x_k]\}$ aligned with respect to $[y]$ itself. This results in a system of interlaced equations defining both the alignment and the barycentric combination. QBS are defined using the alignment map and the logarithmic map in the total space, which is Euclidean. This strategy allows for a computation of weights $w_i \in \mathbb{R}$ in the Euclidean Space X [10]. In the sequel of this paper, we focus only on positive weights to simplify the characterization of the subspaces.

The final goal of barycentric subspace analysis is to capture data variability with subspaces of a given dimension. We define a loss function using QBS Definition 4. Consider a set $\mathcal{Y} = \{[y_1], \ldots, [y_m]\}$ of orbit in the quotient space $X/\mathcal{T}$ and a set $\mathcal{X} = \{[x_1], \ldots, [x_k]\} \subset \mathcal{Y}$ of k reference points selected among our data-points. We want to minimize the following function $\mathcal{L}(\mathcal{X}) = \sum_{i=1}^{m} d_{X/\mathcal{T}}^2([\hat{y}_i], [y_i])$ where $\hat{y}_i \in QBS(\mathcal{X})$ is a projection on the QBS spanned by $\mathcal{X}$.

Equivalence Class   Positive Barycentric Subspaces   Positive Quotient Barycentric Subspaces

Fig. 1. Visualization of the equivalence classes in $X/\mathcal{R}$, the subspaces identified by two reference points, the quotient barycentric subspaces, where only the valid segments defined in Definition 4 are considered. The equidistant set of the red (resp. blue) point orbit is represented by the red (resp. blue) dashed line. (Color figue online)

2.3 Characterization of the Subspaces

Given a set of reference points in X, the subspace identified by the orbits of the reference points results in a set of disjointed subspaces in $X/\mathcal{T}$. In Fig. 1, we show the characterization of the QBS in $X/\mathcal{R}$ described in Example 1. Given a data point, the resulting orbit in $X/\mathcal{R}$ is represented as red dots in Fig. 1 left. Consider two reference orbits (here the red and blue orbits), the possible positive barycentric subspaces are all the possible subspaces of dimension 1 (segments) joining two points of the different orbits - Fig. 1 center. Such segments in the Euclidean space corresponds to barycentric combinations of reference points. Among such subspaces, not all points are valid according to the QBS Definition 4. In Fig. 1 right, the valid QBS are underlined in dark black, showing only the parts of the segments whose points are aligned with both reference points, as stated in Definition 4. Thus, the positive barycentric subspaces are turned into two disjoint valid segments: the *complete segments* if containing the reference orbits and the *incomplete segments* if not containing the reference orbits. The two red and blue dashed lines represent the boundary of the space where the points change alignments with respect to the reference points. In this example the identification of the valid – with respect to Definition 4 – parts of the subspaces is possible in close form due to the low dimension of both the space and the number of reference points. In more general setting, such close form description of the valid parts is not straightforward.

3   Algorithms for the Estimation of QBS

Consider a set of reference orbits $\{[x_1], \ldots, [x_k]\}$ and an orbit $[y]$, the estimation of the projection $\hat{y}$ of such orbit on the barycentric subspace is far from trivial. The complexity of the problem resides in the exploration of the search space. Firstly, we need to explore all the possible subspaces identified by $\{[x_1], \ldots, [x_k]\}$: the number of all the possible subspaces identified by k orbits of cardinality c is c^k. Secondly, we need to find for each subspace the valid part, which has no closed form characterization – as explained in the previous section. Remember that the validity of such parts depends on $[y]$ which is itself an orbit of dimension c. We refer to these two aspects as the combinatorial and the geometrical complexity of the given problem. As a starting point in addressing such complexity, we opted for two different algorithms: *Align to Reference Points* - Algorithm 1 - and *Align to Data Point* - Algorithm 2.

Algorithm 1. Align to Reference Points

Require: data point $y \in [y]$ and reference points $\{[x_1], \ldots, [x_k]\}$ **Return:** $\hat{y}$
  Select the closest reference point $[x_i]$ and its representative x_i optimally aligned with y
  Find $x_j \in [x_j]$ optimally aligned with x_i
  Find $y \in [y]$ optimally aligned with x_i
  Find $\hat{y}$, the projection of y onto $BS(x_1, \ldots, x_k)$ ▷ Orthogonal Projection
  if $\hat{y}$ is not optimally aligned with $x_1, \ldots, x_k$ **then**
      Set $\hat{y} = x_i$, where x_i is the closest reference point to y ▷ Closest Reference Projection
  end if

Algorithm 2. Align to Data Point

Require: data point $y \in [y]$ and reference points $\{[x_1], \ldots, [x_k]\}$ **Return:** $\hat{y}$
  Randomly select a representative $y \in [y]$
  For all i, Find $x_i \in [x_i]$ optimally aligned with y
  Find $\hat{y}$, the projection of y onto $BS(x_1, \ldots, x_k)$ ▷ Orthogonal Projection
  if $\hat{y}$ is not optimally aligned with $x_1, \ldots, x_k$ **then**
      Set $\hat{y} = x_i$, where x_i is the closest reference point to y ▷ Closest Reference Projection
  end if

---

4   Experiments

For the two proposed algorithms, we consider the setting described in Example 1 and we restrict the analysis to positive weights. We run two experiments:

- Experiment 1: fixing the reference orbits $\{[x_1], \ldots, [x_k]\}$ and randomly sampling the data orbit $[y^i]$, $i = 1, \ldots, N$ to project
- Experiment 2: fixing the data orbit $[y]$ and randomly sampling the reference orbits $\{[x_1^i], \ldots, [x_k^i]\}$, $i = 1, \ldots, N$ onto which project the data orbit

Such experiments aim at grasping when the algorithm is able to project a data orbit onto a valid QBS, given different configurations of reference orbits and data orbits to project. In the reflection quotient space $X/\mathcal{R}$, we are able to compute the true projection as we can parameterize the quotient subspaces of dimension 1. For each algorithm, we can compute the projection error as the quotient distance between $[y]$ and the estimated $[\hat{y}]$. To evaluate the performance, we measure the absolute error $\hat{\varepsilon}$ between the algorithm projection error and the true projection error. In Experiment 1, we consider two reference orbits and we sample a set of $N = 10000$ orbits to project. We run several Example of Experiment 1 to better characterize the performance of the two algorithms depending on the relative position of the two reference orbits $[x_1]$ and $[x_2]$.

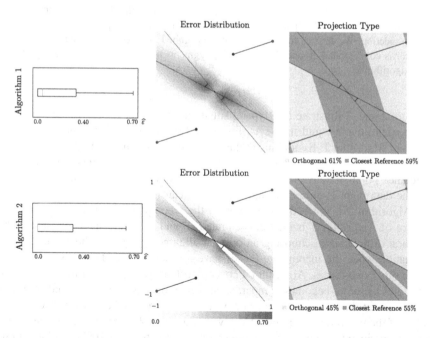

Fig. 2. Experiment 1: Example of two reference points resulting in QBS with small incomplete pieces. We plot the error distribution, the spatial distribution of the error and the spatial distribution of the type of projection. We can see that the error of Algorithm 1 and Algorithm 2 are similar. Algorithm 2 is better performing in projecting onto the incomplete segments, which are short in this example.

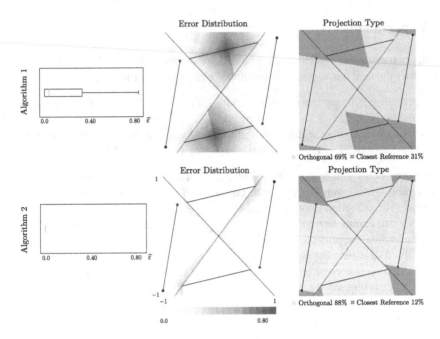

Fig. 3. Experiment 1: Example of two reference points close to each others equidistant sets. We plot the error distribution, the spatial distribution of the error and the spatial distribution of the type of projection. We can see that the error of Algorithm 1 is much higher than Algorithm 2. Such error is concentrated around the incomplete segments in Algorithm 1 and the Equidistant sets in Algorithm 2.

Figure 2 and 3 illustrate two typical situations. In Fig. 3, the reference points lay close to each other's equidistant sets. From the two boxplots, Algorithm 2 performs better than Algorithm 1. By looking at the error distribution in the middle panel, Algorithm 1 is under-performing when $[y]$ is around the incomplete segments. The alignment to reference point strategy is not able to project onto the incomplete segments - even when the data point is closer to such segments - causing an high error. On the other hand, Algorithm 2 is correctly projecting on the incomplete segments when needed. The error of Algorithm 2. is only concentrated around the equidistant sets, caused by the natural difficulties in choosing an optimal alignment. Figure 2 showcase a different situation when the incomplete pieces have a small length. By always projecting onto the other two pieces, Algorithm 2 covers most situations quite similarly to Algorithm 1 and therefore becomes comparable in performances. We understand from this first experiment that the performance of Algorithm 1 strongly depends on the relative position of the two reference points. However, Algorithm 2 is overall better than Algorithm 1. Experiment 2 focuses on changing reference points ($N = 10000$ configurations randomly sampled) fixing the data point $[y]$. We compare the performance of the two algorithms using two boxplots of the error distribution (Fig. 4). The performance of Algorithm 2 is better than for Algorithm 1 even when changing reference points.

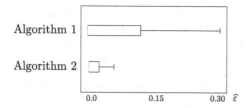

Fig. 4. Experiment 2: Comparison of the performances of Algorithm 1 and Algorithm 2. The boxplot are obtained by simulating a set of different reference points, keeping the data point fix. Algorithm 2 is performing better.

5   Discussion

We address the extension of Exponential Barycentric Subspaces from manifolds to Quotient Spaces obtained by a discrete group action. We define Quotient Barycentric Subspaces (QBS) and we propose two different algorithms for the estimation of the projection of data-points onto QBS. The detailed analysis that we perform shows that the complexity of such projection is not only combinatorial, as could be expected, but also geometrical. The identification of the correct segment onto which project a data point depends on the reciprocal position of reference orbits, the position of the data point with respect to the segment and the orientation of the quotient orbit with respect to the subspace. As a first approach, we showed how algorithm 2 better addresses the identification of the correct subspace for projection. Further developments will include an alignment procedure tailored for subspaces rather then data points (e.g. reference points), which might allow the identification of all valid subspaces, including the incomplete segments.

Acknowledgements. We thank Gstats team for the fruitful discussions. This work was supported by the ERC grant #786854 G-Statistics from the European Research Council under the European Union's Horizon 2020 research and innovation program and by the French government through the 3IA Côte d'Azur Investments ANR-19-P3IA-0002 managed by the National Research Agency.

References

1. Aydın, R., Pataki, G.H., Bullitt, E., Marron, J.S., et al.: A principal component analysis for trees. Annals Appli. Statist. **3**(4), 1597–1615 (2009)
2. Calissano, A., Feragen, A., Vantini, S.: Graph-valued regression: Prediction of unlabelled networks in a non-euclidean graph space. J. Multivar. Anal. **190**, 104950 (2022)
3. Calissano, A., Feragen, A., Vantini, S.: Populations of unlabelled networks: Graph space geometry and generalized geodesic principal components. Biometrika, asad024 (2023)
4. Fletcher, P.T., Joshi, S.: Principal geodesic analysis on symmetric spaces: statistics of diffusion tensors. In: Sonka, M., Kakadiaris, I.A., Kybic, J. (eds.) CVAMIA/MMBIA -2004. LNCS, vol. 3117, pp. 87–98. Springer, Heidelberg (2004). https://doi.org/10.1007/978-3-540-27816-0_8

5. Guo, X., Srivastava, A., Sarkar, S.: A quotient space formulation for generative statistical analysis of graphical data. J. Mathem. Imaging Vis. **63**, 735–752 (2021)

6. Huckemann, S., Hotz, T., Munk, A.: Intrinsic shape analysis: geodesic PCA for Riemannian manifolds modulo isometric Lie group actions. Statist. Sinica **20**(1), 1–58 (2010)

7. Jain, B.J., Obermayer, K.: Structure spaces. J. Mach. Learn. Res. **10**, 2667–2714 (2009)

8. Kendall, D.G.: Shape manifolds, Procrustean metrics, and complex projective spaces. Bull. Lond. Math. Soc. **16**(2), 81–121 (1984)

9. Lee, J.M.: Smooth manifolds. Springer (2013)

10. Pennec, X.: Barycentric subspace analysis on manifolds. Ann. Stat. **46**(6A), 2711–2746 (2018)

11. Severn, K.E., Dryden, I.L., Preston, S.P.: Non-parametric regression for networks. Stat. **10**(1), e373 (2021)

12. Wilker, J.B.: Equidistant sets and their connectivity properties. Proc. Am. Mathem. Soc. **47**(2), 446–452 (1975)

Rethinking the Riemannian Logarithm on Flag Manifolds as an Orthogonal Alignment Problem

Tom Szwagier$^{(\boxtimes)}$ (iD) and Xavier Pennec (iD)

Université Côte D'Azur, and Inria, Epione Project Team, Sophia-Antipolis, France
{tom.szwagier,xavier.pennec}@inria.fr

Abstract. Flags are sequences of nested linear subspaces of increasing dimension. They belong to smooth manifolds generalizing Grassmannians and bring a richer multi-scale point of view to the traditional subspace methods in statistical analysis. Hence, there is an increasing interest in generalizing the formulae and statistical methods already developed for Grassmannians to flag manifolds. In particular, it is critical to compute accurately and efficiently the geodesic distance and the logarithm due to their fundamental importance in geometric statistics. However, there is no explicit expression known in the case of flags. In this work, we exploit the homogeneous quotient space structure of flag manifolds and rethink the geodesic endpoint problem as an alignment of orthogonal matrices on their equivalence classes. The relaxed problem with the Frobenius metric surprisingly enjoys an explicit solution. This is the key to modify a previously proposed algorithm. We show that our explicit alignment step brings drastic improvements in accuracy, speed and radius of convergence, in addition to overcoming the combinatorial issues raised by the non-connectedness of the equivalence classes.

Keywords: Flag manifolds · Riemannian logarithm · Orthogonal alignment · Procrustes analysis

1 Introduction

Flags are sequences of nested linear subspaces of increasing dimension. They are important in statistical analysis [3,9] due to the multi-scale information they provide, compared to traditional subspace methods involving Grassmann manifolds [1]. Flags of a given type form a Riemannian manifold that generalizes Grassmannians. Hence there is a natural interest in generalizing the formulae and statistical methods already developed on Grassmannians.

The geodesic distance and logarithm are central tools in statistics on Riemannian manifolds, as they allow notably to discriminate, interpolate, and optimize [10]. Their explicit formulae are known for Grassmannians, and are related to the problem of finding principal vectors and angles between linear subspaces, which can be solved using Singular Value Decomposition (SVD) [1]. However no

F. Nielsen and F. Barbaresco (Eds.): GSI 2023, LNCS 14071, pp. 375–383, 2023.
https://doi.org/10.1007/978-3-031-38271-0_37

explicit formula is known for flags. One of the main hurdles seems to be the relatively complex structure of the tangent space, that must carry the information of nestedness of several subspaces, compared to Grassmannians which account for only two subspaces, and yield closed-form expressions for the geodesics in terms of sines and cosines of principal angles, like in [1, *Prop. 3.3*].

The question of the distance on flag manifolds is first addressed in [12, *Prop. 10*], but the proposition is actually misleading. Indeed, the proof of this proposition is based on the implicit knowledge of the horizontal tangent vector allowing to shoot from one flag to the other with minimal distance, that is the Riemannian logarithm. However, if this vector is known in the geodesic shooting problem, it is actually unknown in the geodesic endpoint problem, otherwise the geodesic distance would also be known, as it is the norm of the logarithm. Without clearly stating this assumption, the proposition is false. For instance, two orthogonal matrices representing the same flag would have a positive distance. Subsequent works rather focus on algorithms to approximate the Riemannian logarithm and distance, either based on a quotient space point of view and using alternated projections on the horizontal and vertical tangent spaces [3,4], or following an optimization approach [6,7].

In this work, we exploit the homogeneous quotient space structure of flag manifolds and rethink the geodesic endpoint problem (i.e., the search for a geodesic of minimal length joining two flags) as an alignment problem on the orthogonal group, that is the search for orthogonal matrices that are the closest among their equivalence classes. As it does not have an explicit solution, we relax it using the Frobenius metric. Our key result is that this constrained orthogonal Procrustes problem actually enjoys an explicit solution. We show that modifying a previously existing algorithm [4, *Alg. 1*] with this new result drastically improves its accuracy, speed and radius of convergence, in addition to overcoming the combinatorial issues raised by the non-connectedness of the stabilizer subgroup.

2 Flag Manifolds

In this section, we briefly define flag manifolds and the important properties for the next sections. A more complete introduction is given in [12].

Definition 1 (Flag, Signature). *Let $n \geq 2$ and $d_1 < d_2 < \cdots < d_r = n$ be a sequence of strictly increasing natural numbers. A flag of signature $(d_1, \ldots, d_r)$ is a sequence of properly nested linear subspaces of $\mathbb{R}^n$, $E_1 \subset \cdots \subset E_r = \mathbb{R}^n$ of respective dimension $d_1 < \cdots < d_r = n$. Flags with signatures of length n, i.e. $(1, 2, \ldots, n-1, n)$ are called complete.*

Flags can equivalently be defined as a sequence of *incremental orthogonal subspaces* $V_1 \perp \cdots \perp V_r$, by taking the orthogonal complement of one nested subspace into the next one. One then has $E_i = \bigoplus_{j=1}^{i} V_j$ $(i = 1..r)$. The associated sequence of *increments* $\mathcal{I} := (n_1, \ldots, n_r) := (d_1, d_2 - d_1, \ldots, d_r - d_{r-1})$ is called the *type* of the flag. We will use this definition of flags in the rest of the paper.

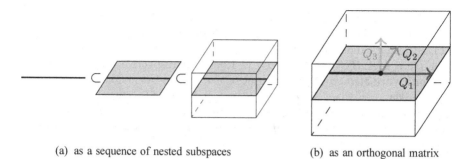

(a) as a sequence of nested subspaces   (b) as an orthogonal matrix

Fig. 1. Different representations of a flag of type $(1, 1, 1)$.

A flag can be represented as an orthogonal matrix whose columns form an orthonormal basis that is *adapted* to the sequence of incremental orthogonal subspaces. More specifically, for a given sequence of subspaces $V_1 \perp \cdots \perp V_r$ such that $\bigoplus_{i=1}^{r} V_i = \mathbb{R}^n$, one can find an orthonormal basis $Q_i := \left[q_i^1 | \ldots | q_i^{n_i} \right] \in \mathbb{R}^{n \times n_i}$ for each V_i, and the concatenation of those bases forms an orthogonal matrix $Q = [Q_1 | \ldots | Q_r] \in \mathcal{O}(n)$ that is a representative of the flag $V_1 \perp \cdots \perp V_r$. A flag of type $(1, 1, 1)$ is represented in Fig. 1, both in terms of nested subspaces and in the orthogonal representation.

Theorem 1 (Flag Manifolds [5]). The set of all flags of type $\mathcal{I}$ is a connected compact smooth manifold. It is noted Flag $(n_1, \ldots, n_r)$, or Flag($\mathcal{I}$) for short.

There exist several orthonormal bases adapted to a flag, hence flags are actually *equivalence classes* of orthogonal matrices. More precisely, let $Q \in \mathcal{O}(n)$ be a representative of a given flag and $R = \text{diag}(R_1, \ldots, R_r) \in \mathcal{O}(n)$, with $R_i \in \mathcal{O}(n_i)$ $(i = 1 .. r)$. Then Q and $QR = [Q_1 R_1 | \ldots | Q_r R_r]$ represent the same flag. Indeed, the right multiplication by the block diagonal orthogonal matrix R on Q only rotates and reflects the incremental orthonormal bases within their subspaces, but do not change their span. Therefore, flag manifolds are *homogeneous quotient spaces* of $\mathcal{O}(n)$ [12, Eq. (8)]

$$\text{Flag}(\mathcal{I}) \cong \mathcal{O}(n) / \left(\mathcal{O}(n_1) \times \cdots \times \mathcal{O}(n_r) \right). \tag{1}$$

For the sake of readability, we will thereafter write $\mathcal{O}(\mathcal{I}) := \mathcal{O}(n_1) \times \cdots \times \mathcal{O}(n_r)$. Therefore one has Flag($\mathcal{I}$) $\cong \mathcal{O}(n)/\mathcal{O}(\mathcal{I})$ and $R \in \mathcal{O}(\mathcal{I})$. Hence, a flag is an equivalence class of orthogonal matrices

$$[Q] := \{ QR, \quad R \in \mathcal{O}(\mathcal{I}) \} := Q \cdot \mathcal{O}(\mathcal{I}) \tag{2}$$

and one defines the canonical projection $\Pi_c \colon Q \in \mathcal{O}(n) \mapsto [Q] \in \mathcal{O}(n)/\mathcal{O}(\mathcal{I})$. One can also embed flags in a product of Grassmannians $\text{Gr}(n_1, n) \times \cdots \times \text{Gr}(n_r, n) := \text{Gr}(\mathcal{I})$, representing them as a sequence of orthogonal projection matrices onto the sequence of incremental orthogonal subspaces of the flag. The embedding map $\Pi_{\text{Gr}(\mathcal{I})} \colon [Q_1 | \ldots | Q_r] \in \mathcal{O}(n) \mapsto \left(Q_1 Q_1^\top, \ldots, Q_r Q_r^\top \right) \in \text{Gr}(\mathcal{I})$ then removes the necessity to work with equivalence classes.

The theory of Riemannian submersions and quotient spaces [8] tells that the Riemannian manifold structure of flags can be deduced from the one of $\mathcal{O}(n)$ and $\mathcal{O}(\mathcal{I})$ with their canonical metric g_Q, given by a multiple of the Frobenius metric: $g_Q(A, B) = \frac{1}{2}\langle A, B\rangle_F := \frac{1}{2}\mathrm{tr}\left(A^\top B\right)$ [2,12]. Their respective tangent spaces at the identity are $\mathfrak{o}(n) := \mathrm{Skew}_n$, the set of $n \times n$ skew-symmetric matrices, and $\mathfrak{o}(\mathcal{I}) := \mathrm{Skew}_I := \mathrm{diag}\left(\mathrm{Skew}_{n_1}, \ldots, \mathrm{Skew}_{n_r}\right)$. In the fiber bundle vocabulary, $\mathfrak{o}(\mathcal{I})$ is referred to as the *vertical space* (noted $\mathrm{Ver}(\mathcal{I})$), and its orthogonal complement in Skew_n as the *horizontal space*, containing skew-symmetric matrices with diagonal zero blocks

$$\mathrm{Hor}(\mathcal{I}) := \left\{ H := \begin{bmatrix} 0_{n_1} & H_{1,2} & \ldots & H_{1,r} \\ -H_{1,2}^\top & 0_{n_2} & \ldots & H_{2,r} \\ \vdots & \vdots & \ddots & \vdots \\ -H_{1,r}^\top & -H_{2,r}^\top & \ldots & 0_{n_r} \end{bmatrix}, \quad H_{ij} \in \mathbb{R}^{n_i \times n_j} \right\}. \quad (3)$$

The tangent space of flag manifolds is then given by

$$T_{[Q]}\left(\mathcal{O}(n)/\mathcal{O}(\mathcal{I})\right) = Q\mathrm{Hor}(\mathcal{I}). \quad (4)$$

The geodesics in the *base space* $\mathcal{O}(n)/\mathcal{O}(\mathcal{I})$ with the canonical metric are inherited from the *horizontal* geodesics in the *top space* $\mathcal{O}(n)$ [12]

$$\exp_{[Q]}(A) = \left[Q\exp((Q^\top A))\right].$$

A common approach for computing the Riemannian logarithm and distance is through solving the *geodesic endpoint problem*

$$\underset{H\in\mathrm{Hor}(\mathcal{I}), R\in\mathcal{O}(\mathcal{I})}{\arg\min} \quad \frac{1}{2}\|H\|_F^2 \quad \text{subject to} \quad P\exp(H) = QR. \quad (5)$$

However, it has a priori no solution for flag manifolds. In the literature, the geodesic endpoint problems are generally optimized using a gradient descent on the tangent space [6,13]. In the following section, we propose a different approach based on the previously described quotient space structure, and reformulate the geodesic endpoint problem as an alignment on the equivalence classes.

3   Alignment on Flag Manifolds

The notion of *alignment* in Riemannian quotient spaces is mentioned in [2, *Def. 5.1.6*]. It refers to the problem of finding a pair of points which minimizes the top space distance within their respective equivalence classes.

Definition 2 (Alignment on Flag Manifolds). *Let $P, Q \in \mathcal{O}(n)$ and $\mathrm{d}_{\mathcal{O}(n)}(P,Q)^2 := \frac{1}{2}\left\|\log(P^\top Q)\right\|_F^2$, the geodesic distance on $\mathcal{O}(n)$. We define the alignment problem on flag manifolds as*

$$\underset{R\in O(\mathcal{I})}{\arg\min} \quad \mathrm{d}_{\mathcal{O}(n)}(P, QR)^2. \quad (6)$$

Given two flags $[P], [Q] \in \mathcal{O}(n)/\mathcal{O}(\mathcal{I})$, finding a point $Q^* := QR^*$ which minimizes the $\mathcal{O}(n)$ geodesic distance to P in the orbit $[Q]$ yields both the geodesic distance between the two flags $\mathrm{d}([P], [Q]) = \mathrm{d}_{\mathcal{O}(n)}(P, Q^*)$ and the logarithm $\log_{[P]}([Q]) = P \log(P^\top Q^*)$, according to [2, *Prop. 5.1.3 & Def. 5.1.6*]. Moreover, [12, *Thm. 4*] ensures that such an alignment always exists on flag manifolds. Hence we get an approach to the computation of the Riemannian logarithm that leverages more information than classical optimization algorithms solving the geodesic endpoint problem, because it takes into consideration the homogeneous quotient space structure of flag manifolds, and takes advantage of the results that already exist on the top space $\mathcal{O}(n)$. Notably, even if the proposed algorithm does not converge, one still has a way to interpolate between two flags using the logarithm in $\mathcal{O}(n)$. Just like the geodesic endpoint problem, the alignment problem (6) does not have either an analytic solution, to the best of our knowledge. However, the embedding of $\mathcal{O}(n)$ in $\mathbb{R}^{n \times n}$ raises the natural idea of relaxing the problem and working with the Frobenius metric.

3.1   Relaxing Orthogonal to Frobenius Distance

Definition 3 ($\mathcal{O}(\mathcal{I})$ Procrustes Problem on Flag Manifolds). *Let $P, Q \in \mathcal{O}(n)$. We define the $\mathcal{O}(\mathcal{I})$-constrained Procrustes problem on flag manifolds as*

$$\underset{R \in O(\mathcal{I})}{\arg\min} \quad \|P - QR\|_F^2. \tag{7}$$

This name refers to the celebrated orthogonal Procrustes problem [11], which also involves the minimization of a Frobenius distance on an orthogonal group. Here, our optimization problem is constrained to the isotropy subgroup $\mathcal{O}(\mathcal{I})$ of flag manifolds in $\mathcal{O}(n)$. We now give the main result of the paper.

Theorem 2 ($\mathcal{O}(\mathcal{I})$ Procrustes Solution on Flag Manifolds).
Let $P := [P_1|\ldots|P_r]$, $Q := [Q_1|\ldots|Q_r] \in \mathcal{O}(n)$ and let us write the SVDs $P_i^\top Q_i := U_i \Sigma_i V_i^\top$ $(i = 1..r)$, with $U_i, V_i \in \mathcal{O}(n_i)$ and $\Sigma_i \in \mathrm{diag}\left(\mathbb{R}_{\geq 0}^{n_i}\right)$. A solution $R^* = \mathrm{diag}\,(R_1^*, \ldots, R_r^*) \in \mathcal{O}(\mathcal{I})$ to the $\mathcal{O}(\mathcal{I})$-constrained Procrustes problem on flag manifolds (7) is given by

$$R_i^* = U_i V_i^\top \quad (i = 1..r). \tag{8}$$

The uniqueness is conditioned on the uniqueness of the SVD.

Proof. We can show that this $\mathcal{O}(\mathcal{I})$ Procrustes problem is equivalent to the independent resolution of classical orthogonal Procrustes problem [11], i.e.

$$\underset{R \in O(\mathcal{I})}{\min} \|P - QR\|_F^2 = \sum_{i=1}^{r} \underset{R_i \in O(n_i)}{\min} \|P_i - Q_i R_i\|_F^2.$$

$\square$

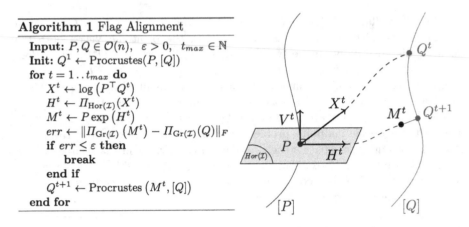

Algorithm 1 Flag Alignment

Input: $P, Q \in \mathcal{O}(n)$, $\varepsilon > 0$, $t_{max} \in \mathbb{N}$
Init: $Q^1 \leftarrow \text{Procrustes}(P, [Q])$
for $t = 1 .. t_{max}$ **do**
$\quad X^t \leftarrow \log\left(P^\top Q^t\right)$
$\quad H^t \leftarrow \Pi_{\text{Hor}(\mathcal{I})}\left(X^t\right)$
$\quad M^t \leftarrow P \exp\left(H^t\right)$
$\quad err \leftarrow \|\Pi_{\text{Gr}(\mathcal{I})}\left(M^t\right) - \Pi_{\text{Gr}(\mathcal{I})}(Q)\|_F$
$\quad$**if** $err \leq \varepsilon$ **then**
$\quad\quad$**break**
$\quad$**end if**
$\quad Q^{t+1} \leftarrow \text{Procrustes}\left(M^t, [Q]\right)$
end for

Fig. 2. The flag alignment Algorithm 1 and a conceptual visualization. The point Q moves in its equivalence class $[Q]$ at each iteration, with the aim that the $\mathcal{O}(n)$ logarithm PX^t becomes horizontal.

3.2 Introducing $\mathcal{O}(\mathcal{I})$ Procrustes into an Existing Algorithm

An algorithm to compute the Riemannian logarithm and distance on flag manifolds is proposed in [4, *Alg. 1*]. This algorithm takes as inputs two special orthogonal matrices $P, Q \in \mathcal{SO}(n)$. It aims at finding a horizontal vector $H \in \text{Hor}(\mathcal{I})$ and a vertical vector $V \in \text{Ver}(\mathcal{I})$ such that one can write $Q = P \exp(H) \exp(V)$. It is based on iterated alternating projections on the horizontal and vertical spaces. One important drawback of this method is that it implicitly assumes that the endpoint Q is in the same connected component in $\mathcal{O}(\mathcal{I})$ as the horizontally aligned point $Q_{aligned} = P \exp(H)$, which is generally not the case. The authors highlight this drawback and explain that the algorithm is actually working on *fully oriented flag manifolds* $\mathcal{SO}(n)/\left(\mathcal{SO}(n_1) \times \cdots \times \mathcal{SO}(n_r)\right)$. To overcome this issue, they first create 2^{r-1} "equivalents" of the endpoint Q in all the connected components, then run 2^{r-1} times their algorithm, and finally take the best outcome among the different runs.

In this work, we take advantage of our new result (Theorem 2), which gives a global minimum on the whole isotropy group $\mathcal{O}(\mathcal{I})$, to overcome this drawback. We reinterpret [4, *Alg. 1*] as an alignment problem on flag manifolds and introduce $\mathcal{O}(\mathcal{I})$ Procrustes alignment both in the initialization and as a substitute for the vertical projection step onto $\mathfrak{o}(\mathcal{I})$. The algorithm and a conceptual visualization of it are given in Fig. 2.

4   Numerical Experiments

We now evaluate and compare Algorithm 1 to the alternating projections algorithm proposed in [4], run on all the connected components of $\mathcal{O}(\mathcal{I})$. The general evaluation process is as follows. First we generate an orthogonal matrix P drawn

Table 1. Evaluation of Algorithm 1 and [4] on $\mathrm{Flag}(1,2,2,15,80)$ averaged over 10 experiments, for different geodesic distances *(from top to bottom: $\frac{\pi}{5}$, $\frac{\pi}{2}$, π).*

$\frac{\pi}{5}/\frac{\pi}{2}/\pi$	endpoint error	distance error	horizontal error	# iter	time
Alg. 1	$(4.0_{\pm.55}) \times 10^{-7}$	$(6.7_{\pm1.6}) \times 10^{-14}$	$(2.9_{\pm.40}) \times 10^{-7}$	$(1.0_{\pm0.0})$	$(5.4_{\pm.22}) \times 10^{-2}$
[4]	$(2.2_{\pm2.2}) \times 10^{-6}$	$(7.8_{\pm14.}) \times 10^{-12}$	$(2.6_{\pm2.2}) \times 10^{-6}$	$(6.1_{\pm1.4})$	$(2.8_{\pm.14}) \times 10^{+1}$
Alg. 1	$(5.8_{\pm3.5}) \times 10^{-6}$	$(7.8_{\pm6.2}) \times 10^{-12}$	$(4.5_{\pm2.7}) \times 10^{-6}$	$(2.4_{\pm.49})$	$(1.0_{\pm.17}) \times 10^{-1}$
[4]	$(9.0_{\pm12.}) \times 10^{-1}$	$(1.9_{\pm2.4}) \times 10^{-01}$	$(1.0_{\pm1.4}) \times 10^{+0}$	$(10._{\pm0.0})$	$(3.1_{\pm.12}) \times 10^{+1}$
Alg. 1	$(9.8_{\pm2.9}) \times 10^{-6}$	$(1.5_{\pm.90}) \times 10^{-11}$	$(1.3_{\pm.41}) \times 10^{-5}$	$(9.9_{\pm.30})$	$(3.4_{\pm.22}) \times 10^{-1}$
[4]	$(7.6_{\pm3.2}) \times 10^{+0}$	$(1.3_{\pm.56}) \times 10^{+00}$	$(7.1_{\pm2.2}) \times 10^{+0}$	$(10._{\pm0.0})$	$(3.0_{\pm.04}) \times 10^{+1}$

from a uniform distribution on $\mathcal{O}(n)$. Second we generate $H \in \mathrm{Hor}(\mathcal{I})$ drawn from a uniform distribution (with a specified norm $\frac{1}{2}\|H\|_F \in \{\frac{\pi}{5}, \frac{\pi}{2}, \pi\}$) and get the aligned endpoint $Q_{aligned} = P \exp(H)$. Third we generate $R \in \mathcal{O}(\mathcal{I})$ drawn from a uniform distribution and get the endpoint $Q = P \exp(H)R$. We run both algorithms with a maximal number of iterations of 10, and repeat the experiment 10 times independently, each time outputting an optimal horizontal vector H^*. The evaluation metrics are: endpoint error $(\|\Pi_{\mathrm{Gr}(\mathcal{I})}(P \exp(H^*)) - \Pi_{\mathrm{Gr}(\mathcal{I})}(Q)\|_F)$, distance error $(\frac{1}{2}\,|\|H^*\|_F - \|H\|_F|)$, horizontal error $(\frac{1}{2}\,\|H^* - H\|_F)$, number of iterations and computing time. The results are reported in Table 1 and the associated learning curves are illustrated in Fig. 3.

It is clear that the introduction of $\mathcal{O}(\mathcal{I})$ Procrustes alignment as an initialization and a substitute for the vertical projection drastically improves [4, *Alg. 1*], in terms of accuracy, speed and radius of convergence, on all the evaluation metrics, as well as overcomes the difficulty of non-connectedness of the orbits. The difference of computing time is particularly important, because Algorithm 1 does not require to try all the connected components of $\mathcal{O}(\mathcal{I})$ and converges faster.

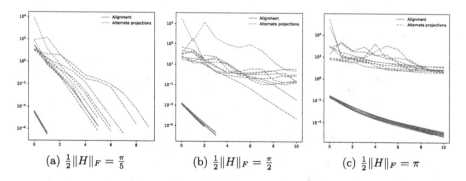

(a) $\frac{1}{2}\|H\|_F = \frac{\pi}{5}$ (b) $\frac{1}{2}\|H\|_F = \frac{\pi}{2}$ (c) $\frac{1}{2}\|H\|_F = \pi$

Fig. 3. Evolution of the endpoint errors along the iterations for different shooting vector lengths $\frac{1}{2}\|H\|_F \in \{\frac{\pi}{5}, \frac{\pi}{2}, \pi\}$. Solid: Algorithm 1, Dashed: [4].

5 Conclusion and Perspectives

In this work, we first gave a quick overview on the state of research in flag manifolds logarithm and distance. Then, rethinking the geodesic endpoint problem as an alignment problem, we proposed a relaxed solution involving orthogonal Procrustes analysis. Finally, we showed how introducing it into [4, *Alg. 1*] drastically improves its accuracy, speed and radius of convergence, as well as overcomes the combinatorial drawback of working on fully oriented flag manifolds. The code is available on GitHub[1].

In future work, we could first improve the computational efficiency of the algorithm, by using linear algebra tricks like the $2k$ Embedding of [3, *Thm. 1*]. Second we could investigate the convergence properties of our algorithm, by understanding the link between the alignment of the target point on the target orbit and the gradient of the geodesic endpoint problem.

Acknowledgements. This work was supported by the ERC grant #786854 G-Statistics from the European Research Council under the European Union's Horizon 2020 research and innovation program and by the French government through the 3IA Côte d'Azur Investments ANR-19-P3IA-0002 managed by the National Research Agency.

References

1. Bendokat, T., Zimmermann, R., Absil, P.A.: A Grassmann Manifold Handbook: Basic Geometry and Computational Aspects (2020). https://doi.org/10.48550/arXiv.2011.13699
2. Guigui, N., Miolane, N., Pennec, X.: Introduction to Riemannian Geometry and Geometric Statistics: From Basic Theory to Implementation with Geomstats. Found. Trends Mach. Learn. **16**(3), 329–493 (Feb 2023). https://doi.org/10.1561/2200000098
3. Ma, X., Kirby, M., Peterson, C.: The Flag Manifold as a Tool for Analyzing and Comparing Sets of Data Sets. In: Proceedings of the IEEE/CVF International Conference on Computer Vision. pp. 4185–4194 (2021). https://www.doi.org/10.1109/ICCVW54120.2021.00465
4. Ma, X., Kirby, M., Peterson, C.: Self-organizing mappings on the flag manifold with applications to hyper-spectral image data analysis. Neural Comput. Appl. **34**(1), 39–49 (2021). https://doi.org/10.1007/s00521-020-05579-y
5. Monk, D.: The geometry of flag manifolds. Proc. London Math. Soc. **s3-9**(2), 253–286 (1959). https://doi.org/10.1112/plms/s3-9.2.253
6. Nguyen, D.: Closed-form Geodesics and Optimization for Riemannian Logarithms of Stiefel and Flag Manifolds. J. Optim. Theor. Appl. **194**(1), 142–166 (2022). https://doi.org/10.1007/s10957-022-02012-3
7. Nijhawan, S., Gupta, A., Appaiah, K., Vaze, R., Karamchandani, N.: Flag manifold-based precoder interpolation techniques for MIMO-OFDM systems. IEEE Trans. Commun. **69**(7), 4347–4359 (2021). https://doi.org/10.1109/TCOMM.2021.3069015

---

[1] Link for the code: https://github.com/tomszwagier/flag-manifold-distance.

8. O'Neill, B.: The fundamental equations of a submersion. Mich. Math. J. **13**(4), 459–469 (1966). https://doi.org/10.1307/mmj/1028999604

9. Pennec, X.: Barycentric subspace analysis on manifolds. Ann. Stat.**46**(6A), 2711–2746 (2018). https://doi.org/10.1214/17-AOS1636, publisher: Institute of Mathematical Statistics

10. Pennec, X., Sommer, S., Fletcher, T.: Riemannian Geometric Statistics in Medical Image Analysis. Academic Press (2020). https://doi.org/10.1016/C2017-0-01561-6

11. Schönemann, P.H.: A generalized solution of the orthogonal procrustes problem. Psychometrika **31**(1), 1–10 (1996). https://doi.org/10.1007/BF02289451

12. Ye, Ke., Wong, Ken Sze-Wai., Lim, Lek-Heng.: Optimization on flag manifolds. Math. Program. **194**, 1–40 (2021). https://doi.org/10.1007/s10107-021-01640-3

13. Zimmermann, R., Hüper, K.: Computing the Riemannian Logarithm on the Stiefel Manifold: Metrics, Methods, and Performance. SIAM J. Matrix Anal. Appl. **43**(2), 953–980 (2022). https://doi.org/10.1137/21M1425426

Characterization of Invariant Inner Products

Yann Thanwerdas[1][(✉)] and Xavier Pennec[2]

[1] Centralesupélec, 3 rue Joliot Curie, 91190 Gif-sur-Yvette, France
yann.thanwerdas@centralesupelec.fr
[2] Université Côte d'Azur and Inria, Epione Project Team, Sophia Antipolis, 2004 route des Lucioles, 06902 Valbonne Cedex, France
xavier.pennec@inria.fr

Abstract. In several situations in differential geometry, one can be interested in determining all inner products on a vector space that are invariant under a given group action. For example, bi-invariant Riemannian metrics on a Lie group G are characterized by $\mathrm{Ad}(G)$-invariant inner products on the Lie algebra $\mathfrak{g}$. Analogously, G-invariant Riemannian metrics on a homogeneous space $\mathcal{M} = G/H$ are characterized by $\mathrm{Ad}(H)$-invariant inner products on the tangent space $T_H\mathcal{M}$. In addition, given a G-equivariant diffeomorphism between a manifold $\mathcal{M}$ and a Euclidean space V, G-invariant log-Euclidean metrics can be defined on $\mathcal{M}$ by pullback of G-invariant inner products on V. There exists a general procedure based on representation theory to find all invariant inner products on a completely reducible Hermitian space. It consists in changing the viewpoint from invariant inner products to equivariant automorphisms. The goal of this work is to diffuse this method to communities of applied mathematics which use differential geometry. Therefore, in this work, we recall this general method that we did not find elsewhere, along with an elementary presentation of the basics of representation theory.

Keywords: Invariant inner product · Invariant Riemannian metric · Group action · Representation theory

1 Introduction

When one looks for appropriate metrics on a given space representing some data, it is natural to require them to be invariant under a certain group action. For example, when data are represented by Symmetric Positive Definite (SPD) matrices, one can use Riemannian metrics that are invariant under the congruence action of the general linear group (affine-invariant metrics [3,8,9,12,13,15]), the orthogonal group [18] (e.g. log-Euclidean [1], Bures-Wasserstein [4,7,10,16], Bogoliubov-Kubo-Mori metrics [11,14]), the group of positive diagonal matrices [5,6,17], the permutation group. In several situations, the question of finding all G-invariant Riemannian metrics on a manifold reduces to finding all H-invariant inner products on a vector space where H is a Lie subgroup of G. For example,

© The Author(s), under exclusive license to Springer Nature Switzerland AG 2023
F. Nielsen and F. Barbaresco (Eds.): GSI 2023, LNCS 14071, pp. 384–391, 2023.
https://doi.org/10.1007/978-3-031-38271-0_38

bi-invariant metrics (or pseudo-metrics) on a Lie group G are characterized by $\mathrm{Ad}(G)$-invariant inner products (or non-degenerate symmetric bilinear forms) on its Lie algebra $\mathfrak{g}$. On a homogeneous space $\mathcal{M} = G/H$, G-invariant metrics are characterized by $\mathrm{Ad}(H)$-invariant inner products on the tangent space $\mathfrak{m} = T_H\mathcal{M}$ at the equivalence class $H \in G/H$.

Another frequent case is when the manifold $\mathcal{M}$ is diffeomorphic to a vector space V. For example, the cone of SPD matrices is diffeomorphic to the vector space of symmetric matrices by the symmetric matrix logarithm log : $\mathrm{Sym}^+(n) \longrightarrow \mathrm{Sym}(n)$. Similarly, the elliptope of full-rank correlation matrices $\mathrm{Cor}^+(n)$ is diffeomorphic to a vector space of dimension $\frac{n(n-1)}{2}$. Hence, given a G-equivariant diffeomorphism $\phi : \mathcal{M} \longrightarrow V$ between the manifold $\mathcal{M}$ and a vector space V, the G-invariant inner products on V provide natural flat G-invariant Riemannian metrics on $\mathcal{M}$ by pullback. These three examples of bi-invariant metrics on Lie groups, Riemannian homogeneous manifolds and "Euclideanized" manifolds motivate to find all G-invariant inner products on a vector space V. Indeed, if one wants to impose an invariance on a space, this requirement defines a family of metrics in general, rarely a unique metric. Therefore, there is no reason a priori to distinguish between all the metrics that satisfy this requirement. Then, it is possible to reduce the choice by requiring other invariance or constraints, or to optimize within the family in function of the data for example.

To answer this question, the central notion is the reducibility or irreducibility of a vector space under a group action. We say that V is G-irreducible when there is no other subvector space than $\{0\}$ and V that is stable under the action of G. It can easily be proved that inner products on an irreducible space are positive scalings of one another. Hence, when V is completely reducible, i.e. V can be expressed into a direct sum of irreducible subspaces, then any positive linear combination of inner products on each irreducible subspace is an invariant inner product. For example, if $G = \{-1, 1\} \times \{-1, 1\}$ acts on $V = \mathbb{R}^2 = \mathbb{R} \oplus \mathbb{R}$ component by component, then one can easily check that all G-invariant inner products are given by $\varphi((x, y), (x, y)) = \alpha x^2 + \beta y^2$ with $\alpha, \beta > 0$.

However, there are other invariant inner products in general. Take another example where $G = \{-1, 1\}$ acts on $V = \mathbb{R}^2 = \mathbb{R} \oplus \mathbb{R}$ globally. It is clear that all the G-invariant inner products are given by $\varphi((x, y), (x, y)) = \alpha x^2 + 2\gamma xy + \beta x^2$ with $\alpha, \beta > 0$ and $\alpha\beta > \gamma^2$. In the first example, G acts differently on each copy of $\mathbb{R}$: for example $(1, -1) \cdot (x, y) = (x, -y)$. In the second example, G acts identically on each copy of $\mathbb{R}$, they are indistinguishable with respect to the action. This is why another coefficient is allowed between the two components. Therefore, in the irreducible decomposition, one has to group all the irreducible spaces on which G acts the same way to find all G-invariant inner products. That is what we is summarized in Theorem 1.

In this work, we assume that V is a complex vector space. We also assume that we know a G-invariant inner product on V and that V is completely reducible, i.e. there exists an irreducible decomposition of V. This is the case when G is finite [2, Maschke's Theorem 10.2.10] and also when the group action is unitary and continuous [2, Corollary 10.3.5]. In particular, when G is finite or compact, it is well known that one can find *one* G-invariant inner product by taking any inner product and averaging over G with the counting measure or the Haar measure μ: $\langle x|y\rangle := \frac{1}{|G|}\sum_{a\in G} ax\cdot ay$ or $\int_G (ax\cdot ay)\mu(da)$. So the two hypotheses are automatically satisfied when G is finite or when G is compact and the action is continuous.

Our method relies on basic representation theory. Indeed, any G-invariant inner product on V is characterized, via the given inner product $\langle\cdot|\cdot\rangle$, by a G-invariant automorphism of V. Then, representation theory allows to find the general form of this automorphism, hence the general form of G-invariant inner products on V. Note that this characterization of invariant inner products is well known in algebra although we could not find it explicitly in the references we read. Yet, it seems to be much less known in the communities of applied mathematics which use differential geometry, although it seems quite useful. That is why we present it with the fewest details on representation theory.

2   Preliminaries on Representation Theory

2.1   Vocabulary of Representation Theory

The following terminology is neither unique in representation theory nor exclusive to this branch of mathematics. We use the term "module" for simplicity, in reference to "G-module" which is frequent.

Definition 1 (Vocabulary of representation theory). *Let G be a group with neutral element e, and V be a complex vector space of finite dimension.*

- *A representation of G on V is a group homomorphism $\rho : G \longrightarrow \mathrm{GL}(V)$. It is equivalent to a linear group action of G on V, i.e. a map $\rho : G \times V \longrightarrow V$ such that for all $a, b \in G$ and all $x \in V$, $\rho(a, \rho(b, x)) = \rho(ab, x)$, $\rho(e, x) = x$ and $\rho_a : x \in V \longmapsto \rho(a, x) \in V$ is linear.*
- *In this work, a (G-)module will designate a vector space V on which G acts linearly and continuously via $\rho : G \times V \longrightarrow V$.*
- *A submodule of V is a ρ-stable subvector space of V.*
- *A module homomorphism (resp. endomorphism, isomorphism, automorphism) is a ρ-equivariant linear map (resp. endormorphism, isomorphism, automorphism) between modules. Given modules V, V', we denote $V \simeq V'$ when there exists a module isomorphism from V to V' and we say that V and V' are isomorphic modules.*
- *A module V is irreducible if it has no other submodules than $\{0\}$ and V. Otherwise, it is reducible.*
- *A module V is completely reducible if it is the direct sum of irreducible modules.*

From now on, let $(V, \langle\cdot|\cdot\rangle)$ be a Hermitian space (i.e. a complex vector space of finite dimension endowed with an inner product), $V \neq \{0\}$. We denote $\|\cdot\|$ the associated norm. The set $\mathrm{U}(V) = \{f \in \mathrm{GL}(V)|\forall x \in V, \|f(x)\| = \|x\|\}$ is the subgroup of isometries of $\mathrm{GL}(V)$. Let $\rho : G \longrightarrow \mathrm{U}(V)$ be a representation of a group G acting isometrically on V.

We assume that V is completely reducible. For example, this is the case when G is finite [2, Maschke's Theorem 10.2.10] or when ρ is unitary and continuous [2, Corollary 10.3.5].

Note that if W is a submodule of V, then $W^\perp$ (where orthogonality refers to the inner product $\langle \cdot | \cdot \rangle$) is also a submodule of V. Indeed, if $x \in W^\perp$ and $a \in G$, then for all $y \in W$, $\rho(a^{-1})(y) \in W$ because W is a module so $\langle \rho(a)(x)|y \rangle = \langle x|\rho(a^{-1})(y) \rangle = 0$, so $\rho(a)(x) \in W^\perp$. Therefore, we can assume that the decomposition of V into irreducible submodules is orthogonal.

2.2   The Big Picture

We split the method into three steps:

1. Transform the problem of finding all ρ-invariant inner products on V into finding all ρ-equivariant automorphisms of V.

Let φ be an inner product on V. We recall the musical isomorphisms of φ:
- "Flat" $\flat_\varphi : x \in V \longmapsto \varphi(x, \cdot) \in V^*$ which is a linear isomorphism by the Riesz representation theorem,
- "Sharp" $\#_\varphi = \flat_\varphi^{-1} : V^* \longrightarrow V$ the inverse linear isomorphism.

We denote $\flat = \flat_{\langle \cdot | \cdot \rangle}$ and $\# = \#_{\langle \cdot | \cdot \rangle}$. Then φ is a ρ-invariant inner product on V if and only if $f_\varphi := \# \circ \flat_\varphi : V \longrightarrow V$ is a ρ-equivariant automorphism of V. We retrieve φ from $f := f_\varphi$ by $\varphi(x, y) = \langle f(x)|y \rangle$ for all $x, y \in V$.

The core of the method is to determine the general form of f.

2. Find all ρ-equivariant automorphisms of V using representation theory.
3. Go back to the initial problem.

Before the general case (V completely reducible), it is natural to start with the particular case where V is irreducible (Sect. 3.1). Then, we explain why the general case does not reduce to the irreducible case but only to an intermediate case (Sect. 3.2). Therefore, we treat this intermediate case (Sect. 3.3) and we conclude with the general case (Sect. 3.4).

3   Find All Equivariant Automorphisms

3.1   The Particular Case: V Irreducible

We start with the simplest case where V is irreducible. The key result to characterize all ρ-equivariant automorphisms of V is Schur's lemma.

Lemma 1 *[2, Schur's Lemma 10.7.6]. Let V, W be irreducible modules.*

1. *A module homomorphism $f : V \longrightarrow W$ is either null or a module isomorphism.*
2. *A module endomorphism $f : V \longrightarrow V$ is a scaling, i.e. there exists $\alpha \in \mathbb{C}$ such that $f = \alpha \, \mathrm{Id}_V$.*

Indeed, the first statement holds because $f(V)$ is a submodule of W so $f(V) = \{0\}$ or W and $\ker(f)$ is a submodule of V so $\ker(f) = V$ or $\{0\}$. The second statement holds because if $\alpha \in \mathbb{C}$ is an eigenvalue of f – it exists because $\mathbb{C}$ is algebraically closed –, then $f - \alpha \operatorname{Id}_V$ is not a module isomorphism so it is null by the first statement.

As a consequence, if V is an irreducible module, then:

- Module automorphisms on V are the non null scalings.
- Module isomorphisms between V and an irreducible module W are unique up to scaling. Indeed, if $f, f_0 : V \longrightarrow W$ are module isomorphisms, then $f_0^{-1} \circ f : V \longrightarrow V$ is a scaling, i.e. there exists $\alpha \in \mathbb{C}^*$ such that $f = \alpha f_0$.
- The ρ-invariant inner products on V are positive multiples of $\langle \cdot | \cdot \rangle$.

3.2   The General Case Does Not Reduce to the Previous Case

We continue with the case where V is completely reducible. Therefore, let $V_1, ..., V_m$ be irreducible submodules of V such that $V = V_1 \overset{\perp}{\oplus} \cdots \overset{\perp}{\oplus} V_m$. From the previous case, it is clear that the map $f : x = x_1 + \cdots + x_m \in V \longmapsto \alpha_1 x_1 + \cdots + \alpha_m x_m \in V$ with $\alpha_1, ..., \alpha_m \in \mathbb{C}^*$ is a module automorphism of V. The question is: are they the only ones? The answer is no and the following consequence of Schur's lemma allows to explain precisely why.

Lemma 2 (Consequence of Schur's lemma on the irreducible decomposition). *We group $V_1, ..., V_m$ by classes $\mathcal{C}^1, ..., \mathcal{C}^p$ of isomorphic irreducible modules. The decomposition becomes $V = \mathcal{V}^1 \overset{\perp}{\oplus} \cdots \overset{\perp}{\oplus} \mathcal{V}^p$ with $\mathcal{V}^k = \overset{\perp}{\bigoplus}_{V_i \in \mathcal{C}^k} V_i$. Let $f : V \longrightarrow V$ be a module automorphism. Then $f(\mathcal{V}^k) = \mathcal{V}^k$ for all $k \in \{1, ..., p\}$.*

Proof. For $i, j \in \{1, ..., m\}$, let $f_{ij} = \operatorname{proj}_{V_j} \circ f_{|V_i} : V_i \longrightarrow V_j$. Let $k \in \{1, ..., p\}$ and let $i \in \{1, ..., n\}$ such that $V_i \in \mathcal{C}^k$. By Schur's lemma, for all $j \in \{1, ..., n\}$, f_{ij} is null or it is an isomorphism. Since f is an isomorphism, there exists $j \in \{1, ..., m\}$ such that f_{ij} is non null, thus an isomorphism. Hence $V_j \in \mathcal{C}^k$ so $V_j \subseteq \mathcal{V}^k$. Therefore, $f(V_i) = V_j \subseteq \mathcal{V}^k$. This inclusion is valid for all $V_i \in \mathcal{C}^k$ so $f(\mathcal{V}^k) \subseteq \mathcal{V}^k$ and $f(\mathcal{V}^k) = \mathcal{V}^k$ by equality of dimensions because f is bijective.

In other words, the study of ρ-invariant automorphisms of V cannot be reduced to the V_i's but only to the $\mathcal{V}^k$'s: there is no reason that $f(V_i) = V_i$ for all $i \in \{1, ..., m\}$ (unless all classes are singletons). So we need to study the case $V = \mathcal{V} \simeq mW$ with W irreducible, where mW is a notation for the direct sum of m irreducible modules isomorphic to W.

3.3   The Intermediate Case: $V \simeq mW$ with W Irreducible

We assume that $V = V_1 \overset{\perp}{\oplus} \cdots \overset{\perp}{\oplus} V_m$ where $V_1 \simeq \cdots \simeq V_m$ are isomorphic irreducible modules. Let W be an irreducible module isomorphic to them, endowed with a G-invariant inner product $(\cdot | \cdot)$. Let $\psi_i : V_i \longrightarrow W$ be the unique module isomorphism which is an isometry. Indeed, the module isomorphism is unique up to scaling and the pullback of the inner product $(\cdot|\cdot)$ onto V_i is necessarily a scaling of the restriction of $\langle\cdot|\cdot\rangle$ to V_i so there is a unique choice of ψ_i such that it is an isometry. For example, W can be taken as one of the V_i's.

Let f be a module automorphism of V. We define the module endomorphisms $f_{ij} = \psi_j \circ \mathrm{proj}_{V_j} \circ f \circ \psi_i^{-1} : W \longrightarrow W$. By Schur's lemma, they are scalings: there exists $S_{ij} \in \mathbb{C}$ such that $f_{ij} = S_{ij}\mathrm{Id}_W$. This defines a matrix $S = (S_{ij})_{1\leqslant i,j\leqslant m} \in \mathrm{Mat}_m(\mathbb{C})$. Then, f writes:

$$f(x) = (\mathrm{Id}_V \circ f)(x) = \left(\sum_{j=1}^m \mathrm{proj}_{V_j}\right) \circ f\left(\sum_{i=1}^m x_i\right) = \sum_{i=1}^m\sum_{j=1}^m (\mathrm{proj}_{V_j} \circ f)(x_i)$$

$$= \sum_{i=1}^m\sum_{j=1}^m (\psi_j^{-1} \circ f_{ij} \circ \psi_i)(x_i) = \sum_{i=1}^m\sum_{j=1}^m S_{ij}\,\psi_j^{-1} \circ \psi_i(x_i). \tag{1}$$

When f comes from an inner product φ as explained in Sect. 2.2, we have for all $x, y \in V$:

$$\varphi(x,y) = \langle f(x)|y\rangle = \sum_{i=1}^m\sum_{j=1}^m S_{ij}\langle\psi_j^{-1}\circ\psi_i(x_i)|y_j\rangle = \sum_{i=1}^m\sum_{j=1}^m S_{ij}(\psi_i(x_i)|\psi_j(y_j)),$$

because the V_i's are orthogonal and $\psi_j : (V_j, \langle\cdot|\cdot\rangle) \longrightarrow (W, (\cdot|\cdot))$ is an isometry. This implies that S is Hermitian Positive Definite (HPD). Indeed, let $w \in W\backslash\{0\}$, $a \in \mathbb{R}^m$ and $x = \sum_{i=1}^k a_i\psi_i^{-1}(w)$. Then:

- if $a = (1, ..., 1)$, then $\varphi(x_i, x_j) = S_{ij}\|w\|^2$ and by conjugate symmetry of φ, we have $\varphi(x_i, x_j) = \overline{\varphi(x_j, x_i)} = \overline{S_{ji}}\|w\|^2$ so $S_{ij} = \overline{S_{ji}}$,
- we have $\varphi(x,x) = \sum_{i,j} S_{ij}\overline{a_i}a_j\|w\|^2$ so for all $a \in \mathbb{R}^m\backslash\{0\}$, we have $\sum_{i,j} S_{ij}\overline{a_i}a_j > 0$.

Conversely, if S is Hermitian positive definite, then $\varphi(x,y) = \langle f(x)|y\rangle$ defines an inner product. It is clearly conjugate symmetric and if $x \neq 0$, there exists $w \in W\backslash\{0\}$ and $a \in \mathbb{R}^m\backslash\{0\}$ such that $x = \sum_{i=1}^k a_i\psi_i^{-1}(w)$ so the equality above proves that $\varphi(x,x) > 0$ (for the existence of w and a, take $i \in \{1, ..., p\}$ such that $x_i \neq 0$ and define $w = \psi_i(x_i)$ and $a_j = \frac{\psi_j(x_j)}{\psi_i(x_i)}$ for $j \in \{1, ..., m\}$).

So f is a module isomorphism of $V = V_1 \overset{\perp}{\oplus} \cdots \overset{\perp}{\oplus} V_m$ if and only if there exists a Hermitian positive definite matrix $S \in \mathrm{Mat}_m(\mathbb{C})$ such that f writes as in Eq. (1). Now we have all the ingredients to state the global result.

3.4   The General Case: V Completely Reducible

Theorem 1 (General form of a ρ-invariant inner product on V). *Let $V = \bigoplus_{k=1}^p \mathcal{V}^k$, with $\mathcal{V}^k = \bigoplus_{i=1}^{m_k} V_i^k$, be an orthogonal decomposition where $V_1^k \simeq \cdots \simeq V_{m_k}^k$ are irreducible modules. For all $k \in \{1, ..., p\}$, let $(W^k, (\cdot|\cdot)_k)$ be a Hermitian space and $\psi_i^k : V_i^k \longrightarrow W^k$ be the unique module isomorphism which* *is an isometry. Then, an inner product φ on V is ρ-invariant if and only if there exist p HPD matrices $S^k \in \mathrm{Mat}_{m_k}(\mathbb{C})$ for $k \in \{1, ..., p\}$ such that for all $x = \sum_{k=1}^p \sum_{i=1}^{m_k} x_i^k \in V$ and $y = \sum_{k=1}^p \sum_{i=1}^{m_k} y_i^k \in V$:*

$$\varphi(x, y) = \sum_{k=1}^p \sum_{1 \leqslant i,j \leqslant m_k} S_{ij}^k (\psi_i^k(x_i^k)|\psi_j^k(y_j^k))_k. \tag{2}$$

The number of parameters is $\sum_{k=1}^p m_k^2$ and the number of positivity constraints is $\sum_{k=1}^p m_k$.

Proof. We assemble the pieces of demonstration of the previous sections together. Let φ be a ρ-invariant inner product on V. Then, the map $f = \#\circ\flat_\varphi :$ $V \longrightarrow V$ is a module automorphism (Sect. 2.2). Hence for all $k \in \{1, ..., p\}$, $f(\mathcal{V}^k) = \mathcal{V}^k$ (Sect. 3.2). Therefore, there exists an HPD matrix $S^k \in \mathrm{Mat}_{m_k}(\mathbb{C})$ such that for all $x^k, y^k \in \mathcal{V}^k$, $\langle f(x^k)|y^k \rangle = \sum_{i=1}^{m_k} \sum_{j=1}^{m_k} S_{ij}^k(\psi_i^k(x_i^k)|\psi_j^k(y_j^k))_k$ (Sect. 3.3). Hence, the inner product φ writes:

$$\varphi(x, y) = \langle f(x)|y \rangle = \sum_{k=1}^p \langle f(x^k)|y^k \rangle = \sum_{k=1}^p \sum_{i=1}^{m_k} \sum_{j=1}^{m_k} S_{ij}^k(\psi_i^k(x_i^k)|\psi_j^k(y_j^k))_k. \tag{3}$$

Conversely, this bilinear form is an inner product since it is the sum of inner products on the $\mathcal{V}^k$'s (Sect. 3.3) which are supplementary. It is ρ-invariant because the orthogonal projections are equivariant (because $\langle \cdot|\cdot \rangle$ is invariant), the ψ_i's are equivariant and the inner products $(\cdot|\cdot)_k$ are invariant. So this bilinear form is a ρ-invariant inner product on V.

Note that the choice of the inner product $(\cdot|\cdot)_k$ on W^k instead of $\lambda(\cdot|\cdot)_k$ with $\lambda > 0$ does not affect the general form of the inner product: it suffices to replace S^k by λS^k. Neither does the choice of the isometric parameterization $\psi_i^k : (V_i^k, \langle \cdot|\cdot \rangle) \longrightarrow (W^k, (\cdot|\cdot)_k)$ instead of $\lambda_i \psi_i^k$ with $\lambda_i > 0$ for all $i \in \{1, ..., m_k\}$: it suffices to replace S^k by $\Lambda S^k \Lambda$ where $\Lambda = \mathrm{Diag}(\lambda_1, ..., \lambda_{m_k})$.

4 Conclusion

We formalized a general method to determine all G-invariant inner products on a completely reducible Hermitian space V. Beyond linear algebra, this characterization is interesting when one wants to characterize invariant Riemannian metrics on Lie groups and homogeneous spaces. The characterization is also interesting when there exists a global diffeomorphism from a manifold to a vector space, such as for symmetric positive definite matrices or full-rank correlation matrices.

An important challenge is to adapt this method to real vector spaces. Moreover, since this method is based on representation theory, it would be interesting to investigate non-linear methods to characterize invariant Riemannian metrics on manifolds when this problem does not reduce to a linear problem.

Acknowledgements. The authors would like to thank the anonymous reviewers for their careful proofreading of this manuscript.

References

1. Arsigny, V., Fillard, P., Pennec, X., Ayache, N.: Log-Euclidean metrics for fast and simple calculus on diffusion tensors. Magn. Reson. Med. **56**(2), 411–421 (2006)
2. Artin, M.: Algebra, 2nd edn. Pearson Prentice Hall (2011)
3. Batchelor, P.G., Moakher, M., Atkinson, D., Calamante, F., Connelly, A.: A rigorous framework for diffusion tensor calculus. Magn. Reson. Med. **53**(1), 221–225 (2005)
4. Bhatia, R., Jain, T., Lim, Y.: On the Bures-Wasserstein distance between positive definite matrices. Expo. Math. **37**(2), 165–191 (2019)
5. David, P., Gu, W.: A Riemannian structure for correlation matrices. Oper. Matrices **13**(3), 607–627 (2019)
6. David, P.: A Riemannian quotient structure for correlation matrices with applications to data science. Ph.D. thesis, Institute of Mathematical Sciences, Claremont Graduate University (2019)
7. Dryden, I.L., Koloydenko, A., Zhou, D.: Non-Euclidean statistics for covariance matrices, with applications to diffusion tensor imaging. Ann. Appl. Stat. **3**(3), 1102–1123 (2009)
8. Fletcher, P.T., Joshi, S.: Riemannian geometry for the statistical analysis of diffusion tensor data. Signal Process. **87**(2), 250–262 (2007)
9. Lenglet, C., Rousson, M., Deriche, R., Faugeras, O.: Statistics on the manifold of multivariate normal distributions: theory and application to diffusion tensor MRI processing. J. Math. Imaging Vision **25**(3), 423–444 (2006)
10. Malagò, L., Montrucchio, L., Pistone, G.: Wasserstein Riemannian geometry of Gaussian densities. Inf. Geom. **1**(2), 137–179 (2018)
11. Michor, P.W., Petz, D., Andai, A.: The curvature of the Bogoliubov-Kubo-Mori scalar product on matrices. Infin. Dimens. Anal. Quantum Probab. Relat. Top. **3**(2), 1–14 (2000)
12. Moakher, M.: A differential geometric approach to the geometric mean of symmetric positive-definite matrices. SIAM J. Matrix Anal. Appl. **26**(3), 735–747 (2005)
13. Pennec, X., Fillard, P., Ayache, N.: A Riemannian framework for tensor computing. Int. J. Comput. Vision **66**(1), 41–66 (2006)
14. Petz, D., Toth, G.: The Bogoliubov inner product in quantum statistics. Lett. Math. Phys. **27**(3), 205–216 (1993)
15. Skovgaard, L.T.: A Riemannian geometry of the multivariate normal model. Scand. J. Stat. **11**(4), 211–223 (1984)
16. Takatsu, A.: Wasserstein geometry of Gaussian measures. Osaka J. Math. **48**(4), 1005–1026 (2011)
17. Thanwerdas, Y., Pennec, X.: Geodesics and curvature of the quotient-affine metrics on full-rank correlation matrices. In: Nielsen, F., Barbaresco, F. (eds.) GSI 2021. LNCS, vol. 12829, pp. 93–102. Springer, Cham (2021). https://doi.org/10.1007/978-3-030-80209-7_11
18. Thanwerdas, Y., Pennec, X.: O(n)-invariant Riemannian metrics on SPD matrices. Linear Algebra Appl. **661**, 163–201 (2023)

Probability and Statistics on Manifolds

Variational Gaussian Approximation of the Kushner Optimal Filter

Marc Lambert[1(✉)], Silvère Bonnabel[2], and Francis Bach[3]

[1] DGA, INRIA, Ecole Normale Supérieure, Paris, France
marc.lambert@inria.fr
[2] MINES Paris PSL, Paris, France
silvere.bonnabel@minesparis.psl.eu
[3] INRIA, Ecole Normale Supérieure, PSL, Paris, France
francis.bach@inria.fr

Abstract. In estimation theory, the Kushner equation provides the evolution of the probability density of the state of a dynamical system given continuous-time observations. Building upon our recent work, we propose a new way to approximate the solution of the Kushner equation through tractable variational Gaussian approximations of two proximal losses associated with the propagation and Bayesian update of the probability density. The first is a proximal loss based on the Wasserstein metric and the second is a proximal loss based on the Fisher metric. The solution to this last proximal loss is given by implicit updates on the mean and covariance that we proposed earlier. These two variational updates can be fused and shown to satisfy a set of stochastic differential equations on the Gaussian's mean and covariance matrix. This Gaussian flow is consistent with the Kalman-Bucy and Riccati flows in the linear case and generalize them in the nonlinear one.

Keywords: filtering · variational inference · information geometry · optimal transport

1 Introduction

We consider the general filtering problem where we aim to estimate the state x_t of a continuous-time stochastic system given noisy observations y_t. If the state follows a Langevin dynamic $f = -\nabla V$ with V a potential function and the observations occur continuously in time, the problem can be described by two stochastic differential equations (SDE) on x_t and z_t, where z_t is related to the observation by the equation $dz_t = y_t dt$:

$$dx_t = -\nabla V(x_t)dt + \sqrt{2\varepsilon}d\beta \tag{1}$$

$$dz_t = h(x_t)dt + \sqrt{R}d\eta. \tag{2}$$

β and η are independent Wiener processes and $Q = 2\varepsilon\mathbb{I}$ and R play the role of covariance matrices of the associated diffusion processes. Many dynamical

© The Author(s), under exclusive license to Springer Nature Switzerland AG 2023
F. Nielsen and F. Barbaresco (Eds.): GSI 2023, LNCS 14071, pp. 395–404, 2023.
https://doi.org/10.1007/978-3-031-38271-0_39

systems can be rewritten in the Langevin canonical form (1), see for instance [6]. In essence (2) means "$y_t = h(x_t) + $ noise", but one has to resort to (2) to avoid problems related to infinitely many observations. The optimal Bayesian filter corresponds to the conditional probability p_t of the state at time t given all past observations. This probability satisfies the Kushner equations which can be split into two parts:

$$dp_t = \mathcal{L}(p_t)dt + d\mathcal{H}(p_t), \tag{3}$$

where $\mathcal{L}$ is defined by the Fokker-Planck partial differential equation (PDE)

$$\mathcal{L}(p_t) = \mathrm{div}[\nabla V p_t] + \varepsilon \Delta p_t, \tag{4}$$

whereas the second term corresponds to the Kushner stochastic PDE (SPDE):

$$d\mathcal{H}(p_t) = (h - \mathbb{E}_{p_t}[h])^T R^{-1}(dz_t - \mathbb{E}_{p_t}[h]dt)p_t,$$

where $\mathbb{E}_{p_t}[h] := \int h(x)p_t(x)dx$ and stochasticity comes from dz_t. These equations cannot be solved in the general case, and we must resort to approximation. In this paper, we consider variational Gaussian approximation, which consists in searching for the Gaussian distribution q_t closest to the optimal one p_t for a particular variational loss. Two variational losses are well suited for our problem.

Jordan-Kinderlehrer-Otto (JKO) [9] showed that the following proximal scheme:

$$\operatorname{argmin} \mathcal{L}^{\delta t}(p) = \operatorname{argmin} \left[KL\left(p \middle\| \pi\right) + \frac{1}{2\delta t}d_w^2(p_t, p) \right], \quad \text{(JKO)} \tag{5}$$

is related to the Fokker-Planck (FP) equation associated to (1) where we denote its stationary distribution $\pi \propto \exp(-V/\varepsilon)$. Indeed, iterating this proximal algorithm yields a curve being solution to FP as $\delta t \to 0$. It is referred to as variational since it is an optimization problem over the function p, and it involves the Kullback-Leibler divergence defined by $KL(p\|\pi) = \int p \log \frac{p}{\pi}$, and the Wasserstein (or optimal transport) distance $d_w^2(p_t, p)$ [2].

The variational loss associated to the Kushner PDE is the Laugesen-Mehta-Meyn-Raginsky (LMMR) proximal scheme [14] defined by:

$$\operatorname{argmin} \mathcal{H}^{\delta t}(p) = \operatorname{argmin} \left[\mathbb{E}_p \frac{1}{2}\|\delta z_t - h(x)\delta t\|_{(R\delta t)^{-1}}^2 + KL(p\|p_t) \right], \quad \text{(LMMR)} \tag{6}$$

where $\delta z_t := z_{t+\delta t} - z_t$ comes from the Euler-Marayama discretization of the observation SDE: $\delta z_t = h(x_t)\delta t + \sqrt{R}\delta\eta$ such that $p(\delta z_t|x_t) = \mathcal{N}(h(x_t)\delta t, R\delta t)$.

For small δt those schemes generate a sequence of probability distributions that converge to the solutions of the corresponding PDE in the limit $\delta t \to 0$. We see the KLs in both schemes play a different role, though. In (5), the proximal scheme shows that the solution to the FP equation follows a gradient of the KL to the stationary distribution π. This gradient is computed with respect to the Wasserstein metric. In (6), the proximal scheme defines a gradient over the state prediction p of the expected prediction error. This gradient is computed in the sense of the metric defined by the KL around its null value, which may be related to the Fisher metric.

To approximate the solutions, we propose to constrain them to lie in the space of Gaussian distributions. That can be done by constraining in the proximal schemes the general distribution p_t to be a Gaussian distribution $q_t = \mathcal{N}(\mu, P)$. The proximal problems become finite-dimensional and boil down to minimizing $\mathcal{L}^{\delta t}$ and $\mathcal{H}^{\delta t}$ over (μ, P). The Gaussian approximation of the JKO scheme yields in the limit a set of ODEs on μ and P as shown in [13]. In this paper, we extend these results showing the Gaussian solution to the LMMR scheme corresponds to the R-VGA solution [11] which yields in the limit a set of SDEs on μ and P. Moreover, using a two-step approach, we can fuse the two Gaussian solutions to approximate the Kushner equation (3). As shall be shown presently, we find the following SDEs for μ and P:

The fully continuous-time variational Kalman filter

$$d\mu_t = b_t dt + P_t dC_t$$

$$dP_t = A_t P_t dt + P_t A_t^T dt + \frac{1}{2} dH_t P_t + \frac{1}{2} P_t dH_t^T + 2\varepsilon \mathbb{I} dt \qquad (7)$$

where $b_t = -\mathbb{E}_{q_t}[\nabla V(x)]; \quad dC_t = \mathbb{E}_{q_t}[\nabla h(x_t)^T R^{-1}(dz_t - h(x_t)dt)]$

$A_t = -\mathbb{E}_{q_t}[\nabla^2 V(x)]; \quad dH_t = \mathbb{E}_{q_t}[(x_t - \mu_t)(dz_t - h(x_t)dt)^T R^{-1} \nabla h(x_t)].$

The equation for P_t can be seen as a generalization of the Riccati equation in the nonlinear case. Indeed, if we replace V and h with linear functions, the ODE on P_t matches the Riccati equations and we recover the Kalman-Bucy filter, known to solve exactly the Kushner equations.

This paper is organized as follows: Sect. 2 is dedicated to related works on the approximation of the optimal nonlinear filter. In Sect. 3 we derive the variational Gaussian approximation of the LMMR scheme. In Sect. 4 we recall the variational Gaussian approximation of the JKO scheme proposed in our previous work. In Sect. 5 we combine these two results to obtain the Continuous Variational Kalman filter equations and show the equivalence with the Kalman-Bucy filter in the linear case.

2   Related Works

In 1967, Kushner proposed a Gaussian assumed density filter to solve his PDE [10]. This filter is derived by keeping only the first two moments of p_t in (3) which can be computed in closed form using the Ito formula. These moments involve integrals under the unknown distribution p_t and the heuristic is to integrate them rather on the current Gaussian approximation q_t leading to a recursive scheme. A more rigorous way to do this approximation was proposed later [4,8] with the projected filter. In this approach, the solution of the Kushner PDE is projected onto the tangent space to Gaussian distributions equipped with the Fisher information metric. This leads to ODEs that are quite different from

(7). A third approach is to linearize the stochastic dynamic process to obtain a McKean Vlasov process that allows for Gaussian propagation [12, Sect. 4.1.1]. The connexion between approximated SDEs and projected filters was analyzed in detail earlier in [3].

The latter approach is the one explored in the current paper, i.e., considering proximal schemes associated with the Kushner PDE where we constrain the solution to be Gaussian. It is equivalent to projecting the exact gradient flow onto the tangent space of the manifold of Gaussian distributions. This approach is preferred since it exhibits the problem's geometric structure and allows convergence guarantees to be proven. Approximation of gradient flows is an active field and several recent papers have followed this direction: the connexion between the propagation part of the Gaussian assumed density filter and the variational JKO scheme [9] was recently studied [13]; the connexion between the update part of the Gaussian assumed density filter and the variational LMMR scheme [14] was studied in [7] where a connexion with a gradient flow was first established but limited to the linear case. To the best of our knowledge, the variational approximation of the LMMR scheme in the nonlinear case has never been addressed. The various ways to obtain the ODEs (7) lead to nice connexions between geometric projection, constrained optimization, and statistical linearization. These different approaches are illustrated in Fig. 1 which addresses only the approximation of dynamics (1) without measurements (i.e., propagation only) for which all methods prove equivalent.

3   Variational Gaussian Approximation of the LMMR Proximal

In this section, we compute the closest Gaussian solution to the LMMR problem (6). The corresponding Gaussian flow is closely related to natural gradient descent used in information geometry. This flow approximates the Kushner optimal filter when the state is static. In the next sections, we will generalize this result to a dynamic state.

3.1   The Recursive Variational Gaussian Approximation

The proximal LMMR problem (6) where we constrain the solution q to be a Gaussian reads (given q_t a current Gaussian distribution at time t):

$$q_{t+\delta t} = \underset{q \in \mathcal{N}(\mu, P)}{\arg\min} \quad \mathbb{E}_q \frac{1}{2} \|\delta z_t - h(x)\delta t\|^2_{(R\delta t)^{-1}} + KL(q\|q_t) \qquad (8)$$

$$= \underset{q \in \mathcal{N}(\mu, P)}{\arg\min} \quad -\int q(x)\log p(\delta z_t|x)dx + KL(q\|q_t) \qquad (9)$$

$$= \underset{q \in \mathcal{N}(\mu, P)}{\arg\min} \quad KL\left(q\,\middle\|\,\frac{1}{Z}p(\delta z_t|x)q_t\right), \qquad (10)$$

where we have introduced a normalization constant Z which does not change the problem.

Distribution p		Gaussian approx. $q(\mu, P)$
$dx_t = -\nabla V(x_t)dt$ $+\sqrt{2\varepsilon}d\beta$	SDE linearization [12]	$dx_t = A(t)(x_t - \mu_t)dt + b(t)dt$ $+\sqrt{2\varepsilon}d\beta$
nonlinear SDE process	$\Rightarrow$	McKeanVlasov process

$\frac{\partial p}{\partial t} = \text{div}(\nabla V p) + \varepsilon \Delta p$	Riemanian projection [8]	$\dot{\mu}_t = b(t)$ $\dot{P}_t = A(t)P_t + P_t A(t)^T + 2\varepsilon \mathbb{I}$
Fokker-Planck	$\Rightarrow$	Variational Gaussian flow

$KL(p\|\pi) + \frac{1}{2\delta t}d_w(p, p_t)^2$	constrained optim. [13]	$KL(q\|\pi) + \frac{1}{2\delta t}d_{bw}(q, q_t)^2$
proximal JKO	$\Rightarrow$	proximal Bures-JKO

$\delta p = -\nabla_w KL(p\|\pi)$	gradient projection [13]	$\delta q = -\nabla_{bw}KL(q\|\pi)$
W2 gradient flow	$\Rightarrow$	Bures-W2 gradient flow

$$\text{where} \quad A(t) = -\mathbb{E}_{q_t}[\nabla^2 V(x)]; \quad b(t) = -\mathbb{E}_{q_t}[\nabla V(x)]$$

Fig. 1. Various equivalent approaches for Gaussian approximation of the SDE (1). We denote $\pi \propto \exp(-V/\varepsilon)$ as the stationary distribution of the associated Fokker-Planck equation. The left column presents equivalent definitions of p whereas the right column corresponds to the approximated solution q in the space of Gaussian distributions. d_w denotes the Wasserstein distance whereas d_{bw} denotes the Bures-Wasserstein distance, which is its restriction to the subset of Gaussian distributions. At the last row, the tangent vector δp, respectively (resp. δq) and the gradient ∇_w (resp. ∇_{bw}) are defined with respect to the Wasserstein metric space of distribution $\left(\mathcal{P}(\mathbb{R}^d), d_w^2\right)$ (resp. the Bures-Wasserstein metric space of Gaussians $\left(\mathcal{N}(\mathbb{R}^d), d_{bw}^2\right)$). These geometries are briefly explained in Sect. 4.2.

Equation (10) falls into the framework of variational Gaussian approximation (R-VGA) [11]. The solution satisfies the following updates [11, Theorem 1]:

$$\mu_{t+\delta t} = \mu_t + P_t \mathbb{E}_{q_{t+\delta t}}[\nabla_x \log p(\delta z_t | x)]$$
$$P_{t+\delta t}^{-1} = P_t^{-1} - \mathbb{E}_{q_{t+\delta t}}[\nabla_x^2 \log p(\delta z_t | x)],$$

where the expectations are under the Gaussian $q_{t+\delta t} \sim \mathcal{N}(\mu_{t+\delta t}, P_{t+\delta t})$ making the updates implict. In the linear case, that is, if we take $h(x) = Hx$, these updates are equivalent to the online Newton algorithm [11, Theorem 2]. Computing the Hessian $\nabla_x^2 \log p$ can be avoided using integration by part:

$$P_{t+\delta t}^{-1} = P_t^{-1} - P_{t+\delta t}^{-1}\mathbb{E}_{q_{t+\delta t}}[(x - \mu_{t+\delta t})\nabla_x \log p(\delta z_t | x)^T]$$

By rearranging the terms and using that P is symmetric (see [12, Sect. 4.2]) we can let appear an update on the covariance:

$$P_{t+\delta t} = P_t + \frac{1}{2}\mathbb{E}_{q_{t+\delta t}}[(x - \mu_t)\nabla_x \log p(\delta z_t|x)^T]P_t + \frac{1}{2}P_t\mathbb{E}_{q_{t+\delta t}}[\nabla_x \log p(\delta z_t|x)(x - \mu_t)^T].$$

Finally, using that $\nabla_x \log p(\delta z_t|x) = \nabla h(x)^T R^{-1}(\delta z_t - h(x)\delta t)$ we obtain:

$$\mu_{t+\delta t} = \mu_t + P_t\delta C_t, \quad P_{t+\delta t} = P_t + \frac{1}{2}\delta H_t P_t + \frac{1}{2}P_t\delta H_t^T, \tag{11}$$

where:

$$\delta C_t = \mathbb{E}_{q_{t+\delta t}}[\nabla h(x)^T R^{-1}(\delta z_t - h(x)\delta t)]$$

$$\delta H_t = \mathbb{E}_{q_{t+\delta t}}[(x - \mu_t)(\delta z_t - h(x)\delta t)^T R^{-1}\nabla h(x)].$$

Letting $\delta t \to 0$, we obtain the following SDE in the sense of Ito:

$$d\mu_t = P_t dC_t, \quad dP_t = \frac{1}{2}dH_t P_t + \frac{1}{2}P_t dH_t^T, \tag{12}$$

where it shall be noted that dH_t is non-deterministic owing to dz_t. Since the LMMR scheme has been proven to converge to the solution of the Kushner SPDE [14], this SDE describes the best Gaussian approximation of the optimal filter when the state is static.

3.2   Information Geometry Interpretation

We show here how the LMMR proximal scheme is related to the Fisher information geometry in the general case. Let's consider a family of densities: $S = \left\{p(.|\theta); \theta \in \Theta; \Theta \subseteq \mathbb{R}^m\right\}$ and let $F(\theta) = \int \nabla_\theta \log p(x|\theta)\nabla_\theta \log p(x|\theta)^T p(x|\theta)dx$, be the Fisher information matrix, where θ regroups all the parameters.

If we consider now the proximal LMMR on S, and if we use the second-order Taylor expansion of the KL divergence between these two distributions, we have:

$$KL(p(x|\theta)||p(x|\theta_t)) = \frac{1}{2}(\theta - \theta_t)^T F(\theta_t)(\theta - \theta_t) + o((\theta - \theta_t)^2).$$

Rather than minimizing the proximal LMMR scheme (6) in the infinite space of distributions, we now search the minimum in the finite space of parameters:

$$\theta_{t+\delta t} = \underset{\theta \in \Theta}{\arg\min} \; \mathbb{E}_{p(x|\theta)}\left[\frac{1}{2}\|\delta z_t - h(x)\delta t\|_{(R\delta t)^{-1}}^2\right] + \frac{1}{2}(\theta - \theta_t)^T F(\theta_t)(\theta - \theta_t).$$

Considering that the minimum must cancel the gradient of the above proximal loss, we obtain:

$$0 = \nabla_\theta\left(\mathbb{E}_{p(x|\theta)}\left[\frac{1}{2}\|\delta z_t - h(x)\delta t\|_{(R\delta t)^{-1}}^2\right]\right)\Big|_{\theta_{t+\delta t}} + F(\theta_t)(\theta_{t+\delta t} - \theta_t)$$

$$\theta_{t+\delta t} = \theta_t - F(\theta_t)^{-1}\nabla_\theta\left(\frac{1}{2}\mathbb{E}_{p(x|\theta)}\left[\|\delta z_t - h(x)\delta t\|_{(R\delta t)^{-1}}^2\right]\right)\Big|_{\theta_{t+\delta t}}, \tag{13}$$

which corresponds to a gradient descent of the averaged stochastic likelihood:

$$\theta_{t+\delta t} = \theta_t - F(\theta_t)^{-1}\nabla_\theta \mathbb{E}_{p(x|\theta)}[-\log p(\delta z_t|x)]\big|_{\theta_{t+\delta t}}. \tag{14}$$

Remarkably, the optimal filer equations with a static state x are given by an implicit Bayesian variant of the natural gradient descent [1]. Indeed here x plays the role of the parameter of the likelihood distribution. The original natural gradient should be a descent with the gradient $\nabla_x - \log p(\delta z_t|x)\big|_{x_t}$.

4 Variational Gaussian Approximation of the JKO Proximal

The canonical Langevin form (1) assumes that the drift term $f = -\nabla V$ derives from a potential V. This potential has a physical meaning in filtering (consider a gravity field for example). The evolution of the state in the filter mimics the true evolution of the physical system. It's not the case in statistical physics, where the potential is constructed such that $V = -\log \pi$ where π is the asymptotic distribution of a variable x which doesn't correspond to a physical system. We used this property in our previous work [13] and simulated a dynamic to approximate the target π with a Gaussian distribution. Here we do not want to estimate a distribution but to propagate a Gaussian through the nonlinear physical dynamic (1).

4.1   The Bures-JKO Proximal

The proximal JKO problem (5) where we constrained the solution q to be a Gaussian distribution writes:

$$\min_{q \in \mathcal{N}(\mu,P)} \quad KL\left(q\middle\|\pi\right) + \frac{1}{2\delta t}d_{bw}(q,q_t)^2,$$

where $d_{bw}(q,q_t)$ is the Bures distance between two Gaussians given by:

$$d_{bw}(q,q_t) = ||\mu - \mu_t||^2 + \mathcal{B}^2(P,P_t), \tag{15}$$

where $\mathcal{B}^2(P,P_t) = \text{Tr}(P + P_t - 2(P^{\frac{1}{2}}P_tP^{\frac{1}{2}})^{\frac{1}{2}})$ is the squared Bures metric [5], which has a derivative available in closed form. After some computation [13, Appendix A] we can obtain implicit equations that the parameters of the optimal Gaussian solution q must satisfy:

$$\mu_{t+\delta t} = \mu_t - \delta t.\mathbb{E}_{q_{t+\delta t}}[\nabla V(x)]$$
$$P_{t+\delta t} = P_t - \delta t.\mathbb{E}_{q_{t+\delta t}}[\nabla^2 V(x)]P_t - \delta t.P_t\mathbb{E}_{q_{t+\delta t}}[\nabla^2 V(x)]^T + 2\varepsilon\delta t.\mathbb{I}, \tag{16}$$

and at the limit $\delta t \to 0$, we obtain the following ODEs:

$$\dot{\mu}_t = -\mathbb{E}_{q_t}[\nabla V(x)] := b_t \tag{17}$$
$$\dot{P}_t = A_tP_t + P_tA_t^T + 2\varepsilon\mathbb{I} \quad \text{where } A_t := -\mathbb{E}_{q_t}[\nabla^2 V(x)].$$

4.2 Wasserstein Geometry Interpretation

The Wasserstein geometry is defined by the metric space of measure endowed with the Wasserstein distance $(\mathcal{P}(\mathbb{R}^d), d_w^2)$. The definition of a tangent vector in this space is tedious because the measure μ must satisfy the conservation of mass $\int \mu(x)dx = 1$. To handle this constraint we can use the continuity equation. This equation allows to represent any regular curves of measures with a continuous flow along a vector field $v_t \in \mathbb{L}^2$. It is closely related to the Fokker-Planck equation as we show now (see [2] for more details). The JKO proximal scheme (5) gives a sequence of distribution that satisfies at the limit the Fokker-Planck equation (4), this equation rewrites as follows:

$$\dot{p}_t = \nabla.(\nabla V p_t) + \varepsilon\nabla.\nabla p_t = \nabla.(\nabla V p_t) + \varepsilon\nabla.(p_t\nabla\log p_t)$$
$$= \nabla.(p_t(\nabla V + \varepsilon\nabla\log p_t)) = -\mathrm{div}(p_t v_t), \tag{18}$$

which is a continuity equation where $v_t \in \mathbb{L}^2(\mathbb{R}^d)$ plays the role of the tangent vector δp_t along the path p_t and satisfies:

$$v_t = -\nabla V - \varepsilon\nabla\log p_t = -\nabla_w KL(p_t||\pi),$$

with $\pi \propto \exp(-V/\varepsilon)$. The last equality comes from variational calculus in the measure space: the Wasserstein gradient of a functional F is given by the Euclidian gradient of the first variation $\nabla_w F(\rho) = \nabla\delta F(\rho)$, see [2, Chapter 10].

Let's sum up what's going on: starting from a stochastic state x_t following the Langevin dynamic (1) with drift $-\nabla V$, we have rewritten the Fokker-Planck equation which describes the evolution of the density $p(x_t)$ as a continuity equation (18) where the diffusion term has disappeared. At this continuity equation correspond a deterministic ODE $\dot{x}_t = -\nabla_w KL(p_t||\pi)$. It's a nice property of the Wasserstein geometry where PDE can be described by a continuity equation that corresponds to a simple gradient flow.

Following the same track, the sequence of Gaussian distributions satisfying the ODE (17) correspond to a Wasserstein gradient flow given by the continuity equation: $\dot{q}_t = -\mathrm{div}(q_t w_t)$, where $w_t = -\nabla_{bw} KL(q_t||\pi)$ is now a gradient with respect to the Bures-Wasserstein distance (15), see [13, Appendix B3] for the analytical expression of this gradient.

5 Variational Gaussian Approximation of the Kushner Optimal Filter

We have tackled the two proximal problems independently but how to solve them jointly? The simplest method to do so is to alternate between propagation through dynamics (1) for a small time δt, and Bayesian update through LMMR in the light of the accumulated observations δz_t, and let $\delta t \to 0$. This is what we do presently.

5.1   The Continuous Variational Kalman Filter

Consider one step of the Euler-Maruyama method with length δt of SDEs (1) and (2). As the Wiener processes β and η are independent, we may write:

$$p(x_t, y_{t+\delta t}, x_{t+\delta t}) = p(y_{t+\delta t}|x_{t+\delta t}, x_t)p(x_{t+\delta t}, x_t) = p(y_{t+\delta t}|x_{t+\delta t})p(x_{t+\delta t}|x_t),$$

denoting $y_{t+\delta t} = \delta z_t$. In other words, we can solve the proximal LMMR update Eq. (10) using as prior $q_t(x) = \mathcal{N}(\mu_{t+\delta t|t}, P_{t+\delta t|t})$, the solution of the proximal JKO. The LMMR/R-VGA discrete-time equations (11) then become:

$$\mu_{t+\delta t} = \mu_{t+\delta t|t} + P_{t+\delta t|t}\delta C_t$$

$$P_{t+\delta t} = P_{t+\delta t|t} + \frac{1}{2}\delta H_t P_{t+\delta t|t} + \frac{1}{2}P_{t+\delta t|t}\delta H_t^T.$$

Replacing $\mu_{t+\delta t|t}$ and $P_{t+\delta t|t}$ by their expressions as the solutions to the JKO scheme (16) and putting in a residual all the terms in δt^2, we obtain:

$$\mu_{t+\delta t} = \mu_t + \delta t b_t + P_t \delta C_t$$

$$P_{t+\delta t} = P_t + \delta t A_t P_t + \delta t P_t A_t^T + \delta t 2\varepsilon\mathbb{I} + \frac{1}{2}\delta H_t P_t + \frac{1}{2}P_t \delta H_t^T + O(\delta t^2).$$

By Ito calculus, we obtain the continuous variational Kalman updates (7).

5.2   The Kalman-Bucy Filter as a Particular Case

Let us consider the linear case where the SDEs (1) and (2) rewrite:

$$dx_t = Fx_t dt + \sqrt{2\varepsilon}d\beta, \quad dz_t = Gx_t dt + \sqrt{R}d\eta.$$

The various expectations that appear in the proposed filter (7) apply either to quantities being independent of x_t or being linear or quadratic in x_t, yielding

$$d\mu_t = F\mu_t dt + P_t G^T R^{-1}(dz_t - G\mu_t dt)$$

$$\frac{d}{dt}P_t = FP_t + P_t F^T - P_t G^T R^{-1}GP_t + 2\varepsilon\mathbb{I}.$$

We see we exactly recover the celebrated Kalman-Bucy filter.

6   Conclusion

We have approximated the Kushner optimal filter by a Gaussian filter based on variational approximations related to the JKO and LMMR proximal discrete schemes related to the Wasserstein and Fisher geometry respectively. As the dynamic and observation processes are assumed independent, we can mix the two variational solutions to form a set of SDEs on the Gaussian parameters generalizing the Riccati equations associated to the linear systems. In the linear case, the proposed filter boils down to the Kalman-Bucy optimal filter. It is still unclear, though, which global variational loss is minimized by the optimal filter.

Acknowledgements. This work was funded by the French Defence procurement agency (DGA) and by the French government under the management of Agence Nationale de la Recherche as part of the "Investissements d'avenir" program, reference ANR-19-P3IA-0001(PRAIRIE 3IA Institute). We also acknowledge support from the European Research Council (grant SEQUOIA 724063).

References

1. Amari, S.I.: Natural gradient works efficiently in learning. Neural Comput. **10**, 251–276 (1998)
2. Ambrosio, L., Gigli, N., Savare, G.: Gradient Flows in Metric Spaces and in the Space of Probability measures. Lectures in Mathematics, ETH Zürich (2005)
3. Brigo, D.: On nonlinear SDE's whose densities evolve in a finite-dimensional family, vol. 23, pp. 11–19. Birkhäuser (1997)
4. Brigo, D., Hanzon, B., Gland, F.: Approximate nonlinear filtering by projection on exponential manifolds of densities. Bernoulli **5** (1999)
5. Bures, D.: An extension of Kakutani's theorem on infinite product measures to the tensor product of semifinite w^*-algebras. Trans. Am. Math. Soc. **135**, 199–212 (1969)
6. Halder, A., Georgiou, T.: Gradient flows in uncertainty propagation and filtering of linear Gaussian systems (2017)
7. Halder, A., Georgiou, T.: Gradient flows in filtering and Fisher-Rao geometry. In: Annual American Control Conference (2018)
8. Hanzon, B., Hut, R.: New results on the projection filter. Serie Research Memoranda **0023**, 1 (1991)
9. Jordan, R., Kinderlehrer, D., Otto, F.: The variational formulation of the Fokker-Planck equation. SIAM J. Math. Anal. **29**, 1–17 (1998)
10. Kushner, H.: Approximations to optimal nonlinear filters. IEEE Trans. Autom. Control **12**(5), 546–556 (1967)
11. Lambert, M., Bonnabel, S., Bach, F.: The recursive variational Gaussian approximation (R-VGA). Stat. Comput. **32**(1) (2022)
12. Lambert, M., Bonnabel, S., Bach, F.: The continuous-discrete variational Kalman filter (CD-VKF). In: Conference on Decision and Control (2022)
13. Lambert, M., Chewi, S., Bach, F., Bonnabel, S., Rigollet, P.: Variational inference via Wasserstein gradient flows. In: Advances in Neural Information Processing Systems (2022)
14. Laugesen, R., Mehta, P.G., Meyn, S.P., Raginsky, M.: Poisson's equation in nonlinear filtering. In: Conference on Decision and Control, pp. 4185–4190 (2014)

Learning with Symmetric Positive Definite Matrices via Generalized Bures-Wasserstein Geometry

Andi Han[1(✉)], Bamdev Mishra[2], Pratik Jawanpuria[2], and Junbin Gao[1]

[1] University of Sydney, Camperdown, Australia
{andi.han,junbin.gao}@sydney.edu.au
[2] Microsoft India, Hyderabad, India
{bamdevm,pratik.jawanpuria}@microsoft.com

Abstract. Learning with symmetric positive definite (SPD) matrices has many applications in machine learning. Consequently, understanding the Riemannian geometry of SPD matrices has attracted much attention lately. A particular Riemannian geometry of interest is the recently proposed Bures-Wasserstein (BW) geometry which builds on the Wasserstein distance between the Gaussian densities. In this paper, we propose a novel generalization of the BW geometry, which we call the GBW geometry. The proposed generalization is parameterized by a symmetric positive definite matrix $\mathbf{M}$ such that when $\mathbf{M} = \mathbf{I}$, we recover the BW geometry. We provide a rigorous treatment to study various differential geometric notions on the proposed novel generalized geometry which makes it amenable to various machine learning applications. We also present experiments that illustrate the efficacy of the proposed GBW geometry over the BW geometry.

Keywords: Riemannian geometry · SPD matrices · Bures-Wasserstein

1 Introduction

Symmetric positive definite (SPD) matrices play a fundamental role in various fields of machine learning, such as metric learning [21], signal processing [8], sparse coding [9,17], computer vision [14,25], and medical imaging [24,29], etc. The set of SPD matrices, denoted as $\mathbb{S}^n_{++}$, is a subset of the Euclidean space $\mathbb{R}^{n(n+1)/2}$. To measure the (dis)similarity between SPD matrices, one needs to assign a metric (an inner product structure on the tangent space) on $\mathbb{S}^n_{++}$, which yields a Riemannian manifold. Consequently, various Riemannian metrics have been studied such as the affine-invariant [3,29], Log-Euclidean [2], and Log-Cholesky [23] metrics, and those induced from symmetric divergences [33,34]. Different metrics lead to different differential structures on the SPD matrices, and therefore, picking the "right" one depends on the application at hand. Indeed, the choice of metric has profound effect on the performance of learning algorithms [16,28,32].

© The Author(s), under exclusive license to Springer Nature Switzerland AG 2023
F. Nielsen and F. Barbaresco (Eds.): GSI 2023, LNCS 14071, pp. 405–415, 2023.
https://doi.org/10.1007/978-3-031-38271-0_40

The Bures-Wasserstein (BW) metric and its geometry for SPD matrices have lately gained popularity, especially in machine learning applications [4,26,36] such as statistical optimal transport [4], computer graphics [31], neural sciences [13], and evolutionary biology [10], among others. It also connects to the theory of optimal transport and the L_2-Wasserstein distance between zero-centered Gaussian densities [4]. More recently, [16] analyzes the BW and the affine-invariant (AI) geometries in SPD learning problems and compare their advantages/disadvantages in various machine learning applications.

In this paper, we propose a natural generalization of the BW metric by scaling SPD matrices with a given parameter SPD matrix $\mathbf{M}$. The introduction of $\mathbf{M}$ gives flexibility to the BW metric. Choosing $\mathbf{M}$ is equivalent to choosing a suitable metric for learning tasks on SPD matrices. For example, a proper choice of $\mathbf{M}$ can lead to faster convergence of algorithms for certain class of optimization problems (see more discussions in Sect. 4). Indeed, when $\mathbf{M} = \mathbf{I}$, the generalized metric reduces to the BW metric for SPD matrices. When $\mathbf{M} = \mathbf{X}$, the proposed metric coincides locally with the AI metric, i.e., around the neighbourhood of a SPD matrix $\mathbf{X}$. The proposed generalized metric allows to connect the BW and AI metrics (locally) with different choices of $\mathbf{M}$. The following are our contributions.

- We propose a novel generalized BW (GBW) metric by generalizing the Lyapunov operation in the BW metric (Sect. 2). In addition, it can also be viewed as a generalized Procrustes distance and also as the Wasserstein distance with Mahalanobis cost metric for Gaussians.
- The GBW metric leads to a Riemannian geometry for SPD matrices. In Sect. 3.1, we derive various Riemannian operations like geodesics, exponential and logarithm maps, Levi-Civita connection. We show that they are also natural generalizations of operations with the BW geometry. Section 3.2 derives Riemannian optimization ingredients under the proposed geometry.
- In Sect. 4, we show the usefulness of the GBW geometry in the applications of covariance estimation and Gaussian mixture models.

2   Generalized Bures-Wasserstein Metric

The Bures-Wasserstein (BW) distance is defined as

$$d_{\text{bw}}(\mathbf{X}, \mathbf{Y}) = \sqrt{\text{tr}(\mathbf{X}) + \text{tr}(\mathbf{Y}) - 2\text{tr}(\mathbf{XY})^{1/2}}, \tag{1}$$

where $\mathbf{X}$ and $\mathbf{Y}$ are SPD matrices, $\text{tr}(\mathbf{X})$ denotes the matrix trace, and $\text{tr}(\mathbf{X})^{1/2}$ denotes the trace of the matrix square root. It has been shown in [4,26] that the BW distance (1) induces a Riemannian metric and geometry on the manifold of SPD matrices. The BW metric that leads to the distance (1) is defined as

$$g_{\text{bw}}(\mathbf{U}, \mathbf{V}) = \frac{1}{2}\text{tr}(\mathcal{L}_{\mathbf{X}}[\mathbf{U}]\mathbf{V}) = \frac{1}{2}\text{vec}(\mathbf{U})^{\top}(\mathbf{X} \otimes \mathbf{I} + \mathbf{I} \otimes \mathbf{X})^{-1}\text{vec}(\mathbf{V}), \tag{2}$$

where $\mathbf{U}, \mathbf{V}$ on $T_{\mathbf{X}}\mathbb{S}_{++}^{n}$ are the symmetric matrices and the Lyapunov operator $\mathcal{L}_{\mathbf{X}}[\mathbf{U}]$ is defined as the solution of the matrix equation $\mathbf{X}\mathcal{L}_{\mathbf{X}}[\mathbf{U}] + \mathcal{L}_{\mathbf{X}}[\mathbf{U}]\mathbf{X} = \mathbf{U}$ for $\mathbf{U} \in \mathbb{S}^{n}$ (which is the set of symmetric matrices of size $n \times n$). Here, $\mathrm{vec}(\mathbf{U})$ and $\mathrm{vec}(\mathbf{V})$ are the vectorization of matrices $\mathbf{U}$ and $\mathbf{V}$, respectively, and $\otimes$ denotes the Kronecker product. Our proposed GBW metric generalizes (2) and is parameterized by a given $\mathbf{M} \in \mathbb{S}_{++}^{n}$ as

$$g_{\mathrm{gbw}}(\mathbf{U}, \mathbf{V}) = \tfrac{1}{2}\mathrm{tr}(\mathcal{L}_{\mathbf{X},\mathbf{M}}[\mathbf{U}]\mathbf{V}) = \tfrac{1}{2}\mathrm{vec}(\mathbf{U})^{\top}(\mathbf{X} \otimes \mathbf{M} + \mathbf{M} \otimes \mathbf{X})^{-1}\mathrm{vec}(\mathbf{V}), \quad (3)$$

where $\mathcal{L}_{\mathbf{X},\mathbf{M}}[\mathbf{U}]$ is the generalized Lyapunov operator, defined as the solution to the linear matrix equation $\mathbf{X}\mathcal{L}_{\mathbf{X},\mathbf{M}}[\mathbf{U}]\mathbf{M} + \mathbf{M}\mathcal{L}_{\mathbf{X},\mathbf{M}}[\mathbf{U}]\mathbf{X} = \mathbf{U}$. Similar to the special Lyapunov operator, the solution is symmetric given that $\mathbf{X}, \mathbf{M} \in \mathbb{S}_{++}^{n}$ and $\mathbf{U} \in \mathbb{S}^{n}$. As we show later that the Riemannian distance associated with the GBW metric is derived as

$$d_{\mathrm{gbw}}(\mathbf{X}, \mathbf{Y}) = \sqrt{\mathrm{tr}(\mathbf{M}^{-1}\mathbf{X}) + \mathrm{tr}(\mathbf{M}^{-1}\mathbf{Y}) - 2\mathrm{tr}(\mathbf{X}\mathbf{M}^{-1}\mathbf{Y}\mathbf{M}^{-1})^{1/2}}, \quad (4)$$

which can be seen as the BW distance (1) between $\mathbf{M}^{-1/2}\mathbf{X}\mathbf{M}^{-1/2}$ and $\mathbf{M}^{-1/2}\mathbf{Y}\mathbf{M}^{-1/2}$. Note that the affine-invariant metric [3] is given by $g_{\mathrm{ai}}(\mathbf{U}, \mathbf{V}) = \mathrm{vec}(\mathbf{U})^{\top}(\mathbf{X} \otimes \mathbf{X})^{-1}\mathrm{vec}(\mathbf{V})$. Clearly, the proposed metric (3) coincides locally with the affine-invariant (AI) metric when $\mathbf{M} = \mathbf{X}$, i.e., around the neighbourhood of $\mathbf{X}$. (Implications of this observation are discussed later in experiments.)

Below, we show that the same GBW distance (4) is realized under various contexts naturally. In those cases, the Euclidean norm, denoted by $\| \cdot \|_{2}$ is replaced with the more general Mahalanobis norm defined as $\|\mathbf{X}\|_{\mathbf{M}^{-1}} := \sqrt{\mathrm{tr}(\mathbf{X}^{\top}\mathbf{M}^{-1}\mathbf{X})}$.

Orthogonal Procrustes Problem: Any SPD matrix $\mathbf{X} \in \mathbb{S}_{++}^{n}$ can be factorized as $\mathbf{X} = \mathbf{P}\mathbf{P}^{\top}$ for $\mathbf{P} \in \mathrm{M}(n)$, the set of invertible matrices. Such a factorization is invariant under the action of the orthogonal group $O(n)$, the set of orthogonal matrices. That is, for any $\mathbf{O} \in O(n)$, $\mathbf{P}\mathbf{O}$ is also a valid parameterization. In [4], the BW distance is verified as the extreme solution of the orthogonal Procrustes problem where $\mathbf{P}$ is set to be $\mathbf{X}^{1/2}$, i.e., $d_{\mathrm{bw}}(\mathbf{X}, \mathbf{Y}) = \min_{\mathbf{O} \in O(n)} \|\mathbf{X}^{1/2} - \mathbf{Y}^{1/2}\mathbf{O}\|_{2}$. We can show that the GBW distance is obtained as the solution to the same orthogonal Procrustes problem in the Mahalanobis norm parameterized by $\mathbf{M}^{-1}$.

Proposition 1. $d_{\mathrm{gbw}}(\mathbf{X}, \mathbf{Y}) = \min_{\mathbf{O} \in O(n)} \|\mathbf{X}^{1/2} - \mathbf{Y}^{1/2}\mathbf{O}\|_{\mathbf{M}^{-1}}$.

Wasserstein Distance and Optimal Transport: To demonstrate the connection of the GBW distance to the Wasserstein distance, recall that the L_{2}-Wasserstein distance between two probability measures μ, ν with finite second moments is $W^{2}(\mu, \nu) = \inf_{\mathbf{x} \sim \mu, \mathbf{y} \sim \nu} \mathbb{E}\|\mathbf{x} - \mathbf{y}\|_{2}^{2} = \inf_{\gamma \sim \Gamma(\mu,\nu)} \int_{\mathbb{R}^{n} \times \mathbb{R}^{n}} \|\mathbf{x} - \mathbf{y}\|_{2}^{2} d\gamma(\mathbf{x}, \mathbf{y})$, where $\Gamma(\mu, \nu)$ is the set of all probability measures with marginals μ, ν. It is well known that the L_{2}-Wasserstein distance between two zero-centered Gaussian distributions is equal to the BW distance between their covariance matrices [4,30,36]. The following proposition shows that the L_{2}-Wasserstein distance between such measures with respect to a Mahalanobis cost metric (which we term as generalized Wasserstein distance) coincides with the GBW distance in (4).

Proposition 2. *Define the generalized Wasserstein distance as* $\tilde{W}^2(\mu, \nu) :=$ $\inf_{\mathbf{x} \sim \mu, \mathbf{y} \sim \nu} \mathbb{E}\|\mathbf{x} - \mathbf{y}\|_{\mathbf{M}^{-1}}^2$, *for any* $\mathbf{M}^{-1} \in \mathbb{S}_{++}^n$. *Suppose* μ, ν *are two Gaussian measures with zero mean and covariances as* $\mathbf{X}, \mathbf{Y} \in \mathbb{S}_{++}^n$ *respectively. Then, we have* $\tilde{W}^2(\mu, \nu) = d_{\mathrm{gbw}}^2(\mathbf{X}, \mathbf{Y})$.

Alternatively, the same distance is recovered by considering two scaled random Gaussian vector $\mathbf{M}^{-1/2}\mathbf{x}, \mathbf{M}^{-1/2}\mathbf{y}$ under the Euclidean distance, i.e., $d^2(\mathbf{X}, \mathbf{Y}) = \inf_{\mathbf{x} \sim \mu, \mathbf{y} \sim \nu} \mathbb{E}\|\mathbf{M}^{-1/2}\mathbf{x} - \mathbf{M}^{-1/2}\mathbf{y}\|_2^2$. For completeness, we also derive the optimal transport plan corresponding to the GBW distance in our extended report [15].

3   Generalized Bures-Wasserstein Riemannian Geometry

In this section, the geometry arising from the GBW metric (3) is shown to have a Riemannian structure for a given $\mathbf{M} \in \mathbb{S}_{++}^n$, which we denote as $\mathcal{M}_{\mathrm{gbw}}$. We show the expressions of the Riemannian distance, geodesic, exponential/logarithm maps, Levi-Civita connection, sectional curvature as well as the geometric mean and barycenter. A summary of the results is presented in Table 1. Additionally, we discuss optimization on the SPD manifold with the proposed GBW geometry. We defer the detailed derivations in this section to our extended report [15].

3.1   Differential Geometric Properties of GBW

To derive the various expressions in Table 1, we provide two strategies, one is by a Riemannian submersion from the general linear group and another is by a Riemannian isometry from the BW Riemannian geometry, $\mathcal{M}_{\mathrm{bw}}$. These claims are formalized in Propositions 3 and 4 respectively.

Perspective from Riemannian Submersion: A *Riemannian submersion* [22] between two manifolds is a smooth surjective map where its differential restricted to the horizontal space is isometric (formally defined in our extended report [15]). The general linear group GL(n) is the set of invertible matrices with the group action of matrix multiplication. When endowed with the standard Euclidean inner product $\langle \cdot, \cdot \rangle_2$, the group becomes a Riemannian manifold, denoted as $\mathcal{M}_{\mathrm{gl}}$. The proposition below introduces a Riemannian submersion from $\mathcal{M}_{\mathrm{gl}}$ to $\mathcal{M}_{\mathrm{gbw}}$.

Proposition 3. *The map* $\pi : \mathcal{M}_{\mathrm{gl}} \to \mathcal{M}_{\mathrm{gbw}}$ *defined as* $\pi(\mathbf{P}) = \mathbf{M}^{1/2}\mathbf{P}\mathbf{P}^\top\mathbf{M}^{1/2}$ *is a Riemannian submersion, for* $\mathbf{P} \in \mathrm{GL}(n)$ *and* $\mathcal{M}_{\mathrm{gbw}}$ *parameterized by* $\mathbf{M} \in \mathbb{S}_{++}^n$ *as in* (3).

Perspective from Riemannian Isometry: A *Riemannian isometry* between two manifolds is a diffeomorphism (i.e., bijective, differentiable, and its inverse

Table 1. Summary of expressions for the proposed generalized Bures-Wasserstein (GBW) Riemannian geometry, which is parameterized by $\mathbf{M} \in \mathbb{S}^d_{++}$.

Metric	$g_{\text{gbw}}(\mathbf{U}, \mathbf{V}) = \frac{1}{2}\text{tr}(\mathcal{L}_{\mathbf{X},\mathbf{M}}[\mathbf{U}]\mathbf{V})$
Distance	$d^2_{\text{gbw}}(\mathbf{X}, \mathbf{Y}) = \text{tr}(\mathbf{M}^{-1}\mathbf{X}) + \text{tr}(\mathbf{M}^{-1}\mathbf{Y}) - 2\text{tr}(\mathbf{X}\mathbf{M}^{-1}\mathbf{Y}\mathbf{M}^{-1})^{1/2}$
Geodesic	$\gamma(t) = ((1-t)\mathbf{X}^{1/2} + t\mathbf{Y}^{1/2}\mathbf{O})((1-t)\mathbf{X}^{1/2} + t\mathbf{Y}^{1/2}\mathbf{O})^\top$ with $\mathbf{O}$ the orthogonal polar factor of $\mathbf{Y}^{1/2}\mathbf{M}^{-1}\mathbf{X}^{1/2}$
Exp	$\text{Exp}_{\mathbf{X}}(\mathbf{U}) = \mathbf{X} + \mathbf{U} + \mathbf{M}\mathcal{L}_{\mathbf{X},\mathbf{M}}[\mathbf{U}]\mathbf{X}\mathcal{L}_{\mathbf{X},\mathbf{M}}[\mathbf{U}]\mathbf{M}$
Log	$\text{Log}_{\mathbf{X}}(\mathbf{Y}) = \mathbf{M}(\mathbf{M}^{-1}\mathbf{X}\mathbf{M}^{-1}\mathbf{Y})^{1/2} + (\mathbf{Y}\mathbf{M}^{-1}\mathbf{X}\mathbf{M}^{-1})^{1/2}\mathbf{M} - 2\mathbf{X}$
Connection	$\nabla_\xi\eta = D_\xi\eta + \{\mathbf{X}\mathcal{L}_{\mathbf{X},\mathbf{M}}[\eta]\mathbf{M}\mathcal{L}_{\mathbf{X},\mathbf{M}}[\xi]\mathbf{M} +$ $\mathbf{X}\mathcal{L}_{\mathbf{X},\mathbf{M}}[\xi]\mathbf{M}\mathcal{L}_{\mathbf{X},\mathbf{M}}[\eta]\mathbf{M}\}_\text{S} - \{\mathbf{M}\mathcal{L}_{\mathbf{X},\mathbf{M}}[\eta]\xi\}_\text{S} - \{\mathbf{M}\mathcal{L}_{\mathbf{X},\mathbf{M}}[\xi]\eta\}_\text{S},$ where $\{\mathbf{A}\}_\text{S} := \frac{1}{2}(\mathbf{A} + \mathbf{A}^\top)$
Min/Max Curvature	$K_{\min}(\pi(\mathbf{P})) = 0$, and $K_{\max}(\pi(\mathbf{P})) = \frac{3}{\sigma_n^2 + \sigma_{n-1}^2}$, where σ_i is the i-th largest singular value of $\mathbf{P}$, and $\pi(\mathbf{P}) = \mathbf{M}^{1/2}\mathbf{P}\mathbf{P}^\top\mathbf{M}^{1/2}$

is differentiable) that pulls back the Riemannian metric from one to another [22]. We show in the following proposition that there exists a Riemannian isometry between the GBW and BW geometries.

Proposition 4. *Define a map as $\tau(\mathbf{D}) = \mathbf{M}^{-1/2}\mathbf{D}\mathbf{M}^{-1/2}$, for $\mathbf{D} \in \mathbb{S}^n$. Then, the GBW metric can be written as $g_{\text{gbw},\mathbf{X}}(\mathbf{U}, \mathbf{V}) = g_{\text{bw},\tau(\mathbf{X})}(\tau(\mathbf{U}), \tau(\mathbf{V}))$, where the subscript $\mathbf{X}, \tau(\mathbf{X})$ indicates the tangent space. Hence, $\tau : \mathcal{M}_{\text{bw}} \to \mathcal{M}_{\text{gbw}}$ is a Riemannian isometry.*

The proofs of the results in Table 1 are in our extended report [15] and derived from the first perspective of Riemannian submersion, taking inspiration from the analysis in [4,26,27]. In our extended version [15], we also include various additional developments on the GBW geometry, such as geometric interpolation and barycenter, connection to robust Wasserstein distance and metric learning.

3.2 Riemannian Optimization with the GBW Geometry

Learning over SPD matrices usually concerns optimizing an objective function with respect to the parameter, which is constrained to be SPD. Riemannian optimization is an elegant approach that converts the constrained optimization into an unconstrained problem on manifolds [1,6]. Among the metrics for the SPD matrices, the affine-invariant (AI) metric is seemingly the most popular choice for Riemannian optimization due to its efficiency and convergence guarantees. Recently, however, in [16], the BW metric is shown to be a promising alternative for various learning problems. Below, we derive the expressions for Riemannian gradient and Hessian of an objective function for the GBW geometry.

Riemannian gradient (and Hessian) are generalized gradient (and Hessian) on the tangent space of Riemannian manifolds. The expressions allow to implement various Riemannian optimization methods, using toolboxes like Manopt [7], Pymanopt [35], ROPTLIB [20], etc.

Proposition 5. *The Riemannian gradient and Hessian on* $\mathcal{M}_{\mathrm{gbw}}$ *is derived as* $\mathrm{grad}f(\mathbf{X}) = 2\mathbf{X}\nabla f(\mathbf{X})\mathbf{M} + 2\mathbf{M}\nabla f(\mathbf{X})\mathbf{X}$ *and* $\mathrm{Hess}f(\mathbf{X})[\mathbf{U}] = 4\{\mathbf{M}\nabla^2 f(\mathbf{X})[\mathbf{U}]\mathbf{X}\}_{\mathrm{S}} + 2\{\mathbf{M}\nabla f(\mathbf{X})\mathbf{U}\}_{\mathrm{S}} + 4\{\mathbf{X}\{\nabla f(\mathbf{X})\mathbf{M}\mathcal{L}_{\mathbf{X},\mathbf{M}}[\mathbf{U}]\}_{\mathrm{S}}\mathbf{M}\}_{\mathrm{S}} - \{\mathbf{M}\mathcal{L}_{\mathbf{X},\mathbf{M}}[\mathbf{U}]\mathrm{grad}f(\mathbf{X})\}_{\mathrm{S}}$, *where* $\nabla f(\mathbf{X}), \nabla^2 f(\mathbf{X})$ *represent the Euclidean gradient and Hessian, respectively.*

In our extended report [15], we discuss *geodesic convexity* of functions on the SPD manifold endowed with the GBW metric, generalizing the results in [16].

4   Experiments

In this section, we perform experiments showing the benefit of the GBW geometry. The algorithms are implemented in Matlab using the Manopt toolbox [7]. The codes are available on https://github.com/andyjm3/GBW.

4.1   Log-Determinant Riemannian Optimization

Problem Formulation: Log-determinant (log-det) optimization is common in statistical machine learning, such as for estimating the covariance, with an objective concerning $\min_{\mathbf{X}\in\mathbb{S}^n_{++}} f(\mathbf{X}) = -\log\det(\mathbf{X})$. From [16], optimization with the BW geometry is less well-conditioned compared to the AI geometry. This is because the Riemannian Hessians at optimality are $\mathrm{Hess}_{\mathrm{ai}}f(\mathbf{X}^*)[\mathbf{U}] = \mathbf{U}$ for the AI geometry and $\mathrm{Hess}_{\mathrm{bw}}f(\mathbf{X}^*)[\mathbf{U}] = 4\{(\mathbf{X}^*)^{-1}\mathbf{U}\}_{\mathrm{S}}$ for the BW geometry. This suggests, under the BW geometry, the condition number of Hessian at optimality depends on the solution $\mathbf{X}^*$, while no dependence on $\mathbf{X}^*$ under the AI geometry. Thus, this leads to a poor performance on BW geometry [16].

Here, we show how the GBW geometry helps to address this issue. Specifically, with the GBW geometry, we see from Proposition 5 that by choosing $\mathbf{M} = \mathbf{X}^*$, the Riemannian Hessian is $\mathrm{Hess}_{\mathrm{gbw}}f(\mathbf{X}^*)[\mathbf{U}] = \mathbf{U}$, which becomes well-conditioned (around the optimal solution). This provides the motivation for a choice of $\mathbf{M}$. As the optimal solution $\mathbf{X}^*$ is unknown in optimization problems, choice of $\mathbf{M}$ is not trivial. In practice, one may choose $\mathbf{M} = \mathbf{X}$ dynamically at every or after a few iterations. This strategy corresponds to modifying the GBW geometry dynamically with iterations.

As an example, we consider the following inverse covariance estimation problem [12,19] as $\min_{\mathbf{X}\in\mathbb{S}^n_{++}} f(\mathbf{X}) = -\log\det(\mathbf{X}) + \mathrm{tr}(\mathbf{C}\mathbf{X})$, where $\mathbf{C} \in \mathbb{S}^n_{++}$ is a given SPD matrix. The Euclidean gradient $\nabla f(\mathbf{X}) = -\mathbf{X}^{-1} + \mathbf{C}$ and the Euclidean Hessian $\nabla^2 f(\mathbf{X})[\mathbf{U}] = \mathbf{X}^{-1}\mathbf{U}\mathbf{X}^{-1}$. From the analysis in our extended report [15], this problem is geodesic convex and the optimal solution

Table 2. Riemannian optimization ingredients for the affine-invariant (AI) and Generalized Bures-Wasserstein (GBW) with $\mathbf{M} = \mathbf{X}$ geometries for log-det optimization.

	AI	GBW (with $\mathbf{M} = \mathbf{X}$)
Exp	$\text{Exp}_{\mathbf{X}}(\mathbf{U}) = \mathbf{X}\exp(\mathbf{X}^{-1}\mathbf{U})$	$\text{Exp}_{\mathbf{X}}(\mathbf{U}) = \mathbf{X} + \mathbf{U} + \frac{1}{4}\mathbf{U}\mathbf{X}^{-1}\mathbf{U}$
Grad	$\text{grad}f(\mathbf{X}) = \mathbf{X}\mathbf{C}\mathbf{X} - \mathbf{X}$	$\text{grad}f(\mathbf{X}) = 4\mathbf{X}\mathbf{C}\mathbf{X} - 4\mathbf{X}$
Hess	$\text{Hess}f(\mathbf{X})[\mathbf{U}] = 2\mathbf{U} + \{\mathbf{U}\mathbf{C}\mathbf{X}\}_S$	$\text{Hess}f(\mathbf{X})[\mathbf{U}] = 2\mathbf{U} + 2\{\mathbf{U}\mathbf{C}\mathbf{X}\}_S$

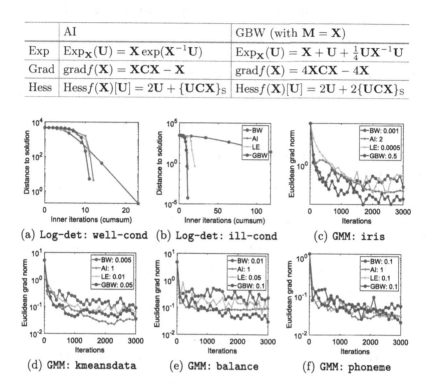

(a) Log-det: well-cond   (b) Log-det: ill-cond   (c) GMM: iris

(d) GMM: kmeansdata   (e) GMM: balance   (f) GMM: phoneme

Fig. 1. Figures (a) & (b): convergence for log-det optimization problem via Riemannian trust region algorithm. Figures (c)–(f): Gaussian mixture model via Riemannian stochastic gradient descent algorithm with optimal initial stepsize. In both the settings, the GBW algorithm outperforms the BW algorithm and performs similar to the AI algorithm. This can be attributed to the choice of $\mathbf{M}$, which offers additional flexibility to the GBW modeling.

is $\mathbf{X}^* = \mathbf{C}^{-1}$, which we seek to estimate as a direct computation is challenging for ill-conditioned $\mathbf{C}$.

Choosing $\mathbf{M} = \mathbf{X}$ and following derivations in Sect. 3.2, the expressions for the exponential map, Riemannian gradient, and Hessian under the GBW geometry are shown in Table 2, where we also draw comparisons to the AI geometry. We see that the choice of $\mathbf{M} = \mathbf{X}$ allows GBW to locally approximate the AI geometry up to some constants. For example, the AI exponential map $\mathbf{X}\exp(\mathbf{X}^{-1}\mathbf{U})$ can by approximated by second-order terms as $\mathbf{X} + \mathbf{U} + \frac{1}{2}\mathbf{U}\mathbf{X}^{-1}\mathbf{U}$. This matches the GBW expression up to an additional term $\frac{1}{4}\mathbf{U}\mathbf{X}^{-1}\mathbf{U}$. Overall, the similarity of optimization ingredients help GBW (with $\mathbf{M} = \mathbf{X}$) perform as similar as the AI geometry, which helps to resolve the poor performance of BW for log-det optimization problems observed in [16].

Experimental Setup and Results: We follow the same settings as in [16] to create problem instances and consider two instances where the condition number of $\mathbf{X}^*$ is 10 (well-conditioned) and 1000 (ill-conditioned). $\mathbf{C}$ is then obtained as $(\mathbf{X}^*)^{-1}$. To compare the convergence performance of optimization methods under the AI, LE, BW, and GBW (with $\mathbf{M} = \mathbf{X}$) geometries, we implement the Riemannian trust region (a second-order solver) with the considered geometries [1,6]. To measure convergence, we use the distance to (theoretical) optimal solution, i.e., $\|\mathbf{X}_t - \mathbf{X}^*\|_2$. We plot this distance against the cumulative inner iterations that the trust region method takes to solve a particular trust region sub-problem at every iteration. The inner iterations are a good measure to show convergence of trust region algorithms [1, Chapter 7].

From Figs. 1(a) and 1(b), we observe the faster convergence with the GBW geometry compared to other geometries regardless of the condition number. In contrast, the BW geometry performs poorly in log-determinant optimization problems as shown in [16]. The GBW geometry effectively resolves the convergence issues with the BW geometry for such settings. Based on our discussion earlier, we see that GBW with $\mathbf{M} = \mathbf{X}$ performs similar to the AI geometry. Empirically, it shows that the GBW geometry effectively bridges the gap between BW and AI geometries for optimization problems.

4.2   Gaussian Mixture Model (GMM)

Problem Formulation: We now consider Gaussian density estimation and mixture model problem. Let $\mathbf{x}_i \in \mathbb{R}^d, i = 1, ..., N$, be the given i.i.d. samples. Following [18], we consider a reformulated GMM problem on augmented samples $\mathbf{y}_i^\top = [\mathbf{x}_i^\top; 1] \in \mathbb{R}^{d+1}$. The density of a GMM is parameterized by the augmented covariance matrix $\Sigma \in \mathbb{R}^{d+1}$. It should be noted that the log-likelihood of Gaussian is geodesic convex under the AI geometry [18] but not under the GBW geometry. However, if we define $\mathbf{S} = \Sigma^{-1}$ [16], the reparameterized log-likelihood $p_\mathcal{N}(\mathbf{Y}; \mathbf{S}) = \sum_{i=1}^N \log\left((2\pi)^{1-d/2} \exp(1/2) \det(\mathbf{S})^{1/2} \exp(-\frac{1}{2}\mathbf{y}_i^\top \mathbf{S}\mathbf{y}_i)\right)$ is geodesic convex on $\mathcal{M}_{\text{gbw}}$. Similar trick was employed in [16] to obtained geodesic convex log-likelihood objective for GMM under the BW geometry. Overall, we solve the GMM problem similar as discussed in [16,18].

Experimental Setup and Results: We consider datasets: iris, kmeansdata, balance, and phoneme from Matlab database and Keel database [11]. For comparisons, we implement the Riemannian stochastic gradient descent method [5] as it is widely used in GMM problems [18]. The batch size is set to 50 and we use a decaying stepsize for all the geometries [16]. As discussed in Sect. 4.1, we set $\mathbf{M} = \mathbf{X}$ at every iteration for optimizing under the GBW geometry. Without access to the optimal solution, the convergence is measured in terms of the Euclidean gradient norm $\|\Sigma_t \nabla L(\Sigma_t)\|_2$ for comparability across geometries.

Figures 1(c)–1(f) show convergence along with the best selected initial stepsize. We observe that convergence under the GBW geometry is competitive and clearly outperforms the BW geometry based algorithm.

Remark 1. For all the experiments in this section, we simply set $\mathbf{M} = \mathbf{X}$. In general, $\mathbf{M}$ can be learned according to the applications. We demonstrate several examples in Appendix.

5 Conclusion

In this paper, we propose a Riemannian geometry that generalizes the recently introduced Bures-Wasserstein geometry for SPD matrices. This generalized geometry has natural connections to the orthogonal Procrustes problem as well as to the optimal transport theory, and still possesses the properties of the Bures-Wasserstein geometry (which is a special case). The new geometry is shown to be parameterized by a SPD matrix $\mathbf{M}$. This offers necessary flexibility in applications. Experiments show that learning of $\mathbf{M}$ leads to better modeling in applications.

References

1. Absil, P.-A., Mahony, R., Sepulchre, R.: Optimization Algorithms on Matrix Manifolds. Princeton University Press (2008)
2. Arsigny, V., Fillard, P., Pennec, X., Ayache, N.: Log-Euclidean metrics for fast and simple calculus on diffusion tensors. Magn Reson. Med. Official J. Int. Soc. Magn. Reson. Med. **56**(2), 411–421 (2006)
3. Bhatia, R.: Positive Definite Matrices. Princeton University Press (2009)
4. Bhatia, R., Jain, T., Lim, Y.: On the Bures-Wasserstein distance between positive definite matrices. Expo. Math. **37**(2), 165–191 (2019)
5. Bonnabel, S.: Stochastic gradient descent on Riemannian manifolds. IEEE Trans. Autom. Control **58**(9), 2217–2229 (2013)
6. Boumal, N.: An introduction to optimization on smooth manifolds, August 2020
7. Boumal, N., Mishra, B., Absil, P.-A., Sepulchre, R.: Manopt, a Matlab toolbox for optimization on manifolds. J. Mach. Learn. Res. **15**(1), 1455–1459 (2014)
8. Brooks, D.A., Schwander, O., Barbaresco, F., Schneider, J.-Y., Cord, M.: Exploring complex time-series representations for Riemannian machine learning of radar data. In: IEEE International Conference on Acoustics, Speech and Signal Processing (2019)
9. Cherian, A., Sra, S.: Riemannian dictionary learning and sparse coding for positive definite matrices. IEEE Trans. Neural Networks Learn. Syst. **28**(12), 2859–2871 (2016)
10. Demetci, P., Santorella, R., Sandstede, B., Noble, W.S., Singh, R.: Gromov-Wasserstein optimal transport to align single-cell multi-omics data, BioRxiv (2020)
11. Derrac, J., Garcia, S., Sanchez, L., Herrera, F.: KEEL data-mining software tool: data set repository, integration of algorithms and experimental analysis framework. J. Mult-.Valued Logic Soft Comput. **17** (2015)
12. Friedman, J., Hastie, T., Tibshirani, R.: Sparse inverse covariance estimation with the graphical lasso. Biostatistics **9**(3), 432–441 (2008)
13. Gramfort, A., Peyré, G., Cuturi, M.: Fast optimal transport averaging of neuroimaging data. In: International Conference on Information Processing in Medical Imaging (2015)

14. Guillaumin, M., Verbeek, J., Schmid, C.: Is that you? Metric learning approaches for face identification. In: International Conference on Computer Vision (2009)
15. Han, A., Mishra, B., Jawanpuria, P., Gao, J.: Learning with symmetric positive definite matrices via generalized Bures-Wasserstein geometry. arXiv:2110.10464 (2021)
16. Han, A., Mishra, B., Jawanpuria, P., Gao, J.: On Riemannian optimization over positive definite matrices with the Bures-Wasserstein geometry. In: Advances in Neural Information Processing Systems (2021)
17. Harandi, M.T., Hartley, R., Lovell, B., Sanderson, C.: Sparse coding on symmetric positive definite manifolds using Bregman divergences. IEEE Trans. Neural Networks Learn. Syst. 27(6), 1294–1306 (2015)
18. Hosseini, R., Sra, S.: An alternative to EM for Gaussian mixture models: batch and stochastic Riemannian optimization. Math. Program. 181(1), 187–223 (2020)
19. Hsieh, C.-J., Dhillon, I., Ravikumar, P., Sustik, M.: Sparse inverse covariance matrix estimation using quadratic approximation. In: Advances in Neural Information Processing Systems (2011)
20. Huang, W., Absil, P.-A., Gallivan, K.A., Hand, P.: ROPTLIB: an object-oriented C++ library for optimization on Riemannian manifolds. ACM Trans. Math. Softw. (TOMS) 44(4), 1–21 (2018)
21. Huang, Z., Wang, R., Shan, S., Li, X., Chen, X.: Log-Euclidean metric learning on symmetric positive definite manifold with application to image set classification. In: International Conference on Machine Learning (2015)
22. Lee, J.M.: Riemannian Manifolds: An Introduction to Curvature, vol. 176. Springer, Cham (2006). https://doi.org/10.1007/b98852
23. Lin, Z.: Riemannian geometry of symmetric positive definite matrices via Cholesky decomposition. SIAM J. Matrix Anal. Appl. 40(4), 1353–1370 (2019)
24. Lotte, F., et al.: A review of classification algorithms for EEG-based brain-computer interfaces: a 10 year update. J. Neural Eng. 15(3), 031005 (2018)
25. Mahadevan, S., Mishra, B., Ghosh, S.: A unified framework for domain adaptation using metric learning on manifolds. In: European Conference on Machine Learning and Knowledge Discovery in Databases (2019)
26. Malagò, L., Montrucchio, L., Pistone, G.: Wasserstein Riemannian geometry of Gaussian densities. Inf. Geom. 1(2), 137–179 (2018). https://doi.org/10.1007/s41884-018-0014-4
27. Massart, E., Hendrickx, J.M., Absil, P.-A.: Curvature of the manifold of fixed-rank positive-semidefinite matrices endowed with the Bures-Wasserstein metric. In: International Conference on Geometric Science of Information (2019)
28. Mishra, B., Sepulchre, R.: Riemannian preconditioning. SIAM J. Optim. 26(1), 635–660 (2016)
29. Pennec, X., Fillard, P., Ayache, N.: A Riemannian framework for tensor computing. Int. J. Comput. Vision 66(1), 41–66 (2006)
30. Peyré, G., Cuturi, M.: Computational optimal transport. Found. Trends Mach. Learn. 11(5-6), 355–607 (2019)
31. Rabin, J., Peyré, G., Delon, J., Bernot, M.: Wasserstein Barycenter and its application to texture mixing. In: Bruckstein, A.M., ter Haar Romeny, B.M., Bronstein, A.M., Bronstein, M.M. (eds.) SSVM 2011. LNCS, vol. 6667, pp. 435–446. Springer, Heidelberg (2012). https://doi.org/10.1007/978-3-642-24785-9_37
32. Shustin, B., Avron, B.: Preconditioned Riemannian optimization on the generalized Stiefel manifold. arXiv:1902.01635 (2019)
33. Sra, S.: A new metric on the manifold of kernel matrices with application to matrix geometric means. In: Advances in Neural Information Processing Systems (2012)

34. Sra, S.: Positive definite matrices and the S-divergence. Proc. Am. Math. Soc. **144**(7), 2787–2797 (2016)
35. Townsend, J., Koep, N., Weichwald, S.: Pymanopt: a Python toolbox for optimization on manifolds using automatic differentiation. J. Mach. Learn. Res. **17**(137), 1–5 (2016)
36. van Oostrum, J.: Bures-Wasserstein geometry. arXiv:2001.08056 (2020)

Fisher-Rao Riemannian Geometry of Equivalent Gaussian Measures on Hilbert Space

Hà Quang Minh$^{(\boxtimes)}$ (ID)

RIKEN Center for Advanced Intelligence Project, Tokyo, Japan
minh.haquang@riken.jp

Abstract. This work presents an explicit description of the Fisher-Rao Riemannian metric on the Hilbert manifold of equivalent centered Gaussian measures on an infinite-dimensional Hilbert space. We show that the corresponding quantities from the finite-dimensional setting of Gaussian densities on Euclidean space, including the Riemannian metric, Levi-Civita connection, curvature, geodesic curve, and Riemannian distance, when properly formulated, directly generalize to this setting. Furthermore, we discuss the connection with the Riemannian geometry of positive definite unitized Hilbert-Schmidt operators on Hilbert space, which can be viewed as a regularized version of the current setting.

Keywords: Fisher-Rao geometry · Gaussian measures · Hilbert space · positive Hilbert-Schmidt operators

1 Introduction

We first briefly review the Fisher-Rao metric, which is an object of central importance in information geometry, for more detail we refer to e.g. [1]. Let $\mathcal{S}$ be a family of probability density functions P_θ on $\mathcal{X} = \mathbb{R}^n$, parametrized by a parameter $\theta = (\theta^1, \ldots, \theta^k) \in \Theta$, where Θ is an open subset in $\mathbb{R}^k$, for some $k \in \mathbb{N}$, that is $\mathcal{S} = \{P_\theta = P(x; \theta) \mid \theta = (\theta^1, \ldots, \theta^k) \in \Theta \subset \mathbb{R}^k\}$, where the mapping $\theta \to P_\theta$ is assumed to be injective. Such an $\mathcal{S}$ is called a k-dimensional *statistical model* or a *parametric model* on $\mathcal{X}$. Assume further that for each fixed $x \in \mathcal{X}$, the mapping $\theta \to P(x; \theta)$ is C^∞, so that all partial derivatives, such as $\frac{\partial P(x;\theta)}{\partial \theta^i}$, $1 \leq i \leq k$, are well-defined and continuous.

A k-dimensional statistical model $\mathcal{S}$ can be considered as a smooth manifold. At each point $\theta \in \Theta$, the *Fisher information matrix* [11] of $\mathcal{S}$ at θ is the $k \times k$ matrix $G(\theta) = [g_{ij}(\theta)]$, $1 \leq i, j \leq k$, with the (i, j)th entry given by

$$g_{ij}(\theta) = \int_{\mathbb{R}^n} \frac{\partial \ln P(x;\theta)}{\partial \theta^i} \frac{\partial \ln P(x;\theta)}{\partial \theta^j} P(x;\theta) dx. \tag{1}$$

Assume that $G(\theta)$ is strictly positive definite $\forall \theta \in \Theta$, then it defines an inner product on the tangent space $T_{P_\theta}(\mathcal{S})$, via the inner product on the basis $\{\frac{\partial}{\partial \theta^j}\}_{j=1}^k$

This research is partially supported by JSPS KAKENHI Grant Number JP20H04250.

© The Author(s), under exclusive license to Springer Nature Switzerland AG 2023
F. Nielsen and F. Barbaresco (Eds.): GSI 2023, LNCS 14071, pp. 416–425, 2023.
https://doi.org/10.1007/978-3-031-38271-0_41

of $T_{P_\theta}(\mathcal{S})$, by $\left\langle \frac{\partial}{\partial \theta^i}, \frac{\partial}{\partial \theta^j} \right\rangle_{P_\theta} = g_{ij}(\theta)$. This inner product defines a Riemannian metric on $\mathcal{S}$, the so-called *Fisher-Rao metric*, or *Fisher information metric* [25], turning $\mathcal{S}$ into a Riemannian manifold.

Gaussian Density Setting. Let $\mathrm{Sym}^{++}(n)$ denote the set of $n \times n$ symmetric, positive definite matrices. Consider the family $\mathcal{S}$ of multivariate Gaussian density functions on $\mathbb{R}^n$ with mean zero

$$\mathcal{S} = \left\{ P(x; \theta) = \frac{1}{\sqrt{(2\pi)^n \det(\Sigma(\theta))}} \exp\left(-\frac{1}{2} x^T \Sigma(\theta)^{-1} x \right), \Sigma(\theta) \in \mathrm{Sym}^{++}(n), \theta \in \mathbb{R}^k \right\}.$$
(2)

Here $k = \frac{n(n+1)}{2}$ and $\theta = [\theta^1, \ldots, \theta^k]$, with the θ^j's corresponding to the upper triangular entries in $\Sigma(\theta)$ according to the following order: $\Sigma(\theta)_{11} = \theta^1$, $\Sigma(\theta)_{12} = \theta^2, \ldots, \Sigma(\theta)_{22} = \theta^{n+1}, \ldots, \Sigma(\theta)_{nn} = \theta^{\frac{n(n+1)}{2}}$. Thus each P_θ in $\mathcal{S}$ is completely characterized by the positive definite covariance matrix $\Sigma(\theta)$. In this case, the Fisher information matrix is given by the following (see e.g [10,17,28])

$$g_{ij}(\theta) = g_{ij}(\Sigma(\theta)) = \frac{1}{2}\mathrm{tr}[\Sigma^{-1}(\partial_{\theta^i}\Sigma)\Sigma^{-1}(\partial_{\theta^j}\Sigma)], \quad 1 \le i,j \le k,$$
(3)

where $\partial_{\theta^i} = \frac{\partial}{\partial \theta^i}$, $\partial_{\theta^j} = \frac{\partial}{\partial \theta^j}$. With the one-to-one correspondence $P_\theta \leftrightarrow \Sigma(\theta)$, we can identify the statistical manifold $\mathcal{S}$ with the manifold $\mathrm{Sym}^{++}(n)$ and the corresponding tangent space $T_{P_\theta}(\mathcal{S})$ with the tangent space $T_{\Sigma(\theta)}(\mathrm{Sym}^{++}(n)) \cong \mathrm{Sym}(n)$. The corresponding Riemannian metric on $\mathrm{Sym}^{++}(n)$ is given by

$$\langle A, B \rangle_\Sigma = \frac{1}{2}\mathrm{tr}(\Sigma^{-1}A\Sigma^{-1}B), \quad A, B \in \mathrm{Sym}(n), \Sigma \in \mathrm{Sym}^{++}(n)$$
(4)

$$= \frac{1}{2}\mathrm{tr}[(\Sigma^{-1/2}A\Sigma^{-1/2})(\Sigma^{-1/2}B\Sigma^{-1/2})].$$
(5)

This is $1/2$ the widely used affine-invariant Riemannian metric, see e.g. [5,21].

Infinite-Dimensional Gaussian Setting. In this work, we generalize the Fisher-Rao metric for the Gaussian density in $\mathbb{R}^n$ to the setting of Gaussian measures on an infinite-dimensional separable Hilbert space $\mathcal{H}$. In the general Gaussian setting, this is not possible since no Lebesgue measure exists on $\mathcal{H}$, hence density functions are not well-defined. Instead, we show the generalization of the Fisher-Rao metric to the set of Gaussian measures *equivalent to a fixed Gaussian measure*, which is a Hilbert manifold, along with the corresponding Riemannian connection, curvature tensor, geodesic, and Riemannian distance, all in closed form expressions. Furthermore, we show that this setting is closely related to the geometric framework of positive definite unitized Hilbert-Schmidt operators in [16], which can be viewed as a regularized version of the equivalent Gaussian measure setting. This work thus provides a link between the framework in [16] with the information geometry of Gaussian measures on Hilbert space.

Related Work in the Infinite-Dimensional Setting. While most work in information geometry is concerned with the finite-dimensional setting, many authors have also considered the infinite-dimensional setting. In [24] and subsequent work [7,12,23], the authors constructed an infinite-dimensional Banach manifold, modeled on Orlicz spaces, for the set of all probability measures equivalent to a given one. In [20], the author constructed a Hilbert manifold of all probability measures equivalent to a given one, with finite entropy, along with the definition of the Fisher-Rao metric, which is generally a pseudo-Riemannian metric. In [2–4], the authors constructed general parametrized measure models and statistical models on a given sample space by utilizing the natural immersion of the set of probability measures into the Banach space of all finite signed measures under the total variation norm. This framework is independent of the reference measure and encompasses that proposed in [24]. The previously mentioned work all deal with highly general settings. Instead, the current work focuses exclusively on the concrete setting of equivalent Gaussian measures on Hilbert space, where a concrete Hilbert manifold structure exists and many quantities of interest can be computed explicitly.

2   Background: Gaussian Measures and Positive Hilbert-Schmidt Operators on Hilbert Space

Throughout the following, let $(\mathcal{H}, \langle, \rangle)$ be a real, separable Hilbert space, with $\dim(\mathcal{H}) = \infty$ unless explicitly stated otherwise. For two separable Hilbert spaces $(\mathcal{H}_i, \langle, \rangle_i)$, $i = 1, 2$, let $\mathcal{L}(\mathcal{H}_1, \mathcal{H}_2)$ denote the Banach space of bounded linear operators from $\mathcal{H}_1$ to $\mathcal{H}_2$, with operator norm $||A|| = \sup_{||x||_1 \leq 1} ||Ax||_2$. For $\mathcal{H}_1 = \mathcal{H}_2 = \mathcal{H}$, we use the notation $\mathcal{L}(\mathcal{H})$. Let $\mathrm{Sym}(\mathcal{H}) \subset \mathcal{L}(\mathcal{H})$ be the set of bounded, self-adjoint linear operators on $\mathcal{H}$. Let $\mathrm{Sym}^+(\mathcal{H}) \subset \mathrm{Sym}(\mathcal{H})$ be the set of self-adjoint, *positive* operators on $\mathcal{H}$, i.e. $A \in \mathrm{Sym}^+(\mathcal{H}) \iff A^* = A, \langle Ax, x \rangle \geq 0 \forall x \in \mathcal{H}$. Let $\mathrm{Sym}^{++}(\mathcal{H}) \subset \mathrm{Sym}^+(\mathcal{H})$ be the set of self-adjoint, *strictly positive* operators on $\mathcal{H}$, i.e. $A \in \mathrm{Sym}^{++}(\mathcal{H}) \iff A^* = A, \langle x, Ax \rangle > 0$ $\forall x \in \mathcal{H}, x \neq 0$. We write $A \geq 0$ for $A \in \mathrm{Sym}^+(\mathcal{H})$ and $A > 0$ for $A \in \mathrm{Sym}^{++}(\mathcal{H})$. If $\gamma I + A > 0$, where I is the identity operator, $\gamma \in \mathbb{R}, \gamma > 0$, then $\gamma I + A$ is also invertible, in which case it is called *positive definite*. In general, $A \in \mathrm{Sym}(\mathcal{H})$ is said to be positive definite if $\exists M_A > 0$ such that $\langle x, Ax \rangle \geq M_A ||x||^2 \forall x \in \mathcal{H}$ - this is equivalent to A being both strictly positive and invertible, see e.g. [22].

The Banach space $\mathrm{Tr}(\mathcal{H})$ of trace class operators on $\mathcal{H}$ is defined by (see e.g. [26]) $\mathrm{Tr}(\mathcal{H}) = \{A \in \mathcal{L}(\mathcal{H}) : ||A||_{\mathrm{tr}} = \sum_{k=1}^{\infty} \langle e_k, (A^*A)^{1/2} e_k \rangle < \infty\}$, for any orthonormal basis $\{e_k\}_{k \in \mathbb{N}} \subset \mathcal{H}$. For $A \in \mathrm{Tr}(\mathcal{H})$, its trace is defined by $\mathrm{tr}(A) = \sum_{k=1}^{\infty} \langle e_k, Ae_k \rangle$, which is independent of choice of basis $\{e_k\}_{k \in \mathbb{N}}$. The Hilbert space $\mathrm{HS}(\mathcal{H}_1, \mathcal{H}_2)$ of Hilbert-Schmidt operators from $\mathcal{H}_1$ to $\mathcal{H}_2$ is defined by (see e.g. [14]) $\mathrm{HS}(\mathcal{H}_1, \mathcal{H}_2) = \{A \in \mathcal{L}(\mathcal{H}_1, \mathcal{H}_2) : ||A||_{\mathrm{HS}}^2 = \mathrm{tr}(A^*A) = \sum_{k=1}^{\infty} ||Ae_k||_2^2 < \infty\}$, for any orthonormal basis $\{e_k\}_{k \in \mathbb{N}}$ in $\mathcal{H}_1$, with inner product $\langle A, B \rangle_{\mathrm{HS}} = \mathrm{tr}(A^*B)$. For $\mathcal{H}_1 = \mathcal{H}_2 = \mathcal{H}$, we write $\mathrm{HS}(\mathcal{H})$.

Equivalence of Gaussian Measures. On $\mathbb{R}^n$, any two Gaussian densities are equivalent, that is they have the same support, which is all of $\mathbb{R}^n$. The situation is drastically different in the infinite-dimensional setting. Let Q, R be two self-adjoint, positive trace class operators on $\mathcal{H}$ such that $\ker(Q) = \ker(R) = \{0\}$. Let $m_1, m_2 \in \mathcal{H}$. A fundamental result in the theory of Gaussian measures is the Feldman-Hajek Theorem [9,13], which states that two Gaussian measures $\mu = \mathcal{N}(m_1, Q)$ and $\nu = \mathcal{N}(m_2, R)$ are either mutually singular or equivalent, that is either $\mu \perp \nu$ or $\mu \sim \nu$. The necessary and sufficient conditions for the equivalence of the two Gaussian measures ν and μ are given by the following.

Theorem 1 ([6], Corollary 6.4.11, [8], Theorems 1.3.9 and 1.3.10). *Let $\mathcal{H}$ be a separable Hilbert space. Consider two Gaussian measures $\mu = \mathcal{N}(m_1, Q)$, $\nu = \mathcal{N}(m_2, R)$ on $\mathcal{H}$. Then μ and ν are equivalent if and only if the following conditions both hold*

1. *$m_2 - m_1 \in \text{range}(Q^{1/2})$.*
2. *There exists $S \in \text{Sym}(\mathcal{H}) \cap \text{HS}(\mathcal{H})$, without the eigenvalue 1, such that $R = Q^{1/2}(I - S)Q^{1/2}$.*

Riemannian Geometry of Positive Definite Hilbert-Schmidt Operators. Each zero-mean Gaussian measure $\mu = \mathcal{N}(0, C)$ corresponds to an operator $C \in \text{Sym}^+(\mathcal{H}) \cap \text{Tr}(\mathcal{H})$ and vice versa. The set $\text{Sym}^+(\mathcal{H}) \cap \text{Tr}(\mathcal{H})$ of positive trace class operators is a subset of the set $\text{Sym}^+(\mathcal{H}) \cap \text{HS}(\mathcal{H})$ of positive Hilbert-Schmidt operators. The affine-invariant Riemannian metric on $\text{Sym}^{++}(n)$ is generalized to this set via extended (unitized) Hilbert-Schmidt operators, as follows.

Extended Hilbert-Schmidt Operators. In [16], the author considered the following set of *extended*, or *unitized*, Hilbert-Schmidt operators

$$\text{HS}_X(\mathcal{H}) = \{A + \gamma I : A \in \text{HS}(\mathcal{H}), \gamma \in \mathbb{R}\}. \tag{6}$$

This set is a Hilbert space under the *extended Hilbert-Schmidt inner product and norm*, under which the Hilbert-Schmidt and scalar operators are orthogonal,

$$\langle A + \gamma I, B + \mu I \rangle_{\text{HS}_X} = \langle A, B \rangle_{\text{HS}} + \gamma \mu, \quad ||A + \gamma I||^2_{\text{HS}_X} = ||A||^2_{\text{HS}} + \gamma^2. \tag{7}$$

Manifold of Positive Definite Hilbert-Schmidt Operators. Consider the following subset of *(unitized) positive definite Hilbert-Schmidt operators*

$$\mathscr{PC}_2(\mathcal{H}) = \{A + \gamma I > 0 : A \in \text{Sym}(\mathcal{H}) \cap \text{HS}(\mathcal{H}), \gamma \in \mathbb{R}\} \subset \text{HS}_X(\mathcal{H}). \tag{8}$$

The set $\mathscr{PC}_2(\mathcal{H})$ is an open subset in the Hilbert space $\text{Sym}(\mathcal{H}) \cap \text{HS}_X(\mathcal{H})$ and is thus a Hilbert manifold. It can be equipped with the following Riemannian metric, generalizing the finite-dimensional affine-invariant metric,

$$\langle A + \gamma I, B + \mu I \rangle_P = \langle P^{-1/2}(A + \gamma I)P^{-1/2}, P^{-1/2}(B + \mu I)P^{-1/2} \rangle_{\text{HS}_X}, \tag{9}$$

for $P \in \mathscr{PC}_2(\mathcal{H})$, $A + \gamma I, B + \mu I \in T_P(\mathscr{PC}_2(\mathcal{H})) \cong \text{Sym}(\mathcal{H}) \cap \text{HS}_X(\mathcal{H}))$. Under this metric, $\mathscr{PC}_2(\mathcal{H})$ becomes a Cartan-Hadamard manifold. There is a unique geodesic joining every pair $(A + \gamma I), (B + \mu I)$, given by

$$\gamma_{AB}(t) = (A + \gamma I)^{1/2} \exp[t \log((A + \gamma I)^{-1/2}(B + \mu I)(A + \gamma I)^{-1/2})](A + \gamma I)^{1/2}. \tag{10}$$

The Riemannian distance between $(A + \gamma I), (B + \mu I) \in \mathscr{PC}_2(\mathcal{H})$ is the length of this geodesic and is given by

$$d_{\text{aiHS}}[(A + \gamma I), (B + \mu I)] = \| \log[(A + \gamma I)^{-1/2}(B + \mu I)(A + \gamma I)^{-1/2}] \|_{\text{HS}_X}. \tag{11}$$

The definition of the extended Hilbert-Schmidt norm $\| \ \|_{\text{HS}_X}$ guarantees that the distance d_{aiHS} is always well-defined and finite on $\mathscr{PC}_2(\mathcal{H})$. We show below that, when restricted to the subset $\text{Sym}^+(\mathcal{H}) \cap \text{Tr}(\mathcal{H})$, this can be viewed as a regularized version of the exact Fisher-Rao distance between two equivalent centered Gaussian measures on $\mathcal{H}$. In the next section, we show that the set of equivalent Gaussian measures on a separable Hilbert space forms a Hilbert manifold, modeled on the Hilbert space of extended Hilbert-Schmidt operators.

3   Fisher-Rao Metric for Equivalent Gaussian Measures

Throughout the following, let $C_0 \in \text{Sym}^+(\mathcal{H}) \cap \text{Tr}(\mathcal{H})$ be fixed, with $\ker(C_0) = 0$. Let $\mu_0 = \mathcal{N}(0, C_0)$ be the corresponding *nondegenerate* Gaussian measure. Consider the set of all zero-mean Gaussian measures on $\mathcal{H}$ equivalent to μ_0, which, by Theorem 1, is given by

$$\text{Gauss}(\mathcal{H}, \mu_0) = \{\mu = \mathcal{N}(0, C), \quad C = C_0^{1/2}(I - S)C_0^{1/2},$$
$$S \in \text{Sym}(\mathcal{H}) \cap \text{HS}(\mathcal{H}), I - S > 0\}. \tag{12}$$

Motivated by Theorem 1, we define the following set

$$\text{SymHS}(\mathcal{H})_{<I} = \{S : S \in \text{Sym}(\mathcal{H}) \cap \text{HS}(\mathcal{H}), I - S > 0\}. \tag{13}$$

This is an open subset in the Hilbert space $\text{Sym}(\mathcal{H}) \cap \text{HS}(\mathcal{H})$ and hence is a Hilbert manifold (for Banach manifolds in general, see e.g. [15]).

Lemma 1. *The set $S \in \text{SymHS}(\mathcal{H})_{<I}$ is a Hilbert manifold with tangent space* $T_S(\text{SymHS}(\mathcal{H})_{<I}) \cong \text{Sym}(\mathcal{H}) \cap \text{HS}(\mathcal{H}) \ \forall S \in \text{SymHS}(\mathcal{H})_{<I}$.

The set $\text{Gauss}(\mathcal{H}, \mu_0)$ corresponds to the following subset of $\text{Sym}^+(\mathcal{H}) \cap \text{Tr}(\mathcal{H})$

$$\text{Tr}(\mathcal{H}, C_0) = \{C \in \text{Sym}^+(\mathcal{H}) \cap \text{Tr}(\mathcal{H}) : C = C_0^{1/2}(I - S)C_0^{1/2}$$
$$\text{for some } S \in \text{SymHS}(\mathcal{H})_{<I}\}. \tag{14}$$

The set $\text{Tr}(\mathcal{H}, C_0)$, equivalently $\text{Gauss}(\mathcal{H}, \mu_0)$, is a Hilbert manifold modeled on $\text{Sym}(\mathcal{H}) \cap \text{HS}(\mathcal{H})$ via the following bijection

$$\varphi : \text{Tr}(\mathcal{H}, C_0) \to \text{SymHS}(\mathcal{H})_{<I}, \quad \varphi(C) = I - C_0^{-1/2}CC_0^{-1/2}. \tag{15}$$

We show that $\text{Tr}(\mathcal{H}, C_0)$ can also be embedded as an open subset in a larger Hilbert space, as follows. Define the following set

$$\text{HS}_X(\mathcal{H}, C_0) = \{C_0^{1/2}(A + \gamma I)C_0^{1/2} : A \in \text{HS}(\mathcal{H}), \gamma \in \mathbb{R}\} \subset \text{Tr}(\mathcal{H}). \tag{16}$$

This is a Hilbert space under the following inner product and norm

$$\langle C_0^{1/2}(A+\gamma I)C_0^{1/2}, C_0^{1/2}(B+\mu I)C_0^{1/2}\rangle_{\mathrm{HS}_X(\mathcal{H},C_0)} = \langle A,B\rangle_{\mathrm{HS}} + \gamma\mu, \qquad (17)$$

$$\|C_0^{1/2}(A+\gamma I)C_0^{1/2}\|_{\mathrm{HS}_X(\mathcal{H},C_0)}^2 = \|A\|_{\mathrm{HS}}^2 + \gamma^2. \qquad (18)$$

$\mathrm{Tr}(\mathcal{H},C_0)$ is a subset of the following subspace of self-adjoint operators in $\mathrm{HS}_X(\mathcal{H},C_0)$

$$\mathrm{SymHS}_X(\mathcal{H},C_0) = \{C_0^{1/2}(A+\gamma I)C_0^{1/2} : A \in \mathrm{Sym}(\mathcal{H}) \cap \mathrm{HS}(\mathcal{H}), \gamma \in \mathbb{R}\}. \qquad (19)$$

Since the set $\{S : S \in \mathrm{Sym}(\mathcal{H}) \cap \mathrm{HS}(\mathcal{H}), I - S > 0\}$ is an open subset in the Hilbert space $\mathrm{Sym}(\mathcal{H}) \cap \mathrm{HS}(\mathcal{H})$ under the $\|\ \|_{\mathrm{HS}}$ norm, it follows that $\mathrm{Tr}(\mathcal{H},C_0)$ is an open subset in the Hilbert space $\mathrm{SymHS}_X(\mathcal{H},C_0)$ under the $\|\ \|_{\mathrm{HS}_X(\mathcal{H},C_0)}$ norm. Thus $\mathrm{Tr}(\mathcal{H},C_0)$ is a Hilbert manifold modeled on $\mathrm{SymHS}_X(\mathcal{H},C_0)$. By the correspondence $\mathcal{N}(0,C) \in \mathrm{Gauss}(\mathcal{H},\mu_0) \iff C \in \mathrm{Tr}(\mathcal{H},C_0)$, the corresponding set of zero-mean Gaussian measures $\mathrm{Gauss}(\mathcal{H},\mu_0)$ is a Hilbert manifold modeled on $\mathrm{SymHS}_X(\mathcal{H},C_0)$. Subsequently, we focus on this manifold structure.

Tangent Space. Let $\Sigma \in \mathrm{Tr}(\mathcal{H},C_0)$ be fixed, with $\Sigma = C_0^{1/2}(I-S)C_0^{1/2}$, $S \in \mathrm{SymHS}(\mathcal{H})_{<I}$. We first specify the tangent space to $\mathrm{Tr}(\mathcal{H},C_0)$ at Σ. Let

$$\mathrm{SymHS}(\mathcal{H},C_0) = \{V = C_0^{1/2}XC_0^{1/2}, X \in \mathrm{Sym}(\mathcal{H}) \cap \mathrm{HS}(\mathcal{H})\}. \qquad (20)$$

Proposition 1. *Let $\Sigma \in \mathrm{Tr}(\mathcal{H},C_0)$ be fixed. The tangent space of the Hilbert manifold $\mathrm{Tr}(\mathcal{H},C_0)$ at Σ is the following Hilbert subspace of $\mathrm{SymHS}_X(\mathcal{H},C_0)$*

$$T_\Sigma(\mathrm{Tr}(\mathcal{H},C_0)) \cong \mathrm{SymHS}(\mathcal{H},C_0) = \mathrm{SymHS}(\mathcal{H},\Sigma). \qquad (21)$$

We now give the formula for the Fisher metric on $\mathrm{Gauss}(\mathcal{H},\mu_0)$, using the abstract framework in [2–4]. Let $\mu = \mathcal{N}(0,\Sigma) \in \mathrm{Gauss}(\mathcal{H},\mu_0)$, with $\Sigma = C_0^{1/2}(I-S)C_0^{1/2}$. Let $\frac{d\mu}{d\mu_0}$ denote its Radon-Nikodym derivative with respect to μ_0, which has an explicit form, see e.g. [19]. We consider the set $\mathrm{Gauss}(\mathcal{H},\mu_0)$ to be parametrized by $S \in \mathrm{SymHS}(\mathcal{H})_{<I}$. For a fixed $S \in \mathrm{SymHS}(\mathcal{H})_{<I}$, the Fisher metric at S is defined to be, for $V_1,V_2 \in T_S(\mathrm{SymHS}(\mathcal{H})_{<I}) \cong \mathrm{Sym}(\mathcal{H}) \cap \mathrm{HS}(\mathcal{H})$,

$$g_S(V_1,V_2) = \int_{\mathcal{H}} D\log\left\{\frac{d\mu}{d\mu_0}(x)\right\}(S)(V_1)D\log\left\{\frac{d\mu}{d\mu_0}(x)\right\}(S)(V_2)d\mu(x). \qquad (22)$$

Here D denotes the Fréchet derivative and the quantity $g_S(V_1,V_2)$ is finite whenever $D\log\left\{\frac{d\mu}{d\mu_0}\right\}(S)(V_j) \in \mathcal{L}^2(\mathcal{H},\mu)$, $j = 1,2$. In the current setting, it can be shown that this is always the case.

Equivalently, the Fisher metric can be obtained by taking the second derivative of the Kullback-Leibler (KL) divergence between two equivalent Gaussian measures on $\mathcal{H}$. Let $\mu = \mathcal{N}(m_1,Q)$ and $\nu = \mathcal{N}(m_2,R)$ on $\mathcal{H}$. If $\mu \perp \nu$, then $\mathrm{KL}(\nu\|\mu) = \infty$. If $\mu \sim \nu$, then we have the following result.

Theorem 2 ([19]). *Let* $\mu = \mathcal{N}(m_1, Q)$, $\nu = \mathcal{N}(m_2, R)$, *with* $\ker(Q) = \ker R = \{0\}$, *and* $\mu \sim \nu$. *Let* $S \in \mathrm{HS}(\mathcal{H}) \cap \mathrm{Sym}(\mathcal{H})$, $I - S > 0$, *be such that* $R = Q^{1/2}(I - S)Q^{1/2}$, *then*

$$\mathrm{KL}(\nu \| \mu) = \frac{1}{2}\|Q^{-1/2}(m_2 - m_1)\|^2 - \frac{1}{2}\log \det{}_2(I - S). \tag{23}$$

Here $\det_2$ is the Hilbert-Carleman determinant, see e.g. [27], with $\det_2(I + A) = \det[(I + A)\exp(-A)]$ for $A \in \mathrm{HS}(\mathcal{H})$, where $\det$ is the Fredholm determinant, given by $\det(I + A) = \prod_{j=1}^{\infty}(1 + \lambda_k(A))$, $A \in \mathrm{Tr}(\mathcal{H})$, $\{\lambda_k(A)\}_{k=1}^{\infty}$ being the eigenvalues of A. The following gives the explicit expression for the Fisher metric.

Theorem 3 (Riemannian metric). *The Fisher metric on* $\mathrm{Gauss}(\mathcal{H}, \mu_0)$ *is given as follows. Let* $S \in \mathrm{SymHS}(\mathcal{H})_{<I}$ *be fixed. Then*

$$g_S(V_1, V_2) = \frac{1}{2}\mathrm{tr}[(I - S)^{-1}V_1(I - S)^{-1}V_2], \quad V_1, V_2 \in \mathrm{Sym}(\mathcal{H}) \cap \mathrm{HS}(\mathcal{H}). \tag{24}$$

Let $\Sigma \in \mathrm{Tr}(\mathcal{H}, C_0)$ *be fixed. The corresponding Riemannian metric on* $\mathrm{Tr}(\mathcal{H}, C_0)$ *is given by, for* $A_1, A_2 \in T_\Sigma(\mathrm{Tr}(\mathcal{H}, C_0)) \cong \mathrm{SymHS}(\mathcal{H}, C_0) = \mathrm{SymHS}(\mathcal{H}, \Sigma)$,

$$\langle A_1, A_2 \rangle_\Sigma = \frac{1}{2}\langle \Sigma^{-1/2}A_1\Sigma^{-1/2}, \Sigma^{-1/2}A_2\Sigma^{-1/2} \rangle_{\mathrm{HS}} \tag{25}$$

$$= \frac{1}{2}\mathrm{tr}(\Sigma^{-1/2}A_1\Sigma^{-1}A_2\Sigma^{-1/2}). \tag{26}$$

Remark 1. In the finite-dimensional setting, if we characterize the Gaussian density P_θ by $S = S(\theta)$, instead of $\Sigma = \Sigma(\theta)$, with $S(\theta)_{11} = \theta^1$, $S(\theta)_{12} = \theta^2$, $\ldots, S(\theta)_{22} = \theta^{n+1}, \ldots, S(\theta)_{nn} = \theta^{\frac{n(n+1)}{2}}$, then we have the following expression for the Fisher metric

$$g_{ij}(\theta) = g_{ij}(S(\theta)) = \frac{1}{2}\mathrm{tr}[(I - S)^{-1}\partial_{\theta_i}S(I - S)^{-1}\partial_{\theta_j}S]. \tag{27}$$

By identifying the basis $\{\partial_{\theta_i}\}$ of $T_{P_\theta}(\mathcal{S})$ with a basis for $\mathrm{Sym}(n)$, we obtain Eq. (24) for $V_1, V_2 \in \mathrm{Sym}(n)$.

We note that Eq. (5) in the finite-dimensional setting directly generalizes to Eqs. (25) and (26) in the current setting of equivalent Gaussian measures on Hilbert space. However, Eq. (4) is generally not well-defined in this setting, since for $A \in \mathrm{SymHS}(\mathcal{H}, \Sigma)$, $\Sigma^{-1}A$ is not necessarily bounded.

Theorem 4 *(Riemannian structures on* $\mathrm{Tr}(\mathcal{H}, C_0)$). *Under the Riemannian metric in Eq. (26),* $\mathrm{Tr}(\mathcal{H}, C_0)$ *is an infinite-dimensional Cartan-Hadamard manifold. Let* $P \in \mathrm{Tr}(\mathcal{H}, C_0)$, *let* X, Y, Z *be smooth vector fields on* $\mathrm{Tr}(\mathcal{H}, C_0)$, *with* $X_P, Y_P, Z_P \in T_P(\mathrm{Tr}(\mathcal{H}, C_0)) \cong \mathrm{SymHS}(\mathcal{H}, C_0) = \mathrm{SymHS}(\mathcal{H}, P)$.

1. The Levi-Civita connection is given by

$$(\nabla_X Y)_P = D(Y)(P)[X_P] - \frac{1}{2}[X_P P^{-1}Y_P + Y_P P^{-1}X_P]. \tag{28}$$

Here $D(Y)(P)$ *denotes the Fréchet derivative of* Y *at* P *in the open subset* $\mathrm{Tr}(\mathcal{H}, C_0)$ *of the Hilbert space* $\mathrm{SymHS}_X(\mathcal{H}, C_0)$.

2. Let $X_P = P^{1/2}\tilde{X}_P P^{1/2}$, $Y_P = P^{1/2}\tilde{Y}_P P^{1/2}$, $Z_P = P^{1/2}\tilde{Z}_P P^{1/2}$, for $\tilde{X}_P, \tilde{Y}_P, \tilde{Z}_P \in \mathrm{Sym}(\mathcal{H}) \cap \mathrm{HS}(\mathcal{H})$. The Riemannian curvature tensor is given by

$$[R(X,Y)Z](P) = -\frac{1}{4}P^{1/2}[[\tilde{X}_P, \tilde{Y}_P], \tilde{Z}_P]P^{1/2}. \tag{29}$$

Here $[,]$ denotes the operator commutator $[A, B] = AB - BA$.

3. The sectional curvature is everywhere nonpositive

$$S_P(X_P, Y_P) = -\frac{\mathrm{tr}[(\tilde{X}_P)^2(\tilde{Y}_P)^2 - (\tilde{X}_P\tilde{Y}_P)^2]}{\mathrm{tr}(\tilde{X}_P)^2\mathrm{tr}(\tilde{Y}_P)^2 - \mathrm{tr}(\tilde{X}_P\tilde{Y}_P)^2} \le 0, \tag{30}$$

where X_P, Y_P are any two linearly independent operators in $\mathrm{SymHS}(\mathcal{H}, P)$.

4. There is a unique geodesic connecting any pair $A \in \mathrm{Tr}(\mathcal{H}, C_0)$, $B = A^{1/2}(I - S)A^{1/2} \in \mathrm{Tr}(\mathcal{H}, C_0)$, given by

$$\gamma_{AB}(t) = A^{1/2}\exp[t\log(I - S)]A^{1/2}. \tag{31}$$

The length of this geodesic is the Riemannian distance between A and B, or equivalently, the Fisher-Rao distance between $\mathcal{N}(0, A)$ and $\mathcal{N}(0, B)$,

$$d(A, B) = \frac{1}{2}\|\log(A^{-1/2}BA^{-1/2})\|_{\mathrm{HS}} = \frac{1}{2}\|\log(I - S)\|_{\mathrm{HS}}. \tag{32}$$

Remark 2. The Levi-Civita connection in Eq. (28) is formulated using the fact that $\mathrm{Tr}(\mathcal{H}, C_0)$ is an open subset in the Hilbert space $\mathrm{SymHS}_X(\mathcal{H}, C_0)$ under the inner product and norm defined in Eqs. (17) and (18).

Remark 3. In the finite-dimensional setting, the Riemannian curvature tensor in Eq. (29) is equivalent to

$$[R(X,Y)Z](P) = -\frac{1}{4}P[[P^{-1}X_P, P^{-1}Y_P], P^{-1}Z_P]. \tag{33}$$

In [15] (Theorem 3.9), this formula was given in the case $P = I$, the identity operator, but without the factor $\frac{1}{4}$. This formula is also valid on the manifold $\mathscr{PC}_2(\mathcal{H})$ under the affine-invariant Riemannian metric defined in Eq. (9) (see [16], Eq. (5)). It is generally not valid in the setting of equivalent infinite-dimensional Gaussian measures, however, since $P^{-1}X_P = P^{-1/2}\tilde{X}_P P^{1/2}$, $\tilde{X}_P \in \mathrm{Sym}(\mathcal{H}) \cap \mathrm{HS}(\mathcal{H})$, is not necessarily bounded.

Connection with the Riemannian Geometry of Positive Definite Unitized Hilbert-Schmidt Operators. The following shows that the Fisher-Rao distance in Eq. (32) can be obtained from the Riemannian distance between positive definite unitized Hilbert-Schmidt operators in Eq. (11) as $\gamma = \mu \to 0$.

Theorem 5. Let $A, B \in \mathrm{Tr}(\mathcal{H}, C_0)$ with $B = A^{1/2}(I - S)B^{1/2}$. Then

$$\lim_{\gamma \to 0^+} \|\log[(A + \gamma I)^{-1/2}(B + \gamma I)(A + \gamma I)^{-1/2}]\|_{\mathrm{HS}} = \|\log(I - S)\|_{\mathrm{HS}}. \tag{34}$$

Thus, for $A, B \in \mathrm{Sym}^+(\mathcal{H}) \cap \mathrm{Tr}(\mathcal{H})$ and $\gamma \in \mathbb{R}, \gamma > 0$, Eq. (11) can be considered as a regularized version of Eq. (32), up to the multiplicative factor $\frac{1}{2}$. The advantage of the distance in Eq. (11) is that it is always finite for all pairs $A, B \in \mathrm{Sym}^+(\mathcal{H}) \cap \mathrm{Tr}(\mathcal{H})$ and all $\gamma \in \mathbb{R}, \gamma > 0$. Furthermore, Eq. (11) can be applied for estimating the distance between the infinite-dimensional Gaussian measures corresponding to measurable Gaussian processes with squared integrable paths, using distances between the corresponding finite-dimensional Gaussian measures, with explicit finite sample complexity, see [18].

Remark 4. The proofs of all results stated above will be presented in the full version of the current work.

References

1. Amari, S., Nagaoka, H.: Methods of Information Geometry. American Mathematical Society (2000)
2. Ay, N., Jost, J., Lê, H.V., Schwachhöfer, L.: Information geometry and sufficient statistics. Prob. Theory Relat. Fields **162**, 327–364 (2015)
3. Ay, N., et al.: Parametrized measure models. Bernoulli **24**(3), 1692–1725 (2018)
4. Ay, N., Jost, J., Lê, H.V., Schwachhöfer, L.: Information Geometry. EMGFASMSM, vol. 64. Springer, Cham (2017). https://doi.org/10.1007/978-3-319-56478-4
5. Bhatia, R.: Positive Definite Matrices. Princeton University Press, Princeton (2007)
6. Bogachev, V.: Gaussian Measures. American Mathematical Society (1998)
7. Cena, A., Pistone, G.: Exponential statistical manifold. Ann. Inst. Stat. Math. **59**, 27–56 (2007)
8. Da Prato, G., Zabczyk, J.: Second Order Partial Differential Equations in Hilbert Spaces, vol. 293. Cambridge University Press, Cambridge (2002)
9. Feldman, J.: Equivalence and perpendicularity of Gaussian processes. Pac. J. Math. **8**(4), 699–708 (1958)
10. Felice, D., Hà Quang, M., Mancini, S.: The volume of Gaussian states by information geometry. J. Math. Phys. **58**(1), 012201 (2017)
11. Fisher, R.A.: On the mathematical foundations of theoretical statistics. Phil. Trans. Roy. Soc. Lond. Ser. A **222**, 309–368 (1922)
12. Gibilisco, P., Pistone, G.: Connections on non-parametric statistical manifolds by orlicz space geometry. Infinite Dimen. Anal. Quant. Prob. Relat. Topics **1**(02), 325–347 (1998)
13. Hájek, J.: On a property of normal distributions of any stochastic process. Czechoslovak Math. J. **08**(4), 610–618 (1958)
14. Kadison, R., Ringrose, J.: Fundamentals of the Theory of Operator Algebras. Volume I: Elementary Theory. Academic Press, Cambridge (1983)
15. Lang, S.: Fundamentals of Differential Geometry, vol. 191. Springer, Heidelberg (2012). https://doi.org/10.1007/978-1-4612-0541-8
16. Larotonda, G.: Nonpositive curvature: a geometrical approach to hilbert-schmidt operators. Differ. Geom. Appl. **25**, 679–700 (2007)
17. Lenglet, C., Rousson, M., Deriche, R., Faugeras, O.: Statistics on the manifold of multivariate normal distributions: theory and application to diffusion tensor MRI processing. J. Math. Imaging Vision **25**(3), 423–444 (2006)

18. Minh, H.Q.: Estimation of Riemannian distances between covariance operators and Gaussian processes (2021). arXiv preprint arXiv:2108.11683
19. Minh, H.: Regularized divergences between covariance operators and Gaussian measures on Hilbert spaces. J. Theor. Prob. **34**, 580–643 (2021)
20. Newton, N.J.: An infinite-dimensional statistical manifold modelled on Hilbert space. J. Funct. Anal. **263**(6), 1661–1681 (2012)
21. Pennec, X., Fillard, P., Ayache, N.: A Riemannian framework for tensor computing. Int. J. Comput. Vision **66**(1), 41–66 (2006)
22. Petryshyn, W.: Direct and iterative methods for the solution of linear operator equations in Hilbert spaces. Trans. Am. Math. Soc. **105**, 136–175 (1962)
23. Pistone, G., Rogantin, M.P.: The exponential statistical manifold: mean parameters, orthogonality and space transformations. In: Bernoulli, pp. 721–760 (1999)
24. Pistone, G., Sempi, C.: An infinite-dimensional geometric structure on the space of all the probability measures equivalent to a given one. In: The Annals of Statistics, pp. 1543–1561 (1995)
25. Rao, C.: Information and accuracy attainable in the estimation of statistical parameters. Bull. Calcutta Math. Soc. **37**(3), 81–91 (1945)
26. Reed, M., Simon, B.: Methods of Modern Mathematical Physics: Functional Analysis. Academic Press, Cambridge (1975)
27. Simon, B.: Notes on infinite determinants of Hilbert space operators. Adv. Math. **24**, 244–273 (1977)
28. Skovgaard, L.T.: A Riemannian geometry of the multivariate normal model. Scand. J. Stat. **11**, 211–223 (1984)

The Gaussian Kernel on the Circle and Spaces that Admit Isometric Embeddings of the Circle

Nathaël Da Costa, Cyrus Mostajeran$^{(\boxtimes)}$, and Juan-Pablo Ortega

Division of Mathematical Sciences, School of Physical and Mathematical Sciences,
Nanyang Technological University, Singapore 637371, Singapore
cyrus.mostajeran@gmail.com

Abstract. On Euclidean spaces, the Gaussian kernel is one of the most widely used kernels in applications. It has also been used on non-Euclidean spaces, where it is known that there may be (and often are) scale parameters for which it is not positive definite. Hope remains that this kernel is positive definite for many choices of parameter. However, we show that the Gaussian kernel is not positive definite on the circle for any choice of parameter. This implies that on metric spaces in which the circle can be isometrically embedded, such as spheres, projective spaces and Grassmannians, the Gaussian kernel is not positive definite for any parameter.

Keywords: kernel methods · Gaussian kernel · positive definite kernels · geodesic exponential kernel · metric spaces · Riemannian manifolds

1 Introduction

In many applications, it is useful to capture the geometry of the data and view it as lying in a non-Euclidean space, such as a metric space or a Riemannian manifold. Examples of such applications include computer vision [15], robot learning [5] and brain-computer interfaces [1]. We are interested in the problem of applying kernel methods on such non-Euclidean spaces.

Kernel methods are prominent in machine learning, with some examples of algorithms including support vector machines [7], kernel principal component analysis [18], solvers for controlled and stochastic differential equations [16], and reservoir computing [11,12]. These algorithms rely on the existence of a reproducing kernel Hilbert space into which the kernel maps the data. This in turn requires the chosen kernel to be positive definite (PD).

One of the most common types of kernel used in applications is the Gaussian kernel. Defined on a Euclidean space, this kernel is PD for any choice of parameter. Moreover, [17] shows that the Gaussian kernel defined on a metric space is PD for all parameters if and only if the metric space can be isometrically embedded into an inner product space. This implies that Euclidean spaces are the only complete Riemannian manifolds for which the Gaussian kernel is

F. Nielsen and F. Barbaresco (Eds.): GSI 2023, LNCS 14071, pp. 426–435, 2023.
https://doi.org/10.1007/978-3-031-38271-0_42

PD for all parameters [9,14]. However, the problem of determining for which parameters the Gaussian kernel is PD on a given metric space is not solved. [19] shows that the Gaussian kernel may be PD for a wide range of parameters even when it is not PD for every parameter. However, we rule out such a possibility for a large class of spaces of interest.

We start by defining positive definite kernels. Then we give a brief review of the literature on the positive definiteness of the Gaussian kernel, and introduce some new notation to study this problem. Finally, we show that the Gaussian kernel is not PD for any choice of parameter on the circle, and consequently for any metric space admitting an isometrically embedded circle.

We should note that since producing the results of this paper, we have discovered that certain general characterisations of positive definite functions on the circle exist in the literature. Using Bochner's theorem, sufficient conditions for non-positive definiteness have been provided, which encompass our result on the circle [10,20]. Our proof, however, is specific to the Gaussian kernel, relies only on elementary analysis, and provides further insight into the extent to which the Gaussian kernel fails to be PD, which may have practical relevance for applications of kernel methods to non-Euclidean data processing.

2 Kernels

Definition 1. *A* kernel *on a set* X *is a symmetric map* $k : X \times X \to \mathbb{R}$. k *is said to be* positive definite (PD) *if for all* $N \in \mathbb{N}$, $x_1, \ldots, x_N \in X$ *and all* $c_1, \ldots, c_N \in \mathbb{R}$, $\sum_{i=1}^{N} \sum_{j=1}^{N} c_i c_j k(x_i, x_j) \geq 0$, *i.e. the matrix* $\left(k(x_i, x_j) \right)_{i,j}$, *which we call the* Gram matrix *of* $x_1, \ldots, x_N$, *is positive semidefinite.*

Proposition 1. *Suppose the* $(k_n)_{n \geq 1}$ *are PD kernels on* X.

(i) $a_1 k_1 + a_2 k_2$ *is a PD kernel on* X *for all* $a_1, a_2 \geq 0$.
(ii) The Hadamard (pointwise) product $k_1 \cdot k_2$ *is a PD kernel on* X.
(iii) If $k_n \to k$ *pointwise as* $n \to \infty$, *then* k *is a PD kernel on* X.
(iv) If $Y \subset X$ *then* $k_1|_Y$ *is a PD kernel on* Y.

Proof. The $N \times N$ symmetric positive semidefinite matrices $\mathrm{Sym}^{0+}(N)$ form a closed convex cone in the space of symmetric matrices $\mathrm{Sym}(N)$, which implies (i) and (iii). $\mathrm{Sym}^{0+}(N)$ is also closed under pointwise multiplication, as shown in [2, Chapter 3 Theorem 1.12.], which implies (ii). Finally, proving (iv) is trivial.

□

Proposition 1 (i), (ii), and (iii) say that PD kernels on X form a convex cone, closed under pointwise convergence and pointwise multiplication.

3 The Gaussian Kernel

In this section, X is a metric space equipped with the metric d. A common type of kernel on such a space is the Gaussian kernel $k(\, \cdot \, , \, \cdot \,) := \exp(-\lambda d(\, \cdot \, , \, \cdot \,)^2)$ where $\lambda > 0$. Write

$$\Lambda_+(X) := \{\lambda > 0 : \text{the Gaussian kernel with parameter } \lambda \text{ is PD}\}.$$

We would like to characterise $\Lambda_+(X)$ in terms of X. In what follows, Propositions 2 and 3 are analogous to Proposition 1 for Gaussian kernels.

Proposition 2. *(i) $\Lambda_+(X)$ is closed under addition.*
(ii) $\Lambda_+(X)$ is topologically closed in $(0, \infty)$.

Proof. (i) and (ii) follow from Proposition 1 (ii) and (iii) respectively.   □

Corollary 1. *(i) If there is $\epsilon > 0$ s.t. $(0, \epsilon) \subset \Lambda_+(X)$ then $\Lambda_+(X) = (0, \infty)$.*
(ii) If there is $\epsilon > 0$ s.t. $\Lambda_+(X) \subset (0, \epsilon)$ then $\Lambda_+(X) = \varnothing$.

Proof. These both follow from Proposition 2 (i).   □

Definition 2. *Let Y be another metric space with metric d'. We say Y isometrically embeds into X, written $Y \hookrightarrow X$ if there is a function $\iota : Y \to X$ such that $d(\iota(\,\cdot\,), \iota(\,\cdot\,)) = d'(\,\cdot\,,\,\cdot\,)$.*

Note that, while the notion of 'isometry' in the context of Riemannian manifolds and in the context of metric spaces correspond (Myers-Steenrod theorem), the notion of 'isometric embedding' is stronger in the context of metric spaces than in the context of Riemannian manifolds. For example, the unit 2-sphere S^2 can be isometrically embedded in $\mathbb{R}^3$ in the sense of Riemannian manifolds, but not in our sense.

Proposition 3. *Let Y be another metric space with metric d'. If $Y \hookrightarrow X$, then $\Lambda_+(X) \subset \Lambda_+(Y)$.*

Proof. Follows from Proposition 1 (iv).   □

As of now, we have only made rather elementary observations about $\Lambda_+(X)$, but now we state the first major result, without proof.

Theorem 1 (due to I.J. Schoenberg [17]). *The following are equivalent:*

1. $\Lambda_+(X) = (0, \infty)$.
2. $X \hookrightarrow \mathcal{V}$ *for some inner product space $\mathcal{V}$.*

Note that, if it exists, the isometric embedding $X \hookrightarrow \mathcal{V}$ is not in general related to the reproducing kernel Hilbert space (RKHS) map for the Gaussian kernel. Given a positive definite kernel k on X, the RKHS map is a set-theoretic map $K : X \to \mathcal{H}$ where $\mathcal{H}$ is a Hilbert space such that $k(\,\cdot\,,\,\cdot\,) = \langle K(\,\cdot\,), K(\,\cdot\,) \rangle$. These are different objects.

Theorem 1 is already very powerful, and guarantees that $\Lambda_+(X) = (0, \infty)$ for many spaces.

Corollary 2. $\Lambda_+(X) = (0, \infty)$ *for the following spaces X:*

1. $\mathbb{R}^n$ *with the Euclidean metric, for $n \geq 1$.*
2. $L^2_\mathbb{R}(\Omega, \mu)$ *for any measure space (Ω, μ).*
3. $\mathrm{Sym}^{++}(n)$ *the space of symmetric $n \times n$ positive definite matrices, with the Frobenius metric, for $n \geq 1$.*

4. $Sym^{++}(n)$ with the log-Euclidean metric $d(A, B) = \| \log(A) - \log(B) \|_F$, for $n \geq 1$, where $\| \cdot \|_F$ denotes the Frobenius norm.

5. $Gr_\mathbb{R}(k, n)$ the real Grassmanian with the projection metric $d([A], [B]) = \| AA^T - BB^T \|_F$, where A, B are the $n \times k$ matrices representing the subspaces $[A], [B]$, respectively, for $1 \leq k < n$.

Proof. Follows directly from Theorem 1. □

Moreover, [9] and [14] deduce from Theorem 1 the following result, which we state without proof.

Theorem 2. *If X is a complete Riemannian manifold, $\Lambda_+(X) = (0, \infty)$ if and only if X is isometric to a Euclidean space.*

While Theorem 1 is powerful, the full characterisation of $\Lambda_+(X)$ is far from solved. $\Lambda_+(X)$ can be non-empty and different from $(0, \infty)$; it is easy to construct finite metric spaces with more complicated $\Lambda_+(X)$. This can also be the case for more complex metric spaces: [19, Theorem 3.10] shows that on the space of symmetric positive definite matrices $X = Sym^{++}(n)$ equipped with the metric of Stein divergence, we have $\Lambda_+(X) = \{\frac{1}{2}, \frac{2}{2}, \ldots, \frac{n-2}{2}\} \cup [\frac{n-1}{2}, \infty)$. While this result gives hope that the Gaussian kernel may be PD for many parameters on many interesting spaces, we show that this is often not the case.

4   The Gaussian Kernel on the Circle

Theorem 3. $\Lambda_+(S^1) = \varnothing$ *where S^1 is the unit circle with its classical intrinsic metric.*

Proof. Let $N \in \mathbb{N}$. Define $x_k = 2\pi k/N$ for $0 \leq k \leq N - 1$. So

$$d(x_k, x_l) = \frac{2\pi}{N} \min\{|k - l + mN| : m \in \mathbb{Z}\}$$

for all $0 \leq k, l \leq N - 1$ (Fig. 1).

So the Gram matrix of $x_0, \ldots, x_{N-1}$ is

$$K = \begin{bmatrix} 1 & \exp(-\lambda(\frac{2\pi}{N})^2) & \exp(-\lambda(\frac{4\pi}{N})^2) & \ldots & \exp(-\lambda(\frac{2\pi}{N})^2) \\ \exp(-\lambda(\frac{2\pi}{N})^2) & 1 & & \ldots & \exp(-\lambda(\frac{4\pi}{N})^2) \\ \vdots & \vdots & \ddots & & \vdots \\ \exp(-\lambda(\frac{2\pi}{N})^2) & \exp(-\lambda(\frac{4\pi}{N})^2) & \exp(-\lambda(\frac{6\pi}{N})^2) & \ldots & 1 \end{bmatrix}.$$

To show that the Gaussian kernel with parameter λ is not PD, all we need to do is to show that we can choose N such that K has a negative eigenvalue.

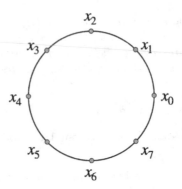

Fig. 1. Points x_k on S^1 for $N = 8$.

K is a circulant matrix, so its eigenvalues are given by the discrete Fourier transform of the first row. Explicitly, these eigenvalues are

$$w_j = \sum_{k=0}^{\lfloor N/2 \rfloor} \exp\left(-\mu \frac{k^2}{N^2}\right) e^{i\frac{2\pi}{N}kj} + \sum_{k=\lfloor N/2 \rfloor+1}^{N-1} \exp\left(-\mu \frac{(N-k)^2}{N^2}\right) e^{i\frac{2\pi}{N}kj}$$

for $0 \leq j \leq N - 1$, where $\mu = \lambda(2\pi)^2$. Taking $N \equiv 0 \bmod 2$, this gives

$$w_j = 1 + \sum_{k=1}^{N/2-1} \exp\left(-\mu \frac{k^2}{N^2}\right) \left(e^{i\frac{2\pi}{N}kj} + e^{i\frac{2\pi}{N}(N-k)j}\right) + \exp(-\mu/4)e^{i\pi j}$$

$$= -1 + 2 \sum_{k=0}^{N/2-1} \exp\left(-\mu \frac{k^2}{N^2}\right) \cos\left(\frac{2\pi}{N}kj\right) + (-1)^j \exp(-\mu/4).$$

Restricting further to $N \equiv 0 \bmod 4$ and $j = N/2$, the sum conveniently becomes alternating:

$$w_{N/2} = -1 + 2 \underbrace{\sum_{k=0}^{N/2-1} (-1)^k \exp\left(-\mu \frac{k^2}{N^2}\right)}_{(*)} + \exp(-\mu/4). \qquad (1)$$

We will show that $w_{N/2}$ is negative for N large enough. For this, we need to estimate the second term of (1). The difficulty lies in the fact that the variable N appears in both the terms and the indices of the sum. To remedy this we define

$$S_r(N) := \sum_{k=0}^{\infty} (-1)^k \exp\left(-\mu \frac{k^2}{N^2} - r\frac{k}{N}\right)$$

for $r \geq 0$. These series are instances of partial theta functions, and below we leverage two facts about them from the literature. But first, let us express $w_{N/2}$

in terms of such series. We have

$$S_0(N) = \sum_{k=0}^{\infty} (-1)^k \exp\left(-\mu \frac{k^2}{N^2}\right)$$

$$= \underbrace{\sum_{k=0}^{N/2-1} (-1)^k \exp\left(-\mu \frac{k^2}{N^2}\right)}_{(*)} + \exp(-\mu/4) - \exp(-\mu/4)\exp\left(-\mu \frac{1}{N^2} - \mu \frac{1}{N}\right)$$

$$+ \underbrace{\sum_{k=N/2+2}^{\infty} (-1)^k \exp\left(-\mu \frac{k^2}{N^2}\right)}_{(**)}.$$

$$\tag{2}$$

We remove the dependency on N from the indices of $(**)$:

$$\sum_{k=N/2+2}^{\infty} (-1)^k \exp\left(-\mu \frac{k^2}{N^2}\right) = \sum_{k=0}^{\infty} (-1)^{k+N/2} \exp\left(-\mu \frac{(k+N/2+2)^2}{N^2}\right)$$

$$= \exp\left(-\mu \frac{(N/2+2)^2}{N^2}\right) \sum_{k=0}^{\infty} (-1)^k \exp\left(-\mu \frac{k^2}{N^2} - \mu \left(1 + \frac{4}{N}\right) \frac{k}{N}\right)$$

$$= \exp(-\mu/4) \exp\left(-\mu \frac{4}{N^2} - \mu \frac{2}{N}\right) S_{\mu(1+4/N)}(N).$$

$$\tag{3}$$

Substituting (3) into (2), and in turn substituting (2) into (1) we get

$$w_{N/2} = -1 + 2S_0(N) + \exp(-\mu/4)\Bigg(-1 + 2\exp\left(-\mu \frac{1}{N^2} - \mu \frac{1}{N}\right)$$

$$- 2\exp\left(-\mu \frac{4}{N^2} - \mu \frac{2}{N}\right) S_{\mu(1+4/N)}(N)\Bigg).$$

Now we use the following lemma.

Lemma 1. $S_r(N) \geq S_0(N)$ for all $r \geq 0$ and for all $N \in \mathbb{N}$.

Proof. This follows from [6, Proposition 14 Eq. 5.8], which makes use of the maximum principle for the heat equation.   □

So

$$w_{N/2} \leq -1 + 2S_0(N) + \exp(-\mu/4)\Bigg(-1 + 2\exp\left(-\mu \frac{1}{N^2} - \mu \frac{1}{N}\right)$$

$$- 2\exp\left(-\mu \frac{4}{N^2} - \mu \frac{2}{N}\right) S_0(N)\Bigg).$$

$$\tag{4}$$

The limit of the RHS of (4) as $N \to \infty$ is 0, so it is not enough just to take the limit. Instead, we will need to take an asymptotic expansion with respect to $1/N$ to the second order. For this, we need a second lemma.

Lemma 2.
$$S_0(N) = \frac{1}{2} + O(1/N^n) \ \text{as} \ N \to \infty$$

for all $n \geq 1$.

Proof. [4, Theorem 1.1 (i)] says that for $n \geq 1$,

$$S_0(N) = \sum_{a=0}^{n} \left(\frac{1}{2\pi i} \frac{\partial}{\partial z}\right)^{2a} \left[\frac{1}{1 - e^{2\pi i z}}\right]_{z=1/2} \frac{(-\mu)^a}{a!} \frac{1}{N^{2a}} + O(1/N^{2n+1}) \ \text{as} \ N \to \infty.$$

Now observe

$$f(z) := \frac{1}{1 - e^{2\pi i(z-1/2)}} - \frac{1}{2} = \frac{i}{2} \tan(\pi z)$$

is odd, so the even terms in the Taylor series of f vanish, and hence

$$\left(\frac{1}{2\pi i} \frac{\partial}{\partial z}\right)^{2a} \left[\frac{1}{1 - e^{2\pi i z}}\right]_{z=1/2} = \begin{cases} \frac{1}{2} & \text{if } a = 0 \\ 0 & \text{if } a \geq 1 \end{cases}$$

which yields the result. ∎

So taking the asymptotic expansion with respect to $1/N$ to second order, (4) simplifies to

$$w_{N/2} \leq \exp(-\mu/4)\frac{2\mu - \mu^2}{N^2} + O(1/N^3) \ \text{as} \ N \to \infty.$$

If $\lambda > \frac{1}{2\pi^2}$ then $\mu > 2$ so $2\mu - \mu^2 < 0$ and hence $w_{N/2}$ is negative for N large enough, with $N \equiv 0$ mod 4. It is possible to improve these inequalities to obtain the result for all λ, although this is unnecessary: Corollary 1 (ii) is enough to conclude the proof. □

Thanks to Proposition 3, Theorem 3 gives us much more than one may suspect at first.

Corollary 3. *If $S^1 \hookrightarrow X$ then $\Lambda_+(X) = \varnothing$. So $\Lambda_+(X) = \varnothing$ for the following spaces, equipped with their classical intrinsic Riemannian metrics:*

1. *S^n the sphere, for $n \geq 1$.*
2. *$\mathbb{RP}^n$ the real projective space, for $n \geq 1$.*
3. *$Gr_{\mathbb{R}}(k, n)$ the real Grassmannian, for $1 \leq k < n$, viewed as the homogeneous space $O(n)/(O(k) \times O(n-k))$.*
4. *$\mathbb{T}^n$ the n-dimensional torus, for $n \geq 1$.*

Proof Sketch. This follows from Theorem 3 and Proposition 3. Now we briefly argue for the specific examples. For 1., $S^1 \hookrightarrow S^n$ (e.g., a 'great circle'). For 2., $1/2 \cdot S^1 \cong \mathbb{RP}^1 \hookrightarrow \mathbb{RP}^n$ where '$\cong$' means isometric and '$1/2 \cdot S^1$' means S^1 rescaled by a factor of $1/2$. This factor does not affect the conclusion. For 3., the metric in question is

$$d([A], [B]) = \left(\sum_{i=1}^{k} \theta_i^2\right)^{1/2}$$

where θ_i is the i-th principal angle between $[A]$ and $[B]$ (see [21] and [22]), $[A], [B] \in Gr_{\mathbb{R}}(k, n)$. Fixing any $[A] \in Gr_{\mathbb{R}}(k, n)$, travelling on $Gr_{\mathbb{R}}(k, n)$ while keeping $\theta_i = 0$ for $i > 1$ and varying θ_1 only, we get an isometric embedding of $1/2 \cdot S^1$ into $Gr_{\mathbb{R}}(k, n)$. For 4., $\mathbb{T}^n := \underbrace{S^1 \times \cdots \times S^1}_{n}$ equipped with the product metric, so $S^1 \hookrightarrow \mathbb{T}^n$ by embedding into one component.   □

It is conceivable that S^1 can be isometrically embedded (in the metric sense from Definition 2) into any compact Riemannian manifold (up to rescaling). We have yet to think of a counterexample. If this is true, then Theorem 3 would solve the problem of characterising $\Lambda_+(X)$ for all compact Riemannian manifolds. However, while the Lyusternik-Fet theorem tells us that any compact Riemannian manifold has a closed geodesic, it appears to be an open question whether any such manifold admits an isometric embedding of S^1.

Fig. 2. Examples of Riemannian manifolds that admit isometric embeddings of S^1. From left to right: a sphere, a torus, a hyperbolic hyperboloid.

Note that non-compact manifolds may also admit isometric embeddings of S^1: consider a hyperbolic hyperboloid. There is precisely one (scaled) isometric embedding of S^1 into it. This example is particularly interesting since, as opposed to the examples above with positive curvature, it has everywhere negative curvature. See Fig. 2.

5   Discussion

In machine learning, most kernel methods rely on the existence of an RKHS embedding. This, in turn, requires the chosen kernel to be positive definite. Theorem 3 shows that the Gaussian kernel defined in this work cannot provide such RKHS embeddings of the circle, spheres, and Grassmannians. It reinforces the conclusion from Theorem 2 that one should be careful when using the Gaussian kernel in the sense defined in this work on non-Euclidean Riemannian manifolds. The authors in [3] propose a different way to generalise the Gaussian kernel from Euclidean spaces to Riemannian manifolds by viewing it as a solution to the heat equation. This produces positive definite kernels by construction.

Nevertheless, perhaps we should not be so fast to altogether reject our version of the Gaussian kernel, which has the advantage of being of particularly simple form. It is worth noting that the proof of Theorem 3 relies on taking $N \to \infty$, where N is the number of data points. [8] lists three open problems regarding the positive definiteness of the Gaussian kernels on metric spaces. It suggests that we should not only look at whether the Gaussian kernel is PD on the whole space but whether there are conditions on the spread of the data such that the Gram matrix of this data is PD. Our proof of Theorem 3 relying on the assumption of infinite data suggests that this may be the case. In general, fixing N data points, the Gram matrix with components $\exp(-\lambda d(\cdot, \cdot)^2)$ tends to the identity as $\lambda \to \infty$, so will be PD for λ large enough. This observation has supported the use of the Gaussian kernel on non-Euclidean spaces, for example, in [13] where it is used on spheres. However, it is important to keep Theorem 3 in mind in applications where the data is not fixed, and we need to be able to deal with new and incoming data, which is often the case.

Acknowledgments. The authors acknowledge financial support from the School of Physical and Mathematical Sciences and the Presidential Postdoctoral Fellowship programme at Nanyang Technological University.

References

1. Barachant, A., Bonnet, S., Congedo, M., Jutten, C.: Riemannian geometry applied to BCI classification. In: Vigneron, V., Zarzoso, V., Moreau, E., Gribonval, R., Vincent, E. (eds.) LVA/ICA 2010. LNCS, vol. 6365, pp. 629–636. Springer, Heidelberg (2010). https://doi.org/10.1007/978-3-642-15995-4_78
2. Berg, C., Christensen, J.P.R., Ressel, P.: Harmonic Analysis on Semigroups, Graduate Texts in Mathematics, vol. 100. Springer, New York (1984). https://doi.org/10.1007/978-1-4612-1128-0
3. Borovitskiy, V., Terenin, A., Mostowsky, P., Deisenroth, M.: Matérn Gaussian processes on Riemannian manifolds. In: Larochelle, H., Ranzato, M., Hadsell, R., Balcan, M., Lin, H. (eds.) Advances in Neural Information Processing Systems, vol. 33, pp. 12426–12437. Curran Associates, Inc. (2020)
4. Bringmann, K., Folsom, A., Milas, A.: Asymptotic behavior of partial and false theta functions arising from Jacobi forms and regularized characters. J. Math. Phys. **58**(1), 011702 (2017). https://doi.org/10.1063/1.4973634
5. Calinon, S.: Gaussians on Riemannian manifolds: applications for robot learning and adaptive control. IEEE Rob. Autom. Maga. **27**(2), 33–45 (2020). https://doi.org/10.1109/MRA.2020.2980548
6. Carneiro, E., Littmann, F.: Bandlimited approximations to the truncated Gaussian and applications. Constr. Approx. **38**(1), 19–57 (2013). https://doi.org/10.1007/s00365-012-9177-8
7. Cristianini, N., Ricci, E.: Support Vector Machines, pp. 2170–2174. Springer, New York (2016). https://doi.org/10.1007/978-0-387-77242-4
8. Feragen, A., Hauberg, S.: Open problem: kernel methods on manifolds and metric spaces. What is the probability of a positive definite geodesic exponential kernel? In: Feldman, V., Rakhlin, A., Shamir, O. (eds.) 29th Annual Conference on Learning Theory. Proceedings of Machine Learning Research, vol. 49, pp. 1647–1650. PMLR, Columbia University (2016). https://proceedings.mlr.press/v49/feragen16.html

9. Feragen, A., Lauze, F., Hauberg, S.: Geodesic exponential kernels: when curvature and linearity conflict. In: 2015 IEEE Conference on Computer Vision and Pattern Recognition (CVPR), pp. 3032–3042 (2015). https://doi.org/10.1109/CVPR.2015.7298922

10. Gneiting, T.: Strictly and non-strictly positive definite functions on spheres. Bernoulli **19**(4), 1327–1349 (2013). https://doi.org/10.3150/12-BEJSP06

11. Gonon, L., Grigoryeva, L., Ortega, J.P.: Reservoir kernels and Volterra series. ArXiv Preprint (2022)

12. Grigoryeva, L., Ortega, J.P.: Dimension reduction in recurrent networks by canonicalization. J. Geom. Mech. **13**(4), 647–677 (2021). https://doi.org/10.3934/jgm.2021028

13. Jaquier, N., Rozo, L.D., Calinon, S., Bürger, M.: Bayesian optimization meets Riemannian manifolds in robot learning. In: Kaelbling, L.P., Kragic, D., Sugiura, K. (eds.) 3rd Annual Conference on Robot Learning, CoRL 2019, Osaka, Japan, 30 October–1 November 2019, Proceedings. Proceedings of Machine Learning Research, vol. 100, pp. 233–246. PMLR (2019). http://proceedings.mlr.press/v100/jaquier20a.html

14. Jayasumana, S., Hartley, R., Salzmann, M., Li, H., Harandi, M.: Kernel methods on Riemannian manifolds with Gaussian RBF kernels. IEEE Trans. Pattern Anal. Mach. Intell. **37**(12), 2464–2477 (2015). https://doi.org/10.1109/TPAMI.2015.2414422

15. Romeny, B.M.H.: Geometry-Driven Diffusion in Computer Vision. Springer, Heidelberg (2013). google-Books-ID: Fr2rCAAAQBAJ

16. Salvi, C., Cass, T., Foster, J., Lyons, T., Yang, W.: The Signature Kernel is the solution of a Goursat PDE. SIAM J. Math. Data Sci. **3**(3), 873–899 (2021)

17. Schoenberg, I.J.: Metric spaces and positive definite functions. Trans. Am. Math. Soc. **44**(3), 522–536 (1938). http://www.jstor.org/stable/1989894

18. Schölkopf, B., Smola, A., Müller, K.R.: Nonlinear component analysis as a kernel eigenvalue problem. Neural Comput. **10**(5), 1299–1319 (1998). https://doi.org/10.1162/089976698300017467

19. Sra, S.: Positive definite matrices and the S-divergence. Proc. Am. Math. Soc. **144**(7), 2787–2797 (2016). https://www.jstor.org/stable/procamermathsoci.144.7.2787

20. Wood, A.T.A.: When is a truncated covariance function on the line a covariance function on the circle? Stat. Prob. Lett. **24**(2), 157–164 (1995). https://doi.org/10.1016/0167-7152(94)00162-2

21. Ye, K., Lim, L.H.: Schubert varieties and distances between subspaces of different dimensions. SIAM J. Matrix Anal. Appl. **37**(3), 1176–1197 (2016). https://doi.org/10.1137/15M1054201

22. Zhu, P., Knyazev, A.: Angles between subspaces and their tangents. J. Numer. Math. **21**(4), 325–340 (2013). https://doi.org/10.1515/jnum-2013-0013

Determinantal Expressions of Certain Integrals on Symmetric Spaces

Salem Said[1] and Cyrus Mostajeran[2(✉)]

[1] CNRS, Laboratoire Jean Kuntzmann (UMR 5224), Grenoble, France
[2] School of Physical and Mathematical Sciences, Nanyang Technological University,
Singapore 637371, Singapore
cyrus.mostajeran@gmail.com

Abstract. The integral of a function f defined on a symmetric space $M \simeq G/K$ may be expressed in the form of a determinant (or Pfaffian), when f is K-invariant and, in a certain sense, a tensor power of a positive function of a single variable. The paper presents a few examples of this idea and discusses future extensions. Specifically, the examples involve symmetric cones, Grassmann manifolds, and classical domains.

Keywords: symmetric space · matrix factorisation · random matrices

1 Introduction

Riemannian symmetric spaces were classified by É. Cartan, back in the 1920 s. A comprehensive account of this classification may be found in the monograph [1]. In the 1960 s, a classification of quantum symmetries led Dyson to introduce three kinds of random matrix ensembles, orthogonal, unitary, and symplectic [2]. These three kinds of ensembles are closely related to the symmetric spaces known as symmetric cones, and also to their compact duals, which provide for so-called circular ensembles. More recently, Dyson's classification of quantum symmetries has been extended to free fermionic systems. It turned out that this extended classification is in on-to-one correspondance with Cartan's old classification of symmetric spaces [3]. This correspondance has motivated the notion that the relationship between random matrices and symmetric spaces extends well beyond symmetric cones, and is of a general nature (for example [4] or [5,6]).

The present submission has a modest objective. It is to show how the integral of a function f, defined on a symmetric space $M \simeq G/K$, can be expressed in the form of a determinant or Pfaffan, when f is K-invariant and satisfies an additional hypothesis, formulated in Sect. 4 below. This is not carried out in a general setting, but through a non-exhaustive set of examples, including symmetric cones, Grassmann manifolds, classical domains, and their duals (for the case of compact Lie groups, yet another example of symmetric spaces, see [7]).

The determinantal expressions obtained here, although elementary, are an analytic pre-requisite to developing the *random matrix theory of Riemannian symmetric spaces*. This long-term goal is the motivation behind the present work.

F. Nielsen and F. Barbaresco (Eds.): GSI 2023, LNCS 14071, pp. 436–443, 2023.
https://doi.org/10.1007/978-3-031-38271-0_43

Unfortunately, due to limited space, no proofs are provided for statements made in the following. These will be given in an upcoming extended version.

2   Integral Formulas

Let M be a Riemannian symmetric space, given by the symmetric pair (G, K). Write $\mathfrak{g} = \mathfrak{k} + \mathfrak{p}$ the corresponding Cartan decomposition, and let $\mathfrak{a}$ be a maximal abelian subspace of $\mathfrak{p}$. Then, denote by Δ a set of positive reduced roots on $\mathfrak{a}$ [1].

Assume that $\mathfrak{g} = \mathfrak{z}(\mathfrak{g}) + \mathfrak{g}_{ss}$ where $\mathfrak{z}(\mathfrak{g})$ is the centre of $\mathfrak{g}$ and $\mathfrak{g}_{ss}$ is semisimple and non-compact ($\mathfrak{g}_{ss}$ is a real Lie algebra). The Riemannian exponential Exp maps $\mathfrak{a}$ isometrically onto a totally flat submanifold of M, and any $x \in M$ is of the form $x = k \cdot \mathrm{Exp}(a)$ where $k \in K$ and $a \in \mathfrak{a}$.

Let $f : M \to \mathbb{R}$ be a K-invariant function, $f(k \cdot x) = f(x)$ for $k \in K$ and $x \in M$. There is no ambiguity in writing $f(x) = f(a)$ where $x = k \cdot \mathrm{Exp}(a)$. With this notation, there exists a constant C_M such that [1]

$$\int_M f(x)\mathrm{vol}(dx) = C_M \int_{\mathfrak{a}} f(a) \prod_{\lambda \in \Delta} \sinh^{m_\lambda} |\lambda(a)| \, da \tag{1}$$

where da is the Lebesgue measure on $\mathfrak{a}$.

The dual $\hat{M}$ of M is a symmetric space given by the symmetric pair (U, K), where U is a compact Lie group, with the Cartan decomposition $\mathfrak{u} = \mathfrak{k} + i\mathfrak{p}$ ($i = \sqrt{-1}$). Now, Exp maps $i\mathfrak{a}$ onto a torus T which is totally flat in $\hat{M}$, and any point $x \in \hat{M}$ is of the form $x = k \cdot \mathrm{Exp}(ia)$ where $k \in K$ and $a \in \mathfrak{a}$.

If $f : \hat{M} \to \mathbb{R}$ is K-invariant, there is no ambiguity in writing $f(x) = f(t)$ where $x = k \cdot t$, $t = \mathrm{Exp}(ia)$. In this notation [1],

$$\int_{\hat{M}} f(x)\mathrm{vol}(dx) = C_M \int_T f(t) \prod_{\lambda \in \Delta} \sin^{m_\lambda} |\lambda(t)| \, dt \tag{2}$$

where dt is the Haar measure on T. Here, $\sin|\lambda(t)| = \sin|\lambda(a)|$ where $t = \mathrm{Exp}(ia)$, and this does not depend on the choice of a.

3   Determinantal Expressions

Let μ be a positive measure on a real interval I. Consider the multiple integrals,

$$z_\beta(\mu) = \frac{1}{N!} \int_I \cdots \int_I |V(u_1, \ldots, u_N)|^\beta \, \mu(du_1) \ldots \mu(du_N) \tag{3}$$

where V denotes the Vandermonde determinant and $\beta = 1, 2$ or 4. Consider also the following bilinear forms,

$$(h, g)_{(\mu, 1)} = \int_I \int_I (h(u)\varepsilon(u - v)g(v)) \, \mu(du)\mu(dv) \tag{4}$$

$$(h, g)_{(\mu,2)} = \int_I h(u)g(u)\, \mu(du) \tag{5}$$

$$(h, g)_{(\mu,4)} = \int_I (h(u)g'(u) - g(u)h'(u))\, \mu(du) \tag{6}$$

Here, ε denotes the unit step function and the prime denotes the derivative. In the following proposition, det denotes the determinant and pf the Pfaffian.

Proposition 1. *The following hold for any probability measure μ as above.*
(a) if N is even,

$$z_1(\mu) = \mathrm{pf}\left\{ \left(u^k, u^\ell\right)_{(\mu,1)} \right\}_{k,\ell=0}^{N-1} \tag{7}$$

(b) on the other hand, if N is odd,

$$z_1(\mu) = \mathrm{pf}\left\{ \begin{matrix} \left(u^k, u^\ell\right)_{(\mu,1)} & \left(1, u^k\right)_{(\mu,2)} \\ -\left(u^\ell, 1\right)_{(\mu,2)} & 0 \end{matrix} \right\}_{k,\ell=0}^{N-1} \tag{8}$$

(c) moreover,

$$z_2(\mu) = \det\left\{ \left(u^k, u^\ell\right)_{(\mu,2)} \right\}_{k,\ell=0}^{N-1} \tag{9}$$

(d) and, finally,

$$z_4(\mu) = \mathrm{pf}\left\{ \left(u^k, u^\ell\right)_{(\mu,4)} \right\}_{k,\ell=0}^{2N-1} \tag{10}$$

On the other hand, if μ is a probability measure on the unit circle S^1, and

$$z_\beta(\mu) = \frac{1}{N!} \int_{S^1} \cdots \int_{S^1} |V(u_1, \ldots, u_N)|^\beta\, \mu(du_1) \ldots \mu(du_N) \tag{11}$$

consider the bilinear form

$$(h, g)_{(\mu,1)} = \int_0^{2\pi} \int_0^{2\pi} (h(e^{ix})\varepsilon(x - y)g(e^{iy}))\, \tilde{\mu}(dx)\tilde{\mu}(dy) \tag{12}$$

where $\tilde{\mu}$ is the pullback of the measure μ through the map that takes x to e^{ix}, and let $(h, g)_{(\mu,2)}$ and $(h, g)_{(\mu,4)}$ be given as in (5) and (6), with integrals over S^1 instead of I.

Proposition 2. *The following hold for any probability measure μ on S^1.*
(a) if N is even,

$$z_1(\mu) = (-i)^{N(N-1)/2} \times \mathrm{pf}\left\{ (g_k, g_\ell)_{(\mu,1)} \right\}_{k,\ell=0}^{N-1} \tag{13}$$

where $g_k(u) = u^{k-(N-1)/2}$.
(b) on the other hand, if N is odd,

$$z_1(\mu) = (-i)^{N(N-1)/2} \times \mathrm{pf}\left\{ \begin{matrix} (g_k, g_\ell)_{(\mu,1)} & (1, g_k)_{(\mu,2)} \\ -(g_\ell, 1)_{(\mu,2)} & 0 \end{matrix} \right\}_{k,\ell=0}^{N-1} \tag{14}$$

with the same definition of $g_k(u)$.

(c) moreover,

$$z_2(\mu) = \det\left\{\left(u^k, u^{-\ell}\right)_{(\mu,2)}\right\}_{k,\ell=0}^{N-1} \tag{15}$$

(d) and, finally,

$$z_4(\mu) = \text{pf}\left\{\left(h_k, h_\ell\right)_{(\mu,4)}\right\}_{k,\ell=0}^{2N-1} \tag{16}$$

where $h_k(u) = u^{k-(N-1)}$.

Both of the above Propositions 1 and 2 are directly based on [8].

4 Main Idea

An additional hypothesis is made on the function $f(a)$ (in (1)) or $f(t)$ (in (2)): that there exists a natural orthonormal basis $(e_j; j = 1, \ldots, r)$ of $\mathfrak{a}$, such that

$$f(a) = \prod_{j=1}^{r} w(a_j) \qquad f(t) = \prod_{j=1}^{r} w(t_j) \tag{17}$$

where w is a positive function of a single variable, and a_j are the components of a in the basis $(e_j; j = 1, \ldots, r)$, while $t_j = \text{Exp}(ia_j e_j)$. In this sense, it may be said that f is the r-th tensor power of w.

What is meant by *natural* is that (17) will imply that the integral (1) or (2) can be transformed into a multiple integral of the form (3) or (11), respectively. Thus, in the case of (1), there exists a measure μ on an interval I, which satisfies

$$\int_M f(x)\text{vol}(dx) = \tilde{C}_M \times z_\beta(\mu) \qquad (\tilde{C}_M \text{ is a new constant})$$

and, in the case of (2), there is a measure μ on S^1, which yields a similar identity. It should be noted that this measure μ will depend on the function w from (17).

Then, Propositions 1 and 2 provide a determinantal (or Pfaffian) expression of the initial integral on the symmetric space M or $\hat{M}$.

At present, this is not a theorem, but a mere idea or observation, supported by the examples in the following section.

5 Examples

5.1 Symmetric Cones

Consider the following Lie groups (in the usual notation, as found in [1]).

β	G_β	U_β	K_β
1	$GL_N(\mathbb{R})$	$U(N)$	$O(N)$
2	$GL_N(\mathbb{C})$	$U(N) \times U(N)$	$U(N)$
4	$GL_N(\mathbb{H})$	$U(2N)$	$Sp(N)$

Then, $M_\beta \simeq G_\beta / K_\beta$ is a Riemannian symmetric space, with dual $\hat{M}_\beta = U_\beta / K_\beta$. In fact, M_β is realised as a so-called symmetric cone: the cone of positive-definite real, complex, or quaternion matrices (according to the value of $\beta = 1, 2$ or 4).

Each $x \in M_\beta$ is of the form $k \lambda k^\dagger$ where $k \in K_\beta$ and λ is a positive diagonal matrix ($\dagger$ denotes the transpose, conjugate-transpose, or quaternion conjugate-transpose). If $f : M_\beta \to \mathbb{R}$ is K_β-invariant, and can be written $f(x) = \prod w(\lambda_j)$,

$$\int_{M_\beta} f(x)\mathrm{vol}(dx) = \tilde{C}_\beta \times z_\beta(\mu) \tag{18}$$

where $\mu(du) = (w(u)u^{-N_\beta})du$, with $N_\beta = (\beta/2)(N-1) + 1$, on the interval $I = (0, \infty)$. The constant $\tilde{C}_\beta$ is known explicitly, but this is irrelevant at present.

The dual $\hat{M}_\beta$ can be realised as the space of symmetric unitary matrices ($\beta = 1$), of unitary matrices ($\beta = 2$), or of antisymmetric unitary matrices with double dimension $2N$, ($\beta = 4$).

If $\beta = 1, 2$, then $x \in \hat{M}_\beta$ is of the form $ke^{i\theta} k^\dagger$ where $k \in K_\beta$ and θ is real diagonal. However, if $\beta = 4$, there is a somewhat different matrix factorisation,

$$x = k \begin{pmatrix} & -e^{i\theta} \\ e^{i\theta} & \end{pmatrix} k^{\mathrm{tr}} \qquad \text{(tr denotes the transpose)} \tag{19}$$

where $k \in Sp(N)$ is considered as a $2N \times 2N$ complex matrix (rather than a $N \times N$ quaternion matrix). If $f : \hat{M}_\beta \to \mathbb{R}$ is K_β-invariant, $f(x) = \prod w(e^{i\theta_j})$,

$$\int_{\hat{M}_\beta} f(x)\mathrm{vol}(dx) = \tilde{C}_\beta \times z_\beta(\mu) \tag{20}$$

where $\mu(du) = w(u)|du|$ on the unit circle S^1 ($|du| = d\varphi$ if $u = e^{i\varphi}$).

Remark: in many textbooks, $\hat{M}_1$ is realised as the space of real structures on $\mathbb{C}^N$, and $\hat{M}_4$ as the space of quaternion structures on $\mathbb{C}^{2N}$. The alternative realisations proposed here seem less well-known, but more concrete, so to speak.

5.2 Grassmann Manifolds

Consider the following Lie groups (again, for the notation, see [1]).

β	G_β	U_β	K_β
1	$O(p,q)$	$O(p+q)$	$O(p) \times O(q)$
2	$U(p,q)$	$U(p+q)$	$U(p) \times U(q)$
4	$Sp(p,q)$	$Sp(p+q)$	$Sp(p) \times Sp(q)$

Then, $M_\beta \simeq G_\beta / K_\beta$ is a Riemannian symmetric space, with dual $\hat{M}_\beta = U_\beta / K_\beta$. The M_β may be realised as follows [9] ($\mathbb{K} = \mathbb{R}, \mathbb{C}$ or $\mathbb{H}$, according to β),

$$M_\beta = \{x : x \text{ is a } p\text{-dimensional and space-like subspace of } \mathbb{K}^{p+q}\} \tag{21}$$

Here, x is space-like if $|\xi_p|^2 - |\xi_q|^2 > 0$ for all $\xi \in x$ with $\xi = (\xi_p, \xi_q)$, where $|\cdot|$ denotes the standard Euclidean norm on $\mathbb{K}^p$ or $\mathbb{K}^q$. Moreover, for each $x \in M_\beta$, $x = k(x_\tau)$ where $k \in K_\beta$ and $x_\tau \in M_\beta$ is spanned by the vectors

$$\cosh(\tau_j)\xi_j + \sinh(\tau_j)\xi_{p+j} \qquad j = 1, \ldots, p$$

with $(\xi_k; k = 1, \ldots, p+q)$ the canonical basis of $\mathbb{K}^{p+q}$, and $(\tau_j; j = 1, \ldots, p)$ real ($p \leq q$ throughout this paragraph).

If $f : M_\beta \to \mathbb{R}$ is K_β-invariant, $f(x) = f(\tau)$, the right-hand side of (1) reads (the positive reduced roots can be found in [10])

$$C_\beta \int_{\mathbb{R}^p} f(\tau) \prod_{j=1}^{p} \sinh^{\beta(q-p)} |\tau_j| \sinh^{\beta-1} |2\tau_j| \prod_{i<j} |\cosh(2\tau_i) - \cosh(2\tau_j)|^\beta \, d\tau \quad (22)$$

and this can be transformed into the form (3), by introducing $u_j = \cosh(2\tau_j)$. This will reappear, with $\beta = 2$ and $p = q$, in the following paragraph.

Now, the duals $\hat{M}_\beta$ are real, complex, or quaternion Grassmann manifolds,

$$\hat{M}_\beta = \{x : x \text{ is a } p\text{-dimensional subspace of } \mathbb{K}^{p+q}\} \qquad (23)$$

For each $x \in \hat{M}_\beta$, $x = k(x_\theta)$ where $k \in K_\beta$ and x_θ is spanned by the vectors

$$\cos(\theta_j)\xi_j + \sin(\theta_j)\xi_{p+j} \qquad j = 1, \ldots, p$$

with $(\theta_j; j = 1, \ldots, p)$ real.

If $f : \hat{M}_\beta \to \mathbb{R}$ is K_β-invariant, $f(x) = f(\theta)$, the right-hand side of (2) reads

$$C_\beta \int_{(0,\pi)^p} f(\theta) \prod_{j=1}^{p} \sin^{\beta(q-p)} |\theta_j| \sin^{\beta-1} |2\theta_j| \prod_{i<j} |\cos(2\theta_i) - \cos(2\theta_j)|^\beta \, d\theta \quad (24)$$

which can be transformed into the form (11), by introducing $u_j = \cos(2\theta_j)$. In [4], this is used to recover the Jacobi ensembles of random matrix theory. **Remark:** the angles θ_j may be taken in the interval $(-\pi/2, \pi/2)$ instead of $(0, \pi)$. In this case, $|\theta_j|$ are the principal angles between x_θ and the subspace x_o spanned by $(\xi_j; j = 1, \ldots, p)$. By analogy, it is natural to think of $|\tau_j|$ as the 'principal boosts' (using the language of special relativity) between x_τ and x_o.

5.3   Classical Domains

Consider, finally, the following Lie groups (again, for the notation, see [1]).

β	G_β	U_β	K_β
1	$Sp(N, \mathbb{R})$	$Sp(N)$	$U(N)$
2	$U(N, N)$	$U(2N)$	$U(N) \times U(N)$
4	$O^*(4N)$	$O(4N)$	$U(2N)$

Then, $M_\beta \simeq G_\beta/K_\beta$ is a Riemannian symmetric space, with dual $\hat{M}_\beta = U_\beta/K_\beta$. The M_β are realised as classical domains, whose elements are $N \times N$ complex matrices (if $\beta = 1, 2$) or $2N \times 2N$ complex matrices (if $\beta = 4$), with operator norm < 1, and which are in addition symmetric ($\beta = 1$) or antisymmetric ($\beta = 4$).

If $\beta = 1, 2$, then any $x \in M_\beta$ may be written

$$x = k_1 (\tanh(\lambda)) k_2 \tag{25}$$

where k_1 and k_2 are unitary ($k_2 = k_1^{\mathrm{tr}}$, in case $\beta = 1$), and λ is real diagonal. However, if $\beta = 4$,

$$x = k \begin{pmatrix} & -\tanh(\lambda) \\ \tanh(\lambda) & \end{pmatrix} k^{\mathrm{tr}} \tag{26}$$

where k is $2N \times 2N$ unitary. If $f : M_\beta \to \mathbb{R}$ is K_β-invariant, and $f(x) = \prod w(\lambda_j)$,

$$\int_{M_\beta} f(x)\mathrm{vol}(dx) = \tilde{C}_\beta \int_{\mathbb{R}^N} \prod_{j=1}^N w(\lambda_j) \sinh|2\lambda_j| \prod_{i<j} |\cosh(2\lambda_i) - \cosh(2\lambda_j)|^\beta \, d\lambda$$

After introducing $u_j = \cosh(2\lambda_j)$, this immediately becomes

$$\int_{M_\beta} f(x)\mathrm{vol}(dx) = \tilde{C}_\beta \times z_\beta(\mu) \tag{27}$$

where $\mu(du) = w(\mathrm{acosh}(u)/2)du$ on the interval $I = (1, \infty)$.

Remark: the domain M_2 is sometimes called the Siegel disk. As an application of (27), consider a random $x \in M_2$ with a Gaussian probability density function

$$p(x|\bar{x}, \sigma) = (Z(\sigma))^{-1} \exp\left[-\frac{d^2(x, \bar{x})}{2\sigma^2} \right] \tag{28}$$

with respect to $\mathrm{vol}(dx)$, where $d(x, \bar{x})$ denotes Riemannian distance and $\sigma > 0$. Then, following the arguments in [6], (27) can be used to obtain

$$Z(\sigma) = \tilde{C}_2 \times \det \{m_{k+\ell}(\sigma)\}_{k,\ell=0}^{N-1} \quad m_j(\sigma) = \int_1^\infty \exp\left(-\mathrm{acosh}^2(u)/8\sigma^2\right) u^j \, du$$

The integrals $m_j(\sigma)$ are quite easy to compute, and one is then left with a determinantal expression of $Z(\sigma)$. The starting point to the study of the random matrix x is the following observation. If x is written as in (25) and $u_j = \cosh(2\lambda_j)$, then the random subset $\{u_j; j = 1, \ldots, N\}$ of $I = (1, \infty)$ is a determinantal point process (see [11]). By writing down its kernel function, one may begin to investigate in detail many of its statistical properties, including asymptotic ones, such as the asymptotic density of the (u_j), or the asymptotic distribution of their maximum, in the limit where $N \to \infty$ (of course, with suitable re-scaling).

6   Future Directions

The present submission developed determinantal expressions for integrals on symmetric spaces on a case-by-case basis, only through a non-exhaustive set of examples. Future work should develop these expressions in a fully general way, by transforming (1) and (2) into (3) or (11), for any system of reduced roots.

The long-term goal is to understand the *random matrix theory of symmetric spaces*. One aspect of this is to understand the asymptotic properties of a joint probability density (in the notation of (1))

$$f(a) \prod_{\lambda \in \Delta} \sinh^{m_\lambda} |\lambda(a)| \, da$$

and analyse how these depend on the set of positive reduced roots Δ. It is worth mentioning that, in previous work [6], it was seen that a kind of universality holds, where different root systems lead to the same asymptotic properties.

Random matrix theory (in its classical realm of orthogonal, unitary, and symplectic ensembles) has so many connections to physics, combinatorics, and complex systems in general. A further important direction is to develop such connections for the random matrix theory of symmetric spaces.

References

1. Helgason, S.: Differential geometry and symmetric spaces. Academic Press (1962)
2. Dyson, F.: The threefold way. Algebraic structures of symmetry groups and ensembles in quantum mechanics. J. Mathemat. Phys. **3**(6), 1199–1215 (1962)
3. Zirnbauer, M.: Symmetry classes. In: Akemann, G., Baik, J., Di Francesco, P. (eds.) The Oxford Handbook of Random Matrix Theory (2018)
4. Edelman, A., Jeong, S.: On the Cartan decomposition for classical random matrix ensembles. J. Mathem. Phys. **63**(6) (2022)
5. Santilli, L., Tierz, M.: Riemannian Gaussian distributions, random matrix ensembles, and diffusion kernels. Nuclear Phys. B **973** (2021)
6. Said, S., Heuveline, S., Mostajeran, C.: Riemannian statistics meets random matrix theory: towards learning from high-dimensional covariance matrices. IEEE Trans. Inf. Theory **69**(1), 472–481 (2023)
7. Meckes, E.S.: The random matrix theory of the classical compact groups. Cambridge University Press (2019)
8. Mehta, M.L.: Random Matrices (Third Edition). Elsevier (2004)
9. Huang, Y.: A uniform description of Riemannian symmetric spaces as Grassmannians using magic square. PhD Thesis, The Chinese University of Hong Kong (2007)
10. Sakai, T.: On cut loci of compact symmetric spaces. Hokkaido Math. J. **6**, 136–161 (1977)
11. Johansson, K.: Random matrices and determinantal processes. arXiv:match-ph/0510038 (2005)

Projective Wishart Distributions

Emmanuel Chevallier[✉]

Aix Marseille Univ, CNRS, Centrale Marseille, Institut Fresnel,
Marseille, France
emmanuel.chevallier@univ-amu.fr

Abstract. We are interested in the distribution of Wishart samples after forgetting their scaling factors. We call such a distribution a projective Wishart distribution. We show that projective Wishart distributions have strong links with the affine-invariant geometry of symmetric positive definite matrices in the real case or Hermitian positive definite matrices in the complex case. First, the Fréchet mean of a projective Wishart distribution is the covariance parameter, up to a scaling factor, of the corresponding Wishart distribution. Second, in the case of 2×2 matrices, the densities have simple expressions in term of the affine-invariant distance.

Keywords: Wishart distributions · positive definite matrices · hyperbolic spaces · Fréchet means

1 Introduction

A Wishart distribution is the law of the empirical second order moment of a set of i.i.d. Gaussian random vectors, see [7] for the original paper of Wishart, or [1,2] for a more modern presentation. These distributions are parametrized by the covariance Σ of the Gaussian and the number n of i.i.d. random vectors. It is well known that when a linear map A acts on the random vectors, a Wishart distribution of parameter Σ is turned into a Wishart distribution of parameter $A\Sigma A^*$.

This equivariance property draws a link between Wishart distributions and the affine-invariant geometry of symmetric positive definite and Hermitian positive definite matrices. We are interested here in explicit links between Wishart distributions and affine invariant distance functions.

The link with affine-invariant geometry plays a role in analysis on symmetric cones, see [4,5]. In [4], the author shows how the Wishart distribution and its normalizing constant relates to certain integrals on symmetric positive definite matrices. These results rely mostly on the equivariance property itself and the relation between Wishart distributions and the distance function is indirect.

Authors of [6] have build an estimator of the parameter Σ based on empirical Fréchet means, defined using an affine-invariant distance. In order to obtain a consistent estimator, they need to introduce a multiplicative correcting factor.

F. Nielsen and F. Barbaresco (Eds.): GSI 2023, LNCS 14071, pp. 444–451, 2023.
https://doi.org/10.1007/978-3-031-38271-0_44

The underlying reason is that the Euclidean mean $n\Sigma$ and the Fréchet mean of a Wishart distribution do not coincide.

The link between affine-invariant distances and Wishart distributions appears more clearly when we consider second order moments up to a scaling factor. To do so, we renormalize the second order moments by their determinant and call the corresponding distribution a projective Wishart distribution. These distributions are formally introduced in Sect. 2. In Sect. 3, we describe the action of invertible linear maps of second order moments and the associated distance function. Section 4 contains the main results. First, the Fréchet mean of a projective Wishart distribution is the covariance parameter, up to a scaling factor, of the corresponding Wishart distribution. Second, in the case of 2×2 matrices, the densities have simple expressions in term of the affine-invariant distance.

2 Wishart Distributions

Let $\mathcal{N}_d(0, \Sigma)$ denote the Gaussian distribution on column vectors of K^d with $K = \mathbb{R}$ or $K = \mathbb{C}$, whose mean is 0 and covariance is Σ. Note $\mathcal{P}_d(K)$ the set of $d \times d$ symmetric positive definite matrices when $K = \mathbb{R}$ and the set of Hermitian positive definite matrices when $K = \mathbb{C}$. In the rest of the paper, $\Sigma \in \mathcal{P}_d(K)$. Let $(Y_i)_{i \in \mathbb{N}}$ be a sequence of i.i.d. random variables with

$$Y_i \sim \mathcal{N}_d(0, \Sigma).$$

For $n \in \mathbb{N}$, consider the random variable

$$X_n = \sum_{i=1}^{i=n} Y_i Y_i^T.$$

The Wishart distribution with parameters Σ and n, noted $\mathcal{W}(\Sigma, n)$, is defined as the law of the random variable X_n. When $n \geq d$, $X_n \in \mathcal{P}_d(K)$ almost surely and the distribution $\mathcal{W}(\Sigma, n)$ is supported on $\mathcal{P}_d(K)$.

Our aim is to study covariance matrices up to a scaling factor. Hence, for $X \in \mathcal{P}_d(K)$, consider the equivalence classes

$$\bar{X} = \{\alpha X, \alpha \in \mathbb{R}_{>0}\},$$

and denote $\mathcal{S} = \mathcal{P}_d(K)/\mathbb{R}_{>0}$ the set of equivalence classes. In the rest of the paper X denotes an element of $\mathcal{P}_d(K)$ and x an element of $\mathcal{S}$. There exists several standard parametrizations of $\mathcal{S}$, such as matrices of $\mathcal{P}_d(K)$ of fixed trace or matrices of fixed determinant. In the rest of the paper, we identify $\mathcal{S}$ with matrices of determinant 1,

$$\mathcal{S} \sim \{X \in \mathcal{P}_d(K), \det(X) = 1\}.$$

Note π the canonical projection

$$\pi(X) = \frac{1}{\det(X)^d} X.$$

We are interested in the law $PW(\Sigma, n)$ of $\pi(X)$ when X follows a Wishart distribution $W(\Sigma, n)$. Distributions PW will be called projective Wishart distributions.

In the next sections, we will study properties of such distributions. In particular, we will show that $\pi(\Sigma)$ is the Frechet mean of $PW(\Sigma, n)$ for the affine-invariant metric on $\mathcal{S}$. We will also show that when $d = 2$, the density of $PW(\Sigma, n)$ evaluated in x has a simple expression depending on the distance between x and $\bar{\Sigma}$.

3   The Geometry of $\mathcal{S}$

The space $\mathcal{P}_d(K)$ is an open cone of the vector space of (Hermitian-)symmetric matrices. This cone is invariant by the following action of invertible matrices

$$G \cdot X = GXG^*,$$

where $G \in GL_d(K)$, $X \in \mathcal{P}_d(K)$, and where A^* refers to the transpose or the conjugate transpose of A.

Let denote $SL_d(K)$ the subset of $GL_d(K)$ of matrices with determinant 1. The action of $G \in SL_d(K)$ preserves the determinant:

$$\det(G \cdot X) = \det(X).$$

Hence $SL_d(K)$ preserves $\mathcal{S}$. The so called affine-invariant on $\mathcal{S}$ is defined up to a multiplicative constant by

$$d(x, y) \propto \| \log(x^{-\frac{1}{2}} y x^{-\frac{1}{2}}) \|, \tag{1}$$

see [3,8,9].

This distance $d(.,.)$ will be used to define the Fréchet mean. In the particular case $d = 2$, $\mathcal{P}_d(K)$ endowed with the distance $d(.,.)$ becomes a hyperbolic space of dimension 2 when $K = \mathbb{R}$ or 3 when $K = \mathbb{C}$. The property of isotropy of hyperbolic spaces enable to express the density of $PW(\Sigma, n)$ as a function of $d(.,.)$.

4   Properties of Projective Wishart Distributions

In order to prove the results on Fréchet mean and on densities, we need to state two lemmas. The first lemma states that the projective Wishart distribution $PW(\Sigma, n)$ is invariant by a certain subgroup, noted H_Σ, of isometries of $d(.,.)$ which fix Σ. The second lemma states that $\bar{\Sigma}$ is the only fixed point of H_Σ in $\mathcal{S}$.

Note H the subset of matrices $R \in SL_d(K)$ with $RR^* = I$. When $K = \mathbb{R}$, H is the special orthogonal group SO_d and when $K = \mathbb{C}$, H is the special unitary group SU_d. Define $H_\Sigma \subset SL_d(K)$ as the set of matrices

$$R_\Sigma = \Sigma^{\frac{1}{2}} R \Sigma^{-\frac{1}{2}}$$

with $R \in H$. Matrices of H_Σ leave Σ stable. Note $R_\Sigma \cdot PW(\Sigma, n)$ the action of R_Σ on the distribution $PW(\Sigma, n)$:

$$R_\Sigma \cdot PW(\Sigma, n)(R_\Sigma \cdot A) = PW(\Sigma, n)(A)$$

for all measurable subset A of $\mathcal{S}$. As R_Σ leaves $\mathcal{S}$ stable, $R_\Sigma \cdot PW(\Sigma, n)$ is still a distribution on $\mathcal{S}$. State the first lemma.

Lemma 1. $\forall R_\Sigma \in H_\Sigma$,

$$R_\Sigma \cdot PW(\Sigma, n) = PW(\Sigma, n).$$

Proof. The invariance for Wishart distributions, namely $R_\Sigma \cdot W(\Sigma, n) = W(\Sigma, n)$, is well known and can easily be checked from their definition, see [1,2]. Projective Wishart distributions are images of Wishart distributions by the projection π. Since the projection π commutes the action of $Gl_d(K)$, the invariance also holds for projective Wishart distributions.

Lemma 2. $\bar{\Sigma} = \pi(\Sigma)$ *is the only fixed point of* H_Σ *in* $\mathcal{S}$.

Proof. Consider first a matrix $X \in \mathcal{P}_d(K)$ fixed by H. Since

$$RXR^* = X \Leftrightarrow RX = XR,$$

the matrix X preserves the set of fixed points of all matrices $R \in H$. Furthermore, for all $u \in K^d$, there exist R_1 and $R_2 \in H$ such that if $R_1(v) = R_2(v) = v$ then $v \in Ku$ (if $K = \mathbb{R}$ and if d is even, there is no $R \in H$ such that Ku is the set of fixed points). Now $R_i(X(u)) = X(R_i(u)) = X(u)$, $i = 1, 2$, hence $X(u) \in Ku$. It follows that X leaves all lines stable, which in turn implies that $X = cI$ for some $c \in \mathbb{R}_{>0}$.

If X is fixed by H_Σ then $\Sigma^{-\frac{1}{2}} \cdot X$ if fixed by H. Hence there is $c \in \mathbb{R}_{>0}$ such that $\Sigma^{-\frac{1}{2}} \cdot X = cI$, and $X = c\Sigma$. This proves that $\bar{\Sigma}$ is the only fixed point in $\mathcal{S}$.

4.1 Fréchet Means

The Fréchet mean of a probability distribution μ on $\mathcal{S}$ can be defined as

$$F(\mu) = \operatorname{argmin}_x \int_{\mathcal{S}} d(x, y)^2 \mathrm{d}\mu(y)$$

[10,11].

Theorem 1. *Let* $\Sigma \in \mathcal{P}_d(K)$. *The Fréchet mean of* $PW(\Sigma, n)$ *is unique and equals to* $\bar{\Sigma}$:

$$F(PW(\Sigma, n)) = \frac{1}{\det \Sigma^d} \Sigma$$

Proof. Prove first the equivariance of the Fréchet mean with respect to isometries. Let φ be an isometry of $d(.,.)$. Let μ be any probability measure on $\mathcal{S}$. Note $\varphi_*\mu$ the pushforward of μ, defined as $\varphi_*\mu(A) = \mu(\varphi^{-1}(A))$. Since

$$\int_\mathcal{S} d(x,y)^2 d\mu(y) = \int_\mathcal{S} d(x,\varphi^{-1}(y))^2 d\varphi_*\mu(y) = \int_\mathcal{S} d(\varphi(x),y)^2 d\varphi_*\mu(y),$$

it can be checked that $\varphi(F(\mu)) = F(\varphi_*\mu)$. Using Lemma 1, the equivariance of the mean, and that $SL_d(K)$ is acting by isometries, we see that

$$\forall R_\Sigma \in H_\Sigma, \; R_\Sigma \cdot F(PW(\Sigma,n)) = F(R_\Sigma \cdot PW(\Sigma,n)) = F(PW(\Sigma,n))$$

from which we deduce with Lemma 2 that $F(PW(\Sigma,n)) \subset \{\bar{\Sigma}\}$: if the mean exists, it is $\bar{\Sigma}$. In order to show that a Fréchet mean exists, we need to show that

$$a(x) = \int_\mathcal{S} d(x,y)^2 d\nu(y)$$

is finite for at least one $x \in \mathcal{S}$. Consider the distribution $PW(I,n)$ and show that $a(I)$ is finite. An equivariance argument can then transfer the reasoning to $PW(\Sigma,n)$ and $a(\bar{\Sigma})$. Note $\lambda_i \in \mathbb{R}_+$ the eigenvalues of $X \in \mathcal{P}_d(K)$. A calculation shows that using the distance of Eq. (1),

$$d^2(\bar{X},I) \propto \sum_i \log^2\left(\frac{\lambda_i}{\prod_j \lambda_j^{\frac{1}{d}}}\right) = \left(\sum_i \log^2(\lambda_i) - \frac{1}{d}\log^2\left(\prod_i \lambda_i\right)\right).$$

From the marginal distribution of the λ_i when $X \sim W(I,n)$, see [1,12], we deduce that

$$a(I) \propto \int_{\mathbb{R}_+^d} \left(\sum_i \log^2(\lambda_i) - \log^2\left(\prod_i \lambda_i\right)\right)\left(\prod_i \lambda_i^{k_1}\right) e^{-\frac{1}{2}\sum_i \lambda_i} \prod_{i>j}|\lambda_i-\lambda_j|^{k_2} d\lambda,$$

where k_1, k_2 are positive constants depending on the case $K = \mathbb{R}$ or $K = \mathbb{C}$. Since the λ_i are positive, the exponential terms dominate the other terms and the integral converges.

4.2   Densities in the 2 × 2 Case

Let us start by defining densities on $\mathcal{S}$. It can be proved that up to a multiplicative factor, there exists a unique volume measure $\mathcal{S}$ invariant by the action of $SL_d(K)$. Note ν such a measure. The density of the probability distribution $PW(\Sigma,n)$ is defined as a function $f_{PW}(.;\Sigma,n) : \mathcal{S} \to \mathbb{R}$ such that for any measurable subset A of $\mathcal{S}$

$$PW(\Sigma,n)(A) = \int_A f_{PW}(x;\Sigma,n)d\nu(x).$$

To simplify notations, we will only write the parameters Σ and n when it is necessary. Since both $PW(\Sigma, n)$ and ν are invariant by the action of H_Σ, it can be checked that the density f is also invariant:

$$\text{for } \nu\text{-almost all } x \in \mathcal{S}, f_{PW}(R_\Sigma \cdot x) = f(x).$$

It is easy to prove that the density f_{PW} can be chosen continuous. In that case, the equality holds for all x.

Hence the density f_{PW} is constant on the orbits of the action of H_Σ. Since H_Σ is acting by isometries which fix $\bar\Sigma$, the orbits are contained in balls of center $\bar\Sigma$ of fixed radius. When $d = 2$, orbits of H_Σ are the full balls. Indeed, as mentioned in Sect. 3, in that case, $\mathcal{S}$ is an hyperbolic space of dimension 2 when $K = \mathbb{R}$ and an hyperbolic space of dimension 3 when $K = \mathbb{C}$. In these cases, $\mathcal{S}$ is not only a symmetric space but also an isotorpic space. We have the additional property that

$$d(x, \bar\Sigma) = d(y, \bar\Sigma) \implies \exists R_\Sigma \in H_\Sigma, R_\Sigma \cdot x = y.$$

Hence, we have the following theorem.

Theorem 2. *The density $f_{PW}(x; \Sigma, n)$ can be factored through a function $h_{\Sigma,n}$:* $\mathbb{R}_+ \to \mathbb{R}_+$,

$$f_{PW}(x; \Sigma, n) = h_{\Sigma,n}(d(x, \bar\Sigma)),$$

where $\bar\Sigma = \frac{1}{\det(\Sigma)^d}\Sigma$.

As we will see later in the explicit calculation of f_{PW}, $h_{\Sigma,n}$ does not depend on Σ. This follows from the transitivity of the action of $GL_d(K)$ on $\mathcal{P}_d(K)$ and the commutation relations $G \cdot W(\Sigma, n) = W(G \cdot \Sigma, n)$ and $G \cdot \pi(X) = \pi(G \cdot X)$.

To compute the density f_{PW} of $PW(\Sigma, n)$, we will integrate the density of $W(\Sigma, n)$ along fibers $\pi^{-1}(x)$ of the projection π. Although we introduced the measure ν on $\mathcal{S}$, we shall need a reference measure on the entire set $\mathcal{P}_d(K)$ to define the density of $W(\Sigma, n)$. Note first that the determinant of a matrix in $X \in \mathcal{P}_d(K)$ is always a positive real number since the eigenvalues of X are real and positive. Consider the following identification between $\mathcal{P}_d(K)$ and the cartesian product $\mathcal{S} \times \mathbb{R}$,

$$\theta : \mathcal{P}_d(K) \to \mathcal{S} \times \mathbb{R}$$
$$X \mapsto \left(\frac{1}{\det(X)^d}X, \log(\det(X))\right),$$

and define the measure ν_{tot} on $\mathcal{P}_d(K)$ as the product measure between ν and the Lebesgue measure on $\mathbb{R}$. It can be checked that ν_{tot} is invariant by the action of $GL_d(K)$.

Let f_W denote the density of the Wishart distribution with parameters Σ and n, with respect to ν_{tot}. From the definition of the projective Wishart distribution and the definition of the reference measure ν_{tot}, we have

$$\int_{A \subset \mathcal{S}} f_{PW} d\nu = \int_{\pi^{-1}(A)} f_W d\nu_{tot} = \int_{x \in A} \int_{\alpha \in \mathbb{R}} f_W(e^{\frac{\alpha}{d}}x) d\alpha d\nu,$$

where $d\alpha$ refers to the Lebesgue measure on $\mathbb{R}$. The factor $\frac{1}{d}$ arise from the d-linearity of the determinant. Hence, with the change of variable $\beta = e^{\frac{\alpha}{d}}$,

$$f_{PW}(x) = \int_{\mathbb{R}} f_W(e^{\frac{\alpha}{d}}x) d\alpha = \int_{\mathbb{R} > 0} f_W(\beta x) \frac{d}{\beta} d\beta.$$

Now, the Wishart density with respect to ν_{tot} is given up to a multiplicative constant by

$$f_W(X) \propto (\det X)^{\frac{kn}{2}} e^{-\frac{1}{2} \operatorname{tr}(\Sigma^{-1} X)},$$

where $k = 1$ when $K = \mathbb{R}$ and $k = 2$ when $K = \mathbb{C}$. The calculation in the real case can be found in [4], and the complex case is obtained by the same reasoning. Hence

$$f_{PW}(x) \propto \int (\beta^d)^{\frac{kn}{2}} e^{-\frac{\beta}{2} \operatorname{tr}(\Sigma^{-1} x)} \frac{d}{\beta} d\beta.$$

Set $\gamma = \frac{\beta}{2} \operatorname{tr}(\Sigma^{-1} x)$. The integral becomes

$$f_{PW}(x) \propto \left(\frac{2}{\operatorname{tr}(\Sigma^{-1} x)}\right)^{\frac{dkn}{2}} \int \gamma^{\frac{dkn}{2}} e^{-\gamma} d\frac{d\gamma}{\gamma} \propto \left(\frac{2}{\operatorname{tr}(\Sigma^{-1} x)}\right)^{\frac{dkn}{2}}. \qquad (2)$$

Note that the formula (2) is valid in for any dimension d. We have $\operatorname{tr}(\Sigma^{-1}x) = \det(\Sigma)^{-\frac{1}{d}} \operatorname{tr}(\bar{\Sigma}^{-\frac{1}{2}} x \bar{\Sigma}^{-\frac{1}{2}})$ and since the matrix $\bar{\Sigma}^{-\frac{1}{2}} x \bar{\Sigma}^{-\frac{1}{2}}$ is in S, there exists $R \in H$ such that

$$R\bar{\Sigma}^{-\frac{1}{2}} x \bar{\Sigma}^{-\frac{1}{2}} R^* = \begin{pmatrix} \lambda & 0 \\ 0 & \frac{1}{\lambda} \end{pmatrix}.$$

We have then $\operatorname{tr}(\bar{\Sigma}^{-1}x) = \lambda + \frac{1}{\lambda}$. By Eq. (1) in Sect. 3,

$$d(x, \bar{\Sigma}) \propto \| \log(\bar{\Sigma}^{-\frac{1}{2}} x \bar{\Sigma}^{-\frac{1}{2}}) \| = \| \log(R\bar{\Sigma}^{-\frac{1}{2}} x \bar{\Sigma}^{-\frac{1}{2}} R^*) \| = \sqrt{2} |\log \lambda|.$$

Since we can choose the multiplicative factor in Eq. (1), suppose that $d(x, \bar{\Sigma}) = |\log \lambda|$. We have then

$$\left(\frac{2}{\operatorname{tr}(\Sigma^{-1} x)}\right)^{\frac{dkn}{2}} = \left(\frac{2(\det \Sigma)^{\frac{1}{2}}}{\lambda + \frac{1}{\lambda}}\right)^{kn} \propto \left(\frac{e^{d(x,\bar{\Sigma})} + e^{-d(x,\bar{\Sigma})}}{2}\right)^{-kn},$$

which leads to the following theorem.

Theorem 3. *The distance $d(.,.)$ can be normalized such that for all $x \in \mathcal{S}$, $\Sigma \in \mathcal{P}_2(K)$ and $n \in \mathbb{N}, n > 2$,*

$$f_{PW}(x; \Sigma, n) \propto \cosh(d(x, \bar{\Sigma}))^{-kn}.$$

As announced, the function $h_{\Sigma,n} = \cosh^{-kn}$ does not depends on Σ.

5 Conclusion

We exhibited simple links between projective Wishart distributions and the affine invariant distance of positive definite matrices of constant determinant. Our future researches will focus on two aspects. Firstly, we will investigate the convergence of the estimation of the parameters of projective Wishart distributions. Secondly, we will investigate the use of the geometric properties of these distributions in signal processing applications.

References

1. Kollo, T., Dietrich Rosen, D.: Advanced Multivariate Statistics with Matrices. Springer, Dordrecht (2005). https://doi.org/10.1007/1-4020-3419-9
2. Muirhead, R.J.: Aspects of Multivariate Statistical Theory. Wiley Series in Probability and Statistics (1982)
3. Bhatia, R.: Positive Definite Matrices. Princeton University Press, Princeton (2007). https://doi.org/10.1515/9781400827787
4. Terras, A.: Harmonic Analysis on Symmetric Spaces and Applications II, 2nd edn. Springer, New York (1988). https://doi.org/10.1007/978-1-4612-3820-1
5. Faraut, J.: Analysis on Symmetric Cones. Oxford Mathematical Monographs (1994)
6. Zhuang, L., Walden, A.T.: Sample mean versus sample Fréchet mean for combining complex Wishart matrices: a statistical study. IEEE Trans. Sig. Process. **65**(17), 4551–4561 (2017). https://doi.org/10.1109/TSP.2017.2713763
7. Wishart, J.: The generalised product moment distribution in samples from a normal multivariate population. Biometrika **20**, 32–52 (1928)
8. Pennec, X., Fillard, P., Ayache, N.: A Riemannian framework for tensor computing. Int. J. Comput. Vis. **66**, 41–66 (2006). https://doi.org/10.1007/s11263-005-3222-z
9. Thanwerdas, Y., Pennec, X.: Is affine-invariance well defined on SPD matrices? A principled continuum of metrics. In: Nielsen, F., Barbaresco, F. (eds.) Geometric Science of Information, GSI 2019. LNCS, vol. 11712, pp. 502–510. Springer, Cham (2019). https://doi.org/10.1007/978-3-030-26980-7_52
10. Afsari, B.: Riemannian L^p center of mass: existence, uniqueness, and convexity. Proc. Am. Math. Soc. **139**(2), 655–673 (2011)
11. Arnaudon, M., Barbaresco, F., Yang, L.: Riemannian medians and means with applications to radar signal processing. IEEE J. Sel. Top. Sig. Process. **7**(4), 595–604 (2013). https://doi.org/10.1109/JSTSP.2013.2261798
12. James, A.T.: Distributions of matrix variates and latent roots derived from normal samples. Ann. Math. Statist. **35**(2), 475–501 (1964). https://doi.org/10.1214/aoms/1177703550

Computing Geometry and Algebraic Statistics

Convex Hulls of Curves: Volumes and Signatures

Carlos Améndola[1]([✉])[iD], Darrick Lee[2][iD], and Chiara Meroni[3][iD]

[1] Technical University of Berlin, Berlin, Germany
amendola@math.tu-berlin.de
[2] University of Oxford, Oxford, UK
darrick.lee@maths.ox.ac.uk
[3] MPI-MiS Leipzig, Leipzig, Germany
chiara.meroni@mis.mpg.de

Abstract. Taking the convex hull of a curve is a natural construction in computational geometry. On the other hand, path signatures, central in stochastic analysis, capture geometric properties of curves, although their exact interpretation for levels larger than two is not well understood. In this paper, we study the use of path signatures to compute the volume of the convex hull of a curve. We present sufficient conditions for a curve so that the volume of its convex hull can be computed by such formulae. The canonical example is the classical moment curve, and our class of curves, which we call *cyclic*, includes other known classes such as d-order curves and curves with totally positive torsion. We also conjecture a necessary and sufficient condition on curves for the signature volume formula to hold. Finally, we give a concrete geometric interpretation of the volume formula in terms of lengths and signed areas.

Keywords: Convex hull · Path signature · Volume · Cyclic polytope

1 Introduction

Taking the convex hull of a curve is a classical geometric construction. Understanding both its computation and properties is important for non-linear computational geometry, with the case of space curves particularly relevant for applications such as geometric modeling [13,16,17]. On the other hand, volumes are a fundamental geometric invariant. Computing volumes of convex hulls leads to interesting isoperimetric problems in optimization [11], and has applications in areas like ecology [3] and spectral imaging [12].

Supported by NCCR-Synapsy Phase-3 SNSF grant number 51NF40-185897 and Hong Kong Innovation and Technology Commission (InnoHK Project CIMDA).

F. Nielsen and F. Barbaresco (Eds.): GSI 2023, LNCS 14071, pp. 455–464, 2023.
https://doi.org/10.1007/978-3-031-38271-0_45

In the recent article [5], the authors show that for curves with *totally positive torsion*, one can compute the volume of their convex hull using a certain integral formula. An example of such a path is the *moment curve*

$$\mathbf{x}(t) = (t, t^2, \ldots, t^d) : [0,1] \to \mathbb{R}^d.$$

From the perspective of discrete geometry, the moment curve is also a canonical example of a larger class of curves called *d-order curves* [18]. Our main contribution is Theorem 3, which extends the integral formulae of [5], for volumes of convex hulls, to the class of *cyclic curves*, which are uniform limits of *d*-order curves. The motivation behind this name arises from the connection to cyclic polytopes, which play a central role in our proof of this generalization.

Our method uses the notion of the *path signature* [2], a powerful tool which is widely used in both stochastic analysis [6] and machine learning [10]. Certain orthogonal invariants of the path signature can be understood as a notion of signed volume [4]. In particular, they show that this orthogonal invariant computes the volume of the convex hull in the specific case of the moment curve. This result inspires our extension to the setting of cyclic curves, which forms a connection between the convex hull formulae of [5] and the path signature.

2 Classes of Curves

Throughout this article, we consider Lipschitz-continuous paths $\mathbf{x} : [0,1] \to \mathbb{R}^d$ and we write $\mathbf{x} \in \mathrm{Lip}([0,1], \mathbb{R}^d)$. The following definition appears in [5].

Definition 1 (Totally positive torsion ($\mathbf{T}^d$)). *A path* $\mathbf{x} \in C^d((0,1), \mathbb{R}^d) \cap C([0,1], \mathbb{R}^d)$ *has totally positive torsion if all the leading principal minors of the matrix of derivatives* $(\mathbf{x}'(t), \ldots, \mathbf{x}^{(d)}(t))$ *are positive for all* $t \in (0,1)$. *The space of totally positive torsion paths is denoted by* $\mathbf{T}^d$.

Note that this definition depends on the parametrization, is not invariant under the action of the orthogonal group, and requires a high regularity of $\mathbf{x}$. On the other hand, the following class of curves is purely geometric and therefore independent of the parametrization.

Definition 2 (d-order path (Ord^d)). *A path* $\mathbf{x} \in \mathrm{Lip}([0,1], \mathbb{R}^d)$ *is d-order if any affine hyperplane in* $\mathbb{R}^d$ *intersects the image of* $\mathbf{x}$ *in at most d points. This occurs exactly when*

$$\det \begin{pmatrix} 1 & 1 & \cdots & 1 \\ \mathbf{x}(t_0) & \mathbf{x}(t_1) & \cdots & \mathbf{x}(t_d) \end{pmatrix} > 0 \qquad \text{for all } 0 \le t_0 < \ldots < t_d \le 1. \tag{1}$$

The space of d-order paths is denoted Ord^d.

These curves have special properties and have been studied in the literature intensively under many different names, for instance as *strictly convex curves*. We mention here [9], where the authors compute volume and Caratheodory number

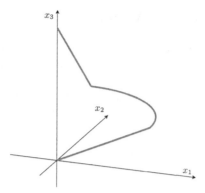

Fig. 1. A curve in $\mathsf{Cyc}^d \setminus \mathsf{Ord}^d$, containing two line segments and an arc which lies in the plane $\{x_3 = \frac{1}{2}\}$.

of convex curves and [15], where a volume formula for their convex hull appears, in the case of closed even dimensional such curves. One of the peculiarities of d-order curves that we will exploit is that the convex hull of every n-tuple of points on the curve is a *cyclic polytope* [20] with those n points as vertices.

By [5, Corollary 2.5], the totally positive torsion property implies that $\mathbf{x}$ is a d-order path, hence $\mathsf{T}^d \subseteq \mathsf{Ord}^d$. It is easy to construct examples proving that this inclusion is strict. We can interpret the condition for T^d to be a *local* condition, in the sense that it is checked simply at individual points on the curve. In contrast, the condition for Ord^d is a *global* condition, in the sense that the condition simultaneously takes any $d + 1$ points of the curve into consideration. We can further extend the class of Ord^d.

Definition 3 (Cyclic paths (Cyc^d)). *A path* $\mathbf{x} \in Lip([0, 1], \mathbb{R}^d)$ *is cyclic if it is a limit, in the Lipschitz (or equivalently, uniform) topology, of d-order curves. The space of cyclic paths is denoted by* Cyc^d.

In particular, cyclic curves include some piecewise linear curves, which are not contained in any of the previous classes. It also allows subsets of the curve to lie in a lower dimensional space. See Fig. 1 for an example. We point out that for a curve $\mathbf{x}$ that spans the whole $\mathbb{R}^d$, cyclicity is the same as the relaxed version of condition (1), where the determinants are required to be non-negative. In the case that $\mathbf{x}$ is contained in some lower dimensional subspace, the determinants in (1) are all zero and thus the relaxed version of (1) puts no condition on the curve, whereas cyclicity does. In conclusion, we obtain the following sequence of strict inclusions, transitioning from local to global properties of curves:

$$\mathsf{T}^d \subset \mathsf{Ord}^d \subset \mathsf{Cyc}^d.$$

3   Path Signatures and Volume

In this section, we introduce the connection between volume formulae for convex hulls of curves and the path signature, a characterization of curves based on iterated integrals [2]. In particular, we show that the volume of the convex hull of a cyclic curve can be written in terms of an antisymmetrization of the path signature. This connection was first considered in [4] for the moment curve.

Definition 4. *Suppose* $\mathbf{x} = (x_1, \ldots, x_d) \in Lip([0, 1], \mathbb{R}^d)$. *The level* k *path signature of* $\mathbf{x}$ *is a tensor* $\sigma^{(k)}(\mathbf{x}) \in (\mathbb{R}^d)^{\otimes k}$, *defined by*

$$\sigma^{(k)}(\mathbf{x}) := \int_{\Delta^k} \mathbf{x}'(t_1) \otimes \cdots \otimes \mathbf{x}'(t_k) \, dt_1 \cdots dt_k \in (\mathbb{R}^d)^{\otimes k} \tag{2}$$

where the integration is over the k-*simplex* $\Delta^k := \{0 \le t_1 < \cdots < t_k \le 1\}$. *Given a multi-index* $I = (i_1, \ldots, i_k) \in [d]^k$, *the path signature of* $\mathbf{x}$ *with respect to* I *is*

$$\sigma_I(\mathbf{x}) := \int_{\Delta^k} x'_{i_1}(t_1) \cdots x'_{i_k}(t_k) \, dt_1 \cdots dt_k \in \mathbb{R}. \tag{3}$$

The *path signature* $\sigma(\mathbf{x})$ of a path $\mathbf{x}$ is the formal power series obtained by summing up all levels $\sigma^{(k)}(\mathbf{x})$. It is well known that the path signature characterizes paths up to *tree-like equivalence* [2,8]; however, the individual entries of the signature $\sigma_I(\mathbf{x})$ are often difficult to understand geometrically. We aim to provide a geometric interpretation of signature terms via antisymmetrization into the exterior algebra $\Lambda(\mathbb{R}^d) := \bigoplus_{k=0}^{d} \Lambda^k \mathbb{R}^d$, where $\Lambda^k \mathbb{R}^d$ is the vector space of alternating tensors of $\mathbb{R}^d$ of degree k. These tensors are indexed using order preserving injections $P : [k] \to [d]$, denoted here by $\mathcal{O}_{k,d}$.

Definition 5. *Suppose* $\mathbf{x} = (x_1, \ldots, x_d) \in Lip([0, 1], \mathbb{R}^d)$. *The level* k *alternating signature of* $\mathbf{x}$ *is a tensor* $\alpha^{(k)}(\mathbf{x}) \in \Lambda^k \mathbb{R}^d$, *defined by*

$$\alpha^{(k)}(\mathbf{x}) := \int_{\Delta^k} \mathbf{x}'(t_1) \wedge \cdots \wedge \mathbf{x}'(t_k) \, dt_1 \cdots dt_k \in \Lambda^k \mathbb{R}^d. \tag{4}$$

Given $P \in \mathcal{O}_{k,d}$, *the alternating signature of* $\mathbf{x}$ *with respect to* P *is*

$$\alpha_P(\mathbf{x}) := \frac{1}{k!} \int_{\Delta^k} \det(x'_{P(1)}(t_1), \ldots, x'_{P(k)}(t_k)) \, dt_1 \cdots dt_k \in \mathbb{R}. \tag{5}$$

We will primarily be interested in the level $k = 2$ and level $k = d$ alternating signature. Let $\mathbf{x} \in Lip([0, 1], \mathbb{R}^d)$ and consider the level $k = 2$. At this level, $\alpha^{(2)}(\mathbf{x})$ is a $d \times d$ antisymmetric matrix whose (i,j)-th entry is $\alpha_{i,j}(\mathbf{x}) = \frac{1}{2}(\sigma_{i,j}(\mathbf{x}) - \sigma_{j,i}(\mathbf{x}))$, which is exactly the *signed area* of the path $\mathbf{x}$ projected to the (e_i, e_j)-plane. In the case of level $k = d$, the exterior power $\Lambda^d \mathbb{R}^d$ is one-dimensional, and we can express the level d alternating signature as

$$\alpha^{(d)}(\mathbf{x}) = \frac{1}{d!} \int_{\Delta^d} \det(\mathbf{x}'(t_1), \ldots, \mathbf{x}'(t_d)) dt_1 \ldots dt_d.$$

In fact, this expression is rotation invariant.

Theorem 1. ([4]).   *The level d alternating signature for paths $\mathbf{x} \in Lip([0,1], \mathbb{R}^d)$ is invariant under the special orthogonal group $\mathrm{SO}(d)$ (where the action acts pointwise over $[0,1]$), i.e., given $V \in \mathrm{SO}(d)$, we have $\alpha^{(d)}(V\mathbf{x}) = \alpha^{(d)}(\mathbf{x})$.*

3.1   Convex Hull Formulae for Cyclic Curves

In [4], the authors interpret the alternating signature $\alpha^{(d)}$ as the *signed-volume* of a curve, and show that $\alpha^{(d)}$ is the volume of the convex-hull of the moment curve. The first part of our generalization extends this to d-order curves. This was initially proved using other methods in [9, Theorem 6.1].

Theorem 2. *If $\mathbf{x} : [0,1] \to \mathbb{R}^d$ is a d-order curve, then $\mathrm{vol}(\mathrm{conv}(\mathbf{x})) = \alpha^{(d)}(\mathbf{x})$.*

Proof. Consider the $n+1$ points $p_0 = \mathbf{x}(t_0), \ldots, p_n = \mathbf{x}(t_n)$ on the d-order curve, with $0 \le t_0 < \ldots < t_n \le 1$. Denote by C_n their convex hull $\mathrm{conv}(p_0, \ldots, p_n)$, which is a cyclic polytope. We can realize a triangulation of C_n by pulling one vertex, as in [4, Lemma 3.29]. Indeed, by Gale evenness criterion, a triangulation of a cyclic polytope is given by simplices with vertices $p_{i_0}, \ldots, p_{i_d}$ satisfying

- for even d: $i_0 = 0$ and $i_{2\ell+1} + 1 = i_{2\ell+2}$ for any $1 \le \ell \le d$;
- for odd d: $i_0 = 0$, $i_d = n$, and $i_{2\ell+1} + 1 = i_{2\ell+2}$ for any $1 \le \ell \le d-1$.

Let us denote by $\mathcal{I}$ the set of all d-tuples of indices that satisfy these conditions. Therefore, the volume of C_n is the sum of the volumes of all these simplices. Because $\mathbf{x}$ is a d-order curve, condition (1) holds. Hence,

$$\mathrm{vol}(C_n) = \frac{1}{d!} \sum_{\{i_0, \ldots, i_d\} \in \mathcal{I}} \det \begin{pmatrix} 1 & 1 & \ldots & 1 \\ p_{i_0} & p_{i_1} & \ldots & p_{i_d} \end{pmatrix}. \tag{6}$$

Since $\mathbf{x}$ has bounded variation, we can take the limit of 3.1 for $n \to \infty$, as in [4, Lemma 3.29]. In particular, we use the continuity of convex hulls from [14, Section 1.8] and the continuity of truncated signatures from [7, Proposition 7.63]. This gives the formula $\mathrm{vol}(\mathrm{conv}(\mathbf{x})) = \alpha^{(d)}(\mathbf{x})$.

Furthermore, the volume formula still holds for limits of d-order curves.

Theorem 3. *Let $\mathbf{x}: [0,1] \to \mathbb{R}^d$ be a cyclic curve. Then $\mathrm{vol}(\mathrm{conv}(\mathbf{x})) = \alpha^{(d)}(\mathbf{x})$.*

Proof. By cyclicity, let $\{\mathbf{x}_k\}$ be a sequence of d-order curves such that $\mathbf{x} = \lim_{k \to \infty} \mathbf{x}_k$. Because conv is a continuous operation [14, Section 1.8], it holds that $\mathrm{conv}(\mathbf{x}) = \lim \mathrm{conv}(\mathbf{x}_k)$. Hence we have

$$\mathrm{vol}(\mathrm{conv}(\mathbf{x})) = \lim_{k \to \infty} \mathrm{vol}(\mathrm{conv}(\mathbf{x}_k)) = \lim_{k \to \infty} \alpha^{(d)}(\mathbf{x}_k) = \alpha^{(d)}(\mathbf{x}), \tag{7}$$

where the first equality is due to the continuity of volume, the second due to Theorem 2 and the last one due to the stability property of the signature [7, Proposition 7.63].

Example 1. Fix d distinct non-negative real numbers $a_1, a_2, \ldots, a_d$ and consider the associated logarithmic curve $\mathbf{x}$ in $\mathbb{R}^d$, or *log-curve*, parameterized by

$$\mathbf{x} : [0, 1] \to \mathbb{R}^d, \quad t \mapsto \begin{pmatrix} \log(1 + a_1 t) \\ \log(1 + a_2 t) \\ \vdots \\ \log(1 + a_d t) \end{pmatrix}. \tag{8}$$

One can prove that log-curves are d-order, and therefore they are cyclic. Therefore, we can compute the volume of their convex hull using the signature formula. Let $G(z_1, \ldots, z_d; 1)$ be a multiple polylogarithm, as defined in [19, Section 8.1]. Then, the volume of the convex hull of the log-curve $\mathbf{x}$ is a combination of multiple polylogarithms:

$$\mathrm{vol}(\mathrm{conv}(\mathbf{x})) = \sum_{\tau \in \Sigma_d} \mathrm{sgn}(\tau) G\left(-\frac{1}{a_{\tau(1)}}, -\frac{1}{a_{\tau(2)}}, \ldots, -\frac{1}{a_{\tau(d)}}; 1 \right), \tag{9}$$

where Σ_d denotes the symmetric group on d elements.

3.2   Towards a Necessary Condition

At this point, a natural question is whether the sufficient condition of being cyclic is also a necessary condition for the volume formula to hold. In other words, we would like for the converse of Theorem 3 to hold. Unfortunately, this is not the case, as the following example shows.

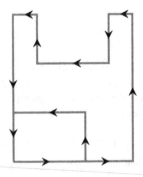

Fig. 2. Non-cyclic curve for which the convex hull volume formula holds.

Example 2. The path $\mathbf{x}$ in Fig. 2 starts (and ends) on the bottom left, traces the small cycle and then goes around the outer loop. Note that the convex hull $\mathrm{conv}(\mathbf{x})$ is a solid rectangle, whose area will equal the sum of the areas enclosed by each of the two loops. The latter is precisely the alternating signature and thus

$$\mathrm{vol}(\mathrm{conv}(\mathbf{x})) = \alpha^{(2)}(\mathbf{x}).$$

However, the curve $\mathbf{x}$ is not cyclic. Indeed, it can be proved that a cyclic curve $\mathbf{x}'$ satisfies $\mathbf{x}' \subset \partial \operatorname{conv}(\mathbf{x}')$, which is not true for $\mathbf{x}$.

Notice that any subpath of a cyclic curve is also cyclic, and thus the volume formula holds for all subpaths of a cyclic curve. We conjecture that cyclic curves are the largest class of curves for which the volume formula holds for all subpaths.

Conjecture 1. Let $\mathbf{x} : [0,1] \to \mathbb{R}^d$ be a Lipschitz path. Then, $\mathbf{x}$ is cyclic if and only if $\operatorname{vol}(\operatorname{conv}(\widetilde{\mathbf{x}})) = \alpha^{(d)}(\widetilde{\mathbf{x}})$ holds for all restrictions $\widetilde{\mathbf{x}} : [a,b] \to \mathbb{R}^d$ of $\mathbf{x}$ to a subinterval $[a,b] \subset [0,1]$.

4 Volumes in Terms of Signed Areas

The individual terms of the path signature are equipped with the *shuffle product*, and when we transfer this to the alternating signature, we obtain the following decomposition of the alternating signature into first and second level terms.

Lemma 1 ([4], Lemma 3.17). *Let $\mathbf{x} \in Lip([0,1], \mathbb{R}^d)$ and $P \in \mathcal{O}_{k,d}$. If k is even, then*

$$\alpha_P(\mathbf{x}) = \frac{1}{k! \left(\frac{k}{2}\right)!} \sum_{\tau \in \Sigma_k} \operatorname{sgn}(\tau) \prod_{r=1}^{k/2} \alpha_{P(\tau(2r-1)), P(\tau(2r))}(\mathbf{x}), \tag{10}$$

and if k is odd, then

$$\alpha_P(\mathbf{x}) = \frac{1}{k! \left(\frac{k-1}{2}\right)!} \sum_{\tau \in \Sigma_k} \operatorname{sgn}(\tau) \sigma_{P(\tau(1))}(\mathbf{x}) \prod_{r=1}^{(k-1)/2} \alpha_{P_i(\tau(2r)), P_i(\tau(2r+1))}(\mathbf{x}). \tag{11}$$

We can use this decomposition to rewrite the volume formula based on projected signed areas. Suppose $\mathbf{x} = (x_1, \ldots, x_d) : [0,1] \to \mathbb{R}^d$, where $d = 2n+1$ and $\mathbf{x}(1) \neq \mathbf{x}(0)$. Without loss of generality (due to the $SO(d)$ invariance of the alternating signature), we suppose that $\mathbf{x}(1) - \mathbf{x}(0)$ is restricted to the $n+1$ coordinate. Then, the odd decomposition in the above lemma is

$$\alpha^{(d)}(\mathbf{x}) = \frac{1}{d!}(x_d(1) - x_d(0)) \cdot \alpha^{(2n)}(\overline{\mathbf{x}}),$$

where $\overline{\mathbf{x}} = (x_1, \ldots, x_{2n}) : [0,1] \to \mathbb{R}^{2n}$. Thus, it remains to interpret the top-level alternating signature of even-dimensional paths.

Consider the even case $d = 2n$. Because $\alpha^{(2)}(\mathbf{x})$ is a antisymmetric matrix, it has conjugate pairs of purely imaginary eigenvalues. We can block-diagonalize the matrix $\alpha^{(2)}$ as follows (see, e.g., [1,21]):

$$\alpha^{(2)}(\mathbf{x}) = Q\Lambda Q^T,$$

where

$$\Lambda = \begin{pmatrix} 0 & \lambda_1 & 0 & 0 & \cdots & 0 & 0 \\ -\lambda_1 & 0 & 0 & 0 & \cdots & 0 & 0 \\ 0 & 0 & 0 & \lambda_2 & \cdots & 0 & 0 \\ 0 & 0 & -\lambda_2 & 0 & \cdots & 0 & 0 \\ \vdots & \vdots & \vdots & \vdots & & \vdots & \vdots \\ 0 & 0 & 0 & 0 & \cdots & 0 & \lambda_n \\ 0 & 0 & 0 & 0 & \cdots & -\lambda_n & 0 \end{pmatrix} \tag{12}$$

and Q is the orthogonal matrix of eigenvectors

$$Q = (\mathbf{v}_1, \mathbf{v}_2, \ldots, \mathbf{v}_{2n-1}, \mathbf{v}_{2n}).$$

Here, $\mathbf{v}_{2k-1}$ and $\mathbf{v}_{2k}$ are the real and imaginary parts of the conjugate pair of eigenvectors for $i\lambda_k$ and $-i\lambda_k$. Moreover, we can choose $Q \in SO(d)$ and by equivariance of the path signature, we have

$$\alpha^{(2)}(Q^T\mathbf{x}) = \Lambda.$$

By the $SO(d)$-invariance of the top-level alternating signature (Theorem 1), the decomposition can be written as

$$\alpha^{(2n)}(\mathbf{x}) = \frac{(-1)^n}{(2n)! \cdot n!} \prod_{k=1}^{n} \lambda_k.$$

Theorem 4. *Let $d = 2n$ and $\mathbf{x} = (x_1, \ldots, x_d) : [0,1] \to \mathbb{R}^d$ a cyclic curve. Then*

$$\mathrm{vol}(\mathrm{conv}(\mathbf{x})) = \frac{(-1)^n}{(2n)! \cdot n!} \prod_{k=1}^{n} \lambda_k, \tag{13}$$

where λ_k are the entries of Λ in (12). If $d = 2n + 1$ and the displacement $\mathbf{x}(1) - \mathbf{x}(0)$ is restricted to the x_d coordinate ($x_k(1) - x_k(0) = 0$ for $k \in [2n]$ and $x_d(1) - x_d(0) > 0$), then

$$\mathrm{vol}(\mathrm{conv}(\mathbf{x})) = \frac{(-1)^n}{(2n+1)! \cdot n!} (x_d(1) - x_d(0)) \prod_{k=1}^{n} \lambda_k, \tag{14}$$

where λ_k are the entries of Λ in the decomposition (12) of $\alpha^{(2)}(\overline{\mathbf{x}})$, corresponding to $\overline{\mathbf{x}} = (x_1, \ldots, x_{2n}) : [0,1] \to \mathbb{R}^{2n}$.

With the above decomposition, we obtain a geometric interpretation of the volume formula for the convex hull of the curve $\mathbf{x}$. More precisely, up to rotation, the volume of the convex hull of a cyclic curve is the product of the distance between the start point and end point of $\mathbf{x}$, if $\mathbf{x} \in \mathbb{R}^d$ with d odd, and $\lfloor d/2 \rfloor$ signed areas of the projections of $\mathbf{x}$ onto respective 2-planes.

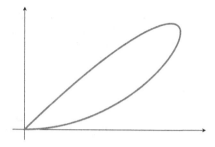

Fig. 3. The projection of the moment curve onto the plane orthogonal to $(1,1,1)$.

Example 3. Let $\mathbf{x} \subset \mathbb{R}^3$ be the moment curve parametrized by (t, t^2, t^3). We can apply the rotation

$$Q = \frac{1}{\sqrt{6}} \begin{pmatrix} 2 & -1 & -1 \\ 0 & \sqrt{3} & -\sqrt{3} \\ \sqrt{2} & \sqrt{2} & \sqrt{2} \end{pmatrix}, \tag{15}$$

which sends the vector $\mathbf{x}(1) = (1,1,1)$ to $(0,0,1)$. Then, the off-diagonal entry λ_1 of the associated matrix Λ is given by

$$\frac{1}{2} \int_0^1 \int_0^{t_2} \frac{2 - 2t_1 - 3t_1^2}{\sqrt{6}} \frac{(2 - 3t_2)t_2}{\sqrt{2}} - \frac{2 - 2t_2 - 3t_2^2}{\sqrt{6}} \frac{(2 - 3t_1)t_1}{\sqrt{2}} \mathrm{d}t_1 \mathrm{d}t_2 = \frac{1}{30\sqrt{3}}. \tag{16}$$

This value is the signed area of the projection of $\mathbf{x}$ onto the plane orthogonal to $(1,1,1)$, the curve shown in Fig. 3. Applying the odd version of Theorem 4 we get that

$$\mathrm{vol}(\mathrm{conv}(\mathbf{x})) = \frac{1}{3!} \cdot \sqrt{3} \cdot \frac{1}{30\sqrt{3}} = \frac{1}{180}.$$

Acknowledgements. We are grateful to Bernd Sturmfels, Anna-Laura Sattelberger, and Antonio Lerario for helpful discussions.

References

1. Baryshnikov, Y., Schlafly, E.: Cyclicity in multivariate time series and applications to functional MRI data. In: 2016 IEEE 55th Conference on Decision and Control (CDC). pp. 1625–1630 (2016)
2. Chen, K.T.: Integration of paths - a faithful representation of paths by noncommutative formal power series. Trans. Amer. Math. Soc. **89**(2), 395–407 (1958)
3. Cornwell, W.K., Schwilk, D.W., Ackerly, D.D.: A trait-based test for habitat filtering: convex hull volume. Ecology **87**(6), 1465–1471 (2006)
4. Diehl, J., Reizenstein, J.: Invariants of multidimensional time series based on their iterated-integral signature. Acta Applicandae Math. **164**(1), 83–122 (2019)
5. de Dios Pont, J., Ivanisvili, P., Madrid, J.: A new proof of the description of the convex hull of space curves with totally positive torsion. arXiv:2201.12932 (2022)

6. Friz, P.K., Hairer, M.: A Course on Rough Paths: With an Introduction to Regularity Structures. Springer International Publishing, second edn, Universitext (2020)
7. Friz, P.K., Victoir, N.B.: Multidimensional Stochastic Processes as Rough Paths: Theory and Applications. Cambridge Studies in Advanced Mathematics, Cambridge University Press (2010)
8. Hambly, B., Lyons, T.: Uniqueness for the signature of a path of bounded variation and the reduced path group. Ann. of Math. **171**(1), 109–167 (2010)
9. Karlin, S., Studden, W.: Tchebycheff Systems: With Applications in Analysis and Statistics. Interscience, Interscience Publishers, New York, Pure and Applied Mathematics (1966)
10. Lyons, T., McLeod, A.D.: Signature Methods in Machine Learning. arXiv:2206.14674 (2022)
11. Melzak, Z.: The isoperimetric problem of the convex hull of a closed space curve. Proc. Am. Math. Soc. **11**(2), 265–274 (1960)
12. Messinger, D., Ziemann, A., Schlamm, A., Basener, B.: Spectral image complexity estimated through local convex hull volume. In: 2nd Workshop on Hyperspectral Image and Signal Processing: Evolution in Remote Sensing. pp. 1–4. IEEE (2010)
13. Ranestad, K., Sturmfels, B.: On the convex hull of a space curve. Adv. Geom. **12**(1), 157–178 (2012)
14. Schneider, R.: Convex Bodies: The Brunn-Minkowski Theory, 2nd edn. Encyclopedia of Mathematics and its Applications, Cambridge University Press (2013)
15. Schoenberg, I.J.: An isoperimetric inequality for closed curves convex in even-dimensional Euclidean spaces. Acta Math. **91**, 143–164 (1954)
16. Sedykh, V.D.: Structure of the convex hull of a space curve. J. Sov. Math. **33**(4), 1140–1153 (1986)
17. Seong, J.K., Elber, G., Johnstone, J.K., Kim, M.S.: The convex hull of freeform surfaces. In: Geometric Modelling, pp. 171–183. Springer (2004). https://doi.org/10.1007/978-3-7091-0587-0_14
18. Sturmfels, B.: Cyclic polytopes and d-order curves. Geom. Dedicata **24**(1), 103–107 (1987)
19. Weinzierl, S.: Feynman Integrals: A Comprehensive Treatment for Students and Researchers. Springer International Publishing, UNITEXT for Physics (2022)
20. Ziegler, G.M.: Lectures on polytopes, Graduate Texts in Mathematics, vol. 152. Springer Verlag, New York (1995), revised edition, 1998; 7th updated printing 2007
21. Zumino, B.: Normal forms of complex matrices. J. Math. Phys. **3**(5), 1055–1057 (1962)

Avoiding the General Position Condition When Computing the Topology of a Real Algebraic Plane Curve Defined Implicitly

Jorge Caravantes[1] , Gema M. Diaz–Toca[2]([⊠]) ,
and Laureano Gonzalez–Vega[3]

[1] Universidad de Alcalá, Madrid, Spain
jorge.caravantes@uah.es
[2] Universidad de Murcia, Murcia, Spain
gemadiaz@um.es
[3] CUNEF Universidad, Madrid, Spain
laureano.gonzalez@cunef.edu

Abstract. The problem of computing the topology of curves has received special attention from both Computer Aided Geometric Design and Symbolic Computation. It is well known that the general position condition simplifies the computation of the topology of a real algebraic plane curve defined implicitly since, under this assumption, singular points can be presented in a very convenient way for that purpose. Here we will show how the topology of cubic, quartic and quintic plane curves can be computed in the same manner even if the curve is not in general position, avoiding thus coordinate changes. This will be possible by applying new formulae, derived from subresultants, which describe multiple roots of univariate polynomials as rational functions of the considered polynomial coefficients. We will also characterize those higher degree curves where this approach can be used and use this technique to describe the curve arising when intersecting two ellipsoids.

Keywords: Topology of curves · Subresultants · Singular points

1 Introduction

The problem of computing the topology of an implicitly defined real algebraic plane curve has received special attention from both Computer Aided Geometric Design and Symbolic Computation, independently. For the Computer Aided Geometric Design community, this problem is a basic subproblem appearing often in practice when dealing with intersection problems. For the Symbolic Computation community, on the other hand, this problem has been the motivation for many achievements in the study of subresultants, symbolic real root counting, infinitesimal computations, etc. By a comparison between the seminal papers and the more renewed works, one can see how the theoretical and practical complexities of the algorithms dealing with this problem have been dramatically improved (see, for example, [4] and [7]).

© The Author(s), under exclusive license to Springer Nature Switzerland AG 2023
F. Nielsen and F. Barbaresco (Eds.): GSI 2023, LNCS 14071, pp. 465–473, 2023.
https://doi.org/10.1007/978-3-031-38271-0_46

Sweeping algorithms to compute the topology of a real algebraic plane curve are greatly simplified if there are no more than one critical point in each vertical line, which is part of the so called general position condition. When this condition is not satisfied, there exists a suitable change of coordinates that moves the initial curve to one in general position. However, this step might be inconvenient in most cases transforming a sparse defined curve into a dense ones, so increasing the complexity of some steps of the algorithm. In [3], the authors introduce an algorithm that does not require the general position condition, mainly by enclosing the potentially irrational critical points inside boxes. In this paper, we deal with some particular cases where the topology of the considered curve around a critical point is easy to compute even if the curve is not in general position (but this requires also such critical point to be represented in an easy way, question we will deal with here too) with *ad hoc* approaches, taking advantage of a manageable presentation of the critical points, that do not require the general method in [3].

The paper is distributed as follows: after a section devoted to the well known features and issues of sweeping algorithms, Sect. 3 introduces a proposal to manage the connection of branches when two critical points of a low degree curve appear in the same vertical line. The following section provides an application and the final one states some conclusions.

2   Sweeping Algorithms for Topology Computation

The characterization of the topology of a curve $\mathcal{C}_P$ presented by the equation $P(x, y) = 0$ follows a sweeping strategy, usually based on the location of the critical points of P with respect to y (i.e. those singular points or points with a vertical tangent), and, on the study of the half-branches of $\mathcal{C}_P$ around these points since, for any other point of $\mathcal{C}_P$, there will be only one half-branch to the left and one half-branch to the right.

Definition 1. *Let $P(x, y) \in \mathbb{R}[x, y]$ and $\mathcal{C}_P = \{(\alpha, \beta) \in \mathbb{R}^2 : P(\alpha, \beta) = 0\}$ the real algebraic plane curve defined by P. Let P_x and P_y the partial derivatives of P. We say that a point $(\alpha, \beta) \in \mathcal{C}_P$ is*

- *a critical point of $\mathcal{C}_P$ if $P_y(\alpha, \beta) = 0$.*
- *a singular point of $\mathcal{C}_P$ if it is critical and $P_x(\alpha, \beta) = 0$.*
- *a regular point of $\mathcal{C}_P$ if it is not critical.*

Non singular critical points are called ramification points. Vertical lines through critical points will be referred as critical lines.

The usual strategy to compute the topology of a real algebraic plane curve $\mathcal{C}$ defined implicitly by a polynomial $P(x, y) \in \mathbb{R}[x, y]$ proceeds in the following way (for details see [1,5]):

1. Find all the real critical lines $x = \alpha_i$ with $\alpha_1 < \alpha_2 < \ldots < \alpha_r$.
2. For every α_i, compute the real roots of $P(\alpha_i, y)$, $\beta_{i,1} < \ldots < \beta_{i,s_i}$ and determining those $\beta_{i,j}$ regular points and those $\beta_{i,j}$ critical points.

3. For every α_i and every $\beta_{i,j}$, compute the number of half-branches to the right and to the left of each point $(\alpha_i, \beta_{i,j})$. Joining them in an appropriate way finishes the topology computation.

First two steps provide the vertices of a graph that will represent the topology of the considered curve (see [5] and [6]). The last step provides the edges. This paper is focused on that last step, which usually requieres general position of the curve (i.e. no real vertical asymptotes and just one critical point in each critical line), attainable through a coordinate change. In this paper, we address the possibility of ignoring the coverticality condition for low degree curves.

3 Avoiding General Position: Branch Computations Around a Critical Point

In order to analyse the topology of $\mathcal{C}_P$, when there is only one critical point in a critical line, it is easy to determine how many half-branches there are to the left and to the right of the considered critical point: each non-critical point gets one half-brach per side and all the remaining correspond to the critical one. Dealing with a critical line with more than one critical point is more complicated but having an explicit description of the considered critical points allows to determine the required information about the half-branches.

Namely, given a critical line defined by $x = \alpha$, with $P(\alpha, \beta) = 0$, (α, β) critical point, there are many cases where an explicit description for β in terms of α, $\beta = \Psi(\alpha)$ can be computed. Probably, the most efficient way to calculate Ψ is by using subresultants. In [2], these explicit expressions are found for degrees 3, 4, and 5 and the cases where this can be achieved are fully characterised. In addition, the multiplicity of β is determined too.

Remark 1. Vertical asymptotes are those lines $x = \alpha$ where α is a root of the leading coefficient of P with respect to y. Simple vertical asymptotes (related to simple roots of the leading coefficient) can be worked, as in [3], by checking the sign of the second coefficient at the critical line. More complicated asymptotes imply that the curve either has a cusp or an inflection point in its vertical infinity point in the difficult cases or it is an arrangement of lines. This paper is devoted to the management of affine critical points, however, so we refer to [3, Subsection 4.5] for vertical asymptotes.

3.1 Ramification Points of $\mathcal{C}_P$

The method here is not far from what one can see in [3]. However, since the authors of this paper could not find a proof or reference to the correctness of the method in [3], a motivation is written here. Recall that $(\alpha, \beta) \in \mathbb{R}^2$ is a critical and non singular point of $\mathcal{C}_P$ when $P(\alpha, \beta) = P_y(\alpha, \beta) = 0$ and $P_x(\alpha, \beta) \neq 0$. Applying the Implicit Function Theorem this means that around (α, β) the curve $\mathcal{C}_P$ can be described as a function $x = \Phi(y)$ such that $\alpha = \Phi(\beta)$. Since

$$\Phi'(\beta) = -\frac{P_y(\alpha, \beta)}{P_x(\alpha, \beta)} = 0$$

we have three possibilities, since $y = \beta$ can be:

- a local minimun of Φ: 2 half-branches to the right of (α, β), 0 to the left; or
- a local maximun of Φ: 0 half-branches to the right of (α, β), 2 to the left; or
- an inflection point of Φ: 1 half-branch to the right of (α, β) and 1 to the left.

Characterising the behaviour of the function $x = \Phi(y)$ at $y = \beta$ requires to evaluate the derivatives $P_{yy}(\alpha, \beta), P_{yyy}(\alpha, \beta), P_{yyyy}(\alpha, \beta), \ldots$ until one of them does not vanishes since: if $\Phi^{(1)}(\beta) = \Phi^{(2)}(\beta) = \cdots = \Phi^{(k-1)}(\beta) = 0$, $\Phi^{(k)}(\beta) \neq 0$, then (applying recursively implicit differentiation to $P(\Phi(y), y) = 0$), we have that $\Phi^{(k)}(\beta) P_x(\alpha, \beta) = -P_{\underset{y \cdots y}{k}}(\alpha, \beta)$ and we can apply the higher order derivative test to determine whether the point is a local maximum, a local minimum, or a flex.

3.2   Singular Points of $\mathcal{C}_P$ When $\deg_y(P) = 4$

Due to the low degree of the curve, the only ambiguity that can arise happens when there are two double roots $\beta_1 < \beta_2$ of the polynomial $P(\alpha, y)$. Otherwise it is locally general position for the critical line $x = \alpha$. Since we work with a quartic, we have, at most 4 branches to each of the sides of our critical line. Since we have found two double roots for $P(\alpha, y)$, we know that the coefficient of y^4 must be nonzero. We suppose it is 1 for simplicity. Moreover, since P is defined over the reals, the number of real branches to each side of the critical line must be even.

If we have four branches to join with the singular points to one of the sides, then it must be two for each due to multiplicity. If there are no real branches to one of the sides, we have no work to do for such branch.

Finally, the remaining case is when we have just two branches to join. These two branches must go to the same point, since the other critical point must attract two conjugate complex branches. First of all, we consider $Q_i(s, t) = P(s + \alpha, t + \beta_i)$. Then the behaviour of $(0, 0)$ as a point for Q_i is the same as the behaviour of (α, β_i) for P. Factoring the lowest homogeneous component of Q_i we have the slopes of the (at most two) tangent lines to $\mathcal{C}_P$ at (α, β_i). Then:

- If one of the points has all slopes to be complex and non real, then it is an isolated point, so the other one takes the branches.
- If one of the points has two different real slopes, then it takes the two arcs since it is a real node.

In the case that there is just one slope for the tangent lines to the curve at the critical points, we will consider the cubic curve given by $P_y(x, y) = 0$. The polynomial $P_y(\alpha, y)$ vanishes in β_1, β_2 and an intermediate point $\gamma \in (\beta_1, \beta_2)$ since it is the derivative of $P(\alpha, y)$. This means that there are three real branches of $\mathcal{C}_{P_y}$ through the vertical line $x = \alpha$. Due to the low degree, the only possibility is what happens in Fig. 1 or the symmetric case, and the relative position of the branches of $\mathcal{C}_P$ and $\mathcal{C}_{P_y}$ determines how to join the half branches.

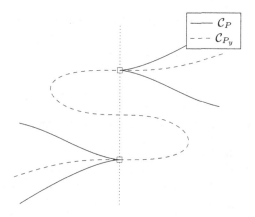

Fig. 1. To the left, the real branches of $\mathcal{C}_P$ go to the critical point below because two branches of $\mathcal{C}_{P_y}$ are above both of them. To the right, we have the complementary situation.

3.3 Singular Points of $\mathcal{C}_P$ When $\deg_y(P) = 5$

We now consider $\deg_y P = 5$. The nontrivial cases here are:

– $P(\alpha, y)$ has two double roots.
– $P(\alpha, y)$ has a triple root (β) and a double root (γ).

Formulae in [2] allow to represent easily these roots of $P(\alpha, y)$. For example, in the second case, we have

$$\beta = -\frac{\mathbf{sres}_{1,0}(\tau_1(y))}{\mathbf{sres}_1(\tau_1(y))}, \quad \gamma = -\frac{\mathbf{sres}_{3,0}(P(\alpha, y))}{\mathbf{sres}_3(P(\alpha, y))\beta^2}$$

where $\tau_1(y) = \mathbf{Sres}_3(P(\alpha, y))$ and $\mathbf{Sres}_k$ denotes the polynomial subresultant and $\mathbf{sres}_k$ the subresultant coefficients of a polynomial and its derivative (with respecto to y in this case) of index k.

We will address each case separately, but first we consider, as before, that P is monic on y (and $\deg_y(P) = 5$, otherwise, we proceed as in lower degree). Then, reasoning in a similar way, we see that the number of real branches between critical lines must be 1, 3 or 5.

$P(\alpha, y)$ has Two Double Roots. This case can be treated as before (degree 4). We have two critical points that take either two or no branches each, and one single point that will take one.

If we have 5 branches to distribute, then each critical point takes two branches and the non-critical point takes one. We distribute the branches to avoid crossings outside the critical line. If we have just one branch, then the non-critical point takes it.

If we have three branches, one goes to the noncritical point and the other two are assigned either checking whether the tangent lines are real and different at the singularities, or considering the curve $\mathcal{C}_{P_y}$ and reasoning as in the degree 4 case:

- If there are at least two branches of $\mathcal{C}_{P_y}$ above at least two of the three branches of $\mathcal{C}_P$, then the critical point below takes two branches.
- Otherwise, the critical point above takes two branches.

$P(\alpha, y)$ has a Triple Root and a Double Root. Here, the triple critical point takes one or three branches, and the double one takes two or none.

If there are five branches to distribute, then three branches go to the critical point corresponding to the triple root, and two branches go to the critical point corresponding to the double root. If there is just one branch, then the critical point corresponding to the triple root takes it.

If there are three branches, we again check the slopes of the tangent lines:

- If one of the critical points has two complex non-real slopes, then the other one takes the until now unassigned branches.
- If one of the critical points has (at least) two real slopes, then it takes the until now unassigned branches.

If we do not have enough data, then we consider again $\mathcal{C}_{P_y}$. It has one real branch through the double point, one real branch passing between the critical points and two possibly non-real branches passing through the triple point.

- If $\mathcal{C}_{P_y}$ has just two real branches, then the double point takes the until now unassigned branches.
- If two of the four real branches of $\mathcal{C}_{P_y}$ lie above the three branches of $\mathcal{C}_P$, then the below critical point takes the until now unassigned branches.
- Otherwise, the above critical point takes the until now unassigned branches.

It is impossible that the three branches of $\mathcal{C}_P$ lie between the four branches of $\mathcal{C}_{P_y}$ with this configuration at the critical line.

3.4   Examples

This section is devoted to testing the ideas introduced in the previous sections with some examples, which serve as an argument against imposing general position when treating with low degree curves.

Example 1. Consider the quartic $\mathcal{C}_P$ given by $P(x, y) = x^4 + 2x^2y^2 + y^4 + x^3 - 3xy^2$ (see Fig. 2). If we apply the sweeping algorithm to P, then we find the critical line $x = 9/16$ with two double critical points: $(9/16, \pm 3\sqrt{15}/16)$. To the left, there are four half branches, so each critical point takes two. To the right, there are no critical branches, so there's no need to join anything. We can make this decision even without considering whether the critical points are singular. While the change $x \mapsto y, y \mapsto x$ gives general position, the general coordinate change transforms this sparse represented curve into a dense one.

Example 2. Consider the quintic C_P given by $P(x,y) = x^5 - y^5 - 4x^4 + 6y^4 + x^3 - 12y^3 + 10x^2 + 8y^2 - 4x - 8$ (see Fig. 2). The restriction of P to the critical line $x = -1$ gives a double root $y = 0$ and a triple root $y = 2$. There are three half-branches on each sides. Since the localization at the critical point $(-1, 0)$, $P(x - 1, y)$, has the lowest homogeneous component equal to $8y^2 - 27x^2$, the curve has two tangent lines at such point. This means, as mentioned in the previous section, that the point $(-1, 0)$ must take two half-branches from each side, which means that the remaining one goes to the point $(-1, 2)$.

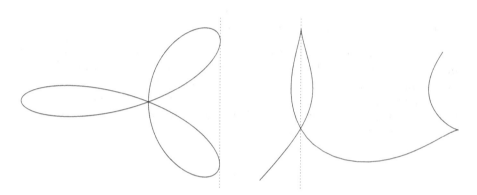

Fig. 2. To the left, the curve in Example 1 with the mentioned critical line. To the right, the curve in Example 2 with the mentioned critical line.

4 Application: Computing the Intersection Curve Between Two Ellipsoids

Given two ellipsoids $\mathcal{A} : XAX^T = 0$ and $\mathcal{B} : XBX^T = 0$, $X = (x, y, z, 1)$, their characteristic equation is defined as $f(\lambda) = \det(\lambda A + B) = \det(A)\lambda^4 + \ldots + \det(B)$ which is a quartic polynomial in λ with real coefficients. The characterization of the relative position of two ellipsoids (separation, externally touching and overlapping) in terms of the sign of the real roots of their characteristic equation was introduced by [9]. However, the root pattern of the characteristic polynomial is not enough to characterize the arrangement of two ellipsoids.

A more in-depth algebraic characterization using the so-called index sequence was introduced in [8] to classify the morphology of the intersection curve of two quadratic surfaces in the the 3D real projective space. The index sequence of a quadric pencil not only includes the root pattern of the characteristic polynomial, but also involves the Jordan form associated to each root and the information between two consecutive roots.

The behaviour of the index function for a pencil of ellipsoids is captured by the eigenvalue curve $\mathcal{S}$ defined by the equation $S(\lambda, \mu) = \det(\lambda A + B - \mu \mathbb{I}_4) = 0$.

This curve has degree four in both λ and μ. Because $\lambda A + B$ is a real symmetric matrix for each $\lambda \in \mathbb{R}$, there are in total four real roots for $S(\lambda, \mu) = 0$, counting multiplicities. For each value λ_0, the index function $\mathrm{Id}(\lambda_0)$ equals to the number of positive real roots of $S(\lambda_0, \mu) = 0$.

Since $S(\lambda, \mu) = 0$ is a very special quartic curve (there are always four real branches (taking into account multiplicities), its analysis is extremely simple. If $\lambda = \alpha$ is a critical line, then $S(\alpha, \mu)$ factorises in the following way:

1. $S(\alpha, \mu) = \tau_4(y - \beta)^4$.
2. $S(\alpha, \mu) = \tau_4(y - \beta)^3(y - \gamma)$ with $\gamma \in \mathbb{R}$.
3. $S(\alpha, \mu) = \tau_4(y - \beta)^2(y - \gamma)^2$ with $\gamma \in \mathbb{R}$.
4. $S(\alpha, \mu) = \tau_4(y - \beta)^2(y - \gamma_1)(y - \gamma_2)$ with $\gamma_1 \neq \gamma_2$.

In [2], we can find formulae showing, for each case and in terms of α, the values of β, γ, γ_1 and γ_2 allowing to determine easily $\mathrm{Id}(\alpha)$. Computing $\mathrm{Id}(\lambda)$ for λ not giving a critical line reduce to apply Descartes' law of signs (see Remark 2.38 in [1]) to the polynomial $S(\lambda, \mu)$ as polynomial in μ. And finally the way the four branches touch every critical line is easily determined by using the techniques described in Sect. 3.2.

5   Conclusions

In this paper (together with the formulae in [2]), we have shown how to avoid the use of the "general position condition" when computing the topology of a real algebraic plane curve defined implicitly. A concrete application has been also described and our next step will be to design a new algorithm computing the topology of an arrangement of quartics and quintics by using the formulae and strategy introduced here.

Acknowledgements. The authors are partially supported by the grant PID2020-113192GB-I00/AEI/ 10.13039/501100011033 (Mathematical Visualization: Foundations, Algorithms and Applications) from the Spanish Agencia Estatal de Investigación (Ministerio de Ciencia e Innovación). J. Caravantes belongs to the Research Group ASYNACS (Ref. CT-CE2019/683).

References

1. Basu, S., Pollack, R., Roy, M.-F.: Algorithms in Real Algebraic Geometry. Algorithms and Computations in Mathematics, vol. 10. Springer, Heidelberg (2003). https://doi.org/10.1007/3-540-33099-2
2. Caravantes, J., Diaz-Toca, G.M., Gonzalez-Vega, L.: Closed formulae for multiple roots of univariate polynomials through subresultants (Preprint 2023). https://doi.org/10.48550/arXiv.2303.02780
3. Diatta, D.N., Diatta, S., Rouillier, F., et al.: Bounds for polynomials on algebraic numbers and application to curve topology. Disc. Comput. Geom. **67**, 631–697 (2022). https://doi.org/10.1007/s00454-021-00353-w

4. Diatta, D.N., Rouillier, F., Roy, M.-F.: On the computation of the topology of plane curves. In: Proceedings of the ISSAC International Symposium on Symbolic and Algebraic Computation, pp. 130–137 (2014)

5. Gonzalez-Vega, L., Necula, I.: Efficient topology determination of implicitly defined algebraic plane curves. Comput. Aided Geom. Des. **19**, 719–743 (2002)

6. Hong, H.: An efficient method for analyzing the topology of plane real algebraic curves. Math. Comput. Simul. **42**, 571–582 (1996)

7. Kobel, A., Sagraloff, M.: On the complexity of computing with planar algebraic curves. J. Complex. **31**, 206–236 (2015)

8. Tu, C., Wang, W., Mourrain, B., Wang, J.: Using signature sequences to classify intersection curves of two quadrics. Comput. Aided Geom. Des. **26**, 317–335 (2009)

9. Wang, W., Wang, J., Kim, M.-S.: An algebraic condition for the separation of two ellipsoids. Comput. Aided Geom. Des. **18**, 531–539 (2001)

Dynamical Geometry and a Persistence K-Theory in Noisy Point Clouds

Sita Gakkhar[(✉)] and Matilde Marcolli

California Institute of Technology, Pasadena, CA 91125, USA
{sgakkhar,matilde}@caltech.edu

Abstract. The question of whether the underlying geometry of a dynamical point cloud is invariant is considered from the perspective of the algebra of trajectories of the points as opposed to their point-set topology. We sketch two approaches to identifying when the geometry remains invariant, one that accounts for a model of stochastic effects as well, and a second that is based on a persistence K-theory. Additional geometric structure in both approaches is made apparent by viewing them as finite noncommutative spaces (spectral triples) embedded inside the Hodge-de Rham spectral triple. A general reconstruction problem for such spaces is posed. The ideas are illustrated in the setting of understanding the dependence of grid cell population activity on environmental input.

Keywords: Grid cell modules · K-theory · Persistence homology · Noncommutative geometry · Stochastic differential geometry · Discrete differential geometry

1 Introduction

A dynamical point cloud is a family of point clouds (D_θ) parameterized by time or other environmental input, $\theta \in \Theta$. For each θ, the data, D_θ, are assumed to be sampled from a compact Riemannian manifold, M_θ. Characterizing the change in geometry and topology defined by the point cloud has important applications in many fields. Towards this, we study the geometry of a dynamic point cloud through discrete differential geometry and the persistence of the K_0 functor. This algebraic approach naturally connects with viewing the point clouds as finite spectral triples embedded inside the Hodge-de Rham spectral triple for M. The connection is provided by the results from [2] on the convergence of point cloud Laplacians to the Laplace-Beltrami operator and a Hodge theory on metric spaces developed by [1]. The point cloud Laplacians also allow for considering a stochastic version of the question with the Laplacian as the generator for the noise process. We begin by putting forward a model describing the case where the geometry is invariant over Θ up to stochastic effects and statistical testing in such a setup. Then we establish a stability theorem for an algebraic persistence theory to complement the topological persistence homology by capturing the dynamics of individual points without the complexity of multidimensional

F. Nielsen and F. Barbaresco (Eds.): GSI 2023, LNCS 14071, pp. 474–483, 2023.
https://doi.org/10.1007/978-3-031-38271-0_47

persistence. Finally, we consider the convergence of discrete Dirac operators for point clouds to the Dirac operator for the ambient Hodge-de Rham spectral triple. This is needed to be able to argue that the discretely sampled trajectories are sufficient to understand the geometry. A general reconstruction question is posed for such embedded finite spectral triples. The underlying motivation is understanding the modulation of grid cell firing by the environment. We start by introducing this illustrative example.

1.1 Modulation of Grid Cell Firing by Environmental Input

In the entorhinal cortex grid cells are cells with spatial firing fields organized on regular grids that form a part of the neural system responsible for navigation and mapping. Grid cells are organized in modules with structured correlations between different cells in the module. The neural code used by grid cell networks can be probed using persistence homology. In [10], Gardner et al, find that the activity of grid cell modules lies on a toroidal manifold that persists across brain states and offers support for continuous attractor models of grid cell activity. They also show evidence for environmental input-driven deformation of the geometry of population activity[1]. This can be thought of as an example of homeostatic plasticity. The question of the degree of stability of population dynamics is interesting, and one would like to relate this deformation to mechanistic models. A first step in this direction is putting forward a statistical test for the simplest case where the geometry is invariant and the point clouds evolve under a diffusion process on this fixed geometry.

We set this question up as follows: suppose that the spike train data from N neurons measured at K spatial locations $x_k^0, k \in [K], x_k \in \mathbb{R}^N$ in environmental conditions E_0 at time t_0. The environmental conditions are then updated to E_1 with firing data $(x_k^1)_{k \in [K]}$. The point cloud $\hat{M}_K^t := \{x_k^t : k \in [K]\} \subset \mathbb{R}^N$ changes with $t \in T$. The question now is of testing if the geometry M_K^t from which the point cloud $\hat{M}_K^t$ is invariant with respect to environmental input over $t \in T$, that is, $M_K^t = M_K^0 := M$, where the sample path of individual points, x_t^k, follows Brownian motion process on the invariant geometry M, that is, the diffusion generated by the Laplace-Beltrami operator, $\triangle_M$. While K is fixed, data from multiple runs of the experiment can be pooled to consider large size limit of the point cloud.

The choice of the process provides a natural null model for testing the presence of non-Markovian dynamics, as well as for testing synchronization in the point cloud. The hypothesis being tested is not just that the point cloud lives on an invariant geometry, that is, it's sampled from $M \times [0, T]$, but also that the time evolution follows Brownian motion on M. One can consider more general diffusion processes for such model testing, with the parameters learned from the time-series data, however, if the geometry is relevant then the Laplace-Beltrami operator is expected to play a role.

---

[1] [10, Tori persist despite grid distortions].

1.2   A Diffusive Model and Random Matrices

As a prelude to introducing L^2 Hodge theory [1], we consider the question of testing the hypothesis that the point cloud $M_K^t = M$ for all t. The Riemannian manifold (M, g), with $\dim M = d$ and metric g, is assumed to be embedded smoothly and isometrically in an ambient space, $\phi : M \hookrightarrow \mathbb{R}^N$, and for each $k \in K$, x_k^t is evolving by $\triangle_M/2$ diffusion on M.

Recalling that on a filtered probability space $(\Omega, F_*, \mathbb{P})$ a M-valued, F_*-adapted, stochastic process (X_t) is a (local-, semi-)-martingale on $[0, \tau)$ if $f(X_t)$ is a real-valued (local-, semi-)-martingale for all $f \in C^\infty(M)$ where τ is a F_* stopping time (see, for instance, [12]). Brownian motion, $X := (X_t)$, on M is the $\triangle_M/2$ generated diffusion process, that is, a F_*-adapted process $X : \Omega \to W(M)$ (where $W(M)$ is the path space on M) such that for all $f \in C^2(M)$, $\omega \in W(M)$, M^f as defined below is a local martingale:

$$M^f(\omega)_t := f(\omega_t) - f(\omega_0) - \frac{1}{2} \int_0^t \triangle_M f(X_s) ds \qquad (1)$$

By the results of Belkin-Niyogi [2], the convergence of empirical estimates of Laplacians on finite metric space to $\triangle_M$ is known. This is formulated as follows: data $X_n = (x_i)_{i \in [n]}$ is n samples form M sampled with respect to uniform measure, μ_M, $\dim M = d$, giving an increasing sequence of metric spaces $X_1 \subset X_2 \subset \ldots X_i \subset X_{i+1} \cdots \subset M$. To each X_n is the associated empirical Laplacian, $\triangle_{t_n, n}$, defined for $p \in M$ by $\triangle_{t_n, n} f(p) := \sum_{i \in [n]} K_{t_n}(p - x_i)(f(x_i) - f(p)) n t_n^{d+2}$ where $K_{t_n}(u) = \exp(-\|u\|^2/4t_n)$ and t_n an appropriate sequence decreasing to 0, $\|\cdot\| = \|\cdot\|_{\mathbb{R}^N}$, c, then we have $\lim_{n \to \infty} \triangle_{t_n, n} f(x)/t_n (4\pi t_n)^{d/2} = \triangle_M f(x)/\text{VOL}(M)$.

An analogous result holds for any probability measure μ_M on M. Now the local-martingale characterization of $\triangle_M/2$-diffusion (Eq. 1) applied to $f_i = \pi_i \circ \psi$, the coordinate functions of the smooth embedding ψ to easily test the question that $M_K^t = M$ for all t and X_k^t follows $\triangle_M/2$ diffusion. This is further simplified by noting that $f_i(X_s)$ is uniformly bounded and therefore a martingale, so the mean at each t is constant. The needed statistical test is just the test for constancy of the mean estimated by averaging data from l repeated experiments and using the control on $\triangle_M f(x)$ from [11] which gives a quantitative version of the convergence of the point cloud Laplacian. This is stronger than testing for stationary, e.g. using the unit root tests, as it's additionally required that the generator is the Laplacian.

Simplicial homology of random configurations and dynamical models for random simplicial complexes have been studied (for example, [6,8]), the simple example here suggests that (co)homology, both with rational coefficients and the α-scale theory of [1] for randomly evolving configurations is also meaningful from an applications perspective as well.

1.3   Discrete Differential Operators with Heat Kernel Weights

On a finite metric space, (X_n, d), with a probability measure μ, the point cloud Laplacian can be realized as Hodge Laplacian of a (co)chain complex [1]. Note that for a finitely supported measure ν on M, the point cloud Laplacian on M is an empirical estimate (via concentration bounds) for the functional approximation to the Laplace-Beltrami operator $\triangle_t f(x) = \int_X (f(x) - f(y)) K_t(x,y) d\nu(y)$. We work in the picture that n point metric space X is n samples from M, d is the distance in ambient euclidean space, d_M the geodesic distance on M, and as n increases we have inclusions $i_n : X_n \to X_{n+1}$, $|X_n| = n$ and $X_{n+1} \setminus X_n$ is the one additional sample from M.

Fix $X_n = X$. Barthodi et al. [1] consider (co)chain complexes on $L^2(X^l)$ using the coboundary map $\delta_{l-1} : L^2(X^l) \to L^2(X^{l+1})$, $[\delta f](z_0, z_1 \ldots z_l) = \sum_{i=0}^{l}(-1)^i \prod_{i \neq j} \sqrt{K(z_i, z_j)}\, f(z_0, \ldots \hat{z}_i \ldots z_l)$ where $X^l = \prod_{i \in [l]} X$, $L^\infty(X^2) \ni$ $K : X^2 \to \mathbb{R}$ is symmetric, nonnegative and measurable; $K := K_t(\cdot, \cdot)$ is taken the t_n scaled heat kernel. The boundary map $\partial_l : L^2(X^{l+1}) \to L^2(X^l)$ is defined by $[\partial g](z_0 \ldots z_{l-1}) = \sum_{i=0}^{l}(-1)^i \int_X \prod_{j=0}^{l-1} \sqrt{K(s, z_j)}$ $g(z_0 \ldots z_{j-1}, s, z_{j+1} \ldots z_{l-1})\, d\mu(s)$ and satisfies $\delta_{l-1}^* = \partial_l$, and the laplacian, $\triangle_l = (\delta_l^* \delta_l + \delta_{l-1} \delta_{l-1}^*)$ can be defined. The constructions and results also hold for $L_a^2(X^l) = \{ f \in L^2(X^l) : f(x_0, \ldots x_l) = (-1)^{\mathrm{sgn}(\sigma)} f(\sigma(x_0), \ldots \sigma(x_l)), \sigma \in \mathcal{S}_{l+1} \}$. In [1], they also establish that for a Riemannian manifold, (X, g, μ), on restricting this construction to a suitable neighborhood of the diagonal, de Rham cohomology of X can be recovered and a Hodge decomposition exists for each $L^2(X^l)$.

Observing that $\triangle_0^t(f(x)) = \int_X (f(x) - f(y)) K_t(x,y) d\mu(y)$, i.e., $\triangle_0|_{L^2(X)}$ is exactly the functional approximation to the Laplace-Beltrami operator which in the large sample-small t limit approaches the Laplace-Beltrami operator, and since on restricting to functions, Hodge-de Rham Laplacian agrees with the Laplace-Beltrami operator up to a sign suggests that in this limit $\delta^{(n)}$ associated to the sequence of n-point metric spaces (X_n) must approach the usual exterior derivative d acting on $\Omega^0(X)$. We give a quick proof using covariant Taylor series with respect to the canonical Riemannian connection ∇.

Theorem 1. *Suppose $U \subset \mathbb{R}^N$ is such that $M \cap U$ is a normal neighborhood of $x \in M$, and for any $y \in M \cap U$, $y \neq x$, $x(t)$ is the unique unit speed geodesic joining x, y, $v := \dot{x}(0)$. Then for $s = d_M(x,y)$ and $K_t(x,y) = \exp(-\|x - y\|_N^2 / 4t)$, $s = t + O(t^2)$ implies $|\delta f(x,y)/t - df_x(v)| = O(t)$.*

Proof. Since $x(t)$ is unit speed geodesic with $x(0) = x$, so $x(s) = y$. Expanding in a covariant Taylor series about $x(0)$, $f(x(t)) = \sum_{n=0}^{\infty} t^n/n! d^n/d\tau^n f(x(\tau))|_{\tau=0}$, with $d/d\tau = \dot{x}^i(\tau) \nabla_i$, gives $f(y) - f(x) = s \cdot df(v) + O(s^2)$ since first order term is $\dot{x}^i(\tau) \nabla_i f|_{\tau=0} = s \cdot g(v, \nabla f(x)) = s \cdot df_x(v)$. We have $\delta f(x,y) = \sqrt{K_t(x,y)}(f(y) - f(x)) = \sqrt{K_t(x,y)} s\, df_x(v) + \sqrt{K_t(x,y)} O(s^2)$. For fixed x, using that there exists $\eta \geq 0$, such that $d_M(x,y)^2 - \|x - y\|_N^2 = \eta(y)$ with $|\eta(y)| \leq C d_M(x,y)^4$ for a constant C on the normal neighborhood U, so $\|x - y\|_N^2 = d_M(x,y) - \eta(y)$. Using $e^\alpha = 1 + O(\alpha e^\alpha)$ for $\alpha > 0$, $1/(1 + \alpha) \leq 1 + O(\alpha)$ yields the following estimate from which the result follows for $s = t + O(t^2)$

$$\left| \sqrt{K_t(x,y)} \frac{s}{t} df(v) - df(v) \right| = \left| \left(e^{\eta(y)} e^{-d_M(x,y)^2/8t} \frac{s}{t} - 1 \right) df(v) \right|$$

$$\leq \left| \left(\frac{s}{t}(1 + O(s^2/t))(1 + O(s^4/t)) - 1 \right) df(v) \right|$$

In the large sample limit as the sampled points get closer s/t approaches identity while $s^k/t, k > 1$ terms vanish, and the exterior derivative is recovered. This observation is the basis for the attempt in Sect. 3 to formalize how sample paths, x_k^t, (from Sect. 1.1) encode the underlying geometry using Hodge-de Rham spectral triples. To warm up to the idea of replacing topological spaces (X_n) by the algebras $\mathcal{C}(X_n)$, we consider the persistence theory K_0 functor and use it towards analyzing dynamical geometry in point clouds.

2 $\mathbb{Q} \otimes K_0$-Persistence

Dynamical point clouds have been studied through persistent homology theories that use multiple persistence parameters for the incomparable space and time dimensions [13]. However, theories that use independent persistence parameters introduce complexity that intuitively is not necessary. Consider the question of detecting synchronization. Suppose in the extreme case, the point cloud completely synchronizes to evolve by rotation, so that the distance matrices $[D_{ij}]_{i,j\in[K]}$, are invariant in time, and persistence homology is constant for every value of space and time persistence parameters. One can detect this synchronization by analyzing the time persistence, but one now needs to test ranges of multiple independently varying persistence parameters to assign statistical confidence.

Since in the setup of the basic question, we are not exploring the development of new structures in relationships between points in time and are only interested in the sample paths of the points themselves, one expects that persistence in time is unnecessary. This intuition can be verified by showing that a persistence theory with only spatial parameters is sufficient in this setting. Furthermore, this theory is shown to be equivalent to a topological persistence theory.

2.1   A Category-Theoretic Formulation of Persistence

In [3] Bubenik and Scott formulate persistence homology abstractly in terms of functor F from a small poset category C into a category D called C-indexed diagram in D. The space of such functors with natural transformations is the category D^C. Composing a diagram in the category of topological spaces TOP indexed by $(\mathbb{R}, \geq)$, $F \in \text{TOP}^{(\mathbb{R}, \geq)}$, $F : (\mathbb{R}, \geq) \to \text{TOP}$ with the k-th homology functor H_k into the category of finite dimensional vector spaces VEC gives a diagram $H_k F \in \text{VEC}^{(\mathbb{R}, \geq)}$. For a topological space X, a map $f : X \to \mathbb{R}$ defines a functor $F \in \text{TOP}^{(\mathbb{R}, \geq)}$ by $F(a) = f^{-1}((-\infty, a])$, and from this data the p-persistent k-th homology group for the topological space X is defined as the image of map $H_k F(a \leq a+p)$ induced on homology by the inclusion $H_k F(a) \hookrightarrow H_k F(a+p)$. The construction of a persistence K-theory is analogous. We first use the functor $\mathcal{C} : \text{TOP} \to \mathcal{C}_1^*$, where $\mathcal{C}_1^*$ is the category of unital $\mathcal{C}^*$ algebras, that associates to compact Hausdorff topological spaces X, Y, the unital $\mathcal{C}^*$-algebras $\mathcal{C}(X), \mathcal{C}(Y)$ and to continuous map $\phi : X \to Y$, the pullback, $\phi^* : \mathcal{C}(Y) \to \mathcal{C}(X), \phi^*(h) = h \circ \phi$. Note that $\mathcal{C}$ reverse the direction of the arrows: for $\epsilon > 0$, the inclusion $i : F(a) \hookrightarrow F(a + \epsilon)$ induces $i^* : \mathcal{C}(F(a + \epsilon)) \to \mathcal{C}(F(A))$, we adjust this by using the opposite category to index, equivalently the diagram $F : (\mathbb{R}, \geq) \to \text{TOP}$ the associated diagram is $\mathcal{C}F : (\mathbb{R}, \geq) \to \mathcal{C}_1^*, -a \to \mathcal{C}(F(a))$.

2.2 The $\mathbb{Q} \otimes K_0$-Functor: Stability and Computation

On diagrams $F_1, F_2 \in D^C$, there exists an extended pseudo-metric, d^{IL}, defined as $d^{IL}(F_1, F_2) = \min\{\epsilon : \epsilon > 0, F_1, F_2 \text{ are } \epsilon \text{ interleaved}\}$ where F_1, F_2 are ϵ-interleaved if there exists natural transformations $\phi_{12} : F_1 \Rightarrow F_2, \phi_{21} : F_2 \Rightarrow F_1$ such that the following diagrams commute for $i, j \in \{1, 2\}, i \neq j$, the horizontal arrows being the inclusions of the diagram:

$$
\begin{array}{ccc}
F_i(a) & \longrightarrow & F_i(b) \\
\downarrow{\scriptstyle \phi_{ij(a)}} & & \downarrow{\scriptstyle \phi_{ij}(b)} \\
F_j(a + \epsilon) & \longrightarrow & F_j(b + \epsilon)
\end{array}
\qquad
\begin{array}{ccc}
F_i(a) & \longrightarrow & F_i(a + 2\epsilon) \\
& \searrow{\scriptstyle \phi_{ij}(a)} & \uparrow{\scriptstyle \phi_{ji}(a+\epsilon)} \\
& & F_j(a + \epsilon)
\end{array}
$$

The K_0-functor is the functor from $\mathcal{C}_1^*$ to the category of abelian groups ABGRP that associates to an unital $\mathcal{C}^*$-algebra its Grothendieck group. We consider the diagrams in $\text{ABGRP}^{(\mathbb{R}, \geq)}$, $K_0 \mathcal{C}F$. The p K_0-persistence is now defined for the diagram $\mathcal{C}F$ as the image of map $K_0 F_{\mathcal{C}}(a \geq a + p)$ induced on K_0-group by the map $K_0 \mathcal{C}F(a) \to K_0 \mathcal{C}F(a + p)$. As for topological persistence, a stability theorem is needed that ensures that similar topological spaces have similar K_0 persistence for their continuous function algebras. We have that $\mathcal{C}$ is contractive with respect to the interleaving distance even though it reverses the arrows. And since by [3, Prop 3.6], for any functor $H : \mathcal{C}_1^* \to E$ to any category $d^{IL}(H\mathcal{C}F_1, H\mathcal{C}F_2) \leq d^{IL}(\mathcal{C}F_1, \mathcal{C}F_2)$. This yields the needed stability theorem analogous to [3, Thm 5.1] as a corollary.

Lemma 1. *For $F_1, F_2 \in \text{TOP}^{(\mathbb{R}, \geq)}$, $d^{IL}(\mathcal{C}F_1, \mathcal{C}F_2) \leq d^{IL}(F_1, F_2)$*

Proof. This follows since if F_1, F_2 are ϵ-interleaved then $\mathcal{C}F_1, \mathcal{C}F_2$ are as well: the associated natural transformation obtained by composing $\phi_{ij} \circ \mathcal{C}$ and the as $\mathcal{C}$ simply reverse the arrows the interleaving relations still hold.

Corollary 1. *If $F_1, F_2 \in \text{TOP}^{(\mathbb{R}, \geq)}$ are such that $F_i(a) = f_i^{-1}((-\infty, a])$, then*

$$d^{IL}(K_0 \mathcal{C}F_1, K_0 \mathcal{C}F_2) \leq d^{IL}(\mathcal{C}F_1, \mathcal{C}F_2) \leq \|f_1 - f_2\|_\infty$$

Proof. From the proof of [3, Thm 5.1], $d^{IL}(F_1, F_2) \leq \|f_1 - f_2\|_\infty$, and the rest follows.

For increasing finite metric spaces arising by sampling from a manifold M, $X_1 \hookrightarrow X_2 \ldots \hookrightarrow M$, the inclusions $X_n \hookrightarrow X_{n+1}$ induce maps $\mathcal{C}(X_{n+1}) \to \mathcal{C}(X_n)$. Recovering the algebra $\mathcal{C}(M)$ in large n limit of such systems is difficult as projective limits of $\mathcal{C}^*$-algebras are more general pro $\mathcal{C}^*$-algebras. Even K_0 may not be continuous under the projective limits. Keeping in mind that the goal is simply a statistical test for the invariance of the underlying geometry, one can use the following observation to derive the test.

Lemma 2. $\mathbb{Q} \otimes K_0(\mathcal{C}(X)) \otimes \cong H^{even}(X, \mathbb{Q})$

Proof. This is obvious from results in topological K-theory [16] : $K^0(X) \otimes \mathbb{Q} \cong H^{even}(X, \mathbb{Q})$ for any topological space X where $K^0(X)$ is topological K^0 group associated to isomorphism classes of vector bundles over X. When X is compact Hausdorff space, as abelian groups $K_0(\mathcal{C}(X)) \cong K^0(X)$, and on taking the tensor product with $\mathbb{Q}$, they are isomorphic as $\mathbb{Q}$-vector spaces.

This reduces the algebraic K-theoretic persistence to the persistence of the even rational cohomology of the topological space X for which the sample paths approximate $\mathcal{C}(X)$. We offer a candidate space next such that the a topological persistence parameter can be obtained from K-persistence parameter.

Notice that if the time evolution is constrained to be by a possibly random isometry, then the hypothesis that the geometry of the point cloud is invariant translates to the null model being that the time evolution of the topological Rips simplicial complex at persistence parameter ϵ is simply the mapping cylinder M_1, formed by gluing $(x_k^t, t) \sim (x_k^{t+1}, t+1)$. Since the evolution is isometric, the maps are simplicial under the null hypothesis, and l-cells in complexes, $X_\epsilon^t, X_\epsilon^{t+1}$, at times $t, t + 1$, can be glued. Confidence in how well the true data conforms to the hypothesis can be quantified by testing the cohomology of the time-evolved complex X^T for actual data against the expected.

If the evolution is not isometric, then picking a single persistence parameter is difficult as distances in various parts of the geometry will change differently. This can be accounted for by using that as in the Brownian motion diffusive model, the generator is the Laplace-Beltrami operator, $\triangle$, which is being approximated by the point cloud Laplacian, $\triangle^{PC}$, the evolution will be isometric in expectation after adjusting for the eigenvalues of $\triangle^{PC}$; we will work with this rescaled metric. The rescaling does not affect the cohomology and allows for using a uniform spatial persistence parameter for the time-evolved complex. The actual data can now be tested against the simulated data or against the expectation to see if the null hypothesis of a diffusive model can be accepted.

The presence of stochastic effects is measured by the distribution of lifetimes of the simplices in this process since if the data is not evolving by a process generated by the Laplacian, then rescaling by the eigenvalues of the point cloud Laplacian will not yield isometric evolution, leading to simplices splitting and merging. At the same time, longer than expected lifetimes for simplices for the unscaled metric indicate likely synchronized sub-populations and possible homeostatic plasticity in the population response to input, which is of interest.

3 Embedded Finite and Hodge-de Rham Spectral Triples

For T large, the data of Brownian motion sample paths $\gamma : [0, T] \to M$ on a finite point cloud, composing with coordinate functions of the embedding $\psi : M \hookrightarrow \mathbb{R}^N$ gives a discretized version of the algebra $\mathcal{C}(M)$ because of the asymptotics of the time taken to get within r of each point, the r-covering time [7]. If this is enough to recover the geometry of M is central to the program we have outlined. This is best viewed as a question in noncommutative geometry: we recall how commutative geometry is encoded in the noncommutative language. The Hodge-de Rham spectral triple, $\mathfrak{A}_M$, for Riemannian manifold (M, g) is the data $(C^\infty(M), \Omega^\bullet(M), d+d^\dagger)$ where $d+d^\dagger$ is the Hodge-de Rham Dirac operator, d the exterior derivative on differential forms $\Omega^\bullet(X)$, $d^\dagger$ the Hodge dual. By Connes' spectral characterization of manifolds [4], (M, g) can be recovered from $\mathfrak{A}_M$. A finite spectral triple is the triple, $\mathfrak{A}_F := (\mathcal{A}_F, H_F, D_F)$, where $\mathcal{A}_F$ is an unital $*$-algebra represented faithfully on a Hilbert space H_F, $\dim H_F$ finite, and D a symmetric operator on H_F subject to some additional requirements. There's a standard representation of a finite metric space as a finite spectral triple.

We instead define an alternative representation using theorem 1 to obtain a finite Dirac operator. Suppose $\ldots X_i \subset X_{i+1} \ldots$ is an increasing sequence of metric space sampled from M, with $X_n = \{x_i : i \in [n]\}$. Then using the L^2-Hodge theory, with a uniform measure on X_n (except weighed multiply if $x_i = x_j, i \neq j$), we associate to it the restriction of the algebra $C^\infty(M)$ and $\Omega^\bullet(M)$. $\Omega^\bullet(M)$ is permissible as the space of co-chains is alternating, that is, $L_a(X_n^\bullet)$. Similar to theorem 1 it's possible to show that for the operator $\delta_{l-1}^{(n)}$ on $L^2(X^l)$, $\delta f(x_0 \ldots x_l)$ converges to $df_{x_0}(v_1 \ldots v_l)$ where v_i is the tangent at x_0 to the unit speed geodesic to x_i, the idea being to fix $l - 1$ of x_i's to get back to 1-cochain setting, although it needs to be checked that this is well defined regardless of order and number of fixed x_i's. This can be achieved using continuity of f as in the limit we restrict to infinitesimal neighborhoods of x_0. From this, the result below follows which for transparency can be roughly stated as –

Theorem 2. *The finite Dirac operators, $D_n := \delta^{(n)} + (\delta^{(n)})^*$, for X_n converges to the Hodge-de Rham Dirac operator $d + d^\dagger$ for M.*

To reconstruct the full Hodge-de Rham spectral triple from finite spectral triples (and (M, g) by [4]) the knowledge of $C^\infty(M)$ and $\Omega^\bullet(M)$ cannot be assumed. For recovering the algebra of the spectral triple, instead of taking the projective limit of $\mathcal{C}(X_n)$, we use a classical result in PL-topology [14]: M being smooth implies there exists a homeomorphism $\phi : K \to M$, where K is a polyhedron with triangulation $\{\sigma_i\}$ and ϕ is a piecewise diffeomorphism on σ_i, and therefore, $\mathcal{C}(M) \cong \mathcal{C}(K)$.

For the polyhedron K, viewed as the geometric realization $|\Sigma|$ of an abstract simplicial complex Σ on the finite vertex set $V_\Sigma = i \in [N]$ for $\{\sigma_i\}$, define $\mathcal{C}_\Sigma^{ab}$ as the abelianization of the universal C^*-algebra generated by positive generators $h_i, i \in V_\Sigma$, $h_{i_1} h_{i_2} \ldots h_{i_k} = 0$ whenever $\{i_j : j \in [k]\} \subset \Sigma$ and for all $m \in V_\Sigma, \sum_{k \in V_\Sigma} i_m i_k = i_m$ with the dense subalgebra generated algebraically on the same generators and relations. Then from [5], $\mathcal{C}_\Sigma^{ab} \cong \mathcal{C}_0(|\Sigma|)$ where $|\Sigma|$ is the geometric realization of Σ. As M, K are compact, $\mathcal{C}_\Sigma^{ab} \cong \mathcal{C}(M)$. The last ingredient needed to recover the Hodge-de Rham spectral triple is how the Dirac operator acts on $\mathcal{C}(K)$, but this is given by the homeomorphism ϕ, although some care is required as ϕ is only a piece-wise diffeomorphism (so the action of Dirac operator is not everywhere defined and we need to restrict to a differentiable subalgebra).

Finally, from d and $C^\infty(M)$, $\Omega^\bullet(M)$ can be constructed. The Dirac operators for the finite spectral triples we have used are weighed by the Euclidean heat kernel of the ambient space and are not standard finite spectral triples. This spectral triple with the Dirac operator coming from the L^2 Hodge theory is defined as an *embedded* finite spectral triple. The details of convergence to the Hodge-de Rham spectral triple[2] are developed in forthcoming work [9].

We end this article by posing the question of computationally reconstructing the Hodge-de Rham spectral triple, that is, recovering K and ϕ from the point cloud data, (X_n), in the large n limit. In particular, one does not expect to have access to the Euclidean embedding $\psi : M \to \mathbb{R}^N$, but can only construct the simplicial complex from sampled points, and the discrepancy of the action of Dirac operator on constructed simplex and M needs to be bound in terms of the geometry (e.g. M's maximum sectional curvature).

References

1. Bartholdi, L., Schick, T., Smale, N., Smale, S.: Hodge theory on metric spaces. Found. Comput. Math. **12**, 1–48 (2012)
2. Belkin, M., Niyogi, P.: Towards a theoretical foundation for Laplacian-based manifold methods. J. Comput. Syst. Sci. **74**(8), 1289–1308 (2008)
3. Bubenik, P., Scott, J.A.: Categorification of persistent homology. Discrete Comput. Geom. **51**(3), 600–627 (2014)
4. Connes, A.: On the spectral characterization of manifolds. J. Noncommutative Geom. **7**(1), 1–82 (2013)
5. Cuntz, J.: Noncommutative simplicial complexes and the Baum-Connes conjecture. Geom. Funct. Anal. GAFA **12**(2), 307–329 (2002)
6. Decreusefond, L., Ferraz, E., Randriambololona, H., Vergne, A.: Simplicial homology of random configurations. Adv. Appl. Probab. **46**(2), 325–347 (2014)
7. Dembo, A., Peres, Y., Rosen, J.: Brownian motion on compact manifolds: cover time and late points. Electron. J. Probab. **8** (2003)
8. Fountoulakis, N., Iyer, T., Mailler, C., Sulzbach, H.: Dynamical models for random simplicial complexes. Ann. Appl. Probab. **32**(4), 2860–2913 (2022)
9. Gakkhar, S., Marcolli, M.: Embedded finite and Hodge-de Rham spectral triples (2023, in preparation)
10. Gardner, R.J., et al.: Toroidal topology of population activity in grid cells. Nature, 1–6 (2022)
11. Giné, E., Koltchinskii, V.: Empirical graph Laplacian approximation of Laplace-Beltrami operators: large sample results. IN: Lecture Notes-Monograph Series, pp. 238–259 (2006)
12. Hsu, E.P.: Stochastic Analysis on Manifolds, No. 38, American Mathematical Soc. (2002)

---

[2] The very recent article [15], considers a similar approach for convergence of finite spectral triples without the heat kernel weights, building on the Hodge theory for the non-commutative laplacian. Using the L^2 Hodge theory complements their ideas, and in particular, gives a canonical choice for both the Dirac operator and Laplacian regardless of the distribution on the finite spaces, making progress on an open research direction they cite.

13. Kim, W., Mémoli, F.: Spatiotemporal persistent homology for dynamic metric spaces. Discrete Comput. Geom. **66**, 831–875 (2021)
14. Lurie, J.: Topics in geometric topology (18.937) (2009)
15. Tageddine, D., Nave, J.C.: Statistical fluctuation of infinitesimal spaces. arXiv preprint arXiv:2304.10617 (2023)
16. Wegge-Olsen, N.E.: K-theory and C^*-algebras (1993)

An Approximate Projection onto the Tangent Cone to the Variety of Third-Order Tensors of Bounded Tensor-Train Rank

Charlotte Vermeylen[1]([✉]) [ID], Guillaume Olikier[2] [ID], and Marc Van Barel[1] [ID]

[1] Department of Computer Science, KU Leuven, Heverlee, Belgium
charlotte.vermeylen@kuleuven.be
[2] ICTEAM Institute, UCLouvain, Louvain-la-Neuve, Belgium

Abstract. An approximate projection onto the tangent cone to the variety of third-order tensors of bounded tensor-train rank is proposed and proven to satisfy a better angle condition than the one proposed by Kutschan (2019). Such an approximate projection enables, e.g., to compute gradient-related directions in the tangent cone, as required by algorithms aiming at minimizing a continuously differentiable function on the variety, a problem appearing notably in tensor completion. A numerical experiment is presented which indicates that, in practice, the angle condition satisfied by the proposed approximate projection is better than both the one satisfied by the approximate projection introduced by Kutschan and the proven theoretical bound.

Keywords: Projection · Tangent cone · Angle condition · Tensor-train decomposition

1 Introduction

Tangent cones play an important role in constrained optimization to describe admissible search directions and to formulate optimality conditions [9, Chap. 6]. In this paper, we focus on the set

$$\mathbb{R}^{n_1 \times n_2 \times n_3}_{\leq (k_1, k_2)} := \{ X \in \mathbb{R}^{n_1 \times n_2 \times n_3} \mid \mathrm{rank}_{\mathrm{TT}}(X) \leq (k_1, k_2) \}, \tag{1}$$

where $\mathrm{rank}_{\mathrm{TT}}(X)$ denotes the tensor-train rank of X (see Sect. 2.2), which is a real algebraic variety [4], and, given $X \in \mathbb{R}^{n_1 \times n_2 \times n_3}_{\leq (k_1, k_2)}$, we propose an *approximate projection* onto the tangent cone $T_X \mathbb{R}^{n_1 \times n_2 \times n_3}_{\leq (k_1, k_2)}$, i.e., a set-valued mapping $\tilde{\mathcal{P}}_{T_X \mathbb{R}^{n_1 \times n_2 \times n_3}_{\leq (k_1, k_2)}} : \mathbb{R}^{n_1 \times n_2 \times n_3} \multimap T_X \mathbb{R}^{n_1 \times n_2 \times n_3}_{\leq (k_1, k_2)}$ such that there exists $\omega \in (0, 1]$ such that, for all $Y \in \mathbb{R}^{n_1 \times n_2 \times n_3}$ and all $\tilde{Y} \in \tilde{\mathcal{P}}_{T_X \mathbb{R}^{n_1 \times n_2 \times n_3}_{\leq (k_1, k_2)}} Y$,

$$\langle Y, \tilde{Y} \rangle \geq \omega \| \mathcal{P}_{T_X \mathbb{R}^{n_1 \times n_2 \times n_3}_{\leq (k_1, k_2)}} Y \| \| \tilde{Y} \|, \tag{2}$$

© The Author(s), under exclusive license to Springer Nature Switzerland AG 2023
F. Nielsen and F. Barbaresco (Eds.): GSI 2023, LNCS 14071, pp. 484–493, 2023.
https://doi.org/10.1007/978-3-031-38271-0_48

where $\langle \cdot, \cdot \rangle$ is the inner product on $\mathbb{R}^{n_1 \times n_2 \times n_3}$ given in [1, Example 4.149], $\| \cdot \|$ is the induced norm, and the set

$$\mathcal{P}_{T_X \mathbb{R}^{n_1 \times n_2 \times n_3}_{\leq (k_1, k_2)}} Y := \underset{Z \in T_X \mathbb{R}^{n_1 \times n_2 \times n_3}_{\leq (k_1, k_2)}}{\mathrm{argmin}} \|Z - Y\|^2 \qquad (3)$$

is the projection of Y onto $T_X \mathbb{R}^{n_1 \times n_2 \times n_3}_{\leq (k_1, k_2)}$. By [10, Definition 2.5], inequality (2) is called an *angle condition*; it is well defined since, as $T_X \mathbb{R}^{n_1 \times n_2 \times n_3}_{\leq (k_1, k_2)}$ is a closed cone, all elements of $\mathcal{P}_{T_X \mathbb{R}^{n_1 \times n_2 \times n_3}_{\leq (k_1, k_2)}} Y$ have the same norm (see Sect. 3). Such an approximate projection enables, e.g., to compute a gradient-related direction in $T_X \mathbb{R}^{n_1 \times n_2 \times n_3}_{\leq (k_1, k_2)}$, as required in the second step of [10, Algorithm 1] if the latter is used to minimize a continuously differentiable function $f : \mathbb{R}^{n_1 \times n_2 \times n_3} \to \mathbb{R}$ on $\mathbb{R}^{n_1 \times n_2 \times n_3}_{\leq (k_1, k_2)}$, a problem appearing notably in tensor completion; see [11] and the references therein.

An approximate projection onto $T_X \mathbb{R}^{n_1 \times n_2 \times n_3}_{\leq (k_1, k_2)}$ satisfying the angle condition (2) with $\omega = \frac{1}{6\sqrt{n_1 n_2 n_3}}$ was proposed in [5, §5.4]. If X is a singular point of the variety, i.e., $(r_1, r_2) := \mathrm{rank}_{\mathrm{TT}}(X) \neq (k_1, k_2)$, the approximate projection proposed in this paper ensures (see Theorem 1)

$$\omega = \sqrt{\max \left\{ \frac{k_1 - r_1}{n_1 - r_1}, \frac{k_2 - r_2}{n_3 - r_2} \right\}}, \qquad (4)$$

which is better, and can be computed via SVDs (see Algorithm 1). We point out that no general formula to project onto the closed cone $T_X \mathbb{R}^{n_1 \times n_2 \times n_3}_{\leq (k_1, k_2)}$, which is neither linear nor convex (see Sect. 2.3), is known in the literature.

This paper is organized as follows. Preliminaries are introduced in Sect. 2. Then, in Sect. 3, we introduce the proposed approximate projection and prove that it satisfies (2) with ω as in (4) (Theorem 1). Finally, in Sect. 4, we present a numerical experiment where the proposed approximate projection preserves the direction better than the one from [5, §5.4].

2   Preliminaries

In this section, we introduce the preliminaries needed for Sect. 3. In Sect. 2.1, we recall basic facts about orthogonal projections. Then, in Sect. 2.2, we review the tensor-train decomposition. Finally, in Sect. 2.3, we review the description of the tangent to $\mathbb{R}^{n_1 \times n_2 \times n_3}_{\leq (k_1, k_2)}$ given in [4, Theorem 2.6].

2.1   Orthogonal Projections

Given $n, p \in \mathbb{N}$ with $n \geq p$, we let $\mathrm{St}(p, n) := \{U \in \mathbb{R}^{n \times p} \mid U^\top U = I_p\}$ denote the Stiefel manifold. For every $U \in \mathrm{St}(p, n)$, we let $P_U := UU^\top$ and $P_U^\perp := I_n - P_U$ denote the orthogonal projections onto the range of U and its orthogonal complement, respectively. The proof of Theorem 1 relies on the following basic result.

Lemma 1. *Let $A \in \mathbb{R}^{n \times m}$ have rank r. If $\hat{A} = \hat{U}\hat{S}\hat{V}^\top$ is a truncated SVD of rank s of A, with $s < r$, then, for all $U \in \mathrm{St}(s,n)$ and all $V \in \mathrm{St}(s,m)$,*

$$\|P_{\hat{U}}A\| \geq \|P_U A\|, \qquad\qquad \|P_{\hat{U}}A\|^2 \geq \frac{s}{r}\|A\|^2, \qquad (5)$$

$$\|AP_{\hat{V}}\| \geq \|AP_V\|, \qquad\qquad \|AP_{\hat{V}}\|^2 \geq \frac{s}{r}\|A\|^2. \qquad (6)$$

Proof. By the Eckart–Young theorem, $\hat{A}$ is a projection of A onto

$$\mathbb{R}^{n \times m}_{\leq s} := \{X \in \mathbb{R}^{n \times m} \mid \mathrm{rank}(X) \leq s\}.$$

Thus, since $\mathbb{R}^{n \times m}_{\leq s}$ is a closed cone, the same conditions as in (14) hold. Moreover, since $\hat{S}\hat{V}^\top = \hat{U}^\top A$ and thus $\hat{A} = \hat{U}\hat{U}^\top A = P_{\hat{U}}A$, it holds that

$$\|P_{\hat{U}}A\|^2 = \max\left\{\|A_1\|^2 \mid A_1 \in \mathbb{R}^{n \times m}_{\leq s}, \langle A_1, A\rangle = \|A_1\|^2\right\}.$$

Furthermore, for all $U \in \mathrm{St}(s,n)$, $\langle P_U A, A\rangle = \langle P_U A, P_U A + P_U^\perp A\rangle = \|P_U A\|^2$. Hence,

$$\{P_U A \mid U \in \mathrm{St}(s,n)\} \subseteq \{A_1 \in \mathbb{R}^{n \times m}_{\leq s} \mid \langle A_1, A\rangle = \|A_1\|^2\}.$$

Thus, $\|P_{\hat{U}}A\|^2 = \max_{U \in \mathrm{St}(s,n)}\|P_U A\|^2$. The left inequality in (5) follows, and the one in (6) can be obtained similarly.

By orthogonal invariance of the Frobenius norm and by definition of $\hat{A}$,

$$\|A\|^2 = \sum_{i=1}^r \sigma_i^2, \qquad\qquad \|\hat{A}\|^2 = \sum_{i=1}^s \sigma_i^2,$$

where $\sigma_1, \ldots, \sigma_r$ are the singular values of A in decreasing order. Moreover, either $\sigma_s^2 \geq \frac{1}{r}\sum_{i=1}^r \sigma_i^2$ or $\sigma_s^2 < \frac{1}{r}\sum_{i=1}^r \sigma_i^2$. In the first case, we have

$$\|\hat{A}\|^2 = \sum_{i=1}^s \sigma_i^2 \geq s\sigma_s^2 \geq s\frac{\sum_{i=1}^r \sigma_i^2}{r} = \frac{s}{r}\|A\|^2.$$

In the second case, we have

$$\|\hat{A}\|^2 = \sum_{i=1}^r \sigma_i^2 - \sum_{i=s+1}^r \sigma_i^2 \geq \|A\|^2 - (r-s)\sigma_s^2 > \|A\|^2 - (r-s)\frac{\sum_{i=1}^r \sigma_i^2}{r} = \frac{s}{r}\|A\|^2.$$

Thus, in both cases, the second inequality in (5) holds. The second inequality in (6) can be obtained in a similar way.   □

2.2   The Tensor-Train Decomposition

In this section, we review basic facts about the tensor-train decomposition (TTD) that are used in Sect. 3; we refer to the original paper [8] and the subsequent works [3,11,12] for more details.

A *tensor-train decomposition* of $X \in \mathbb{R}^{n_1 \times n_2 \times n_3}$ is a factorization

$$X = X_1 \cdot X_2 \cdot X_3, \tag{7}$$

where $X_1 \in \mathbb{R}^{n_1 \times r_1}$, $X_2 \in \mathbb{R}^{r_1 \times n_2 \times r_2}$, $X_3 \in \mathbb{R}^{r_2 \times n_3}$, and '$\cdot$' denotes the contraction between a matrix and a tensor. The minimal (r_1, r_2) for which a TTD of X exists is called the *TT-rank* of X and is denoted by $\text{rank}_{\text{TT}}(X)$. By [2, Lemma 4], the set

$$\mathbb{R}^{n_1 \times n_2 \times n_3}_{(k_1, k_2)} := \{ X \in \mathbb{R}^{n_1 \times n_2 \times n_3} \mid \text{rank}_{\text{TT}}(X) = (k_1, k_2) \} \tag{8}$$

is a smooth embedded submanifold of $\mathbb{R}^{n_1 \times n_2 \times n_3}$.

Let $X^{\text{L}} := [X]^{n_1 \times n_2 n_3} := \text{reshape}(X, n_1 \times n_2 n_3)$ and $X^{\text{R}} = [X]^{n_1 n_2 \times n_3} := \text{reshape}(X, n_1 n_2 \times n_3)$ denote respectively the left and right unfoldings of X. Then, $\text{rank}_{\text{TT}}(X) = (\text{rank}(X^{\text{L}}), \text{rank}(X^{\text{R}}))$ and the minimal rank decomposition can be obtained by computing two successive SVDs of unfoldings; see [8, Algorithm 1]. The contraction interacts with the unfoldings according to the following rules:

$$X_1 \cdot X_2 \cdot X_3 = \left[X_1 (X_2 \cdot X_3)^{\text{L}} \right]^{n_1 \times n_2 n_3}, \quad X_2 \cdot X_3 = \left[X_2^{\text{R}} X_3 \right]^{r_1 \times n_2 n_3}.$$

For every $i \in \{1, 2, 3\}$, if $U_i \in \text{St}(r_i, n_i)$, then the mode-$i$ vectors of $[P_{U_1} X^{\text{L}} (P_{U_3} \otimes P_{U_2})]^{n_1 \times n_2 n_3}$ are the orthogonal projections onto the range of U_i of those of X. A similar property holds for X^{R}. The tensor X is said to be *left-orthogonal* if $n_1 \leq n_2 n_3$ and $(X^{\text{L}})^{\top} \in \text{St}(n_1, n_2 n_3)$, and *right-orthogonal* if $n_3 \leq n_1 n_2$ and $X^{\text{R}} \in \text{St}(n_3, n_1 n_2)$.

As a TTD is not unique, certain orthogonality conditions can be enforced, which can improve the numerical stability of algorithms working with TTDs. Those used in this work are given in Lemma 2.

2.3   The Tangent Cone to the Low-Rank Variety

In [4, Theorem 2.6], a parametrization of the tangent cone to $\mathbb{R}^{n_1 \times n_2 \times n_3}_{\leq (k_1, k_2)}$ is given and, because this parametrization is not unique, corresponding orthogonality conditions are added. The following lemma recalls this parametrization however with slightly different orthogonality conditions which make the proofs in the rest of the paper easier, and the numerical computations more stable because they enable to avoid matrix inversion in Algorithm 1.

Lemma 2. *Let* $X \in \mathbb{R}^{n_1 \times n_2 \times n_3}_{(r_1, r_2)}$ *have* $X = X_1 \cdot X_2'' \cdot X_3'' = X_1' \cdot X_2' \cdot X_3$ *as TTDs, where* $(X_2''^{\text{L}})^{\top} \in \text{St}(r_1, n_2 r_2)$, $X_3''^{\top} \in \text{St}(r_2, n_3)$, $X_1' \in \text{St}(r_1, n_1)$, *and* $X_2'^{\text{R}} \in \text{St}(r_2, r_1 n_2)$. *Then,* $T_X \mathbb{R}^{n_1 \times n_2 \times n_3}_{\leq (k_1, k_2)}$ *is the set of all* G *such that*

$$G = \begin{bmatrix} X_1' & U_1 & W_1 \end{bmatrix} \cdot \begin{bmatrix} X_2' & U_2 & W_2 \\ 0 & Z_2 & V_2 \\ 0 & 0 & X_2'' \end{bmatrix} \cdot \begin{bmatrix} W_3 \\ V_3 \\ X_3'' \end{bmatrix} \tag{9}$$

with $U_1 \in \text{St}(s_1, n_1)$, $W_1 \in \mathbb{R}^{n_1 \times r_1}$, $U_2 \in \mathbb{R}^{r_1 \times n_2 \times s_2}$, $W_2 \in \mathbb{R}^{r_1 \times n_2 \times r_2}$, $Z_2 \in \mathbb{R}^{s_1 \times n_2 \times s_2}$, $V_2 \in \mathbb{R}^{s_1 \times n_2 \times r_2}$, $W_3 \in \mathbb{R}^{r_2 \times n_3}$, $V_3^\top \in \text{St}(s_2, n_3)$, $s_i = k_i - r_i$ *for all* $i \in \{1, 2\}$, *and*

$$U_1^\top X_1' = 0, \quad W_1^\top X_1' = 0, \quad (U_2^R)^\top X_2'^R = 0,$$
$$W_3 X_3''^\top = 0, \quad V_3 X_3''^\top = 0, \quad V_2^L (X_2''^L)^\top = 0. \tag{10}$$

Proof. By [4, Theorem 2.6], $T_X \mathbb{R}^{n_1 \times n_2 \times n_3}_{\leq (k_1, k_2)}$ is the set of all $G \in \mathbb{R}^{n_1 \times n_2 \times n_3}$ that can be decomposed as

$$G = \begin{bmatrix} X_1' & \dot{U}_1 & \dot{W}_1 \end{bmatrix} \cdot \begin{bmatrix} X_2' & \dot{U}_2 & \dot{W}_2 \\ 0 & \dot{Z}_2 & \dot{V}_2 \\ 0 & 0 & X_2' \end{bmatrix} \cdot \begin{bmatrix} \dot{W}_3 \\ \dot{V}_3 \\ X_3 \end{bmatrix},$$

with the orthogonality conditions

$$\dot{U}_1^\top X_1' = 0, \quad \left(\dot{V}_2 \cdot X_3\right)^L \left(\left(X_2' \cdot X_3\right)^L\right)^\top = 0, \quad \dot{V}_3 X_3^\top = 0, \tag{11}$$
$$\dot{W}_1^\top X_1' = 0, \quad \left(\dot{W}_2^R\right)^\top X_2'^R = 0, \quad \left(\dot{U}_2^R\right)^\top X_2'^R = 0.$$

The following invariances hold for all $B \in \mathbb{R}^{s_2 \times s_2}, C \in \mathbb{R}^{r_2 \times r_2}, Q \in \mathbb{R}^{s_1 \times s_1}$, and $R \in \mathbb{R}^{r_1 \times r_1}$:

$$G = \begin{bmatrix} X_1' & \dot{U}_1 Q^{-1} & \dot{W}_1 R^{-1} \end{bmatrix} \cdot \begin{bmatrix} X_2' & \dot{U}_2 \cdot B & \dot{W}_2 \cdot C \\ 0 & Q \cdot \dot{Z}_2 \cdot B & Q \cdot \dot{V}_2 \cdot C \\ 0 & 0 & R \cdot X_2' \cdot C \end{bmatrix} \cdot \begin{bmatrix} \dot{W}_3 \\ B^{-1} \dot{V}_3 \\ C^{-1} X_3' \end{bmatrix}.$$

Then, if we define $U_1 := \dot{U}_1 Q^{-1}$, $W_1 := \dot{W}_1 R^{-1}$, $U_2 := \dot{U}_2 \cdot B$, $Z_2 := Q \cdot \dot{Z}_2 \cdot B$, $V_2 := Q \cdot \dot{V}_2 \cdot C$, $X_2'' := R \cdot X_2' \cdot C$, $V_3 := B^{-1} \dot{V}_3$, and $X_3'' := C^{-1} X_3'$, the matrices B, C, Q, and R can be chosen such that X_2'' is left-orthogonal, $X_3''^\top \in \text{St}(r_2, n_3)$, $V_3^\top \in \text{St}(s_2, n_3)$, and $U_1 \in \text{St}(s_1, n_1)$, e.g., using SVDs. Additionally, $\dot{W}_3$ can be decomposed as $\dot{W}_3 = \dot{W}_3 X_3''^\top X_3'' + W_3$. The two terms involving $\dot{W}_3$ and $\dot{W}_2 \cdot C$ can then be regrouped as

$$X_1' \cdot X_2' \cdot \dot{W}_3 + X_1' \cdot \dot{W}_2 \cdot C X_3'' = X_1' \cdot W_2 \cdot X_3'' + X_1' \cdot X_2' \cdot W_3,$$

where we have defined $W_2 := X_2' \cdot \dot{W}_3 X_3''^\top + \dot{W}_2 \cdot C$, obtaining the parametrization (9) satisfying (10). □

Expanding (9) yields, by (10), a sum of six mutually orthogonal TTDs:

$$G = W_1 \cdot X_2'' \cdot X_3'' + X_1' \cdot X_2' \cdot W_3 + X_1' \cdot W_2 \cdot X_3''$$
$$+ U_1 \cdot V_2 \cdot X_3'' + X_1' \cdot U_2 \cdot V_3 + U_1 \cdot Z_2 \cdot V_3. \tag{12}$$

Thus, the following holds:

$$W_1 = G^L \left(\left(X_2'' \cdot X_3''\right)^L\right)^\top, \quad W_2 = X_1'^\top \cdot G \cdot X_3''^\top, \quad W_3 = \left(\left(X_1'' \cdot X_2''\right)^R\right)^\top G^R,$$
$$U_2 = X_1'^\top \cdot G \cdot V_3^\top, \quad V_2 = U_1^\top \cdot G \cdot X_3''^\top, \quad Z_2 = U_1^\top \cdot G \cdot V_3^\top. \tag{13}$$

The first three terms in (12) form the tangent space $T_X \mathbb{R}^{n_1 \times n_2 \times n_3}_{(r_1, r_2)}$, the projection onto which is described in [7, Theorem 3.1 and Corollary 3.2].

3 The Proposed Approximate Projection

In this section, we prove Proposition 1 and then use it to prove Theorem 1. Both results rely on the following observation. By [6, Proposition A.6], since $T_X \mathbb{R}^{n_1 \times n_2 \times n_3}_{\leq (k_1, k_2)}$ is a closed cone, for all $Y \in \mathbb{R}^{n_1 \times n_2 \times n_3}$ and $\hat{Y} \in \mathcal{P}_{T_X \mathbb{R}^{n_1 \times n_2 \times n_3}_{\leq (k_1, k_2)}} Y$, it holds that $\langle Y - \hat{Y}, \hat{Y} \rangle = 0$ or, equivalently, $\langle Y, \hat{Y} \rangle = \|\hat{Y}\|^2$. Thus, all elements of $\mathcal{P}_{T_X \mathbb{R}^{n_1 \times n_2 \times n_3}_{\leq (k_1, k_2)}} Y$ have the same norm and (3) can be rewritten as

$$\mathcal{P}_{T_X \mathbb{R}^{n_1 \times n_2 \times n_3}_{\leq (k_1, k_2)}} Y = \underset{\substack{Z \in T_X \mathbb{R}^{n_1 \times n_2 \times n_3}_{\leq (k_1, k_2)} \\ \langle Y, Z \rangle = \|Z\|^2}}{\operatorname{argmax}} \|Z\| = \underset{\substack{Z \in T_X \mathbb{R}^{n_1 \times n_2 \times n_3}_{\leq (k_1, k_2)} \\ \langle Y, Z \rangle = \|Z\|^2}}{\operatorname{argmax}} \left\langle Y, \frac{Z}{\|Z\|} \right\rangle. \quad (14)$$

Proposition 1. *Let X be as in Lemma 2. For every $Y \in \mathbb{R}^{n_1 \times n_2 \times n_3}$ and every $\hat{Y} \in \mathcal{P}_{T_X \mathbb{R}^{n_1 \times n_2 \times n_3}_{\leq (k_1, k_2)}} Y$, if U_1 and V_3 are the parameters of $\hat{Y}$ in (12), then the parameters W_1, W_2, W_3, U_2, V_2, and Z_2 of $\hat{Y}$ can be written as*

$$W_1 = P^\perp_{X_1'} \left(Y \cdot X_3''^\top \right)^{\mathrm{L}} \left(X_2''^{\mathrm{L}} \right)^\top, \qquad\qquad W_3 = X_2'^{\mathrm{R}} \left(X_1'^\top \cdot Y \right)^{\mathrm{R}} P^\perp_{X_3''^\top},$$

$$U_2 = \left[P^\perp_{X_2^{\mathrm{R}}} \left(X_1'^\top \cdot Y \right)^{\mathrm{R}} \right]^{r_1 \times n_2 \times n_3} \cdot V_3^\top, \qquad W_2 = X_1'^\top \cdot Y \cdot X_3''^\top, \qquad (15)$$

$$V_2 = U_1^\top \cdot \left[\left(Y \cdot X_3''^\top \right)^{\mathrm{L}} P^\perp_{\left(X_2''^{\mathrm{L}} \right)^\top} \right]^{n_1 \times n_2 \times r_2}, \qquad Z_2 = U_1^\top \cdot Y \cdot V_3^\top.$$

Furthermore, $Y_\|(U_1, V_3)$ defined as in (12) with the parameters from (15) is a feasible point of (14) for all U_1 and all V_3.

Proof. Straightforward computations show that $\langle Y, \hat{Y} \rangle = \langle Y_\|(U_1, V_3), \hat{Y} \rangle$ and $\langle Y_\|(U_1, V_3), Y - Y_\|(U_1, V_3) \rangle = 0$. Thus, $Y_\|(U_1, V_3)$ is a feasible point of (14). Since $\hat{Y}$ is a solution to (14), $\|Y_\|(U_1, V_3)\| \leq \|\hat{Y}\|$. Therefore, if $\hat{Y} = 0$, then $Y_\|(U_1, V_3) = 0$ and consequently all parameters in (15) are zero because of (13). Otherwise, by using the Cauchy–Schwarz inequality, we have

$$\|\hat{Y}\|^2 = \langle Y, \hat{Y} \rangle = \langle Y_\|(U_1, V_3), \hat{Y} \rangle \leq \|Y_\|(U_1, V_3)\| \|\hat{Y}\| \leq \|\hat{Y}\|^2, \quad (16)$$

where the last inequality holds because $\hat{Y}$ is a solution to (14). It follows that the Cauchy–Schwarz inequality is an equality and hence there exists $\lambda \in (0, \infty)$ such that $Y_\|(U_1, V_3) = \lambda \hat{Y}$. By (16), $\lambda = 1$. Thus, because of (13), the parameters in (15) are those of $\hat{Y}$. $\qquad\square$

Theorem 1. *Let X be as in Lemma 2 with $(r_1, r_2) \neq (k_1, k_2)$. The approximate projection that computes the parameters U_1 and V_3 of $\hat{Y} \in \hat{\mathcal{P}}_{T_X \mathbb{R}^{n_1 \times n_2 \times n_3}_{\leq (k_1, k_2)}} Y$ in (12) with Algorithm 1 and the parameters W_1, W_2, W_3, U_2, V_2, and Z_2 with (15) satisfies (2) with ω as in (4) for all ε and all $i_{\max}$ in Algorithm 1.*

Proof. Let $(s_1, s_2) := (k_1 - r_1, k_2 - r_2)$ and $\hat{Y} \in \mathcal{P}_{T_X \mathbb{R}^{n_1 \times n_2 \times n_3}_{\leq(k_1,k_2)}} Y$. Thus, $s_1 + s_2 > 0$. Because W_1, W_2, W_3, U_2, V_2, and Z_2 are as in (15), it holds that $\tilde{Y} = Y_{\|}(U_1, V_3)$, thus $\tilde{Y}$ is a feasible point of (14), and hence (2) is equivalent to $\|\tilde{Y}\| \geq \omega \|\hat{Y}\|$. To compare the norm of $\tilde{Y}$ with the norm of $\hat{Y}$, $\hat{U}_1$ and $\hat{V}_3$ are defined as the parameters of $\hat{Y}$. From (15), and because all terms are mutually orthogonal, we have that

$$\|\tilde{Y}\|^2 = \left\| \mathcal{P}_{T_X \mathbb{R}^{n_1 \times n_2 \times n_3}_{(r_1,r_2)}} Y \right\|^2 + \left\| P_{U_1} (Y \cdot X_3''^{\top})^{\mathrm{L}} P_{(X_2''^{\mathrm{L}})^{\top}}^{\perp} (X_3'' \otimes I_{n_2}) \right\|^2$$
$$+ \left\| \left((P_{U_1} \cdot Y)^{\mathrm{R}} + (I_{n_2} \otimes X_1') P_{X_2'^{\mathrm{R}}}^{\perp} (X_1'^{\top} \cdot Y)^{\mathrm{R}} \right) P_{V_3^{\top}}^{\perp} \right\|^2.$$

Now, assume that $s_2/(n_3 - r_2) > s_1/(n_1 - r_1)$ and consider the first iteration of Algorithm 1. Because in the second step V_3 is obtained by a truncated SVD of $\left((P_{U_1} \cdot Y)^{\mathrm{R}} + (I_{n_2} \otimes X_1') P_{X_2'^{\mathrm{R}}}^{\perp} (X_1'^{\top} \cdot Y)^{\mathrm{R}} \right) P_{X_3''^{\top}}^{\perp}$ and, by using (6),

$$\|\tilde{Y}\|^2 \geq \left\| \mathcal{P}_{T_X \mathbb{R}^{n_1 \times n_2 \times n_3}_{(r_1,r_2)}} Y \right\|^2 + \left\| P_{U_1} (Y \cdot X_3''^{\top})^{\mathrm{L}} P_{(X_2''^{\mathrm{L}})^{\top}}^{\perp} (X_3'' \otimes I_{n_2}) \right\|^2$$
$$+ \frac{s_2}{n_3 - r_2} \left\| \left((P_{U_1} \cdot Y)^{\mathrm{R}} + (I_{n_2} \otimes X_1') P_{X_2'^{\mathrm{R}}}^{\perp} (X_1'^{\top} \cdot Y)^{\mathrm{R}} \right) P_{X_3''^{\top}}^{\perp} \right\|^2.$$

Furthermore, because in the first step U_1 is obtained from the truncated SVD of $P_{X_1'}^{\perp} \left((Y \cdot P_{X_3''^{\top}}^{\perp})^{\mathrm{L}} + (Y \cdot X_3''^{\top})^{\mathrm{L}} P_{X_2''^{\mathrm{L}}}^{\perp} (X_3'' \otimes I_{n_2}) \right)$ and by using (5),

$$\|\tilde{Y}\|^2 \geq \left\| \mathcal{P}_{T_X \mathbb{R}^{n_1 \times n_2 \times n_3}_{(r_1,r_2)}} Y \right\|^2 + \frac{s_2}{n_3 - r_2} \left\| P_{\hat{U}_1} (Y \cdot X_3''^{\top})^{\mathrm{L}} P_{(X_2''^{\mathrm{L}})^{\top}}^{\perp} (X_3'' \otimes I_{n_2}) \right\|^2$$
$$+ \frac{s_2}{n_3 - r_2} \left\| \left((P_{\hat{U}_1} \cdot Y)^{\mathrm{R}} + (I_{n_2} \otimes X_1') P_{X_2'^{\mathrm{R}}}^{\perp} (X_1'^{\top} \cdot Y)^{\mathrm{R}} \right) P_{X_3''^{\top}}^{\perp} \right\|^2,$$

where we have used that a multiplication with $\frac{s_2}{n_3 - r_2}$ can only decrease the norm. The same is true for a multiplication with $P_{\hat{V}_3^{\top}}$ and thus

$$\|\tilde{Y}\|^2 \geq \frac{s_2}{n_3 - r_2} \left(\left\| \mathcal{P}_{T_X \mathbb{R}^{n_1 \times n_2 \times n_3}_{(r_1,r_2)}} Y \right\|^2 + \left\| P_{\hat{U}_1} (Y \cdot X_3''^{\top})^{\mathrm{L}} P_{(X_2''^{\mathrm{L}})^{\top}}^{\perp} (X_3'' \otimes I_{n_2}) \right\|^2 \right.$$
$$\left. + \left\| \left((P_{\hat{U}_1} \cdot Y)^{\mathrm{R}} + (I_{n_2} \otimes X_1') P_{X_2'^{\mathrm{R}}}^{\perp} (X_1'^{\top} \cdot Y)^{\mathrm{R}} \right) P_{\hat{V}_3^{\top}} \right\|^2 \right) = \frac{s_2}{n_3 - r_2} \|\hat{Y}\|^2.$$

In Algorithm 1, the norm of the approximate projection increases monotonously. Thus, this lower bound is satisfied for any ε and $i_{\max}$. A similar derivation can be made if $s_2/(n_3 - r_2) \leq s_1/(n_1 - r_1)$. □

This section ends with three remarks on Algorithm 1. First, the instruction "$[U, S, V] \leftarrow \mathrm{SVD}_s(A)$" means that $USV^{\top}$ is a truncated SVD of rank s of A. Since those SVDs are not necessarily unique, Algorithm 1 can output several (U_1, V_3) for a given input, and hence the approximate projection is set-valued.

Second, the most computationally expensive operation in Algorithm 1 is the truncated SVD. The first step of the first iteration requires to compute either

s_1 singular vectors of a matrix of size $n_1 \times n_2 n_3$ or s_2 singular vectors of a matrix of size $n_1 n_2 \times n_3$. All subsequent steps are computationally less expensive since each of them merely requires to compute either s_1 singular vectors of

$$A := P_{X_1'}^\perp \left[(Y \cdot V_3^\top)^{\mathrm{L}}, (Y \cdot X_3''^\top)^{\mathrm{L}} P_{X_2''^{\mathrm{L}\top}}^\perp \right] \in \mathbb{R}^{n_1 \times n_2(r_2 + s_2)} \text{ or } s_2 \text{ singular vectors}$$

of $B := \left[(U_1^\top \cdot Y)^{\mathrm{R}}; P_{X_2'^{\mathrm{R}}}^\perp (X_1'^\top \cdot Y)^{\mathrm{R}} \right] P_{X_3''^\top}^\perp \in \mathbb{R}^{(r_1 + s_1)n_2 \times n_3}$, and, in general, $s_1 + r_1 \ll n_1$ and $s_2 + r_2 \ll n_3$. This is because

$$P_{X_1'}^\perp \left((Y \cdot P_{V_3^\top})^{\mathrm{L}} + (Y \cdot X_3''^\top)^{\mathrm{L}} P_{X_2''^{\mathrm{L}\top}}^\perp (X_3'' \otimes I_{n_2}) \right) = A \begin{bmatrix} V_3 \otimes I_{n_2} \\ X_3'' \otimes I_{n_2} \end{bmatrix},$$

and the rightmost matrix, being in $\mathrm{St}(n_2(s_2 + r_2), n_2 n_3)$ by Lemma 2, does not change the left singular vectors (and values). The argument for B is similar. The MATLAB implementation of Algorithm 1 that is used to perform the numerical experiment in Sect. 4 computes a subset of singular vectors (and singular values) using the svd function with 'econ' flag.

Third, studying the numerical stability of Algorithm 1 would require a detailed error analysis, which is out of the scope of the paper. Nevertheless, the modified orthogonality conditions improve the stability compared to the approximate projection described in [5, §5.4.4] because Algorithm 1 uses only orthogonal matrices to project onto vector spaces (it uses no Moore–Penrose inverse).

Algorithm 1. Iterative method to obtain U_1 and V_3 of $\tilde{\mathcal{P}}_{T_X \mathbb{R}^{n_1 \times n_2 \times n_3}_{\leq(k_1, k_2)}} Y$

Input: $Y \in \mathbb{R}^{n_1 \times n_2 \times n_3}$, $X = X_1' \cdot X_2' \cdot X_3 = X_1 \cdot X_2'' \cdot X_3'' \in \mathbb{R}^{n_1 \times n_2 \times n_3}_{(r_1, r_2)}$, $\varepsilon > 0$, $i_{\max}, s_1, s_2 \in \mathbb{N} \setminus \{0\}$

  Initialize: $i \leftarrow 0$, $V_3 \leftarrow P_{X_3''^\top}^\perp$, $U_1 \leftarrow P_{X_1'}^\perp$, $\eta_1 \leftarrow 0$, $\eta_{\text{new}} \leftarrow \infty$

  if $s_2/(n_3 - r_2) > s_1/(n_1 - r_1)$ **then**

    while $i < i_{\max}$ **and** $|\eta_{\text{new}} - \eta_1| \leq \varepsilon$ **do**

      $\eta_1 \leftarrow \eta_{\text{new}}$, $i \leftarrow i + 1$

      $[U_1, \sim, \sim] \leftarrow \mathrm{SVD}_{s_1} \left(P_{X_1'}^\perp \left((Y \cdot P_{V_3^\top})^{\mathrm{L}} + (Y \cdot X_3''^\top)^{\mathrm{L}} P_{X_2''^{\mathrm{L}\top}}^\perp (X_3'' \otimes I_{n_2}) \right) \right)$

      $[\sim, S, V_3^\top] \leftarrow \mathrm{SVD}_{s_2} \left(\left((P_{U_1} \cdot Y)^{\mathrm{R}} + (I_{n_2} \otimes X_1') P_{X_2'^{\mathrm{R}}}^\perp (X_1'^\top \cdot Y)^{\mathrm{R}} \right) P_{X_3''^\top}^\perp \right)$

      $\eta_{\text{new}} \leftarrow \|S\|^2 + \|P_{U_1} (Y \cdot X_3''^\top)^{\mathrm{L}} P_{X_2''^{\mathrm{L}\top}}^\perp (X_3'' \otimes I_{n_2})\|^2$

  else

    while $i < i_{\max}$ **and** $|\eta_{\text{new}} - \eta_1| \leq \varepsilon$ **do**

      $\eta_1 \leftarrow \eta_{\text{new}}$, $i \leftarrow i + 1$

      $[\sim, \sim, V_3^\top] \leftarrow \mathrm{SVD}_{s_2} \left(\left((P_{U_1} \cdot Y)^{\mathrm{R}} + (I_{n_2} \otimes X_1') P_{X_2'^{\mathrm{R}}}^\perp (X_1'^\top \cdot Y)^{\mathrm{R}} \right) P_{X_3''^\top}^\perp \right)$

      $[U_1, S, \sim] \leftarrow \mathrm{SVD}_{s_1} \left(P_{X_1'}^\perp \left((Y \cdot P_{V_3^\top})^{\mathrm{L}} + (Y \cdot X_3''^\top)^{\mathrm{L}} P_{X_2''^{\mathrm{L}\top}}^\perp (X_3'' \otimes I_{n_2}) \right) \right)$

      $\eta_{\text{new}} \leftarrow \|S\|^2 + \|(I_{n_2} \otimes X_1') P_{X_2'^{\mathrm{R}}}^\perp (X_1'^\top \cdot Y)^{\mathrm{R}} P_{V_3^\top}\|^2$

Output: U_1, V_3.

---

4 A Numerical Experiment

To compute gradient-related directions in the tangent cone, the input tensor for Algorithm 1 would be the gradient of the continuously differentiable function that is considered. Since, in general, such tensors are dense, we consider in this section randomly generated pairs of dense tensors (X, Y) with $X \in \mathbb{R}^{n_1 \times n_2 \times n_3}_{\leq (k_1, k_2)}$ and $Y \in \mathbb{R}^{n_1 \times n_2 \times n_3}$, and compare the values of $\left\langle \frac{\tilde{Y}}{\|\tilde{Y}\|}, \frac{Y}{\|Y\|} \right\rangle$ obtained by computing the approximate projection $\tilde{Y}$ of Y onto $T_X \mathbb{R}^{n_1 \times n_2 \times n_3}_{\leq (k_1, k_2)}$ using Theorem 1, the tensor diagrams from [5, §5.4.4], which we have implemented in MATLAB, and the point output by the built-in MATLAB function `fmincon` applied to (3). The latter can be considered as a benchmark for the exact projection. Since $\|Y\| \geq \|\mathcal{P}_{T_X \mathbb{R}^{n_1 \times n_2 \times n_3}_{\leq (k_1, k_2)}} Y\|$, (2) is satisfied if $\left\langle \frac{\tilde{Y}}{\|\tilde{Y}\|}, \frac{Y}{\|Y\|} \right\rangle \geq \omega$.

For this experiment, we set $(k_1, k_2) := (3, 3)$ and generate fifty random pairs (X, Y), where $X \in \mathbb{R}^{5 \times 5 \times 5}_{(2,2)}$ and $Y \in \mathbb{R}^{5 \times 5 \times 5}$, using the built-in MATLAB function `randn`. For such pairs, the ω from (4) equals $\frac{1}{3}$. We use Algorithm 1 with $\varepsilon := 10^{-16}$, which implies that $i_{\max}$ is used as stopping criterion. In the left subfigure of Fig. 1, the box plots for this experiment are shown for two values of $i_{\max}$. As can be seen, for both values of $i_{\max}$, the values of $\left\langle \frac{\tilde{Y}}{\|\tilde{Y}\|}, \frac{Y}{\|Y\|} \right\rangle$ obtained by the proposed approximate projection are close to those obtained by `fmincon` and are larger than those obtained by the approximate projection from [5, §5.4]. We observe that $\left\langle \frac{\tilde{Y}}{\|\tilde{Y}\|}, \frac{Y}{\|Y\|} \right\rangle$ is always larger than $\frac{1}{3}$, which suggests that (4) is a pessimistic estimate. The middle subfigure compares ten of the fifty pairs. For one of these pairs, the proposed method obtains a better result than `fmincon`. This is possible since the `fmincon` solver does not necessarily output a global solution because of the nonconvexity of (3). An advantage of the proposed approximate projection is that it requires less computation time than the `fmincon` solver (a fraction of a second for the former and up to ten seconds for the latter). In

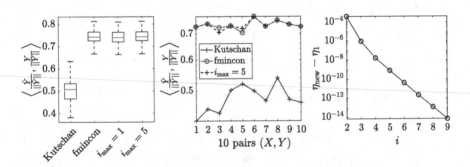

Fig. 1. A comparison of $\left\langle \frac{\tilde{Y}}{\|\tilde{Y}\|}, \frac{Y}{\|Y\|} \right\rangle$ for fifty randomly generated pairs (X, Y), with $X \in \mathbb{R}^{5 \times 5 \times 5}_{(2,2)}$, $Y \in \mathbb{R}^{5 \times 5 \times 5}$, and $(k_1, k_2) := (3, 3)$, for the approximate projection defined in Theorem 1, the one from [5, §5.4], and the one output by `fmincon`. On the rightmost figure, the evolution of $\eta_{\text{new}} - \eta_1$ is shown for one of the fifty pairs.

the rightmost subfigure, the evolution of $\eta_{\text{new}} - \eta_1$ is shown for one of the fifty pairs. This experiment was run on a laptop with a AMD Ryzen 7 PRO 3700U processor (4 cores, 8 threads) having 13.7 GiB of RAM under Kubuntu 20.04. The MATLAB version is R2020a. The code is publicly available.[1]

References

1. Hackbusch, W.: Tensor Spaces and Numerical Tensor Calculus. Springer Series in Computational Mathematics, 2nd edn., vol. 56. Springer, Cham (2019). https://doi.org/10.1007/978-3-642-28027-6

2. Holtz, S., Rohwedder, T., Schneider, R.: On manifolds of tensors of fixed TT-rank. Numer. Math. **120**(4), 701–731 (2012)

3. Kressner, D., Steinlechner, M., Uschmajew, A.: Low-rank tensor methods with subspace correction for symmetric eigenvalue problems. SIAM J. Sci. Comput. **36**(5), A2346–A2368 (2014)

4. Kutschan, B.: Tangent cones to tensor train varieties. Linear Algebra Appl. **544**, 370–390 (2018)

5. Kutschan, B.: Convergence of gradient methods on hierarchical tensor varieties. Ph.D. thesis, TU Berlin (2019)

6. Levin, E., Kileel, J., Boumal, N.: Finding stationary points on bounded-rank matrices: a geometric hurdle and a smooth remedy. Math. Program. (2022)

7. Lubich, C., Oseledets, I.V., Vandereycken, B.: Time integration of tensor trains. SIAM J. Numer. Anal. **53**(2), 917–941 (2015)

8. Oseledets, I.V.: Tensor-train decomposition. SIAM J. Sci. Comput. **33**(5), 2295–2317 (2011)

9. Rockafellar, R.T., Wets, R.J.B.: Variational Analysis, Grundlehren der mathematischen Wissenschaften, vol. 317. Springer, Heidelberg (1998). Corrected 3rd printing 2009

10. Schneider, R., Uschmajew, A.: Convergence results for projected line-search methods on varieties of low-rank matrices via Łojasiewicz inequality. SIAM J. Optim. **25**(1), 622–646 (2015)

11. Steinlechner, M.: Riemannian optimization for high-dimensional tensor completion. SIAM J. Sci. Comput. **38**(5), S461–S484 (2016)

12. Steinlechner, M.: Riemannian optimization for solving high-dimensional problems with low-rank tensor structure. Ph.D. thesis, EPFL (2016)

---

[1] URL: https://github.com/golikier/ApproxProjTangentConeTTVariety.

Toric Fiber Products in Geometric Modeling

Eliana Duarte[1,2]📍, Benjamin Hollering[1]📍, and Maximilian Wiesmann[1(✉)]📍

[1] Max-Planck-Institute for Mathematics in the Sciences, Leipzig, Germany
{benjamin.hollering,wiesmann}@mis.mpg.de
[2] Centro de Matemática, Universidade do Porto, Porto, Portugal
eliana.gelvez@fc.up.pt

Abstract. An important challenge in Geometric Modeling is to classify polytopes with rational linear precision. Equivalently, in Algebraic Statistics one is interested in classifying scaled toric varieties, also known as discrete exponential families, for which the maximum likelihood estimator can be written in closed form as a rational function of the data (rational MLE). The toric fiber product (TFP) of statistical models is an operation to iteratively construct new models with rational MLE from lower dimensional ones. In this paper we introduce TFPs to the Geometric Modeling setting to construct polytopes with rational linear precision and give explicit formulae for their blending functions. A special case of the TFP is taking the Cartesian product of two polytopes and their blending functions.

Keywords: Toric variety · Exponential family · Blending function

1 Introduction

A discrete statistical model with m outcomes is a subset $\mathcal{M}$ of the open probability simplex $\Delta_{m-1}^\circ = \{(p_1, \dots, p_m) : p_i > 0, \sum p_i = 1\}$. Each point in Δ_{m-1}° specifies a probability distribution for a random variable X with outcome space $[m] := \{1, \dots, m\}$ by setting $p_i = P(X = i)$. Given an i.i.d. sample $\mathcal{D} = \{X_1, \dots, X_N\}$ of X, let u_i be the number of times the outcome i appears in $\mathcal{D}$ and set $u = (u_1, \dots, u_m)$. The maximum likelihood estimator of the model $\mathcal{M}$ is the function $\Phi : \mathbb{N}^m \to \mathcal{M}$ that assigns to u the point in $\mathcal{M}$ that maximizes the log-likelihood function $\ell(u|p) := \sum_i u_i \log(p_i)$. For discrete regular exponential families, the log-likelihood function is concave, and under certain genericity conditions on $u \in \mathbb{N}^m$, existence and uniqueness of the maximum likelihood estimate $\Phi(u)$ is guaranteed [9]. This does not mean that the MLE is given in closed form but rather that it can be computed using iterative proportional scaling [5].

In Algebraic Statistics, discrete exponential families are studied from an algebro-geometric perspective using the fact that the Zariski closure of any such family is a scaled projective toric variety, we refer to these as toric varieties from this point forward. In this setting, the complexity of maximum likelihood estimation for a model $\mathcal{M}$, or more generally any algebraic variety, is measured in

F. Nielsen and F. Barbaresco (Eds.): GSI 2023, LNCS 14071, pp. 494–503, 2023.
https://doi.org/10.1007/978-3-031-38271-0_49

terms of its maximum likelihood degree (ML degree). The ML degree of $\mathcal{M}$ is the number of critical points of the likelihood function over the complex numbers for generic u and it is an invariant of $\mathcal{M}$ [11]. If a model has ML degree one it means that the coordinate functions of Φ are rational functions in u, thus the MLE has a closed form expression which is in fact determined completely in terms of a Horn matrix as explained in [6,10]. It is an open problem in Algebraic Statistics to characterize the class of toric varieties with ML degree one and their respective Horn matrices.

The toric fiber product (TFP), introduced by Sullivant [14], is an operation that takes two toric varieties $\mathcal{M}_1, \mathcal{M}_2$ and, using compatibility criteria determined by a multigrading $\mathcal{A}$, creates a higher dimensional toric variety $\mathcal{M}_1 \times_{\mathcal{A}} \mathcal{M}_2$. This operation is used to construct a Markov basis for $\mathcal{M}_1 \times_{\mathcal{A}} \mathcal{M}_2$ by using Markov bases of $\mathcal{M}_1$ and $\mathcal{M}_2$. Interestingly, the ML degree of a TFP is the product of the ML degrees of its factors, therefore the TFP of two models with ML degree one yields again a model with ML degree one [2]. The Cartesian product of two statistical models is an instance of a TFP. Another example is the class of decomposable graphical models, each of these models has ML degree one and can be constructed iteratively from lower dimensional ones using TFPs [13,14].

In Geometric Modeling, it is an open problem to classify polytopes in dimension $d \geq 3$ having rational linear precision [3]. Remarkably, a polytope has rational linear precision if and only if its corresponding toric variety has ML degree one [8]. Inspired by Algebraic Statistics, it is our goal in this article to introduce the toric fiber product construction to Geometric Modeling. In statistics, the interest is in the closed form expression for the MLE; in Geometric Modeling, the interest is in explicitly writing blending functions defined on the polytope that satisfy the property of linear precision. Our main Theorem 2 gives an explicit formula for the blending functions defined on the toric fiber product of two polytopes that have rational linear precision.

2 Preliminaries

In this section we provide background on blending functions, rational linear precision, scaled projective toric varieties and toric fiber products. For a friendly introduction to Algebraic Statistics, we refer the reader to the book by Sullivant [15], in particular to Chapter 7 on maximum likelihood estimation. To the readers looking for more background on toric geometry we recommend the book by Cox, Little and Schenck [4].

2.1 Blending Functions

Let $P \subset \mathbb{R}^d$ be a lattice polytope with facet representation $P = \{\mathbf{p} \in \mathbb{R}^d : \langle \mathbf{p}, n_i \rangle \geq a_i, \forall i \in [R]\}$, where n_i is a primitive inward facing normal vector to the facet F_i. Without loss of generality, we will always assume that P is full-dimensional inside $\mathbb{R}^d$. The lattice distance of a point $\mathbf{p} \in \mathbb{R}^d$ to F_i is $h_i(\mathbf{p}) :=$

$\langle \mathbf{p}, n_i \rangle + a_i$, $i \in [R]$. Set $\mathcal{B} := P \cap \mathbb{Z}^d$, so $\mathcal{B}$ is the set of lattice points in P and let $w = (w_\mathbf{b})_{\mathbf{b} \in \mathcal{B}}$ be a vector of positive weights. To each $\mathbf{b} \in \mathcal{B}$ we associate the rational functions $\beta_\mathbf{b}, \beta_w, \beta_{w,\mathbf{b}} : P \to \mathbb{R}$ defined by

$$\beta_\mathbf{b}(\mathbf{p}) := \prod_{i=1}^{R} h_i(\mathbf{p})^{h_i(\mathbf{b})}, \quad \beta_w(\mathbf{p}) := \sum_{\mathbf{b} \in \mathcal{B}} w_\mathbf{b} \beta_\mathbf{b}(\mathbf{p}), \quad \text{and} \quad \beta_{w,\mathbf{b}} := w_\mathbf{b} \beta_\mathbf{b} / \beta_w. \quad (1)$$

The functions $\beta_{w,\mathbf{b}}$, $\mathbf{b} \in \mathcal{B}$, are the *toric blending functions* of the pair (P, w), introduced by Krasauskas [12] as generalizations of Bézier curves and surfaces to more general polytopes. Blending functions usually satisfy additional properties that make them amenable for computation, see for instance [12]. Given a set of control points $\{Q_\mathbf{b}\}_{\mathbf{b} \in \mathcal{B}}$, a *toric patch* is defined by the rule $F(\mathbf{p}) := \sum_{\mathbf{b} \in \mathcal{B}} \beta_\mathbf{b}(\mathbf{p}) Q_\mathbf{b}$.

The *scaled projective toric variety* $X_{\mathcal{B},w}$ is the Zariski closure of the image of the map $(\mathbb{C}^*)^d \to \mathbb{P}^{|\mathcal{B}|-1}$ defined by $\mathbf{t} \mapsto [w_\mathbf{b} \mathbf{t}^\mathbf{b}]_{\mathbf{b} \in \mathcal{B}}$. Here $\mathbf{t} = (t_1, \ldots, t_d)$, $\mathbf{b} = (b_1, \ldots, b_d)$ and $\mathbf{t}^\mathbf{b} = \prod_{i \in [d]} t_i^{b_i}$. The image of $X_{\mathcal{B},w}$ under the map $\mathbb{P}^{|\mathcal{B}|-1} \to \mathbb{C}^{|\mathcal{B}|}$, $[x_1 : \cdots : x_{|\mathcal{B}|}] \mapsto \frac{1}{x_1 + \cdots + x_{|\mathcal{B}|}}(x_1, \cdots, x_{|\mathcal{B}|})$ intersected with the positive orthant defines a discrete regular exponential family $\mathcal{M}_{\mathcal{B},w}$ inside $\Delta^\circ_{|\mathcal{B}|-1}$. In the literature these are also called log-linear models. In this construction we require that the vector of ones is in the rowspan of the matrix whose columns are the points in $\mathcal{B}$. If this is not the case, we add the vector of ones to this matrix.

Definition 1. *The pair (P, w) has* rational linear precision *if there is a set of rational functions $\{\hat{\beta}_\mathbf{b}\}_{\mathbf{b} \in \mathcal{B}}$ on $\mathbb{C}^d$ satisfying:*

1. *$\sum_{\mathbf{b} \in \mathcal{B}} \hat{\beta}_\mathbf{b} = 1$.*
2. *The functions $\{\hat{\beta}_\mathbf{b}\}_{\mathbf{b} \in \mathcal{B}}$ define a rational parametrization*

$$\hat{\beta} : \mathbb{C}^d \dashrightarrow X_{\mathcal{B},w} \subset \mathbb{P}^{|\mathcal{B}|-1}, \quad \hat{\beta}(\mathbf{t}) = (\hat{\beta}_\mathbf{b}(\mathbf{t}))_{\mathbf{b} \in \mathcal{B}}.$$

3. *For every $\mathbf{p} \in \text{Relint}(P) \subset \mathbb{C}^d$, $\hat{\beta}_\mathbf{b}(\mathbf{p})$ is defined and is a nonnegative real number.*
4. *Linear precision: $\sum_{\mathbf{b} \in \mathcal{B}} \hat{\beta}_\mathbf{b}(\mathbf{p})\mathbf{b} = \mathbf{p}$ for all $\mathbf{p} \in P$.*

The property of rational linear precision does not hold for arbitrary toric patches but it is desirable because the blending functions "provide barycentric coordinates for general control point schemes" [8]. A deep relation to Algebraic Statistics is provided by the following statement.

Theorem 1 ([8]). *The pair (P, w) has rational linear precision if and only if $X_{\mathcal{B},w}$ has ML degree one.*

Remark 1. Henceforth, to ease notation, we drop the usage of a vector of weights w for the blending functions $\beta_{w,\mathbf{b}}$ and the scaled projective toric variety $X_{\mathcal{B},w}$. Although we will not in general write them explicitly in the proofs, the weights play an important role in determining whether the toric variety has ML degree one or, equivalently, if the polytope has rational linear precision. A deep dive into the study of these scalings for toric varieties by using principal A-determinants is presented in [1].

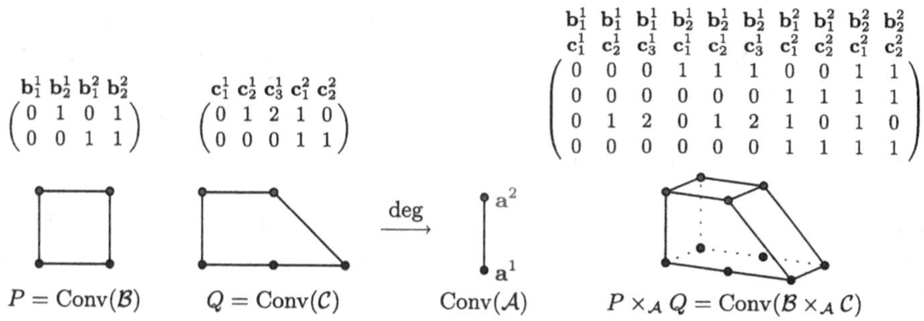

$$
\begin{array}{cccc}
\mathbf{b}_1^1\ \mathbf{b}_2^1\ \mathbf{b}_1^2\ \mathbf{b}_2^2 \\
\begin{pmatrix} 0 & 1 & 0 & 1 \\ 0 & 0 & 1 & 1 \end{pmatrix}
\end{array}
\qquad
\begin{array}{c}
\mathbf{c}_1^1\ \mathbf{c}_2^1\ \mathbf{c}_3^1\ \mathbf{c}_1^2\ \mathbf{c}_2^2 \\
\begin{pmatrix} 0 & 1 & 2 & 1 & 0 \\ 0 & 0 & 0 & 1 & 1 \end{pmatrix}
\end{array}
$$

$$
\begin{array}{c}
\mathbf{b}_1^1\ \mathbf{b}_1^1\ \mathbf{b}_1^1\ \mathbf{b}_2^1\ \mathbf{b}_2^1\ \mathbf{b}_2^1\ \mathbf{b}_1^2\ \mathbf{b}_1^2\ \mathbf{b}_2^2\ \mathbf{b}_2^2 \\
\mathbf{c}_1^1\ \mathbf{c}_2^1\ \mathbf{c}_3^1\ \mathbf{c}_1^2\ \mathbf{c}_2^2\ \mathbf{c}_3^1\ \mathbf{c}_1^2\ \mathbf{c}_2^2\ \mathbf{c}_1^2\ \mathbf{c}_2^2 \\
\begin{pmatrix}
0 & 0 & 0 & 1 & 1 & 1 & 0 & 0 & 1 & 1 \\
0 & 0 & 0 & 0 & 0 & 0 & 1 & 1 & 1 & 1 \\
0 & 1 & 2 & 0 & 1 & 2 & 1 & 0 & 1 & 0 \\
0 & 0 & 0 & 0 & 0 & 0 & 1 & 1 & 1 & 1
\end{pmatrix}
\end{array}
$$

$P = \mathrm{Conv}(\mathcal{B})$ $Q = \mathrm{Conv}(\mathcal{C})$ $\xrightarrow{\ \deg\ }$ $\mathrm{Conv}(\mathcal{A})$ $P \times_{\mathcal{A}} Q = \mathrm{Conv}(\mathcal{B} \times_{\mathcal{A}} \mathcal{C})$

Fig. 1. Toric fiber product of the point configurations $\mathcal{B}$ and $\mathcal{C}$ in Example 2. Each point configuration is displayed as a matrix with its corresponding convex hull below. The blue vertices in each polytope have degree e_1 while the red vertices in each polytope have degree e_2 in the associated multigrading $\mathcal{A}$. The degree map is $\deg(\mathbf{b}_j^i) = \deg(\mathbf{c}_k^i) = \mathbf{a}^i$. (Color figure online)

Example 1. Consider the point configurations $\mathcal{B} = \{(0,0),(1,0),(0,1),(1,1)\}$, $\mathcal{C} = \{(0,0),(1,0),(2,0),(1,1),(0,1)\}$ and set $P = \mathrm{Conv}(\mathcal{B})$, $Q = \mathrm{Conv}(\mathcal{C})$; these are displayed in Fig. 1. The facet presentation of P is

$$P = \{(x_1,x_2) \in \mathbb{R}^2 : x_1 \geq 0, x_2 \geq 0, 1 - x_1 \geq 0, 1 - x_2 \geq 0\}.$$

The lattice distance functions of a point $(x_1, x_2) \in \mathbb{R}^2$ to the facets of P are

$$h_1 = x_1, \ h_2 = x_2, \ h_3 = 1 - x_1, \ h_4 = 1 - x_2.$$

Therefore the toric blending functions of P with weights $w = (1,1,1,1)$ are:

$$\beta_{\binom{0}{0}} = (1-x_1)(1-x_2), \ \beta_{\binom{1}{0}} = x_2(1-x_1), \ \beta_{\binom{0}{1}} = x_1(1-x_2), \ \beta_{\binom{1}{1}} = x_1 x_2. \quad (2)$$

These toric blending functions satisfy the conditions in Definition 1; when this is the case, P is said to have *strict linear precision*. The polytope Q has rational linear precision for the vector of weights $w = (1,2,1,1,1)$. In this case, the toric blending functions do not satisfy condition 4 in Definition 1, however, as explained in [3], the following functions do:

$$\tilde{\beta}_{\binom{0}{0}} = \frac{(1-y_2)(2-y_1-y_2)^2}{(2-y_2)^2}, \ \tilde{\beta}_{\binom{1}{0}} = \frac{2y_1(1-y_2)(2-y_1-y_2)}{(2-y_2)^2}, \ \tilde{\beta}_{\binom{2}{0}} = \frac{y_1^2(1-y_2)}{(2-y_2)^2},$$

$$\tilde{\beta}_{\binom{0}{1}} = \frac{y_2(2-y_1-y_2)}{2-y_2}, \qquad \tilde{\beta}_{\binom{1}{1}} = \frac{y_1 y_2}{2-y_2}.$$

2.2 Toric Fiber Products of Point Configurations

Let $r \in \mathbb{N}$ and $s_i, t_i \in \mathbb{N}$ for $1 \leq i \leq r$. Fix integral point configurations $\mathcal{A} = \{\mathbf{a}^i : i \in [r]\} \subseteq \mathbb{Z}^d$, $\mathcal{B} = \{\mathbf{b}_j^i : i \in [r], j \in [s_i]\} \subseteq \mathbb{Z}^{d_1}$ and $\mathcal{C} = \{\mathbf{c}_k^i : i \in$

$[r], k \in [t_i]\} \subseteq \mathbb{Z}^{d_2}$. For any point configuration $\mathcal{P}$, we use $\mathcal{P}$ interchangeably to denote a set of points or the matrix whose columns are the points in $\mathcal{P}$; the symbol $|\mathcal{P}|$ will be used to denote the indexing set of $\mathcal{P}$. For each $i \in |\mathcal{A}|$, set $\mathcal{B}^i := \{\mathbf{b}_j^i : j \in [s_i]\}$ and $\mathcal{C}^i = \{\mathbf{c}_k^i : k \in [t_i]\}$. The indices i, j, k are reserved for elements in $|\mathcal{A}|, |\mathcal{B}^i|$ and $|\mathcal{C}^i|$, respectively.

Throughout this paper, we assume linear independence of $\mathcal{A}$ and the existence of a $\overline{w} \in \mathbb{Q}^d$ such that $\overline{w}\mathbf{a}^i = 1$ for all i; the latter condition ensures that if an ideal is homogeneous with respect to a multigrading in $\mathcal{A}$ it is also homogeneous in the usual sense. Sullivant introduces the TFP as an operation on toric ideals which are multigraded by $\mathcal{A}$; such condition, as explained in [7], is equivalent to the existence of linear maps $\pi_1 : \mathbb{Z}^{d_1} \to \mathbb{Z}^r$ and $\pi_2 : \mathbb{Z}^{d_2} \to \mathbb{Z}^r$ such that $\pi_1(\mathbf{b}_j^i) = \mathbf{a}^i$ for all i and j, and $\pi_2(\mathbf{c}_k^i) = \mathbf{a}^i$ for all i and k. We use deg to denote the projections π_1, π_2.

The *toric fiber product* of $\mathcal{B}$ and $\mathcal{C}$ is the point configuration $\mathcal{B} \times_{\mathcal{A}} \mathcal{C}$ given by

$$\mathcal{B} \times_{\mathcal{A}} \mathcal{C} = \{(\mathbf{b}_j^i, \mathbf{c}_k^i) : i \in |\mathcal{A}|, j \in |\mathcal{B}^i|, k \in |\mathcal{C}^i|\}.$$

In terms of toric varieties, introduced in Sect. 2.1, the toric fiber product of $X_{\mathcal{B}}$ and $X_{\mathcal{C}}$ is the toric variety $X_{\mathcal{B} \times_{\mathcal{A}} \mathcal{C}}$ associated to $\mathcal{B} \times_{\mathcal{A}} \mathcal{C}$ which is given in the following way. Let $X_{\mathcal{B}}$ and $X_{\mathcal{C}}$ have coordinates x_j^i and y_k^i respectively. Then $X_{\mathcal{B}} \times_{\mathcal{A}} \mathcal{C} = \phi(X_{\mathcal{B}} \times X_{\mathcal{C}})$ where ϕ is the monomial map

$$\phi : \mathbb{C}^{|\mathcal{B}|} \times \mathbb{C}^{|\mathcal{C}|} \to \mathbb{C}^{|\mathcal{B} \times_{\mathcal{A}} \mathcal{C}|}$$

$$(x_j^i, y_k^i) \mapsto x_j^i y_k^i = z_{jk}^i.$$

Furthermore, if $w, \tilde{w}$ are weights for $\mathcal{B}, \mathcal{C}$, respectively, then the vector of weights for $\mathcal{B} \times_{\mathcal{A}} \mathcal{C}$ is $w_{\mathcal{B} \times_{\mathcal{A}} \mathcal{C}} := (w_j^i \tilde{w}_k^i)_{(j,k) \in |\mathcal{B}^i \times \mathcal{C}^i|}^{i \in |\mathcal{A}|}$. We end this section with an example illustrating this operation.

Example 2. Consider the point configurations $\mathcal{B}$ and $\mathcal{C}$ in Example 1 and let $\mathcal{A} = \{e_1, e_2\}$ consist of two standard basis vectors. The construction of a degree map and the corresponding toric fiber product $\mathcal{B} \times_{\mathcal{A}} \mathcal{C}$ is explained in Fig. 1. The convex hull of the polytope $\mathcal{B} \times_{\mathcal{A}} \mathcal{C}$ is not full-dimensional in $\mathbb{R}^4$, it has dimension three. The 3D polytope in Fig. 1 is unimodularly equivalent to $\mathrm{Conv}(\mathcal{B} \times_{\mathcal{A}} \mathcal{C})$.

3   Blending Functions of Toric Fiber Products

In this section we show that the blending functions of the toric fiber product of two polytopes with rational linear precision can be constructed from the blending functions of the original polytopes and give an explicit formula for them. Throughout this section we use the setup for the toric fiber product introduced in Sect. 2.2. We let $P = \mathrm{Conv}(\mathcal{B})$ and $Q = \mathrm{Conv}(\mathcal{C})$ be polytopes with rational linear precision and denote their blending functions satisfying Definition 1 by $\{\beta_j^i\}_{j \in |\mathcal{B}^i|}^{i \in |\mathcal{A}|}$ and $\{\beta_k^i\}_{k \in |\mathcal{C}^i|}^{i \in |\mathcal{A}|}$, respectively.

Theorem 2. *If P and Q are polytopes with rational linear precision for weights w, $\tilde{w}$, respectively, then the toric fiber product $P\times_{\mathcal{A}}Q$ has rational linear precision with vector of weights $w_{\mathcal{B}\times_{\mathcal{A}}\mathcal{C}}$. Moreover, blending functions with rational linear precision for $P\times_{\mathcal{A}}Q$ are given by*

$$\beta_{j,k}^i(\mathbf{p},\mathbf{q}) = \frac{\beta_j^i(\mathbf{p})\beta_k^i(\mathbf{q})}{\sum_{j'\in|\mathcal{B}^i|}\beta_{j'}^i(\mathbf{p})} = \frac{\beta_j^i(\mathbf{p})\beta_k^i(\mathbf{q})}{\sum_{k'\in|\mathcal{C}^i|}\beta_{k'}^i(\mathbf{q})} \tag{3}$$

where $(\mathbf{p},\mathbf{q})\in P\times_{\mathcal{A}}Q$.

Remark 2. The two expressions on the right hand side of Eq. (3) are well defined on $\mathrm{Relint}(P\times_{\mathcal{A}}Q)$. The morphism $\beta_{j,k}^i$ extends to a rational function $\beta_{j,k}^i\colon \mathbb{C}^d \dashrightarrow \mathbb{C}$ where $d = \dim(P\times_{\mathcal{A}}Q)$. By abuse of notation, we will sometimes write $\beta_{j,k}^i(\mathbf{t}) = \frac{1}{N^i(\mathbf{t})}\beta_j^i(\mathbf{t})\beta_k^i(\mathbf{t})$ where $\mathbf{t}\in\mathbb{C}^d$ and $N^i(\mathbf{t})$ denotes the denominator as in (3).

The following example illustrates the construction in Theorem 2.

Example 3. Consider the polytopes P and Q from Example 1, with their vectors of weights. By Theorem 2, the blending functions for $P\times_{\mathcal{A}}Q$ are $\frac{\beta_j^i\tilde{\beta}_k^i}{\sum_j\beta_j^i} = \frac{\beta_j^i\tilde{\beta}_k^i}{\sum_k\tilde{\beta}_k^i}$. For example, the blending function corresponding to the point $\begin{pmatrix}\mathbf{b}_2^1 & \mathbf{c}_3^1\end{pmatrix}^T$ is

$$\beta_{2,3}^1 = \frac{\beta_2^1\tilde{\beta}_3^1}{\beta_1^1+\beta_1^2} = \frac{x_1(1-x_2)y_1^2(1-y_2)}{(1-x_2)(2-y_2)^2} = \frac{\beta_2^1\tilde{\beta}_3^1}{\tilde{\beta}_1^1+\tilde{\beta}_2^1+\tilde{\beta}_3^1} = \frac{x_1(1-x_2)y_1^2(1-y_2)}{(1-y_2)(2-y_2)^2}.$$

Note that while the denominators are not the same, the two expressions above are equal at all points in $\mathrm{Relint}(P\times_{\mathcal{A}}Q)$.

Before proving Theorem 2 we will prove two lemmas which will be used in the final proof. Our first lemma demonstrates how the blending functions behave on certain faces of P and Q. The second lemma shows that the two parametrizations in Eq. (3) yield the same MLE for a generic data point u.

Lemma 1. *Let P^i be the subpolytope defined by $P^i = \mathrm{Conv}\{\mathbf{b}_j^i : j\in|\mathcal{B}^i|\}$. Then, for $\mathbf{p}\in P^i$, we have*

$$\sum_{j\in|\mathcal{B}^i|}\beta_j^i(\mathbf{p}) = 1.$$

Proof. By assumption, $\beta\colon\mathbb{C}^{d_1}\dashrightarrow X_{\mathcal{B}}$, $\beta(\mathbf{t}) = \left(\beta_j^i(\mathbf{t})\right)_{j\in|\mathcal{B}^i|}^{i\in|\mathcal{A}|}$ is a rational parametrization of $X_{\mathcal{B}}$. Let $X_{\mathcal{B}}^i$ be the toric variety associated to P^i; we claim that $X_{\mathcal{B}}^i$ is parametrized by $\left(\beta_j^i(\mathbf{t})\right)_{j\in|\mathcal{B}^i|}$ and setting all other coordinates of β to zero. Indeed, consider the linear map

$$\deg\colon P\to\mathrm{Conv}(\mathcal{A}),\quad \mathbf{b}_j^i\mapsto\mathbf{a}^i.$$

As $\mathcal{A}$ is linearly independent, $\mathbf{a}^i$ is a vertex of $\mathrm{Conv}(\mathcal{A})$. Note that $P^i = \deg^{-1}(\mathbf{a}^i)$; as preimages of faces under linear maps are again faces, P^i is a face of P. The claim then follows from the Orbit-Cone Correspondence [4, Theorem 3.2.6]. We know that $\sum_{(i,j)\in|\mathcal{B}|} \beta_j^i = 1$. On P^i, all $\beta_j^{i'}$ for $i' \neq i$ vanish, so we must have $\sum_{j\in|\mathcal{B}^i|} \beta_j^i(\mathbf{p}) = 1$ for $\mathbf{p} \in P^i$. $\square$

We record the following fact as a consequence from the proof above.

Corollary 1. *Let P be a polytope equipped with a linearly independent multigrading $\mathcal{A}$. Then $P^i = \mathrm{Conv}(\mathbf{b}_j^i \mid j \in |\mathcal{B}^i|)$ is a face of P.*

Example 4. For the polytope Q in Fig. 1, we have $Q^1 = \mathrm{Conv}(\mathbf{c}_1^1, \mathbf{c}_2^1, \mathbf{c}_3^1)$ and $Q^2 = \mathrm{Conv}(\mathbf{c}_1^2, \mathbf{c}_2^2)$. The projection deg is illustrated in Fig. 1. To illustrate the result of Lemma 1, note that the sum of the blending functions associated to the lattice points in Q^1 is equal to $1 - y_2$.

Lemma 2. *Let P and Q be polytopes with rational linear precision and β_1, β_2 be two rational functions defined by*

$$\beta_1(\mathbf{t}) = \left(\frac{\beta_j^i(\mathbf{t})\beta_k^i(\mathbf{t})}{\sum_{j'\in|\mathcal{B}^i|} \beta_{j'}^i(\mathbf{t})} \right)^{i\in|\mathcal{A}|}_{(j,k)\in|\mathcal{B}^i\times\mathcal{C}^i|} , \quad \beta_2(\mathbf{t}) = \left(\frac{\beta_j^i(\mathbf{t})\beta_k^i(\mathbf{t})}{\sum_{k'\in|\mathcal{C}^i|} \beta_{k'}^i(\mathbf{t})} \right)^{i\in|\mathcal{A}|}_{(j,k)\in|\mathcal{B}^i\times\mathcal{C}^i|} .$$

For $u = \left(u_{j,k}^i \right)^{i\in|\mathcal{A}|}_{(j,k)\in|\mathcal{B}^i\times\mathcal{C}^i|}$, set $\mathbf{p} = \sum_{(i,j,k)\in|\mathcal{B}\times_{\mathcal{A}}\mathcal{C}|} \frac{u_{j,k}^i}{u_{+,+}^i} \mathbf{m}_{j,k}^i \in \mathbb{C}^d$. Then the maximum likelihood estimate for $X_{\mathcal{B}\times_{\mathcal{A}}\mathcal{C}}$ is

$$\beta_1(\mathbf{p}) = \beta_2(\mathbf{p}) = \left(\hat{p}_{j,k}^i \right)^{i\in|\mathcal{A}|}_{(j,k)\in|\mathcal{B}^i\times\mathcal{C}^i|} .$$

Proof. As P and Q have rational linear precision, by [3, Proposition 8.4] we have $\beta_j^i(\mathbf{p}) = (\hat{p}_{\mathcal{B}})_j^i$ and $\beta_k^i(\mathbf{p}) = (\hat{p}_{\mathcal{C}})_k^i$. Furthermore, by [2, Theorem 5.5], the MLE of the toric fiber product is given by $\hat{p}_{j,k}^i = \frac{(\hat{p}_{\mathcal{B}})_j^i(\hat{p}_{\mathcal{C}})_k^i}{(\hat{p}_{\mathcal{A}})^i}$. From the proof of [2, Lemma 5.10], as a consequence of Birch's Theorem, it follows that $(\hat{p}_{\mathcal{B}})_+^i = \frac{u_{+,+}^i}{u_{+,+}^i} = (\hat{p}_{\mathcal{A}})^i$, and analogously $(\hat{p}_{\mathcal{C}})_+^i = (\hat{p}_{\mathcal{A}})^i$. Therefore,

$$\sum_{j'\in|\mathcal{B}^i|} \beta_{j'}^i(\mathbf{p}) = (\hat{p}_{\mathcal{B}})_+^i = \sum_{k'\in|\mathcal{C}^i|} \beta_{k'}^i(\mathbf{p}) = (\hat{p}_{\mathcal{C}})_+^i = (\hat{p}_{\mathcal{A}})^i$$

and the desired statement follows. $\square$

We are now ready to prove Theorem 2.

Proof. Having rational linear precision is equivalent to having ML degree one by Theorem 1. Then the first statement is a direct consequence of the multiplicativity of the ML degree under toric fiber products [2, Theorem 5.5].

We first show that both expressions in (3) define rational parametrizations

$$\beta_1(\mathbf{t}) = \left(\frac{\beta_j^i(\mathbf{t})\beta_k^i(\mathbf{t})}{\sum_{j'\in|\mathcal{B}^i|} \beta_{j'}^i(\mathbf{t})} \right)^{i\in|\mathcal{A}|}_{(j,k)\in|\mathcal{B}^i\times\mathcal{C}^i|} , \quad \beta_2(\mathbf{t}) = \left(\frac{\beta_j^i(\mathbf{t})\beta_k^i(\mathbf{t})}{\sum_{k'\in|\mathcal{C}^i|} \beta_{k'}^i(\mathbf{t})} \right)^{i\in|\mathcal{A}|}_{(j,k)\in|\mathcal{B}^i\times\mathcal{C}^i|}$$

of $X_{\mathcal{B} \times_{\mathcal{A}} \mathcal{C}}$. To do this, we first show that the products $\beta_j^i \beta_k^i$ parametrize $X_{\mathcal{B} \times_{\mathcal{A}} \mathcal{C}}$ and the result then follows since β_1 and β_2 are equivalent to $\beta_j^i \beta_k^i$ under the torus action associated to the multigrading $\mathcal{A}$. Let $\phi : \mathbb{C}^{|\mathcal{B}|} \times \mathbb{C}^{|\mathcal{C}|} \to \mathbb{C}^{|\mathcal{B} \times_{\mathcal{A}} \mathcal{C}|}$ be the map given by

$$\phi(\mathbf{x}, \mathbf{y}) = (x_j^i y_k^i)_{(j,k) \in |\mathcal{B}^i \times \mathcal{C}^i|}^{i \in |\mathcal{A}|} .$$

Then the toric fiber product $X_{\mathcal{B} \times_{\mathcal{A}} \mathcal{C}}$ is precisely given by $\phi(X_{\mathcal{B}} \times X_{\mathcal{C}})$. Since the blending functions β_j^i and β_k^i parametrize $X_{\mathcal{B}}$ and $X_{\mathcal{C}}$, respectively, and $\beta_j^i \beta_k^i = \phi \circ (\beta_j^i, \beta_k^i)$, we immediately get that the $\beta_j^i \beta_k^i$ parametrize $X_{\mathcal{B} \times_{\mathcal{A}} \mathcal{C}}$. Now observe that the multigrading $\mathcal{A}$ induces an action of the torus $T_{\mathcal{A}} = (\mathbb{C}^*)^{|\mathcal{A}|}$ via

$$T_{\mathcal{A}} \times X_{\mathcal{B} \times_{\mathcal{A}} \mathcal{C}} \to X_{\mathcal{B} \times_{\mathcal{A}} \mathcal{C}}, \ (t^1, \ldots, t^{|\mathcal{A}|}) . (x_{j,k}^i)_{(j,k) \in |\mathcal{B}^i \times \mathcal{C}^i|}^{i \in |\mathcal{A}|} = (t^i x_{j,k}^i)_{(j,k) \in |\mathcal{B}^i \times \mathcal{C}^i|}^{i \in |\mathcal{A}|} .$$

Define $\tau : \mathbb{C}^{d_1} \to T_{\mathcal{A}}$ by

$$\tau = (\tau^1, \ldots, \tau^{|\mathcal{A}|}), \ \tau^i(\mathbf{t}) = \begin{cases} \left(\sum_{j \in |\mathcal{B}^i|} \beta_j^i(\mathbf{t}) \right)^{-1} & \text{if } \sum_{j \in |\mathcal{B}^i|} \beta_j^i(\mathbf{t}) \neq 0 \\ 1 & \text{else.} \end{cases}$$

Note that $\tau(\mathbf{t}) \in T_{\mathcal{A}}$ and $\tau(\mathbf{x}).(\beta_j^i(\mathbf{x}) \beta_k^i(\mathbf{x}))_{j \in |\mathcal{B}^i|, k \in |\mathcal{C}^i|}^{i \in |\mathcal{A}|} = \beta_1(\mathbf{x})$ for all $\mathbf{x} \in P \times_{\mathcal{A}} Q$, showing that $\beta_j^i(\mathbf{x}) \beta_k^i(\mathbf{x})$ and $\beta_1(\mathbf{x})$ lie in the same $T_{\mathcal{A}}$-orbit. A similar argument shows the same for $\beta_2(\mathbf{x})$, thus both β_1 and β_2 parametrize $X_{\mathcal{B} \times_{\mathcal{A}} \mathcal{C}}$.

We will now show the two expressions in Eq. 3 are equal. Let us define a new $\tau : \mathbb{C}^{d_1 + d_2} \to T_{\mathcal{A}}$ by

$$\tau = (\tau^1, \ldots, \tau^{|\mathcal{A}|}), \ \tau^i(\mathbf{t}) = \begin{cases} \frac{\sum_{j \in |\mathcal{B}^i|} \beta_j^i(\mathbf{t})}{\sum_{k \in |\mathcal{C}^i|} \beta_k^i(\mathbf{t})} & \text{if } \sum_{j \in |\mathcal{B}^i|} \beta_j^i(\mathbf{t}) \neq 0 \neq \sum_{k \in |\mathcal{C}^i|} \beta_k^i(\mathbf{t}) \\ 1 & \text{else.} \end{cases}$$

Clearly, $\tau(\mathbf{t}) \in T_{\mathcal{A}}$; we claim that $\tau(\mathbf{x}).\beta_1(\mathbf{x}) = \beta_2(\mathbf{x})$ for $\mathbf{x} \in P \times_{\mathcal{A}} Q$. First consider the case $\mathbf{x} \in P^i \times Q^i$, with P^i and Q^i defined as in Lemma 1. By the Orbit-Cone Correspondence applied to the $T_{\mathcal{A}}$-action, all coordinates in $\beta_1(\mathbf{x})$ and $\beta_2(\mathbf{x})$ vanish except for those graded by $\mathbf{a}^i$. By Lemma 1, $\sum_{j \in |\mathcal{B}^i|} \beta_j^i(\mathbf{x}) = \sum_{k \in |\mathcal{C}^i|} \beta_k^i(\mathbf{x}) = 1$, so in particular the claim holds. Now consider the case where $\mathbf{x} \notin \bigcup_{i \in |\mathcal{A}|} P^i \times Q^i$. Then, again by the Orbit-Cone Correspondence applied to the $T_{\mathcal{A}}$-action, for each $i \in |\mathcal{A}|$ there exist $j \in |\mathcal{B}^i|$ and $k \in |\mathcal{C}^i|$ such that $\beta_j^i(\mathbf{x}), \beta_k^i(\mathbf{x}) \neq 0$. Thus, by definition, $\tau(\mathbf{x}).\beta_1(\mathbf{x}) = \beta_2(\mathbf{x})$. We conclude that for all $\mathbf{x} \in P \times_{\mathcal{A}} Q$, $\beta_1(\mathbf{x})$ and $\beta_2(\mathbf{x})$ lie in the same $T_{\mathcal{A}}$-orbit. Equality of β_1 and β_2 then follows once there exists at least one point in each orbit where the two parametrizations agree. This is indeed the case: for the maximal orbit this is the point given in Lemma 2, for smaller orbits corresponding to faces of $P^i \times Q^i$ we can pick a point as in Lemma 1. It now remains to show that the $\beta_{j,k}^i$ sum to one and satisfy condition 4. in Definition 1; these are straightforward computations using the two different forms of $\beta_{j,k}^i$. Firstly, we have

$$\sum_{(i,j,k) \in |\mathcal{B} \times_{\mathcal{A}} \mathcal{C}|} \beta_{j,k}^i = \sum_{i \in |\mathcal{A}|, k \in |\mathcal{C}^i|} \beta_k^i \sum_{j \in |\mathcal{B}^i|} \frac{\beta_j^i}{\sum_{j' \in |\mathcal{B}^i|} \beta_{j'}^i} = \sum_{i \in |\mathcal{A}|, k \in |\mathcal{C}^i|} \beta_k^i = 1.$$

Finally, we compute

$$
\sum_{(i,j,k)\in|\mathcal{B}\times_{\mathcal{A}}\mathcal{C}|} \beta^i_{j,k}(\mathbf{p})\mathbf{m}^i_{j,k} = \sum_{i\in|\mathcal{A}|,j\in|\mathcal{B}^i|} \beta^i_j(\mathbf{p}) \sum_{k\in|\mathcal{C}^i|} \frac{\beta^i_k(\mathbf{p})}{\sum_{k'\in|\mathcal{C}^i|}\beta^i_{k'}(\mathbf{p})}(\mathbf{b}^i_j,0)
$$

$$
+ \sum_{i\in|\mathcal{A}|,k\in|\mathcal{C}^i|} \beta^i_k(\mathbf{p}) \sum_{j\in|\mathcal{B}^i|} \frac{\beta^i_j(\mathbf{p})}{\sum_{j'\in|\mathcal{B}^i|}\beta^i_{j'}(\mathbf{p})}(0,\mathbf{c}^i_k)
$$

$$
= \left(\sum_{i\in|\mathcal{A}|,j\in|\mathcal{B}^i|} \beta^i_j(\mathbf{p})\mathbf{b}^i_j , \sum_{i\in|\mathcal{A}|,k\in|\mathcal{C}^i|} \beta^i_k(\mathbf{p})\mathbf{c}^i_k \right) = \mathbf{p}.
$$

Therefore, the $\beta^i_{j,k}$ constitute blending functions with rational linear precision. □

Acknowledgements. Eliana Duarte was supported by the FCT grant 2020. 01933. CEECIND, and partially supported by CMUP under the FCT grant UIDB/00144/ 2020.

References

1. Améndola, C., et al.: The maximum likelihood degree of toric varieties. J. Symb. Comput. **92**, 222–242 (2019). https://doi.org/10.1016/j.jsc.2018.04.016
2. Améndola, C., Kosta, D., Kubjas, K.: Maximum likelihood estimation of toric Fano varieties. Algebraic Stat. **11**(1), 5–30 (2020). https://doi.org/10.2140/astat.2020. 11.5
3. Clarke, P., Cox, D.A.: Moment maps, strict linear precision, and maximum likelihood degree one. Adv. Math. **370**, 107233 (2020). https://doi.org/10.1016/j.aim. 2020.107233
4. Cox, D.A., Little, J.B., Schenck, H.K.: Toric Varieties, vol. 124. American Mathematical Society (2011)
5. Darroch, J.N., Ratcliff, D.: Generalized iterative scaling for log-linear models. Ann. Math. Stat. **43**(5), 1470–1480 (1972). https://doi.org/10.1214/aoms/1177692379
6. Duarte, E., Marigliano, O., Sturmfels, B.: Discrete statistical models with rational maximum likelihood estimator. Bernoulli **27**(1), 135–154 (2021). https://doi.org/ 10.3150/20-BEJ1231
7. Engström, A., Kahle, T., Sullivant, S.: Multigraded commutative algebra of graph decompositions. J. Algebraic Combin. **39**(2), 335–372 (2013). https://doi.org/10. 1007/s10801-013-0450-0
8. Garcia-Puente, L.D., Sottile, F.: Linear precision for parametric patches. Adv. Comput. Math. **33**(2), 191–214 (2010). https://doi.org/10.1007/s10444-009-9126-7
9. Haberman, S.J.: Log-linear models for frequency tables derived by indirect observation: maximum likelihood equations. Ann. Stat. **2**(5), 911–924 (1974). https:// doi.org/10.1214/aos/1176342813
10. Huh, J.: Varieties with maximum likelihood degree one. J. Algebraic Stat. **5**(1), 1–17 (2014). https://doi.org/10.52783/jas.v5i1.25
11. Huh, J., Sturmfels, B.: Likelihood geometry. Comb. Algebraic Geom. **2108**, 63–117 (2014). https://doi.org/10.1007/978-3-319-04870-3

12. Krasauskas, R.: Toric surface patches. Adv. Comput. Math. **17**(1), 89–113 (2002). https://doi.org/10.1023/A:1015289823859
13. Lauritzen, S.L.: Graphical Models, Oxford Statistical Science Series, vol. 17. The Clarendon Press, Oxford University Press, New York (1996). Oxford Science Publications
14. Sullivant, S.: Toric fiber products. J. Algebra **316**(2), 560–577 (2007). https://doi.org/10.1016/j.jalgebra.2006.10.004
15. Sullivant, S.: Algebraic Statistics, Graduate Studies in Mathematics, vol. 194. American Mathematical Society, Providence, RI (2018). https://doi.org/10.1090/gsm/194

Geometric and Analytical Aspects of Quantization and Non-Commutative Harmonic Analysis on Lie Groups

Twirled Products and Group-Covariant Symbols

Paolo Aniello[1,2]([✉])[iD]

[1] Dipartimento di Fisica "Ettore Pancini", Università di Napoli "Federico II", Complesso Universitario di Monte S. Angelo, via Cintia, 80126 Napoli, Italy
paolo.aniello@na.infn.it
[2] Istituto Nazionale di Fisica Nucleare, Sezione di Napoli, Complesso Universitario di Monte S. Angelo, via Cintia, 80126 Napoli, Italy

Abstract. A quantum stochastic product is a binary operation on the convex set of states (density operators) of a quantum system preserving the convex structure. We review the notion of twirled product of quantum states, a group-theoretical construction yielding a remarkable class of group-covariant associative stochastic products. In the case where the relevant group is abelian, we then realize the twirled product in terms of the covariant symbols associated with the quantum states involved in the product. Finally, the special case of the twirled product associated with the group of phase-space translations is considered.

Keywords: Quantum state · Twirled product · Group-covariant symbol · Square integrable representation

1 Introduction

In classical statistical mechanics, physical states are realized as probability measures on phase space — usually, as probability distributions w.r.t. the Liouville measure — and the experimentally observable quantities (as well as other physically relevant quantities like, say, the entropy) are associated with suitably regular real functions on phase space [1,2]. In the quantum setting, both classes of fundamental objects — states and observables — are realized by means of (suitable classes of) linear operators on a separable complex Hilbert space $\mathcal{H}$ [3]. In an alternative approach to quantum mechanics [2,4–6], states and observables are realized, instead, as phase-space functions — that, in the case of physical states, are often called *quasi-probability distributions* [7] — so obtaining a mathematical formalism sharing several common features and intriguing analogies with the classical case. From a technical point of view, this formalism relies on the peculiar properties of a class of *irreducible projective representations* of the group of translations on phase space, i.e., the *Weyl systems* [8]. These group representations are *square integrable* [6,9,10], hence, they generate 'resolutions of the identity' in the carrier Hilbert space and canonical maps transforming Hilbert space operators into phase-space functions [6,11–14]. Such a map is called the *Wigner transform* and admits a remarkable generalization where phase-space translations are

F. Nielsen and F. Barbaresco (Eds.): GSI 2023, LNCS 14071, pp. 507–515, 2023.
https://doi.org/10.1007/978-3-031-38271-0_50

replaced with an abstract *locally compact group* G — typically, a Lie group — admitting square integrable representations [6,13–15]. In this generalization, the (real-valued) Wigner quasi-probability distribution (more precisely, its Fourier transform, the *quantum characteristic function* [16,17]) is replaced with a complex function on G, the *generalized Wigner transform* or *group-covariant symbol* [6,13–18]. In a recent series of papers [19–21], we have introduced a notion of *stochastic product* of quantum states, a binary operation on the convex set $\mathcal{D}(\mathcal{H})$ of quantum states (the density operators on the Hilbert space $\mathcal{H}$) that preserves the convex structure of $\mathcal{D}(\mathcal{H})$. We have also shown that a class of associative stochastic products — the *twirled products* — can be obtained via a group-theoretical construction. Extending a twirled product from $\mathcal{D}(\mathcal{H})$ to the full Banach space $\mathcal{T}(\mathcal{H})$ of trace class operators, one gets a *Banach algebra*. This structure may be regarded as a quantum counterpart of the 'classical' convolution algebra on a locally compact group [22], and, in the case where the group of phase-space translations is involved, it has been called the *quantum convolution* [21]. In a forthcoming paper [23], we express the twirled products in terms of the covariant symbols associated with some square integrable representation. Here, we anticipate the result in the case where the relevant group is abelian (e.g., the group of phase-space translations). Specifically, in Sect. 2, we introduce the notion of stochastic product and review the construction of the twirled products. In Sect. 3, we study the problem of expressing the twirled products via group-covariant symbols. In Sect. 4, a few conclusions are drawn.

2   The Twirled Products as Covariant Stochastic Products

Using the notations introduced in the previous section, a *convex-linear* application $\mathfrak{S} \colon \mathcal{D}(\mathcal{H}) \to \mathcal{D}(\mathcal{H})$ is called a *quantum stochastic map*. Such a map admits a unique (trace-preserving, positive) *linear extension* $\mathfrak{S}_{\mathrm{ext}} \colon \mathcal{T}(\mathcal{H}) \to \mathcal{T}(\mathcal{H})$ [19], a *linear stochastic map* on $\mathcal{T}(\mathcal{H})$. Accordingly, a *quantum stochastic product* is a binary operation $(\cdot) \odot (\cdot) \colon \mathcal{D}(\mathcal{H}) \times \mathcal{D}(\mathcal{H}) \to \mathcal{D}(\mathcal{H})$, that is *convex-linear w.r.t. both its arguments*; namely, for all $\rho, \sigma, \tau, \upsilon \in \mathcal{D}(\mathcal{H})$ and all $\alpha, \epsilon \in [0,1]$, we assume that

$$(\alpha \rho + (1-\alpha)\sigma) \odot (\epsilon \tau + (1-\epsilon)\upsilon) = \alpha \epsilon \, \rho \odot \tau + \alpha(1-\epsilon) \, \rho \odot \upsilon$$
$$+ (1-\alpha)\epsilon \, \sigma \odot \tau + (1-\alpha)(1-\epsilon) \, \sigma \odot \upsilon. \quad (1)$$

A binary operation $(\cdot) \boxdot (\cdot)$ on $\mathcal{T}(\mathcal{H})$ is said to be *state-preserving* if it is such that $\mathcal{D}(\mathcal{H}) \boxdot \mathcal{D}(\mathcal{H}) \subset \mathcal{D}(\mathcal{H})$.

Proposition 1 ([19]). *Every quantum stochastic product on $\mathcal{D}(\mathcal{H})$ is continuous w.r.t. the topology inherited from $\mathcal{T}(\mathcal{H})$; i.e., denoting by $\|\cdot\|_1 \equiv \|\cdot\|_{\mathrm{tr}}$ the standard trace norm on $\mathcal{T}(\mathcal{H})$, w.r.t. the topology on $\mathcal{D}(\mathcal{H})$ and on $\mathcal{D}(\mathcal{H}) \times \mathcal{D}(\mathcal{H})$ induced by the distance functions*

$$\mathsf{d}_1(\rho,\sigma) := \|\rho - \sigma\|_1 \quad and \quad \mathsf{d}_{1,1}((\rho,\tau),(\sigma,\upsilon)) := \max\{\|\rho-\sigma\|_1, \|\tau - \upsilon\|_1\}. \quad (2)$$

For every quantum stochastic product $(\cdot) \odot (\cdot)$: $\mathcal{D}(\mathcal{H}) \times \mathcal{D}(\mathcal{H}) \to \mathcal{D}(\mathcal{H})$, *there exists a unique bilinear stochastic map* $(\cdot) \boxdot (\cdot)$: $\mathcal{T}(\mathcal{H}) \times \mathcal{T}(\mathcal{H}) \to \mathcal{T}(\mathcal{H})$ — *i.e., a unique state-preserving bilinear map on* $\mathcal{T}(\mathcal{H})$ — *such that* $\rho \odot \sigma = \rho \boxdot \sigma$, *for all* $\rho, \sigma \in \mathcal{D}(\mathcal{H})$.

Definition 1. $\mathcal{T}(\mathcal{H})$, *endowed with a map* $(\cdot) \boxdot (\cdot)$: $\mathcal{T}(\mathcal{H}) \times \mathcal{T}(\mathcal{H}) \to \mathcal{T}(\mathcal{H})$ *which is bilinear, state-preserving and associative, is called a* stochastic algebra.

Note that, by Proposition 1, every quantum stochastic product $(\cdot) \odot (\cdot)$ on $\mathcal{D}(\mathcal{H})$ may be thought of as the restriction of a uniquely determined bilinear stochastic map $(\cdot) \boxdot (\cdot)$ on $\mathcal{T}(\mathcal{H})$, that is *associative* iff $(\cdot) \odot (\cdot)$ is associative. Therefore, a stochastic algebra can also be defined as a Banach space of trace class operators $\mathcal{T}(\mathcal{H})$, together with an *associative* quantum stochastic product $(\cdot) \odot (\cdot)$: $\mathcal{D}(\mathcal{H}) \times \mathcal{D}(\mathcal{H}) \to \mathcal{D}(\mathcal{H})$. Let $\mathsf{BL}(\mathcal{H})$ denote the complex vector space of all bounded bilinear maps on $\mathcal{T}(\mathcal{H})$, which becomes a Banach space once endowed with the norm $\|\cdot\|_{(1)}$ defined as follows. For every $\beta(\cdot, \cdot)$: $\mathcal{T}(\mathcal{H}) \times \mathcal{T}(\mathcal{H}) \to \mathcal{T}(\mathcal{H})$ in $\mathsf{BL}(\mathcal{H})$, we define its norm as

$$\|\beta(\cdot, \cdot)\|_{(1)} := \sup\{\|\beta(A, B)\|_1 : \|A\|_1, \|B\|_1 \le 1\}. \tag{3}$$

Proposition 2 ([19]). *Every bilinear stochastic map* $(\cdot)\boxdot(\cdot)$: $\mathcal{T}(\mathcal{H}) \times \mathcal{T}(\mathcal{H}) \to \mathcal{T}(\mathcal{H})$ *is bounded and its norm is such that* $\|(\cdot)\boxdot(\cdot)\|_{(1)} \le 2$, *while its restriction* $(\cdot) \boxminus (\cdot)$ *to a bilinear map on the real Banach space* $\mathcal{T}(\mathcal{H})_s$ *of all selfadjoint trace class operators on* $\mathcal{H}$ *is such that* $\|(\cdot) \boxminus (\cdot)\|_{(1)} = 1$. *Hence, whenever a stochastic product* $(\cdot)\boxdot(\cdot)$ *on* $\mathcal{T}(\mathcal{H})$ *is associative, the pair* $(\mathcal{T}(\mathcal{H})_s, (\cdot)\boxminus(\cdot))$ *is a real Banach algebra, because, for all* $A, B \in \mathcal{T}(\mathcal{H})_s$, $\|A \boxminus B\|_1 \le \|A\|_1 \|B\|_1$.

Here, it is worth observing that

1. The restriction of a stochastic product $(\cdot) \boxdot (\cdot)$ on $\mathcal{T}(\mathcal{H})$ to a bilinear map on $\mathcal{T}(\mathcal{H})_s$ is well defined, since the fact that the map $(\cdot) \boxdot (\cdot)$ is bilinear and state-preserving implies that it is also *adjoint-preserving*.
2. The inequality $\|(\cdot) \boxdot (\cdot)\|_{(1)} \le 2$ may *not* be saturated. E.g., the algebra $(\mathcal{T}(\mathcal{H}), (\cdot) \boxdot (\cdot))$ may be a Banach algebra too; see Theorem 2 below.

We now outline a group-theoretical construction yielding a class of quantum stochastic products. Let us first summarize our main notations and assumptions:

- We suppose that G is a *unimodular* locally compact, second countable, Hausdorff topological group (in short, a unimodular l.c.s.c. group), which admits square integrable representations [10]. Let $\mathscr{B}(G)$ denote the Borel σ-algebra of G, and $\mathscr{P}(G)$ the set of all Borel probability measures on G, endowed with the standard *convolution* product $(\cdot) \circledast (\cdot)$: $\mathscr{P}(G) \times \mathscr{P}(G) \to \mathscr{P}(G)$ [22].
- Next, we pick a *square integrable projective representation* $U \colon G \to \mathcal{U}(\mathcal{H})$, where $\mathcal{U}(\mathcal{H})$ is the unitary group of $\mathcal{H}$; in particular, U is supposed to be irreducible. The assumption of square-integrability for U cannot be dispensed with. It ensures the validity (of the subsequent relations (4) and) of the final claim of Theorem 1 below, a fundamental ingredient of our construction.

- It is convenient to suppose henceforth that the (both left and right invariant) Haar measure μ_G on G is normalized in such a way that the *orthogonality relations* [9,10,13] for the representation U are precisely the form

$$\int_G d\mu_G(g) \, \langle \eta, U(g) \, \phi \rangle \, \langle U(g) \, \psi, \chi \rangle = \langle \eta, \chi \rangle \langle \psi, \phi \rangle, \quad \forall \eta, \chi, \psi, \phi \in \mathcal{H}, \quad (4)$$

where $\langle \cdot, \cdot \rangle \colon \mathcal{H} \times \mathcal{H} \to \mathbb{C}$ is the scalar product in $\mathcal{H}$.
- The representation U induces an *isometric* representation $G \ni g \mapsto S_U(g)$ in the Banach space $\mathcal{T}(\mathcal{H})$, namely,

$$S_U(g) \, A := U(g) \, A \, U(g)^*, \quad A \in \mathcal{T}(\mathcal{H}), \quad (5)$$

and $G \times \mathcal{D}(\mathcal{H}) \ni (g, \rho) \mapsto S_U(g) \rho \in \mathcal{D}(\mathcal{H})$ is the *symmetry action* [24,25] of G on $\mathcal{D}(\mathcal{H})$. Although U is, in general, *projective* — $U(gh) = \gamma(g,h) \, U(g) \, U(h)$, where $\gamma \colon G \times G \to \mathbb{T}$ is a Borel function, the *multiplier* [26] associated with U — note that S_U behaves as a group *homomorphism*: $S_U(gh) = S_U(g) \, S_U(h)$.
- We also pick a *fiducial state* $\upsilon \in \mathcal{D}(\mathcal{H})$ and a *probability measure* $\varpi \in \mathscr{P}(G)$. E.g., $\varpi = \delta \in \mathscr{P}(G)$ is the Dirac (point mass) measure at the identity $e \in G$.

Theorem 1 ([19])**.** *For every probability measure $\mu \in \mathscr{P}(G)$, the linear map*

$$\mu[U] \colon \mathcal{T}(\mathcal{H}) \ni A \mapsto \int_G d\mu(g) \, (S_U(g) \, A) \in \mathcal{T}(\mathcal{H}) \quad (6)$$

is positive and trace-preserving. Therefore, one can define the quantum stochastic map

$$\mathcal{D}(\mathcal{H}) \ni \rho \mapsto \mu[U] \, \rho \in \mathcal{D}(\mathcal{H}). \quad (7)$$

Moreover, for every density operator $\rho \in \mathcal{D}(\mathcal{H})$, the mapping

$$\nu_{\rho,\upsilon} \colon \mathscr{B}(G) \ni \mathcal{E} \mapsto \int_{\mathcal{E}} d\mu_G(g) \, \mathrm{tr}\big(\rho \, (S_U(g) \upsilon)\big) \in \mathbb{R}^+ \quad (8)$$

belongs to $\mathscr{P}(G)$.

We next define a *binary operation* on $\mathcal{D}(\mathcal{H})$, associated with the triple (U, υ, ϖ):

$$\rho \overset{\upsilon}{\underset{\varpi}{\odot}} \sigma := \big((\nu_{\rho,\upsilon} \circledast \varpi)[U]\big) \, \sigma = \int_G d(\nu_{\rho,\upsilon} \circledast \varpi)(g) \, (S_U(g)\sigma), \quad \forall \rho, \sigma \in \mathcal{D}(\mathcal{H}). \quad (9)$$

Here, we have considered the fact that $\nu_{\rho,\upsilon}$ is a probability measure associated with ρ and υ (final assertion of Theorem 1). Then, we have taken the convolution $\mu \equiv \nu_{\rho,\upsilon} \circledast \varpi \in \mathscr{P}(G)$ of $\nu_{\rho,\upsilon}$ with the probability measure ϖ. Eventually, we have applied the stochastic map $\mu[U] = (\nu_{\rho,\upsilon} \circledast \varpi)[U]$, see Theorem 1, to the state σ.

We can express this product in a more explicit form:

$$\rho \overset{\upsilon}{\underset{\varpi}{\odot}} \sigma = \int_G d\mu_G(g) \int_G d\varpi(h) \, \mathrm{tr}\big(\rho \, (S_U(g)\upsilon)\big) \, (S_U(gh)\sigma). \quad (10)$$

Definition 2. *We call the binary operation defined by* (9) *the twirled product generated by the triple* (U, υ, ϖ). *Here,* U *is called the* inducing representation *of the product; the states* ρ, υ *and* σ *are called the* input, *the* probe *and the* whirligig, *respectively; finally,* $\varpi \in \mathscr{P}(G)$ *is called the* smearing measure.

In particular, with $\varpi = \delta$ (Dirac measure) in (10), we find

$$\rho \overset{\upsilon}{\odot} \sigma \equiv \rho \overset{\upsilon}{\underset{\delta}{\odot}} \sigma = \int_G \mathrm{d}\mu_G(g)\, \mathrm{tr}\big(\rho\,(\mathsf{S}_U(g)\,\upsilon)\big)\,(\mathsf{S}_U(g)\,\sigma), \tag{11}$$

the *un-smeared twirled product* generated by the pair (U, υ). The twirled product has an interesting interpretation within quantum measurement theory and information, in terms of quantum *channels, observables* and *instruments* [19–21].

Theorem 2 ([19]). *The twirled product generated by the triple* (U, υ, ϖ) *is an associative quantum stochastic product, which is left-covariant w.r.t. the representation* U, *namely, such that*

$$\rho_g \overset{\upsilon}{\underset{\varpi}{\odot}} \sigma = \left(\rho \overset{\upsilon}{\underset{\varpi}{\odot}} \sigma \right)_g, \quad \forall g \in G,\ \forall \rho, \sigma \in \mathcal{D}(\mathcal{H}),\ \textit{where}\ \rho_g \equiv \mathsf{S}_U(g)\,\rho. \tag{12}$$

Extending this product to a state-preserving bilinear map on $\mathcal{T}(\mathcal{H})$, *one obtains a Banach algebra — i.e., a stochastic Banach algebra — that, in the case where the l.c.s.c. group* G *is abelian, is commutative.*

3   Twirled Products via Group-Covariant Symbols

With the same notations and assumptions adopted for our construction of twirled products (Sect. 2) — and denoting by $\mathcal{S}(\mathcal{H}) \supset \mathcal{T}(\mathcal{H})$ the Hilbert space of all Hilbert-Schmidt operators on $\mathcal{H}$, and by $\|\cdot\|_{\mathrm{HS}}$ the associated norm — we can define a *linear isometry* $\mathfrak{D}\colon \mathcal{S}(\mathcal{H}) \to \mathrm{L}^2(G) \equiv \mathrm{L}^2(G, \mu_G; \mathbb{C})$, determined by putting $\check{A}(g) \equiv (\mathfrak{D}A)(g) = \mathrm{tr}(U(g)^*A)$, for all $A \in \mathcal{T}(\mathcal{H})$. Here, we are using the fact that $\mathcal{T}(\mathcal{H})$ is a $\|\cdot\|_{\mathrm{HS}}$-dense subspace of $\mathcal{S}(\mathcal{H})$; see [15, 18] for the details. The isometry $\mathfrak{D}$ may be thought of as a *dequantization map*, which is directly related to the Wigner transform in the case where G is the group of translations on phase space [13–15,18]. We will call the bounded Borel function $\check{A} = \mathfrak{D}A \in \mathrm{L}^2(G)$ the (group-covariant) *symbol* of the operator $A \in \mathcal{T}(\mathcal{H})$ associated with the representation U. The operator A can be easily re-constructed from its symbol [15]. In the case where G is the group of phase-space translations and U is the *Weyl system*, for $A \equiv \rho \in \mathcal{D}(\mathcal{H})$ the symbol $\check{A} \equiv \check{\rho}$ is also called the *quantum characteristic function* of the state ρ (essentially, the Fourier transform of the Wigner distribution of ρ), in a natural analogy with the 'classical' characteristic function of a probability measure on a locally compact abelian group [13,16,17]. Note that, for every $\rho \in \mathcal{D}(\mathcal{H})$, $\check{\rho}(e) = \mathrm{tr}(\rho) = 1$, where e is the identity in G.

We will henceforth suppose that G is an *abelian* l.c.s.c. group.

It will be useful to introduce the following function $\gamma_\diamond \colon G \times G \to \mathbb{T}$ associated with the multiplier γ of the projective representation U:

$$\gamma_\diamond(g, h) := \gamma(g, h)\, \overline{\gamma(h, g)}. \tag{13}$$

We have that

$$\gamma_\diamond(g, h) = \overline{\gamma_\diamond(h, g)}, \quad \gamma_\diamond(g, g^{-1}) = 1, \tag{14}$$

where the second relation descends from a well known property [26] — namely, $\gamma(g, g^{-1}) = \gamma(g^{-1}, g)$ — of a group multiplier. Note that, if the multiplier γ is *trivial* [26] — i.e., of the form $\gamma(g, h) = \beta(g)\,\beta(h)\,\overline{\beta(gh)} = \gamma(h, g)$ (G being abelian), for some Borel function $\beta \colon G \to \mathbb{T}$ — then $\gamma_\diamond \equiv 1$. Accordingly, $\gamma_\diamond$ depends on the *similarity class* [26] of γ only. By Lemma 7.1 of [27], we have:

Proposition 3. *The function* $\gamma_\diamond \colon G \times G \to \mathbb{T}$ *is a (continuous) skew-symmetric bicharacter; i.e., in addition to the the first of relations* (14), *we also have that*

$$\gamma_\diamond(gh, \tilde{g}) = \gamma_\diamond(g, \tilde{g})\, \gamma_\diamond(h, \tilde{g}), \quad \forall g, h, \tilde{g} \in G. \tag{15}$$

Remark 1. Let $\widehat{G}$ be the *Pontryagin dual* of G, and denote by $[\![\,\cdot\,,\cdot\,]\!] \colon G \times \widehat{G} \to \mathbb{T}$ the *pairing* map [22]. Then, $\mathsf{h}_\gamma \colon G \ni g \mapsto \gamma_\diamond(\,\cdot\,, g) \in \widehat{G}$ is a continuous homomorphism. We say that the multiplier γ is *nondegenerate* if the homomorphism h_γ is injective; namely, if the closed subgroup $G_\gamma := \{g \in G \colon \gamma_\diamond(h, g) = 1, \ \forall\, h \in G\}$ of G is trivial, i.e., $G_\gamma = \ker(\mathsf{h}_\gamma) = \{e\}$. It is easy to check that γ is nondegenerate iff $\mathsf{h}_\gamma(G)$ is a *dense* subgroup of $\widehat{G}$. Indeed, observe that the *closure* $\mathrm{cl}(\mathsf{h}_\gamma(G))$ of the subgroup $\mathsf{h}_\gamma(G)$ of $\widehat{G}$, is such that

$$\mathrm{cl}(\mathsf{h}_\gamma(G))^\perp \equiv \left\{ h \in G \colon [\![h, \widehat{g}]\!] \equiv \widehat{g}(h) = 1, \ \forall\, \widehat{g} \in \mathrm{cl}(\mathsf{h}_\gamma(G)) \subset \widehat{G} \right\}$$

$$= \left\{ h \in G \colon [\![h, \widehat{g}]\!] = 1, \ \forall\, \widehat{g} \in \mathsf{h}_\gamma(G) \right\}$$

$$= \left\{ h \in G \colon [\![h, \mathsf{h}_\gamma(g)]\!] = \gamma_\diamond(h, g) = 1 = \gamma_\diamond(g, h), \ \forall\, g \in G \right\} = G_\gamma.$$

Moreover, by Theorem 4.39 of [22], $\big(\widehat{G}/\mathrm{cl}(\mathsf{h}_\gamma(G))\big)^{\widehat{\ }}$ is isomorphic, as a topological group, to $\mathrm{cl}(\mathsf{h}_\gamma(G))^\perp = G_\gamma$. Thus, $G_\gamma = \{e\}$ iff $\widehat{G} = \mathrm{cl}(\mathsf{h}_\gamma(G))$. In particular, in the case where G is *selfdual* — i.e., $G \simeq \widehat{G}$ — the (nondegenerate) multiplier γ is called *regular* if $\mathsf{h}_\gamma(G) = \widehat{G}$; i.e., if h_γ is an isomorphism of topological groups.

Now, given a complex (Radon) measure ν on G — let us denote by $\mathcal{M}(G)$ the Banach space of all such measures — we can define its *Fourier-Stieltjes transform* [22] $\widehat{\nu} \colon \widehat{G} \to \mathbb{C}$ (the classical characteristic function of ν), i.e.,

$$\widehat{\nu}(\widehat{g}) := \int_G \mathrm{d}\nu(h)\, \overline{[\![h, \widehat{g}]\!]}, \quad [\![h, \widehat{g}]\!] \equiv \widehat{g}(h), \ \widehat{g} \in \widehat{G}, \tag{16}$$

which is a bounded continuous function. We then define the (classical) *symbol* of ν associated with the multiplier γ; namely, the bounded continuous function $\widecheck{\nu} \equiv \widecheck{\nu}_\gamma \colon G \to \mathbb{C}$ — depending on the similarity class of γ only — defined by

$$\widecheck{\nu}(g) := \widehat{\nu} \circ \mathsf{h}_\gamma(g)$$

$$= \int_G \mathrm{d}\nu(h)\, \overline{[\![h, \mathsf{h}_\gamma(g)]\!]} = \int_G \mathrm{d}\nu(h)\, \overline{\gamma_\diamond(h, g)} = \int_G \mathrm{d}\nu(h)\, \gamma_\diamond(g, h). \tag{17}$$

Proposition 4. *The mapping* $\mathcal{M}(G) \ni \nu \mapsto \breve{\nu} \equiv \breve{\nu}_\gamma$ *is injective if the multiplier* γ *is nondegenerate.*

Proof. The mapping $\mathcal{M}(G) \ni \nu \mapsto \hat{\nu}$ is injective ('Fourier uniqueness theorem'; see Theorem 4.33 of [22]). Moreover, if γ is nondegenerate, then $\mathsf{h}_\gamma(G)$ is a dense subgroup of $\hat{G}$. Therefore, given $\nu_1, \nu_2 \in \mathcal{M}(G)$, if $\breve{\nu}_1 = \breve{\nu}_2$, then $\hat{\nu}_1(\hat{g}) = \hat{\nu}_2(\hat{g})$, for all $\hat{g}$ in the dense subset $\mathsf{h}_\gamma(G)$ of $\hat{G}$; hence, actually, for all $\hat{g} \in \hat{G}$, because $\hat{\nu}_1$ and $\hat{\nu}_2$ are continuous functions. Thus, the mapping $\nu \mapsto (\hat{\nu} \mapsto) \breve{\nu}$ is injective.

Under suitable assumptions, the twirled product can be expressed via the symbols of the relevant density operators and probability measures [23]:

Theorem 3. *Let the abelian group* G *be* selfdual, *and let the multiplier* γ *of the square integrable representation* U *be (nodegenerate and)* regular. *For every triple of states* $\rho, \upsilon, \sigma \in \mathcal{D}(\mathcal{H})$ *and every probability measure* ϖ *on* G, *we have:*

$$\left(\rho \underset{\varpi}{\overset{\upsilon}{\odot}} \sigma\right)(g) = \overline{\gamma(g, g^{-1})} \, \breve{\varpi}(g) \, \breve{\rho}(g) \, \breve{\upsilon}(g^{-1}) \, \breve{\sigma}(g) = \breve{\varpi}(g) \, \breve{\rho}(g) \, \overline{\breve{\upsilon}(g)} \, \breve{\sigma}(g). \tag{18}$$

Let us consider the case of the group of *translations on phase space* (with, say, n position variables) — $G = \mathbb{R}^n \times \mathbb{R}^n$ — and of the phase-space stochastic product generated by the *Weyl system* ($\hbar \equiv 1$), i.e., by the projective representation $(q, p) \mapsto U(q, p) = \mathrm{e}^{-\mathrm{i}\, q \cdot p/2} \, \mathrm{e}^{\mathrm{i}\, p \cdot \hat{q}} \, \mathrm{e}^{-\mathrm{i}\, q \cdot \hat{p}}$, where $\hat{q}$, $\hat{p}$ are the position and momentum operators. The multiplier of U is of the form $\gamma(q, p; \tilde{q}, \tilde{p}) = \exp(\mathrm{i}(q \cdot \tilde{p} - p \cdot \tilde{q})/2)$ so that $\gamma(q, p; -q, -p) \equiv 1$ and $\gamma_\diamond(q, p; \tilde{q}, \tilde{p}) = \exp(\mathrm{i}(q \cdot \tilde{p} - p \cdot \tilde{q}))$. Thus, the bicharacter $\gamma_\diamond$ involves the standard symplectic form on $\mathbb{R}^n \times \mathbb{R}^n$, and γ is a regular multiplier since we can identify G with its dual $\hat{G}$ via the *symplectic pairing*

$$G \times \hat{G} \equiv G \times G \ni (q, p; \tilde{q}, \tilde{p}) \mapsto [\![q, p; \tilde{q}, \tilde{p}]\!] = \gamma_\diamond(q, p; \tilde{q}, \tilde{p}) \in \mathbb{T}. \tag{19}$$

Moreover, for every probability measure ϖ on G, we can identify its Fourier-Stieltjes transform $\hat{\varpi} \colon \hat{G} \to \mathbb{C}$, the characteristic function of ϖ, with the function

$$\breve{\varpi}(q, p) := \int_{\mathbb{R}^n \times \mathbb{R}^n} \mathrm{d}\varpi(\tilde{q}, \tilde{p}) \, \exp(\mathrm{i}(q \cdot \tilde{p} - p \cdot \tilde{q})). \tag{20}$$

By Theorem 3, the phase-space stochastic product generated by the triple (U, υ, ϖ) — expressed in terms of the characteristic function $\breve{\varpi}$ and of the symbols $\breve{\rho} := \mathrm{tr}(U(q, p)^* \rho)$, $\breve{\upsilon}$, $\breve{\sigma}$ of the states ρ, υ, σ — has the manifestly commutative form

$$\left(\rho \underset{\varpi}{\overset{\upsilon}{\odot}} \sigma\right)(q, p) = \breve{\varpi}(q, p) \, \breve{\rho}(q, p) \, \breve{\upsilon}(-q, -p) \, \breve{\sigma}(q, p)$$

$$= \breve{\varpi}(q, p) \, \breve{\rho}(q, p) \, \overline{\breve{\upsilon}(q, p)} \, \breve{\sigma}(q, p). \tag{21}$$

4   Conclusions

We have reviewed the notion of *stochastic product*, a binary operation on the set of density operators preserving the convex structure. By a group-theoretical approach, one finds a class of associative stochastic products, the *twirled products*, that exist in every Hilbert space dimension and admit an interesting interpretation within quantum measurement theory and information science [19–21]. If the relevant group is *abelian*, one obtains a *commutative* stochastic product that can be expressed via *group-covariant symbols*. Work is in progress on this topic [23].

Acknowledgements. We thank J.-P. Gazeau and P. Bieliavsky for their kind invitation to participate in the session "Geometric and Analytical Aspects of Quantization and Non-Commutative Harmonic Analysis on Lie Groups" (GSI'23).

References

1. Wehrl, A.: General properties of entropy. Rev. Mod. Phys. **50**(2), 221–260 (1978)
2. Hillery, M., O'Connell, R.F., Scully, M.O., Wigner, E.P.: Distribution functions in physics: fundamentals. Phys. Rep. **106**(3), 121–167 (1984)
3. Heinosaari, T., Ziman, M.: The Mathematical Language of Quantum Theory. Cambridge University Press, Cambridge (2012)
4. Wigner, E.: On the quantum correction for thermodinamic equilibrium. Phys. Rev. **40**, 749–759 (1932)
5. Folland, G.B.: Harmonic Analysis in Phase Space. Princeton University Press, Princeton, NJ (1989)
6. Ali, S.T., Antoine, J.-P., Gazeau, J.-P.: Coherent States. Wavelets and Their Generalizations. Second Edn. Springer, New York (2014). https://doi.org/10.1007/978-1-4614-8535-3
7. Aniello, P., Man'ko, V., Marmo, G., Solimeno, S., Zaccaria, F.: On the coherent states, displacement operators and quasidistributions associated with deformed quantum oscillators. J. Opt. B: Quantum Semiclass. Opt. **2**, 718–725 (2000)
8. Aniello, P.: On the notion of Weyl system. J. Russ. Laser Res. **31**(2), 102–116 (2010)
9. Aniello, P., Cassinelli, G., De Vito, E., Levrero, A.: Square-integrability of induced representations of semidirect products. Rev. Math. Phys. **10**(3), 301–313 (1998)
10. Aniello, P.: Square integrable projective representations and square integrable representations modulo a relatively central subgroup. Int. J. Geom. Meth. Mod. Phys. **3**(2), 233–267 (2006)
11. Aniello, P., Cassinelli, G., De Vito, E., Levrero, A.: Wavelet transforms and discrete frames associated to semidirect products. J. Math. Phys. **39**(8), 3965–3973 (1998)
12. Aniello, P., Cassinelli, G., De Vito, E., Levrero, A.: On discrete frames associated with semidirect products. J. Fourier Anal. and Appl. **7**(2), 199–206 (2001)
13. Aniello, P.: Square integrable representations, an invaluable tool. In: Antoine, J.-P., Bagarello, F., Gazeau, J.-P. (eds.) Coherent States and Their Applications: a Contemporary Panorama. Springer Proceedings in Physics, vol. 205, pp. 17–40. Springer, Cham (2018). https://doi.org/10.1007/978-3-319-76732-1_2
14. Aniello, P.: Discovering the manifold facets of a square integrable representation: from coherent states to open systems. J. Phys: Conf. Ser. **1194**, 012006 (2019)
15. Aniello, P.: Star products: a group-theoretical point of view. J. Phys. A: Math. Theor. **42**, 475210 (2009)
16. Aniello, P.: Playing with functions of positive type, classical and quantum. Phys. Scr. **90**, 074042 (2015)
17. Aniello, P.: Functions of positive type on phase space, between classical and quantum, and beyond. J. Phys: Conf. Ser. **670**, 012004 (2016)

18. Aniello, P., Man'ko, V.I., Marmo, G.: Frame transforms, star products and quantum mechanics on phase space. J. Phys. A: Math. Theor. **41**, 285304 (2008)

19. Aniello, P.: A class of stochastic products on the convex set of quantum states. J. Phys. A: Math. Theor. **52**, 305302 (2019)

20. Aniello, P.: Covariant stochastic products of quantum states. J. Phys: Conf. Ser. **1416**, 012002 (2019)

21. Aniello, P.: Quantum stochastic products and the quantum convolution. Geometry Integrability Quantization **22**, 64–77 (2021)

22. Folland, G.: A Course in Abstract Harmonic Analysis. CRC Press, Boca Raton (1995)

23. Aniello, P.: Group-theoretical stochastic products: the twirled products and their realization via covariant symbols. Manuscript in preparation

24. Bargmann, V.: Note on Wigner's theorem on symmetry operations. J. Math. Phys. **5**(7), 862–868 (1964)

25. Aniello, P., Chruściński, D.: Symmetry witnesses. J. Phys. A: Math. Theor. **50**, 285302 (2017)

26. Varadarajan, V.S.: Geometry of Quantum Theory, 2nd edn. Springer, New York (1985). https://doi.org/10.1007/978-0-387-49386-2

27. Kleppner, A.: Mutipliers on abelian groups. Math. Annalen **158**, 11–34 (1965)

Coherent States and Entropy

Tatyana Barron[1]([✉]) and Alexander Kazachek[1,2]

[1] Department of Mathematics, University of Western Ontario,
London, ON N6A 5B7, Canada
{tatyana.barron,akazache}@uwo.ca
[2] Department of Applied Mathematics, University of Waterloo,
Waterloo, ON N2L 3G1, Canada

Abstract. Let H_k, $k \in \mathbb{N}$, be the Hilbert spaces of geometric quantization on a Kähler manifold M. With two points in M we associate a Bell-type state $b_k \in H_k \otimes H_k$. When M is compact or when M is $\mathbb{C}^n$, we provide positive lower bounds for the entanglement entropy of b_k (asymptotic in k, as $k \to \infty$).

Keywords: entanglement · Hilbert spaces · asymptotics

1 Introduction and Main Results

Let H_k, $k = 1, 2, 3, \ldots$ be the Hilbert spaces of Kähler quantization on a Kähler manifold M. Let $p, q \in M$ and $\Theta_p^{(k)} \in H_k$ and $\Theta_q^{(k)} \in H_k$ be the coherent states at p and q respectively. With the pair p, q we associate the Bell-type pure state

$$w_k = w_k(p, q) = \frac{1}{||\Theta_p^{(k)}||^2} \Theta_p^{(k)} \otimes \Theta_p^{(k)} + \frac{1}{||\Theta_q^{(k)}||^2} \Theta_q^{(k)} \otimes \Theta_q^{(k)} \in H_k \otimes H_k. \quad (1)$$

It is entangled. The two theorems below are for the cases when M is compact and when M is $\mathbb{C}^n$ ($n \in \mathbb{N}$), respectively. We address the question how the entanglement entropy E_k [1] of

$$b_k = b_k(p, q) = \frac{1}{||w_k||} w_k \in H_k \otimes H_k \quad (2)$$

depends on the quantum parameter k and on the distance between p and q, $\text{dist}(p, q)$. The theorems provide positive lower bounds on $E_k(b_k)$. In quantum information theory, when quantum systems are used for communication, one interpretation of entanglement entropy is the amount of information that can be transmitted. Bell states are maximally entangled (e.g.

$$\frac{1}{\sqrt{2}} (e_0 \otimes e_0 + e_1 \otimes e_1) \quad (3)$$

is one of the standard Bell states).

Partially supported by a 2022 Bridge grant, University of Western Ontario, proposal ID 53917.

F. Nielsen and F. Barbaresco (Eds.): GSI 2023, LNCS 14071, pp. 516–523, 2023.
https://doi.org/10.1007/978-3-031-38271-0_51

Our state (1) is constructed from the coherent vectors $\Theta_p^{(k)}$ and $\Theta_q^{(k)}$ which are typically not orthogonal to each other (unlike e_0 and e_1 in the Bell state (3)), although $\langle \Theta_p^{(k)}, \Theta_q^{(k)} \rangle \to 0$ as $k \to \infty$.

To provide some background, in quantum information theory, given a nonzero vector v in the tensor product of two separable Hilbert spaces V_1 and V_2, its entanglement entropy $E(v)$ characterizes "how nondecomposable" (or, in other words, how entangled) this vector is. It is defined as follows. For the purposes of this paper we only need the case $V_1 = V_2 = V$. Let $\langle ., . \rangle$ be the inner product in V. Let $\{e_i\}$ be an orthonormal basis in V. Let $Tr_2(A) \in End(V)$ denote the partial trace of a density matrix A. It is defined by

$$\langle x, Tr_2(A)y \rangle = \sum_i \langle x \otimes e_i, A(y \otimes e_i) \rangle$$

for every $x, y \in V$. The entanglement entropy is

$$E(v) = -Tr(\rho \ln \rho) = -\sum_i \langle (\rho \ln \rho)e_i, e_i \rangle$$

where

$$\rho = Tr_2(P_v)$$

and P_v is the rank 1 orthogonal projection onto the 1-dimensional linear subspace of $V \otimes V$ spanned by v. The operator $\rho \ln \rho$ is defined via the continuous functional calculus. When V is finite-dimensional, $E(v)$ is a real number in the interval $[0, \ln dim(V)]$. When V is infinite-dimensional, $E(v)$ is a nonnegative real number or $+\infty$. The value of $E(v)$ does not depend on the choice of the basis $\{e_i\}$.

Kähler quantization deals with asymptotic analysis on Kähler manifolds in the context of classical mechanics and quantum mechanics. Let (M, ω) be an integral Kähler manifold. Let $L \to M$ be a holomorphic line bundle whose first Chern class is represented by ω. One can consider (M, ω) as a classical phase space, i.e. a space that parametrizes position and momentum of a classical particle. Classical mechanics on M is captured in ω and a choice of a Hamiltonian (a smooth function on M). The symplectic form defines a Poisson bracket on $C^\infty(M)$. Dirac's correspondence principle seeks a linear map from $C^\infty(M)$ to linear operators on the Hilbert space of quantum mechanical wave functions that takes the Poisson bracket of functions to the commutator of operators. In geometric quantization or Kähler quantization, a standard choice of the Hilbert space V is the space of holomorphic sections of L^k, where the positive integer k is interpreted (philosophically) as $1/\hbar$. If M is compact, then V is finite-dimensional. If M is noncompact, then V is infinite-dimensional.

The motivation to bring techniques from quantum information theory to geometric quantization was to obtain new insights in the interplay between geometry and analysis on Kähler manifolds. It would be interesting to investigate if there is a meaningful relationship between the information-theoretic entropy and other concepts of entropy. In the opposite direction, some geometric intuition may be useful in information transmission problems.

1.1   Compact Case

Let $L \to M$ be a positive hermitian holomorphic line bundle on a compact n-dimensional complex manifold M. Denote by ∇ the Chern connection in L. Equipped with the 2-form $\omega = i \operatorname{curv}(\nabla)$, M is a Kähler manifold. Denote by $d\mu$ the measure on M associated with the volume form $\dfrac{\omega^n}{n!}$. As before, let k be a positive integer. For p, q in M, let $\Theta_p^{(k)}, \Theta_q^{(k)} \in H_k = H^0(M, L^k)$ be Rawnsley coherent states at p and q (see e.g. [2]).

Let us recall the definition of $\Theta_p^{(k)}$ for $p \in M$ and $k \in \mathbb{N}$. Choose a unit vector $\xi \in L$. Then by Riesz representation theorem there is a unique vector $\Theta_p^{(k)}$ in the Hilbert space $H^0(M, L^k)$ with the property

$$\langle s, \Theta_p^{(k)} \rangle = \langle s(p), \xi^{\otimes k} \rangle$$

for every $s \in H^0(M, L^k)$.

The $k \to \infty$ asymptotics of the norms $||\Theta_p^{(k)}||$, $||\Theta_q^{(k)}||$ are determined by the asymptotics of the Bergman kernels for L^k on the diagonal. We take these asymptotics from [5]. Asymptotic bounds for the inner products $\langle \Theta_p^{(k)}, \Theta_q^{(k)} \rangle$ can be obtained from the off-diagonal estimates on the Bergman kernels [4].

Theorem 1. *Suppose $L \to M$ is a positive hermitian holomorphic line bundle on a compact n-dimensional complex manifold M. Let $p, q \in M$. Let $k \in \mathbb{N}$. Then there is the following (positive) lower bound for the entanglement entropy of the pure state $b_k(p, q)$ (2). There are positive constants C_1 and C_2 that depend on M and ω such that as $k \to \infty$*

$$E_k(b_k) \geq \frac{1}{2}\left(1 - C_1 e^{-C_2 \sqrt{k} \, \operatorname{dist}(p,q)}\right)^4.$$

1.2 $M = \mathbb{C}^n$

Let $M = \mathbb{C}^n$, $n \geq 1$. We will use the notations

$$z^T \bar{w} = z_1 \bar{w}_1 + \ldots + z_n \bar{w}_n$$

or

$$\langle z, w \rangle = z_1 \bar{w}_1 + \ldots + z_n \bar{w}_n$$

and

$$|z| = \sqrt{z^T \bar{z}}$$

or

$$||z|| = \sqrt{z^T \bar{z}}$$

for $z, w \in \mathbb{C}^n$.

For $k \in \mathbb{N}$ let H_k be the Segal-Bargmann space that consists of holomorphic functions on M with the inner product

$$\langle f, g \rangle = \left(\frac{k}{\pi}\right)^n \int_M f(z)\overline{g(z)} e^{-k|z|^2} dV(z),$$

where dV is the Lebesgue measure on $\mathbb{R}^{2n}$. It is a reproducing kernel Hilbert space. For $p \in \mathbb{C}^n$ the coherent vector at p is $\Theta_p^{(k)} \in H_k$ defined by

$$\Theta_p^{(k)}(z) = e^{kz^T \bar{p}}. \tag{4}$$

It is defined by the property

$$\langle f, \Theta_p^{(k)} \rangle = f(p) \tag{5}$$

for every $f \in H_k$. Similarly for $q \in M$ the coherent vector at q is

$$\Theta_q^{(k)}(z) = e^{kz^T \bar{q}}. \tag{6}$$

Theorem 2. *Let $p, q \in \mathbb{C}^n$. Let $k \in \mathbb{N}$. Let $\Theta_p^{(k)}$ and $\Theta_p^{(k)}$ be the coherent states at p and q respectively (4), (6). Then there is the following (positive) lower bound for the entanglement entropy of the pure state $b_k(p, q)$ (2): as $k \to \infty$*

$$E_k(b_k) \geq \frac{1}{2}(1 - e^{-k|p-q|^2})^4.$$

Remark 1. The change from $e^{-C\sqrt{k}\,\mathrm{dist}(p,q)}$ to $e^{-Ck\,\mathrm{dist}^2(p,q)}$ in Theorem 2 reflects the fact that the latter appears in the Bergman asymptotics for real analytic metrics (see the discussion in [3]). In the proof of Theorem 1 we used the Bergman kernel expansion for smooth metrics.

2   Proofs

2.1   General Lower Bound

Theorem 3. *Let H be a separable Hilbert space, with the inner product $\langle ., . \rangle$. Let u and v be nonzero vectors in H, such that u is not a multiple of v. Let*

$$w = \frac{1}{||u||^2} u \otimes u + \frac{1}{||v||^2} v \otimes v \in H \otimes H.$$

There is the following (positive) lower bound on the entanglement entropy $E(b)$ of the vector $b = \frac{1}{||w||} w$

$$E(b) \geq 2\frac{(||u||^2 ||v||^2 - |\langle u, v \rangle|^2)^2}{(2||u||^2||v||^2 + \langle u, v \rangle^2 + \langle v, u \rangle^2)^2}. \tag{7}$$

Proof. Let e_1, e_2 be an orthonormal basis of the 2-dimensional complex linear subspace spanned by u and v, defined as follows:

$$e_1 = \frac{1}{||u||} u,$$

e_2 is the unit vector in the direction of $v - \langle v, e_1 \rangle e_1$

$$e_2 = \frac{1}{||v - \langle v, e_1 \rangle e_1||}(v - \langle v, e_1 \rangle e_1) = \frac{1}{\sqrt{||v||^2 - |\langle v, e_1 \rangle|^2}}(v - \langle v, e_1 \rangle e_1)$$

We get:

$$w = \frac{||v||^2 + \langle v, e_1 \rangle^2}{||v||^2} e_1 \otimes e_1 + \frac{\langle v, e_1 \rangle \sqrt{||v||^2 - |\langle v, e_1 \rangle|^2}}{||v||^2}(e_1 \otimes e_2 + e_2 \otimes e_1)$$

$$+ \frac{||v||^2 - |\langle v, e_1 \rangle|^2}{||v||^2} e_2 \otimes e_2$$

$$||w|| = \frac{\sqrt{2||v||^2 + \langle v, e_1 \rangle^2 + \langle e_1, v \rangle^2}}{||v||}$$

Denote

$$\beta = \sqrt{2||v||^2 + \langle v, e_1 \rangle^2 + \langle e_1, v \rangle^2}.$$

We get:

$$b = \frac{1}{\beta ||v||}\Big((||v||^2 + \langle v, e_1 \rangle^2)e_1 \otimes e_1 +$$

$$\langle v, e_1 \rangle \sqrt{||v||^2 - |\langle v, e_1 \rangle|^2}(e_1 \otimes e_2 + e_2 \otimes e_1) + (||v||^2 - |\langle v, e_1 \rangle|^2)e_2 \otimes e_2 \Big)$$

Let

$$A = \frac{1}{\beta ||v||}\begin{pmatrix} ||v||^2 + \langle v, e_1 \rangle^2 & \langle v, e_1 \rangle \sqrt{||v||^2 - |\langle v, e_1 \rangle|^2} \\ \langle v, e_1 \rangle \sqrt{||v||^2 - |\langle v, e_1 \rangle|^2} & ||v||^2 - |\langle v, e_1 \rangle|^2 \end{pmatrix}.$$

Then $A^*A =$

$$\frac{1}{\beta^2}\begin{pmatrix} ||v||^2 + \langle v, e_1 \rangle^2 + \langle e_1, v \rangle^2 + |\langle v, e_1 \rangle|^2 & \sqrt{||v||^2 - |\langle v, e_1 \rangle|^2}(\langle v, e_1 \rangle + \langle e_1, v \rangle) \\ \sqrt{||v||^2 - |\langle v, e_1 \rangle|^2}(\langle v, e_1 \rangle + \langle e_1, v \rangle) & ||v||^2 - |\langle v, e_1 \rangle|^2 \end{pmatrix}$$

The equation for the eigenvalues of A^*A is

$$\lambda^2 - \lambda + \frac{(||v||^2 - |\langle v, e_1 \rangle|^2)^2}{\beta^4} = 0.$$

The eigenvalues of A^*A are

$$\lambda_{1,2} = \frac{1}{2} \pm \frac{1}{2}\frac{(\langle v, e_1 \rangle + \langle e_1, v \rangle)\sqrt{4||v||^2 + (\langle v, e_1 \rangle - \langle e_1, v \rangle)^2}}{\beta^2} =$$

$$\frac{1}{2} \pm \frac{1}{2}\frac{(\langle u, v \rangle + \langle v, u \rangle)\sqrt{4||u||^2||v||^2 + (\langle u, v \rangle - \langle v, u \rangle)^2}}{2||u||^2||v||^2 + \langle u, v \rangle^2 + \langle v, u \rangle^2}. \tag{8}$$

The singular values of A are the square roots of the eigenvalues of A^*A. The entanglement entropy of b equals

$$E(b) = -\lambda_1 \ln \lambda_1 - \lambda_2 \ln \lambda_2. \tag{9}$$

Since for $0 < x < 1$

$$-\ln x > 1 - x,$$

we have:

$$E(b) \geq \lambda_1(1 - \lambda_1) + \lambda_2(1 - \lambda_2) = 1 - \lambda_1^2 - \lambda_2^2. \tag{10}$$

Now, (7) is obtained by plugging (8) into (10).

Remark 2. Since u and v in Theorem 3 are linearly independent, it follows that the vector b is entangled, i.e. $E(b) > 0$. This follows from the fact that the right hand side of the inequality (7) is positive. Another way to see it is to refer to (9) and to observe that in (8)

$$\left| \frac{(\langle u, v \rangle + \langle v, u \rangle)\sqrt{4||u||^2||v||^2 + (\langle u, v \rangle - \langle v, u \rangle)^2}}{2||u||^2||v||^2 + \langle u, v \rangle^2 + \langle v, u \rangle^2} \right| < 1$$

(it is straightforward to check that this inequality is equivalent to

$$(||u||^2||v||^2 - \langle u, v \rangle \langle v, u \rangle)^2 > 0),$$

hence $0 < \lambda_1 < 1$ and $0 < \lambda_2 < 1$.

2.2   Proof of Theorem 1

Proof. It follows from [5] that there is a constant $A_0 > 0$ such that as $k \to \infty$, $||\Theta_p^{(k)}||^2$ and $||\Theta_q^{(k)}||^2$ are asymptotic to

$$A_0 k^n + O(k^{n-1}). \tag{11}$$

It follows from [4] that there are constants $A_1 > 0$, $A_2 > 0$ such that as $k \to \infty$

$$|\langle \Theta_p^{(k)}, \Theta_q^{(k)} \rangle| \leq A_1 k^n e^{-A_2\sqrt{k}\ \mathrm{dist}(p,q)}.$$

From (7) in Theorem 3 we get:

$$E_k(b_k) \geq \frac{1}{2} \frac{(1 - \frac{|\langle \Theta_p^{(k)}, \Theta_q^{(k)} \rangle|^2}{||\Theta_p^{(k)}||^2||\Theta_q^{(k)}||^2})^2}{(1 + \frac{\langle \Theta_p^{(k)}, \Theta_q^{(k)} \rangle^2 + \langle \Theta_q^{(k)}, \Theta_p^{(k)} \rangle^2}{2||\Theta_p^{(k)}||^2||\Theta_q^{(k)}||^2})^2}.$$

As $k \to \infty$,

$$E_k(b_k) \geq \frac{1}{2}(1 - \frac{|\langle \Theta_p^{(k)}, \Theta_q^{(k)} \rangle|^2}{||\Theta_p^{(k)}||^2||\Theta_q^{(k)}||^2})^2(1 - \frac{\langle \Theta_p^{(k)}, \Theta_q^{(k)} \rangle^2 + \langle \Theta_q^{(k)}, \Theta_p^{(k)} \rangle^2}{2||\Theta_p^{(k)}||^2||\Theta_q^{(k)}||^2})^2 \geq$$

$$\frac{1}{2}(1 - \frac{A_1^2 k^{2n} e^{-2A_2\sqrt{k}\ \mathrm{dist}(p,q)}}{||\Theta_p^{(k)}||^2||\Theta_q^{(k)}||^2})^4.$$

The conclusion now follows from (11).

2.3   Proof of Theorem 2

Proof. Using (7) in Theorem 3 we get:

$$E_k(b_k) \geq 2\frac{(\|\Theta_p^{(k)}\|^2\|\Theta_q^{(k)}\|^2 - |\langle\Theta_p^{(k)},\Theta_q^{(k)}\rangle|^2)^2}{(2\|\Theta_p^{(k)}\|^2\|\Theta_q^{(k)}\|^2 + \langle\Theta_p^{(k)},\Theta_q^{(k)}\rangle^2 + \langle\Theta_q^{(k)},\Theta_p^{(k)}\rangle^2)^2}.$$

By the reproducing property (5)

$$\|\Theta_p^{(k)}\|^2 = \Theta_p(p) = e^{k|p|^2}$$

$$\|\Theta_q^{(k)}\|^2 = \Theta_q(q) = e^{k|q|^2}$$

$$\langle\Theta_p^{(k)},\Theta_q^{(k)}\rangle = \Theta_p^{(k)}(q) = e^{kq^T\bar{p}}.$$

Therefore

$$E_k(b_k) \geq \frac{1}{2}\frac{(1 - e^{kq^T\bar{p}+kp^T\bar{q}-k|p|^2-k|q|^2})^2}{(1 + \frac{e^{2kq^T\bar{p}}+e^{2kp^T\bar{q}}}{2e^{k|p|^2+k|q|^2}})^2}.$$

As $k \to \infty$

$$E_k(b_k) \geq \frac{1}{2}(1 - e^{kq^T\bar{p}+kp^T\bar{q}-k|p|^2-k|q|^2})^2(1 - \frac{e^{2kq^T\bar{p}} + e^{2kp^T\bar{q}}}{2e^{k|p|^2+k|q|^2}})^2.$$

and the conclusion follows.

3   Example for 1.1

Let $M = \mathbb{CP}^1$ and let $L \to M$ be the hyperplane bundle with the standard hermitian metric. The Kähler form is the Fubini-Study form on M. The Hilbert space $V = H^0(M, L)$ is isomorphic to the space of polynomials in 1 and z, with the inner product

$$\langle f, g\rangle = \frac{2}{\pi}\int_{\mathbb{C}} \frac{f(z)\overline{g(z)}}{(1 + |z|^2)^3}dx\,dy.$$

The monomials $e_0 = 1$ and $e_1 = z$ form an orthonormal basis in V. For $p \in \mathbb{C}$ (an affine chart of M) let Θ_p be the unique vector in V defined by the property

$$\langle f, \Theta_p\rangle = f(p)$$

for all $f \in V$. It is immediate that

$$\Theta_p(z) = 1 + z\bar{p}$$

and

$$\|\Theta_p\| = \sqrt{1 + |p|^2}.$$

Let us consider two particles at $p = x + i0$ and $q = -x + i0$, $x > 0$. The associated state (2) is

$$b_1 = b_1(p, q) = \frac{1}{\sqrt{1 + x^4}} e_0 \otimes e_0 + \frac{x^2}{\sqrt{1 + x^4}} e_1 \otimes e_1.$$

The Schmidt coefficients are $\alpha_0 = \frac{1}{\sqrt{1+x^4}}$ and $\alpha_1 = \frac{x^2}{\sqrt{1+x^4}}$. The entanglement entropy of b_1 equals

$$E(x) = -\alpha_1^2 \ln(\alpha_1^2) - \alpha_2^2 \ln(\alpha_2^2) = \frac{(1 + x^4) \ln(1 + x^4) - x^4 \ln(x^4)}{1 + x^4}. \tag{12}$$

The graph of E is shown in Fig. 1. From (12), we observe that $E(x) \to 0$ as $x \to \infty$, and the maximum value of $E(x)$ is attained at $x = 1$.

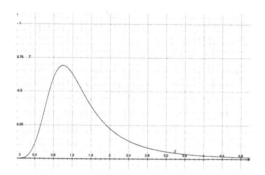

Fig. 1. The graph of $E(x)$, $x > 0$.

References

1. Araki, H., Lieb, E.: Entropy inequalities. Commun. Math. Phys. **18**, 160–170 (1970)
2. Berceanu, S., Schlichenmaier, M.: Coherent state embeddings, polar divisors and Cauchy formulas. J. Geom. Phys. **34**(3–4), 336–358 (2000)
3. Hezari, H., Lu, Z., Xu, H.: Off-diagonal asymptotic properties of Bergman kernels associated to analytic Kähler potentials. Int. Math. Res. Not. IMRN **8**, 2241–2286 (2020)
4. Ma, X., Marinescu, G.: Exponential estimate for the asymptotics of Bergman kernels. Math. Ann. **362**, 1327–1347 (2015)
5. Zelditch, S.: Szegö kernels and a theorem of Tian. Internat. Math. Res. Notices **6**, 317–331 (1998)

A Non-formal Formula
for the Rankin-Cohen Deformation
Quantization

Pierre Bieliavsky[1] and Valentin Dendoncker[2($\boxtimes$)]

[1] Université de Louvain, Ottignies-Louvain-la-Neuve, Belgium
[2] ICHEC Brussels Management School, Woluwe-Saint-Pierre, Belgium
valentin.dendoncker@gmail.com

Abstract. Don Zagier's superposition of Rankin-Cohen brackets on the Lie group $SL_2(\mathbb{R})$ defines an associative formal deformation of the algebra of modular forms on the hyperbolic plane [9]. This formal deformation has been used in [6] to establish strong connections between the theory of modular forms and that of regular foliations of co-dimension one. Alain Connes and Henri Moscovici also proved that Rankin-Cohen's deformation gives rise to a formal universal deformation formula (UDF) for actions of the group $ax + b$. In a joint earlier work [5], the first author, Xiang Tang and Yijun Yao proved that this UDF is realized as a truncated Moyal star-product. In the present work, we use a method to explicitly produce an equivariant intertwiner between the above mentioned truncated Moyal star-product (i.e. Rankin-Cohen deformation) and a non-formal star-product on $ax + b$ defined by the first author in an earlier work [2]. The specific form of the intertwiner then yields an oscillatory integral formula for Zagier's Rankin-Cohen UDF, answering a question raised by Alain Connes.

In this paper, we will study equivalences between two particular star-products on the symplectic manifold underlying the group $ax + b$. We start by recalling that for a symplectic manifold (M, ω), a *star-product* on M is a bilinear map $\star_\nu :$ $C^\infty(M) \times C^\infty(M) \to C^\infty(M)[[\nu]] : (f, g) \mapsto f \star_\nu g := \sum_{k=0}^{+\infty} \nu^k C_k(f, g)$, where ν is a formal parameter called *deformation parameter* and with $C^\infty(M)[[\nu]] :=$ $\left\{ \sum_{k=0}^{+\infty} \nu^k f_k \mid f_k \in C^\infty(M) \; \forall k \in \mathbb{N} \right\}$, such that (i) for each $k \in \mathbb{N}\backslash\{0\}$, the map $C_k : C^\infty(M) \times C^\infty(M) \to C^\infty(M)$ defines a bidifferential operator on M; (ii) the law $\star_\nu$, extended $\mathbb{C}[[\nu]]$-linearly to the space of formal power series $C^\infty(M)[[\nu]]$, gives an associative product on $C^\infty(M)[[\nu]]$; (iii) $C_0(f, g) = fg$ and $C_1(f, g) - C_1(g, f) = c\{f, g\}$ for some constant $c \in \mathbb{C}$ depending on the chosen normalization and where $\{\cdot, \cdot\}$ means the Poisson structure associated with ω; (iv) $1 \star_\nu f = f \star_\nu 1 = f$. In what follows, we will assume that $\nu \in i\mathbb{R}_0$. Given a star-product $\star_\nu$ on (M, ω), a *derivation* D of $\star_\nu$ is a linear operator on $C^\infty(M)[[\nu]]$ of the form $D = \sum_{k=0}^{+\infty} \nu^k D_k$, where for every k, D_k is a $\mathbb{C}[[\nu]]$-linear differential operator, with D satisfying the following relation for every $f, g \in C^\infty(M)[[\nu]]$: $D(f \star_\nu g) = Df \star_\nu g + f \star_\nu Dg$. The space of derivations of $\star_\nu$ is denoted by $\mathrm{Der}(\star_\nu)$. Also, for any $\varphi, \psi \in C^\infty(M)[[\nu]]$, their $\star_\nu$-*commutator*

© The Author(s), under exclusive license to Springer Nature Switzerland AG 2023
F. Nielsen and F. Barbaresco (Eds.): GSI 2023, LNCS 14071, pp. 524–532, 2023.
https://doi.org/10.1007/978-3-031-38271-0_52

is defined by $[\varphi, \psi]_{\star_\nu} := \varphi \star_\nu \psi - \psi \star_\nu \varphi$. A star-product $\star_\nu$ on a (M, ω) is said to be *invariant* under an action of $\mathfrak{g} = \text{Lie}(G)$ on M (that is the data of a homomorphism $\mathfrak{g} \to \Gamma^\infty(TM) : X \mapsto X^\star$) if each $X^\star$ is a derivation of $\star_\nu$. In that case, we will say that $\star_\nu$ is $\mathfrak{g}$-invariant. Suppose now that $\star_\nu$ and $\star_\nu'$ are two star-products on the same symplectic manifold (M, ω). There are said to be *equivalent* if there exists a linear formal operator T on $C^\infty(M)[[\nu]]$ of the form $T = \sum_{k=0}^{+\infty} \nu^k T_k$, such that (i) $T_0 = \text{Id}$; (ii) T_k is a $\mathbb{C}[[\nu]]$-linear differential operator for all $k \in \mathbb{N}$; (iii) T *intertwines* $\star_\nu$ and $\star_\nu'$, that is, for $f, g \in C^\infty(M)[[\nu]]$, $T(f \star_\nu g) = Tf \star_\nu' Tg$ (this property will be denoted shortly as: $T(\star_\nu) = \star_\nu'$). For this last reason, such an equivalence is also called an *intertwiner*. Let us now introduce the two star-products that we will link in the Sect. 2 by studying their intertwiners.

1 Rankin-Cohen and Bieliavsky-Gayral Star-Products

1.1 Rankin-Cohen Brackets and Formal Deformation Quantization

The first star-product that we consider is due to Zagier, Connes and Moscovici, but based on the researches of Rankin and Cohen concerning the modular forms. Let us consider the Hopf algebra $\mathcal{H}_1$ described by Connes and Moscovici in [6] in the particular case where all the δ_n are trivial. In this case, $\mathcal{H}_1$ reduces to the universal enveloping algebra $\mathcal{U}(\mathfrak{s})$ of the Lie algebra $\mathfrak{s}$ with basis $\{X, Y\}$ and bracket $[Y, X] = X$. Note that $\mathfrak{s} := \text{Lie}(\mathbb{S})$, i.e. the Lie algebra associated with the $ax + b$ Lie group that we denote $\mathbb{S}$. Authors in [6] then show that the superposition of (a reduced form of) Rankin-Cohen brackets (denoted as RC_n^{red} for n a positive integer) gives rise to a formal universal deformation formula (UDF) $\star_\nu^{RC} := \sum_{n=0}^{+\infty} \nu^n RC_n^{red}$ for actions of the group $\mathbb{S}$. It is worth mentioning that the Lie group $\mathbb{S}$ corresponds to the Iwasawa component of $SL_2(\mathbb{R})$. Setting $\mathfrak{sl}_2(\mathbb{R}) := \text{Lie}(SL_2(\mathbb{R})) = \text{span}_\mathbb{R}\{H, E, F\}$ with brackets $[H, E] = 2E$, $[H, F] = -2F$ and $[E, F] = H$, one has $\mathfrak{s} = \text{span}_\mathbb{R}\{H, E\}$ and thus $\mathbb{S} = \exp(\text{span}_\mathbb{R}\{H, E\})$ (it means that $X = E$ and $Y = \frac{H}{2}$ within notations in [6]). This leads to the following global coordinate system on $\mathbb{S}$:

$$\mathfrak{s} \simeq \mathbb{R}^2 \to \mathbb{S} : (a, \ell) \mapsto \exp(aH)\exp(\ell E). \tag{1}$$

We endow this space $\mathbb{S}$ with the symplectic form $\omega^\mathbb{S} = da \wedge d\ell$. As noticed in [5], the deformation $\star_\nu^{RC}$ defines a *left invariant* star-product on the symplectic manifold $(\mathbb{S}, \omega^\mathbb{S})$, meaning that for all $x \in \mathbb{S}$, the left action L_x of $\mathbb{S}$ is a *symmetry* of $\star_\nu^{RC}$, i.e. it satisfies the identity $L_x^\star \varphi \star_\nu^{RC} L_x^\star \psi = L_x^\star(\varphi \star_\nu^{RC} \psi)$, for all $\varphi, \psi \in C^\infty(\mathbb{S})[[\nu]]$. Hereafter, the star-product $\star_\nu^{RC}$ on the symplectic manifold $(\mathbb{S}, \omega^\mathbb{S})$ will be called the *Rankin-Cohen star-product* (or *RC star-product* in short). Let us now consider the symplectic vector space $(\mathbb{R}^2 = \{(q, p)\}, \omega := dq \wedge dp)$ and $\mathfrak{h} := \text{span}_\mathbb{R}\{Q, P\} \oplus \mathbb{R}Z$ the corresponding Heisenberg algebra with brackets given by $[Q, P] = \omega(Q, P)Z = Z$, $[Y, Z] = 0 \,\forall\, Y \in \text{span}_\mathbb{R}\{Q, P\}$. We form the semi-direct product $\mathfrak{g}_{RC} := \mathfrak{sl}_2(\mathbb{R}) \ltimes \mathfrak{h}$. In [5], authors show that $\star_\nu^{RC}$ is $\mathfrak{g}_{RC}$-invariant, through the action of $\mathfrak{g}_{RC}$ on $(\mathbb{S}, \omega^\mathbb{S})$ associated with the fundamental vector fields. Note that considering the natural action through matrix multiplication of $\mathbb{S} \subset SL_2(\mathbb{R})$ on $(\mathbb{R}^2, \omega)$, we can form the orbit $\mathcal{O} = \mathbb{S}.(0,1) = \{(\ell e^a, e^{-a}) \mid a, \ell \in \mathbb{R}\}$ (within coordinates (1)), corresponding to the upper half-plane of $\mathbb{R}^2$. As shown in [5], Proposition 3.4, the star-product $\star_\nu^{RC}$ realized on $\mathcal{O} \subset \mathbb{R}^2$ coincides with the restriction to $\mathcal{O}$ of the Moyal star-product on $(\mathbb{R}^2, \omega)$. A last property of $\star_\nu^{RC}$ is given below.

Proposition 1 ([4], **Lemma 5.3**). *The star-product $\star_\nu^{RC}$ is (up to a redefinition of the deformation parameter) the only $\mathfrak{s} \ltimes \mathfrak{h}$-invariant formal star-product on $(\mathbb{S}, \omega^{\mathbb{S}})$.*

1.2 Bieliavsky-Gayral Star-Product and Non-formal Deformation Quantization

The second star-product that we consider constitutes a particular example of a construction presented in [3]. In what follows, the Poincaré algebra $\mathbb{R}H \ltimes (\mathbb{R}E \oplus \mathbb{R}F)$, with table given by $[H, E] = 2E$, $[H, F] = -2F$, $[E, F] = 0$, will be denoted as $\mathfrak{g}_{BG}$. The method used in [3] consists in twisting the two-dimensional Moyal star-product $\star_\nu^M$ (in coordinates (1)) by means of an intertwiner T, in such a way that $T\left(\star_\nu^M\right)$ yields a left invariant star-product on $(\mathbb{S}, \omega^{\mathbb{S}})$. As pointed out by the authors, there exists a family (indexed by functions ϑ) of operators T leading to a *convergent form* of the corresponding star-product (as in the case of Moyal-Weyl). Indeed, by considering for instance the operator T_{ν, ϑ_0} associated with $\vartheta := \vartheta_0$ given by $T_{\nu, \vartheta_0} := \mathcal{F}^{-1} \circ \mathcal{M}_{\exp(\vartheta_0)} \circ \left(\Psi_\nu^{-1}\right)^\star \circ \mathcal{F}$, where $\mathcal{F}$ is the partial Fourier transform in the second variable, $\mathcal{M}_{\exp(\vartheta_0)}$ is the multiplication by $\exp(\vartheta_0)$ whith $\vartheta_0(\xi) := -\frac{1}{4} \log\left(1 - 4\nu^2 \xi^2\right)$ and where $\Psi_\nu(a, \xi) = \left(a, \frac{\sinh(2i\nu\xi)}{2i\nu}\right)$, the star-product $\star_{\nu, \vartheta_0}^{BG} := T_{\nu, \vartheta_0}\left(\star_\nu^M\right)$ on the space $\mathcal{E}_{\nu, \vartheta_0}^{BG} := T_{\nu, \vartheta_0}\left(\mathcal{S}(\mathbb{S})\right)$ (where $\mathcal{S}(\mathbb{S})$ is the Euclidean Schwartz space on $\mathbb{S} \simeq \mathbb{R}^2$ with respect to (1)) enjoys the following properties.

Theorem 1 ([3], **Theorem 4.5**). *Within the previous notations,*

(i) $\left(\mathcal{E}_{\nu, \vartheta_0}^{BG}, \star_{\nu, \vartheta_0}^{BG}\right)$ *is an associative algebra, endowed with the Fréchet algebra structure transported from $\mathcal{S}(\mathbb{S}) \simeq \mathcal{S}\left(\mathbb{R}^2\right)$ under the operator T_{ν, ϑ_0}.*

(ii) *For all compactly supported functions $\varphi, \psi \in \mathcal{D}(\mathbb{S}) \subset \mathcal{E}_{\nu, \vartheta_0}^{BG}$, the expression $\varphi \star_{\nu, \vartheta_0}^{BG} \psi$ admits an oscillatory integral representation.*

(iii) *The star-product $\star_{\nu, \vartheta_0}^{BG}$ is $\mathfrak{g}_{BG}$-invariant, through the action of $\mathfrak{g}_{BG}$ on $(\mathbb{S}, \omega^{\mathbb{S}})$ associated with the fundamental vector fields. In particular, $\star_{\nu, \vartheta_0}^{BG}$ is also left invariant on $(\mathbb{S}, \omega^{\mathbb{S}})$.*

The star-product given in Theorem 1 will be called the *strongly-closed Bieliavsky-Gayral star-product* and will be denoted as $\underline{\star}^{BG}$. It is worth mentioning that the choice $\vartheta \equiv 0$ leading to the associative algebra $\left(\mathcal{E}_{\nu,0}^{BG}, \star_{\nu,0}^{BG}\right)$ (where $\star_{\nu,0}^{BG} = T_{\nu,0}\left(\star_\nu^M\right)$) corresponds to that of [2], Theorem 6.13, where $\star_{\nu,0}^{BG}$ is also $\mathfrak{g}_{BG}$-invariant (and in particular left invariant). The star-product $\star_\nu^{BG} := \star_{\nu,0}^{BG}$ will be called the *Bieliavsky-Gayral star-product* (or *BG star-product* in short). Note that we can easily show that $\underline{T}_\nu \left(\star_\nu^{BG}\right) = \underline{\star}_\nu^{BG}$, where $\underline{T}_\nu := \mathcal{F}^{-1} \circ \mathcal{M}_{\exp(\vartheta_0)} \circ \mathcal{F}$. At this point, we can naturally ask whether the RC star-product also admits a convergent expression like the BG star-product (on a suitable subspace of $C^\infty(\mathbb{S})$). This is precisely the subject of the next section, where we start by describing a method to obtain the expressions of intertwiners between RC and BG star-products.

2 Non-formal Rankin-Cohen Star-Product

2.1 Guessing the Intertwiners Between RC and BG Star-Products

Consider the symplectic manifold $(\mathbb{S}, \omega^{\mathbb{S}})$. Thanks to the particular invariances of $\star_\nu^{BG}$ and $\star_\nu^{RC}$ mentioned previously, we get the following injections: $\mathfrak{g}_{BG} \hookrightarrow \mathrm{Der}\left(\star_\nu^{BG}\right) : Y \mapsto Y^\star$ and $\mathfrak{g}_{RC} \hookrightarrow \mathrm{Der}\left(\star_\nu^{RC}\right) : Y \mapsto Y^\star$, where $Y^\star$ denotes the fundamental vector field associated with Y. Note that $\mathfrak{s}$ is a subalgebra of both $\mathfrak{g}_{BG}$ and $\mathfrak{g}_{RC}$. By virtue of [1], Theorem 4.1, one knows that the set of $\mathbb{S}$-equivalence classes of left invariant star-products on $\mathbb{S}$ is canonically parametrized by the set of sequences of elements belonging to the second de Rham cohomology space $H^2_{dR}(\mathbb{S})^\mathbb{S}$ of the $\mathbb{S}$-invariant de Rham complex on $\mathbb{S}$. Thanks to the fact that every Chevalley-Eilenberg 2-cocycle on $\mathfrak{s}$ is a coboundary, it turns out that there exists at least one intertwiner T on $C^\infty(\mathbb{S})[[\nu]]$ between $\star_\nu^{BG}$ and $\star_\nu^{RC}$ that is $\mathbb{S}$-*equivariant*, i.e. such that for all $x \in \mathbb{S}$, $L_x^\star T = T L_x^\star$. At the level of the space $\mathcal{D}(\mathbb{S})$ of smooth compactly supported functions, we can then express T as a (formal) convolution operator by means of the Schwartz kernel theorem. Consequently, there exists a distribution u such that for any $\varphi \in \mathcal{D}(\mathbb{S})$ and $x \in \mathbb{S}$, $T\varphi(x) = \int_{\mathbb{S}} u(x,y)\varphi(y)d^L_\mathbb{S}(y)$, where $d^L_\mathbb{S}$ denotes a left invariant Haar measure on $\mathbb{S}$. Using the $\mathbb{S}$-equivariance of T and setting $u_T(x) := u\left(e, x^{-1}\right)$ (for $e \in \mathbb{S}$ the neutral element), we get the following particular form: $T\varphi(x) = \int_{\mathbb{S}} u_T \left(y^{-1}x\right)\varphi(y)d^L_\mathbb{S}(y)$. This yields in turn the following algebra morphism: $D : \mathfrak{g}_{RC} \to \mathrm{Der}\left(\star_\nu^{BG}\right) : Y \mapsto D_Y := T^{-1}Y^\star T$, extending to $\mathfrak{g}_{RC}$ the injection $\mathfrak{s} \hookrightarrow \mathrm{Der}\left(\star_\nu^{BG}\right)$. Since $H^1_{dR}(\mathbb{S}) = \{0\}$ thanks to the Poincaré lemma, we deduce from [8], Theorem 8.2 that for any $Y \in \mathfrak{g}_{RC}$, the associated derivation D_Y is inner, i.e. $D_Y = \frac{1}{2\nu}[\Phi_Y, \cdot]_{\star_\nu^{BG}}$ for some $\Phi_Y \in C^\infty(\mathbb{S})[[\nu]]$. For instance, in the basis $\{H, E\}$ of $\mathfrak{s}$, $\Phi_H(a, \ell) = \ell$ and $\Phi_E(a, \ell) = \frac{1}{2}e^{-2a}$ (within (1)). Now, remark that the algebra $\mathfrak{g}_{RC}$ decomposes (as vector space) as $\mathfrak{s} \oplus V_{RC}$, with $V_{RC} := \mathrm{span}_\mathbb{R}\{F, Q, P, Z\}$. For any element X_k in the basis of $\mathfrak{s}$ and any element Y_l in the basis of V_{RC}, one has: $[X_k, Y_l] = \sum_a C^a_{kl}X_a + \sum_b \underline{C}^b_{kl}Y_b$, with $a \in \{1, 2\}$, $b \in \{1, \ldots, 4\}$ and where $C^a_{kl}, \underline{C}^b_{kl}$ are some real constants. The fact that D is a Lie algebra morphism then leads to the following ODE: $X_k^\star(\Phi_{Y_l}) = \sum_a C^a_{kl}\Phi_{X_a} + \sum_b \underline{C}^b_{kl}\Phi_{Y_b}$, up to additive formal constants. Once the solution Φ_{Y_l} is obtained for any Y_l in the basis of V_{RC}, we can deduce the expression of D_Y for any $Y \in \mathfrak{g}_{RC}$ by means of linear extension. Regarding the convolution kernel, recall first that $d^L_\mathbb{S}(y)$ is preserved under $Y^\star$ for all $Y \in \mathfrak{g}_{RC}$. Since the inverse T^{-1} of T is also a convolution operator with associated kernel

v, for $x, x_0 \in \mathbb{S}$ and $Y \in \mathfrak{g}_{RC}$, one has: $D_Y|_{x_0} v\left(x^{-1}x_0\right) = -Y^*|_x v\left(x^{-1}x_0\right)$, where $|_{x_0}$ (respectively $|_x$) means that the operator is applied by considering the function on its right as depending on the x_0-variable (resp. x-variable) only. A straightforward computation then leads to the observation that any intertwiner T such that $T\left(\star_\nu^{BG}\right)$ is a $\mathfrak{g}_{RC}$-invariant star-product admits as an inverse a convolution operator $T^{-1} = v \times \cdot$, with v a solution of the following two equations:

$$D_Q|_{x_0} v\left(x^{-1}x_0\right) = -Q^*|_x v\left(x^{-1}x_0\right) \ , \ D_P|_{x_0} v\left(x^{-1}x_0\right) = -P^*|_x v\left(x^{-1}x_0\right),$$
(2)

for $D_Q|_{(a,\ell)} = \frac{1}{2\nu}\left[K_1 e^{-a}, \cdot\right]_{\star_\nu^{BG}}$, $D_P|_{(a,\ell)} = \frac{1}{2\nu}\left[(K_2 - K_1\ell)e^a, \cdot\right]_{\star_\nu^{BG}}$, $Q^*|_{(a,\ell)} = -e^{-a}\partial_\ell$ and $P^*|_{(a,\ell)} = e^a\left(\partial_a - \ell\partial_\ell\right)$ (within the chart (1)), for some parameters $K_1, K_2 \in \mathbb{C}[[\nu]]$. We set $\boldsymbol{K} := (K_1, K_2)$. Solving the previous equations then shows that to the corresponding intertwiner $T_{\nu,\boldsymbol{K}} = v \times \cdot$ is formally given by:

$$T_{\nu,\boldsymbol{K}}\,\varphi(a_0, \ell_0) =$$

$$\frac{1}{2\pi} \int_{\mathbb{R}^2} e^{i\left[\xi\left(\ell - K_1^2\eta_\nu^2(r)\ell_0\right) - \frac{K_2}{2i\nu K_1}\operatorname{arcsinh}(2i\nu\xi)\right]} \underline{\eta_\nu}(\xi)\varphi\left(a_0 - \log(K_1\eta_\nu(\xi)), \ell\right) d\ell d\xi,$$
(3)

where $\eta_\nu : \mathbb{R} \to \mathbb{R}_0^+ : \xi \mapsto \sqrt{\frac{2}{1+\sqrt{1-4\nu^2\xi^2}}}$ and $\underline{\eta_\nu} := \eta_\nu \exp(2\vartheta_0)$. From now on, we will always assume that K_1 is a strictly positive real number and K_2 a real number, satisfying the limit condition $\lim_{\nu\to 0}(K_1, K_2) = (1, 0)$. The next section is devoted to an analytical study of the intertwiner $T_{\nu,\boldsymbol{K}}$ as well as the formula $\natural_{\nu,\boldsymbol{K}} := T_{\nu,\boldsymbol{K}}\left(\star_\nu^{BG}\right)$.

2.2   Towards a Non-formal RC Star-Product

In order to give rise to an oscillatory integral formula for the RC star-product, we will focus on the intertwiner between $\star_\nu^{BG}$ (the strongly-closed BG star-product) and $\star_\nu^{RC}$, rather than the one between $\star_\nu^{BG}$ and $\star_\nu^{RC}$. The reason is that Proposition 4.10 in [3] exhibits explicitly a space $\mathcal{S}^{BG}(\mathbb{S})$ of Schwartz-type functions such that, endowed with the multiplication $\star_\nu^{BG}$, $\left(\mathcal{S}^{S_{\nu,\vartheta_0}^{BG}}(\mathbb{S}), \star_\nu^{BG}\right)$ becomes an associative Fréchet algebra. It turns out that $\mathcal{S}^{BG}(\mathbb{S})$ corresponds to the usual Schwartz space in coordinates $(x, y) = (\sinh(2a), \ell)$. In what follows, the idea is then to transport that space under the operator $\underline{T}_{\nu,\boldsymbol{K}} := T_{\nu,\boldsymbol{K}} \circ \underline{T}_\nu^{-1}$.

Proposition 2. Let $\varphi \in \mathcal{S}^{BG}(\mathbb{S})$. Then, $\underline{T}_{\nu,\boldsymbol{K}}\varphi \in \mathcal{S}(\mathbb{S})$.

Proof. Let $\varphi \in \mathcal{S}^{BG}(\mathbb{S})$. Note first that $\mathcal{S}^{BG}(\mathbb{S}) \subset \mathcal{S}(\mathbb{S})$. Also, the element $\mathbf{E} := e^{i\left[r\left(z - K_1^2\eta_\nu^2(r)\ell_0\right) - \frac{K_2}{2i\nu K_1}\operatorname{arcsinh}(2i\nu r)\right]}$ satisfies $\frac{1}{1+r^2}\left(\operatorname{Id} - \partial_z^2\right)\mathbf{E} = \mathbf{E}$. Thanks to the self-adjointness of the differential operator $D := \operatorname{Id} - \partial_z^2$ (w.r.t. the $L^2(\mathbb{R})$-inner product), we get the following expression for $\underline{T}_{\nu,\boldsymbol{K}}$:

$$\underline{T}_{\nu,\boldsymbol{K}}\,\varphi(a_0, \ell_0) = \frac{1}{2\pi} \int_{\mathbb{R}^2} \mathbf{E}\,\frac{\tilde{\eta}_\nu(r)}{1+r^2}\,D\varphi\left(a_0 - \log(K_1\eta_\nu(r)), z\right) dr dz,$$
(4)

where $\tilde{\eta}_\nu : \mathbb{R} \to \mathbb{R}_0^+ : r \mapsto \eta_\nu(r)\exp(\vartheta_0(r))$. The Lebesgue dominated convergence theorem and the mean-value theorem then imply that $\underline{T}_{\nu,K}\varphi$, $\partial_1\underline{T}_{\nu,K}\varphi$ and $\partial_2\underline{T}_{\nu,K}\varphi$ (where ∂_j denotes the partial derivative w.r.t. the jth-variable) are continuous, which means that $\underline{T}_{\nu,K}\varphi \in C^1(\mathbb{S})$. Denoting by $f_\varphi(a_0,\ell_0,r,z)$ the integrand in (4), remark now that for $m,n \in \mathbb{N}$, $\partial_1^m\partial_2^n f_\varphi(a_0,\ell_0,r,z) = \left(-iK_1^2 r\eta_\nu^2(r)\right)^n \mathbf{E}\frac{\tilde{\eta}_\nu(r)}{1+r^2}\partial_1^m(D\varphi)(a_0 - \log(K_1\eta_\nu(r)),z)$. An induction on the degree of the partial derivatives then shows that $\underline{T}_{\nu,K}\varphi \in C^\infty(\mathbb{S})$. Now, thanks to the Fubini theorem, notice that performing the integral w.r.t. the z-variable within the expression of $\underline{T}_{\nu,K}\varphi$ leads to

$$\underline{T}_{\nu,K}\varphi(a,\ell) = \int_\mathbb{R} e^{i\left[-K_1^2 r\eta_\nu^2(r)\ell - \frac{K_2}{2i\nu K_1}\mathrm{arcsinh}(2i\nu r)\right]}\tilde{\eta}_\nu(r)\hat{\varphi}\left(a - \log(K_1\eta_\nu(r)),r\right)dr \tag{5}$$

with $\hat{\varphi} := \mathcal{F}^{-1}\varphi$. An obvious but important fact about η_ν, $\log\eta_\nu$, $\tilde{\eta}_\nu$ and arcsinh is that those functions belong to the space $\mathcal{O}_M(\mathbb{R})$ of Schwartz multipliers on $\mathbb{R}$. In particular, $\tilde{\eta}_\nu\hat{\varphi}\left(a - \log(K_1\eta_\nu(r)),r\right)$ is a Schwartz function on $\mathbb{R}$ (for fixed a). That being said, we set $\mu(r) := r\eta_\nu^2(r)$, $\mathbf{E} := e^{-i\mu(r)\ell}$, $\tilde{\mathbf{E}} := e^{-\frac{K_2}{2\nu K_1}\mathrm{arcsinh}(2i\nu r)}$ and $A := i\left(\frac{d}{dr}\mu(r)\right)^{-1}\partial_r$. Note that $\frac{1}{1+\ell^2}\left[\mathrm{Id} + A^2\right](\mathbf{E}) = \mathbf{E}$. Remark also that $\left(\frac{d}{dr}\mu\right)^{-1}, \frac{d^2}{dr^2}\mu \in \mathcal{O}_M(\mathbb{R})$. Denoting by $A^\dagger$ the adjoint operator of A w.r.t. the $L^2(\mathbb{R})$-inner product, notice that the coefficients of the differential operator $\mathrm{Id} + \left(A^\dagger\right)^2$ as well as the element $\tilde{\mathbf{E}}$ belong to $\mathcal{O}_M(\mathbb{R})$ (w.r.t. the r-variable). Using now an integration by parts argument (combined with the Lebesgue dominated convergence and the mean-value theorem) leads to the observation that $\sup_{(a,\ell)\in\mathbb{S}}\left|a^k\ell^p\partial_a^m\partial_\ell^n(\underline{T}_{\nu,K}\varphi)(a,\ell)\right| < +\infty$, which concludes the proof.

For a given function f, let $\mathrm{supp}(f)$ denotes its support. Note that using (1), a direct computation leads to the following interesting result.

Lemma 1. *For any $\psi \in \underline{T}_{\nu,K}\left(\mathcal{S}^{BG}(\mathbb{S})\right)$, $\mathrm{supp}(\mathcal{F}\psi) \subset \mathbb{R} \times \left[-\frac{K_1^2}{|i\nu|}, \frac{K_1^2}{|i\nu|}\right]$.*

Following notations due to Gel'fand and Shilov, for a given $\sigma \in \mathbb{R}^+$, we set $\mathcal{S}_\sigma := \left\{f \in \mathcal{S}^{BG}(\mathbb{S}) \mid \mathrm{supp}(f) \subset \mathbb{R} \times [-\sigma,\sigma]\right\}$, $\underline{\mathcal{S}}_\sigma := \cup_{0\le\epsilon<\sigma}\mathcal{S}_\epsilon$ and $\mathcal{S}^\sigma$ (respectively $\underline{\mathcal{S}}^\sigma$) the space $\mathcal{F}^{-1}(\mathcal{S}_\sigma)$ (resp. $\mathcal{F}^{-1}(\underline{\mathcal{S}}_\sigma)$).

Proposition 3. *The operator $\underline{T}_{\nu,K}$ defined by*

$$\underline{T}_{\nu,K}\psi(a_0,\ell_0) =$$
$$\frac{K_1^2}{2\pi}\int_{\mathbb{R}^2} e^{i\left[r(\ell_0 - K_1^2\eta_\nu^2(r)z) - \frac{K_2}{2i\nu K_1}\mathrm{arcsinh}(2i\nu r)\right]}\tilde{\eta}_\nu(r)\psi\left(a_0 + \log(K_1\eta_\nu(r)),z\right)dr\,dz \tag{6}$$

is an inverse for $\underline{T}_{\nu,K}$. More precisely, the following identities hold:

$$\left(\underline{T}_{\nu,K}\circ\underline{T}_{\nu,K}\right)\big|_{\mathcal{S}^{BG}(\mathbb{S})} = Id_{\mathcal{S}^{BG}(\mathbb{S})} \quad and \quad \left(\underline{T}_{\nu,K}\circ\underline{T}_{\nu,K}\right)\big|_{\underline{\mathcal{S}}^{K_1^2/|i\nu|}} = Id\big|_{\underline{\mathcal{S}}^{K_1^2/|i\nu|}}.$$

Proof. Let $\gamma_{\nu,K}$ be the smooth map defined by $\gamma_{\nu,K} : \mathbb{R}^2 \to \mathbb{R}^2 : (a,\xi) \mapsto (a + \log(K_1\eta_\nu(\xi)), K_1^2 r\eta_\nu^2(\xi))$ and let $\mathcal{M}_{\nu,K}$ be the operator defined for any smooth function f on $\mathbb{R}^2$ by $\mathcal{M}_{\nu,K}(f) := \left[(a,\xi) \mapsto K_1^2 \tilde{\eta}_\nu(\xi)e^{-\frac{K_2}{2\nu K_1}\operatorname{arcsinh}(2i\nu\xi)}f(a,\xi)\right]$. A direct computation shows that $\underline{\mathcal{T}}_{\nu,K}$ given by (6) admits the following decompostion: $\underline{\mathcal{T}}_{\nu,K} = \mathcal{F}^{-1} \circ \mathcal{M}_{\nu,K} \circ \gamma_{\nu,K}^\star \circ \mathcal{F}$. The first identity is then a consequence of Proposition 2 combined with the decomposition above. For the second identity, let us choose a function $\psi \in \mathcal{S}^\sigma$, for $0 \le \sigma < \frac{K_1^2}{|i\nu|}$ (i.e. $\psi \in \underline{\mathcal{S}}^{K_1^2/|i\nu|}$). It means that there exists $\psi_0 \in \mathcal{S}_\sigma$ such that $\psi = \mathcal{F}^{-1}\psi_0$. Thanks to the decomposition of $\underline{\mathcal{T}}_{\nu,K}$, we have $\underline{\mathcal{T}}_{\nu,K}\psi = \mathcal{F}^{-1}\left(\mathcal{M}_{\nu,K}\left(\gamma_{\nu,K}^\star\psi_0\right)\right)$. Note that the function $\mathcal{M}_{\nu,K}\left(\gamma_{\nu,K}^\star\psi_0\right)(a,\xi)$ vanishes identically for ξ outside an interval $[-\tilde{\sigma},\tilde{\sigma}]$, for a certain $0 \le \tilde{\sigma} < +\infty$. Moreover $\partial_1\mathcal{M}_{\nu,K}\left(\gamma_{\nu,K}^\star\psi_0\right) = \mathcal{M}_{\nu,K}\left(\gamma_{\nu,K}^\star\partial_1\psi_0\right)$. It implies that $\mathcal{M}_{\nu,K}\left(\gamma_{\nu,K}^\star\psi_0\right) \in \mathcal{S}(\mathbb{R}^2)$, which in turn implies that $\underline{\mathcal{T}}_{\nu,K}\psi \in \mathcal{S}(\mathbb{S})$. Plugging the function $\varphi := \underline{\mathcal{T}}_{\nu,K}\psi$ into the expression (5) then concludes the proof.

We can now provide an oscillatory integral formula for the Rankin-Cohen deformation by transporting the space $\mathcal{S}^{BG}(\mathbb{S})$ under the operator $\underline{\mathcal{T}}_{\nu,K}$.

Theorem 2. *Let $\mathcal{E}_{\nu,K} := \underline{\mathcal{T}}_{\nu,K}\left(\mathcal{S}^{BG}(\mathbb{S})\right)$.*

(i) *The following inclusions hold: $\underline{\mathcal{T}}_{\nu,K}(\mathcal{D}(\mathbb{S})) \subset \underline{\mathcal{S}}^{K_1^2/|i\nu|} \subset \mathcal{E}_{\nu,K} \subset \mathcal{S}(\mathbb{S})$.*

(ii) *For $\varphi,\psi \in \mathcal{E}_{\nu,K}$, the formula $\varphi\natural_{\nu,K}\psi := \underline{\mathcal{T}}_{\nu,K}\left(\underline{\mathcal{T}}_{\nu,K}\varphi\star_\nu^{BG}\underline{\mathcal{T}}_{\nu,K}\psi\right)$ defines an associative product on $\mathcal{E}_{\nu,K}$. The algebra $(\mathcal{E}_{\nu,K},\natural_{\nu,K})$ is then endowed with the Fréchet algebra structure transported under $\underline{\mathcal{T}}_{\nu,K}$ from $\mathcal{S}^{BG}(\mathbb{S})$.*

(iii) *The star-product $\natural_{\nu,K}$ is $\mathfrak{g}_{RC}$-invariant (and, up to a redefinition of the deformation parameter, it coincides with $\star_\nu^{RC}$). In particular, $\natural_{\nu,K}$ is left invariant.*

(iv) *Denoting by R the right action of $\mathbb{S}$, the star-product $\natural_{\nu,K}$ admits the following integral representation at the level of functions belonging to $\underline{\mathcal{S}}^{K_1^2/|i\nu|}$:*

$$\varphi\natural_{\nu,K}\psi =$$

$$\frac{K_1^4}{(2\pi i\nu)^2}\int_{\mathbb{R}^4}\frac{1}{\cosh(a_1)\cosh(a_2)\cosh(a_1-a_2)}e^{\frac{K_1^2}{\nu}[\tanh(a_2)\ell_1-\tanh(a_1)\ell_2]}$$

$$\times R^\star_{\left(a_1+\log\frac{\cosh(a_1-a_2)}{\cosh(a_2)},\ell_1\right)}\varphi R^\star_{\left(a_2+\log\frac{\cosh(a_1-a_2)}{\cosh(a_1)},\ell_2\right)}\psi\, da_1 d\ell_1 da_2 d\ell_2.$$

(v) *Let $\mathbf{K}_0 := (1,0)$. For every $\varphi,\psi \in \underline{\mathcal{T}}_{\frac{\theta}{4},K_0}(\mathcal{D}(\mathbb{S}))$ and for every $x \in \mathbb{S}$, the map $\mathbb{R} \to \mathbb{C} : \theta \mapsto \left(\varphi\natural_{\frac{\theta}{4},K_0}\psi\right)(x)$ is smooth. Its Taylor series at 0 defines an associative star-product on $(\mathbb{S},\omega^{\mathbb{S}})$ which coincides with $\star_{\frac{\theta}{4}}^{RC}$.*

Proof. We start by item (i). For the first inclusion, it comes from the fact that the range of the map $\left[B \to \mathbb{R} : r \mapsto K_1^2 r\eta_\nu^2(r)\right]$ is included in an interval $[-\epsilon,\epsilon]$ for a certain $0 \le \epsilon < \frac{K_1^2}{|i\nu|}$. Choosing $\psi = \underline{\mathcal{T}}_{\nu,K}\varphi$ with $\varphi \in \mathcal{D}(\mathbb{S}) \subset \mathcal{S}^{BG}(\mathbb{S})$, a direct computation of $\mathcal{F}\left(\underline{T}_{\nu,K}\,\varphi\right)$ shows that $\psi \in \underline{\mathcal{S}}^{K_1^2/|i\nu|}$. The second inclusion is straightforward and the last one is a consequence of Proposition 2. The item (ii) comes from the associativity of $\underline{\star}_\nu^{BG}$ (see Theorem 1). The item (iii) comes from the construction of $\natural_{\nu,K}$ itself (cf. the method used in Sect. 2.1) and Proposition 1. Moreover, notice that for $\varphi =: \underline{T}_{\nu,K}\,\varphi_0$ and $\psi =: \underline{T}_{\nu,K}\,\psi_0$ belonging to $\underline{T}_{\nu,K}\left(\mathcal{S}^{BG}(\mathbb{S})\right)$, we have $\varphi\natural_{\nu,K}\,\psi := \underline{T}_{\nu,K}\left(\varphi_0 \underline{\star}_\nu^{BG}\psi_0\right)$. The left invariance of $\natural_{\nu,K}$ is then a direct consequence of that of $\underline{\star}_\nu^{BG}$ (see Theorem 1, item (ii)) and $\underline{T}_{\nu,K}$. For the item (iv), we start by choosing σ such that $\sigma < \frac{K_1^2}{|i\nu|}$ and two functions $\varphi, \psi \in \mathcal{S}^\sigma$. Using the expression of $\underline{\star}_\nu^{BG}$ given in [3] and performing the change of variables $\ell_1 \mapsto \ell_1 + \ell\exp\left[-2\left(a_1 + \log\frac{\cosh(a_1-a_2)}{\cosh(a_2)}\right)\right]$ and $\ell_2 \mapsto \ell_2 + \ell\exp\left[-2\left(a_2 + \log\frac{\cosh(a_1-a_2)}{\cosh(a_1)}\right)\right]$ into the integral expression corresponding to $\frac{(2\pi i\nu)^2}{K_1^4}\underline{T}_{\nu,K}\left(\underline{T}_{\nu,K}\,\varphi\underline{\star}_\nu^{BG}\underline{T}_{\nu,K}\,\psi\right)(a,\ell)$, we get the announced integral formula in the item (iv), thanks to the fact the right action R on $\mathbb{S}$ reads $R_{(a,\ell)}(a',\ell') = \left(a' + a, \ell + \ell'e^{-2a}\right)$. Regarding the item (v), we choose $K_0 := (1,0)$ without any loss of generality (indeed, formula in item (iv) is the same for any K_1 up to a redefinition of ν and independent from K_2). Note first that using similar arguments to that of the proof of Proposition 2, we can see that both functions $\left[(x,\theta) \mapsto \underline{T}_{\frac{\theta}{\tau},K}\,\varphi(x)\right]$ and $\left[(x,\theta) \mapsto \left(\psi_1\natural_{\frac{\theta}{\tau},K_0}\psi_2\right)(x)\right]$ are elements of $C^\infty(\mathbb{S} \times \mathbb{R})$. Theorem 40.1 in [7] then implies that for every $\varphi \in \mathcal{D}(\mathbb{S})$, the function of $\theta \in \mathbb{R}$ given by $\underline{T}_{\frac{\theta}{\tau},K_0}\,\varphi$ belongs to $C^\infty\left(\mathbb{R}, C^\infty(\mathbb{S})\right)$ and for every $\varphi, \psi \in \underline{T}_{\frac{\theta}{\tau},K_0}\left(\mathcal{D}(\mathbb{S})\right)$, the function of $\theta \in \mathbb{R}$ given by $\varphi\natural_{\frac{\theta}{\tau},K_0}\psi$ belongs to $C^\infty\left(\mathbb{R}, C^\infty(\mathbb{S})\right)$. Moreover, by virtue of the associativity of $\underline{\star}_{\frac{\theta}{\tau}}^{BG}$, the Taylor series at 0 of the map $\left[\mathbb{R} \to \mathbb{C} : \theta \mapsto \left(\varphi\natural_{\frac{\theta}{\tau},K_0}\psi\right)(x)\right]$ yields an associative star-product which is $\mathfrak{g}_{RC}$-invariant by construction. We then deduce from Proposition 1 that the formal version of $\natural_{\nu,K_0}$ coincides with $\star_\nu^{RC}$.

References

1. Bertelson, M., Bieliavsky, P., Gutt, S.: Parametrizing equivalence classes of invariant star products. Lett. Math. Phys. **46**, 339–345 (1998)
2. Bieliavsky, P.: Strict quantization of solvable symmetric spaces. J. Symplectic Geo. **1**, 269–320 (2002)
3. Bieliavsky, P., Gayral, V.: Deformation Quantization for Actions of Kählerian Lie groups, vol. 236. Memoirs of the American Mathematical Society (2015)
4. Bieliavsky, P., de Goursac, A., Maeda, Y., Spinnler, F.: Non-formal star-exponential on contracted one-shetted hyperboloids. arXiv:1501.07491v1 [math.QA] (2015)
5. Bieliavsky, P., Tang, X., Yao, Y.: Rankin-Cohen brackets and formal quantization. arXiv:0506506v4 [math.QA] (2008)
6. Connes, A., Moscovici, H.: Rankin-Cohen brackets and the Hopf algebra of transverse geometry. Moscow Mathem. J. **4**(1), 111–130 (2004)
7. Trèves, F.: Topological vector spaces, distributions and kernels. Academic Press, New York (1967)

8. Xu, P.: Fedosov ⋆-products and quantum momentum maps. Commun. Math. Phys. **197**, 167–197 (1998)
9. Zagier, D.: Modular forms and differential operators. Proc. Math. Sci. **104**(1), 57–75 (1994). https://doi.org/10.1007/BF02830874

Equivalence of Invariant Star-Products: The "Retract" Method

Pierre Bieliavsky[1], Valentin Dendoncker[2(✉)], and Stéphane Korvers[3]

[1] Université de Louvain, Ottignies-Louvain-la-Neuve, Belgium
[2] ICHEC Brussels Management School, Woluwe-Saint-Pierre, Belgium
valentin.dendoncker@gmail.com
[3] Gevers Patents, Brussels, Belgium

Abstract. In this article, we present a general method for enlarging the group of symmetries (symplectomorphisms) of a given star-product (or deformation quantization) on a symplectic homogeneous space. We call this method the "retract method".

Keywords: Deformation quantization · star-product · sympletic homogeneous spaces

1 Introduction

Let $M = G/K$ be a homogeneous space of a Lie group G with closed isotropy subgroup K. Let us assume that M is equipped with a G-invariant symplectic structure ω. Assume moreover that there exists a Lie subgroup $\mathbb{S}$ of G whose action on M is simply transitive.

An important example of such a situation is the case where G is a real non-compact simple Lie group and K its maximal compact subgroup. The symplectic condition in this case amounts to require that the center of K is non-discrete. Given an Iwasawa decomposition $G = ANK$, the solvable Iwasawa factor $\mathbb{S} := AN$ then simply transitively acts on $M = G/K$.

Generally, these requirements can be slightly relaxed by letting the G-action to be only local in the sense that one has a finite dimensional Lie algebra $\mathfrak{g}$ of symplectic vector fields on a symplectic manifold (M, ω) which contains a Lie subalgebra $\mathfrak{s}$ generated by complete vector fields. A theorem due to Palais then guarantees that $\mathfrak{s}$ exponentiates to a Lie group of symplectic transformations of (M, ω), which we assume to be simply transitive.

This relaxed situation allows to consider examples such as $M := \{(q, p) \in \mathbb{R}^2 \,|\, p > 0\}$ and $\omega := dp \wedge dq$. The affine group

$$\mathbb{S} := \left\{ \begin{pmatrix} e^a & n \\ 0 & e^{-a} \end{pmatrix} \right\}_{(a,n)\in\mathbb{R}^2}$$

then simply transitively acts on (M, ω) by linear sympletic transformations. The full affine symplectic group $G = Sp(1, \mathbb{R}) \ltimes \mathbb{R}^2$ does not act on M but

F. Nielsen and F. Barbaresco (Eds.): GSI 2023, LNCS 14071, pp. 533–539, 2023.
https://doi.org/10.1007/978-3-031-38271-0_53

its lie algebra does by restricting to M every fundamental vector field $X_x^\star :=$ $\frac{d}{dt}\big|_0 \exp(-tX).x$ with $X \in \mathfrak{g}$. This is the situation which is consider in [4] in the context of the modular form algebra.

Within this context, we now consider a formal star-product $\star$ on the symplectic manifold (M,ω) which we assume to be $\mathbb{S}$-invariant. The question we are addressing here is whether it is possible to explicitly describe an intertwiner between $\star$ and a star-product $\sharp$ on (M,ω) that is not only $\mathbb{S}$-invariant but also $\mathfrak{g}$-invariant (or G-invariant in the global case).

2   Star-Products on Symplectic Manifolds

Definition 1. *Let (M,ω) be a symplectic manifold with associated symplectic Poisson bracket denoted by $\{\,,\,\}^\omega$. Let us denote by $C^\infty(M)[[\nu]]$ the space of formal power series with coefficients in $C^\infty(M)$ in the formal parameter ν. A star-product on (M,ω) is a $\mathbb{C}[[\nu]]$-bilinear associative algebra structure on $C^\infty(M)[[\nu]]$:*

$$\star_\nu : C^\infty(M)[[\nu]] \times C^\infty(M)[[\nu]] \to C^\infty(M)[[\nu]]$$

such that, viewing $C^\infty(M)$ embedded in $C^\infty(M)[[\nu]]$ as the zero order coefficients, and writing for all $u,v \in C^\infty(M)$:

$$u \star_\nu v =: \sum_{k=0}^{\infty} \nu^k C_k(u,v)\,,$$

one has

(1) $C_0(u,v) = uv$ (pointwise multiplication of functions).
(2) $C_1(u,v) - C_1(v,u) = \{u,v\}^\omega$.
(3) For every k, the coefficient $C_k : C^\infty(M) \times C^\infty(M) \to C^\infty(M)$ is a bi-differential operator. These are called the cochains of the star-product.

A star-product is called natural *when C_2 is a second order bi-differential operator.*

Definition 2. *Two such star-products $\star_\nu^j$ ($j=1,2$) are called* equivalent *if there exists a formal power series of differential operators of the form*

$$T = \mathrm{id} + \sum_{k=1}^{\infty} \nu^k T_k : C^\infty(M) \to C^\infty(M)[[\nu]] \tag{1}$$

such that for all $u,v \in C^\infty(M)$:

$$T(u) \star_\nu^2 T(v) = T(u \star_\nu^1 v)\,, \tag{2}$$

where we denote by T the $\mathbb{C}[[\nu]]$-linear extension of (1) to $C^\infty(M)[[\nu]]$.

Since the zeroth order of such an equivalence operator T is the identity, it admits an inverse that we denote by

$$T^{-1} : C^\infty(M)[[\nu]] \to C^\infty(M)[[\nu]] \ .$$

We the adopt the following notation.

Definition 3. *Let $\star_\nu^j$ $(j = 1, 2)$ be two star-products on (M, ω) that are equivalent to each other under an equivalence operator T as in Definition 2. We encode equation (2) by the notation*

$$\star_\nu^2 \ = \ T(\star_\nu^1) \ .$$

Let now G be a group of transformations of M.

Definition 4. *A star-product $\star_\nu$ on (M, ω) is called G-invariant if its cochains are G-invariant bi-differential operators; for every $g \in G$, we write:*

$$g^\star(u \star v) \ = \ g^\star(u) \star_\nu g^\star(v) \quad (u, v \in C^\infty(M)) \ .$$

Two G-invariant star-products $\star_\nu^j$ $(j = 1, 2)$ are called G-equivariantly equivalent if there exists a G-commuting equivalence T between them:

$$T(\star_\nu^1) \ = \ \star_\nu^2 \quad \text{and} \quad T(g^\star u) \ = \ g^\star T(u)$$

for all $g \in G$ and $u \in C^\infty(M)$.

A slightly more general framework, as explained in the Introduction, is the one where only a Lie algebra acts on M. This means that one has an injective Lie algebra homomorphism

$$\mathfrak{g} \to \Gamma^\infty(T(M)) : X \mapsto X^\star \ .$$

In the case of a Lie group action G with Lie algebra $\mathfrak{g}$, this corresponds to the action by fundamental vector fields

$$X_x^\star \ := \ \left.\frac{\mathrm{d}}{\mathrm{d}t}\right|_0 \exp(-tX)x \ . \tag{3}$$

Definition 5. *A derivation of a star-product $\star$ on (M, ω) is a formal series*

$$D \ = \ \sum_{k=0}^{\infty} \nu^k D_k$$

of differential operators $\{D_k\}$ on $C^\infty(M)$ such that for all $u, v \in C^\infty(M)$, one has

$$D(u \star v) \ = \ (Du) \star v + u \star (Dv) \ .$$

The (Lie algebra) of derivations of $\star$ is denoted by $\mathfrak{Der}(\star)$.

We end this section with the following statement, which is a straightforward application of the main result in [1].

Proposition 1. *On a simply connected symplectic Lie group $\mathbb{S}$ with Lie algebra $\mathfrak{s}$, the left-equivariant equivalence classes of left-invariant (symplectic) star-products are parametrized by the sequences of elements of the second Chevalley cohomology space $H_{Chev}^2(\mathfrak{s})$ associated to the trivial representation of $\mathfrak{s}$ on $\mathbb{R}$.*

3   Equivariant Equivalences – The "Retract" Method

As explained in the Introduction, we now consider a homogeneous symplectic space G/K containing $\mathbb{S}$ as a Lie subgroup and such that the action of $\mathbb{S}$ on this space is simply transitive. Identifying G/K with $\mathbb{S}$ through the action, we consider a G-invariant star-product on $(\mathbb{S}, \omega) \star$ where ω denotes the transported G-invariant symplectic structure. We will assume $\mathbb{S}$ to be simply connected and *solvable*. This hypothesis of solvability is in fact not essential. However, it brings proofs that are more Lie theoretic.

Proposition 2. [Relevance of the retract method] *Let* $\mathrm{Hom}_\mathfrak{s}(\mathfrak{g}, \mathfrak{Der}(\star))^1$ *the space of Lie algebra homomorphisms from* $\mathfrak{g}$ *to the Lie algebra* $\mathfrak{Der}(\star)$ *of derivations of* $(C^\infty(\mathbb{S})[[\nu]], \star)$ *that are* $\mathfrak{s}$-*relative in the sense that every element* D *of* $\mathrm{Hom}_\mathfrak{s}(\mathfrak{g}, \mathfrak{Der}(\star))$ *restricts to* $\mathfrak{s}$ *as the identity:*

$$D_X = X^*$$

for every X *in* $\mathfrak{s}$.
Then, $\mathrm{Hom}_\mathfrak{s}(\mathfrak{g}, \mathfrak{Der}(\star))$ *is finite dimensional over the formal field* $\mathbb{R}[[\nu]]$.

We will adopt the following notation in accordance with the distributions that are defined by locally summable functions:

$$T^{-1}\varphi(x) := \int_\mathbb{S} u(x^{-1}y)\,\varphi(y)\,dy =: \tau_u(\varphi)(x)$$

where dy denotes a left-invariant Haar measure on $\mathbb{S}$.

Proposition 3. *Consider such an intertwiner* T *with distributional kernel* u. *Then.*

(i) the map

$$D^T : \mathfrak{g} \to \mathfrak{Der}(\star) : X \mapsto T^{-1}X^*T$$

belongs to $\mathrm{Hom}_\mathfrak{s}(\mathfrak{g}, \mathfrak{Der}(\star))$.
(ii) The kernel u *is a (weak) joint solution of the following evolution equations:*

$$-X_y^*[u(x^{-1}y)] = D_X^T|_x[u(x^{-1}y)] \quad (\forall X \in \mathfrak{g}).$$

Proof. With the notations introduced above and based on the previous propositions, we have

$$T^{-1}X^*(\varphi)(x) = \int u(x^{-1}y)\,X_y^*(\varphi)\,dy = -\int X_y^*[u(x^{-1}y)]\,\varphi(y)\,dy$$

$$= D_X^T|_x \int u(x^{-1}y)\,\varphi(y)\,dy = \int D_X^T|_x u(x^{-1}y)\,\varphi(y)\,dy$$

because φ is compactly supported and, being symplectic, X^* preserves the (Liouville) Haar measure.

---

[1] This space can be interpreted as the space of flat $\mathfrak{g}$-invariant connections on a non-commutative space modelled on the infinite dimensional automorphism group of the star-product (c.f. Vinberg's description of invariant affine connections on a homogeneous space).

The above Proposition leads us to the following definition.

Definition 6. *Let D be an element in $\mathrm{Hom}_s(\mathfrak{g}, \mathfrak{Der}(\star))$. An element $u \in \mathcal{D}'(\mathbb{S})[[\nu]]$ such that*

(a) it is a joint (weak) solution of

$$D_Z u = 0 \quad \forall Z \in \mathfrak{k},$$

(b) its singular support is reduced to the unit element:

$$\mathrm{supp}(u) = \{e\},$$

(c) it satisfies the following semi-classical condition:

$$u \bmod \nu = \delta_e,$$

is called a D-retract based over $\star$.

The following statement is classical.

Proposition 4. *The space of left-invariant differential operators on $\mathbb{S}$ (canonically isomorphic to $\mathcal{U}(\mathfrak{s})$) identifies with the sub-space of distributions in $\mathcal{D}'(\mathbb{S})$ supported at the unit e.*

Corollary 1. *Let u be a D-retract. Then, the convolution operator τ_u is differential and satisfies the semi-classical condition:*

$$\tau_u = \mathrm{Id} \bmod \nu.$$

It therefore extends to an invertible endomorphism

$$\tau_u : C^\infty(\mathbb{S})[[\nu]] \to C^\infty(\mathbb{S})[[\nu]]$$

called inverse retract operator. *Its formal inverse*

$$T_u := \tau_u^{-1}$$

is called direct retract operator.

An argument similar to the one in the proof of Proposition 3 then yields the following theorem.

Theorem 1. *(i) Let u be a D-retract with associated direct retract operator $T_u : C^\infty(\mathbb{S})[[\nu]] \to C^\infty(\mathbb{S})[[\nu]]$. Then,*

$$T_u(\star) =: \sharp^u$$

is a formal star-product on $(\mathbb{S}, \omega)$ that is $\mathfrak{g}$-invariant with respect to the infinitesimal action (3).

(ii) Given any symplectic action of $\mathfrak{g}$ on $(\mathbb{S}, \omega)$ that restricts to $\mathfrak{s}$ as the regular one, every $\mathfrak{g}$-invariant star-product that is $\mathbb{S}$-equivariantly equivalent to $\star$ is of the form $\sharp^u$ for a certain D-retract u.

4 The Complex Unit Ball

In [5], one finds a generalisation to the complex domain $SU(1,n)/U(n)$ of the result of [2] obtained in the 2-dimensional case.

We consider the Lie group $G := SU(1,n)$. The Iwasawa factor $\mathbb{S}$ has the following structure. Its Lie algebra $\mathfrak{s}$ is a semi-direct product between a one dimensional Lie algebra $\mathfrak{a} := \mathbb{R}.H$ with a Heisenberg Lie algebra $\mathfrak{N} := V \oplus \mathbb{R}.Z$ where (V, Ω) is a $2n$-dimensional symplectic vector space with Lie bracket given by $[(v,z),\,(v',z')] := \Omega(v,v')Z$. The extension homomorphism $\rho : \mathfrak{a} \to \mathfrak{Der}(\mathfrak{N})$ is defined as $\rho(H)(v,z) := (v,2z)$.

The exponential mapping on $\mathfrak{N}$ is the identity. We consider the following global coordinate system on $\mathbb{S}$:

$$\mathfrak{a} \oplus V \oplus \mathbb{R}.Z \to \mathbb{S} : (a,v,z) \mapsto \exp(a\,\rho(H))(v,z) \, .$$

It is a global Darboux chart in the sense that the 2-form $\omega := \mathrm{d}a \wedge \mathrm{d}z + \Omega$ is left-invariant and corresponds to the Kähler two-form on G/K where $K := U(n)$ under the $\mathbb{S}$-equivariant diffeomorphsim $\mathbb{S} \to G/K : x \mapsto xK$.

In [3], the authors define the following $\mathbb{S}$-equivariant non-formal star-product which will be the source of the retract.

Theorem 2. *Set*

$$S(x_1,x_2) := \sinh(2a_1)t_2 - \sinh(2a_2)t_1 + \cosh a_1 \cosh a_2\,\omega_0(v_1,v_2) \, ,$$

where $x_j = (a_j, v_j, t_j)$ *(j = 1, 2), and*

$$A(x_1,x_2) :=$$
$$\big(\cosh a_1 \cosh a_2 \cosh(a_1 - a_2) \big)^d \big(\cosh 2a_1 \cosh 2a_2 \cosh 2(a_1 - a_2) \big)^{1/2} \, .$$

Let θ *be a non-zero real number. Then the following formula*

$$\varphi_1 \star_\theta \varphi_2(x_0) := \frac{1}{\theta^{2n+2}} \int_{\mathbb{S}\times\mathbb{S}} A(x_1,x_2)\, e^{\frac{i}{\theta}S(x_1,x_2)}\, \varphi_1(x_0.x_1)\, \varphi_2(x_0.x_2)\, \mathrm{d}x_1\, \mathrm{d}x_2$$

defines a left-invariant associative product on $L^2(\mathbb{S})$. *Here* $x.x'$ *denotes the group multiplication on* $\mathbb{S}$ *and the integration is with respect to the left-invariant Haar measure* $\mathrm{d}x$.

Moreover, when φ_1 *and* φ_2 *are smooth and compactly supported the map* $\theta \mapsto \varphi_1 \star_\theta \varphi_2(x_0)$ *smoothly extends to* $\mathbb{R}$. *Its Taylor formula around* $\theta = 0$ *defines a formal star-product on* $(\mathbb{S}, \omega)$. *We denote by* $\star$ *this formal star-product.*

Theorem 3. *Up to redefining the deformation parameter ν, every D-retract based over $\star$ has the form*

$$u_\nu(a, v, z) :=$$

$$\nu^2 \int_{-\infty}^{\infty} d\xi \, \text{sign}(\xi) \, e^{-2a + i\xi z} \int_{-\infty}^{\infty} d\gamma \, \left(1 + \gamma^2\right)^{\frac{n-3}{2}}$$

$$\Gamma_\nu \left[-4\nu^2 \, \text{sign}(\xi) \, e^{-2a} \left(\frac{1}{\gamma^2 + 1} - \cosh^2\left(\frac{1}{2} arcsinh(i\nu\xi)\right) \right) \right]$$

$$\exp \left[-\frac{1}{\nu} arccotan(\gamma) + \frac{\gamma}{\nu} \left(\frac{e^{-2a}}{\gamma^2 + 1} + \cosh^2\left(\frac{1}{2} arcsinh(i\nu\xi)\right) \right) ||v||^2 \right]$$

where $\Gamma_\nu \in \mathcal{D}'(\mathbb{R})[[\nu]]$ is an arbitrary formal power series with coefficients in the distributions $\mathcal{D}'(\mathbb{R})$ on $\mathbb{R}$.

References

1. Bertelson, M., Bieliavsky, P., Gutt, S.: Parametrizing equivalence classes of invariant star products. Lett. Math. Phys. **46**, 339–345 (1998)
2. Bieliavsky, P., Detournay, S., Spindel, P.: The deformation quantizations of the hyperbolic plane. Commun. Math. Phys. **289**(2), 529–559 (2009)
3. Bieliavsky, P., Gayral, V.: Deformation Quantization for Actions of Kählerian Lie groups, vol. 236. Memoirs of the American Mathematical Society (2015)
4. Dendoncker, V.: Non-formal Rankin-Cohen deformations. Ph.D. thesis, Université catholique de Louvain (2018)
5. Korvers, S.: Quantifications par déformations formelles et non formelles de la boule unité de $\mathbb{C}^n$. Ph.D. thesis, Université catholique de Louvain (2014)

Deep Learning: Methods, Analysis and Applications to Mechanical Systems

DL4TO: A Deep Learning Library for Sample-Efficient Topology Optimization

David Erzmann[1]($\boxtimes$), Sören Dittmer[1,2], Henrik Harms[1], and Peter Maaß[1]

[1] Center for Industrial Mathematics, University of Bremen, Bremen, Germany
erzmann@uni-bremen.de
[2] Cambridge Image Analysis, Centre for Mathematical Sciences,
University of Cambridge, Cambridge, UK

Abstract. We present and publish the DL4TO software library – a Python library for three-dimensional topology optimization. The framework is based on PyTorch and allows easy integration with neural networks. The library fills a critical void in the current research toolkit on the intersection of deep learning and topology optimization. We present the structure of the library's main components and how it enabled the incorporation of physics concepts into deep learning models.

Keywords: Topology optimization · Deep learning · Software library

1 Introduction

We begin by briefly introducing the problem of Topology Optimization (TO) and the recent development of applying deep learning to it.

1.1 Classical Topology Optimization

The computational discipline of topology optimization (TO) aims to optimize mechanical structures. Since its development in 1988 [4], TO is a powerful tool widely adopted by engineers in a variety of fields, e.g., fluid [6] and solid mechanics [15], acoustics [14,29], and heat transfer [10].

The *Solid Isotropic Material with Penalization* (SIMP) method [5] is widely regarded as the most significant classical approach used in TO. SIMP involves a density-based setup where the density takes values between 0 and 1 over a given design domain. The density represents to which degree material is present in different places. SIMP first voxelizes the domain and density and then employs an iterative optimization scheme to improve structural performance by adjusting voxel densities. In the case of linear elasticity, one evaluates the integrity of the resulting structure via *von Mises stresses*, whose computation involves the corresponding partial differential equation (PDE). The specified objective function and constraints may vary depending on the user's needs. The most common

F. Nielsen and F. Barbaresco (Eds.): GSI 2023, LNCS 14071, pp. 543–551, 2023.
https://doi.org/10.1007/978-3-031-38271-0_54

setup for mechanical problems is *compliance minimization* [7], where we minimize a compliance objective subject to volume and possibly stress constraints. See Algorithm 1 for a pseudo-code representation of our SIMP implementation. For more details on SIMP see [5].

Algorithm 1. Our implementation of the *Solid Isotropic Material with Penalization* (SIMP) algorithm for compliance minimization with volume and stress constraints. We typically initialize θ_0 to be 0.5 everywhere, where it is not enforced otherwise. For the filtering we apply a smoothed Heaviside function $H_\beta(\theta) := 1 - \exp(-\beta\theta) + \theta\exp(-\beta)$ with a smoothing factor $\beta > 0$ which we gradually steepen over the iterations. The SIMP exponent p is commonly chosen as $p = 3$ to further discourage non-binary solutions.

Require: F, σ_{ys}, N, λ, μ ▷ Forces, yield stress, #iterations and loss weights
Initialize: θ_0 ▷ Start with an initial density
Set: $p = 3$ ▷ Set the SIMP exponent to its typical value
for $i = 0, \ldots, N - 1$ **do**
   $\theta_i \leftarrow \text{project}(\theta_i)$ ▷ Project density values into the unit interval
   $\theta_i \leftarrow \text{smooth}(\theta_i)$ ▷ Avoid checkerboard patterns via smoothing
   $\theta_i \leftarrow \text{filter}(\theta_i)$ ▷ Encourage binary densities via filtering
   u, $\sigma_{vM} = \text{pde\_solver}(F, \theta_i{}^p)$ ▷ Solve PDE for current density and exponent p
   $\text{loss\_compl} = F^T u$ ▷ Compliance with forces F and displacements u
   $\text{loss\_vol} = \|\theta_i\|_1$ ▷ Compute volume loss term
   $\text{loss\_stress} = \text{softplus}(\sigma_{vM} - \sigma_{ys})$ ▷ Compute stress constraint loss term
   $\text{loss} = \text{loss\_compl} + \lambda \cdot \text{loss\_vol} + \mu \cdot \text{loss\_stress}$ ▷ Sum up and weight losses
   $\theta_{i+1} \leftarrow \text{gradient\_step}(\theta_i, \text{loss})$ ▷ Update density via gradient descent
end for
return θ_N ▷ Return final density distribution

1.2   Neural Networks for Topology Optimization

The iterative nature of density-based methods like SIMP requires repeated solving of the governing PDE. This becomes computationally prohibitive for high voxel mesh resolutions, leading to practical limitations [1]. Recent Deep Learning (DL) research has explored overcoming this challenge. One can broadly classify the advances into four categories [33]:

1. Reduce SIMP iterations: Neural networks map from intermediate SIMP iterations to the final structure, technically performing a deblurring task [2,3,26,28].
2. Eliminate SIMP iterations: Neural networks directly predict the final density distribution without performing any SIMP iterations [21,30,32].
3. Substitute for PDE solver: One replaces classical PDE solvers with neural networks, removing the primary bottleneck [8,20,24].
4. Neural reparameterization: One uses neural networks to reparameterize the density function [11,17,31,33]. However, one usually still requires computationally demanding PDE evaluations for the training.

As advances in computational power and DL have only recently brought the application of DL to TO in the realm of the possible, the literature on it is still in its infancy. As a result, the authors are not aware of any public software framework for TO using DL, requiring every researcher to write their code from scratch.

We want to address this issue by presenting DL4TO, a flexible and easy-to-use python library for three-dimensional TO. The library is open source and based on PyTorch [22], allowing for easy integration of DL and TO methods.

2   The DL4TO Framework

2.1   Motivation

In this section, we give a basic overview of the DL4TO[1] library. The primary motivation for developing DL4TO is the need for a flexible and easy-to-use basis to conduct DL experiments for TO in Python. The library focuses on linear elasticity on structured three-dimensional grids. DL4TO comes with its own PDE solver, a SIMP implementation, and various objective functions for classical and learned TO. The PyTorch [22]-based implementation smoothly connects the world of TO with the world of DL. To our knowledge, only two Python libraries for TO [16,18] exist, and neither allows for easy integration with neural networks.

2.2   Core Classes

In the following we give an overview of how our framework works. Below, we introduce the three main classes that form the core of our library.

- **Problem**: An important novelty of our framework is how TO problems are defined and processed. This is done via the **Problem** class, which contains all information of the underlying TO problem one intends to solve. Since we perform optimization on structured grids, all information is either in scalar or in tensor form. This makes data compatible with DL applications since it allows for a shape-consistent tensor representation. Let (n_x, n_y, n_z) be the number of voxels in each spacial direction. We can create a uniquely characterized problem object via

    problem = Problem(E, ν, σ_ys, h, F, Ω_dirichlet, Ω_design).
    Here,

    - E, ν and σ_ys denote scalar material properties, namely Young's modulus, Poisson's ratio and yield stress.
    - h is a three-dimensional vector that defines the voxel sizes in meters in each direction.

---

[1] The DL4TO library is publicly available at https://github.com/dl4to/dl4to.

- F is a $(3 \times n_x \times n_y \times n_z)$-tensor which encodes external forces given in N/m^3. The three channels correspond to the force magnitudes in each spacial dimension.
- $\Omega\_dirichlet$ is a binary $(3 \times n_x \times n_y \times n_z)$-tensor which we use to encode the presence of directional homogeneous Dirichlet boundary conditions for every voxel. 1s indicate the presence, and 0s the absence of homogeneous Dirichlet boundary conditions. Currently, we do not support non-homogeneous Dirichlet boundary conditions since we believe that they are not required for most TO tasks.
- Ω_design is a $(1 \times n_x \times n_y \times n_z)$-tensor containing values $\in \{0, 1, -1\}$ that we use to encode design space information. We use 0s and 1s to constrain voxel densities to be 0 or 1, respectively. Entries of -1 indicate a lack of density constraints, which signifies that the density in that voxel can be freely optimized. For voxels that have loads assigned to them we automatically enforce the corresponding density value to be 1.
- TopoSolver: This parent class provides different methods for solving TO problems. SIMP, as well as learned methods, are child classes. The initialization arguments slightly differ, depending on the method used. For instance, a SIMP solver for compliance minimization with a volume constraint can be initialized via

```
criterion = Compliance() + λ * VolumeConstraint(vol_fract)
topo_solver = SIMP(criterion, p, n_iters, lr)
```

with some arbitrary scalar choice of volume fraction `vol_fract` and optimization weight λ. The other arguments of SIMP denote the SIMP exponent choice p (by default, $p = 3$), the number of iterations `n_iters` and the learning rate `lr`.

Alternatively, for volume minimization with a stress constraint we could set the optimization criterion as follows:

```
criterion = VolumeFraction() + λ * StressConstraint().
```

By default, our framework uses a built-in finite differences method (FDM) solver whenever the PDE for linear elasticity is solved. This is attributed to the regular grid structure, which makes the FDM a suitable and intuitive approach. It is however also possible to include custom PDE solvers, e.g., learned PDE solvers.

In order to apply a `topo_solver` to a predefined **problem** object, we can simply call it via

```
solution = topo_solver(problem),
```

which returns a `solution` object. Note that this also works with a list of problems, in which case `topo_solver` likewise returns a list of solutions.

For learned solvers the procedure is similar, with the exception that the initialization of the `topo_solver` object additionally requires a preprocessing as input. This determines how a `problem` object should be converted to neural network compatible input tensors when calling the solver (see Sect. 4). The `topo_solver` is trained via the built-in `train` function:

topo_solver.train(dataloader_train, dataloader_val, epochs),

where `dataloader_train` and optionally `dataloader_val` are dataloaders for the training and validation dataset.
– Solution: Objects of this class define solutions to TO problems. They usually result from calling a `topo_solver` with a `problem` object, but can also be instantiated manually by passing a problem and a density distribution:

solution = Solution(problem, θ).

Here, θ is a $(1 \times n_x \times n_y \times n_z)$-tensor that defines a three-dimensional density distribution that solves `problem`. The `Solution` class provides several useful functionalities like logging and plotting.

3 Datasets

DL4TO is compatible with the SELTO datasets [13] introduced in [12] and publicly available at https://doi.org/10.5281/zenodo.7034898. We want to give a short overview of the two SELTO datasets containing samples of mechanical mounting brackets. Each dataset consists of tuples (`problem`, `solution`), where `solution` is a ground truth density distribution for `problem`. The two datasets are called *disc* and *sphere*, referring to the shape of the corresponding design spaces; see Table 1 for an overview of both datasets and Fig. 1 for example samples.

Table 1. Overview of our datasets, called *disc* and *sphere*, with the names referring to the shape of their design spaces. Both datasets are split into a training and a validation subset.

dataset	shape	# samples
disc	$39 \times 39 \times 4$ voxels	9246 (8846 train, 400 val)
sphere	$39 \times 39 \times 21$ voxels	602 (530 train, 72 val)

(a) Disc dataset   (b) Sphere dataset

Fig. 1. Ground truth examples from the disc and sphere dataset [13]. The densities are defined on a voxel grid and smoothed for visualization using Taubin smoothing [27].

4 Model Pipeline

We now present how DL4TO enables efficient setup of data pipelines. DL4TO provides different models and loss functions, e.g., UNets [25] and the weighted binary cross-entropy. Due to the integration with PyTorch, optimizers like Adam [19] can be used plug-and-play.

We begin with two critical components of DL4TO's model pipeline: physics-based preprocessing and equivariant architectures, see [12].

- Preprocessing: A suitable input preprocessing strategy is crucial for DL. Following [12], the library provides two preprocessing strategies, easily combined via channel-wise concatenation.
    1. Trivial preprocessing: For each problem object, the input of the neural network is a 7-channel tensor which results from the channel-wise concatenation of Ω_dirichlet, Ω_design, and normalized loads F.
    2. PDE preprocessing [12,32]: We set up a density distribution that is 1 wherever allowed by problem. We compute normalized von Mises stresses for that density by solving the PDE for linear elasticity. We use the von Mieses stresses as a 1-channel input to the neural network.

- Equivariance: Equivariance is the property of a function to commute with the actions of a symmetry group. For a given transformation group G, a function $f : X \to Y$ is (G-)equivariant if

$$f(T_g^X(x)) = T_g^Y(f(x)) \qquad \forall g \in G,\ x \in X,$$

where T_g^X and T_g^Y denote linear *group actions* in the corresponding spaces X and Y [9]. As shown by [12], mirror and rotation equivariance can drastically improve model performance on TO tasks. We implement equivariance via *group averaging* [23] by defining an equivariance wrapper F_G^f via

$$F_G^f(x) := \frac{1}{|G|} \sum_{g \in G} T_{g^{-1}}^Y \left[f(T_g^X(x)) \right].$$

The plug-and-play nature of F_G^f allows effortless applicability to any finite transformation group G and any model f.

5 Experiments

Generating large datasets is costly; therefore, reducing the required training samples, e.g., by modifying the DL model design, is highly beneficial. Using the DL4TO library, [12] investigates the *sample efficiency* of models, i.e., the model's performance when trained on a few training samples. They visualize the sample efficiency of a model via so-called *sample efficiency curves* (SE curves). Each SE curve uses separate instances of a given model setup trained on subsets of the original training dataset of increasing size. One then determines the performance of these models on a fixed validation dataset, e.g., via *Intersection over Union* (IoU) and *fail%* (the percentage of failing model predictions). Figure 2 shows [12]'s results and presents dramatic boosts in the UNet's performance when incorporating physics via equivariance and trivial+PDE preprocessing. These improvements are especially visible for low numbers of training samples. For a more in-depth analysis and comparison of the proposed modeling approaches, we refer to [12].

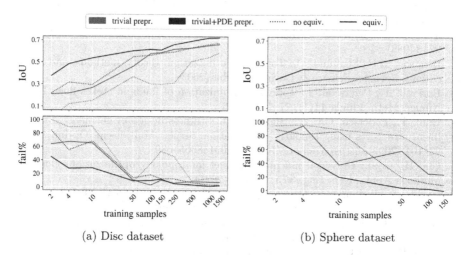

(a) Disc dataset   (b) Sphere dataset

Fig. 2. Sample efficiency curves from [12], trained and evaluated via DL4TO on the SELTO datasets *disc* (left) and *sphere* (right). The horizontal-axis shows the size of the training dataset for the different models on a logarithmic scale. The vertical-axis shows the performance of the criteria IoU and fail%. Each of the plots shows four different models based on trivial preprocessing (red), trivial+PDE preprocessing (blue), equivariance (solid line), and no equivariance (dashed line). (Color figure online)

6   Conclusion

We presented the DL4TO library. The Python library enables research in the intersection of topology optimization and deep learning. Seamlessly integrating with PyTorch, DL4TO enables a smooth interaction between deep learning models and established algorithms like SIMP. Further, DL4TO provides concepts like UNets, SIMP, equivariance, differentiable physics via finite difference analysis, integration with the SELTO datasets [13], as well as three-dimensional interactive visualization. Our library is especially useful for data scientists who want to apply deep learning to topology optimization, as it provides a flexible yet easy-to-use framework. DL4TO will continue to be expanded, and the community is welcome to contribute. Documentation and tutorials can be found at https://dl4to.github.io/dl4to/, providing a guide on how to use the library and its features.

References

1. Aage, N., Andreassen, E., Lazarov, B.S.: Topology optimization using petsc: An easy-to-use, fully parallel, open source topology optimization framework. Struct. Multidiscip. Optim. **51**(3), 565–572 (2015)
2. Abueidda, D.W., Koric, S., Sobh, N.A.: Topology optimization of 2d structures with nonlinearities using deep learning. Comput. Structures **237**, 106283 (2020)
3. Banga, S., Gehani, H., Bhilare, S., Patel, S., Kara, L.: 3d topology optimization using convolutional neural networks. arXiv preprint arXiv:1808.07440 (2018)
4. Bendsøe, M.P., Kikuchi, N.: Generating optimal topologies in structural design using a homogenization method. Comput. Methods Appl. Mech. Eng. **71**(2), 197–224 (1988)
5. Bendsoe, M.P., Sigmund, O.: Topology optimization: theory, methods, and applications. Springer Science & Business Media (2003)
6. Borrvall, T., Petersson, J.: Topology optimization of fluids in stokes flow. Int. J. Numer. Meth. Fluids **41**(1), 77–107 (2003)
7. Buhl, T., Pedersen, C.B.W., Sigmund, O.: Stiffness design of geometrically nonlinear structures using topology optimization. Struct. Multidiscip. Optim. **19**(2), 93–104 (2000). https://doi.org/10.1007/s001580050089
8. Chi, H., et al.: Universal machine learning for topology optimization. Comput. Methods Appl. Mech. Eng. **375**, 112739 (2021)
9. Cohen, T., Welling, M.: Group equivariant convolutional networks. In: International Conference on Machine Learning, pp. 2990–2999. PMLR (2016)
10. Dede, E.M.: Multiphysics topology optimization of heat transfer and fluid flow systems. In: Proceedings of the COMSOL Users Conference, vol. 715 (2009)
11. Deng, H., To, A.C.: Topology optimization based on deep representation learning (drl) for compliance and stress-constrained design. Comput. Mech. **66**(2), 449–469 (2020)
12. Dittmer, S., Erzmann, D., Harms, H., Maass, P.: Selto: Sample-efficient learned topology optimization. arXiv preprint arXiv:2209.05098 (2022)
13. Dittmer, S., Erzmann, D., Harms, H., Falck, R., Gosch, M.: Selto dataset (2023). https://doi.org/10.5281/zenodo.7034898
14. Dühring, M.B., Jensen, J.S., Sigmund, O.: Acoustic design by topology optimization. J. Sound Vib. **317**(3–5), 557–575 (2008)
15. Eschenauer, H.A., Olhoff, N.: Topology optimization of continuum structures: a review. Appl. Mech. Rev. **54**(4), 331–390 (2001)
16. Ferguson, Z.: Topopt - topology optimization in python (2019). https://github.com/zfergus/topopt
17. Hoyer, S., Sohl-Dickstein, J., Greydanus, S.: Neural reparameterization improves structural optimization. arXiv preprint arXiv:1909.04240 (2019)
18. Hunter, W., et al.: Topy - topology optimization with python (2017). https://github.com/williamhunter/topy
19. Kingma, D.P., Ba, J.: Adam: A method for stochastic optimization. arXiv preprint arXiv:1412.6980 (2014)
20. Lee, S., Kim, H., Lieu, Q.X., Lee, J.: Cnn-based image recognition for topology optimization. Knowl.-Based Syst. **198**, 105887 (2020)
21. Nie, Z., Lin, T., Jiang, H., Kara, L.B.: Topologygan: Topology optimization using generative adversarial networks based on physical fields over the initial domain. J. Mech. Design **143**(3) (2021)
22. Paszke, A., et al.: Automatic differentiation in pytorch (2017)

23. Puny, O., Atzmon, M., Ben-Hamu, H., Smith, E.J., Misra, I., Grover, A., Lipman, Y.: Frame averaging for invariant and equivariant network design. arXiv preprint arXiv:2110.03336 (2021)
24. Qian, C., Ye, W.: Accelerating gradient-based topology optimization design with dual-model artificial neural networks. Struct. Multidiscip. Optim. **63**(4), 1687–1707 (2021)
25. Ronneberger, O., Fischer, P., Brox, T.: U-Net: convolutional networks for biomedical image segmentation. In: Navab, N., Hornegger, J., Wells, W.M., Frangi, A.F. (eds.) MICCAI 2015. LNCS, vol. 9351, pp. 234–241. Springer, Cham (2015). https://doi.org/10.1007/978-3-319-24574-4_28
26. Sosnovik, I., Oseledets, I.: Neural networks for topology optimization. Russ. J. Numer. Anal. Math. Model. **34**(4), 215–223 (2019)
27. Taubin, G.: Curve and surface smoothing without shrinkage. In: Proceedings of IEEE International Conference on Computer Vision, pp. 852–857. IEEE (1995)
28. Xue, L., Liu, J., Wen, G., Wang, H.: Efficient, high-resolution topology optimization method based on convolutional neural networks. Front. Mech. Eng. **16**(1), 80–96 (2021). https://doi.org/10.1007/s11465-020-0614-2
29. Yoon, G.H., Jensen, J.S., Sigmund, O.: Topology optimization of acoustic-structure interaction problems using a mixed finite element formulation. Int. J. Numer. Meth. Eng. **70**(9), 1049–1075 (2007)
30. Yu, Y., Hur, T., Jung, J., Jang, I.G.: Deep learning for determining a near-optimal topological design without any iteration. Struct. Multidiscip. Optim. **59**(3), 787–799 (2019)
31. Zehnder, J., Li, Y., Coros, S., Thomaszewski, B.: Ntopo: Mesh-free topology optimization using implicit neural representations. Adv. Neural. Inf. Process. Syst. **34**, 10368–10381 (2021)
32. Zhang, Y., Peng, B., Zhou, X., Xiang, C., Wang, D.: A deep convolutional neural network for topology optimization with strong generalization ability. arXiv preprint arXiv:1901.07761 (2019)
33. Zhang, Z., Li, Y., Zhou, W., Chen, X., Yao, W., Zhao, Y.: Tonr: An exploration for a novel way combining neural network with topology optimization. Comput. Methods Appl. Mech. Eng. **386**, 114083 (2021)

Learning Hamiltonian Systems
with Mono-Implicit Runge-Kutta Methods

Håkon Noren[(✉)]

Department of Mathematical Sciences, Norwegian University of Science
and Technology, Trondheim, Norway
hakon.noren@ntnu.no

Abstract. Numerical integrators could be used to form interpolation
conditions when training neural networks to approximate the vector field
of an ordinary differential equation (ODE) from data. When numerical one-step schemes such as the Runge–Kutta methods are used to
approximate the temporal discretization of an ODE with a known vector field, properties such as symmetry and stability are much studied.
Here, we show that using mono-implicit Runge–Kutta methods of high
order allows for accurate training of Hamiltonian neural networks on
small datasets. This is demonstrated by numerical experiments where the
Hamiltonian of the chaotic double pendulum in addition to the Fermi–
Pasta–Ulam–Tsingou system is learned from data.

Keywords: Inverse problems · Hamiltonian systems · Mono-implicit
Runge–Kutta · Deep neural networks

1 Introduction

In this paper, we apply backward error analysis [11] to motivate the use of numerical integrators of high order when approximating the vector field of ODEs with
neural networks. We particularly consider mono-implicit Runge–Kutta (MIRK)
methods [1,3], a class of one-step methods that are explicit when solving inverse
problems. Such methods can be constructed to have high order with relatively
few stages, compared to explicit Runge–Kutta methods, and attractive properties such as symmetry. Here, we perform numerical experiments learning two
Hamiltonian systems with MIRK methods up to order $p = 6$. To the best of our
knowledge, this is the first demonstration of the remarkable capacity of numerical integrators of order $p > 4$ to facilitate the training of Hamiltonian neural
networks [9] from sparse datasets, to do accurate interpolation and extrapolation
in time.

Recently, there has been a growing interest in studying neural networks
through the lens of dynamical systems. This is of interest both to accelerate
data-driven modeling and for designing effective architectures for neural networks [8,10,16]. Considering neural network layers as the flow of a dynamical

Supported by the Research Council of Norway, through the project DynNoise: Learning
dynamical systems from noisy data. (No. 339389).

F. Nielsen and F. Barbaresco (Eds.): GSI 2023, LNCS 14071, pp. 552–559, 2023.
https://doi.org/10.1007/978-3-031-38271-0_55

system is the idea driving the study of so-called neural ODEs [4] and its discretized counter-part, residual neural networks.

Hamiltonian mechanics provide an elegant formalism that allows a wide range of energy preserving dynamical systems to be described as first order ODEs. Hamiltonian neural networks [9] aim at learning energy-preserving dynamical systems from data by approximating the Hamiltonian using neural networks. A central issue when studying neural networks and dynamical systems is which method to use when discretizing the continuous time dynamics. Several works use backward error analysis to argue for the importance of using symplectic integrators for learning the vector field of Hamiltonian systems [5,14,17]. Using Taylor expansions to derive the exact form of the inverse modified vector field allows for the construction of a correction term that cancels the error stemming from the temporal discretization, up to arbitrary order [6,14].

2 Inverse ODE Problems of Hamiltonian Form

We consider a first-order ODE

$$\frac{d}{dt} y(t) = f(y(t)), \quad y(t) : [0, T] \to \mathbb{R}^n, \tag{1}$$

and assume that the vector field f is unknown, whereas samples $S_N = \{y(t_n)\}_{n=0}^N$ of the solution are available, with constant step size h. Then the inverse problem aims at deriving an approximation $f_\theta \approx f$ where θ is a set of parameters to be chosen. The inverse problem can be formulated as the following optimization problem:

$$\arg\min_\theta \sum_{n=0}^{N-1} \left\| y(t_{n+1}) - \Phi_{h,f_\theta}(y(t_n)) \right\|, \tag{2}$$

where f_θ is a neural network approximation of f with parameters θ, and Φ_{h,f_θ} is a one-step integration method with step size h such that $y_{n+1} = \Phi_{h,f}(y_n)$. In particular, we assume that (1) is a Hamiltonian system, meaning that

$$f(y) = J\nabla H(y(t)), \quad J := \begin{bmatrix} 0 & I \\ -I & 0 \end{bmatrix} \in \mathbb{R}^{2d \times 2d}. \tag{3}$$

We follow the idea of Hamiltonian neural networks [9] aiming at approximating the Hamiltonian, $H : \mathbb{R}^{2d} \to \mathbb{R}$, such that H_θ is a neural network and f is approximated by $f_\theta(y) := J\nabla H_\theta(y)$. It thus follows that the learned vector field f_θ by construction is Hamiltonian.

3 Mono-Implicit Runge–Kutta for Inverse Problems

Since the solution is known point-wise, $S_N = \{y(t_n)\}_{n=0}^N$, the points y_n and y_{n+1} can be substituted by $y(t_n)$ and $y(t_{n+1})$ when computing the next step of a one-step integration method. We denote this substitution as the *inverse injection*, and note that this yields an interpolation condition: $y(t_{n+1}) - \Phi_{h,f_\theta}(y(t_n)) = 0$, for f_θ for each n. If we let Φ_{h,f_θ} in (2) be the so-called implicit midpoint method, we get the following expression to be minimized:

$$\left\| y(t_{n+1}) - \left(y(t_n) + hf_\theta\left(\frac{y(t_n) + y(t_{n+1})}{2}\right) \right) \right\|, \quad n = 0, \ldots, N-1. \quad (4)$$

For the midpoint method, the inverse injection bypasses the computationally costly problem of solving a system of equations within each training iteration, since $y(t_{n+1})$ is known. More generally, mono-implicit Runge–Kutta (MIRK) methods constitute the class of all Runge–Kutta methods that form explicit methods under this substitution. Given vectors $b, v \in \mathbb{R}^s$ and a strictly lower triangular matrix $D \in \mathbb{R}^{s \times s}$, a MIRK method is a Runge–Kutta method where $A = D + vb^T$ are the stage coefficients $a_{ij} = [A]_{ij}$, and is thus given by

$$
\begin{aligned}
y_{n+1} &= y_n + h\sum_{i=1}^{s} b_i k_i, \\
k_i &= f\left(y_n + v_i(y_{n+1} - y_n) + h\sum_{j=1}^{s} d_{ij} k_j\right).
\end{aligned}
\quad (5)
$$

Let us denote $\hat{y}_{n+1}$ and $\hat{k}_i$ as the next time-step and the corresponding stages of a MIRK method when substituting y_n, y_{n+1} by $y(t_n), y(t_{n+1})$ on the right-hand side of (5).

Theorem 1. *Let y_{n+1} be given by a MIRK scheme (5) of order p and $\hat{y}_{n+1}$ be given by the same method under the inverse injection. Assume that only one integration step is taken from a known initial value $y_n = y(t_n)$ and that the vector field f is sufficiently smooth. Then*

$$\hat{y}_{n+1} = y_{n+1} + \mathcal{O}(h^{p+2}) \quad (6)$$

$$\text{and} \quad \hat{y}_{n+1} = y(t_{n+1}) + \mathcal{O}(h^{p+1}). \quad (7)$$

Proof. Since the method (5) is of order p we have that

$$
\begin{aligned}
\hat{k}_1 &= f\left(y(t_n) + v_1(y(t_{n+1}) - y(t_n))\right) \\
&= f\left(y_n + v_1(y_{n+1} - y_n)\right) + \mathcal{O}(h^{p+1}) = k_1 + \mathcal{O}(h^{p+1})
\end{aligned}
$$

The same approximation could be made for $\hat{k}_2, \ldots, \hat{k}_s$, since D is strictly lower triangular, yielding $\hat{k}_i = k_i + \mathcal{O}(h^{p+1})$ for $i = 1, \ldots, s$. In total, we find that

$$\hat{y}_{n+1} = y(t_n) + h \sum_{i=1}^{s} b_i \hat{k}_i$$

$$= y_n + h \sum_{i=1}^{s} b_i k_i + \mathcal{O}(h^{p+2})$$

$$= y_{n+1} + \mathcal{O}(h^{p+2})$$

$$= y(t_{n+1}) + \mathcal{O}(h^{p+1}) + \mathcal{O}(h^{p+2})$$

$$= y(t_{n+1}) + \mathcal{O}(h^{p+1}).$$

For the numerical experiments, we will consider the optimal MIRK methods derived in [12]. The minimal number of stages required to obtain order p is $s = p-1$ for MIRK methods [1]. In contrast, explicit Runge–Kutta methods need $s = p$ stages to obtain order p for $1 \leq p \leq 4$ and $s = p+1$ stages for $p = 5, 6$ [2], meaning that the MIRK methods have significantly lower computational cost for a given order. As an example, a symmetric, A-stable MIRK method with $s = 3$ stages and of order $p = 4$ is given by

$$k_1 = f(y_n), \quad k_2 = f(y_{n+1}),$$

$$k_3 = f\left(\frac{1}{2}(y_n + y_{n+1}) + \frac{h}{8}(k_1 - k_2)\right),$$

$$y_{n+1} = y_n + \frac{h}{6}(k_1 + k_2 + 4k_3).$$

4 Backward Error Analysis

Let $\varphi_{h,f} : \mathbb{R}^n \to \mathbb{R}^n$ be the h-flow of an ODE such that $\varphi_{h,f}(y(t_0)) := y(t_0 + h)$ for an initial value $y(t_0)$. With this notation, the vector field $f_h(y)$ solving the optimization problem (2) exactly must satisfy

$$\varphi_{h,f}(y(t_n)) = \Phi_{h,f_h}(y(t_n)), \quad n = 0, \ldots, N - 1. \tag{8}$$

For a given numerical one-step method Φ, the *inverse modified vector field* [18] f_h could be computed by Taylor expansions. However, since their convergence is not guaranteed, truncated approximations are usually considered. This idea builds on backward error analysis [11, Ch. IX], which is used in the case of forward problems (f is known and $y(t)$ is approximated) and instead computes the modified vector field $\tilde{f}_h$ satisfying $\varphi_{h,\tilde{f}_h}(y(t_n)) = \Phi_{h,f}(y(t_n))$.

An important result, Theorem 3.2 in [18], which is very similar to Theorem 1.2 in [11, Ch. IX], states that if the method $\Phi_{h,f}$ is of order p, then the inverse modified vector field is a truncation of the true vector field, given by

$$f_h(y) = f(y) + h^p f_p(y) + h^{p+1} f_{p+1}(y) + \ldots \tag{9}$$

Furthermore, by the triangle inequality, we can express the objective function of the optimization problem (2) in a given point $y(t_n)$ by

$$\left\| \Phi_{h,f_h}(y(t_n)) - \Phi_{h,f_\theta}(y(t_n)) \right\| \leq \left\| \varphi_{h,f}(y(t_n)) - \Phi_{h,f_h}(y(t_n)) \right\|$$
$$+ \left\| y(t_{n+1}) - \Phi_{h,f_\theta}(y(t_n)) \right\|$$

In the case of formal analysis where we do not consider convergence issues and truncated approximations, the first term is zero by the definition of f_h in (8). Thus it is evident that the approximated vector field will approach the inverse modified vector field as the optimization objective tends to zero. Then, by Equation (9) it is clear that $f_\theta(y)$ will learn an approximation of $f(y)$ up to a truncation $\mathcal{O}(h^p)$, which motivates using an integrator of high order.

5   Numerical Experiments

In this section, MIRK methods of order $2 \leq p \leq 6$, denoted by MIRKp in the plots, in addition to the classic fourth-order Runge–Kutta method (RK4), is utilized for the temporal discretization in the training of Hamiltonian neural networks. We train on samples $y(t_n)$, for $t_n \in [0, 20]$, from solutions of the double pendulum (DP) problem with the Hamiltonian

$$H(y_1, y_2, y_3, y_4) = \frac{\frac{1}{2}y_3^2 + y_4^2 - y_3 y_4 \cos(y_1 - y_2)}{1 + \sin^2(y_1 - y_2)} - 2\cos(y_1) - \cos(y_2).$$

In addition, we consider the highly oscillatory Fermi–Pasta–Ulam–Tsingou (FPUT) problem with $m = 1$, meaning $y(t) \in \mathbb{R}^4$, and $\omega = 2$ as formulated in [11, Ch. I.5]. For both Hamiltonian systems, the data $S_N = \{y(t_n)\}_{n=0}^N$ is found by integrating the system using $DOP853$ [7] with a tolerance of 10^{-15} for the following step sizes and number of steps: $(h, N) = (2, 10), (1, 20), (0.5, 40)$. The initial values used are $y_0^{DP} = [-0.1, 0.5, -0.3, 0.1]^T$ and $y_0^{FPUT} = [0.2, 0.4, -0.3, 0.5]^T$. The results for $[y(t)]_3$ are illustrated in Fig. 1.

After using the specified integrators in training, approximated solutions $\tilde{y}_n$ are computed for each learned vector field f_θ again using DOP853, but now with step size and number of steps given by $(h_{\text{test}}, N_{\text{test}}) = (\frac{h}{20}, 4 \cdot 20N)$, enabling the computation of the interpolation and extrapolation error:

$$e^l(\tilde{y}) = \frac{1}{M+1} \sum_{n=0}^{M} \|\tilde{y}_n - y(t_n)\|_2, \quad t_n \in Q^l, \quad M = |Q^l| - 1. \tag{10}$$

Here $\tilde{y}_{n+1} := \Phi_{h,f_\theta}(\tilde{y}_n)$, and $l \in \{i, e\}$ denotes interpolation or extrapolation: $Q^i = \{hn : 0 \leq hn \leq 20, n \in \mathbb{Z}_+\}$ and $Q^e = \{hn : 20 \leq hn \leq 80, n \in \mathbb{Z}_+\}$, with $h = h_{\text{test}}$. In addition, the error of the learned Hamiltonian is computed along the true trajectory $y(t_n)$ by

Double pendulum   Fermi–Pasta–Ulam–Tsingou

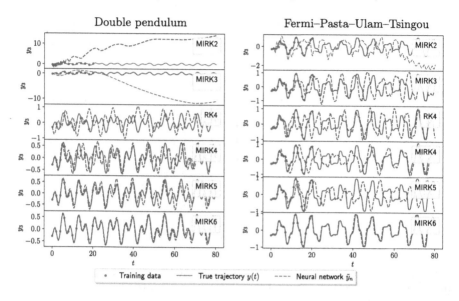

Fig. 1. Result when integrating over the learned vector fields when training on data from the double pendulum (left, $N = 10$) and the Fermi–Pasta–Ulam–Tsingou (right, $N = 20$) Hamiltonian.

$$\overline{e(H_\theta)} = \frac{1}{M+1} \sum_{n=0}^{M} H(y(t_n)) - H_\theta(y(t_n)),$$

$$e(H_\theta) = \frac{1}{M+1} \sum_{n=0}^{M} \left| H(y(t_n)) - H_\theta(y(t_n)) - \overline{e(H_\theta)} \right|, \tag{11}$$

for $t_n \in Q_i \cup Q_e$. The mean is subtracted since the Hamiltonian is only trained by its gradient ∇H_θ. The error terms are shown in Fig. 2. For both test problems the Hamiltonian neural networks have 3 layers with a width of 100 neurons and $\tanh(\cdot)$ as the activation function. Experiments are implemented using PyTorch [15] and the optimization problem is solved using the quasi-Newton L-BFGS algorithm [13] for 100 epochs without batching (all samples used simultaneously).

The implementation of the experiments could be found in the following repository github.com/hakonnoren/learning_hamiltonian_mirk.

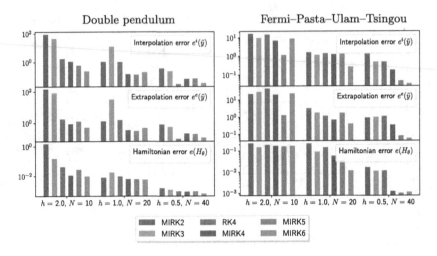

Fig. 2. Errors in interpolation, extrapolation and the Hamiltonian for the double pendulum (left) and the Fermi–Pasta–Ulam–Tsingou problem (right).

6   Conclusion

The mono-implicit Runge–Kutta methods enable the combination of high order and computationally efficient training of Hamiltonian neural networks. The importance of high order is demonstrated by the remarkable capacity of MIRK6 in learning a trajectory of the chaotic double pendulum and the Fermi–Pasta–Ulam–Tsingou Hamiltonian systems from just 11 and 21 points, see Fig. 1. In most cases the error, displayed in Fig. 2, is decreasing when increasing the order. Additionally MIRK4 displays superior performance comparing with the explicit method RK4 of same order. Even though the numerical experiments show promising results, the theoretical error analysis in this work is rudimentary at best. Future work should consider this in greater detail, perhaps along the lines of [18]. Other questions worth investigating are the impact on accuracy if there is noise in the data in addition to how properties of the integrators such as symmetry, symplecticity and stability is transferred to the learned vector fields.

Acknowledgment. The author wishes to express gratitude to Elena Celledoni and Sølve Eidnes for constructive discussions and helpful suggestions while working on this paper.

References

1. Burrage, K., Chipman, F., Muir, P.H.: Order results for mono-implicit Runge–Kutta methods. SIAM J. Numer. Anal. **31**(3), 876–891 (1994)
2. Butcher, J.C.: Numerical methods for ordinary differential equations. John Wiley & Sons (2016)

3. Cash, J.R.: A class of implicit Runge-Kutta methods for the numerical integration of stiff ordinary differential equations. J. ACM (JACM) **22**(4), 504–511 (1975)

4. Chen, R.T., Rubanova, Y., Bettencourt, J., Duvenaud, D.K.: Neural ordinary differential equations. In: Advances in Neural Information Processing Systems, vol. 31 (2018)

5. Chen, Z., Zhang, J., Arjovsky, M., Bottou, L.: Symplectic recurrent neural networks. In: International Conference on Learning Representations (2020). https://openreview.net/forum?id=BkgYPREtPr

6. David, M., Méhats, F.: Symplectic learning for Hamiltonian neural networks. arXiv preprint arXiv:2106.11753 (2021)

7. Dormand, J., Prince, P.: A family of embedded Runge-Kutta formulae. J. Comput. Appl. Math. **6**(1), 19–26 (1980)

8. E, W.: A proposal on machine learning via dynamical systems. Commun. Math. Stat. **5**(1), 1–11 (2017). https://doi.org/10.1007/s40304-017-0103-z

9. Greydanus, S., Dzamba, M., Yosinski, J.: Hamiltonian neural networks. CoRR abs/1906.01563 (2019). https://arxiv.org/abs/1906.01563

10. Haber, E., Ruthotto, L.: Stable architectures for deep neural networks. Inverse Prob. **34**(1), 014004 (2017)

11. Hairer, E., Lubich, C., Wanner, G.: Geometric Numerical Integration: Structure-Preserving Algorithms for Ordinary Differential Equations; 2nd ed. Springer, Dordrecht (2006). https://doi.org/10.1007/3-540-30666-8

12. Muir, P.H.: Optimal discrete and continuous mono-implicit Runge-Kutta schemes for BVODEs. Adv. Comput. Math. **10**(2), 135–167 (1999)

13. Nocedal, J., Wright, S.J.: Numerical optimization. Springer (1999). https://doi.org/10.1007/978-0-387-40065-5

14. Offen, C., Ober-Blöbaum, S.: Symplectic integration of learned Hamiltonian systems. Chaos: An Interdisc. J. Nonlinear Sci. **32**(1), 013122 (2022)

15. Paszke, A., et al.: PyTorch: an imperative style, high-performance deep learning library. Adv. Neural. Inf. Process. Syst. **32**, 8026–8037 (2019)

16. Ruthotto, L., Haber, E.: Deep neural networks motivated by partial differential equations. J. Math. Imag. Vision **62**(3), 352–364 (2020)

17. Zhu, A., Jin, P., Tang, Y.: Deep Hamiltonian networks based on symplectic integrators. arXiv preprint arXiv:2004.13830 (2020)

18. Zhu, A., Jin, P., Zhu, B., Tang, Y.: On numerical integration in neural ordinary differential equations. In: International Conference on Machine Learning, pp. 27527–27547. PMLR (2022)

A Geometric View on the Role of Nonlinear Feature Maps in Few-Shot Learning

Oliver J. Sutton[1]($^{(\boxtimes)}$), Alexander N. Gorban[2], and Ivan Y. Tyukin[1]

[1] King's College London, London, UK
{oliver.sutton,ivan.tyukin}@kcl.ac.uk
[2] University of Leicester, Leicester, UK
a.n.gorban@leicester.ac.uk

Abstract. We investigate the problem of successfully learning from just a few examples of data points in a binary classification problem, and present a brief overview of some recent results on the role of nonlinear feature maps in this challenging task. Our main conclusion is that successful learning and generalisation may be expected to occur with high probability, despite the small training sample, when the nonlinear feature map induces certain fundamental geometric properties in the mapped data.

Keywords: Few-shot learning · High dimensional data · Nonlinear feature maps

1 Introduction

Recent advances in Machine Learning and Artificial Intelligence have undeniably demonstrated the potential impact of learning algorithms for a sweeping range of practically relevant tasks. In many ways, the explosion of growth in the field is due to the onset of the 'big data era', and part of the recent success is simply due to the growing availability of the data and raw computational power required to train models featuring millions of parameters on vast data sets.

There remain areas, however, where obtaining large data sets is infeasible, either because of the costs involved, the availability of data, or privacy and ethical concerns (consider medical data, for instance). Instead, one must undertake the formidable task of *learning from few examples*: extracting algorithmic insight from just a handful of data points from each class in a way which will successfully generalise to unseen data. Various approaches have been proposed to tackle this problem, prominent amongst which are Matching Networks [14] and Prototypical Networks [11], and we refer particularly to the recent review paper by Wang and

The authors are grateful for financial support by the UKRI and EPSRC (UKRI Turing AI Fellowship ARaISE EP/V025295/1). I.Y.T. is also grateful for support from the UKRI Trustworthy Autonomous Systems Node in Verifiability EP/V026801/1.

F. Nielsen and F. Barbaresco (Eds.): GSI 2023, LNCS 14071, pp. 560–568, 2023.
https://doi.org/10.1007/978-3-031-38271-0_56

co-authors [15] for a full overview of the algorithms which are commonly deployed for this task. Despite this, a full theoretical justification of how such algorithms are able to learn from just a few examples is still largely absent.

Classical results from Statistical Learning Theory on Vapnik-Chervonenkis (VC) dimensionality suggest that training a neural network with millions of parameters requires exponentially larger data sets to ensure good generalisation performance. In these restricted data settings large neural networks may therefore not be a good choice, as the risk of failing to generalise is simply too high. Recent developments in the theory of benign [1] and tempered [9] over-fitting suggest one possible scenario in which the data is well conditioned for learning: a combination of low dimensional structure and high dimensional noise, controlled by the spectral properties of the data covariance matrix, can enable over-parameterised models to successfully generalise despite over-fitting to under-sampled data.

Here, we offer a complementary, geometric, perspective on the theory of learning from few examples, broadly summarising our recent results from [12]. The philosophy underlying our results is twofold. Firstly, recent theoretical work has shown that genuinely high dimensional data possesses traits which are highly desirable for learning: concentration of measure phenomena [8] and quasi-orthogonality properties [5–7] ensure that points may be separated from one another and even arbitrary subsets using simple linear separators with high probability [4]. Using these *blessings of dimensionality*, it was shown in recent work [13] that under certain mild assumptions on the data distribution, such high dimensional data can be successfully used to train a simple classifier with strong guarantees on the probability of successfully learning and generalising from just a few examples. Our second motivation is to understand the interplay between nonlinearity and dimensionality: specifically, we wish to understand in what scenarios nonlinear mappings embedding data into higher dimensional spaces are capable of providing the same blessings of dimensionality. This is inspired by the structure of many typical large neural networks, which take the form of sequences of (learned) nonlinear mappings to high dimensional feature spaces, where simpler classifiers are then applied.

Our main result (Theorem 2) shows that if the nonlinear *feature map* used to embed the data to a surface in a higher dimensional space satisfies certain geometric properties, then a simple linear classifier trained on this mapped data may be expected to successfully learn and generalise from just a few data samples. In this sense, the theorem describes exactly the scenario in which nonlinearity can be used to accelerate the onset of the blessing of dimensionality. The properties of the feature map which we focus on are geometric in nature, and measure the interplay between the data distribution and the way in which the nonlinear feature mapping modifies space. We show that, together, they provide symmetric upper and lower bounds on the probability of successfully learning and generalising from small data sets. Our results may therefore be interpreted as a mathematical definition of the properties required from a 'good' feature map, through which the learning problem becomes tractable.

A core thread of narrative from the review article [15] on few-shot learning focuses around the injection of '*a priori* knowledge' into learning schemes, through which the (limited) training data is augmented using some description of extrinsic knowledge to guide the learning algorithm towards a generalisable solution. Although no mathematical definition of this nebulous concept is given in [15], the authors conclude that it seems to be a pervading trait of a wide class of algorithms which successfully learn from few examples. Our results can be interpreted as capturing exactly the essence of this infusion of extra knowledge: if the nonlinear feature map is selected in such a way that it offers beneficial geometric properties, then successful few-shot learning can be expected to occur. Since the (family of the) feature map may be fixed *a priori*, this provides a general mechanism for incorporating prior knowledge into the algorithm.

Incorporating prior knowledge like this is, to some degree, a departure from conventional distribution-agnostic approaches to learning, which fundamentally assume no knowledge of the data other than the training set. However, it is starting to become apparent in this and other areas [2, 3] that assuming no knowledge of the distribution can be actively harmful to the success and stability of learning algorithms. In this context, our results show that providing knowledge of only certain (nonlinear) functionals of the data distribution, encoding geometric properties, can be beneficial, even if complete knowledge remains unknown. Although it is motivated by a very different class of problems, this approach may appear reminiscent of the approach taken by Kernel Mean Embeddings [10], which use nonlinear kernels to embed families of data distributions into a Hilbert space structure. Our key focus here, however, is on using just a few *targeted* properties to describe the situation of interest.

We mathematically formulate the few-shot learning problem and the setting in which we study it in Sect. 2. Our main results are summarised in Sect. 3, and we offer some discussion of them in Sect. 4. Finally, we offer some conclusions in Sect. 5.

2   Problem Setting

We consider the problem of learning from few examples, in a standard binary classification framework. We assume that the data are drawn from two data distributions, one of which has labels in the set $\mathcal{L}_{old}$ and other of which is associated with the label ℓ_{new}. In particular, we assume that there exists a previously trained classification function $F : \mathbb{R}^d \to \mathcal{L}_{old}$ assigning one of the labels in $\mathcal{L}_{old}$ to each point in $\mathbb{R}^d$. In this setting, F models an existing legacy system which has been trained, a possibly expensive task, and needs to be updated to also classify data from the new class ℓ_{new}. The task we consider is to build a simple and computationally cheap new classifier which retains the performance of F on the legacy data classes, while also successfully classifying data points from the new class ℓ_{new}:

Problem 1 (Few-shot learning). Consider a classifier $F : \mathbb{R}^d \to \mathcal{L}_{old}$, trained on a sample $\mathcal{Z}$ drawn from some probability distribution P_Z on $\mathbb{R}^d$, and let $\mathcal{X}$ be a new sample that is drawn from the probability distribution P_X and whose cardinality satisfies $|\mathcal{X}| \ll d$. Let $p_e, p_n \in (0, 1]$ be given positive numbers determining the desired quality of learning.

Find an algorithm $\mathcal{A}(\mathcal{X})$ producing a new classification map $F_{\text{new}} : \mathbb{R}^d \rightarrow \mathcal{L}_{\text{old}} \cup \{\ell_{\text{new}}\}$, such that examples of class ℓ_{new} are correctly learned with probability at least p_n, i.e.

$$P\big(F_{\text{new}}(x) = \ell_{\text{new}}\big) \geq p_n, \tag{1}$$

for x drawn from P_X, while F_{new} remembers the previous classifier F elsewhere with probability at least p_e, i.e.

$$P\big(F_{\text{new}}(x) = F(x)\big) \geq p_e, \tag{2}$$

for x drawn from the distribution P_Z.

3 Theoretical Results

We now present our theoretical results. For further discussion of these results and their proofs, we refer to [12].

Let $\phi : \mathbb{R}^d \rightarrow \mathbb{H}$ be a fixed nonlinear feature map, where $\mathbb{H}$ denotes a (possibly infinite-dimensional) Hilbert space. To concisely state our results, we introduce some functions measuring the probabilities of certain geometric events.

Definition 1 (Geometric probability functions). *Let c_X and c_Z be arbitrary but fixed points in the feature space $\mathbb{H}$. We define the following shorthand notations for probabilities:*

– *Let $p : \mathbb{R}_{\geq 0} \rightarrow [0, 1]$ denote the* projection probability function, *given by*

$$p(\delta) = P(x, y \sim P_X : \langle \phi(x) - c_X, \phi(y) - c_X \rangle \leq \delta),$$

measuring the probability of sampling two data points from the new data distribution P_X which can be separated from one another by a plane which is normal to either $\phi(x)$ or $\phi(y)$ with margin which is a function of δ, $\|\phi(x) - c_X\|$ and $\|\phi(y) - c_X\|$.
– *Let $\lambda_X, \lambda_X : \mathbb{R} \rightarrow [0, 1]$ denote the* localisation probability functions *for P_X and P_Z, given by*

$$\lambda_X(r) = P(x \sim P_X : \|\phi(x) - c_X\| \leq r),$$

and

$$\lambda_Z(r) = P(z \sim P_Z : \|\phi(z) - c_Z\| \leq r).$$

These measure the probability of sampling a point which falls within a ball of radius r of the (arbitrary but fixed) centre point of each distribution.

- *Let $s_X, s_Z : \mathbb{R} \to [0,1]$ denote the class separation probability functions for P_X and P_Z, where*

$$s_X(\delta) = P(x \sim P_X : (\phi(x) - c_X, c_Z - c_X) \leq \delta),$$

and

$$s_Z(\delta) = P(z \sim P_Z : (\phi(z) - c_Z, c_X - c_Z) \leq \delta),$$

which measure the probability of separating points from the two data distributions, and provides a measure of the well-posedness of the classification problem.

Although these probabilities clearly depend on the choice of ϕ and the points c_X and c_Z, we omit this from the notation for brevity.

Our first result shows that these geometric quantities provide upper and lower bounds on the probability of the empirical mean of the training sample falling close to the (arbitrary) centre point c_X. In particular, if the point c_X is such that (i) points sampled from P_X are typically close to c_X or (ii) points sampled from P_X are generally orthogonal about c_X, then the empirical mean of the sampled points will be close to c_X, even for small data samples. The appearance of condition (i) is somewhat natural. Condition (ii), however, is due to the role of the parameter δ in the estimate: the same estimate can be obtained when the sampled points are *less* localised around the centre c_X if they may be expected to be highly orthogonal about c_X. For the proof of this result we refer to [12].

Theorem 1 (Convergence of the empirical mean [12]**).** *Let $s > 0$, let $\{x_i\}_{i=1}^k \subset \mathbb{R}^d$ be independent samples from the distribution P_X, and define $\mu = \frac{1}{k}\sum_{i=1}^k \phi(x_i)$. Then,*

$$P(\{x_i\}_{i=1}^k : \|\mu - c_X\| \leq s) \geq 1 - \inf_{\delta \in \mathbb{R}} \left[k(1 - \lambda_X(r(s,\delta))) + k(k-1)(1 - p(\delta)) \right], \tag{3}$$

where $r(s,\delta) = \{ks^2 - (k-1)\delta\}_+^{1/2}$ and $\{t\}_+ = \max\{t,0\}$. This estimate is symmetric in the sense that

$$P(\{x_i\}_{i=1}^k : \|\mu - c_X\| \leq s) \leq \inf_{\delta \in \mathbb{R}} \left[k\lambda_X(r(s,\delta)) + k(k-1)p(\delta) \right]. \tag{4}$$

Our main result builds on this to show that the geometric quantities in Definition 1 also provide both upper and lower bounds on the probability of successfully learning and generalising from few examples. Once again, we state the result in the form of a supremum over various 'floating' parameters, which naturally appear in the analysis and describe the trade-off which is possible by balancing the various terms. The precise connection between these variables is given in the definition of η and ξ. The general behaviour which this theorem describes, however, is that a small data sample is sufficient for successfully learning and generalising when the problem is well posed (in the sense that the data classes may indeed be separated by a linear separator), the data are reasonably well localised around their central points, and are highly orthogonal around the central point. For the proof of this result we refer to [12].

Theorem 2 (Few shot learning [12]**).** *Suppose that we are in the setting of the few-shot learning problem specified in Problem 1, and let $\{x_i\}_{i=1}^k \subset \mathbb{R}^d$ be independent training samples from the new data distribution P_X. For $\theta \in \mathbb{R}$, construct the classifier*

$$F_{\text{new}}(x) = \begin{cases} \ell_{\text{new}} & \text{for } (\phi(x) - \mu, \mu - c_Z) \geq \theta, \\ F(x) & \text{otherwise,} \end{cases} \tag{5}$$

where $\mu = \frac{1}{k} \sum_{i=1}^k \phi(x_i)$ is the mean of the new class training samples in feature space. Then,

(i) With respect to samples x and $\{x_i\}_{i=1}^k$ drawn independently from the distribution P_X, the probability $P(F_{\text{new}}(x) = \ell_{\text{new}})$ that this classifier has correctly learned and will generalise well to the new class is at least

$$P(F_{\text{new}}(x) = \ell_{\text{new}}) \geq \sup_{a,b,\gamma,\epsilon > 0} P(\|\mu - c_X\| \leq a)\{\lambda_X(b) + s_X(\eta) - 1\}_+, \tag{6}$$

where $\{t\}_+ = \max\{t, 0\}$ and $\eta = -\theta - \frac{1}{2\gamma}\|c_X - c_Z\|^2 - \frac{\epsilon + \gamma + 2}{2}a^2 - \frac{b^2}{2\epsilon}$, and this estimate is symmetric in the sense that the same terms provide an upper bound on the probability, i.e.

$$P(F_{\text{new}}(x) = \ell_{\text{new}}) \leq 1 - \sup_{a,b,\gamma,\epsilon > 0} (1 - P(\|\mu - c_X\| \leq a))\{1 - \lambda_X(b) - s_X(\eta)\}_+. \tag{7}$$

(ii) With respect to samples z drawn from P_Z and $\{x_i\}_{i=1}^k$ drawn independently from P_X, the probability $P(F_{\text{new}}(z) = F(z))$ that the classifier can correctly distinguish the original classes and so will retain its previous learning is at least

$$P(F_{\text{new}}(z) = F(z)) \geq \sup_{a,\beta,\gamma,\epsilon > 0} P(\|\mu - c_X\| \leq a)\{\lambda_Z(\beta) + s_Z(\xi) - 1\}_+, \tag{8}$$

where $\xi = \theta + \left(1 - \frac{1}{\gamma}\right)\|c_X - c_Z\|^2 - \frac{\epsilon + \gamma - 2}{2}a^2 - \frac{\beta^2}{2\epsilon}$, and this estimate is symmetric in the sense that

$$P(F_{\text{new}}(z) = F(z)) \leq 1 - \sup_{a,\beta,\gamma,\epsilon > 0} (1 - P(\|\mu - c_X\| \leq a))\{1 - \lambda_Z(\beta) - s_Z(\xi)\}_+. \tag{9}$$

4   Behaviour of Common Kernels

We can investigate the behaviour of the quantities from Defintion 1 for certain common feature maps, derived through standard nonlinear kernels $\kappa : \mathbb{R}^d \times \mathbb{R}^d \to \mathbb{R}_{\geq 0}$. In particular, we recall the family of polynomial kernels, of the form

$$\kappa(x, y) = (b^2 + x \cdot y)^p$$

for some bias $b > 0$ and polynomial degree $p > 1$, and the Gaussian kernels

$$\kappa(x, y) = \exp(-\frac{1}{2\sigma}\|x - y\|^2)$$

depending on a variance parameter $\sigma > 0$. A key property of these kernels is the existence of a feature map $\phi : \mathbb{R}^d \to \mathbb{H}$, where $\mathbb{H}$ denotes some (typically higher and possibly infinite dimensional) Hilbert space, such that $\kappa(x, y) = (\phi(x), \phi(y))$, where $(\cdot, \cdot)$ denotes the inner product in $\mathbb{H}$.

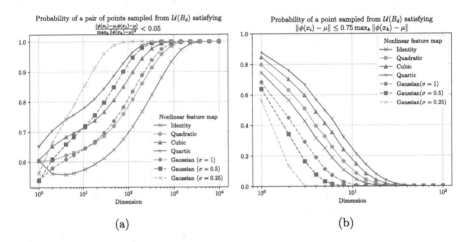

(a)   (b)

Fig. 1. Empirical demonstration of the effect of the choice of nonlinear kernel on the degree of (a) orthogonality (reproduced from [12]) and (b) localisation of data points in the associated feature space. In each case, ϕ denotes the feature map associated with the kernel, and $\|\cdot\|$ denotes the norm induced by $(\cdot, \cdot)$. The data was computed by sampling $N = 5,000$ independent uniformly distributed points $\{x_i\}_{i=1}^N$ from the unit ball in $\mathbb{R}^d$, and $\mu = \frac{1}{N}\sum_{i=1}^N \phi(x_i)$ denotes the empirical mean in feature space.

For various instances of these kernels, we compute the orthogonality and localisation properties of the image of a uniform distribution in the unit ball in $\mathbb{R}^d$ under the associated feature maps. The results of this are plotted in Fig. 1a and Fig. 1b respectively. As a general rule, we observe from Fig. 1a that either increasing the polynomial degree or the data dimension, or decreasing the Gaussian variance of the kernel appears to increase the degree of orthogonality of sampled and mapped points. This implies that the kernels may indeed be capable of accelerating the onset of one aspect of high dimensionality.

On the other hand, however, we see from Fig. 1b that increasing the data dimension reduces the proportion of points which are 'close' to the centre of the mapped distribution, as may be expected from classical concentration of measure results [8], and Gaussian kernels with small variance seem to accelerate this. The use of polynomial kernels, however, appears to hold off this trend (which is undesirable in light of Theorem 2) to some degree.

5 Conclusion

Successfully learning from just a few examples undoubtedly remains a highly challenging task. We have shown how nonlinear feature maps can be beneficial in this area, specifically by offering the geometric properties of localising data sampled from each class, and orthogonalising them about a central point. This provides a mathematical basis for the notion of *prior knowledge* which is often invoked in the literature on few-shot learning [15], and these properties can therefore be used as a design target when crafting bespoke feature mappings for individual datasets. It remains an open problem to be able to assess *in advance* whether a given feature map will be suitable for future few-shot learning tasks on unseen data classes.

References

1. Bartlett, P., Long, P., Lugosi, G., Tsigler, A.: Benign overfitting in linear regression. Proc. Natl. Acad. Sci. **117**(48), 30063–30070 (2020)
2. Bastounis, A., et al.: The crisis of empirical risk in verifiable, robust and accurate learning (2023, in preparation)
3. Bastounis, A., Hansen, A.C., Vlačić, V.: The mathematics of adversarial attacks in AI - why deep learning is unstable despite the existence of stable neural networks (2021). https://doi.org/10.48550/ARXIV.2109.06098. https://arxiv.org/abs/2109.06098
4. Gorban, A.N., Tyukin, I.Y.: Stochastic separation theorems. Neural Netw. **94**, 255–259 (2017)
5. Gorban, A., Tyukin, I., Prokhorov, D., Sofeikov, K.: Approximation with random bases: pro et contra. Inf. Sci. **364–365**, 129–145 (2016)
6. Kainen, P.C., Kůrková, V.: Quasiorthogonal dimension. In: Kosheleva, O., Shary, S.P., Xiang, G., Zapatrin, R. (eds.) Beyond Traditional Probabilistic Data Processing Techniques: Interval, Fuzzy etc. Methods and Their Applications. SCI, vol. 835, pp. 615–629. Springer, Cham (2020). https://doi.org/10.1007/978-3-030-31041-7_35
7. Kainen, P.C., Kurková, V.: Quasiorthogonal dimension of Euclidean spaces. Appl. Math. Lett. **6**(3), 7–10 (1993)
8. Ledoux, M.: The Concentration of Measure Phenomenon. No. 89, American Mathematical Soc. (2001)
9. Mallinar, N., Simon, J.B., Abedsoltan, A., Pandit, P., Belkin, M., Nakkiran, P.: Benign, tempered, or catastrophic: a taxonomy of overfitting. arXiv preprint arXiv:2207.06569 (2022)
10. Smola, A., Gretton, A., Song, L., Schölkopf, B.: A Hilbert space embedding for distributions. In: Hutter, M., Servedio, R.A., Takimoto, E. (eds.) ALT 2007. LNCS (LNAI), vol. 4754, pp. 13–31. Springer, Heidelberg (2007). https://doi.org/10.1007/978-3-540-75225-7_5
11. Snell, J., Swersky, K., Zemel, R.: Prototypical networks for few-shot learning. In: Advances in Neural Information Processing Systems, pp. 4077–4087 (2017)
12. Sutton, O.J., Gorban, A.N., Tyukin, I.Y.: Towards a mathematical understanding of learning from few examples with nonlinear feature maps (2022). https://doi.org/10.48550/ARXIV.2211.03607. https://arxiv.org/abs/2211.03607

13. Tyukin, I.Y., Gorban, A.N., Alkhudaydi, M.H., Zhou, Q.: Demystification of few-shot and one-shot learning. In: 2021 International Joint Conference on Neural Networks (IJCNN), pp. 1–7. IEEE (2021)
14. Vinyals, O., Blundell, C., Lillicrap, T., Kavukcuoglu, K., Wierstra, D.: Matching networks for one shot learning. In: Advances in Neural Information Processing Systems, pp. 3630–3638 (2016)
15. Wang, Y., Yao, Q., Kwok, J.T., Ni, L.M.: Generalizing from a few examples: a survey on few-shot learning. ACM Comput. Surv. **53**(3) (2020). https://doi.org/10.1145/3386252

Learning Discrete Lagrangians for Variational PDEs from Data and Detection of Travelling Waves

Christian Offen$^{(\boxtimes)}$ (iD) and Sina Ober-Blöbaum

Paderborn University, Warburger Street 100, 33098 Paderborn, Germany
christian.offen@uni-paderborn.de
https://www.uni-paderborn.de/en/person/85279

Abstract. The article shows how to learn models of dynamical systems from data which are governed by an unknown variational PDE. Rather than employing reduction techniques, we learn a discrete field theory governed by a discrete Lagrangian density L_d that is modelled as a neural network. Careful regularisation of the loss function for training L_d is necessary to obtain a field theory that is suitable for numerical computations: we derive a regularisation term which optimises the solvability of the discrete Euler–Lagrange equations. Secondly, we develop a method to find solutions to machine learned discrete field theories which constitute travelling waves of the underlying continuous PDE.

Keywords: System identification · discrete Lagrangians · travelling waves

1 Introduction

In data-driven system identification, a model is fitted to observational data of a dynamical system. The quality of the learned model can greatly be improved when prior geometric knowledge about the dynamical system is taken into account such as conservation laws [2,5,9,14,15], symmetries [7,8,10], equilibrium points [18], or asymptotic behaviour of its motions.

One of the most fundamental principles in physics is the variational principle: it says that motions constitute stationary points of an action functional. The presence of variational structure is related to many qualitative features of the dynamics such as the validity of Noether's theorem: symmetries of the action functional are in correspondence with conservation laws. To guarantee that these fundamental laws of physics hold true for learned models, Greydanus et al. propose to learn the action functional from observational data [5] (Lagrangian neural network) and base prediction on numerical integrations of Euler–Lagrange equations. Quin proposes to learn a discrete action instead [17]. An ansatz of a discrete model has the advantage that it can be trained with position data of motions only. In contrast, learning a continuous theory typically requires information about higher derivatives (corresponding to velocity, acceleration, momenta, for instance) which are typically not observed but only approximated. Moreover, the

© The Author(s), under exclusive license to Springer Nature Switzerland AG 2023
F. Nielsen and F. Barbaresco (Eds.): GSI 2023, LNCS 14071, pp. 569–579, 2023.
https://doi.org/10.1007/978-3-031-38271-0_57

discrete action functional (once it is learned) can naturally be used to compute motions numerically.

However, in [14] the authors demonstrate that care needs to be taken when learned action functionals are used to compute motions: even if the data-driven action functional is perfectly consistent with the training data (i.e. the machine learning part of the job is successfully completed), minimal errors in the initialisation process of numerical computations get amplified. As a remedy, the authors develop Lagrangian Shadow Integrators which mitigate these amplified numerical errors based on a technique called backward error analysis. Moreover, using backward error analysis they relate the discrete quantities to their continuous analogues and show how to analyse qualitative aspects of the machine learned model. Action functionals are not uniquely determined by the motions of a dynamical system. Therefore, regularisation is needed to avoid learning degenerate theories. While in [14] the authors develop a regularisation strategy when the action functional is modelled as a Gaussian Process, Lishkova et al. develop a corresponding regularisation technique for artificial neural networks in [10] in the context of ordinary differential equations (ODEs).

In this article we show how to learn a discrete action functional from discrete data which governs solutions to partial differential equations (PDEs) using artificial neural networks extending the regularisation strategy which we have developed in [10]. Our technique to learn (discrete) densities of action functionals can be contrasted to approaches where a spatial discretisation of the problem is considered first, followed by structure-preserving model reduction techniques (data-driven or analytical) [3,4] and then a model for the reduced system of ODEs is learned from data [1,12,19].

Travelling waves solutions of PDEs are of special interest due to their simple structure. When a discrete field theory for a continuous process described by a PDE is learned, they typically "get lost" because the mesh of the discrete theory is incompatible with certain wave speeds. In this article, we introduce a technique to find the solutions of data-driven discrete theories that correspond to travelling waves in the underlying continuous dynamics (shadow travelling waves). The article contains the following novelties:

- We transfer our Lagrangian ODE regularising strategy [10] to data-driven discrete field theories in a PDE setting and provide a justification using numerical analysis.
- The development of a technique to detect travelling waves in data-driven discrete field theories.

The article proceeds with a review of variational principles (Sect. 2), an introduction of our machine learning architecture and derivation of the regularisation strategy (Sect. 3). In Sect. 4, we define the notion of *shadow travelling waves* and show how to find them in data-driven models. The article concludes with numerical examples relating to the wave equation (Sect. 5).

2 Discrete and Continuous Variational Principles

Continuous Variational Principles. Many differential equations describing physical phenomena such as waves, the state of a quantum system, or the evolution of a relativistic fields are derived from a variational principle: solutions are characterised as critical points of a (non-linear) functional S defined on a suitable space of functions $u\colon X \to \mathbb{R}^d$ and has the form

$$S(u) = \int_X L(\boldsymbol{x}, u(\boldsymbol{x}), u_{x_0}(\boldsymbol{x}), u_{x_1}(\boldsymbol{x}), \dots, u_{x_n}(\boldsymbol{x})) \mathrm{d}\boldsymbol{x}, \tag{1}$$

where $\boldsymbol{x} = (x_0, x_1, \dots, x_n) \in X$ and u_{x_j} denote partial derivatives of u. This variational principle can be referred to as a *first-order field theory*, since only derivatives to the first order of u appear. In many applications the free variable x_0 corresponds to time and is denoted by $x_0 = t$. The functional S is stationary at u with respect to all variations $\delta u\colon X \to \mathbb{R}$ vanishing at the boundary (or with the correct asymptotic behaviour) if and only if the Euler–Lagrange equations

$$0 = \mathrm{EL}(L) = \frac{\partial L}{\partial u} - \sum_{j=0}^{n} \frac{\mathrm{d}}{\mathrm{d}x_j} \frac{\partial L}{\partial u_{x_j}} \tag{2}$$

are fulfilled on X.

Example 1. The wave equation

$$u_{tt}(t, x) - u_{xx}(t, x) + \nabla V(u(t, x)) = 0 \tag{3}$$

is the Euler–Lagrange equation $0 = \mathrm{EL}(L)$ to the Lagrangian

$$L(u, u_t, u_x) = \frac{1}{2}(u_t^2 - u_x^2) - V(u). \tag{4}$$

Here ∇V denotes the gradient of a potential V.

Remark 1. Lagrangians are not uniquely determined by the motions of a dynamical system: two first order Lagrangians L and $\tilde{L}$ yield equivalent Euler–Lagrange equations if $sL - \tilde{L}$ ($s \in \mathbb{R} \setminus \{0\}$) is a total divergence $\nabla_{\boldsymbol{x}} \cdot F(\boldsymbol{x}, u(\boldsymbol{x})) = \frac{\partial}{\partial x_1}(F^1(\boldsymbol{x}, u(\boldsymbol{x}))) + \dots + \frac{\partial}{\partial x_n}F^n(\boldsymbol{x}, u(\boldsymbol{x}))$ for $F = (F^1, \dots, F^n)\colon X \times \mathbb{R}^d \to \mathbb{R}^n$.

Discrete Variational Principle. For simplicity, we consider the two dimensional compact case: let $X = [0, T] \times [0, l]/\{0, l\}$ with $T, l > 0$. Here $[0, l]/\{0, l\}$ is the real interval $[0, l]$ with identified endpoints (periodic boundary conditions). Consider a uniform, rectangular mesh X_Δ on X with mesh widths $\Delta t = \frac{T}{N}$ and $\Delta x = \frac{l}{M}$ for $N, M \in \mathbb{N}$. A discrete version of the action functional (1) is

$$S_d\colon (\mathbb{R}^d)^{(N-1) \times M} \to \mathbb{R}^d, \quad S_d(U) = \Delta t \Delta x \sum_{i=1}^{N-1} \sum_{j=0}^{M-1} L_d(u_j^i, u_j^{i+1}, u_{j+1}^i)$$

for a discrete Lagrangian density $L_d\colon (\mathbb{R}^d)^3 \to \mathbb{R}$ together with temporal boundary conditions for $u_j^0 \in \mathbb{R}^d$ and $u_j^N \in \mathbb{R}^d$ for $j = 0, \ldots, M - 1$. Above U denotes the values $(u_j^i)_{j=0,\ldots N-1}^{i=1,\ldots M-1}$ on inner mesh points. We have $u_M^i = u_0^i$ by the periodicity in space. Solutions of the variational principle are $U \in (\mathbb{R}^d)^{(N-1)\times M}$ such that U is a critical point of S_d. This is equivalent to the condition that for all $i = 1, \ldots, N - 1$ and $j = 0, \ldots, M - 1$ the discrete Euler–Lagrange equations

$$\frac{\partial}{\partial u_j^i}\left(L_d(u_j^i, u_j^{i+1}, u_{j+1}^i) + L_d(u_j^{i-1}, u_j^i, u_{j+1}^{i-1}) + L_d(u_{j-1}^i, u_{j-1}^{i+1}, u_j^i)\right) = 0 \tag{5}$$

are fulfilled. The expression on the left of (5) is abbreviated as $\mathrm{DEL}(L_d)_j^i(U)$ in the following.

Remark 2. Instead of periodic boundary conditions in space, S_d can be adapted to other types of boundary conditions such as Dirichlet- or Neumann conditions.

Example 2. The discretised wave equation

$$\frac{(u_j^{i-1} - 2u_j^i + u_j^{i+1})}{\Delta t^2} - \frac{(u_{j-1}^i - 2u_j^i + u_{j-1}^i)}{\Delta x^2} + \nabla V(u_j^i) = 0 \tag{6}$$

is the discrete Euler–Lagrange equations to the discrete Lagrangian

$$L_d((u_j^i, u_j^{i+1}, u_{j+1}^i) = \frac{1}{2}\left(\frac{u_j^{i+1} - u_j^i}{\Delta t^2}\right)^2 - \frac{1}{2}\left(\frac{u_{j+1}^i - u_j^i}{\Delta x^2}\right)^2 - V(u_j^i).$$

Remark 3. In analogy to Remark 1, notice that L_d and $\tilde{L}_d$ yield the same discrete Euler–Lagrange Eqs. (5) if

$$L_d(a,b,c) - s\tilde{L}_d(a,b,c) = \chi_1(a) - \chi_1(b) + \chi_2(a) - \chi_2(c) + \chi_3(b) - \chi_3(c) \tag{7}$$

for differentiable functions $\chi_1, \chi_2, \chi_3\colon X \to \mathbb{R}$ and $s \in \mathbb{R} \setminus \{0\}$.

Remark 4. If $\mathrm{DEL}(L_d)_j^i(U) = 0$ (see (5)) and if $\frac{\partial^2 L_d}{\partial u_j^i \partial u_j^{i+1}}(u_j^i, u_j^{i+1}, u_{j+1}^i)$ is of full rank, then (5) is solvable for u_j^{i+1} as a function of u_j^i, u_{j+1}^i, u_j^{i-1}, u_{j+1}^{i-1}, u_{j-1}^i, u_{j-1}^{i+1} by the implicit function theorem locally around a solution of (5). All of these points correspond to mesh points that either lie to the left or below the point with indices (i, j). If u_j^1 is known for $0 \leq j \leq M - 1$, then utilising the boundary conditions $u_j^0 \in \mathbb{R}^d$ and $u_0^i = u_M^i$ we can compute U by subsequently solving (5). This corresponds to the computation of a time propagation.

The following Proposition analyses the convergence of Newton-Iterations when solving (5) for u_j^{i+1}, as is required to compute time propagations. It introduces a quantity ρ^* that relates to how well the iterations converge. We will make use of this quantity in the design of our machine learning framework.

Proposition 1. *Let u_j^i, u_j^{i+1}, u_{j+1}^i, u_j^{i-1}, u_{j+1}^{i-1}, u_{j-1}^i, u_{j-1}^{i+1} such that (5) holds. Let $O \subset \mathbb{R}^d$ be a convex, neighbourhood of $u^* = u_j^{i+1}$, $\|\cdot\|$ a norm of $\mathbb{R}^d$ inducing an operator norm on $\mathbb{R}^{d \times d}$. Define $p(u) := \frac{\partial^2 L_d}{\partial u_j^i \partial u}(u_j^i, u, u_{j+1}^i)$ and let θ and $\bar{\theta}$ be Lipschitz constants on O for p and for inv $\circ$ p, respectively, where inv denotes matrix inversion. Let*

$$\rho^* := \|\mathrm{inv}(p(u^*))\| = \left\| \left(\frac{\partial^2 L_d}{\partial u_j^i \partial u^*}(u_j^i, u^*, u_{j+1}^i) \right)^{-1} \right\| \tag{8}$$

and let $f(u^{(n)})$ denote the left hand side of (5) with u_j^{i+1} replaced by $u^{(n)}$. If $\|u^{(0)} - u^*\| \leq \min\left(\frac{\rho^*}{\bar{\theta}}, \frac{1}{2\theta\rho^*} \right)$ for $u^{(0)} \in \mathcal{O}$, then the Newton Iterations $u^{(n+1)} :=$ $u^{(n)} - \mathrm{inv}(p(u^{(n)}))f(u^{(n)})$ converge quadratically against u^*, i.e.

$$\|u^{(n+1)} - u^*\| \leq \rho^* \theta \|u^{(n)} - u^*\|^2. \tag{9}$$

Proof. The statement follows from an adaption of the standard estimates for Newton's method (see [6, Sect. 4], for instance) to the considered setting. A detailed proof of the Proposition is contained in the Appendix (Preprint/ArXiv version only). □

Remark 5. The assumptions formulated in Proposition 1 are sufficient but not sharp. The main purpose of the proposition is to identify quantities that are related to the efficiency of our numerical solvers and to use this knowledge in the design of machine learning architectures.

3 Machine Learning Architecture for Discrete Field Theories

We model a discrete Lagrangian L_d as a neural network and fit its parameters

- such that the discrete Euler–Lagrange Eqs. (5) for the learned L_d are consistent with observed solutions $U = (u_j^i)$ of (5)
- and such that (5) is easily solvable for u_j^{i+1} using iterative numerical methods, so that we can use the discrete field theory to predict solutions via forward propagation of initial conditions (see Remark 4).

For given observations $U^{(1)}, \ldots, U^{(K)}$ with $U^{(k)} = (u_j^{i(k)})$ on the interior mesh X_Δ, we consider the loss function $\ell = \ell_{\mathrm{DEL}} + \ell_{\mathrm{reg}}$ consisting of a data consistency term ℓ_{DEL} and a regularising term ℓ_{reg}. We have

$$\ell_{\mathrm{DEL}} = \sum_{k=1}^{K} \sum_{i=1}^{N-1} \sum_{j=0}^{M-1} \mathrm{DEL}(L_d)_j^i (U^{(k)})^2 \tag{10}$$

with $\mathrm{DEL}(L_d)_j^i$ from (5). ℓ_{DEL} measures how well L_d fits to the training data.

Since a discrete Lagrangian L_d is not uniquely determined by the system's motions by Remark 3 (indeed, $L_d \equiv$ const is consistent with any observed dynamics), careful regularisation is required. Indeed, in [14] we demonstrate in an ode setting that if care is not taken, then machine learned models for L_d can be unsuitable for numerical purposes and amplify errors of numerical integration schemes. In view of Proposition 1, we aim to minimize ρ^* (see (8)) and define the regularisation term

$$\ell_{\text{reg}} = \sum_{k=1}^{K} \sum_{i=1}^{N-1} \sum_{j=0}^{M-1} \left\| \left(\frac{\partial^2}{\partial u_j^i \partial u_j^{i+1}} L_d \left(u_j^{i\,(k)}, u_j^{i+1\,(k)}, u_{j+1}^{i\,(k)} \right) \right)^{-1} \right\|^2. \tag{11}$$

In our experiments, we use the spectral norm in (11), which is the operator norm induced by the standard Eucledian vector norm on $\mathbb{R}^d$. Let $A_{\text{reg}}^{i,j,k} := \frac{\partial^2}{\partial u_j^i \partial u_j^{i+1}} L_d \left(u_j^{i\,(k)}, u_j^{i+1\,(k)}, u_{j+1}^{i\,(k)} \right)$, i.e. the summands of ℓ_{reg} are $\|(A_{\text{reg}}^{i,j,k})^{-1}\|^2 = \frac{1}{\lambda_{\min}^2}$, where $\lambda_{\min}$ is the singular value $\lambda_{\min}$ of $A_{\text{reg}}^{i,j,k}$ with the smallest absolute norm. If $u_j^i \in \mathbb{R}^d$ with $d = 1$, then $\|(A_{\text{reg}}^{i,j,k})^{-1}\|$ can be evaluated without problems. Otherwise, $\lambda_{\min}^2$ is computed as the smallest eigenvalue of the symmetric matrix $(A_{\text{reg}}^{i,j,k})^\top A_{\text{reg}}^{i,j,k}$. The eigenvalue can be approximated by inverse matrix vector iterations [6, Sect. 5] or computed exactly if the dimension d is small.

4   Periodic Travelling Waves

For simplicity, we continue within the two-dimensional space time domain $X = [0, T] \times [0, l]/\{0, l\}$ with periodic boundary conditions in space introduced in Sect. 2. A periodic travelling wave (TW) of a pde on X is a solution of the form $u(t, x) = f(x - ct)$ for $c \in \mathbb{R}$ and with $f: [0, l]/\{0, l\} \to \mathbb{R}^d$ defined on the periodic spatial domain. Due to their simple structure, TWs are important solutions to pdes. While the defining feature of a TW is its symmetry $u(t + s, x + sc) = u(t, x)$ for $s \in \mathbb{R}$, evaluated on a mesh X_Δ, no such structure is evident unless the quotient $c\Delta t/\Delta x$ is rational and T sufficiently large. However, after a discrete field theory is learned defined by its discrete Lagrangian L_d, it is of interest, whether the underlying continuous PDE has TWs. As in [13] we define *shadow travelling waves* (TWs) of (5) as solutions to the functional equation

$$0 = \partial_1 L_d(f(\xi), f(\xi - c\Delta t), f(\xi + \Delta x))$$
$$+ \partial_2 L_d(f(\xi + c\Delta t), f(\xi), f(\xi + c\Delta t + \Delta x)) \tag{12}$$
$$+ \partial_3 L_d(f(\xi - \Delta x), f(\xi - c\Delta t - \Delta x), f(\xi))$$

where $\partial_j L_d$ denotes the partial derivative of L_d with respect to its jth slot.

Example 3. A Fourier series ansatz for f reveals that the discrete wave Eq. (6) with potential $V(u) = \frac{1}{2}u^2$ away from resonant cases TWs are $u(t, x) = f(x - c_n t)$ with

$$f(\xi) = \alpha \sin(\kappa_n \xi) + \beta \sin(\kappa_n \xi), \quad \kappa_n = \frac{2\pi n}{l}, \ n \in \mathbb{Z}, \alpha, \beta \in \mathbb{R} \tag{13}$$

and with wave speed c_n a real solution of

$$\cos(\kappa_n c_n \Delta t) = 1 - \frac{\Delta t^2}{2} + \frac{\Delta t^2}{\Delta x^2}(\cos(\kappa_n \Delta x)) - 1). \tag{14}$$

A contour plot for $n = 1$ is shown to the left of Fig. 2.

Remark 6. The TW Eq. (12) inherits variational structure from the underlying PDE: an application of Palais' principle of criticality [16] of the action of $(\mathbb{R}, +)$ on the Sobolev space $H^1(X_c, \mathbb{R})$ with $X_c = [0, l/c] \times [0, l]$ defined by $(s.u)(t, x) := u(t+s, x+cs)$ to the functional $S(u) = \int_{X_c} L_d(u(t, x), u(t+\Delta t, x), u(t, x+\Delta x))\mathrm{d}t$ reveals that (12) is governed by a formal 1st order variational principle. This is investigated more closely in [13].

To identify TWs in a machine learned model of a discrete field theory, we make an ansatz of a discrete Fourier series $f(\xi) = \sum_{m=-\frac{M-1}{2}}^{-\frac{M}{2}} \hat{f}_m \exp(m\frac{2\pi i}{l}\xi)$, where bounds of the sum are rounded such that we have M summands. To locate a TW, the loss function $\ell_{\mathrm{TW}}(c, \hat{f}) + \ell_{\mathrm{TW}}^{\mathrm{reg}}(c, \hat{f})$ is minimised with

$$\ell_{\mathrm{TW}}(c, \hat{f}) = \sum_{i=1}^{N-1} \sum_{j=0}^{M-1} \|\mathrm{DEL}_j^i(U)\|^2, \qquad U = \left(f(i\Delta t - cj\Delta x)\right)_{0 \le i \le N}^{0 \le j \le M-1} \tag{15}$$

and regularisation $\ell_{\mathrm{TW}}^{\mathrm{reg}} = \exp(-100\|U\|_{l^2}^2)$ with discrete l^2-norm $\|\cdot\|_{l^2}$ to avoid trivial solutions. Here $\hat{f} = (\hat{f}_m)_m$.

5 Experiment

Creation of Training Data. We use the space-time domain X (Sect. 2) with $T = 0.5$, $l = 1$, $\Delta x = 0.05$, $\Delta t = 0.025$. To obtain training data that behaves like discretised smooth functions, we compute $K = 80$ solutions to the discrete wave equation (Example 2) with potential $V(u) = \frac{1}{2}u$ on the mesh X_Δ from initial data $\boldsymbol{u}^0 = (u_j^0)_{0 \le j \le M-1}$ and $\boldsymbol{u}^1 = (u_j^1)_{0 \le j \le M-1}$. To obtain $\boldsymbol{u}^0$ we sample r values from a standard normal distribution. Here r is the dimension of the output of a real discrete Fourier transformation of an M-dimensional vector. These are weighted by the function $m \mapsto M \exp(-2j^4)$, where $m = 0, \ldots, r-1$ is the frequency number. The vector $\boldsymbol{u}^0$ is then obtained as the inverse real discrete Fourier transform of the weighted frequencies. To obtain $\boldsymbol{u}^1$ an initial velocity field $\boldsymbol{v}^0 = (v_j^0)_{0 \le j \le M-1}$ is sampled from a standard normal distribution. Then we proceed as in a variational discretisation scheme [11] applied to the Lagrangian density L of the continuous wave equation (Example 1): to compute conjugate momenta we set $L_\Sigma(\boldsymbol{u}, \boldsymbol{v}) = \sum_{j=0}^{M-1} \Delta x L(u_j, v_j)$ and compute $\boldsymbol{p}^0 = \frac{\partial L_\Sigma}{\partial v^0}(\boldsymbol{u}^0, \boldsymbol{v}^0)$. Then $\boldsymbol{p}^0 = -L_\Sigma(\boldsymbol{u}^1, (\boldsymbol{u}^1 - \boldsymbol{u}^0)/\Delta t)$ is solved for $\boldsymbol{u}^1$. A plot of an element of the training data set is displayed in Fig. 1. (TWs are *not* part of the training data.)

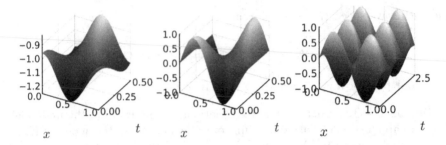

Fig. 1. Left: Element of training data set. Centre: Predicted solution to unseen initial values. Right: Continued solution from centre plot outside training domain

Training and Evaluation. A discrete Lagrangian $L_d \colon \mathbb{R} \times \mathbb{R} \times \mathbb{R} \to \mathbb{R}$ is modelled as a three layer feed-forward neural network, where the interior layer has 10 nodes (160 parameters in total). It is trained on the aforementioned training data set and loss function $\ell = \ell_{\mathrm{DEL}} + \ell_{\mathrm{reg}}$ using the optimiser adam. We perform 1320 epochs of batch training with batch size 10. For the trained model we have $\ell_{\mathrm{DEL}} \approx 8.6 \cdot 10^{-8}$ and $\ell_{\mathrm{reg}} \approx 1.4 \cdot 10^{-7}$. To evaluate the performance of the trained model for L_d, we compute solutions to initial data by forward propagation (Remark 4) and compare with solutions to the discrete wave equation (Example 2). For initial data u^0, u^1 *not* seen during training, the model recovers the exact solution up to an absolute error $\|U - U_{\mathrm{ref}}\|_\infty < 0.012$ on X_Δ and up to $\|U - U_{\mathrm{ref}}\|_\infty < 0.043$ on an extended grid with $T_{\mathrm{ext}} = 2.5$ (Fig. 1).

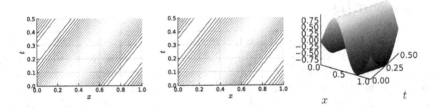

Fig. 2. Left: Reference TW. Centre and Right: Identified TW in learned model

We have $\max_{i,j} \|\mathrm{DEL}(L_d)_{i,j}(U_{\mathrm{ref}}^{\mathrm{TW}})\|^2 < 0.004$, where $U_{\mathrm{ref}}^{\mathrm{TW}}$ is the TW from Example 3 ($n = 1$). This shows that the exact TW is a solution of the learned discrete field theory. This is remarkable since TWs are not part of the training data. However, Remark 6 hints that the ansatz of an autonomous L_d favours TWs as it contains the right symmetries. Using the method of Sect. 4, a TW U^{TW} and speed c can be found numerically: with $(c_1, U_{\mathrm{ref}}^{\mathrm{TW}})$ as an initial guess with normally distributed random noise ($\sigma = 0.5$) added to the Fourier coefficients of $U_{\mathrm{ref}}^{\mathrm{TW}}$ and to c_1, we find U^{TW} and c for the learned L_d with errors $\|U^{\mathrm{TW}} - U_{\mathrm{ref}}^{\mathrm{TW}}\|_\infty < 0.12$ and $|c - c_1| < 0.001$ (using 10^4 epochs of adam) (Fig. 2).

6 Conclusion and Future Work

We present an approach to learn models of dynamical systems that are governed by an (a priori unknown) variational PDE from data. This is done by learning a neural-network model of a discrete Lagrangian such that the discrete Euler–Lagrange equations (DELs) are fulfilled on the training data. As DELs are local, the model can be efficiently trained and used in simulations. Even though the underlying system is infinite dimensional, model order reduction is not required. It would be interesting to relate the implicit locality assumption of our data-driven model to the more widely used approach to fit a dynamical system on a low-dimensional latent space that is identified using model order reduction techniques [1, 3, 4, 12, 19].

Our approach fits in the context of physics-informed machine learning because the data-driven model has (discrete) variational structure by design. However, our model is numerical analysis-informed as well: since our model is discrete by design, it can be used in simulations without an additional discretisation step. Based on an analysis of Newton's method when used to solve DELs, we develop a regulariser that rewards numerical regularity of the model. The regulariser is employed during the training phase. It plays a crucial role to obtain a non-degenerate discrete Lagrangian.

Our work provides a proof of concept illustrated on the wave equation. It is partly tailored to the hyperbolic character of the underlying PDE. It is future work to adapt this approach to dynamical systems of fundamentally different character (such as parabolic or elliptic behaviour) by employing discrete Lagrangians and regularisers that are adapted to the information flow within such PDEs.

Finally, we clarify the notion of travelling waves (TWs) in discrete models and show how to locate TWs in data-driven models numerically. Indeed, in our numerical experiment the data-driven model contains the correct TWs even though the training data does not contain any TWs. In future work it would be interesting to develop techniques to identify more general highly symmetric solutions in data-driven models and use them to evaluate qualitative aspects of learned models of dynamical systems.

Source Code. https://github.com/Christian-Offen/LagrangianDensityML

Acknowledgements. C. Offen acknowledges the Ministerium für Kultur und Wissenschaft des Landes Nordrhein-Westfalen and computing time provided by the Paderborn Center for Parallel Computing (PC2).

References

1. Allen-Blanchette, C., Veer, S., Majumdar, A., Leonard, N.E.: LagNetViP: a Lagrangian neural network for video prediction. In: AAAI 2020 Symposium on Physics Guided AI (2020). https://doi.org/10.48550/ARXIV.2010.12932

2. Bertalan, T., Dietrich, F., Mezić, I., Kevrekidis, I.G.: On learning Hamiltonian systems from data. Chaos Interdisc. J. Nonlinear Sci. **29**(12), 121107 (2019). https://doi.org/10.1063/1.5128231

3. Buchfink, P., Glas, S., Haasdonk, B.: Symplectic model reduction of Hamiltonian systems on nonlinear manifolds and approximation with weakly symplectic autoencoder. SIAM J. Sci. Comput. **45**(2), A289–A311 (2023). https://doi.org/10.1137/21M1466657

4. Carlberg, K., Tuminaro, R., Boggs, P.: Preserving Lagrangian structure in nonlinear model reduction with application to structural dynamics. SIAM J. Sci. Comput. **37**(2), B153–B184 (2015). https://doi.org/10.1137/140959602

5. Cranmer, M., Greydanus, S., Hoyer, S., Battaglia, P., Spergel, D., Ho, S.: Lagrangian neural networks (2020). https://doi.org/10.48550/ARXIV.2003.04630

6. Deuflhard, P., Hohmann, A.: Numerical Analysis in Modern Scientific Computing. Springer, New York (2003). https://doi.org/10.1007/978-0-387-21584-6

7. Dierkes, E., Flaßkamp, K.: Learning Hamiltonian systems considering system symmetries in neural networks. IFAC-PapersOnLine **54**(19), 210–216 (2021). https://doi.org/10.1016/j.ifacol.2021.11.080

8. Dierkes, E., Offen, C., Ober-Blöbaum, S., Flaßkamp, K.: Hamiltonian neural networks with automatic symmetry detection (to appear). Chaos **33** (2023). https://doi.org/10.1063/5.0142969

9. Greydanus, S., Dzamba, M., Yosinski, J.: Hamiltonian Neural Networks. In: Wallach, H., Larochelle, H., Beygelzimer, A., d'Alché Buc, F., Fox, E., Garnett, R. (eds.) Advances in Neural Information Processing Systems, vol. 32. Curran Associates, Inc. (2019), https://proceedings.neurips.cc/paper/2019/file/26cd8ecadce0d4efd6cc8a8725cbd1f8-Paper.pdf

10. Lishkova, Y., et al.: Discrete Lagrangian neural networks with automatic symmetry discovery. In: Accepted Contribution to 22nd World Congress of the International Federation of Automatic Control, Yokohama, Japan, 9–14 July 2023. IFAC-PapersOnLine (2023). https://doi.org/10.48550/ARXIV.2211.10830

11. Marsden, J.E., West, M.: Discrete mechanics and variational integrators. Acta Numerica **10**, 357–514 (2001). https://doi.org/10.1017/S096249290100006X

12. Mason, J., Allen-Blanchette, C., Zolman, N., Davison, E., Leonard, N.: Learning interpretable dynamics from images of a freely rotating 3D rigid body (2022). https://doi.org/10.48550/ARXIV.2209.11355

13. McLachlan, R.I., Offen, C.: Backward error analysis for variational discretisations of pdes. J. Geo. Mech. **14**(3), 447–471 (2022). https://doi.org/10.3934/jgm.2022014

14. Ober-Blöbaum, S., Offen, C.: Variational learning of Euler-Lagrange dynamics from data. J. Comput. Appl. Math. **421**, 114780 (2023). https://doi.org/10.1016/j.cam.2022.114780

15. Offen, C., Ober-Blöbaum, S.: Symplectic integration of learned Hamiltonian systems. Chaos Interdisc. J. Nonlinear Sci. **32**(1), 013122 (2022). https://doi.org/10.1063/5.0065913

16. Palais, R.S.: The principle of symmetric criticality. Comm. Math. Phys. **69**(1), 19–30 (1979). https://projecteuclid.org:443/euclid.cmp/1103905401

17. Qin, H.: Machine learning and serving of discrete field theories. Sci. Rep. **10**(1) (2020). https://doi.org/10.1038/s41598-020-76301-0

18. Ridderbusch, S., Offen, C., Ober-Blobaum, S., Goulart, P.: Learning ODE models with qualitative structure using Gaussian Processes. In: 2021 60th IEEE Conference on Decision and Control (CDC). IEEE (2021). https://doi.org/10.1109/cdc45484.2021.9683426

19. Sharma, H., Kramer, B.: Preserving Lagrangian structure in data-driven reduced-order modeling of large-scale dynamical systems (2022). https://doi.org/10.48550/ARXIV.2203.06361

Stochastic Geometric Mechanics

Gauge Transformations in Stochastic Geometric Mechanics

Qiao Huang[1]([✉])[iD] and Jean-Claude Zambrini[2][iD]

[1] Division of Mathematical Sciences, School of Physical and Mathematical Sciences,
Nanyang Technological University, 21 Nanyang Link,
Singapore 637371, Singapore
qiao.huang@ntu.edu.sg
[2] Group of Mathematical Physics (GFMUL), Department of Mathematics,
Faculty of Sciences, University of Lisbon, Campo Grande, Edifício C6,
1749-016 Lisboa, Portugal
jczambrini@fc.ul.pt

Abstract. In the framework of the dynamical solution of Schrödinger's 1931 problem, we compare key aspects of its Lagrangian and Hamiltonian formalisms. This theory is regarded as a stochastic regularization of classical mechanics, in analogy with Feynman's (informal) path integral approach to quantum mechanics. The role of our counterpart of quantum gauge invariance, in the above stochastic framework, will be described. It establishes, in particular, new dynamical relations between classes of diffusion processes, when the theory is restricted to a natural generalization of Schrödinger's original problem, and illustrates differences between the stochastic Lagrangian and Hamiltonian formulations.

Keywords: Gauge transformations · Canonical transformations · Stochastic geometric mechanics

1 Schrödinger's 1931 Problem in a Nutshell

Let $\mathcal{H} = -\frac{\hbar^2}{2}\Delta + V$ be the n-dimensional (self-adjoint) Hamiltonian operator associated, in classical statistical physics, with the motion of diffusive particles in the potential field V, where $\hbar$ is a positive constant. At initial time t_0, their probability distribution $\rho_{t_0}(q)dq$ is known. Assume that, later on (at time t_1), another arbitrary probability $\rho_{t_1}(q)dq$ is observed. Schrödinger [14] asked what the most likely (stochastic) evolution $\rho_t(q)dq$ (or its associated process X_t) is in-between ($t_0 \leq t \leq t_1$)? The solution of a (slightly generalized) version of this problem can be summarized as the following (cf. [1,3,15], [2, Part II]):

Proposition 1. *Let V be continuous and bounded below. Then if ρ_{t_0} and ρ_{t_1} are strictly positive on $\mathbb{R}^n$, there is an unique diffusion X solving Schrödinger's problem. It solves two Itô's stochastic differential equations, $t_0 \leq t \leq t_1$,*

$$dX_t = \hbar\nabla(\log\eta)(X_t, t)dt + \sqrt{\hbar}dW_t,$$
$$d_*X_t = -\hbar\nabla(\log\eta^*)(X_t, t)dt + \sqrt{\hbar}d_*W_t^*,$$

© The Author(s), under exclusive license to Springer Nature Switzerland AG 2023
F. Nielsen and F. Barbaresco (Eds.): GSI 2023, LNCS 14071, pp. 583–591, 2023.
https://doi.org/10.1007/978-3-031-38271-0_58

where W and W^* are (forward and backward) Wiener processes w.r.t. increasing filtration $\{\mathcal{P}_t : t \geq t_0\}$ and decreasing $\{\mathcal{F}_t : t \leq t_1\}$ respectively, η and η^* are positive solutions of

$$\hbar\frac{\partial \eta}{\partial t} = \mathcal{H}\eta, \quad t \leq t_1, \tag{1}$$

and

$$-\hbar\frac{\partial \eta^*}{\partial t} = \mathcal{H}\eta^*, \quad t \geq t_0, \tag{2}$$

such that $\rho_t(q)dq = (\eta\eta^*)(q,t)dq$ is integrable and is the probability distribution of X_t, for $t_0 \leq t \leq t_1$.

The "Euclidean" transformation $t \to -it$ changes the first parabolic equation into Schrödinger's equation and the second into its complex conjugate. Since η_{t_1}, $\eta_{t_0}^*$ are unrelated, the density ρ_t becomes an L^2 scalar product. This was the whole point of Schrödinger in 1931. Notice that he was referring to an unorthodox stochastic boundary value problem. We are interested in the geometric dynamics of processes solving such problems.

2   Classical (Local) Gauge Symmetry

Let us start with a reminder of classical mechanical systems. Let M be an n-dimensional configuration manifold. Poincaré 1-form on the cotangent bundle T^*M is $\omega_0 = p_i dq^i$, where (q^i) are local coordinates on M and (p_i) parametrize the basis $(dq^1, \cdots, dq^n)$ of the fiber T_q^*M dual to the one $(\frac{\partial}{\partial q^1}, \cdots, \frac{\partial}{\partial q^n})$ of the tangent space T_qM.

Consider a Lagrangian L_0 on TM. If the Legendre transformation $TM \to T^*M$, $(q, \dot{q}) \mapsto (q, p = \frac{\partial L_0}{\partial \dot{q}})$ is diffeomorphic (in this case the Lagrangian is called hyperregular), the actions in phase space and configuration space coincide:

$$S = \int_{\{q(t): t_0 \leq t \leq t_1\}} \omega_0 - H_0 dt = \int_{t_0}^{t_1} L_0(q(t), \dot{q}(t))dt,$$

where $H_0(q, p) = p\dot{q} - L_0(q, \dot{q})$ is the associated Hamiltonian, for $\dot{q}$ solvable by $p = \frac{\partial L_0}{\partial \dot{q}}(q, \dot{q})$. Then, the following Euler-Lagrange (EL) and Hamilton's canonical equations are equivalent:

$$\frac{d}{dt}\left(\frac{\partial L_0}{\partial \dot{q}}\right) - \frac{\partial L_0}{\partial q} = 0,$$

and

$$\dot{q} = \frac{\partial H_0}{\partial p}, \quad \dot{p} = -\frac{\partial H_0}{\partial q}.$$

The respective notions of states are such that those equations can be uniquely solved, but the Lagrangian (Hamilton's) variational principle for S suggests boundary value problems more tricky than Hamiltonian Cauchy ones.

Given a smooth scalar function F on $M \times \mathbb{R}$, we define a new Lagrangian:

$$\tilde{L}_0(q, \dot{q}, t) = L_0(q, \dot{q}) + \frac{d}{dt}[F(q, t)] = L_0(q, \dot{q}) + \frac{\partial F}{\partial t} + \dot{q}\frac{\partial F}{\partial q}. \tag{3}$$

By Hamilton's principle for fixed boundary conditions $q(t_0), q(t_1)$, the action S turns into

$$\tilde{S} = S + F(q(t_1), t_1) - F(q(t_0), t_0).$$

Namely the transformation (3) leaves the Lagrangian formalism "almost" invariant, by affecting only the boundary conditions of the Euler-Lagrange equation. Such a transformation is called gauge transformation. On the other hand, the new conjugate momentum is, by Legendre transform, $\tilde{p} = \frac{\partial \tilde{L}_0}{\partial \dot{q}} = p + \frac{\partial F}{\partial q}$. The corresponding new Hamiltonian $\tilde{H}_0$ is

$$\tilde{H}_0(q, \tilde{p}, t) = \tilde{p}\dot{q} - \tilde{L}_0 = H_0(q, p) - \frac{\partial F}{\partial t}.$$

Clearly, the change of coordinates $(q, p) \to (q, \tilde{p})$ can also be induced by the type two generating function $G_2(q, p, t) = q\tilde{p} + F(q, t)$ [6]. Thus, gauge transformations form a subset of canonical transformations.

3 Quantum and Euclidean Quantum Counterparts

The quantization of the above classical system starts with the one of its state, now ψ, a ray in the complex Hilbert space $L^2(M)$, equipped with scalar product $\langle \cdot | \cdot \rangle_2$. Each classical observable a is associated with a densely defined self-adjoint operator A. For instance, the quantum counterpart of the classical Hamiltonian $H_0(q, p) = \frac{1}{2}|p|^2 + V(q)$ becomes

$$\mathcal{H}(Q, P) = \frac{1}{2}P^2 + V(Q) = -\frac{\hbar^2}{2}\Delta + V(q),$$

with $P = -i\hbar\nabla$ the momentum operator. The state dynamics is given by Schrödinger's equation

$$i\hbar\frac{\partial \psi}{\partial t} = \mathcal{H}\psi. \tag{4}$$

By Born's "probabilistic" interpretation, the modulus squared of the amplitude of ψ represents a probability density ρ, i.e.,

$$\rho(q, t) = |\psi(q, t)|^2. \tag{5}$$

What is a local gauge symmetry in this context? Define a (time-dependent) unitary transformation by [7]

$$\psi \mapsto \tilde{\psi} = U_t\psi, \quad U_t\psi(q) := e^{\frac{i}{\hbar}F(q, t)}\psi(q)$$

for a given smooth $F : M \times \mathbb{R} \to \mathbb{R}$. This transformation will give the same probability, since $|\psi(q,t)|^2 = |\tilde{\psi}(q,t)|^2$. Now $\tilde{\psi}$ solves

$$i\hbar\frac{\partial\tilde{\psi}}{\partial t} = -\frac{\hbar^2}{2}\left(\nabla - \frac{i}{\hbar}\nabla F\right)^2\tilde{\psi} + \left(V - \frac{\partial F}{\partial t}\right)\tilde{\psi},$$

i.e., the new (quantum) Hamiltonian is $\tilde{\mathcal{H}}(Q,\tilde{P},t) = \frac{1}{2}\tilde{P}^2 + V(Q) - \frac{\partial F}{\partial t}(Q)$ with new momentum operator $\tilde{P} = P - \nabla F$. The expectation of any quantum observable A is indeed invariant, since $\langle\tilde{\psi}|\tilde{A}\tilde{\psi}\rangle_2 = \langle\psi|A\psi\rangle_2$, where $\tilde{A} = U_t A U_t^{-1}$.

Born and Feynman suggested that the density ρ in (5), for the Hamiltonian $\mathcal{H}$, should be the probability density of a certain diffusion X, i.e. $\mathbf{P}(X_t \in dq) = \rho(q,t)dq$. A fundamental problem is that X cannot be a diffusion like the one suggested by Feynman. All rigorous quantum probabilistic results, reasonably consistent with regular quantum theory as a whole, are "Euclidean", i.e., follow from $t \to \pm it$ in Schrödinger's Eq. (4) and consider instead two heat equations (1) or (2). Proposition 1 shows that there exist well-defined diffusion processes, as solutions of forward and backward Itô's SDEs, solving Schrödinger's 1931 "statistical" problem.

The classical but Euclidean Hamiltonian for (1) and (2) has a sign opposite to the quantum case, namely,

$$H_0(q,p) = \frac{1}{2}|p|^2 - V(q). \tag{6}$$

For the two heat equations, we have an Euclidean counterpart of the gauge invariance:

$$\eta \mapsto \tilde{\eta} := e^{\frac{1}{\hbar}F}\eta, \quad \eta^* \mapsto \tilde{\eta}^* := e^{-\frac{1}{\hbar}F}\eta^*, \tag{7}$$

for a smooth F. This does not change the distribution of X_t since $\eta\eta^* = \tilde{\eta}\tilde{\eta}^*$. Then $\tilde{\eta}$ solves

$$\hbar\frac{\partial\tilde{\eta}}{\partial t} = -\frac{\hbar^2}{2}\left(\nabla - \frac{1}{\hbar}\nabla F\right)^2\tilde{\eta} + \left(V + \frac{\partial F}{\partial t}\right)\tilde{\eta} = \tilde{\mathcal{H}}\tilde{\eta}.$$

and analogously for $\tilde{\eta}^*$.

For instance, the one-dimensional harmonic oscillators, for classical and Euclidean cases, have potentials of the quadratic form $V(q) = \frac{\beta^2}{2}q^2$ with constant β. The quantum harmonic oscillator bears the Hamiltonian operator $\mathcal{H} = -\frac{\hbar^2}{2}\frac{d^2}{dq^2} + \frac{\beta^2}{2}q^2$ as the quantization of $H_0(q,p) = \frac{1}{2}|p|^2 + \frac{\beta^2}{2}q^2$, while the Euclidean classical counterpart becomes

$$H_0(q,p) = \frac{1}{2}|p|^2 - \frac{\beta^2}{2}q^2. \tag{8}$$

4   Stochastic Geometric Mechanics

In contrast with deterministic smooth trajectories describing the evolution of classical Lagrangian and Hamiltonian systems, diffusion processes, solutions of

Ito's SDEs have more degrees of freedom both probabilistically and geometrically. According to Schwartz and Meyer [4], the "second-order" counterparts of classical geometric structures arise to unveil such probabilistic content.

The second-order extension of the classical phase space T^*M is denoted by $T^{S*}M$, called the second-order cotangent bundle [4]. In addition to the classical coordinates (q,p), the new freedoms in $T^{S*}M$ are represented by coordinates (o_{ij}) which indicate the stochastic deformation. Now the second-order analog of Poincaré form is $\omega = p_i d^2 q^i + \frac{1}{2} o_{ij} dq^i \cdot dq^j$ defined on $T^{S*}M$, where $(d^2 q^i, dq^j \cdot dq^k : 1 \leq i \leq n, 1 \leq j \leq k \leq n)$ form a basis of $T^{S*}M$.

For a second-order Hamiltonian function $H(q,p,o)$ on the extended phase space $T^{S*}M$, the stochastic Hamiltonian equations hold [9]:

$$
\begin{cases}
D^i X = \dfrac{\partial H}{\partial p_i}, \quad Q^{ij} X = 2 \dfrac{\partial H}{\partial o_{ij}}, \\[2mm]
D_i p = -\dfrac{\partial H}{\partial q^i}, \quad o_{ij} = \dfrac{\partial p_i}{\partial q^j} = \dfrac{\partial p_j}{\partial q^i},
\end{cases}
\tag{9}
$$

with unknowns $X = (X_t)$ a diffusion process on M, and (p,o) a time-dependent "section" of $T^{S*}M$, i.e., $p_i = p_i(t,q)$, $o_{ij} = o_{ij}(t,q)$. Here, D and Q are the so called mean derivative and quadratic mean derivative, respectively, which evaluate the drift and diffusive tensor of a diffusion process via limits of conditional expectations. Note that the last equation of (9) implies that the second-order Poincaré form ω is the "differential" of the classical ω_0, i.e., $\omega = dp = d\omega_0$, where d is the differential operator that maps 1-forms to second-order forms [4]. Also observe that Eqs. (9) are invariant under the canonical change of coordinates on $T^{S*}M$ induced by a change of coordinates on M [9, Section 6.2]. Indeed, for example, $(D^i X)$ does not transfrom as a vector because its transformation formula involves second-order term, nor does $(\frac{\partial H}{\partial p_i})$ since the transformation of (o_{ij}) involves (p_i) [9, Lemma 5.2].

When M is equipped with a Riemannian metric g and the associated Levi-Civita connection ∇, a second-order Hamiltonian can be constructed from the classical H_0 by

$$
H_\hbar(q,p,o) = H_0(q,p) + \frac{\hbar}{2} g^{ij}(q)(o_{ij} - \Gamma_{ij}^k p_k).
$$

In this case, the solution of (9) satisfies

$$
Q^{ij} X = \hbar g^{ij}(X),
\tag{10}
$$

i.e., the diffusive tensor coincides with the Riemannian tensor. If the Legendre transform is diffeomorphic, H_0 is associated to the Lagrangian $L_0(q, D_\nabla q) = p D_\nabla q - H_0(q,p)$, and the stochastic Hamiltonian Eqs. (9) are equivalent to the stochastic Euler-Lagrange equation:

$$
\frac{\mathbf{D}}{dt}[(d_{\dot{q}} L_0)(X_t, D_\nabla X_t)] = (d_q L_0)(X_t, D_\nabla X_t),
\tag{11}
$$

where D_∇ is the ∇-compensation of the mean derivative D so that D_∇ takes values as vector fields, $\frac{\overline{D}}{dt}$ is the "damped" mean covariant derivative $\frac{\partial}{\partial t} + \nabla_{D_\nabla X} +$ $\frac{\hbar}{2}\Delta_{LD}$ with Δ_{LD} the Laplace-de Rahm. The stochastic Euler-Lagrange Eq. (11) also solves the stochastic least action principle under conditional expectation $\mathbf{E}_{t,q} = \mathbf{E}(\cdot|X_t = q)$: $\delta\,\mathcal{S} = 0$ with

$$\mathcal{S} = \mathbf{E}_{t,q} \int_t^{t_1} L_0(X_s, D_\nabla X_s)ds$$
$$= \mathbf{E}_{t,q} \int_t^{t_1} p \circ dX_s - H_\hbar ds = \mathbf{E}_{t,q} \int_{\{X_s : t \le s \le t_1\}} \omega - H_\hbar ds, \qquad (12)$$

subject to the constraint (10), where $\circ d$ is Stratonovich's symmetric stochastic differential, and the integral of ω along X is understood as the stochastic line integral [4, Definition 7.3]. These are parts of the framework of stochastic geometric mechanics developed in [9]. Readers may also refer to the proceeding [8] for a short summary.

Given a smooth scalar function F on $M \times \mathbb{R}$, we define a new Lagrangian:

$$\tilde{L}_0(q, D_\nabla q, t) = L_0(q, D_\nabla q) + \mathbf{D}_t[F(q,t)]$$
$$= L_0(q, D_\nabla q) + \frac{\partial F}{\partial t} + D_\nabla^i q \frac{\partial F}{\partial q^i} + \frac{\hbar}{2} g^{ij} \left(\frac{\partial^2 F}{\partial q^i \partial q^j} - \Gamma_{ij}^k \frac{\partial F}{\partial q^k} \right). \qquad (13)$$

By Dynkin's formula, the stochastic action $\mathcal{S}$ turns into $\tilde{\mathcal{S}} = \mathcal{S} + \mathbf{E}_{t,q}[F(X_{t_1}, t_1)] - F(q,t)$. The new conjugate momentum is

$$\tilde{p} = \frac{\partial \tilde{L}_0}{\partial D_\nabla q} = p + \frac{\partial F}{\partial q} \qquad (14)$$

and the new stochastic deformation variable is $\tilde{o}_{ij} = o_{ij} + \frac{\partial^2 F}{\partial q^i \partial q^j}$. The corresponding new second-order Hamiltonian $\tilde{H}$ is

$$\tilde{H}(q, \tilde{p}, \tilde{o}, t) = \tilde{p}D_\nabla q - \tilde{L}_0 + \frac{\hbar}{2} g^{ij}(\tilde{o}_{ij} - \Gamma_{ij}^k \tilde{p}_k) = H(q, p, o) - \frac{\partial F}{\partial t}.$$

Euclidean Quantum Correspondence

Let us go back to the Hamiltonian $H_0(q, p) = \frac{1}{2}|p|^2 - V(q)$ associated with the Euclidean quantum case (6). For this case, the optimal drift of (12) is [3]

$$D_\nabla X_t = \hbar\nabla(\log\eta)(X_t, t),$$

where $\eta > 0$ solves the "retrograde" heat Eq. (1). Thus, the solution of (9) or (11) is just Schrödinger's process of Proposition 1. The Legendre transform indicates that the momentum of (9) is $p = \hbar d(\log\eta)$, where d is de Rham's exterior differential on M. Hence, after the Euclidean quantum gauge transformation (7), the new momentum

$$\tilde{p} = \hbar d(\log\tilde{\eta}) = p + dF,$$

coincides with (14).

In fact, the Cole-Hopf transformation $G = -\hbar \log \eta$, known in PDE theories, yields a second-order Hamiltonian-Jacobi-Bellman (HJB) equation, which plays a central role in the second-order counterpart of the classical Hamiltonian-Jacobi theory [9]. Moreover, as indicated by Schrödinger's problem, there is naturally a time-reversed formulation for the whole stochastic geometric mechanics, where the optimal (backward) drift is $D_\nabla^* X_t = -\hbar \nabla \log \eta^*(X_t, t)$, with η^* solving the heat Eq. (2) and D_∇^* the backward version of the compensated mean derivative D_∇. This also indicates a time-symmetric discretization in the stochastic action (12), as follows,

$$\mathcal{S} = \mathbf{E}_{t,q} \int_t^{t_1} \left[p \frac{D_\nabla X_s + D_\nabla^* X_s}{2} - H_\hbar \right] ds,$$

Transformations from Euclidean Harmonic Oscillator to Free Schrödinger's Motion

The Euclidean harmonic oscillator on the flat real line $\mathbb{R}$ is described by the Euclidean Hamiltonian (8) with potential $V(q) = \frac{\beta^2}{2} q^2$. Its second-order Hamiltonian is

$$H_\hbar(q, p, o) = H_0(q, p) + \frac{\hbar}{2} o = \frac{1}{2} |p|^2 - \frac{\beta^2}{2} q^2 + \frac{\hbar}{2} o$$

and Lagrangian $L_0(q, D_\nabla q) = \frac{1}{2} |D_\nabla q|^2 + \frac{\beta^2}{2} q^2$. The equations of motion on the configuration space are $DDX = \beta^2 X$, $QX = \hbar$. Consider the gauge transformation with $F(q, t) = \frac{\beta}{2}(q^2 - \hbar t)$. The new Lagrangian, by (13), becomes

$$\tilde{L}_0(q, D_\nabla q) = \frac{1}{2} |D_\nabla q + \beta q|^2.$$

and the new Hamiltonian

$$\tilde{H}_\hbar(q, \tilde{p}, \tilde{o}) = \tilde{H}_0(q, \tilde{p}) + \frac{\hbar}{2} \tilde{o} = \frac{1}{2} |\tilde{p}|^2 - \beta q \tilde{p} + \frac{\hbar}{2} \tilde{o},$$

with new conjugate momentum and stochastic deformation variable, respectively,

$$\tilde{p} = p + \beta q, \quad \tilde{o} = o + \beta.$$

The gauge transformation does not change the equations of motion on configuration space.

Next, we consider the following time-dependent canonical transformation [9]

$$\hat{q} = e^{\beta t} q, \quad \hat{p} = e^{-\beta t} \tilde{p}, \quad \hat{t} = \frac{1}{2\beta}(e^{2\beta t} - 1). \tag{15}$$

Clearly, the change of coordinates $(q, \tilde{p}) \mapsto (\hat{q}, \hat{p})$ is induced by the type three generating function $G_3(\hat{q}, \tilde{p}, t) = -e^{-\beta t} \hat{q} \tilde{p}$ via relations $q = -\frac{\partial G_3}{\partial \tilde{p}}$ and $\hat{p} = -\frac{\partial G_3}{\partial \hat{q}}$ [6]. The relation between the latent coordinates o and O is

$$\hat{o} = \frac{\partial \hat{p}}{\partial \hat{q}} = e^{-\beta t} \frac{\partial \tilde{p}}{\partial q} \frac{\partial q}{\partial \hat{q}} = e^{-2\beta t} \tilde{o}.$$

The new 2nd-order Hamiltonian $K_\hbar$ is given by the HJB equation $K_\hbar \frac{d\hat{t}}{dt} - \tilde{H}_\hbar = \frac{\partial G_3}{\partial t}$, i.e.,

$$K_\hbar(\hat{q}, \hat{p}, \hat{o}) = K_0(\hat{q}, \hat{p}) + \frac{\hbar}{2}\hat{o} = \frac{1}{2}|\hat{p}|^2 + \frac{\hbar}{2}\hat{o}.$$

The equations of motion for $K_\hbar$ on the configuration space are $DDY = 0$, $QY = \hbar$.

One can also carry out the above chain of transformations in terms of stochastic actions, denoting $f(q,t) = \mathbf{E}_{t,q}[F(X_{t_1}, t_1)] - F(q,t)$:

$$\mathcal{S} = \frac{1}{2}\mathbf{E}_{t,x} \int_t^{t_1} \left(|DX_s|^2 + \beta^2 X_s^2\right) ds$$

$$= \frac{1}{2}\mathbf{E}_{t,q} \int_t^{t_1} \left[\left(|DX_s|^2 + \beta^2 X_s^2\right) ds + 2dF(X_s, s)\right] - f(q,t)$$

$$= \frac{1}{2}\mathbf{E}_{t,q} \int_t^{t_1} |DX_s + \beta X_s|^2 ds - f(q,t)$$

$$= \frac{1}{2}\mathbf{E}_{\hat{t},\hat{q}} \int_{\hat{t}}^{t_1} |DY_{\hat{s}}|^2 d\hat{s} - f(q,t) = \hat{\mathcal{S}} - f(q,t),$$

since $DY_{\hat{s}} = e^{-\beta s}(DX_s + \beta X_s)$ and $d\hat{s} = e^{2\beta s} ds$.

Consequently, after a gauge transformation and a time-dependent canonical one, the Euclidean harmonic oscillator X is transformed into a free motion of Schrödinger Y. Note that the retrograde heat Eq. (1) for the Hamiltonians $\tilde{H}_0$ and K_0 are the Kolmogorov backward equations of Ornstein–Uhlenbeck (OU) process and free Brownian motion respectively. The relation between X and Y due to (15),

$$X_t = e^{-\beta t} Y_{\frac{1}{2\beta}(e^{2\beta t}-1)},$$

recovers, in particular, Doob's relation between OU process and Brownian motion. See also [12, Theorem 4.1.(2)].

5 Epilogue: Is Schrödinger's Problem Classical Statistical or Quantum Physics?

Clearly, what guided Schrödinger's 1931 cryptic observation at the end of [14] was his scepticism about the foundations of the recently discovered quantum mechanics. Back then, Brownian motion was already familiar but in Statistical Physics (mathematically constructed by N. Wiener in 1923–24), without conceptual difficulties regarding its probabilistic role, in contrast with Quantum theory, its mysterious probabilistic content and interpretation. With the advantage of a retroactive scientific vision, Schrödinger's problem can be seen as the first Euclidean approach to quantum probabilities, very influential in 1970s. But the diffusion solving his problem are also as close as possible to those used by R. Feynman in his Path Integral approach [5]. Those diffusions, known today as Bernstein, or reciprocal, processes for historical reasons [11,13], are generally time-dependent since the data of $\rho_{t_1}(q)$ is "arbitrary". This property makes them natural candidates for nonequilibrium Statistical Physics, as observed in [10]. It should be stressed that Schrödinger was interested in probability distribution of the product form given by Proposition 1, namely Markovian processes. General Bernstein processes are not Markovian, however, and should be relevant not only in nonequilibrium Quantum Statistical Physics, but also in Mass Transportation theory [13].

Acknowledgements. The second author is grateful for the support of FCT, Portugal project UIDB/00208/2020.

References

1. Albeverio, S., Yasue, K., Zambrini, J.C.: Euclidean quantum mechanics: analytical approach. In: Annales de l'IHP Physique théorique, vol. 50, pp. 259–308 (1989)
2. Chung, K., Zambrini, J.C.: Introduction to Random Time and Quantum Randomness, vol. 1. World Scientific Publishing, Singapore (2003)
3. Cruzeiro, A., Zambrini, J.C.: Malliavin calculus and Euclidean quantum mechanics. I. Functional calculus. J. Funct. Anal. **96**(1), 62–95 (1991)
4. Emery, M.: Stochastic Calculus in Manifolds. Springer, Heidelberg (1989). https://doi.org/10.1007/978-3-642-75051-9
5. Feynman, R.: Space-time approach to non-relativistic quantum mechanics. Rev. Mod. Phys. **20**(2), 367 (1948)
6. Goldstein, H.: Classical Mechanics, 2nd edn. Addison-Wesley Publishing Company, Boston (1980)
7. Hall, B.: Quantum Theory for Mathematicians, vol. 267. Springer, Heidelberg (2013). https://doi.org/10.1007/978-1-4614-7116-5
8. Huang, Q., Zambrini, J.C.: Hamilton-Jacobi-Bellman equations in stochastic geometric mechanics. Phys. Sci. Forum **5**(1), 37 (2022)
9. Huang, Q., Zambrini, J.C.: From second-order differential geometry to stochastic geometric mechanics. J. Nonlinear Sci. **33**, 67 (2023)
10. Huang, Q., Zambrini, J.C.: Stochastic geometric mechanics in nonequilibrium thermodynamics: Schrödinger meets Onsager. J. Phys. A: Math. Theor. **56**(13), 134003 (2023)
11. Léonard, C., Rœlly, S., Zambrini, J.C.: Reciprocal processes: a measure-theoretical point of view. Prob. Surv. **11**, 237–269 (2014)
12. Lescot, P., Zambrini, J.C.: Probabilistic deformation of contact geometry, diffusion processes and their quadratures. In: Seminar on Stochastic Analysis, Random Fields and Applications, vol. 59, pp. 203–226. Springer (2007). https://doi.org/10.1007/978-3-7643-8458-6_12
13. Mikami, T.: Stochastic Optimal Transportation: Stochastic Control with Fixed Marginals. Springer, Heidelberg (2021). https://doi.org/10.1007/978-981-16-1754-6
14. Schrödinger, E.: Sur la théorie relativiste de l'électron et l'interprétation de la mécanique quantique. In: Annales de l'institut Henri Poincaré, vol. 2, pp. 269–310 (1932)
15. Zambrini, J.C.: Variational processes and stochastic versions of mechanics. J. Math. Phys. **27**(9), 2307–2330 (1986)

Couplings of Brownian Motions on $SU(2, \mathbb{C})$

Magalie Bénéfice$^{(\boxtimes)}$, Marc Arnaudon, and Michel Bonnefont

Univ. Bordeaux, CNRS, Bordeaux INP, IMB, UMR 5251, 33400 Talence, France
magalie.benefice@math.u-bordeaux.fr

Abstract. We study couplings of Brownian motions on the subRiemannian manifold $SU(2, \mathbb{C})$, that is diffusion processes having the subLaplacian operator as infinitesimal generator. Using some interesting geometric interpretations of the Brownian motion on this space, we present some basic examples of co-adapted couplings. Inspiring ourselves with works on the Heisenberg group, we also construct two successful couplings on $SU(2, \mathbb{C})$: one co-adapted and one not co-adapted but having a good coupling rate. Finally, using the similar structure of $SU(2, \mathbb{C})$ and $SL(2, \mathbb{R})$ we generalise some of our results for the case of this second subRiemannian manifold.

Keywords: SubRiemannian manifolds · Brownian motion · Coupling · Successful coupling

1 Introduction

We consider μ and ν two probability laws on $M \times M$. We call coupling of μ and ν any measure π such that μ is its first marginal distribution and ν its second one. Coupling two Brownian motions $(\mathbb{B}_t)_t$ and $(\mathbb{B}'_t)_t$ consists in coupling their probability laws. This topic has been studied a lot on Riemannian manifolds in particular to obtain inequalities involving the heat semi-group just like Poincaré or Sobolev inequalities [6, 11].

In particular, there is a big interest in constructing successful couplings, that is couplings for which the first meeting time $\tau := inf\{t > 0 | \mathbb{B}_t = \mathbb{B}'_t\}$ of the Brownian motions is almost surely finite. The classical Liouville theorem says that every harmonic bounded function on $\mathbb{R}^2$ (and even on $\mathbb{R}^n$) is constant. As seen in [13], this stays true in every Riemannian manifold with Ricci curvature bounded bellow, if and only if there exists a successful coupling. An other use of successful couplings is the study of the total variation distance between the probability laws $\mathcal{L}(\mathbb{B}_t)$ and $\mathcal{L}(\mathbb{B}'_t)$ of two Brownian motions $\mathbb{B}_t$ and $\mathbb{B}'_t$ starting from different points:

$$d_{TV}(\mathcal{L}(\mathbb{B}_t), \mathcal{L}(\mathbb{B}'_t)) := \sup_{A \text{ measurable}} \{\mathbb{P}(\mathbb{B}_t \in A) - \mathbb{P}(\mathbb{B}'_t \in A)\}.$$

The Aldous inequality (also called Coupling inequality, see [1], chapter VII) says that for every coupling of Brownian motions $(\mathbb{B}_s, \mathbb{B}'_s)_s$ and every $t > 0$:

$$\mathbb{P}(\tau > t) \geq d_{TV}(\mathcal{L}(\mathbb{B}_t), \mathcal{L}(\mathbb{B}'_t)).$$

© The Author(s), under exclusive license to Springer Nature Switzerland AG 2023
F. Nielsen and F. Barbaresco (Eds.): GSI 2023, LNCS 14071, pp. 592–600, 2023.
https://doi.org/10.1007/978-3-031-38271-0_59

In particular, couplings that change this inequality into an equality are called maximal couplings. If it has been proved that such couplings always exist in the case of continuous processes on Polish spaces, they can be very difficult to study (simulation, estimation of a coupling rate) as their construction often needs some knowledge of the future of one of the Brownian motions. To avoid this difficulty one can try to work with co-adapted couplings. This just means that the two considered Brownian motions are adapted to the same filtration. Except in some cases, in $\mathbb{R}^n$ for example, co-adapted couplings are not maximal.

In this paper we study successful couplings on subRiemannian manifolds. In the case of the Heisenberg group $\mathbb{H}$, Kendall has proposed one successful co-adapted coupling [10] and another has been given by Banerjee, Gordina, Mariano [2]. This last one is non co-adapted but has a better coupling rate. Our goal is to use these two previous works to construct successful couplings on $SU(2,\mathbb{C})$ using the fact that its geometric structure has some similarities with the Heisenberg group. Note that the complete proofs of these results will appear in [7,8].

2 Geometric Interpretation of Brownian Motion on $SU(2,\mathbb{C})$

On the Heisenberg group, the Brownian motion can be seen as a two-dimensional real Brownian motion together with its swept area. We have a similar interpretation in the case of $SU(2,\mathbb{C})$. We first recall the subRiemannian structure of this manifold.

The $SU(2,\mathbb{C})$ group is the group of the unitary two-dimensional matrices with complex coefficients and determinant 1. Its group law is the one induced by the multiplication of matrices. As this law is smooth for the usual topology, $SU(2,\mathbb{C})$ is a Lie group. If we denote $X = \frac{1}{2}\begin{pmatrix} 0 & 1 \\ -1 & 0 \end{pmatrix}$, $Y = \frac{1}{2}\begin{pmatrix} 0 & i \\ i & 0 \end{pmatrix}$ and $Z = \frac{1}{2}\begin{pmatrix} i & 0 \\ 0 & -i \end{pmatrix}$ the Pauli matrices, (X,Y,Z) is a basis of the associated Lie algebra $\mathfrak{su}(2,\mathbb{C})$ satisfying

$$[X,Y] = Z \ , \ [Y,Z] = X \text{ and } [Z,X] = Y. \tag{1}$$

We note that every matrice of $SU(2,\mathbb{C})$ can be written on the form:

$$\exp\left(\varphi(\cos(\theta)X + \sin(\theta)Y)\right)\exp(zZ) = \begin{pmatrix} \cos\left(\frac{\varphi}{2}\right)e^{i\frac{z}{2}} & e^{i(\theta - \frac{z}{2})}\sin\left(\frac{\varphi}{2}\right) \\ -e^{-i(\theta - \frac{z}{2})}\sin\left(\frac{\varphi}{2}\right) & \cos\left(\frac{\varphi}{2}\right)e^{-i\frac{z}{2}} \end{pmatrix}$$

This way, we obtain a system of coordinates (φ, θ, z) of $SU(2,\mathbb{C})$ with $\varphi \in [0,\pi]$, $z \in] - 2\pi, 2\pi]$ and $\theta \in [0, 2\pi[$ called cylindrical coordinates. We denote $\bar{X}$, $\bar{Y}$ and $\bar{Z}$ the left-invariant vector fields associated to X, Y, Z. The horizontal plane is defined by $\mathcal{H} = Vect\langle \bar{X}, \bar{Y}\rangle$. Because of (1), the Hörmander condition is satisfied and the induced Carnot-Caratheodory distance is finite. Thus the subRiemannian structure is well defined. Let us remark that, using the Hopf fibration, we have a natural projection Π from our Lie group to the sphere S^2, sending (φ, θ, z) to the point of the sphere described by the spherical coordinates (φ, θ).

We define the subLaplacian operator by $L = \frac{1}{2}\left(\bar{X}^2 + \bar{Y}^2\right)$. Using the cylindrical coordinates we obtain:

$$L = \frac{1}{2}\left(\partial^2_{\varphi,\varphi} + \frac{1}{\sin^2(\varphi)}\partial^2_{\theta,\theta} + \tan^2\left(\frac{\varphi}{2}\right)\partial^2_{z,z} + \frac{1}{\cos^2\left(\frac{\varphi}{2}\right)}\partial^2_{\theta,z} + \cot(\varphi)\partial_\varphi\right).$$

We can now define the Brownian motion $\mathbb{B}_t$ in $SU(2,\mathbb{C})$ as the diffusion process with infinitesimal generator L: using cylindrical coordinates, there exists φ_t, θ_t and z_t three real diffusion processes, such that

$$\mathbb{B}_t = \exp(\varphi_t(\cos(\theta_t)X + \sin(\theta_t)Y))\exp(z_tZ) \text{ and } \begin{cases} \langle d\varphi_t, d\varphi_t\rangle = dt \\ \langle d\theta_t, d\theta_t\rangle = \frac{1}{\sin^2(\varphi_t)}dt \\ \langle dz_t, dz_t\rangle = \tan^2\left(\frac{\varphi_t}{2}\right)dt \\ \langle d\theta_t, dz_t\rangle = \frac{1}{2\cos^2\left(\frac{\varphi_t}{2}\right)}dt \\ \langle d\varphi_t, d\theta_t\rangle = \langle d\varphi_t, dz_t\rangle = 0 \\ Drift(d\varphi_t) = cot(\varphi_t)dt \\ Drift(d\theta_t) = 0 \\ Drift(dz_t) = 0. \end{cases}$$

Then, taking B^1_t and B^2_t two real Brownian motions, we get:

$$\begin{cases} d\varphi_t = dB^1_t + \cot(\varphi_t)dt \\ d\theta_t = \frac{1}{\sin(\varphi_t)}dB^2_t \\ dz_t = \tan\left(\frac{\varphi_t}{2}\right)dB^2_t \end{cases} \tag{2}$$

In particular we note that the infinitesimal generator of the diffusion (φ_t, θ_t) is the Laplace Beltrami operator on the sphere S^2 in spherical coordinates. Studying more in detail the third coordinate z_t, we obtain the following geometric interpretation:

Theorem 1. *The Brownian motion $\mathbb{B}_t$ on $SU(2,\mathbb{C})$ can be described by $\Pi(\mathbb{B}_t)$ a Brownian motion on S^2 and z_t the area swept by $\Pi(\mathbb{B}_t)$ modulo 4π, up to a sign. Here the swept area is described according to the North pole of the sphere.*

Note that $\mathbb{B}_t$ is, in fact, described by its projection $\Pi(\mathbb{B}_t)$ on the sphere. Let $\mathbb{B}^a_t$ and $\mathbb{B}^{a'}_t$ two Brownian motions on $SU(2,\mathbb{C})$ starting from a and a' respectively. As we want to study couplings, we have to deal with the distance between these two processes. It has been proven by Baudoin and Bonnefont in [3–5], that $d^2_{cc}(0, (\varphi, \theta, z))$ is equivalent to $\varphi^2 + |z|$, denoting (φ, θ, z) an element of $SU(2,\mathbb{C})$ in cylindrical coordinates. Using calculations on matrices and spherical trigonometry, we obtain the following result.

Proposition 1. *We consider R_t a continuous signed distance between $\Pi(\mathbb{B}_t^a)$ and $\Pi(\mathbb{B}_t^{a'})$ and A_t the signed swept area between these two paths, up to a constant. Note that if the paths $(\Pi(\mathbb{B}_s^a))_{s \leq t}$ and $(\Pi(\mathbb{B}_s^{a'}))_{s \leq t}$ are crossing at time t, the sign of dA_t changes. We denote $\tilde{A}_t \in]-2\pi, 2\pi]$ such that $\tilde{A}_t \equiv A_t \mod (4\pi)$. Then we have the equivalence:*

$$d_{cc}^2(\mathbb{B}_t^a, \mathbb{B}_t^{a'}) \sim R_t^2 + |\tilde{A}_t|$$

Remark 1. Introducing cylindrical coordinates on $SL(2, \mathbb{R})$, we also have a similar geometric interpretation of the Brownian motion: it can be seen as a Brownian motion on the hyperbolic plane together with its signed swept area.

3   General Construction of Couplings

According to the previous results, to define a coupling on $SU(2, \mathbb{C})$ we just need to define one on S^2. In this section we will use Itô depiction of the Brownian motions in some well chosen frame. The notion of a moving frame along a semimartingale as well as an introduction of the Itô depiction in this frame are taken from Emery [9].

Remark 2. If the results and examples of this section are obtained in studying S^2, that is for a curvature $k = 1$, we can generalise them to the case of the hyperbolic plane, taking $k = -1$ and also to the case of $\mathbb{R}^2$ by passing at the limit for $k = 0$. We then have analogous results on $SL(2, \mathbb{R})$ and $\mathbb{H}$ (for this last one, the results has already been described in [6,10]). Thus we will denote k the curvature of the Riemannian manifold in order to keep generality in our results.

We consider a coupling (X_t, Y_t) of Brownian motions on S^2 and we define the corresponding R_t and A_t as previously. For $0 < |R_t| < \pi$, π being the injectivity radius of S^2, we define:

- $e_1^X(t) := \frac{\exp_{X_t}^{-1}(Y_t)}{R_t}$, with $\exp_{X_t}^{-1}(Y_t) = \dot{\gamma}(0)$, γ the unique geodesic such that $\gamma(0) = X_t$ and $\gamma(1) = Y_t$;
- $e_2^X(t)$ such that $(e_1^X(t), e_2^X(t))$ is a direct orthonormal basis on $T_{X_t}S^2$;
- $(e_1^Y(t), e_2^Y(t))$ the direct orthonormal basis on $T_{Y_t}S^2$ obtained by parallel transport of $(e_1^X(t), e_2^X(t))$ along the geodesic γ joining X_t and Y_t.

Then we can describe (X_t, Y_t) using two dimensional Brownian motions $U = \begin{pmatrix} U_1 \\ U_2 \end{pmatrix}$ and $V = \begin{pmatrix} V_1 \\ V_2 \end{pmatrix}$ with the Itô equation:

$$d^\nabla X_t = dU_1(t)e_1^X(t) + dU_2(t)e_2^X(t) \text{ and } d^\nabla Y_t = dV_1(t)e_1^Y(t) + dV_2(t)e_2^Y(t).$$

In particular, this coupling is co-adapted if and only if U and V form a co-adapted coupling of Brownian motions in $\mathbb{R}^2$.

Calculating the differentials and Hessians of the distance ρ and then, of the signed swept area between two paths, we can use Itô's formula to obtain general stochastic equations for R_t and A_t. We present here some examples of couplings that we will use in the construction of our first successful coupling on $SU(2, \mathbb{C})$. In all these examples, U and V are chosen adapted to the same filtration, that is co-adapted. In what follows C_t and $\tilde{C}_t$ will be two independent real Brownian motions.

Example 1. **Synchronous coupling**
We take $dV_1(t) = dU_1(t)$ and $dV_2(t) = dU_2(t)$. We get R_t deterministic and decreasing and A_t a martingale such that:

$$R_t = \frac{2}{\sqrt{k}} \arcsin\left(e^{-\frac{kt}{2}} \sin\left(\frac{\sqrt{k}R_0}{2}\right)\right) \text{ and } A_t = A_0 + \frac{2}{\sqrt{k}} \int_0^t \tan\left(\frac{\sqrt{k}R_s}{2}\right) d\tilde{C}_s.$$

Example 2. **Reflection coupling**
We take $dV_1(t) = -dU_1(t)$ and $dV_2(t) = dU_2(t)$. With this coupling R_t is a Brownian motion with a negative drift and A_t is still a martingale:

$$R_t = R_0 + 2C_t - \sqrt{k} \int_0^t \tan\left(\frac{\sqrt{k}R_s}{2}\right) ds \text{ and } A_t = A_0 + \frac{2}{\sqrt{k}} \int_0^t \tan\left(\frac{\sqrt{k}R_s}{2}\right) d\tilde{C}_s.$$

Note that, for $k \in \{0, 1\}$, R_t hits 0 at an almost surely finite time. The reflection coupling is a successful coupling on the sphere and on $\mathbb{R}^2$. Moreover, the coupling rate, that is $\mathbb{P}(T > t)$ with T the first meeting time of the processes, is exponentially decreasing for $k = 1$ and of order $\frac{1}{\sqrt{t}}$ for t large for $k = 0$. That is not the case for $k = -1$. In fact, using [13], as the hyperbolic plane doesn't satisfy the Liouville Theorem, we can't construct any successful coupling on it.

Example 3. **Perverse coupling**
We take $dV_1(t) = dU_1(t)$ and $dV_2(t) = -dU_2(t)$. We get R_t deterministic and increasing satisfying:

$$R_t = \frac{2}{\sqrt{k}} \arccos\left(e^{-\frac{kt}{2}} \cos\left(\frac{\sqrt{k}R_0}{2}\right)\right).$$

For this coupling, A_t is constant.

We can also add a noise to these couplings in order to remove the drift part. In particular, this provides a coupling with constant distance between the Brownian motions.

Example 4. **Synchronous coupling with noise/ fixed-distance coupling**
Taking W a real Brownian motion independent of U we let:

$$dV_1(t) = dU_1(t) \text{ and } dV_2(t) = \cos(\sqrt{k}R_t)dU_2(t) + \sin(\sqrt{k}R_t)dW(t).$$

We get R_t constant and A_t a Brownian motion up to a multiplicative constant:

$$A_t = A_0 + \frac{2}{\sqrt{k}} \sin\left(\frac{\sqrt{k}R_0}{2}\right) \tilde{C}_s.$$

Remark 3. In [12], Pascu and Popescu showed that on the sphere, there exists a co-adapted coupling of Brownian motions of deterministic distance function R_t if and only if R_t is continuous and satisfies the inequality:

$$-\tan\left(\frac{R_t}{2}\right) dt \leq dR_t \leq \cot\left(\frac{R_t}{2}\right) dt.$$

The two couplings realizing the extrema of this inequality are the synchronous coupling and the perverse coupling.

Remark 4. These couplings are co-adapted but also Markovian in the sense that the couple $(X_t, Y_t)_t$ is a Markov process.

4   One Co-adapted Successful Coupling

We now lay out a co-adapted successful method on $SU(2, \mathbb{C})$. This method is based on the idea of Kendall for coupling Brownian motions on the Heisenberg group [10]. The original idea is to switch between reflection and synchronous couplings, using reflection coupling to make R_t decrease, and synchronous coupling to keep the swept area comparable to R_t^2 and deacreasing as well.

Let κ, $\epsilon > 0$ such that $0 < \epsilon < \kappa$. We denote τ the first time of coupling of the Brownian motions in S^2 together with their swept areas that is

$$\tau = \inf\{t > 0 \mid X_t = Y_t \text{ and } A_t = 0\}.$$

Using perverse or synchronous coupling if needed, we can suppose that $0 < R_0 < \pi$. Using constant fixed-distance coupling if necessary, we can also suppose that $A_0 = 0$ without changing the value of R_0. We construct the coupling as in [10]:

1. We use reflection coupling until the process $\frac{|A_t|}{R_t^2}$ starting at 0 takes the value κ.
2. While the process $\frac{|A_t|}{R_t^2}$ starting at κ satisfies $\frac{|A_t|}{R_t^2} > \kappa - \epsilon$, we use fixed-distance coupling. Note that, as R_t stays constant during this step, $\frac{|A_t|}{R_t^2}$ is a Brownian motion, up to a multiplicative constant, and will hit $\kappa - \epsilon$ in an almost surely finite time.
3. While the process $\frac{|A_t|}{R_t^2}$, starting at $\kappa - \epsilon$ satisfies $\frac{|A_t|}{R_t^2} < \kappa$, we use reflection coupling.

We iterate steps 2 and 3 until $R_t = 0$. Because of this construction, when it occurs, $A_t = 0$. As the distance stays constant during each fixed-distance coupling step, if we omit these times, R_t is moving as in a simple reflection coupling and so will hit 0 in an almost surely finite time. To prove that this coupling is successful, we then need to show that we don't have too many switchings between the two coupling steps. As the different coupling steps are co-adapted, our final coupling is co-adapted too.

One difference with the case of Heisenberg is that if R_t is too close to π, it is possible to switch too fast from reflection coupling to fixed-distance coupling as the quadratic variation of the swept area will be quite big. To avoid this possibility, an idea is to interrupt the coupling when R_t is too close of π.

Theorem 2. *Let $\eta > 0$, such that $\pi - 2\eta > R_0 > 0$. We denote $\tau_\eta = \inf\{t > 0 | R_t \geq \pi - \eta\}$. Then the co-adapted coupling described below satisfies $\tau \wedge \tau_\eta < +\infty$ a.s. Moreover, we have $\mathbb{P}(\tau > \tau_\eta) < 1$.*

Proof. See [8].

If $\tau_\eta < \tau$, we can use synchronous coupling to decrease R_t until it takes the initial value R_0 and reiterate the method of theorem (2). These iterations are independent and $\mathbb{P}(\tau > \tau_\eta) < 1$, thus the coupling is successful.

Note that, to obtain couplings in $SU(2, \mathbb{C})$, we only need to have $A_t \equiv 0(4\pi)$ instead of $A_t = 0$. Thus we only need to stop reflection coupling if A_t enter intervals of the form

$$\left[4n\pi + \kappa R_t^2, 4(n+1)\pi - \kappa R_t^2\right], \ k \in \mathbb{Z}.$$

For a good choice of κ, this doesn't occur for R_t too close of π. Then we don't need to introduce τ_η to obtain a successful coupling.

Remark 5. Note that this coupling is co-adapted but not Markovian: in fact, if $W_t \in]\kappa - \epsilon, \kappa[$ and we don't know the past of the strategy, we can't choose if we need to continue with step 2 or step 3.

5   One Non Co-adapted Successful Coupling

As explained in the introduction, co-adapted couplings are not, in general, maximal. Moreover the rate of convergence of the previous coupling is not easy to estimate. We introduce here another coupling, not co-adapted, but such that $\mathbb{P}(\tau > t)$ is at most exponentially decreasing. This coupling is inspired by the works of Banerjee, Gordina and Mariano [2], using Brownian bridges couplings. This time we are going to directly couple the cylindrical coordinates of the Brownian motion in $SU(2, \mathbb{C})$:

$$\begin{cases} d\varphi_t = dB_t^1 + \cot(\varphi_t)dt \\ d\theta_t = \frac{1}{\sin(\varphi_t)}dB_t^2 \\ dz_t = \tan\left(\frac{\varphi_t}{2}\right)dB_t^2 \end{cases} \quad \text{and} \quad \begin{cases} d\varphi_t' = dB_t^{1'} + \cot(\varphi_t')dt \\ d\theta_t' = \frac{1}{\sin(\varphi_t')}dB_t^{2'} \\ dz_t' = \tan\left(\frac{\varphi_t'}{2}\right)dB_t^{2'} \end{cases}.$$

Theorem 3. *1. We denote by ρ_0 the distance between (φ_0, θ_0) and (φ_0', θ_0'). Supposing that $\rho_0 = 0$, there exists a successful coupling, not co-adapted, such that $\mathbb{P}(\tau > t)$ has an exponential decay in t.*
2. If $\rho \neq 0$, using first a reflection coupling to obtain $\rho = 0$ and then the coupling above, we obtain the same result.

Proof. The details of the proof will be found in [7]. Supposing that $(\varphi_0, \theta_0) = (\varphi_0', \theta_0')$, the main idea of this coupling is to divide the time in intervals of constant length and to define $((\varphi_t, \theta_t), (\varphi_t', \theta_t'))$ such that the two processes are equal at the end of each interval. Let us explain the method on the first interval $[0, T]$.

- We take $B_t^1 = B_t^{1'}$, this way φ_t and φ_t' will always be equal.
- We can now consider θ_t and θ_t' as two time-changed Brownian motions $C_{\sigma(t)}$ and $C_{\sigma(t)}'$ with $\sigma(t) = \int_0^t \frac{1}{\sin^2(\varphi_s)} ds$. We can decompose C_t (resp C_t') using a Brownian bridge B^{br} (resp $B^{br'}$) on $[0, \sigma(T)]$ and G (resp G') an independent Gaussian variable with mean zero and variance $\sigma(T)$:

$$C_\sigma = B_\sigma^{br} + \frac{\sigma}{\sigma(T)} G \text{ and } C_\sigma' = B_\sigma^{br'} + \frac{\sigma}{\sigma(T)} G'.$$

Choosing $G = G'$, we obtain $C_{\sigma(T)} = C_{\sigma(T)}'$ and so (φ_t, θ_t) and (φ_t', θ_t') meet at time T.

Then we have to choose a good coupling for the Brownian bridges $\left(B_\sigma^{br}, B_\sigma^{br'}\right)$ so that $z_T = z_T' \mod (4\pi)$ with a non zero probability. This way, reproducing this coupling on each interval of time, the Brownian motions on $SU(2, \mathbb{C})$ finally meet at the end of one of the interval which lead to an exponential decreasing coupling rate.

We note that we need to know the path of φ on all the interval $[0, T]$ to know the value of the change of time $\sigma(T)$ and to construct the coupling $\left(B_\sigma^{br}, B_\sigma^{br'}\right)$. Thus, the obtained coupling is not co-adapted.

Remark 6. The first part of theorem (3) is still true in the case of $SL(2, \mathbb{R})$. However, as there is no successful coupling of Brownian motions on the hyperbolic plane, neither on $SL(2, \mathbb{R})$, the second part is false.

Remark 7. Note that the above coupling on $SU(2, \mathbb{C})$ is not a direct generalisation of the coupling of Banerjee, Gordina and Mariano on the Heisenberg group [2]: in their work, Brownian bridges are used to couple Cartesian coordinates on $\mathbb{R}^2$. Our strategy, which uses spherical coordinates of the sphere, can also be directly performed on the Heisenberg group using polar coordinates and leads to a slightly different coupling.

References

1. Asmussen, S.R.: Applied probability and queues, Applications of Mathematics. In: Stochastic Modelling and Applied Probability, vol. 51, 2nd edn. Springer, New York (2003). https://doi.org/10.1007/b97236
2. Banerjee, S., Gordina, M., Mariano, P.: Coupling in the Heisenberg group and its applications to gradient estimates. Ann. Prob. **46**(6), 3275–3312 (2018). https://doi.org/10.1214/17-AOP1247
3. Baudoin, F., Bonnefont, M.: The subelliptic heat kernel on SU(2): representations, asymptotics and gradient bounds. Math. Z. **263**(3), 647–672 (2009). https://doi.org/10.1007/s00209-008-0436-0
4. Bonnefont, M.: Functional inequalities for subelliptic heat kernels. Theses, Université Paul Sabatier - Toulouse III (2009). https://theses.hal.science/tel-00460624
5. Bonnefont, M.: The subelliptic heat kernels on SL(2, $\mathbb{R}$) and on its universal covering SL(2, $\mathbb{R}$): integral representations and some functional inequalities. Potent. Anal. **36**(2), 275–300 (2012). https://doi.org/10.1007/s11118-011-9230-4

6. Bonnefont, M., Juillet, N.: Couplings in L^p distance of two Brownian motions and their Lévy area. Ann. Inst. Henri Poincaré Probab. Stat. **56**(1), 543–565 (2020). https://doi.org/10.1214/19-AIHP972

7. Bénéfice, M.: Non co-adapted couplings of Brownian motions on subRiemannian manifolds. In: Construction (2023)

8. Bénéfice, M.: Couplings of brownian motions on $SU(2, \mathbb{C})$ and $SL(2, \mathbb{R})$ (2023)

9. Émery, M.: Stochastic calculus in manifolds. Universitext. Springer, Berlin (1989). https://doi.org/10.1007/978-3-642-75051-9. with an appendix by P.-A. Meyer

10. Kendall, W.S.: Coupling all the Lévy stochastic areas of multidimensional Brownian motion. Ann. Prob. **35**(3), 935–953 (2007). https://doi.org/10.1214/009117906000001196

11. Kuwada, K.: Duality on gradient estimates and Wasserstein controls. J. Funct. Anal. **258**(11), 3758–3774 (2010). https://doi.org/10.1016/j.jfa.2010.01.010

12. Pascu, M.N., Popescu, I.: Couplings of Brownian motions of deterministic distance in model spaces of constant curvature. J. Theor. Prob. **31**(4), 2005–2031 (2017). https://doi.org/10.1007/s10959-017-0781-1

13. Wang, F.Y.: Liouville theorem and coupling on negatively curved Riemannian manifolds. Stochastic Process. Appl. **100**, 27–39 (2002). https://doi.org/10.1016/S0304-4149(02)00121-7

A Finite Volume Scheme for Fractional Conservation Laws Driven by Lévy Noise

Neeraj Bhauryal[(✉)] [ID]

Grupo de Física Matemática, Universidade de Lisboa, Lisboa, Portugal
nsbhauryal@fc.ul.pt

Abstract. In this article, we study a semi-discrete finite volume scheme for fractional conservation laws perturbed with Lévy noise. With the help of bounded variation estimates and Kružkov's theory we provide a rate of convergence result.

Keywords: Fractional conservation laws · Finite volume scheme · Lévy noise

1 Introduction

We are interested in analyzing a finite difference scheme for the following Cauchy problem

$$du(t,x) - \operatorname{div} f(u(t,x))\, dt + \mathcal{L}_\lambda[u(t,\cdot)](x)\, dt$$
$$= \sigma(u(t,x))\, dW(t) + \int_{|z|>0} \psi(u(t,x); z)\, \widetilde{N}(dz, dt), \text{ in } \Pi_T, \quad (1)$$

whose well-posedness was established in [3]. Here $u(0,x) = u_0(x)$, $u_0 : \mathbb{R}^d \mapsto \mathbb{R}$ is the given initial function, $\Pi_T := (0,T) \times \mathbb{R}^d$ with $T > 0$ fixed, $f : \mathbb{R} \mapsto \mathbb{R}^d$ is a given flux function. We refer the reader to Sect. 2 for the list of assumptions. The operator $\mathcal{L}_\lambda[u]$ denotes the fractional Laplacian $(-\Delta)^\lambda[u]$ of order $\lambda \in (0,1)$, which is defined as $\mathcal{L}_\lambda[\varphi](x) := c_\lambda \, \text{P.V.} \int_{|z|>0} \frac{\varphi(x)-\varphi(x+z)}{|z|^{d+2\lambda}}\, dz$, for some constant $c_\lambda > 0$ and a sufficiently regular function φ. Here P.V. stands for principal value. Furthermore, $W(t)$ is a real-valued Brownian noise and $\widetilde{N}(dz,dt) = N(dz,dt) - \nu(dz)dt$, where N is a Poisson random measure on $\mathbb{R} \times (0,\infty)$ with intensity measure $\nu(dz)$ on $|z| > 0$.

Conservation laws, such as transport type equations or the most celebrated example of Euler equations play an important role in stochastic geometric mechanics. They provide fundamental principles that govern the dynamics of physical systems. It is also important to study the behaviour of systems that have random fluctuations in energy, momentum and angular momentum. Non-local integro differential PDEs as the ones considered here are known for their

Supported by FCT project no. UIDB/00208/2020.

applications in different areas of research, for e.g., in mathematical finance [5], where it is necessary to study efficient numerical schemes to approximate problem (1).

Finite volume schemes (FVSs) for stochastic conservation laws have been analyzed by many authors see [1,8] and with Lévy noise by [2,7]. Recently, for fractional conservation laws with Brownian noise, authors in [6] gave a rate of convergence result for semi-discrete FVS. This article aims to generalize their result for Lévy noise. We propose a semi-discrete FVS and show that the numerical solution converges to entropy solution of (1) in L^1 norm and establish a rate of convergence result depending on the fractional parameter λ. This matches with the result in deterministic setup [4]. We first introduce the notions and scheme in Sect. 2 and prove *a priori* estimates on numerical solution in Sect. 3 which finally helps use to attain rate of convergence result whose proof can be found in Sect. 4.

2 Technical Framework and Statement of the Main Result

The letter C denotes generic constant and for each $r > 0$, we write $\mathcal{L}_\lambda[\varphi] :=$ $\mathcal{L}_{\lambda,r}[\varphi] + \mathcal{L}_\lambda^r[\varphi]$, where $\mathcal{L}_{\lambda,r}[\varphi](x) := c_\lambda$ P.V. $\int_{|z| \le r} \frac{\varphi(x) - \varphi(x+z)}{|z|^{d+2\lambda}}\, dz$, $\mathcal{L}_\lambda^r[\varphi](x) :=$ $c_\lambda \int_{|z|>r} \frac{\varphi(x) - \varphi(x+z)}{|z|^{d+2\lambda}}\, dz$. The function $\eta(x)$ denotes a C^2 approximation of $|x|$ and $\phi(t, x)$ denotes a test function. In the rest of the paper, we consider the following assumptions:

A.1 The initial function u_0 is a deterministic function in $L^2(\mathbb{R}^d) \cap L^\infty(\mathbb{R}^d) \cap BV(\mathbb{R}^d)$.

A.2 The flux function $f = (f_1, f_2, \cdots, f_d) : \mathbb{R} \mapsto \mathbb{R}^d$ is a Lipschitz continuous function with $f_k(0) = 0$, for all $1 \le k \le d$.

A.3 The map σ is Lipschitz and have compact support, i.e. $\exists\, M > 0$ s.t. $\sigma(u) = 0$ for $|u| > M$ with $\sigma(0) = 0$.

A.4 The map ψ has compact support with $\psi(u) = 0$ for $|u| > M$. There exists $\kappa > 0$, s.t., for all $u, v, \in \mathbb{R}$

$$|\psi(u; z) - \psi(v; z)| \le \kappa |u - v|(|z| \wedge 1) \text{ and } \psi(0; z) = 0 \text{ for all } z \in \mathbb{R}.$$

The measure ν is singular at $z = 0$ and satisfies $\int_{|z|>0}(1 \wedge |z|^2)\nu(dz) < \infty$.

Let Δx denote the spatial discretization parameter and set $x_\alpha = \alpha \Delta x$, for $\alpha \in \mathbb{Z}^d$. Following [4], let us define the spatial grid cells as $R_0 = \left[- \Delta x/2, \Delta x/2 \right)^d$, $R_\alpha = x_\alpha + R_0$, where $x_{j\pm 1/2} = x_j \pm \frac{\Delta x}{2}$, for $j \in \mathbb{Z}$ and $\alpha = (\alpha_1, \alpha_2, \cdots, \alpha_d) \in \mathbb{Z}^d$.

For $(u_\alpha)_{\alpha \in \mathbb{Z}^d}$, set $u_{\Delta x} = \sum_{\alpha \in \mathbb{Z}^d} u_\alpha \mathbb{1}_{R_\alpha}$ ($\mathbb{1}_A$ denotes the characteristic function of the set A). Furthermore, the BV semi-norm is defined by

$$|u_{\Delta x}(t, \cdot)|_{BV} = (\Delta x)^{d-1} \sum_{\alpha \in \mathbb{Z}^d} \sum_{i=1}^d |u_{\alpha+e_i}(t) - u_\alpha(t)|,$$

where $\{e_1, e_2, \cdots e_d\}$ denotes the standard basis of $\mathbb{R}^d$. Finally, we denote by $D_k^{\pm}$ the discrete forward and backward differences in space, *i.e.*, $D_k^{\pm} u_{\Delta x} = \sum_{\alpha \in \mathbb{Z}^d} D_k^{\pm} u_\alpha \mathbb{1}_{R_\alpha} = \sum_{\alpha \in \mathbb{Z}^d} \pm \frac{u_{\alpha \pm e_k} - u_\alpha}{\Delta x} \mathbb{1}_{R_\alpha}$. We denote by maximum (resp. minimum) of two reals a and b by $a \vee b$ (resp. $a \wedge b$). Following the techniques in [4,6], we propose the following FVS

$$du_\alpha(t) + \sum_{k=1}^d D_k^- F_k(u_\alpha, u_{\alpha + e_k}) dt + \hat{\mathcal{L}}_\lambda [u_{\Delta x}(t, \cdot)]_\alpha dt$$

$$= \sigma(u_\alpha(t))\, dW(t) + \int_{|z| > 0} \psi(u_\alpha(t); z)\, \tilde{N}(dz, dt), \quad t > 0, \alpha \in \mathbb{Z}^d, \quad (2)$$

where $u_\alpha(0) = \frac{1}{(\Delta x)^d} \int_{R_\alpha} u_0(x) dx$, $\alpha \in \mathbb{Z}^d$, F_i denotes a monotone numerical flux corresponding to f_i, for $i = 1, 2, \cdots, d$. The discretization of the non-local term is as follows for any $u_{\Delta x} = \sum_{\alpha \in \mathbb{Z}^d} u_\alpha \mathbb{1}_{R_\alpha}$, $\hat{\mathcal{L}}_\lambda[u_{\Delta x}] = \sum_{\alpha \in \mathbb{Z}^d} \hat{\mathcal{L}}_\lambda[u_{\Delta x}]_\alpha \mathbb{1}_{R_\alpha} := \frac{1}{(\Delta x)^d} \sum_{\alpha, \beta \in \mathbb{Z}^d} G_{\alpha, \beta} u_\beta \mathbb{1}_{R_\alpha}$, where,

$$\hat{\mathcal{L}}_\lambda[u_{\Delta x}]_\alpha := \frac{1}{(\Delta x)^d} \int_{R_\alpha} \mathcal{L}_\lambda^{\frac{\Delta x}{2}}[u_{\Delta x}](x)\, dx$$

$$= c_\lambda \frac{1}{(\Delta x)^d} \int_{R_\alpha} \left[\int_{|z| > \frac{\Delta x}{2}} \frac{u_{\Delta x}(x) - u_{\Delta x}(x + z)}{|z|^{d + 2\lambda}} dz \right] dx$$

$$= \sum_{\beta \in \mathbb{Z}^d} u_\beta \frac{1}{(\Delta x)^d} \underbrace{\left[\int_{R_\alpha} c_\lambda \int_{|z| > \frac{\Delta x}{2}} \frac{\mathbb{1}_{R_\beta}(x) - \mathbb{1}_{R_\beta}(x + z)}{|z|^{d + 2\lambda}} dz dx \right]}_{:= G_{\alpha, \beta}} = \sum_{\beta \in \mathbb{Z}^d} \frac{1}{(\Delta x)^d} G_{\alpha, \beta} u_\beta.$$

Remark 1. Thanks to the assumptions on the data (**A.**1)-(**A.**3), the solvability of (2) follows from a classical argument of stochastic differential equations with Lipschitz non-linearities (see [9, Chapter 9]).

Remark 2. (Discrete entropy inequality [3, Definition 1.3])

$$0 \leq \int_{\mathbb{R}^d} \eta(u_{\Delta x}(0, x))\, \phi(0, x)\, dx + \int_{\Pi_T} \eta(u_{\Delta x}(t, x))\, \partial_t \phi(t, x)\, dt\, dx \quad (3)$$

$$- \int_{\Pi_T} \eta'(u_{\Delta x}(t, x)) \frac{1}{\Delta x} \sum_{i=1}^d \Big[F_i(u_{\Delta x}(t, x), u_{\Delta x}(t, x + \Delta x e_i))$$

$$- F_i(u_{\Delta x}(t, x - \Delta x e_i), u_{\Delta x}(t, x)) \Big] \phi(t, x)\, dt\, dx$$

$$- \int_{\Pi_T} \eta(u_{\Delta x}(t, x)) \mathcal{L}_{\lambda, r}^{\frac{\Delta x}{2}}[\overline{\phi}(t, \cdot)](x) + \mathcal{L}_\lambda^r [u_{\Delta x}(t, \cdot)](x) \overline{\phi}(t, x)\, \eta'(u_{\Delta x}(t, x))\, dt\, dx$$

$$+ \int_{\Pi_T} \sigma(u_{\Delta x}(t, x)) \eta'(u_{\Delta x}(t, x)) \phi(t, x) dW(t)\, dx$$

$$+ \frac{1}{2} \int_{\Pi_T} \sigma^2(u_{\Delta x}(t, x)) \eta''(u_{\Delta x}(t, x)) \phi(t, x)\, dt\, dx$$

$$+ \int_{\Pi_T} \int_{|z| > 0} \int_0^1 \psi(u_{\Delta x}(t, x); z) \eta'(u_{\Delta x}(t, x) + \tau \psi(u_{\Delta x}(t, x); z)) \phi(t, x) d\tau \tilde{N}(dz, dt) dx$$

$$+ \int_{\Pi_T} \int_{|z|>0} \int_0^1 (1-\tau)\psi^2((u_{\Delta x}); z)\eta''(u_{\Delta x}(t, x) + \tau\psi(u_{\Delta x}(t, x); z))\phi(t, x)\mathrm{d}\tau\nu(\mathrm{d}z)\mathrm{d}t\mathrm{d}x$$

where $\overline{\phi} = \frac{1}{\Delta x} \sum_{\alpha \in \mathbb{Z}^d} \left(\int_{R_\alpha} \psi(t, x)\mathrm{d}x \right) \mathbb{1}_{R_\alpha}(x)$.

Theorem 1. *(Main Theorem) Let $u(t, x)$ be a BV entropy solution to the problem (1), then there exists a constant C, independent of Δx, s.t. for all $t \in (0, T]$,*

$$\mathbb{E}\left[\int_{\mathbb{R}^d} |u_{\Delta x}(t, x) - u(t, x)| \, dx \right] \leq C \begin{cases} \sqrt{\Delta x}, & \text{if } \lambda < \frac{1}{2} \\ \sqrt{\Delta x}|\log \Delta x|, & \text{if } \lambda = \frac{1}{2} \\ (\Delta x)^{1-\lambda}, & \text{if } \lambda > \frac{1}{2} \end{cases}$$

provided the initial error satisfies $\mathbb{E}\|u_{\Delta x}(0, x) - u_0(x)\|_{L^1(\mathbb{R}^d)} = \mathcal{O}(\sqrt{\Delta x})$.

Remark 3. The convergence of approximate solutions $u_{\Delta x}$ to the entropy solution u of (1) in the Young measure sense gives an alternate existence proof of solution for (1).

3   Estimates on Approximating Sequence $u_{\Delta x}$

In this section, we derive a priori estimates for $u_{\Delta x}$ which play an essential role in establishing the proof of the main theorem.

Definition 1. *A pair (η, ζ) is called an entropy-entropy flux pair if $\eta \in C^2(\mathbb{R})$, $\eta \geq 0$ and ζ is a vector field satisfying $\zeta'(r) = \beta'(r)f'(r)$ for all r. This entropy-entropy flux pair (η, ζ) is called convex if $\eta'' \geq 0$.*

3.1   Uniform Moment Estimates

Lemma 1. *(Discrete entropy inequality) For a given convex entropy-entropy flux pair (η, ζ), where η is C^2 and even convex function, the approximate solution $u_\alpha(t)$ generated by the finite volume scheme (2) satisfies the following cell entropy inequality*

$$d\eta(u_\alpha(t)) + \frac{1}{\Delta x} \sum_{\beta \in \mathbb{Z}^d} \eta'(u_\alpha(t))G_{\alpha, \beta} u_\beta(t) \, dt$$

$$+ \sum_{i=1}^d \frac{|\eta'(u_\alpha(t) - k)|}{\Delta x} ((F_i(u_\alpha(t) \vee k, u_{\alpha+e_i} \vee k) - F_i(u_{\alpha-e_i}(t) \vee k, u_\alpha \vee k))$$

$$- (F_i(u_\alpha(t) \wedge k, u_{\alpha+e_i} \wedge k) - F_i(u_{\alpha-e_i}(t) \wedge k, u_\alpha \wedge k))) \, \mathrm{d}t$$

$$\leq \sigma(u_\alpha(t))\eta'(u_\alpha(t)) \, dW(t) + \frac{1}{2}\sigma^2(u_\alpha(t))\eta''(u_\alpha(t)) \, dt$$

$$+ \int_{|z|>0} (\eta(u_\alpha(t) + \psi(u_\alpha(t); z)) - \eta(u_\alpha(t))) \, \widetilde{N}(dz, dt)$$

$$+ \int_{|z|>0} \int_{\tau=0}^1 (1 - \tau)\psi^2(u_\alpha(t); z)\eta''(u_\alpha(t) + \tau\psi(u_\alpha(t); z))\mathrm{d}\tau\nu(\mathrm{d}z)dt,$$

for all $\alpha \in \mathbb{Z}^d$ and almost all $\omega \in \Omega$.

Proof. This proof follows from an application of Itô-Lévy formula to $\eta(u_\alpha(t) - k)$ and estimating the flux term as in [6, Lemma 3.1]. □

We are now in a position to prove uniform moment estimates. In light of this, we have the following lemma:

Lemma 2. *The approximate solutions $u_{\Delta x}(t, x)$ generated by the finite difference scheme* (2) *satisfy the following uniform moment estimates for $p \in \mathbb{N}$ and $t \geq 0$*

$$\sup_{\Delta x > 0} \sup_{0 \leq t \leq T} \mathbb{E}\left[||u_{\Delta x}(t, \cdot)||_p^p\right] \leq M^{\frac{p-1}{p}} ||u_0||_{L^1(\mathbb{R}^d)}^{1/p},$$

for $p = \infty$, we have $||u_{\Delta x}||_{L^\infty(\mathbb{R}^d)} \leq M$.

Proof. We put $k = 0$ in Lemma 1 and take expectation to write

$$\mathbb{E}[\eta(u_\alpha(t))] - \eta(u_\alpha(0)) + \frac{1}{\Delta x} \int_0^t \sum_{\beta \in \mathbb{Z}^d} \mathbb{E}\left[\eta'(u_\alpha(s))G_{\alpha,\beta}\, u_\beta(s)\right] ds$$

$$+ \sum_{i=1}^d \frac{|\eta'(u_\alpha(s))|}{\Delta x} \int_0^t \mathbb{E}\left(\left[F_i(u_\alpha^+(s), u_{\alpha+e_i}^+(s)) - F_i(u_{\alpha-e_i}^+(s), u_\alpha^+(s))\right]\right.$$

$$\left. - \left[F_i(-u_\alpha^-(s), -u_{\alpha+e_i}^-(s)) - F_i(-u_{\alpha-e_i}^-(s), -u_\alpha^-(s))\right]\right) ds$$

$$\leq \frac{1}{2} \int_0^t \mathbb{E}\left[\sigma^2(u_\alpha(s))\eta''(u_\alpha(s))\right] ds$$

$$+ \int_0^t \int_{|z|>0} \int_{\lambda=0}^1 \mathbb{E}(1 - \lambda)\psi^2(u_\alpha(s); z)\eta''(u_\alpha(s) + \lambda\psi(u_\alpha(s); z))d\lambda\nu(dz)ds,$$

$$(4)$$

where the martingale terms vanish due to expectation. We now multiply the inequality (4) *by Δx^d and sum over all $\alpha \in \mathbb{Z}^d$ to write*

$$\sum_{\alpha \in \mathbb{Z}^d} (\mathbb{E}[\eta(u_\alpha(t))] - \eta(u_\alpha(0))) + \frac{1}{\Delta x} \int_0^t \sum_{\alpha \in \mathbb{Z}^d} \sum_{\beta \in \mathbb{Z}^d} \mathbb{E}[\eta'(u_\alpha(s))G_{\alpha,\beta}\, u_\beta(s)] ds$$

$$+ \frac{1}{\Delta x} \sum_{i=1}^d \sum_{\alpha \in \mathbb{Z}^d} \mathbb{E}\left[\int_0^t |\eta'(u_\alpha(s))| \left(\left(F_i[u_\alpha^+(s), u_{\alpha+e_i}^+(s)] - F_i[u_{\alpha-e_i}^+(s), u_\alpha^+(s)]\right)\right.\right.$$

$$\left.\left. - \left(F_i[-u_\alpha^-(s), -u_{\alpha+e_i}^-(s)] - F_i[-u_{\alpha-e_i}^-(s), -u_\alpha^-(s)]\right)\right) ds\right]$$

$$\leq \frac{1}{2} \int_0^t \sum_{\alpha \in \mathbb{Z}^d} \mathbb{E}[\sigma^2(u_\alpha(s))\eta''(u_\alpha(s))] ds$$

$$+ \int_0^t \int_{|z|>0} \int_{\lambda=0}^1 \sum_{\alpha \in \mathbb{Z}^d} \mathbb{E}(1 - \lambda)\psi^2(u_\alpha(s); z)\eta''(u_\alpha(s) + \lambda\psi(u_\alpha(s); z))d\lambda\nu(dz)ds.$$

Recall that $\eta = \eta_\delta$ is the convex even function satisfying $\eta_\delta(0) = 0$ and $\eta'_\delta(x) = \min(1, \frac{x}{\delta})$ for $x > 0$. We pass to the limit as δ goes to zero, one gets for any t,

$$\sum_{\alpha \in \mathbb{Z}^d} \mathbb{E}|u_\alpha(t)| + \frac{1}{\Delta x} \sum_{i=1}^d \sum_{\alpha \in \mathbb{Z}^d} \mathbb{E} \int_0^t \left\{ \left(F_i[u_\alpha^+(s), u_{\alpha+e_i}^+(s)] - F_i[u_{\alpha-e_i}^+(s), u_\alpha^+(s)] \right) \right.$$
$$\left. - \left(F_i[-u_\alpha^-(s), -u_{\alpha+e_i}^-(s)] - F_i[-u_{\alpha-e_i}^-(s), -u_\alpha^-(s)] \right) \right\} ds \leq \sum_{\alpha \in \mathbb{Z}^d} |u_\alpha(0)|.$$

Multiplying this by $(\Delta x)^d$ gives $(\Delta x)^d \sum_{\alpha \in \mathbb{Z}^d} \mathbb{E}|u_\alpha(t)| \leq (\Delta x)^d \sum_{\alpha \in \mathbb{Z}^d} |u_\alpha(0)|$ and thus $\sup_{\Delta x > 0} \sup_{0 \leq t \leq T} \mathbb{E}\left[\|u_{\Delta x}(t, \cdot)\|_{L^1(\mathbb{R}^d)} \right] \leq \|u_0\|_{L^1(\mathbb{R}^d)}$. A similar estimate holds for $p \geq 2$ by choosing η_δ converging to $\frac{|\cdot|^p}{p}$ instead and the general case $p \geq 1$ then follows from an interpolation argument. Finally, the maximum principle follows from [6, Lemma 3.2] and [7, Lemma 3.3]. □

3.2 Spatial Bounded Variation

Lemma 3. *The finite volume approximations $u_{\Delta x}(t, x)$ satisfies*
$$\mathbb{E}\left[|u_{\Delta x}(t, \cdot)|_{BV(\mathbb{R}^d)} \right] \leq C\, \mathbb{E}\left[|u_0(\cdot)|_{BV(\mathbb{R}^d)} \right] \text{ for all } t > 0.$$

Proof. Making use of Eqn. (2) and Itô-Lévy's formula, we estimate $\eta(u_{\alpha+e_j} - u_\alpha)$ and write after taking expectation

$$\mathbb{E}\left[\eta(u_{\alpha+e_j} - u_\alpha)(t) \right] - \mathbb{E}\left[\eta(u_{\alpha+e_j} - u_\alpha)(0) \right]$$
$$+ \frac{1}{(\Delta x)^d} \sum_{\beta \in \mathbb{Z}^d} \mathbb{E}\left[\int_0^t \eta'(u_{\alpha+e_j} - u_\alpha)(s) G_{\alpha,\beta}[u_{\beta+e_j} - u_\beta](s) ds \right]$$
$$= -\sum_{i=1}^d \mathbb{E}\left[\int_0^t \frac{\eta'(u_{\alpha+e_j} - u_\alpha)(s)}{\Delta x} [F_i(u_{\alpha+e_j}, u_{\alpha+e_i+e_j}) - F_i(u_{\alpha-e_i+e_j}, u_{\alpha+e_j}) \right.$$
$$\left. - F_i(u_\alpha, u_{\alpha+e_i}) + F_i(u_{\alpha-e_i}, u_\alpha)](s)\, ds \right]$$
$$+ \frac{1}{2} \mathbb{E}\left[\int_0^t \eta''(u_{\alpha+e_j} - u_\alpha)(s)[\sigma(u_{\alpha+e_j}) - \sigma(u_\alpha)](s) ds \right]$$
$$+ \mathbb{E}\left[\int_0^t \int_{|z|>0} \int_0^1 (1-\tau)\eta''((u_{\alpha+e_j} - u_\alpha)(s) + \tau(\psi(u_{\alpha+e_j}) - \psi(u_\alpha)))(s; z) \right.$$
$$\left. \times \left(\psi(u_{\alpha+e_j}(s); z) - \psi(u_\alpha(s); z) \right)^2 d\tau \nu(dz) ds \right]$$

Using Lebesgue's theorem, the last two terms coming from two noises would vanish as $\eta_\delta \to |\cdot|$, so one can pass to the limit in η_δ, and the above equality holds for the limit of η_δ. After summing over $\alpha \in \mathbb{Z}^d$ one notices (see [6, Lemma 3.3]) that the flux term can be written as a telescopic sum and therefore will *vanish. The fractional term will have no contribution as well using the properties of $G_{\alpha,\beta}$. Finally, after summing over j, we have*

$$\mathbb{E}\left[\sum_{j=1}^{d}\sum_{\alpha\in\mathbb{Z}^d}|u_{\alpha+e_j}(t)-u_\alpha(t)|\right] \leq \mathbb{E}\left[\sum_{j=1}^{d}\sum_{\alpha\in\mathbb{Z}^d}|u_{\alpha+e_j}(0)-u_\alpha(0)|\right].$$

This allows us to conclude that, $\mathbb{E}\left[TV_x(u_{\Delta x}(t))\right] \leq \mathbb{E}\left[TV_x(u_0)\right]$, for all $t > 0$. Along with the L^1 estimate from Lemma 2, we conclude the proof. □

4   Proof of the Theorem

Proof. For $\varepsilon > 0$, we consider the following parabolic perturbation of (1) (for existence-uniqueness result we refer the reader to [3, Theorem 2.5])

$$du_\varepsilon(t,x) - \varepsilon\Delta u_\varepsilon(t,x)\,dt + \mathcal{L}_\lambda[A(u_\varepsilon(t,\cdot))](x)\,dt - \operatorname{div}f(u_\varepsilon(t,x))\,dt$$
$$= \sigma(u_\varepsilon(t,x))\,dW(t). \tag{5}$$

Let ρ and ϱ be the standard mollifiers on $\mathbb{R}$ and $\mathbb{R}^d$ respectively such that $\operatorname{supp}(\rho) \subset [-1,0]$ and $\operatorname{supp}(\varrho) = \overline{B_1}(0)$. We define $\rho_{\delta_0}(r) = \frac{1}{\delta_0}\rho(\frac{r}{\delta_0})$ and $\varrho_\delta(x) = \frac{1}{\delta^d}\varrho(\frac{x}{\delta})$, where δ and δ_0 are two positive parameters. Given a non-negative test function $\psi \in C_c^{1,2}([0,\infty)\times\mathbb{R}^d)$ and two positive constants δ and δ_0, we define

$$\phi_{\delta,\delta_0}(t,x,s,y) = \rho_{\delta_0}(t-s)\,\varrho_\delta(x-y)\,\phi(t,x). \tag{6}$$

Define another mollifier ς on $\mathbb{R}$ with support in $[-1,1]$ and $\varsigma_l(r) = \frac{1}{l}\varsigma(\frac{r}{l})$, for $l > 0$. Our aim is to estimate the L^1 difference between $u_{\Delta x}(t,x)$ and the entropy solution $u(t,x)$, and for this we use Kružkov's doubling argument.

First, we multiply the entropy inequality obtained (with test function (6)) by applying Itô-Lévy formula to (5) by $\varsigma(u_{\Delta y}(s,y) - k)$. Next, we multiply (3) by $\varsigma(u_\varepsilon(t,x) - k)$ and add these two inequalities and integrate the final result wrt all the variables involved. We then follow the analysis done in [6, Sec. 4] and [7, Sec. 4] to pass to the limits in δ_0, η and l to observe that the contribution from the noise terms is zero and the other terms contribute the following

$$0 \leq \mathbb{E}\left[\int_{\mathbb{R}^d}\int_{\mathbb{R}^d}|u_{\Delta y}(0,y) - u_0(x)|\,\varrho_\delta(x-y)\,\phi(0,x)\,dy\,dx\right]$$
$$+ \mathbb{E}\left[\int_{\Pi_T}\int_{\mathbb{R}^d}|u(t,x) - u_{\Delta y}(t,y)|\,\varrho_\delta(x-y)\,\partial_t\phi(t,x)\,dy\,dx\,dt\right]$$
$$+ \mathbb{E}\left[\int_{\Pi_T}\int_{\mathbb{R}^d}F(u(t,x),u_{\Delta y}(t,y))\cdot\nabla\phi(t,x)\varrho_\delta(x-y)\,dy\,dx\,dt\right] + C\frac{\Delta y}{\delta}$$
$$+ \mathbb{E}\left[\int_{\Pi_T}\int_{\mathbb{R}^d}|u_{\Delta y}(t,y) - u(t,x)|\mathcal{L}_\lambda^{\Delta y}[\phi(t,\cdot)](x)\varrho_\delta(x-y)\,dy\,dx\,dt\right]$$

$$+ C\frac{(\Delta y)^{2-2\lambda}}{\delta} + \frac{C_\lambda}{\delta}\|\phi\|_\infty\|u_0\|_{BV}\begin{cases}\Delta y, & \text{if } \lambda < 1/2 \\ \Delta y|\ln \Delta y|, & \text{if } \lambda = 1/2 \\ \Delta y^{2(1-\lambda)}, & \text{if } \lambda > 1/2\end{cases}$$

$$+ \|u_0\|_{BV}\int_0^T\left(\|\nabla\phi(t,\cdot)\|_\infty + \frac{C}{\delta}\|\phi(t,\cdot)\|_\infty\right)\int_{|z|\leq\Delta y}|z|^2\,d\mu_\lambda(z)\,dt.$$

With a particular choice of $\phi = \psi_h^t(s)\,(\psi_R \star \rho)(x)$ for some mollifier ρ and where

$$\psi_h^t(s) = \begin{cases}1, & \text{if } s \leq t, \\ 1 - \frac{s-t}{h}, & \text{if } t \leq s \leq t+h, \\ 0, & \text{if } s \geq t+h.\end{cases} \qquad \psi_R(x) = min\left(1, \frac{R^a}{|x|^a}\right)$$

we pass to the limits as R goes to infinity and h goes to zero to conclude

$$\mathbb{E}\left[\int_{\mathbb{R}^d}\left|u(t,y) - u_{\Delta y}(t,y)\right|dy\right] \leq C\begin{cases}\sqrt{\Delta y}, & \text{if } \lambda < 1/2, \\ \sqrt{\Delta y}|\ln \Delta y|, & \text{if } \lambda = 1/2, \\ \Delta y^{1-\lambda}, & \text{if } \lambda > 1/2,\end{cases}$$

given that the initial error satisfies $\mathbb{E}\|u_{\Delta y}(0,y) - u_0(y)\|_{L^1(\mathbb{R}^d)} \leq \sqrt{\Delta y}$.   □

References

1. Bauzet, C., Charrier, J., Gallouët, T.: Convergence of monotone finite volume schemes for hyperbolic scalar conservation laws with multiplicative noise. Stochastics Partial Diff. Equa. Anal. Comput. **4**(1), 150–223 (2015). https://doi.org/10.1007/s40072-015-0052-z
2. Behera, S., Majee, A.K.: On rate of convergence of finite difference scheme for degenerate parabolic-hyperbolic pde with Lévy noise. https://arxiv.org/abs/2212.12846 (2022)
3. Bhauryal, N., Koley, U., Vallet, G.: The Cauchy problem for fractional conservation laws driven by Lévy noise. Stochastic Process. Appl. **130**(9), 5310–5365 (2020). https://doi.org/10.1016/j.spa.2020.03.009
4. Cifani, S., Jakobsen, E.R.: On numerical methods and error estimates for degenerate fractional convection–diffusion equations. Numer. Math. **127**(3), 447–483 (2013). https://doi.org/10.1007/s00211-013-0590-0
5. Cont, R., Tankov, P.: Financial modelling with jump processes. Chapman & Hall/CRC Financial Mathematics Series, Chapman & Hall/CRC, Boca Raton, FL (2004)
6. Koley, U., Vallet, G.: On the rate of convergence of a numerical scheme for Fractional conservation laws with noise. IMA J. Numer. Anal. (2023). drad015
7. Koley, U., Majee, A.K., Vallet, G.: A finite difference scheme for conservation laws driven by Lévy noise. IMA J. Numer. Anal. **38**(2), 998–1050 (2018). https://doi.org/10.1093/imanum/drx023

8. Kröker, I., Rohde, C.: Finite volume schemes for hyperbolic balance laws with multiplicative noise. Appl. Numer. Math. **62**(4), 441–456 (2012). https://doi.org/10.1016/j.apnum.2011.01.011

9. Peszat, S., Zabczyk, J.: Stochastic partial differential equations with Lévy noise, Encyclopedia of Mathematics and its Applications, vol. 113. Cambridge University Press, Cambridge (2007). https://doi.org/10.1017/CBO9780511721373, An evolution equation approach

The Total Variation-Wasserstein Problem: A New Derivation of the Euler-Lagrange Equations

Antonin Chambolle[1,2], Vincent Duval[1,2], and João Miguel Machado[1,2(✉)]

[1] CEREMADE, Université Paris-Dauphine, PSL University, 75016 Paris, France
joao-miguel.machado@ceremade.dauphine.fr
[2] Inria Paris (Mokaplan), 75012 Paris, France

Abstract. In this work we analyze the Total Variation-Wasserstein minimization problem. We propose an alternative form of deriving optimality conditions from the approach of [8], and as result obtain further regularity for the quantities involved. In the sequel we propose an algorithm to solve this problem alongside two numerical experiments.

Keywords: Total variation · Optimal transport · Image analysis

1 Introduction

The Wasserstein gradient flow of the total variation functional has been studied in a series of recent papers [2,4,8], for applications in image processing. In the present paper, we revisit the work of Carlier & Poon [8] and derive Euler-Lagrange equations for the problem: given $\Omega \subset \mathbb{R}^d$ open, bounded and convex, $\tau > 0$ and an absolutely continuous probability measure $\rho_0 \in \mathcal{P}(\Omega)$

$$\inf_{\rho \in \mathcal{P}(\Omega)} \mathrm{TV}(\rho) + \frac{1}{2\tau} W_2^2(\rho_0, \rho), \qquad \text{(TV-W)}$$

where τ is interpreted as a time discretization parameter for an implicit Euler scheme, as we shall see below.

The *total variation functional* of a Radon measure $\rho \in \mathcal{M}(\Omega)$ is defined as

$$\mathrm{TV}(\rho) = \sup \left\{ \int_\Omega \mathrm{div}\, z \, d\rho : z \in C_c^1\left(\Omega; \mathbb{R}^N\right), \|z\|_\infty \leq 1 \right\}, \qquad \text{(TV)}$$

which is not to be mistaken in this paper with the *total variation measure* $|\mu|$ of a Radon measure μ or its *total variation norm* $|\mu|(\Omega)$. We call $\mathrm{BV}(\Omega)$ the subspace of functions $u \in L^1(\Omega)$ whose weak derivative Du is a *finite Radon measure*. It can also be characterized as the L^1 functions such that $\mathrm{TV}(u) < \infty$, where $\mathrm{TV}(u)$ should be understood as in (TV) with the measure $u\mathcal{L}^d \llcorner \Omega$, and it holds that $\mathrm{TV}(u) = |Du|(\Omega)$. As $\mathrm{BV}(\mathbb{R}^d) \hookrightarrow L^{\frac{d}{d-1}}(\mathbb{R}^d)$, solutions to (TV-W) are also absolutely continuous w.r.t. the Lebesgue measure. Therefore, w.l.o.g. we

© The Author(s), under exclusive license to Springer Nature Switzerland AG 2023
F. Nielsen and F. Barbaresco (Eds.): GSI 2023, LNCS 14071, pp. 610–619, 2023.
https://doi.org/10.1007/978-3-031-38271-0_61

can minimize on $L^{\frac{d}{d-1}}(\Omega)$, which is a reflexive Banach space. In addition, a function ρ will have finite energy only if $\rho \in \mathcal{P}(\Omega)$.

The data term is given by the Wasserstein distance, defined through the value of the optimal transportation problem (see [19])

$$W_2^2(\mu, \nu) \stackrel{\text{def.}}{=} \min_{\gamma \in \Pi(\mu, \nu)} \int_{\Omega \times \Omega} |x - y|^2 \mathrm{d}\gamma = \sup_{\substack{\varphi, \psi \in C_b(\Omega) \\ \varphi \oplus \psi \leq |x-y|^2}} \int_\Omega \varphi \mathrm{d}\mu + \int_\Omega \psi \mathrm{d}\nu, \quad (1)$$

where the minimum is taken over all the probability measures on $\Omega \times \Omega$ whose marginals are μ and ν. An optimal pair (φ, ψ) for the dual problem is referred to as Kantorovitch potentials.

Using total variation as regularization was suggested in [18] with a L^2 data term for the *Rudin-Osher-Fatemi problem*

$$\inf_{u \in L^2(\Omega)} \mathrm{TV}(u) + \frac{1}{2\lambda} \|u - g\|_{L^2(\Omega)}^2, \quad \text{(ROF)}$$

see [6] for an overview. Other data terms were considered to better model the oscillatory behavior of the noise [15,17]. More recently Wasserstein energies have shown success in the imaging community [12], the model (TV-W) being used for image denoising in [2,4].

Existence and uniqueness of solutions for (TV-W) follow from the direct method in the calculus of variations, and the strict convexity of $W_2^2(\rho_0, \cdot)$ whenever ρ_0 is absolutely continuous, see [19, Prop. 7.19]. However, it is not easy to compute the subdifferential of the sum, which makes the derivation of the Euler-Lagrange equations not trivial.

In [8], the authors studied the gradient flow scheme defined by the successive iterations of (TV-W), and following the seminal work [14] they showed that, in dimension 1 as the parameter $\tau \to 0$, the discrete scheme converges to the solution of a fourth order PDE. They used an entropic regularization approach, followed by a Γ-convergence argument, to derive an Euler-Lagrange equation, which states that there exists a Kantorovitch potential ψ_1 coinciding with some $\mathrm{div}\, z \in \partial\mathrm{TV}(\rho_1)$ in the set $\{\rho_1 > 0\}$. On $\{\rho_1 = 0\}$, these quantities are related through a bounded Lagrange multiplier β associated with the nonnegativity constraint $\rho_1 \geq 0$.

In this work we propose an alternative way to derive the Euler-Lagrange equations which relies on the well established properties of solutions of (ROF) and shows further regularity of the quantities $\mathrm{div}\, z, \beta$.

Theorem 1. *For any $\rho_0 \in L^1(\Omega) \cap \mathcal{P}(\Omega)$, let ρ_1 be the unique minimizer of* (TV-W). *The following hold.*

1. *There is a vector field $z \in L^\infty(\Omega; \mathbb{R}^d)$ with $\mathrm{div}\, z \in L^\infty(\Omega)$ and a bounded Lagrange multiplier $\beta \geq 0$ such that*

$$\begin{cases} \operatorname{div} z + \dfrac{\psi_1}{\tau} = \beta, \quad a.e. \ in \ \Omega \\ z \cdot \nu = 0, \quad on \ \partial\Omega \\ \beta\rho_1 = 0, \quad a.e. \ in \ \Omega \\ z \cdot D\rho_1 = |D\rho_1|, \ \|z\|_\infty \leq 1, \end{cases} \qquad \text{(TVW-EL)}$$

where ψ_1 is a Kantorovitch potential associated with ρ_1.

2. The Lagrange multiplier β is the unique solution to (ROF) with $\lambda = 1$ and $g = \psi_1/\tau$.

3. The functions $\operatorname{div} z, \psi_1$ and β are Lipschitz continuous.

2   The Euler-Lagrange Equation

Let X and X^* be duality-paired spaces and $f : X \to \mathbb{R} \cup \{\infty\}$ be a convex function, the subdifferential of f on X is given by

$$\partial_X f(u) \stackrel{\text{def.}}{=} \{p \in X^* : f(v) \geq f(u) + \langle p, v - u \rangle, \text{ for all } v \in X\}. \qquad (2)$$

In order to derive optimality conditions for (TV-W) we will need some properties of the subdifferential of TV and of (ROF).

Proposition 1. *[3,6,16] If $u \in \mathrm{BV}(\Omega) \cap L^2(\Omega)$, then the subdifferential of TV in $L^2(\Omega)$ at u assumes the form*

$$\partial_{L^2} \mathrm{TV}(u) = \left\{ p \in L^2(\Omega) : \begin{array}{l} p = -\operatorname{div} z, \ z \in H_0^1(\operatorname{div}; \Omega), \\ \|z\|_\infty \leq 1, \ |Du| = z \cdot Du \end{array} \right\}$$

If in addition u solves (ROF), then

1. u^+ solves (ROF) with the constraint $u \geq 0$;
2. it holds that

$$0 \in \frac{u - g}{\lambda} + \partial_{L^2} \mathrm{TV}(u), \qquad (3)$$

   and conversely, if u satisfies (3), u minimizes (ROF);
3. *for Ω convex, if g is uniformly continuous with modulus of continuity ω, then u has the same modulus of continuity.*

In the previous proposition, we recall that $H_0^1(\operatorname{div}; \Omega)$ denotes the closure of $C_c^\infty(\Omega; \mathbb{R}^d)$ with respect to the norm $\|z\|_{H^1(\operatorname{div})}^2 = \|z\|_{L^2(\Omega)}^2 + \|\operatorname{div} z\|_{L^2(\Omega)}^2$.

Unless otherwise stated, we consider in the sequel $X = L^{\frac{d}{d-1}}(\Omega)$, $X^* = L^d(\Omega)$ and we drop the index X in the notation ∂_X. Under certain regularity conditions, one can see the Kantorovitch potentials as the first variation of the Wasserstein distance, [19]. As a consequence, Fermat's rule $0 \in \partial \left(W_2^2(\rho_0, \cdot) + \mathrm{TV}(\cdot)\right)(\rho_1)$ assumes the following form.

Lemma 1. *Let ρ_1 be the unique minimizer of (TV-W), then there exists a Kantorovitch potential ψ_1 associated to ρ_1 such that*

$$-\frac{\psi_1}{\tau} \in \partial \left(\mathrm{TV} + \chi_{\mathcal{P}(\Omega)}\right)(\rho_1). \qquad (4)$$

Proof. For simplicity, we assume $\tau = 1$. Take $\rho \in BV(\Omega) \cap \mathcal{P}(\Omega)$ and define $\rho_t \stackrel{\text{def.}}{=} \rho + t(\rho_1 - \rho)$. Since $\overline{\Omega}$ is compact, the sup in (1) admits a maximizer [19, Prop. 1.11]. Let φ_t, ψ_t denote a pair of Kantorovitch potentials between ρ_0 and ρ_t. From the optimality of ρ_1 it follows

$$\frac{1}{2} W_2^2(\rho_0, \rho_1) + \mathrm{TV}(\rho_1) \le \int_\Omega \varphi_t \mathrm{d}\rho_0 + \int_\Omega \psi_t \mathrm{d}\rho_t + \mathrm{TV}(\rho_t)$$

$$\le \int_\Omega \varphi_t \mathrm{d}\rho_0 + \int_\Omega \psi_t \mathrm{d}\rho_1 + \mathrm{TV}(\rho_1) + (1-t) \left(\int_\Omega \psi_t \mathrm{d}(\rho - \rho_1) + \mathrm{TV}(\rho) - \mathrm{TV}(\rho_1) \right)$$

$$\le \frac{1}{2} W_2^2(\rho_0, \rho_1) + \mathrm{TV}(\rho_1) + (1-t) \left(\int_\Omega \psi_t \mathrm{d}(\rho - \rho_1) + \mathrm{TV}(\rho) - \mathrm{TV}(\rho_1) \right).$$

Hence, $-\psi_t \in \partial \left(\mathrm{TV} + \chi_{\mathcal{P}(\Omega)} \right)(\rho_1)$ for all $t \in (0, 1)$. Notice that as the optimal transport map from ρ_0 to ρ_t is given by $T_t = \mathrm{id} - \nabla \psi_t$ and assumes values in the bounded set Ω, the family $(\psi_t)_{t \in [0,1]}$ is uniformly Lipschitz so that by Arzelà-Ascoli's Theorem ψ_t converges uniformly to ψ_1 as t goes to 1 (see also [19, Thm. 1.52]). Therefore, $-\psi_1 \in \partial \left(\mathrm{TV} + \chi_{\mathcal{P}(\Omega)} \right)(\rho_1)$. ∎

With these results we can prove Theorem 1.

Proof (of Theorem 1). Here, to simplify, we still assume $\tau = 1$. The subdifferential inclusion (4) is conceptually the Euler-Lagrange equation for (TV-W), however it can be difficult to verify the conditions for direct sum between subdifferentials and give a full characterization. Therefore, for some arbitrary $\rho \in \mathcal{M}_+(\Omega)$ and $t > 0$, set

$$\rho_t = \frac{\rho_1 + t(\rho - \rho_1)}{1 + t\alpha}, \quad \text{where } \alpha = \int_\Omega \mathrm{d}(\rho - \rho_1).$$

Now ρ_t is admissible for the subdifferential inequality and using the positive homogeneity of TV we can write

$$\mathrm{TV}(\rho_1) - \int_\Omega \psi_1 \mathrm{d}(\rho_t - \rho_1) \le \frac{\mathrm{TV}(\rho_1) + t(\mathrm{TV}(\rho) - \mathrm{TV}(\rho_1))}{1 + t\alpha}.$$

After a few computations we arrive at $\mathrm{TV}(\rho) \ge \mathrm{TV}(\rho_1) + \int_\Omega (C - \psi_1) \mathrm{d}(\rho - \rho_1)$, where $C = \mathrm{TV}(\rho_1) + \int_\Omega \psi_1 \mathrm{d}\rho_1$. Notice that $(\phi + C, \psi - C)$ remains an optimal potential. So we can replace ψ_1 by $\psi_1 - C$, and obtain that for all $\rho \ge 0$ the following holds

$$\mathrm{TV}(\rho) \ge \mathrm{TV}(\rho_1) + \int_\Omega -\psi_1 \mathrm{d}(\rho - \rho_1), \quad \text{with } \mathrm{TV}(\rho_1) = \int_\Omega -\psi_1 \mathrm{d}\rho_1. \quad (5)$$

In particular, this means $-\psi_1 \in \partial \left(\mathrm{TV} + \chi_{\mathcal{M}_+(\Omega)} \right)(\rho_1)$ and ρ_1 is optimal for

$$\inf_{\rho \ge 0} \mathcal{E}(\rho) := \mathrm{TV}(\rho) + \int_\Omega \psi_1(x)\rho(x)\mathrm{d}x. \quad (6)$$

This suggests a penalization with an L^2 term *e.g.*

$$\inf_{u \in L^2(\Omega)} \mathcal{E}_t(u) := \mathrm{TV}(u) + \int_\Omega \psi_1(x)u(x)\mathrm{d}x + \frac{1}{2t} \int_\Omega |u - \rho_1|^2 \mathrm{d}x \quad (7)$$

which is a variation of (ROF) with $g = \rho_1 - t\psi_1$. In order for (7) to make sense, we need $\rho_1 \in L^2(\Omega)$, which is true if ρ_0 is L^∞ since then [8, Thm. 4.2] implies $\rho_1 \in L^\infty$. Suppose for now that ρ_0 is a bounded function.

Let u_t denote the solution of (7), from Prop. 1 if u_t solves (7), then u_t^+ solves the same problem with the additional constraint that $u \geq 0$, see [5, Lemma A.1]. As $\rho_1 \geq 0$ we can compare the energies of u_t^+ and ρ_1 and obtain the following inequalities

$$\mathcal{E}(\rho_1) \leq \mathcal{E}(u_t^+) \text{ and } \mathcal{E}_t(u_t^+) \leq \mathcal{E}_t(\rho_1).$$

Summing both inequalities yields

$$\int_\Omega |u_t^+ - \rho_1|^2 \mathrm{d}x \leq 0, \text{ therefore } u_t^+ = \rho_1 \text{ a.e. on } \Omega. \tag{8}$$

In particular, we also have that $u_t \leq \rho_1$. But as u_t solves a (ROF) problem, the optimality conditions from Prop. 1 give

$$\beta_t - \psi_1 \in \partial_{L^2} \mathrm{TV}(u_t), \text{ where } \beta_t \overset{\text{def.}}{=} \frac{\rho_1 - u_t}{t} \geq 0. \tag{9}$$

Notice from the characterization of $\partial_{L^2} \mathrm{TV}(\cdot)$ that $\partial_{L^2} \mathrm{TV}(u) \subset \partial_{L^2} \mathrm{TV}(u^+)$. Since $u_t^+ = \rho_1$, we have that

$$\beta_t - \psi_1 \in \partial_{L^2} \mathrm{TV}(\rho_1), \tag{10}$$

which proves (TVW-EL).

Now we move on to study the family $(\beta_t)_{t>0}$. Since $\rho_1 = u_t^+$, by definition $\beta_t = u_t^-/t$ and using the fact that $\partial_{L^2} \mathrm{TV}(u) \subset \partial_{L^2} \mathrm{TV}(u^-)$ in conjunction with Eq. (9), it holds that

$$\psi_1 - \beta_t \in \partial_{L^2} \mathrm{TV}(\beta_t). \tag{11}$$

But then, from Prop. 1, β_t solves (ROF) with $g = \psi_1$ and $\lambda = 1$. As this problem has a unique solution, the family $\{\beta_t\}_{t>0} = \{\beta\}$ is a singleton.

Since Ω is convex, and we know that the Kantorovitch potentials are Lipschitz continuous, cf. [19], so β, as a solution of (ROF) with Lipschitz data $g = \psi_1$, is also Lipschitz continuous with the same constant, following [16, Theo. 3.1].

But from (10) and the characterization of the subdifferential of TV, there is a vector field z such that $z \cdot D\rho_1 = |D\rho_1|$ such that

$$\beta - \psi_1 = \mathrm{div}\, z,$$

and as a consequence $\mathrm{div}\, z$ is also Lipschitz continuous, with constant at most twice the constant of ψ_1.

In the general case of $\rho_0 \in L^1(\Omega)$, define $\rho_{0,N} \overset{\text{def.}}{=} c_N(\rho_0 \wedge N)$ for $N \in \mathbb{N}$, where c_N is a renormalizing constant. Then $\rho_{0,N} \in L^\infty(\Omega)$ and $\rho_{0,N} \xrightarrow[N\to\infty]{L^1} \rho_0$.

Let $\rho_{1,N}$ denote the unique minimizer of (TV-W) with data term $\rho_{0,N}$, we can assume that $\rho_{1,N}$ w-$\star$ converges to some $\tilde{\rho}$. Then for any $\rho \in \mathcal{P}(\Omega)$ we have

$$TV(\rho_{1,N}) + \frac{1}{2\tau}W_2^2(\rho_{0,N}, \rho_{1,N}) \le TV(\rho) + \frac{1}{2\tau}W_2^2(\rho_{0,N}, \rho).$$

Passing to the limit on $N \to \infty$ we have that $\tilde{\rho}$ is a minimizer and from uniqueness it must hold that $\tilde{\rho} = \rho_1$.

Hence, consider the functions $z_N, \psi_{1,N}, \beta_N$ that satisfy (TVW-EL) for $\rho_{1,N}$. Up to a subsequence, we may assume that z_N converges weakly-$\star$ to some $z \in L^\infty(\Omega; \mathbb{R}^d)$. Since $\psi_{1,N}$, β_N and div z_N are Lipschitz continuous with the same Lipschitz constant for all N, by Arzelà-Ascoli, we can assume that $\psi_{1,N}, \beta_N$ and div z_N converge uniformly to Lipschitz functions ψ_1, β, div $z = \beta - \psi_1$. In addition, passing to the limit in (11), we find that β solves (ROF) for $\lambda = 1$ and $g = \psi_1$.

Since β_N converges uniformly and $\rho_{1,N}$ converges w-$\star$ we have

$$0 = \lim_{N \to \infty} \int_\Omega \beta_N \rho_{1,N} \mathrm{d}x = \int_\Omega \beta \rho_1 \mathrm{d}x,$$

and hence $\beta \rho_1 = 0$ a.e. in Ω since both are nonnegative. In addition, ψ_1 is a Kantorovitch potential associated to ρ_1 from the stability of optimal transport (see [19, Thm. 1.52]). From the optimality of $\rho_{1,N}$ it holds that

$$TV(\rho_{1,N}) + \frac{1}{2\tau}W_2^2(\rho_{0,N}, \rho_{1,N}) \le TV(\rho_1) + \frac{1}{2\tau}W_2^2(\rho_{0,N}, \rho),$$

so that $\lim TV(\rho_{1,N}) \le TV(\rho_1)$. Changing the roles of ρ_1 and $\rho_{1,N}$ we get an equality. So it follows that

$$\int_\Omega (\beta - \psi_1)\rho_1 \mathrm{d}x = \lim_{N \to \infty} \int_\Omega (\beta_N - \psi_{1,N})\rho_{1,N}\mathrm{d}x = \lim_{N \to \infty} TV(\rho_{1,N}) = TV(\rho_1),$$

Since TV is 1-homogeneous we conclude that $\beta - \psi_1 \in \partial TV(\rho_1)$. ∎

We say E is a set of finite perimeter if the indicator function $\mathbb{1}_E$ is a BV function, and we set $\mathrm{Per}(E) = TV(\mathbb{1}_E)$. As a byproduct of the previous proof we conclude that the level sets $\{\rho_1 > s\}$ are all solutions to the same prescribed curvature problem.

Corollary 1. *The following properties of the level sets of ρ_1 hold.*

1. *For $s > 0$ and ψ_1 in (TVW-EL)*

$$\{\rho_1 > s\} \in \underset{E \subset \Omega}{\mathrm{argmin}}\, \mathrm{Per}(E; \Omega) + \frac{1}{\tau} \int_E \psi_1 \mathrm{d}x$$

2. *$\partial\{\rho_1 > s\} \setminus \partial^*\{\rho_1 > s\}$ is a closed set of Hausdorff dimension at most $d - 8$, where ∂^* denotes the reduced boundary of a set, see [1]. In addition, $\partial^*\{\rho_1 > s\}$ is locally the graph of a function of class $W^{2,q}$ for all $q < +\infty$.*

Proof. For simplicity take $\tau = 1$. Inside the set $\{\rho_1 > s\}$, for $s > 0$, we have $-\psi_1 = \mathrm{div}\, z$, so from the definition of the perimeter we have

$$\int_{\{\rho_1 > s\}} -\psi_1 \mathrm{d}x = \int_{\{\rho_1 > s\}} \mathrm{div}\, z \mathrm{d}x \leq \mathrm{Per}\left(\{\rho_1 > s\}\right).$$

So using the fact that $\mathrm{TV}(\rho_1) = \int_\Omega -\psi_1 \mathrm{d}x$, the coarea formula and Fubini's Theorem give

$$\int_0^{+\infty} \mathrm{Per}(\mathbb{1}_{\{\rho_1 > s\}}) \mathrm{d}s = \int_\Omega -\psi_1 \int_0^{\rho_1(x)} \mathrm{d}s \mathrm{d}x = \int_0^{+\infty} \int_{\{\rho_1 > s\}} -\psi_1 \mathrm{d}x \mathrm{d}s.$$

Hence, $\mathrm{Per}(\{\rho_1 > s\}) = \int_{\{\rho_1 > s\}} -\psi_1 \mathrm{d}x$ for *a.e.* $s > 0$. But as $\beta\psi_1 = 0$ a.e., we have $-\psi_1 = \mathrm{div}\, z$ in $\{\rho_1 > s\}$, so that $-\psi_1 \in \partial \mathrm{TV}(\mathbb{1}_{\{\rho_1 > s\}})$ for *a.e.* $s > 0$; and by a continuity argument, for all $s > 0$. The subdifferential inequality with $\mathbb{1}_E$ gives

$$\{\rho_1 > s\} \in \operatorname*{argmin}_{E \subset \Omega} \mathrm{Per}(E) + \int_E \psi_1(x) \mathrm{d}x. \tag{12}$$

Item (2) follows directly from the properties of (ROF), see [6], since $\rho_1 = u^+$, where u solves a problem (ROF). $\quad\square$

3   Numerical Experiments

We solve (TV-W) for an image denoising application using a Douglas-Rachford algorithm [9] with Halpern acceleration [11], see Table 1. For this we need subroutines to compute the prox operators defined, for a given $\lambda > 0$, as

$$\mathrm{prox}_{\lambda \mathrm{TV}}(\bar{\rho}) \stackrel{\mathrm{def.}}{=} \operatorname*{argmin}_{\rho \in L^2(\Omega)} \mathrm{TV}(\rho) + \frac{1}{2\lambda} \|\rho - \bar{\rho}\|_{L^2(\Omega)}^2, \tag{13}$$

$$\mathrm{prox}_{\lambda W_2^2}(\bar{\rho}) \stackrel{\mathrm{def.}}{=} \operatorname*{argmin}_{\rho \in L^2(\Omega)} \frac{1}{2\tau} W_2^2(\rho_0, \rho) + \frac{1}{2\lambda} \|\rho - \bar{\rho}\|_{L^2(\Omega)}^2. \tag{14}$$

We implemented the prox of TV with the algorithm from [10], modified to account for Dirichlet boundary conditions. From [7, Theo. 2.4] it is consistent with the continuous total variation. The prox of W_2^2 is computed by expanding the L^2 data term as

$$\mathrm{prox}_{\lambda W_2^2}(\bar{\rho}) = \operatorname*{argmin}_{\rho \in L^2(\Omega)} \frac{1}{2\tau} W_2^2(\rho_0, \rho) + \frac{1}{2\lambda} \int_\Omega \rho^2 \mathrm{d}x + \int_\Omega \rho \underbrace{\left(-\frac{\bar{\rho}}{\lambda}\right) \mathrm{d}x}_{=V} + \underbrace{\frac{1}{2\lambda} \bar{\rho}^2 \mathrm{d}x}_{cst}$$

$$= \operatorname*{argmin}_{\rho \in L^2(\Omega)} \frac{1}{2\tau} W_2^2(\rho_0, \rho) + \frac{1}{2\lambda} \int_\Omega \rho^2 \mathrm{d}x + \int_\Omega \rho V \mathrm{d}x,$$

which is one step of the Wasserstein gradient flow of the porous medium equation $\partial_t \rho_t = \lambda^{-1} \Delta(\rho_t^2) + \mathrm{div}\,(\rho_t \nabla V)$, where the potential is $V = -\bar{\rho}/\lambda$, see [13,19]. To compute it we have used the back-n-forth algorithm from [13].

Algorithm 1. Halpern accelerated Douglas-Rachford algorithm

$\beta_0 \leftarrow 0$

$x_0 \leftarrow$ Initial Image

while $n \geq 0$ **do**

$\quad y_n \leftarrow \text{prox}_{\lambda\,\text{TV}}(x_n)$

$\quad \lambda_n \in [\varepsilon, 2 - \varepsilon]$

$\quad z_n \leftarrow x_n + \lambda_n \left(\text{prox}_{\lambda W_2^2}(2y_n - x_n) - y_n \right)$

$\quad \beta_n \leftarrow \frac{1}{2}\left(1 + \beta_{n-1}^2\right)$ ▷ Optimal constants for Halpern acceleration from [11]

$\quad x_{n+1} \leftarrow (1 - \beta_n)x_0 + \beta_n z_n$

end while

3.1 Evolution of Balls

Following [8], in dimension 1, whenever the initial measure is uniformly distributed over a ball, the solutions remain balls. In $\mathbb{R}^d$, one can prove this remains true. If ρ_0 is uniformly distributed over a ball of radius r_0, then the solution to (TV-W) is uniformly distributed in a ball of radius r_1 solving the following polynomial equation for r_1

$$r_1^2(r_1 - r_0) = r_0^2(d + 2)\tau.$$

This theoretical predictions are corroborated by the numerical experiments found in Fig. 1.

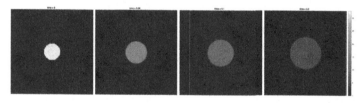

Fig. 1. Evolution of circles: from left to right initial condition and solutions for $\tau = 0.05, 0.1, 0.2$. The red circles correspond to the theoretical radius. (Color figure online)

3.2 Reconstruction of Dithered Images

In this experiment we use model (TV-W) to reconstruct dithered images. In $\mathcal{P}(\mathbb{R}^2)$ the dithered image is a sum of Dirac masses, so the model (TV-W) outputs a new image which is close in the Wasserstein topology, but with small total variation. In Fig. 2 below, we compared the result with the reconstruction given by (ROF), both with a parameter $\tau = 0.2$. Although the classical (ROF) model was able to create complex textures, these remain granulated, whereas the (TV-W) model is able to generate both smooth and complex textures.

Fig. 2. Dithering reconstruction problem. From left to right: Dithered image, TV-Wasserstein and ROF results.

4   Conclusion

In this work we revisited the TV-Wasserstein problem. We showed how it can be related to the classical (ROF) problem and how to exploit this to derive the Euler-Lagrange equations, obtaining further regularity. We proposed a Douglas-Rachford algorithm to solve it and presented two numerical experiments: the first one being coherent with theoretical predictions and the second being an application to the reconstruction of dithered images.

Acknowledgements. The second author acknowledges support from the ANR CIPRESSI project, grant 19-CE48-0017-01 of the French Agence Nationale de la Recherche. The third author acknowledges the financial support from FMJH concerning his master thesis, when this work took place.

References

1. Ambrosio, L., Fusco, N., Pallara, D.: Functions of Bounded Variation and Free Discontinuity Problems, 2nd edn. Oxford University Press, New York (2000)
2. Benning, M., Calatroni, L., Düring, B., Schönlieb, C.-B.: A primal-dual approach for a total variation wasserstein flow. In: Nielsen, F., Barbaresco, F. (eds.) GSI 2013. LNCS, vol. 8085, pp. 413–421. Springer, Heidelberg (2013). https://doi.org/10.1007/978-3-642-40020-9_45
3. Bredies, K., Holler, M.: A pointwise characterization of the subdifferential of the total variation functional. arXiv preprint arXiv:1609.08918 (2016)
4. Burger, M., Franek, M., Schonlieb, C.B.: Regularized regression and density estimation based on optimal transport. Appl. Math. Res. eXpress **2012**(2), 209–253 (2012)
5. Chambolle, A.: An algorithm for mean curvature motion. Interf. Free Bound. **6**(195–218), 2 (2004)
6. Chambolle, A., Caselles, V., Cremers, D., Novaga, M., Pock, T.: An introduction to total variation for image analysis. Theor. Found. Numer. Methods Sparse Rec. **9**(263–340), 227 (2010)

7. Chambolle, A., Pock, T.: Learning consistent discretizations of the total variation. SIAM J. Imaging Sci. **14**(2), 778–813 (2021)
8. Carlier, G., Poon, C.: On the total variation Wasserstein gradient flow and the TV-JKO scheme. ESAIM: COCV **25**(41) (2019)
9. Combettes, P.L., Pesquet, J.C.: Proximal splitting methods in signal processing. In: Bauschke, H., Burachik, R., Combettes, P., Elser, V., Luke, D., Wolkowicz, H. (eds.) Fixed-Point Algorithms for Inverse Problems in Science and Engineering. Springer Optimization and Its Applications, vol. 49. Springer, New York (2011). https://doi.org/10.1007/978-1-4419-9569-8_10
10. Condat, L.: Discrete total variation: new definition and minimization. SIAM J. Imaging Sci. **10**(3), 1258–1290 (2017)
11. Contreras, J.P., Cominetti, R.: Optimal error bounds for non-expansive fixed-point iterations in normed spaces. Math. Program. **199**, 343–374 (2022)
12. Cuturi, M., Peyré, G.: Semidual regularized optimal transport. SIAM Rev. **60**(4), 941–965 (2018)
13. Jacobs, M., Lee, W., Léger, F.: The back-and-forth method for Wasserstein gradient flows. ESAIM Control Optim. Calc. Var. **27**, 28 (2021)
14. Jordan, R., Kinderlehrer, D., Otto, F.: The variational formulation of the Fokker-Planck equation. SIAM J. Math. Anal. **29**(1), 1–17 (1998)
15. Lieu, L.H., Vese, L.A.: Image restoration and decomposition via bounded total variation and negative hilbert-sobolev spaces. Appl. Math. Optim. **58**, 167–193 (2008)
16. Mercier, G.: Continuity results for TV-minimizers. Indiana Univ. Math. J., 1499–1545 (2018)
17. Meyer, Y.: Oscillating patterns in image processing and nonlinear evolution equations: the fifteenth Dean Jacqueline B. Lewis memorial lectures, vol. 22. American Mathematical Society (2001)
18. Rudin, L.I., Osher, S., Fatemi, E.: Nonlinear total variation based noise removal algorithms. Physica D: Nonlinear Phenomena **60**(4), 259–268 (1992)
19. Santambrogio, F.: Optimal Transport for Applied Mathematicians, 1st edn. Birkhauser, New York (2015)

Correction to: Geometric Science of Information

Frank Nielsen⬤ and Frédéric Barbaresco⬤

Correction to:
F. Nielsen and F. Barbaresco (Eds.): *Geometric Science*
of Information, **LNCS 14071,**
https://doi.org/10.1007/978-3-031-38271-0

The original version of the book was inadvertently published with a typo in the frontmatter. In the headline of page xiii it should read "GSI'23 Keynote Speakers" instead of "GSI'21 Keynote Speakers" and a typo sponsor's section. Instead of "European COST CaLISTA" it should read "European Horizon CaLIGOLA". This has been corrected.

The updated original version of the book can be found at
https://doi.org/10.1007/978-3-031-38271-0

Author Index